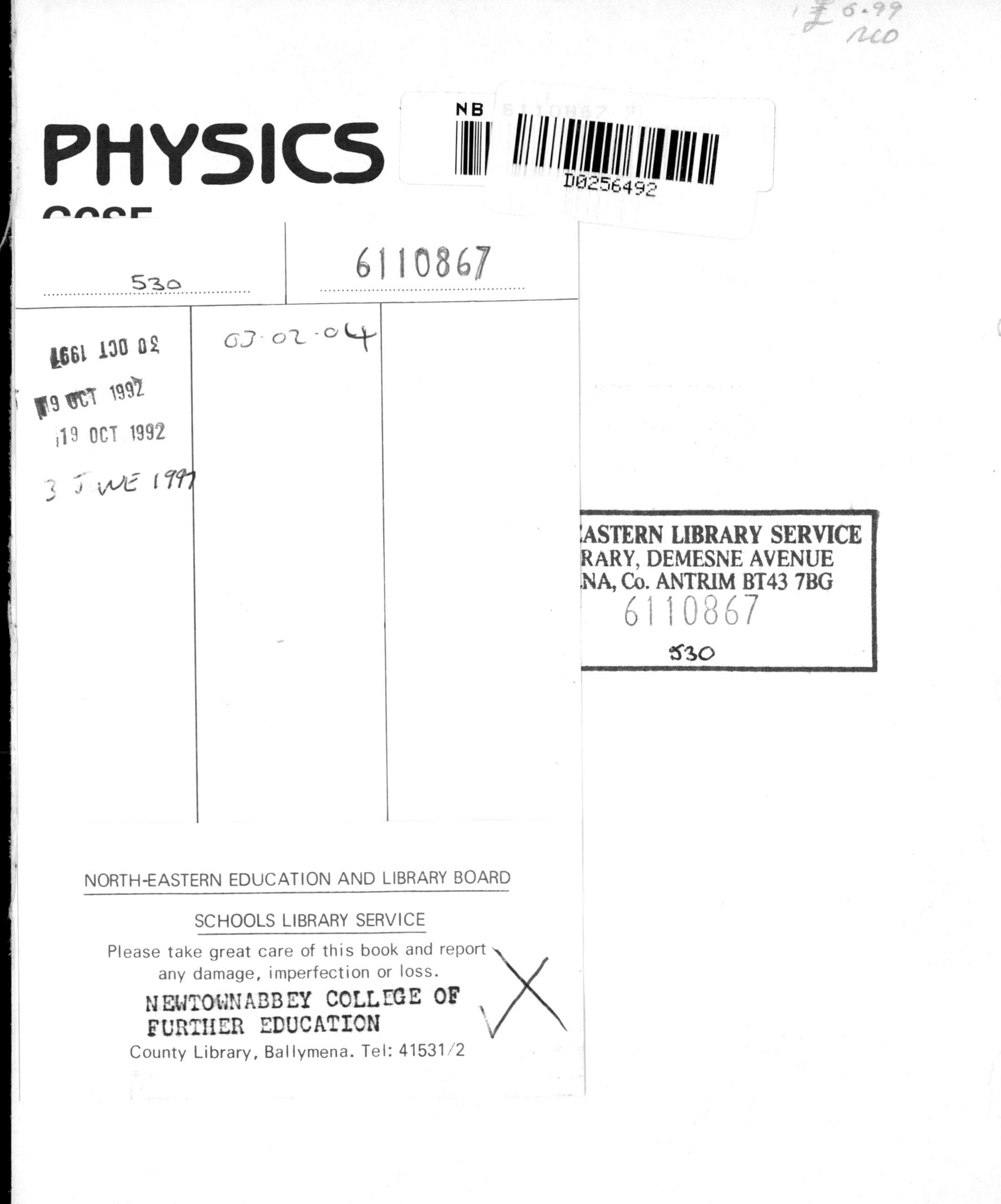

Stanley Thornes & Hulton

First published in Great Britain 1984
by Hulton Educational Publications Ltd
Old Station Drive, Leckhampton, Cheltenham GL53 0DN

Reprinted 1985
Reprinted 1986
Reprinted 1988
GCSE Edition 1989

British Library Cataloguing in Publication Data
Hanson, Edward L. (Edward Latham)
Physics GCSE. – 2nd ed.
1. Physics
I. Title
530

ISBN 1–871402–15–8

Typeset by Hunters Armley Ltd, Leeds & London.
Printed and bound in Great Britain by Butler & Tanner Ltd, Frome.

Contents

Preface

This new edition of ***Physics: an examination course*** is intended specifically for students taking GCSE papers in physics set by one of the six Examining Groups which conduct the GCSE examination in England, Northern Ireland and Wales. As with the first edition, students will find that ***Physics GCSE: an examination course*** will give them a concise, straightforward account of the fundamental principles of physics which they need for their immediate examination goal or as a basis for further studies.

While there is reasonable agreement between the Examining Groups about the basic body of knowledge of elementary physics appropriate for this level of study, when we come to what we may call the 'fringe' topics there is wide variation. For example, only two syllabuses contain a reference to combining forces and only one to resolving forces. One syllabus demands an elementary knowledge of five different logic gates, another a knowledge of four, another of three, and two Groups ignore them altogether! Four Groups require a knowledge of the use of a transistor as a switch and two do not. And so on.

This book covers *all* the topics set by all the Examining Groups, together with brief references to topics not asked for by any examination but which it was felt sensible to include to give a more complete picture of physics. It is clear, therefore, that students will need guidance from their teachers or from the examination syllabus itself as to what particular topics they are likely to be questioned on.

An elementary knowledge of algebra and trigonometry is assumed, as is the ability to construct and interpret graphs.

The exercises at the end of each section are graded so that the early questions (many of them multiple choice) are 'core' level and the later questions are more relevant to the 'extension' level for candidates seeking higher grades.

My thanks are due to many colleagues, teachers and students who over the years have helped me with the simplifying and presentation of physical concepts.

E.L. Hanson
Wimbledon, 1989

INTRODUCTION

It is convenient to divide the vast body of knowledge which we call **science** into two large sections – **physical sciences** and **biological sciences**. The latter deal with the behaviour of living things and the processes taking place in them. The physical sciences are concerned with our knowledge of non-living material and its phenomena. It is further convenient to sub-divide the physical sciences into **physics** and **chemistry**. Chemistry studies the composition of materials and the changes which can take place in this composition. It is concerned with what things are made of. Physics is concerned with how things behave and in discovering the general principles which explain natural phenomena. Physics itself is divided into a number of sections, e.g. mechanics, heat, light, sound, electricity, etc. and this pattern is followed in this book.

However, it would be a great mistake to think that these divisions and sub-divisions of science are completely separate. Processes in living things obey the laws of both physics and chemistry and an understanding of these processes is impossible without the combination of chemical and physical ideas. Different divisions of physics show overlapping relationships: electricity can produce heating, chemical and magnetic effects; light can set free the electrons of atoms in the photoelectric effect. Underneath all our studies are the important ideas of energy and the behaviour of atoms and molecules. It is hoped that your understanding of such relationships will become clearer as your study of physics proceeds.

Physics and experiment

Scientific discovery depends on making observations and doing experiments. From the results of observations and experiments conclusions are drawn. When the same conclusion is reached after performing many experiments, a large generalisation called a **scientific law** can be made. A scientific law is a short statement expressing the fact that the same thing has been observed to take place in the same set of circumstances so many times that it is reasonable to expect it always to happen. Sometimes new observations turn up which 'break the law'. These are checked carefully and if there is no doubt that these new facts do not follow the statement of the law, then the law has to be scrapped or modified.

When many facts are discovered an attempt is made to explain them by producing a working **hypothesis** or possible explanation. The hypothesis should suggest further experiments to test its truth, and on the result of these new experiments the hypothesis stands or falls. When a hypothesis has stood the test of many experiments designed to check its truth, it becomes a **theory**.

These ideas can be illustrated by a look at how our knowledge of the behaviour of gases grew up. The precise study of gases began in the seventeenth century with Galileo's and Torricelli's observations about the pressure of the atmosphere. Robert Boyle studied the compression of gases and concluded that all gases behaved in the same way and obeyed Boyle's law (the pressure of a fixed mass of gas at constant temperature is inversely proportional to the volume). Early in the eighteenth century Jacques Charles and Joseph Louis Gay-Lussac studied the effect of temperature changes on the pressure and volume of a gas. Again, all gases seemed to behave in the same sort of way and the general gas equation (see p.143) was produced. How could these facts be explained? Dalton's idea about atoms and molecules grew out of his interest in the behaviour of gases; a possible explanation of this behaviour could be that gases consisted of particles which were in constant and random movement. This picture is called a **model** and from this model developed the kinetic molecular theory of gases which is discussed in Chapter 16. The theory worked well. It led to other interesting ideas, such as the absolute zero of temperature, the composition and stability of the earth's atmosphere compared with other planets which have lost their atmospheres, and the laws of the diffusion of gases, which were used in the separation of the uranium isotopes to produce the first atomic bomb.

In the development of the scientific method in any particular branch of physics, we should be able to identify some or all of the following steps:

- The recognition of a particular problem.
- The collection of observations and experimental facts.

- The organisation of the facts into scientific laws.
- The invention of a working hypothesis to explain the facts.
- The testing of the hypothesis with its possible justification, overthrow or modification.
- The growth of the hypothesis into a scientific theory.
- The application of the theory to new problems and the making of new discoveries.

Physics and measurement

When we make an observation such as 'a metal bar expands when it is heated', we are making a **qualitative** observation. We are simply stating what happens. If we say 'this metal bar expands 2.6 millimetres when its temperature is raised by 200 degrees C', we are making a **quantitative** statement. A quantitative statement is expressed in terms of measurement and is much more precise and informative. Compare the qualitative and quantitative statements such as 'the man was tall and heavily built' and 'the man was 185 centimetres tall and weighed 89 kilogram weight'. The words 'tall' and 'heavily built' can be interpreted by different people in different ways, but the second statement means the same thing to everybody.

Lord Kelvin, one of the great physicists of the nineteenth century, said, 'When you can measure what you are speaking about and express it in numbers, you know something about it; when you cannot express it in numbers, your knowledge is of a meagre and unsatisfactory kind; it may be the beginning of knowledge, but you have scarcely in your thoughts advanced to the state of science'.

All measurements are subject to some error, either because the measuring instruments are not precise enough or because the human experimenters make mistakes when using and reading the instruments. As instruments get better and the techniques of using them improve, the errors become less, but they can never be entirely eliminated.

All measurement involves a number and a unit. When we say a piece of wood is 3.62 metres long we mean it is 3.62 times as long as a standard length called the metre. The metre is the unit of length which can be exactly defined and exactly reproduced. Just as different currencies are used in different countries of the world, so there are all kinds of units for measuring the same quantity. At different times lengths have been measured in spans, cubits, feet, fathoms, yards, leagues, miles – to name but a few! Now, in order to avoid confusion, scientists use the International System of Units (SI units), which was recommended by the General Conference on Weights and Measures in 1960.

The definitions of the base SI units are given here for reference. A list of some of the units derived from the base units is also given in table 1. Reference can be made to this as the various sections of the book are studied.

The unit of **length** is the **metre**. The metre was originally defined as one ten-millionth part of the meridian between the North Pole and the equator and passing through Paris; it then became the distance between two lines engraved on a platinum–iridium bar kept in Paris; it is now defined in terms of a particular wavelength of light emitted by a krypton-86 atom (1 metre = 1 650 763.73 wavelengths).

The unit of **mass** is the **kilogram**. Originally made equal to the mass of 1000 cubic centimetres of pure water at 4 °C, it is now equal to the mass of the international standard platinum kilogram kept in Paris.

The unit of **time** is the **second**, originally defined in terms of the solar day, that is, the time that elapses between two consecutive passages of the sun across the meridian. It is now defined in terms of the vibration frequency of the radiation emitted by a caesium-133 atom (1 second = 9 192 631 770 periods of vibration).

The unit of **electric current** is the **ampere** which is defined on p. 272.

The unit of **temperature** is the **kelvin** which is explained on pp. 96 and 142.

The unit of **luminous intensity** is the **candela** which is defined in terms of the amount of light emitted by unit area of platinum at its melting point.

Other units are derived from the base units. For example, speed, which is found by dividing length by time, is measured in metres per second, written m/s or m s^{-1}.

For quantities very much larger or smaller than the standard unit, multiples or sub-multiples are used. These go up or down by factors of a thousand. They are given in table 2.

Some units convenient to use do not fit into this plan, for example

the centimetre (10^{-2} m)
the litre (1000 cm^3)
the tonne (1000 kg)

Physics and technology

A distinction has to be drawn between pure science and technology. Obviously a lot of discoveries in physics have had very practical applications. The researches in electricity in the nineteenth century have led to the development of the dynamo and the electric motor. The discovery of the electron has led to enormous developments in the field of electronics. Pure science and technology serve different ends, but in many cases they run along parallel lines and are complementary in their work and discoveries. Pure science aims to acquire as complete a knowledge as

table 1 SI units and symbols. The first six are called **base units**

Quantity	Symbol for quantity	Unit	Symbol for unit
length	*l*	metre	m
mass	*m*	kilogram	kg
time	*t*	second	s
electric current	*I*	ampere	A
temperature	*T*	kelvin	K
luminous intensity	*I*	candela	cd
area	*A*	square metre	m^2
volume	*V*	cubic metre	m^3
velocity	*v*	metre per second	m/s
acceleration	*a*	metre per second squared	m/s^2
frequency	*f*	hertz	Hz
density	*ρ*	kilogram per cubic metre	kg/m^3
force	*F*	newton	N
pressure	*p*	pascal	Pa
energy, work	*E, W*	joule	J
power	*P*	watt	W
electric charge	*Q*	coulomb	C
potential difference	*V*	volt	V
resistance	*R*	ohm	Ω
capacitance	*C*	farad	F

table 2 Decimal multiples and their prefixes

Factor	Multiple	Prefix	Symbol
one thousand million times	10^9	giga-	G
one million times	10^6	mega-	M
one thousand times	10^3	kilo-	k
one thousandth	10^{-3}	milli-	m
one millionth	10^{-6}	micro-	μ
one thousand millionth	10^{-9}	nano-	n
one million millionth	10^{-12}	pico-	p

possible of the material universe. Technology uses this knowledge to serve our needs directly.

When, about 1880, Swan in England and Edison in the USA set out to develop practical forms of electric lamp, the heating effect of an electric current and the laws governing the production of heat were already known. The pure science had been done. Swan and Edison's problems concerned such things as the best kind of filament to use, the type of container to put it in, how to seal metal wires into glass, how to produce a good vacuum. These were technological problems and Swan and Edison solved them and their work had a feedback into pure science because it led indirectly to the discovery of the electron.

The idea that electromagnetic action might travel in the form of waves was suggested by Michael Faraday in 1845. By 1864 James Clerk Maxwell had worked out the mathematics of the problem and showed that the waves travelled at the speed of light. The existence of the waves was demonstrated in the laboratory by Heinrich Hertz in 1886. But it was the technologist, Guglielmo Marconi, who hit upon the idea of using the waves to transmit messages and sent the first radio signal. The whole history of science shows this interplay between pure and applied physics.

Scientific investigations have to be followed for their own sake, without any thought of what 'use' the results will be; otherwise, new fields of knowledge would not be explored at all. In the nineteenth century the standard medical technique of finding out if there was any shrapnel still left in a gunshot wound was to probe the wound with a metal probe. If the problem of developing better methods of carrying out this operation had been passed to the technologists, they would doubtless have produced more sensitive probes, but they would never have discovered X-rays. X-rays were discovered almost by chance by Wilhelm Konrad Röntgen in 1895 when he was carrying out experiments to solve an entirely different problem. He said, 'I have discovered something interesting, but I do not know whether or not my observations are correct'. Neither he nor anyone else at that time could have had the slightest idea where his discovery was going to lead, or how X-ray investigation was going to become a standard medical technique.

1 SPEED, VELOCITY AND ACCELERATION

When a body moves, it travels through a certain distance in a certain time.

Speed is rate of change of distance

$$\textbf{average speed} = \frac{\textbf{distance travelled}}{\textbf{time taken}} \qquad \text{units } \textbf{m/s, km/h}$$

If we wish to know where the body will be after a given time, we need to know not only its speed but also the direction in which it is travelling. This information is given by the **velocity**.

Velocity is the distance travelled in unit time in a given direction

The distance travelled in a given direction is called the **displacement** of the body, so

Velocity = rate of change of displacement

Quantities, such as velocity, which involve the idea of direction are called **vector quantities**. Other vector quantities are acceleration and force.

Quantities which are completely described when their size is known and do not involve the idea of direction are **scalar quantities**. Examples of scalars are length, speed and energy.

If the velocity of a body is not constant, i.e. it is 'speeding up' or 'slowing down', it is travelling with an acceleration.

Acceleration is rate of change of velocity

$$\textbf{acceleration} = \frac{\textbf{increase in velocity}}{\textbf{time taken}} \qquad \text{units } \mathbf{m/s^2} \text{ or } \mathbf{m\ s^{-2}}$$

When a body is slowing down its negative acceleration is often called a **retardation**.

If the velocity changes by equal amounts in equal times, the body is travelling with **uniform acceleration**.

An object falling freely under the gravitational attraction of the earth is moving with uniform acceleration. The value of the **acceleration due to gravity** (g) varies for different points on the earth's surface, but an average value is 9.8 m/s^2. For many calculations it is sufficiently accurate to take $g = 10\ m/s^2$.

The rate of fall of a body at a given place is independent of the body's weight. This was realised by Galileo in the sixteenth century. In air a coin falls more rapidly than a feather or a piece of paper because of air resistance, which is greater on the object of larger area. In an evacuated tube from which all the air has been pumped, the feather and the coin will fall at the same speed, figure 1.1.

fig. 1.1 Two versions of Galileo's experiment. The stroboscopic photograph shows clearly that the heavy ball falls at the same speed as the light one

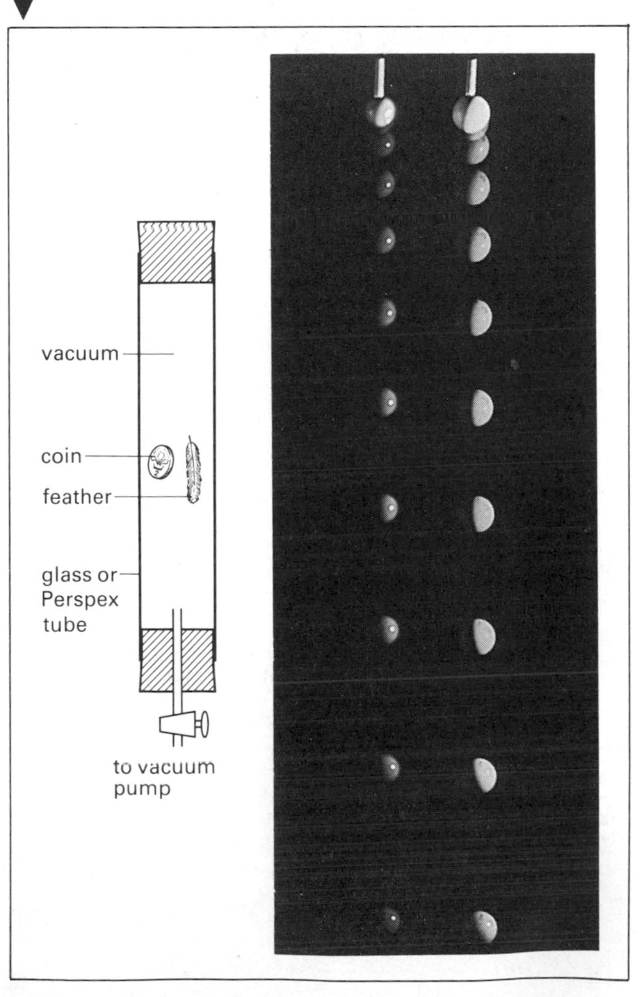

Displacement–time graphs

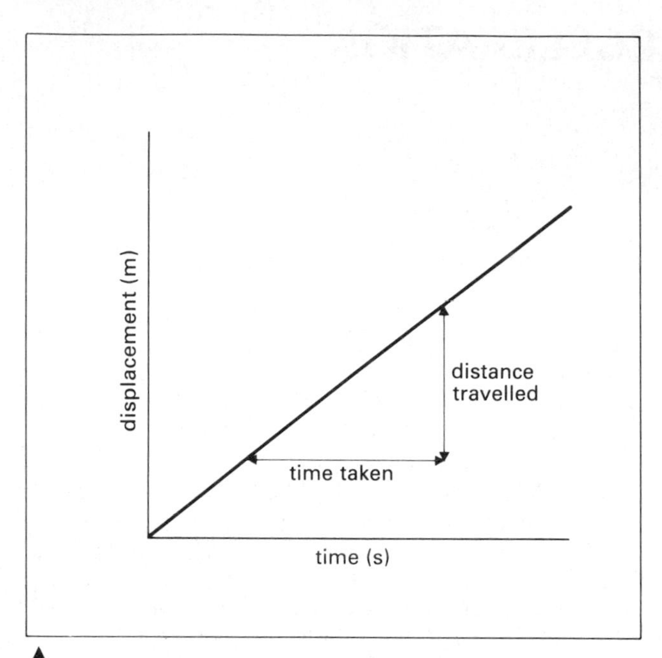

▲
fig. 1.2 If equal distances are travelled in equal times, the graph is a straight line. The **slope** or **gradient** of the graph gives the **velocity**, which in this case is constant

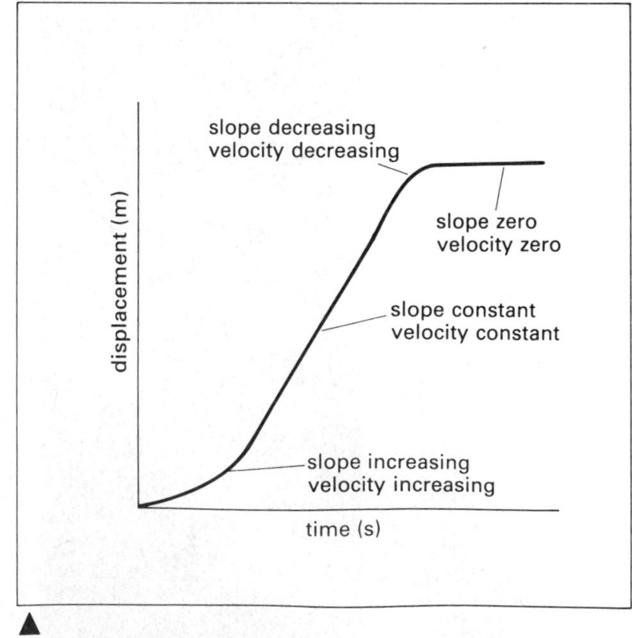

▲
fig. 1.3 A train moves from rest, travels at constant speed and then slows down to stop at a station

Velocity–time graphs

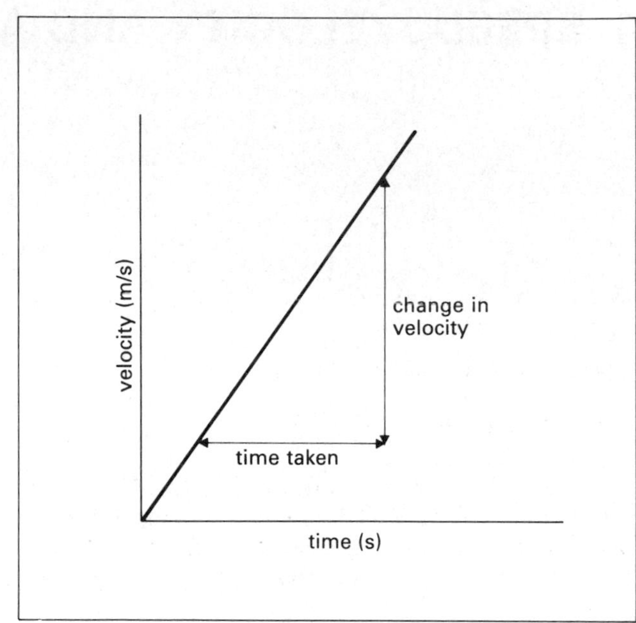

▲
fig. 1.4 The slope represents the **acceleration**. The graph shows a car moving from rest with uniform acceleration

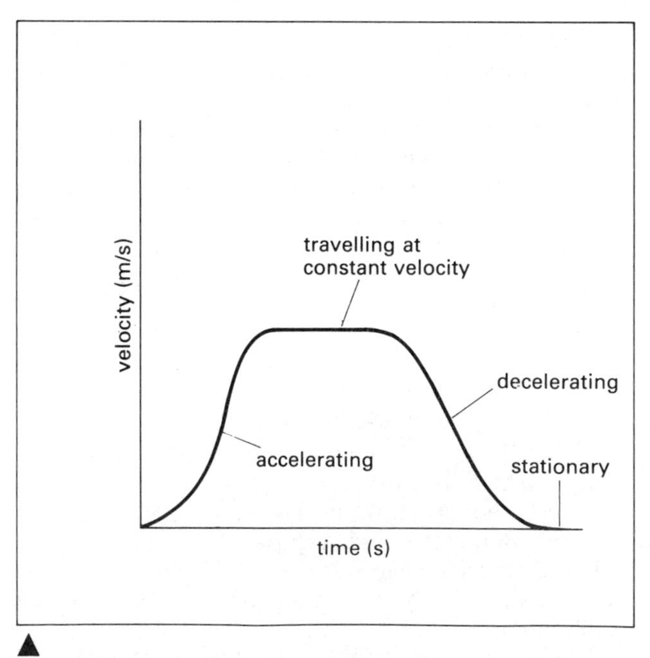

▲
fig. 1.5 A train starts from rest, travels at constant speed and then slows down to stop at a station

The velocity–time graph for a body moving at constant velocity is a straight line parallel to the time axis, figure 1.6.

distance travelled = velocity × time
= 8 × 5
= 40 m

The distance travelled is equal to the area under the graph and up to the required time coordinate.

If the body is moving with constant acceleration the velocity–time graph is a straight line with a gradient equal to the acceleration, figure 1.7.

In the graph shown in figure 1.7, the average velocity is $\frac{1}{2}(6 + 10) = 8$ m/s.

distance travelled = average velocity × time
= 8 × 4
= 32 m

Once again the distance travelled can be shown to be equal to the area under the graph.

area = rectangle PUVO + triangle PUQ
= $4 \times 6 + \frac{1}{2} \times 4 \times 4$
= $24 + 8$
= 32 square units

The area under the velocity–time graph is equal to the distance travelled

This holds good even if the acceleration is not uniform and the velocity–time graph is curved.

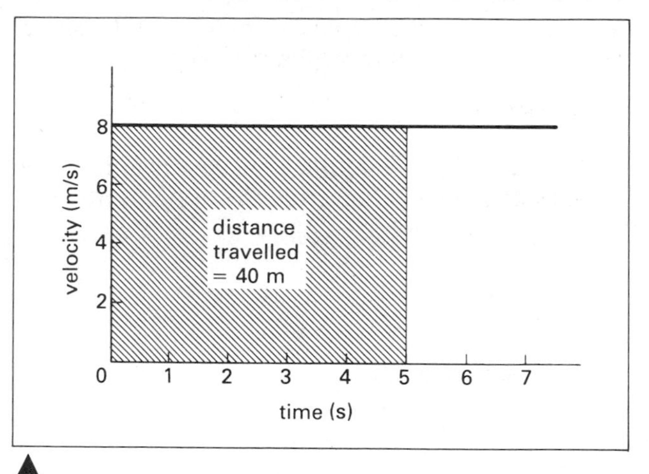

fig. 1.6 Constant velocity

fig. 1.7 Constant acceleration

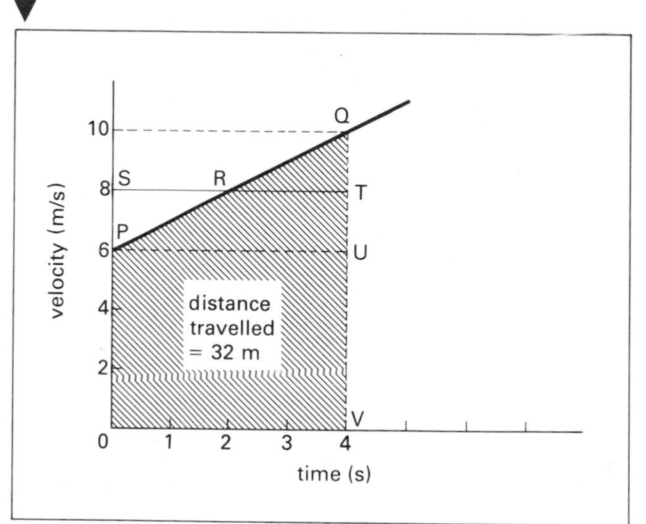

The equations of motion for a uniformly accelerated body

Many problems on motion can be solved from first principles or graphically. However, it is useful to have a set of general formulae which can be applied to all problems of uniformly accelerated motion.

Let a body have an initial velocity $= u$
a final velocity $= v$
an acceleration $= a$
and travel a distance $= d$
in a time $= t$

From the definition of acceleration,

$$\text{acceleration} = \frac{\text{change in velocity}}{\text{time taken}}$$

$$a = \frac{v - u}{t}$$

$$at = v - u$$

$$\boldsymbol{v = u + at} \qquad (1)$$

$$\text{average velocity} = \frac{\text{distance travelled}}{\text{time taken}} = \frac{d}{t}$$

Since the acceleration is uniform,

$$\text{average velocity} = \frac{u + v}{2}$$

therefore,

$$\frac{d}{t} = \frac{u + v}{2} \quad \text{so} \quad d = \frac{u + v}{2} \times t$$

Substituting the value of v from equation (1) gives

$$\boldsymbol{d = ut + \tfrac{1}{2}at^2} \qquad (2)$$

These are the two basic equations, but a third can be found by substituting the value for t from equation (1) in equation (2). This gives

$$\boldsymbol{v^2 - u^2 = 2ad} \qquad (3)$$

Remember that the equations only apply to objects which are accelerating *uniformly*.

EXAMPLES

1 Find the height of a bridge above a river if a stone takes 2 seconds to fall from the bridge to the water.

$u = 0$
$a = 10\ \text{m/s}^2$
$t = 2\ \text{s}$

Using equation (2)

$d = ut + \frac{1}{2}at^2$
$d = \frac{1}{2} \times 10 \times 4 = 20\ \text{m}$

The bridge is 20 m above the water.

2 A body moves from rest with uniform acceleration and in 10 seconds acquires a speed of 14 m/s. What is its acceleration and how far has the body travelled in the 10 seconds?

$u = 0$
$v = 14\ \text{m/s}$
$t = 10\ \text{s}$

Using equation (1)

$v = u + at$
$14 = 10a$
$a = 1.4\ \text{m/s}^2$

Using equation (2)

$d = ut + \frac{1}{2}at^2$
$= \frac{1}{2} \times 1.4 \times 100$
$= 70\ \text{m}$

The acceleration of the body is 1.4 m/s² and it travels 70 m.

Measurement of velocities and accelerations using a timer

The determination of velocities and accelerations involves the measurement of lengths and time intervals. Small time intervals can be measured in the laboratory using a ticker-tape timer, figure 1.8. This device makes use of the controlled frequency of the alternating mains current as do mains-energised electric clocks. The soft steel strip is magnetised. Its poles change every time the direction of the current changes and so the strip vibrates with the frequency of the mains, i.e. 50 cycles per second (50 Hz). As the paper tape is moved under the vibrator, a dot is made on the paper every $\frac{1}{50}$ second. The distance between two dots is thus the distance moved by the tape in $\frac{1}{50}$ second.

The appearance of the dots on the tape gives immediate information about the movement of the tape.

fig. 1.8 Ticker-tape timer ▼

fig. 1.9 Measuring *g* using ticker tape ▼

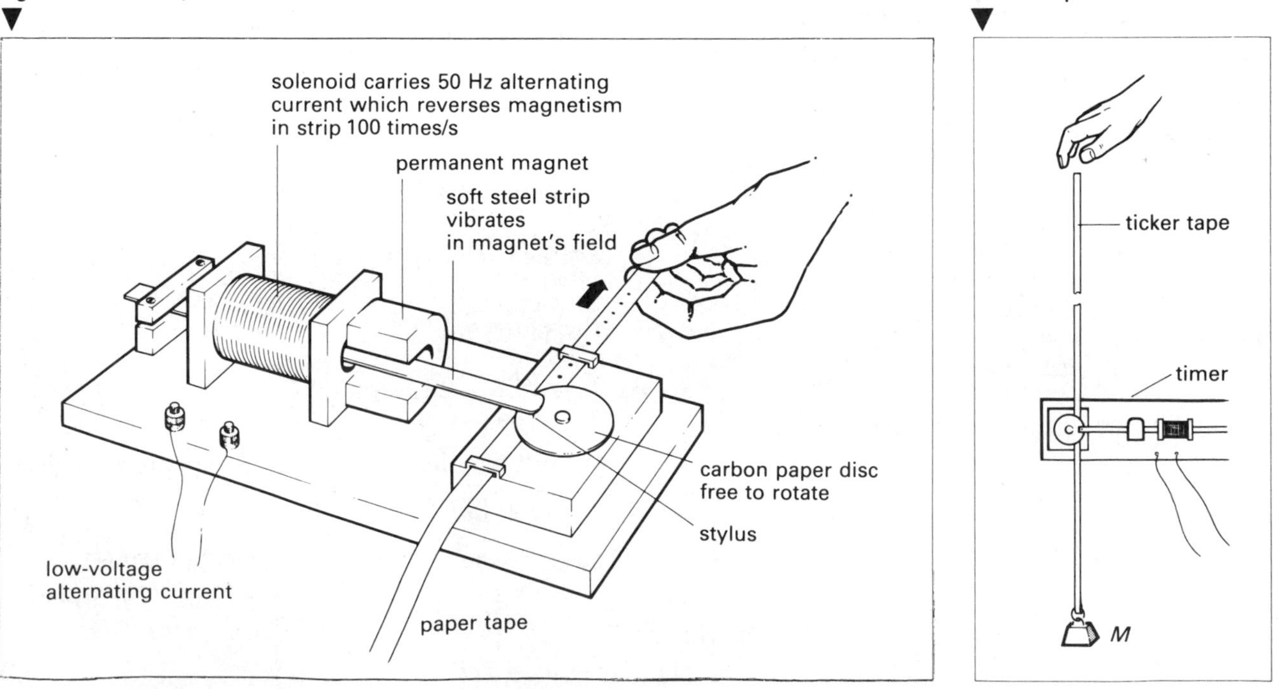

Evenly-spaced dots show that equal distances are travelled in equal times, so the tape is moving with constant velocity:

When the distance between the dots increases, the tape is accelerating:

When the distance between the dots decreases, the tape is decelerating:

Describe the motion of the tape below:

Distances travelled can be measured directly on the tape and velocities and changes in velocity found, or the tape may be cut into pieces of equal time interval and used to make a block graph. The graph produced, figure 1.10, is a velocity–time graph.

Each small strip represents the distance travelled in $5 \times \frac{1}{50} = \frac{1}{10}$ second.

What is the total distance travelled?
What is the total time taken?
What is the average velocity?
What is the velocity for strip A?
What is the velocity for strip B?
What is the change in velocity from A to B?
What is the time taken for this change?
What is the acceleration from A to B?

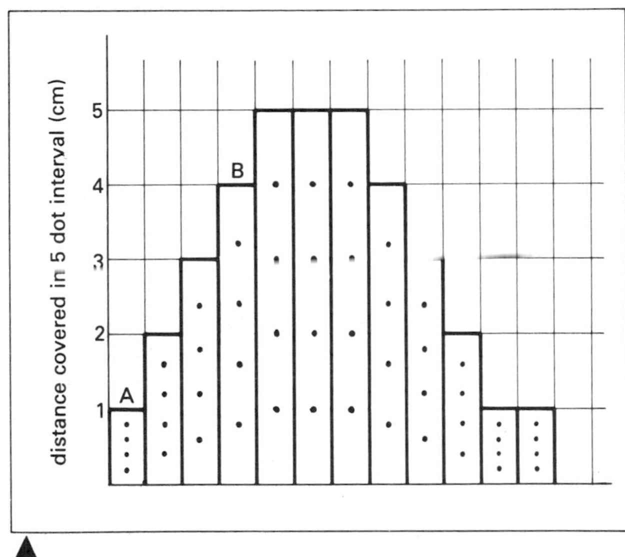

fig. 1.10

Measuring the acceleration due to gravity

Method 1

A mass M is fastened to a length of tape about a metre long. The tape is allowed to fall through a timer supported on its side, figure 1.9. If we ignore the air resistance, the upthrust of the air and the friction, we shall assume the acceleration of the mass is constant and equal to g.

Choose two places PQ and RS ten or twenty intervals apart. Measure the distances PQ and RS and find the velocities over PQ and RS, figure 1.11.

fig. 1.11 A typical ticker-tape record when measuring g

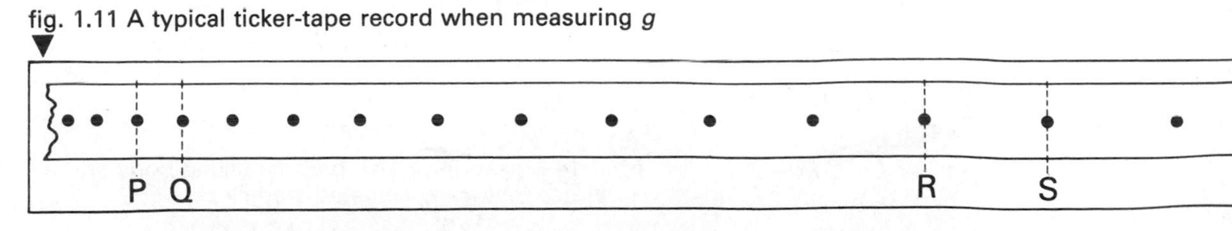

The acceleration between PQ and RS is the change in velocity divided by the time taken from the mid-point of PQ to the mid-point of RS.

Use different positions for PQ and RS along the tape and find an average value for g.

Method 2

The time taken for a freely-falling body to cover a certain distance can be accurately measured using either an electric stop clock or a digital timer, figures 1.12 and 1.13. Both these instruments will measure accurately to 0.01 second.

Figure 1.14 shows a steel ball bearing supported by an electromagnet. When the switch is opened, the ball drops and the clock starts. When the ball hits the hinged switch, it knocks it open, breaks the circuit and stops the clock. The time to fall through a distance h is thus known. The experiment is repeated several times and an average time found. If t is the time to fall a distance h, then

$$g = \frac{2h}{t^2}$$

fig. 1.12 Electric stop clock

fig. 1.13 Digital timer

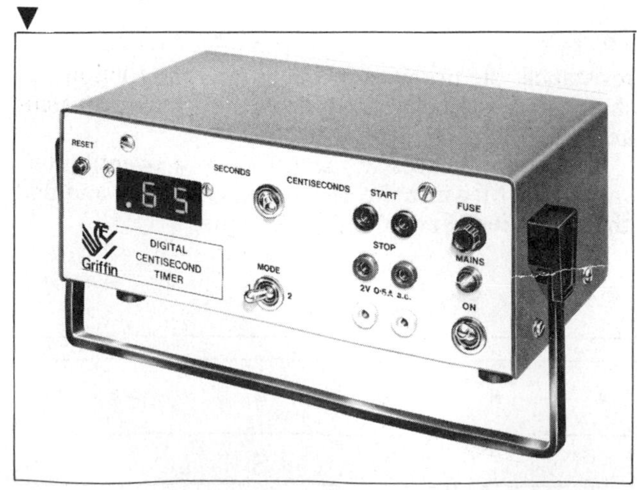

Measuring velocities and accelerations using stroboscopic photography

Another method of investigating the behaviour of a moving object is to take photographs of it at fixed time intervals. This can be done using a **stroboscope** which produces flashes of light at fixed times. The simplest form of stroboscope is a rotating disc with one or more slits or circular holes in it, figure 1.15. When the disc is rotated in front of a lamp, light is emitted each time a hole is opposite the lamp.

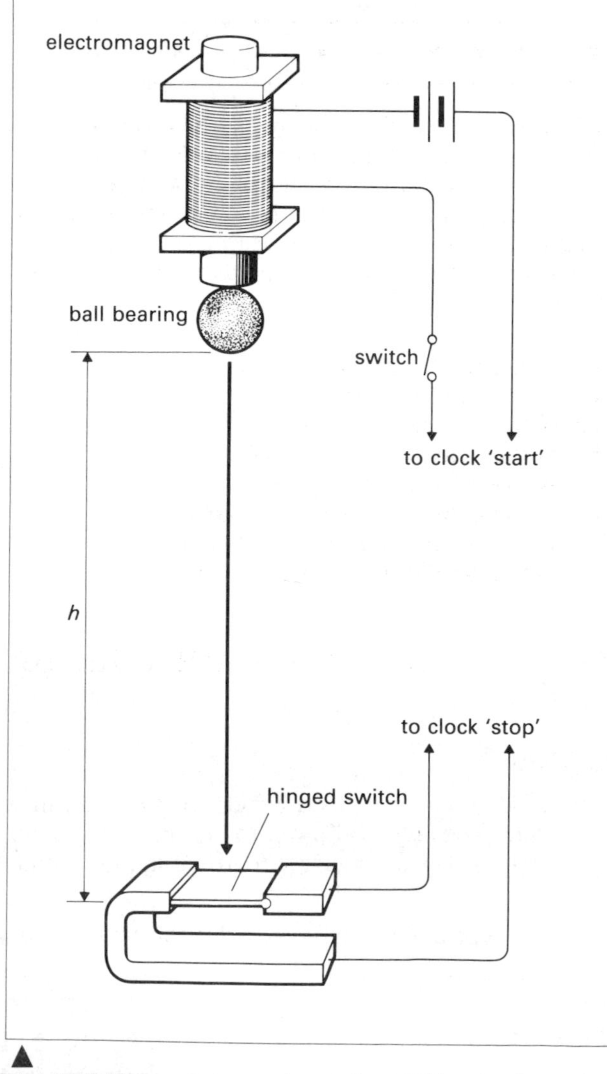

fig. 1.14 Measuring g using a free-falling body and an accurate electrically actuated stop-clock

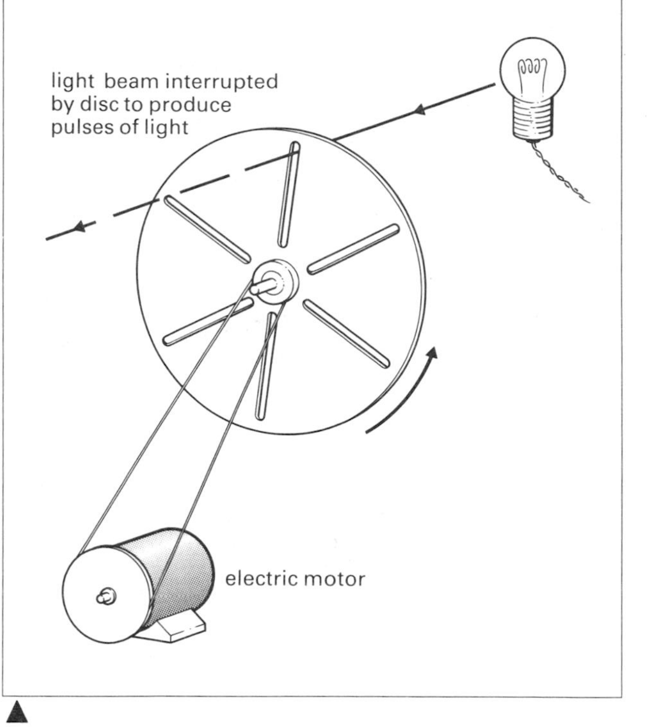

fig. 1.15 Stroboscope disc and motor

If a disc with four holes in it is rotating 5 times per second, there will be $4 \times 5 = 20$ flashes of light per second or the frequency of the stroboscope is 20 Hz. The time interval between the flashes is $\frac{1}{20}$ second.

In a more complicated stroboscope, figure 1.16, an electronic circuit produces an electric discharge in a tube filled with xenon gas. The discharge produces a flash of light of high intensity which lasts for about a hundred thousandth of a second. The frequency of the flashes can be varied from 1 to 250 Hz.

When a photograph is taken using stroboscopic illumination the motion is 'frozen' at fixed time intervals so the movement can be studied, figure 1.17. Since the time intervals are constant, the distance moved between photographs is a measure of the speed. The greater the distance between photographs, the greater the speed, so we can tell whether the speed is increasing or decreasing.

Another method of obtaining stroboscopic photographs is to rotate the stroboscopic disc in front of the camera lens with the camera shutter kept open. Such an experiment is shown in figure 1.18. Each time a slit in the disc comes opposite the lens, a photograph of the rolling ball is taken and the positions of the ball at fixed time intervals are recorded.

Since the distances travelled by the ball in fixed time intervals are obviously increasing, the ball is accelerating. The average velocities over the different time intervals can be worked out. The frequency of the disc is 10 Hz so each time interval is 0.1 second. Table 1.1 gives some typical results.

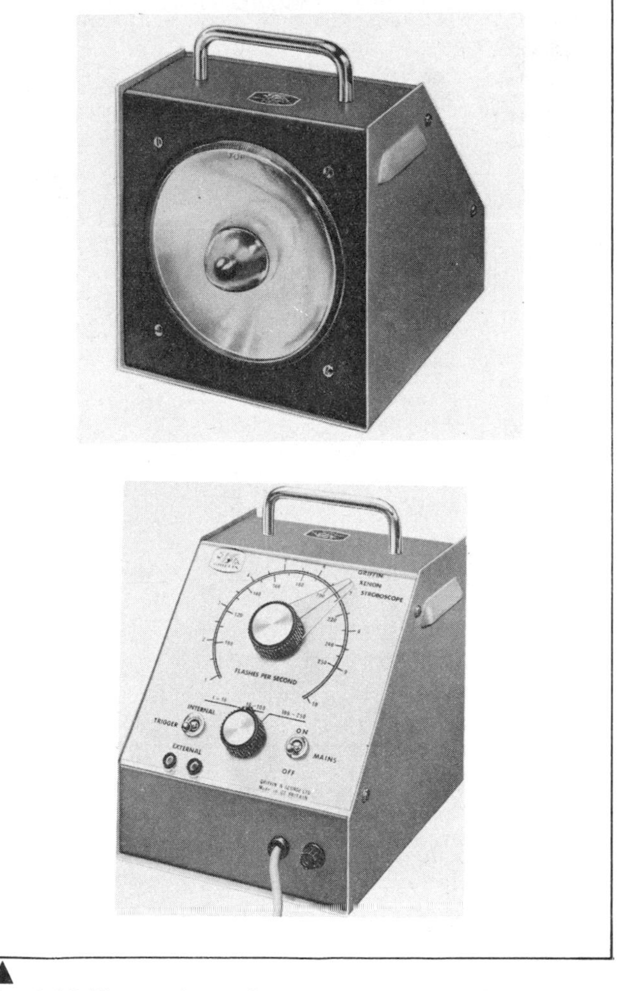

fig. 1.16 Electronic stroboscope

fig. 1.17 The flight of this bird was photographed every 0.1 s

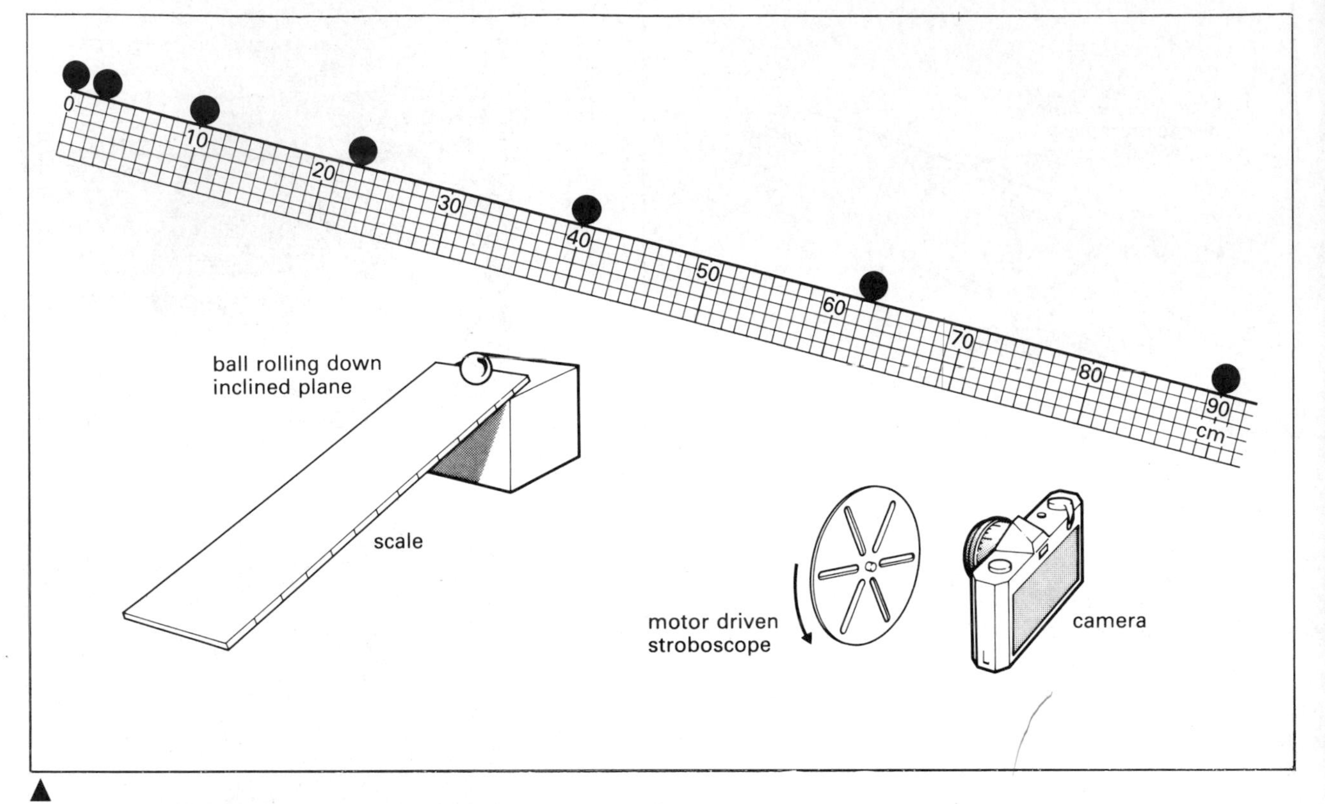

fig. 1.18 Stroboscopic photography for a ball on an inclined plane

fig. 1.19 Velocity–time graph for a ball rolling down an inclined plane

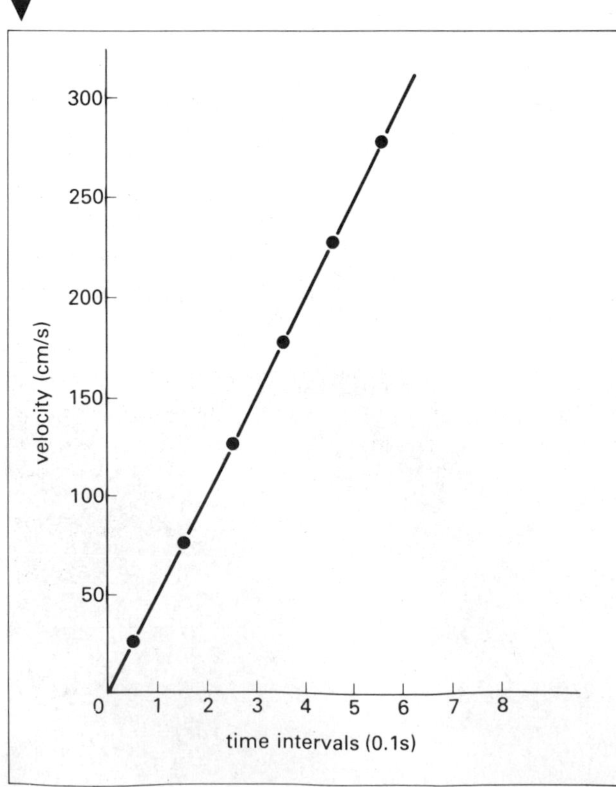

Time interval	Distance travelled	Average velocity $= \frac{\text{distance}}{0.1}$
1	2.5 cm	25 cm/s
2	7.5 cm	75 cm/s
3	12.5 cm	125 cm/s
4	17.5 cm	175 cm/s
5	22.5 cm	225 cm/s
6	27.5 cm	275 cm/s

table 1.1

A velocity–time graph can now be drawn, figure 1.19. Since we are dealing with average velocities, the values are plotted at the mid-points of the time intervals. The velocity–time graph is a straight line showing that the acceleration of the ball is uniform. The value of the acceleration can be obtained by calculating the slope of the graph.

$$\text{acceleration} = \text{slope of graph}$$

$$= \frac{\text{increase in velocity}}{\text{increase in time}}$$

$$= \frac{275}{0.55}$$

$$= 500 \text{ cm/s}^2 = 5 \text{ m/s}^2$$

If it is known in any particular case that we are dealing with a body moving with uniform acceleration, then the equations of motion can be used to solve problems about the motion.

In the case of the ball rolling down the plane:

$$a = \frac{v - u}{t}$$

Substituting

$u = 25$ cm/s
$v = 275$ cm/s
$t = 0.5$ s

we obtain

$$a = \frac{275 - 25}{0.5}$$

$$= 500 \text{ cm/s}^2 = 5 \text{ m/s}^2$$

EXERCISE 1

In questions 1–4 select the most suitable answer.

1 A vehicle travels a distance of 160 km in 5 hours. The average speed is
A 32 km/h (km h^{-1})
B 40 km/h
C 80 km/h
D 165 km/h
E 800 km/h

2 Which of the following is a scalar quantity?
A velocity
B retardation
C speed
D displacement
E acceleration

3 A ball is dropped from rest. What is its speed after 3 seconds?
A 15 m/s
B 25 m/s
C 30 m/s
D 45 m/s
E 90 m/s

4 A stone is dropped from rest from the top of a tall building. The fraction

$$\frac{\text{(distance fallen in the first 4 seconds)}}{\text{(distance fallen in the first 2 seconds)}}$$

is approximately
A 1/4
B 1/2
C 2/1
D 4/1
E 16/1

5 The graph in figure 1.20 shows the distance–time relationship for an object which moves in a straight line past a fixed point.
a Find the distance travelled after 7 seconds.
b Find the time to travel 15 metres.
c Find the average speed in metres/second.

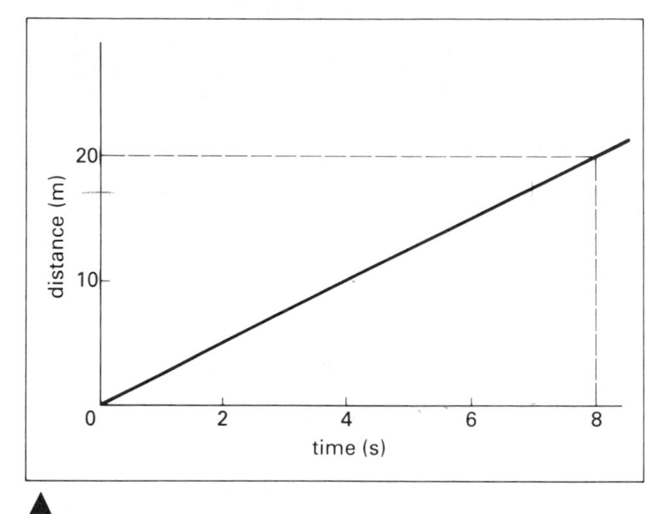

fig. 1.20

6 Figure 1.21 represents the motion of somebody walking along a straight road.
a Between which two points on the graph is the walker travelling with uniform speed away from his starting point? Calculate the value of the uniform speed at this stage.
b Between which two points is the walker travelling with uniform speed towards his starting point? Calculate the speed at this stage.
c Between which points is the walker stationary? For how long a time over the whole motion shown is the walker stationary?
d Between which points is the walker travelling with non-uniform speed?
e How far does the walker go in the first 50 seconds?
f How far does he go in the second 50 seconds?
g If he had continued at this rate, how long would he have taken to get 1000 metres from his starting point?

fig. 1.21

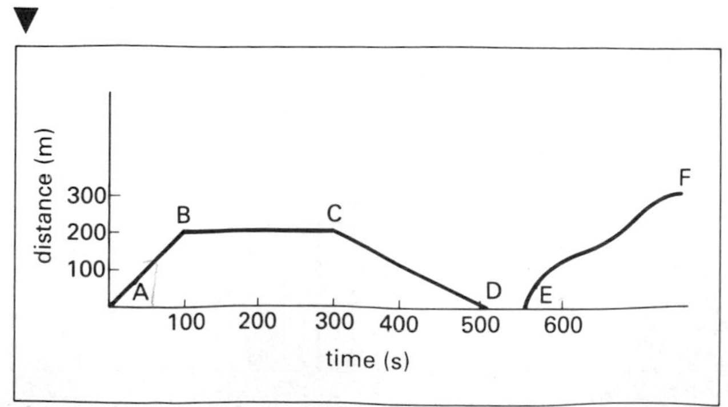

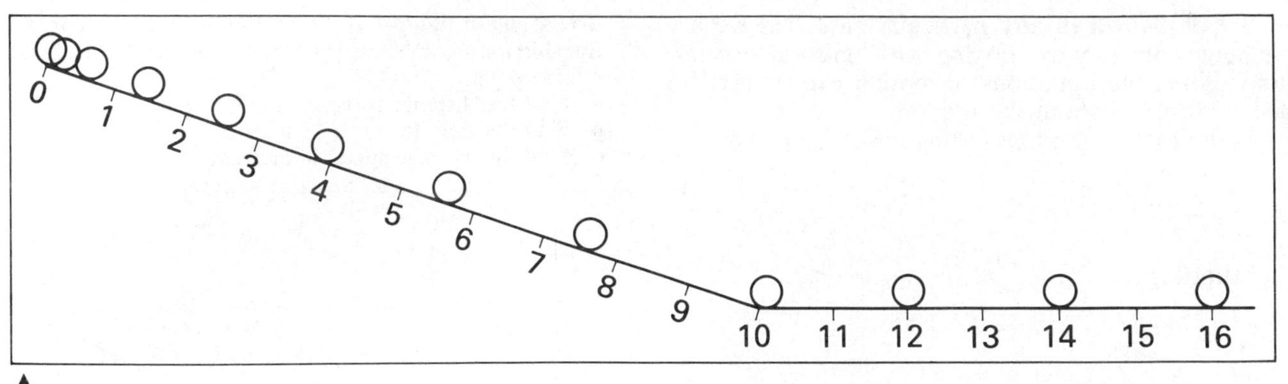

fig. 1.22

7 An object falls down a slope on to a friction-free horizontal surface in the way shown in figure 1.22. The position of the object is shown every $\frac{1}{8}$ second and the distances are marked in centimetres from the start.

a The object is accelerating whilst on the slope. Explain how you can see from the diagram that this is so.
b What happens to the speed of the object once it has reached the horizontal surface?
c Taking readings from the diagram, calculate the speed of the object on the horizontal surface.
d Calculate the approximate value of the maximum speed reached during the whole journey.
e Sketch a graph to show how the velocity of the object varies with time for the whole 16 seconds.

8 An object was moved from rest and its speed recorded every second. The following results were obtained:

Time (s)	1.0	2.0	3.0	4.0	5.0	6.0	7.0	8.0	9.0
Speed (m/s)	0.28	0.53	0.78	1.06	1.33	1.35	1.34	1.31	1.33

a Plot a graph of the speed (y–axis) against time (x–axis).
b Use the graph to find (i) the time at which the acceleration changed, (ii) the average speed 7.5 s from the start of the motion.
c State the types of motion before and after the change in acceleration.
d Use the graph to calculate the average acceleration 3.5 s from the start of the motion.
e Outline any one experiment which could be used to find data similar to those shown in the table.

9 Graphs A, B and C in figure 1.23 show successive ten-tick lengths of ticker tapes stuck side by side for the motion of a trolley along a runway on three different occasions.

a Describe briefly the kind of motion which occurred on each occasion.
b (i) Which ten-tick length of tape reached the greatest velocity? (ii) What was that velocity in centimetres per ten ticks? (iii) What was that velocity in centimetres per second if the vibrator tapped out 50 ticks per second?
c Why is it likely that the fastest velocity reached was more than that calculated in **b** (iii)?
d How far did the trolley move in 100 ticks on each occasion?
e (i) What is the total time taken for the movement of the trolley in graph C? (ii) What is the average velocity in centimetres per second for this trolley?
f Did the trolley in graph C ever have its actual velocity equal to its average velocity? If so, in which tape or tapes?

10 A brick drops from a platform on a building site and reaches the ground in 2 seconds (g, the acceleration due to gravity, is 10 m/s²).

a At what speed is the brick travelling after 1 second?
b At what speed is the brick travelling just before it hits the ground?
c How high is the platform above the ground?

fig. 1.23

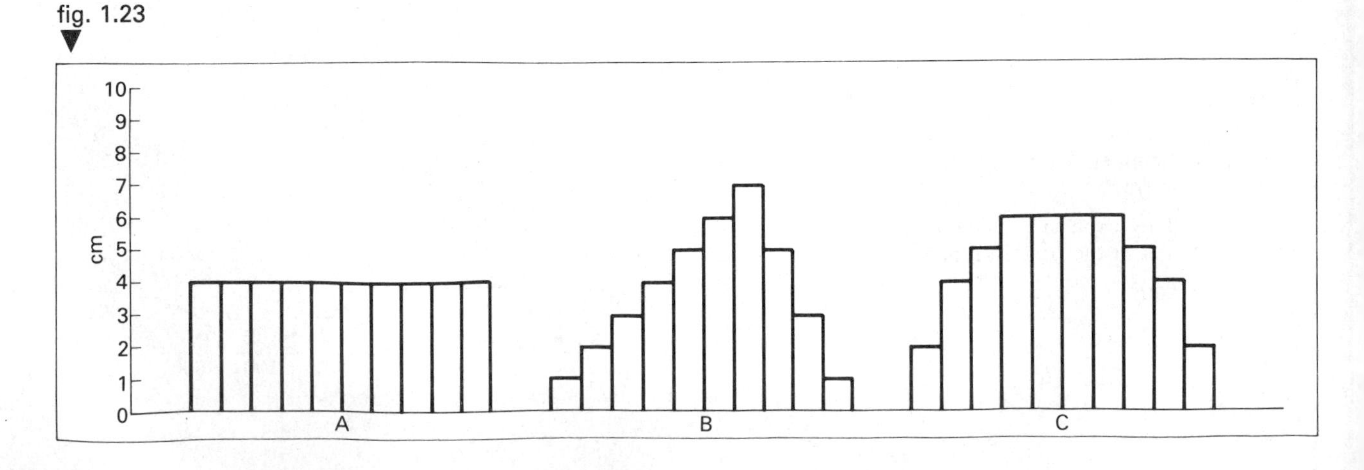

2 FORCES

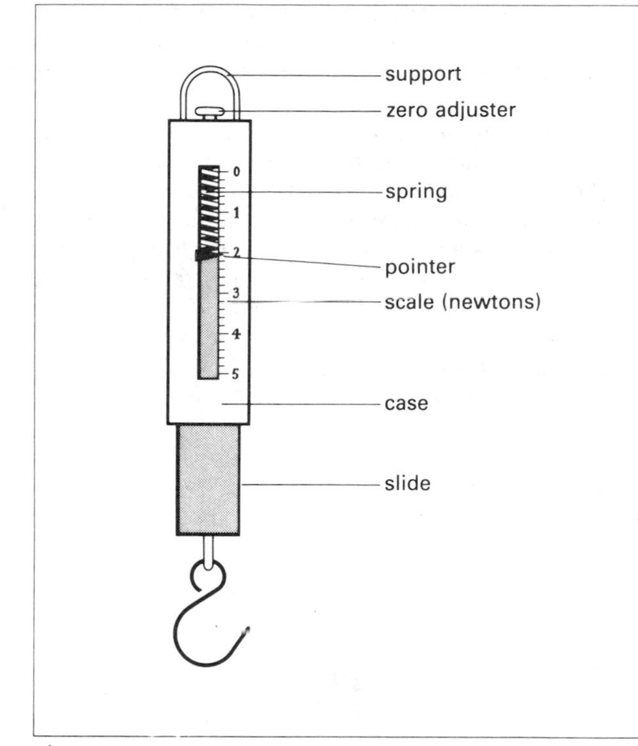

▲
fig. 2.1 Spring balance

fig. 2.2 Force on a free-falling body
▼

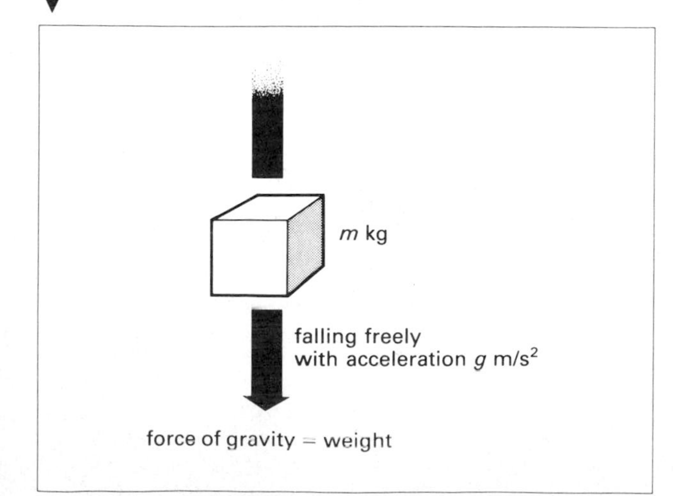

In everyday language, forces push or pull or twist. In mechanics, we define force more precisely.

Force is that which changes or tends to change the state of rest or uniform motion of a body

This is really another way of stating Newton's **first law of motion**, which is

A body remains in a state of rest or of uniform motion in a straight line unless it is acted on by an external force

Force is given precisely by the equation

force = mass × acceleration

$$F = m \times a$$

The **mass** of a body is the amount of material it contains.

Force is a vector quantity

The unit of force, the **newton** (N), is the force which produces an acceleration of one metre per second per second in a mass of one kilogram. For example

1 N produces an acceleration 1 m/s^2 in mass 1 kg
1 N produces an acceleration 2 m/s^2 in mass $\frac{1}{2}$ kg
10 N produce an acceleration 5 m/s^2 in mass 2 kg

This relationship between force, mass and acceleration will be discussed in more detail in Chapter 3 which deals with Newton's laws and their applications.

The pull of the earth or the force of gravity on a body is called the **weight** of a body. Weight is measured on a **spring balance** which is really a force measurer, figure 2.1. To measure forces the balance is marked in newtons.

The mass of an object is constant, but the weight varies with its position relative to the earth. The further away from the earth, the less the gravitational attraction and the less the weight.

When a body falls freely under the force of gravity, it has an acceleration of g m/s^2, figure 2.2.

If a mass of m kilograms falls freely, the force acting on it is mg (mass × acceleration).

Since the force acting is the gravitational attraction or weight, we have

weight = ***mg***

For a mass of 1 kilogram and taking $g = 10\ \text{m/s}^2$, we have

1 kilogram weight = 10 newtons

One newton is just about equal to the weight of 100 grams. More accurately, of course, 1 kilogram weight = 9.8 newtons.

EXAMPLES

force	=	**mass**	×	**acceleration**
(newtons)		(kilograms)		(metres/sec²)

1 What force is necessary to cause a train of mass 500 000 kg to accelerate at $0.04\ \text{m/s}^2$?

$$\begin{aligned}\text{force} &= \text{mass} \times \text{acceleration}\\ &= 500\,000 \times 0.04\\ &= 20\,000\ \text{N}\end{aligned}$$

2 An electric railway locomotive of mass 50 000 kg starts from rest and after 20 seconds has accelerated to a speed of 25 m/s. Calculate
a the acceleration of the locomotive,
b the horizontal driving force,
c the distance travelled in 20 seconds.

a Using
$$\begin{aligned}v &= u + at\\ 25 &= 0 + 20a\\ a &= \frac{25}{20}\\ &= 1.25\ \text{m/s}^2\end{aligned}$$

b
$$\begin{aligned}F &= m \times a\\ &= 50\,000 \times 1.25\\ &= 62\,500\ \text{N}\end{aligned}$$

c Using
$$\begin{aligned}d &= ut + \tfrac{1}{2}at^2\\ &= 0 + \tfrac{1}{2} \times 1.25 \times 20^2\\ &= 250\ \text{m}\end{aligned}$$

3 A railway engine of mass 20 000 kg pulls a truck of mass 5000 kg and both accelerate at $2\ \text{m/s}^2$. Find the force provided by the engine and the tension in the coupling to the truck. Friction and air resistance can be neglected.

total mass moved is 20 000 + 5000 = 25 000 kg

$$\begin{aligned}\text{force exerted by the engine} &= 25\,000 \times 2\\ &= 50\,000\ \text{N}\end{aligned}$$

The force acting on the truck and making it accelerate is the pull or tension in the coupling.

tension in the coupling = 5000 × 2 = 10 000 N

Note: 40 000 N out of the 50 000 N exerted by the engine are being used to accelerate the engine itself.

Balanced forces

When a body is stationary or is not accelerating, we can conclude either that there is no force acting on it or that several forces are acting but they just balance each other so that there is no resultant force to produce acceleration. A body is said to be in **equilibrium** under the action of several forces when it has zero acceleration.

A book lying on the table does not move although two forces are acting on it, figure 2.3a. These are the **weight** of the book and the push of the table on the book, called the **reaction** of the table. These two forces just balance. If the weight were too great, the table would collapse; if the reaction of the table were too great, the book would jump up!

Even if the table is tilted and a new force, the force of **friction** is brought into play, the book may still not move because the forces may still be in equilibrium, figure 2.3b.

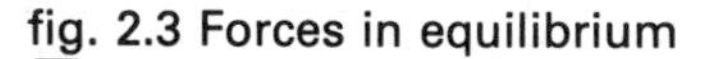

fig. 2.3 Forces in equilibrium

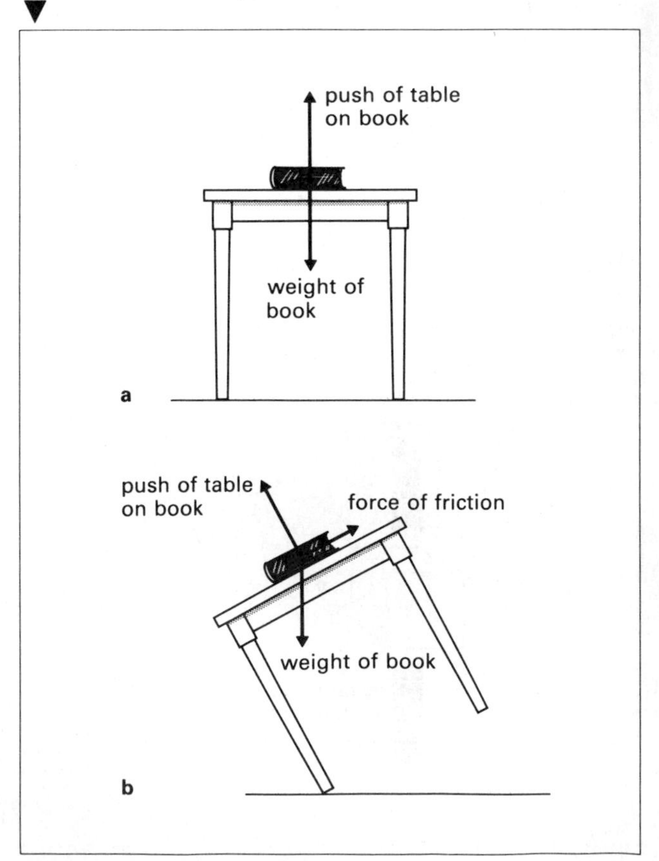

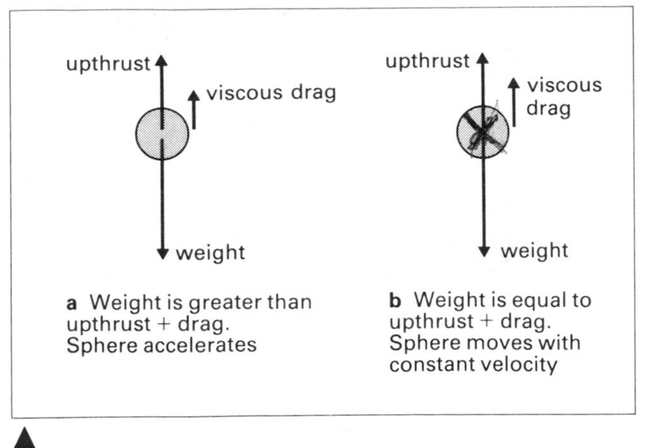

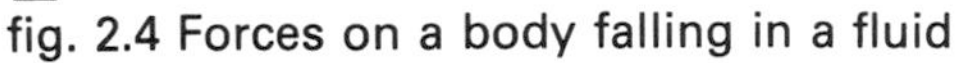
fig. 2.4 Forces on a body falling in a fluid

The falling of a small body through a liquid or gas is an interesting case of forces in equilibrium, figures 2.4a and b. Consider a ball bearing falling through glycerine. The forces acting on the bearing are (1) its weight, (2) the upthrust or buoyancy of the glycerine and (3) the frictional resistance or 'viscous drag' of the glycerine as the bearing moves through it. At first the weight is greater than the upthrust and the drag, so the bearing accelerates. But as it moves more quickly the viscous drag increases until the weight is just equal to the upthrust and drag. The forces are now balanced. There is no acceleration and the bearing falls at a constant rate. This constant velocity is called the **terminal velocity.** Raindrops reach a terminal velocity; so does a parachutist.

Combining forces

Since a force is a vector quantity it can be represented by a line with an arrow on it. The length of the line represents the magnitude of the force and the arrow shows in which direction it is acting.

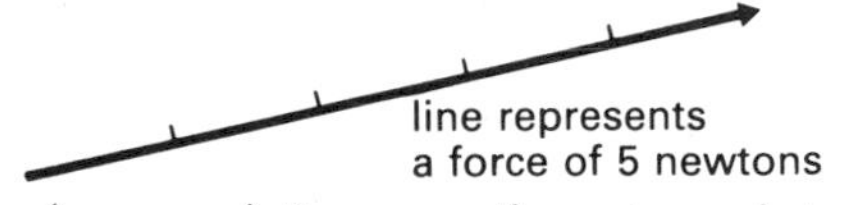

Two (or more) forces acting at a point can be combined into a single force which produces the same effect as the two forces. This force is known as the **resultant**. If the forces are acting in the same straight line, they can be added or subtracted depending on whether they are in the same or opposite sense.

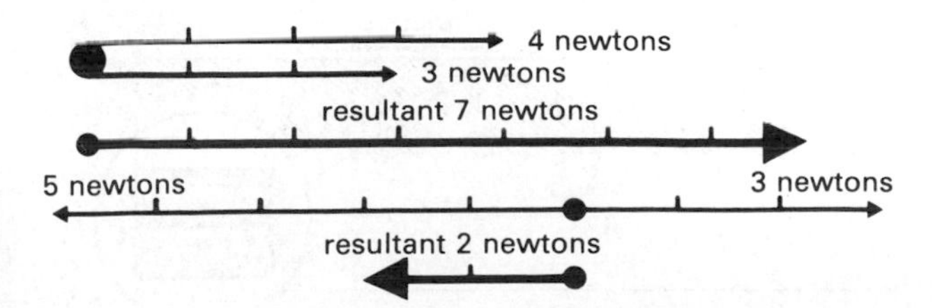

If two forces acting at a point make an angle with each other, they can be combined using the **parallelogram of forces**.

If two forces acting at a point are represented in magnitude and direction by the two adjacent sides of a parallelogram, their resultant is represented in magnitude and direction by the diagonal of the parallelogram from the point

In figure 2.5 the two forces of 3 N and 5 N acting in directions OA and OB are equivalent to the resultant force of 7 N acting in direction OR.

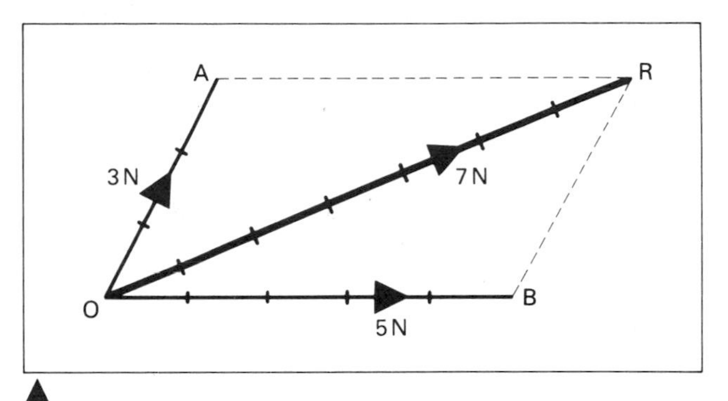

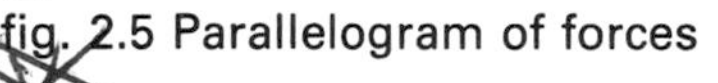
fig. 2.5 Parallelogram of forces

In figure 2.6 the two forces OA and OB can be combined into a single resultant OR. If a force equal and opposite to OR were added to the two original forces, it would produce a net force of zero. The forces OA, OB and OE would be in equilibrium and the acceleration of the point O would be zero. A force equal and opposite to the resultant is called the **equilibrant**.

fig. 2.6 Resultant and equilibrant of two forces

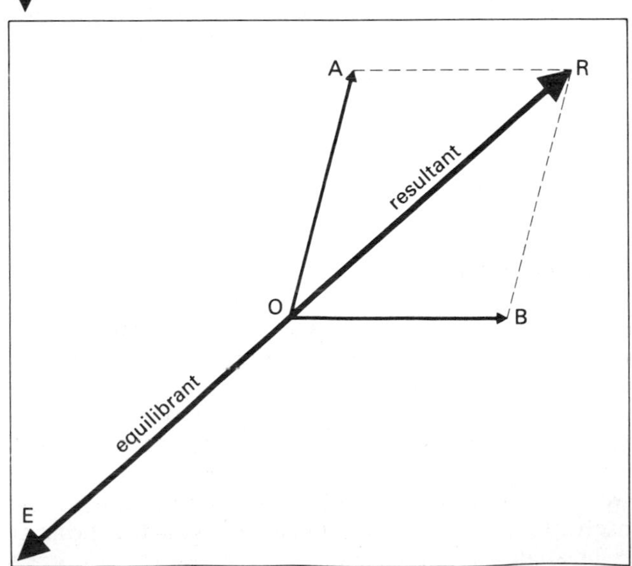

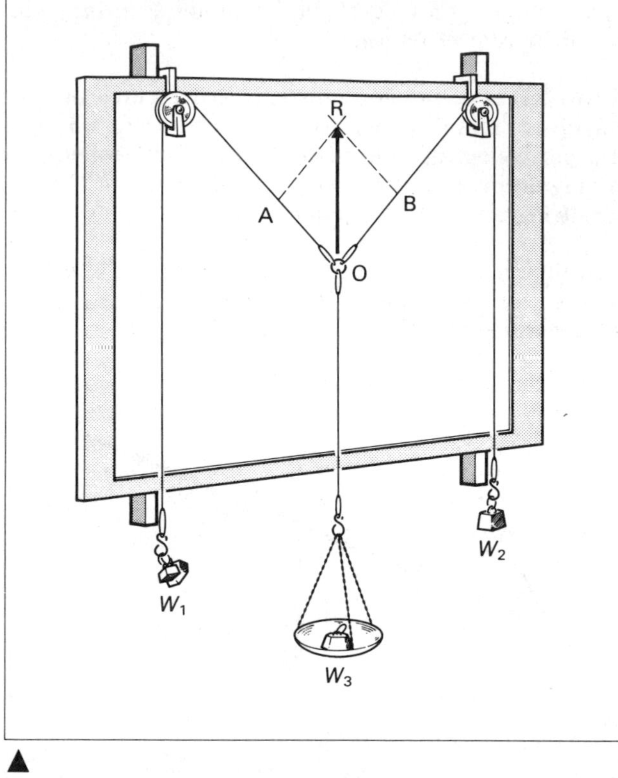

fig. 2.7 Verifying the parallelogram of forces law

The parallelogram of forces can be investigated using the apparatus shown in figure 2.7.

When the strings have settled into an equilibrium position, mark their lines on the paper. Displace the weights and let them settle down again. They should return to the original position, but sometimes friction at the pulleys throws them out a bit. It may be necessary to take an average of the lines.

Along OA mark a length to represent the weight W_1. Along OB mark a length to represent W_2. Complete the parallelogram OARB. OR should be in the same line as W_3 and the length OR should represent the magnitude of W_3.

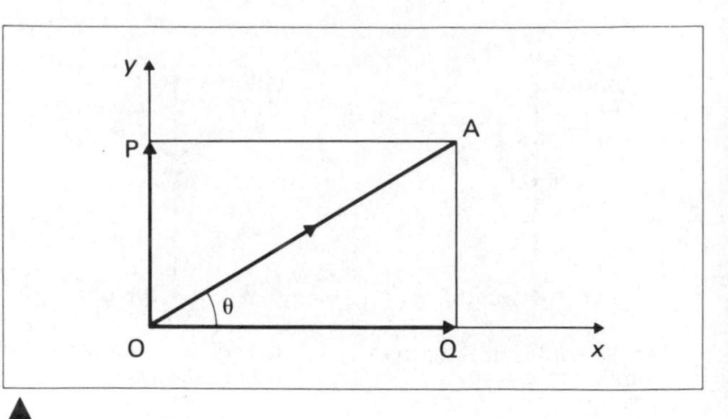

fig. 2.8 Resolving a force into two separate forces

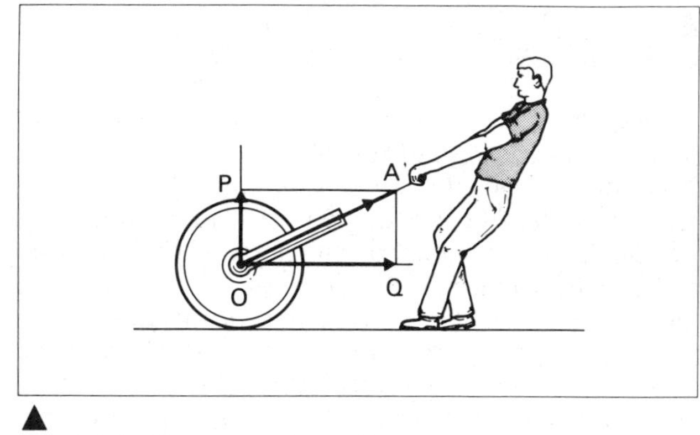

fig. 2.9 Pulling a garden roller

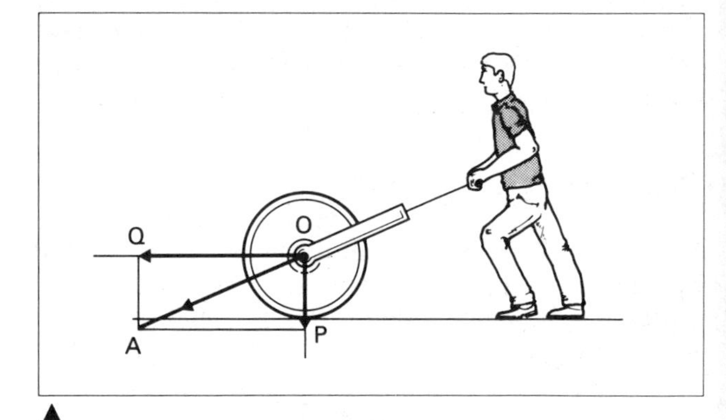

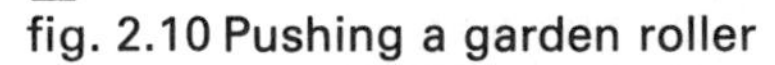
fig. 2.10 Pushing a garden roller

fig. 2.11 Force of friction and its origin

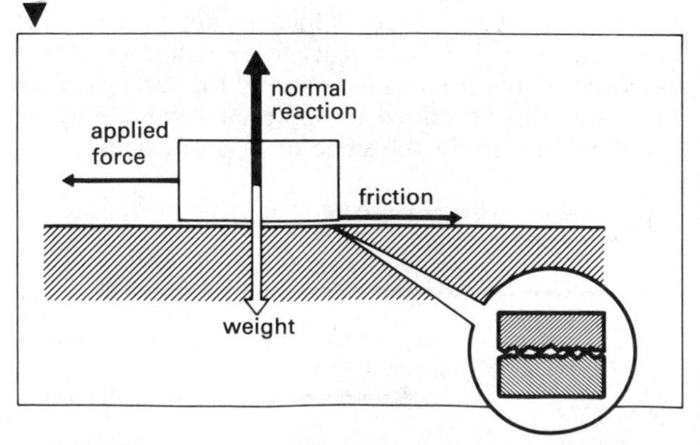

Resolving forces

The parallelogram law enables us to combine two (or more) forces into a single resultant force. The idea can also be used in reverse to convert a single force into two components. The process is known as the **resolution** or **resolving** of a force into two components. There is an infinite number of directions in which a force can be resolved but it is generally most useful to resolve a force in two chosen directions at right angles to each other.

The resolved part of the force tells us the effect of the force in that direction. In figure 2.8 the force OA has components

OP in direction Oy

OQ in direction Ox

The magnitudes of the components can be found by a scale diagram or by calculation:

$$OQ = OA \cos \theta$$

$$OP = OA \sin \theta$$

When a garden roller is being pulled along by a force represented by OA, the horizontal and vertical components of the force are represented by OQ and OP, figure 2.9. The horizontal component OQ is the part of the force which is effective in moving the roller forward. The vertical component OP acts against the weight of the roller and reduces the force which the roller exerts on the ground.

If the roller is being pushed with the same force at the same angle, the vertical component pushes the roller into the ground, figure 2.10. It is easier to pull a roller than to push it, but it does not roll the ground so effectively.

The force of friction

Friction is the resistance which must be overcome when one surface moves over another. Friction always acts in the direction opposite to the movement and so always opposes you when you are trying to do mechanical work.

If a force is applied to a body resting on a surface, and the magnitude of the force gradually increased from zero, there comes a point at which the body just begins to move. Up to this point the frictional force on the body has just balanced the applied force and the body has been in equilibrium, figure 2.11. When the applied force is greater than the frictional force the body will accelerate.

Although surfaces may appear to be smooth, they are, in fact, very rough and when 'in contact' only really touch at a very few places. When an attempt is made to move two surfaces relative to each other the hills interlock with the valleys, figure 2.11. If the pressure between the surfaces is great, the points of contact may actually fuse together. These interlocking and fusion forces are the cause of the friction between the two surfaces.

The friction between two surfaces depends on how tightly they are pressed together and this can be investigated experimentally as follows.

A block of wood is placed on the table and the force just necessary to make it move is found. The

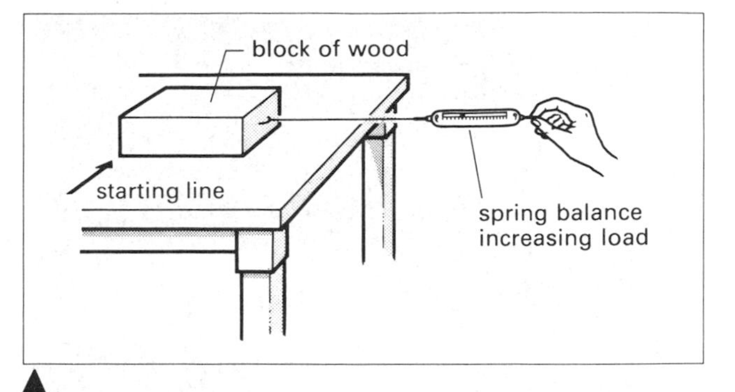

fig. 2.12 Measuring static friction

force can be measured using a spring balance, figure 2.12 or by gently putting weights in a scale pan, figure 2.13. It is important to start at the same point each time and to take the average of several attempts. A weight is placed on the block and the test repeated. The weight is increased and so on until about eight readings have been obtained.

normal reaction R = total weight acting downwards
= weight of block + weights on top

frictional force F = force which causes block to start moving

A graph of F against R is a straight line showing that the friction force is directly proportional to the normal reaction, or

$$F = \mu R$$

where μ is the coefficient of friction.

The force necessary to start the block moving is greater than the force needed to keep the block moving at constant speed, so it is useful to distinguish between **static** and **sliding** friction.

Friction is a great nuisance in machines, causing

fig. 2.13 Another method of measuring static friction

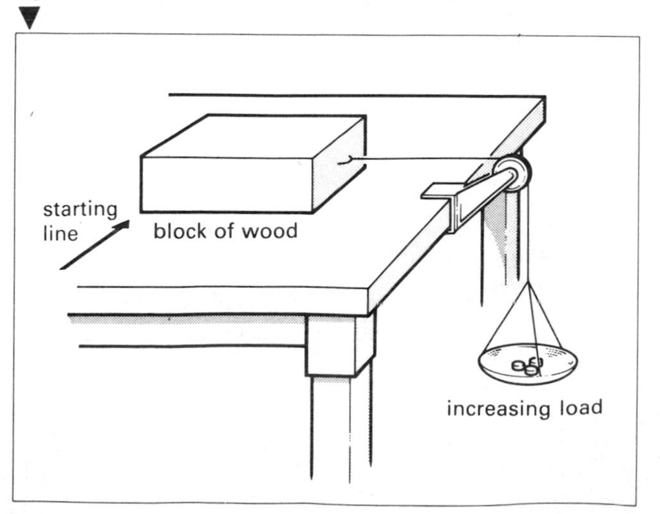

fig. 2.14 A water chute is smooth and lubricated with water

them to move stiffly and so waste some of the energy put into them in moving parts of the machine itself. It also causes wear and makes the machine hot. Kinetic energy is converted into heat energy. Sliding surfaces must be very smooth and well lubricated. A lubricant is a substance such as water, oil, grease or graphite which fills up the irregularities of the surfaces, figures 2.14 and 2.15. When things roll there is much less friction than when they slide.

Friction is also useful. If there were no friction between our feet and the ground, we should not be able to walk. If the wheels of a car could not get a grip on the road, the car could not move forward. Without friction nails would not hold in wood or corks in bottles. Friction between threads makes it possible to weave cloth. We keep some of our clothes in position by friction!

fig. 2.15 The piston slide on an engine is lubricated with oil

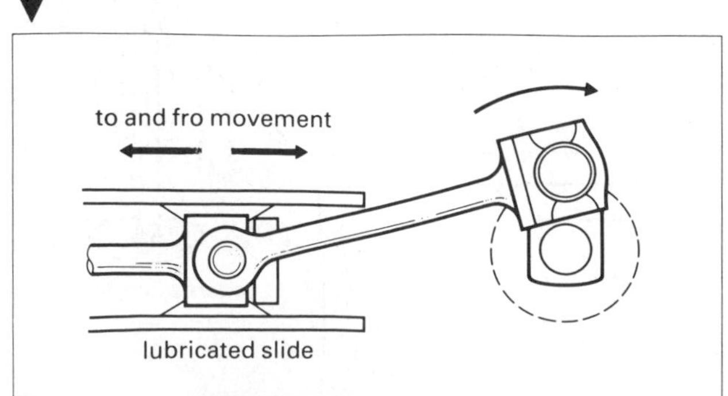

Elasticity and Hooke's law

The use of a spring balance to measure forces depends on two properties of springs. A spring has the property of **elasticity**. When the force stretching the spring is removed it returns to its original size and shape. A piece of rubber stretches when pulled and contracts again when the pull is removed. The rubber is behaving elastically.

Some substances do not return to their original size and shape when the deforming force is removed. Such substances, e.g. putty and plasticine, are said to be **plastic.**

The second property of the spring which is useful in the spring balance is that it obeys **Hooke's law** which states that the **change in length is proportional to the force applied**, provided the force is not too great.

If a large force is applied to an elastic body, the body may stretch disproportionately. The force at which this takes place is the **elastic limit**. Even greater forces may produce permanent deformation of the body so that it does not return to its original size.

The validity of Hooke's law can be tested with the apparatus in figure 2.16. Experiment with an elastic band in place of the spring and see how it behaves.

From the graph, figure 2.17, the force to produce any given extension can be found.

Expressed mathematically

extension = k × load

where k is a constant (in **m/N**) for any given spring.

fig. 2.16 Verifying Hooke's law

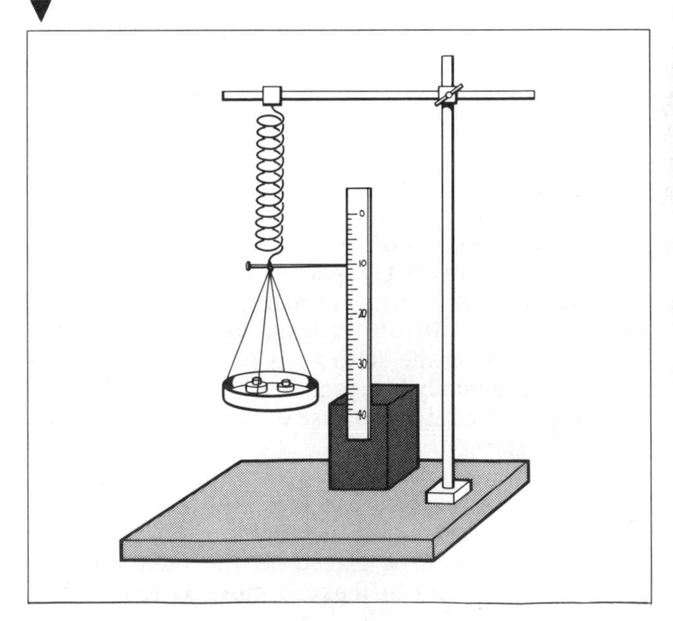

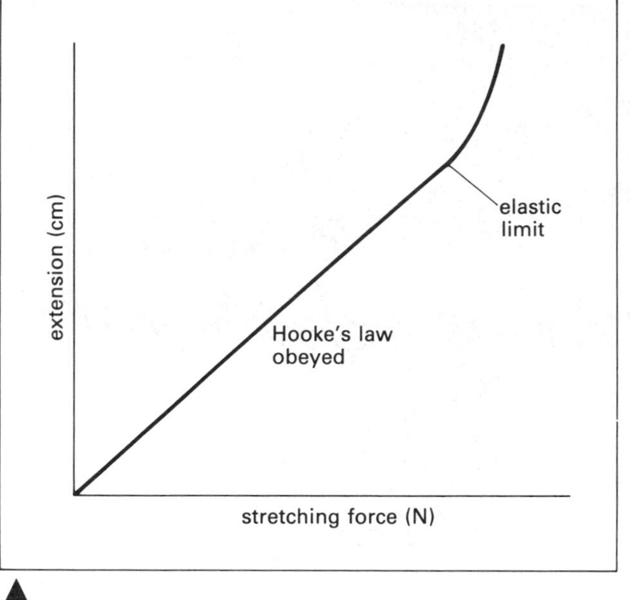

▲
fig 2.17 Typical graph for a stretched spring

The experiment should be repeated with other materials such as rubber, nylon, a strip of polythene and a steel or copper wire. In the case of the wires a specially accurate device (a vernier) will have to be used to measure the extension. Two results are shown in figure 2.18. What do the graphs tell about the behaviour on stretching of the two materials?

fig. 2.18 Typical graphs for steel and rubber
▼

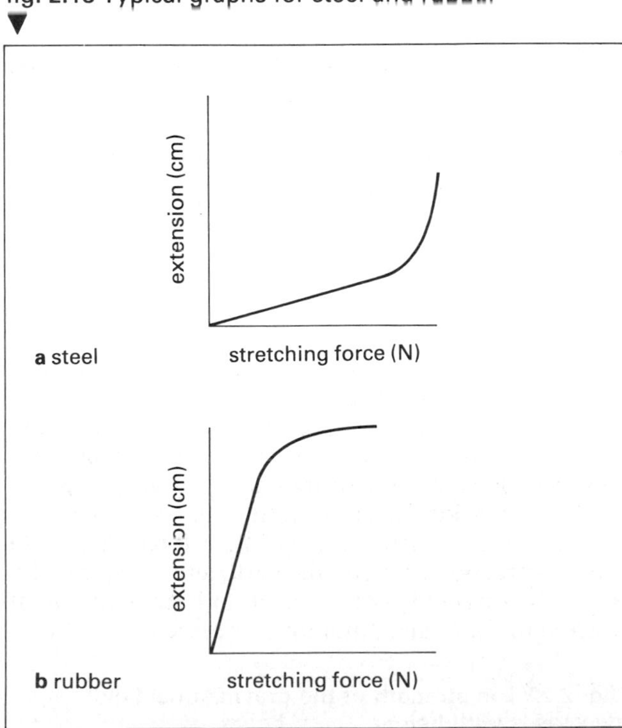

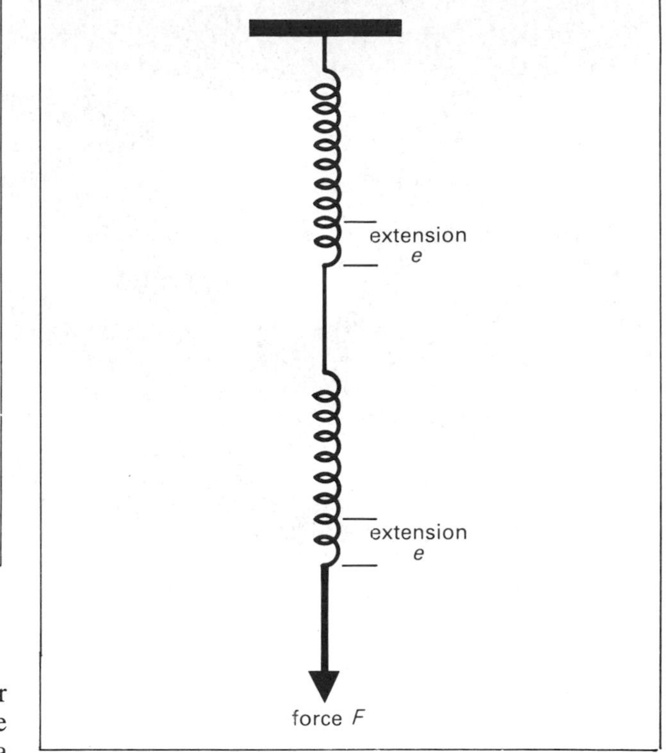

▲
fig. 2.19 Extension of two springs in series

Suppose a force F applied to a given spring produces an extension e. For this spring the constant $k = e/F$. If two such springs are connected end to end (in series) and a force F applied, the tension in each spring is F, the extension in each spring is e and the total extension is $2e$, figure 2.19.

If the two springs are connected side by side (in parallel) and the force F applied, the force applied to each spring is $\frac{1}{2}F$ and the extension of each spring is $\frac{1}{2}e$. That is, the extension is halved, figure 2.20.

fig. 2.20 Extension of two springs in parallel
▼

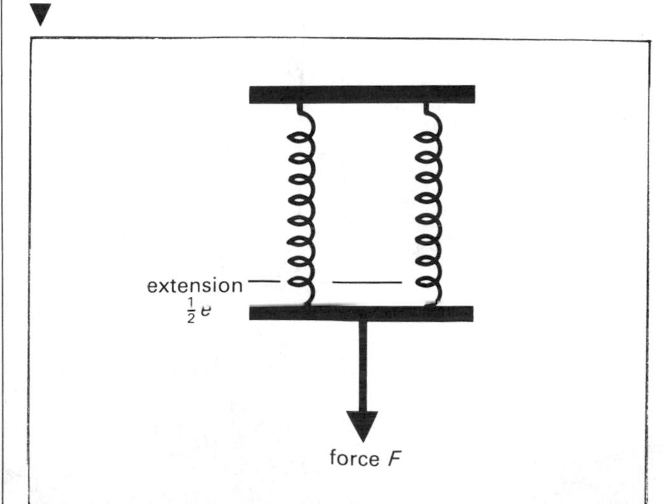

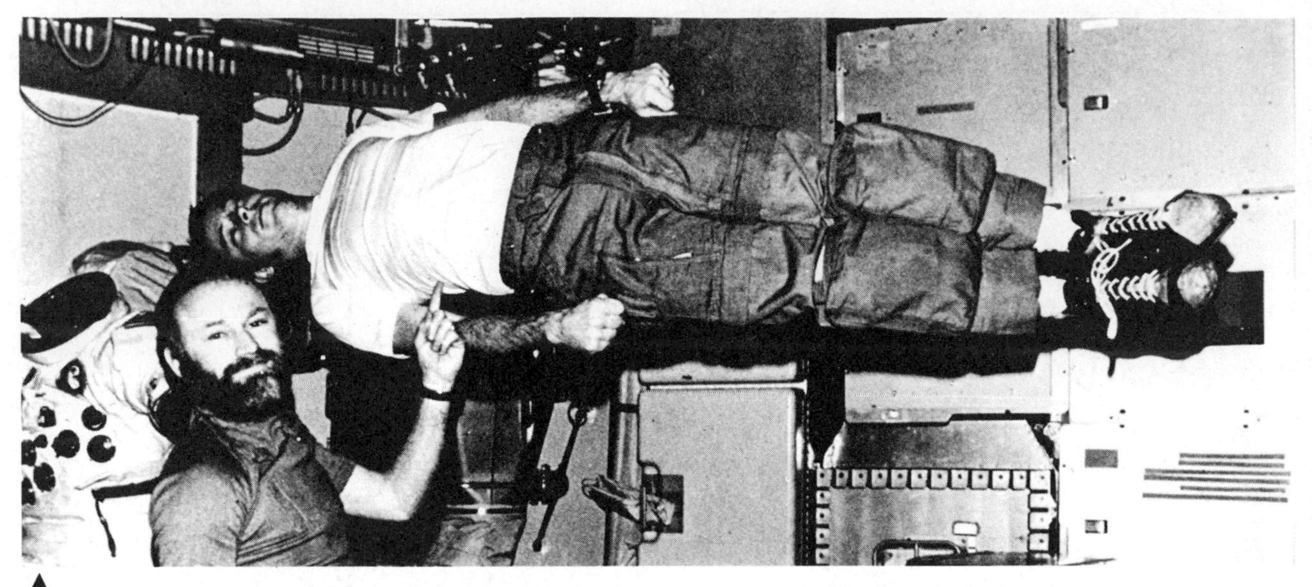

▲ fig. 2.21 Skylab 4 astronauts demonstrating weightlessness

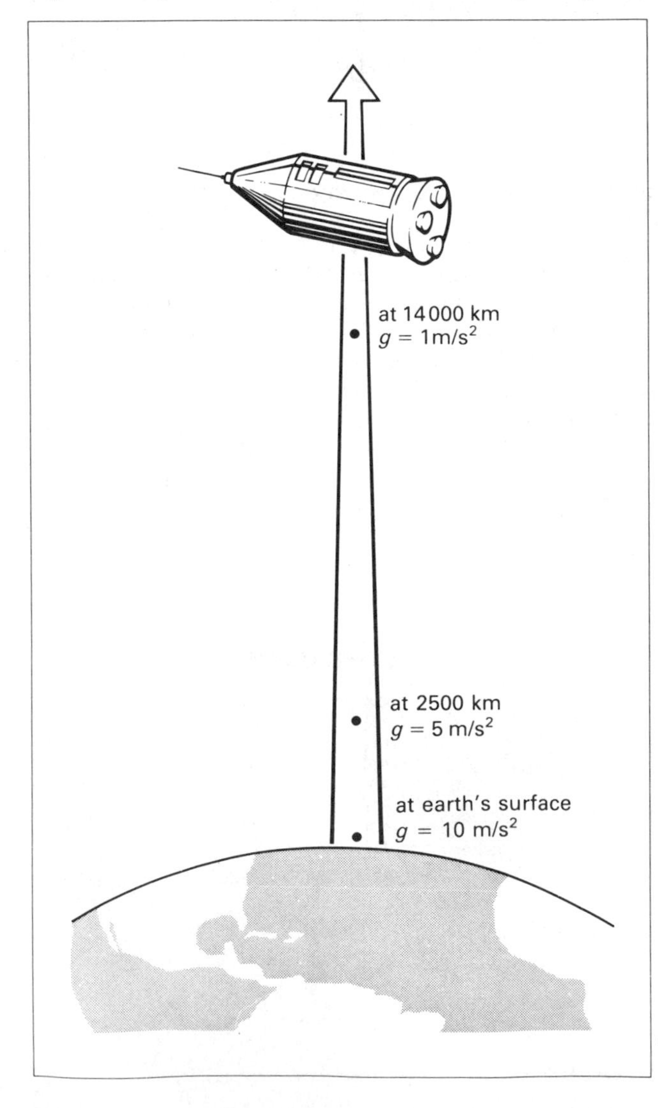

◀ fig. 2.22 The strength of the gravitational field decreases with height

Gravitation

We have already mentioned the force of gravity in connection with the weight of a body and also the acceleration due to gravity. These are only illustrations of a much more general idea. Sir Isaac Newton put forward his theory of universal gravitation in 1666 to explain the movements and orbits of the planets. Newton showed that the behaviour of the planets could be explained if he assumed that

> 'every particle of matter attracts every other particle of matter with a force which varies directly as the product of the masses and inversely as the square of the distance between them'.

This is Newton's law of universal gravitation. Expressed mathematically

$$F = G\frac{m_1 \times m_2}{d^2}$$

where F is the force in newtons, m_1 and m_2 are the masses in kilograms and d is the distance between the masses in metres. G is the gravitation constant and has the very small value of 6.67×10^{-11} N m^2/kg^2. This means that the force of attraction only becomes appreciable when one of the masses is very large.

If we consider a stone a few metres above the surface of the earth there will be a force of gravitational attraction between the earth and the stone. The stone is attracted to the earth and the earth is attracted to the stone. Although the same force is acting

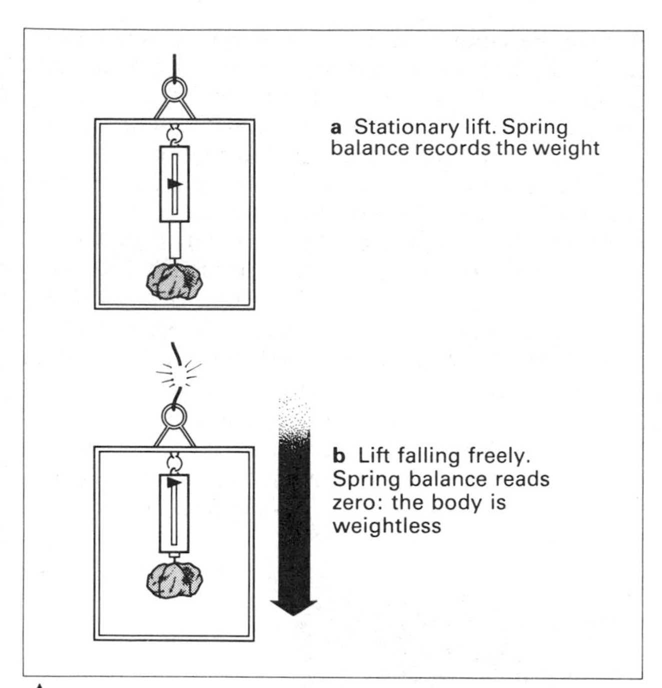

fig. 2.23 Weightlessness in a free-falling lift

on both the stone and the earth, because the mass of the earth is so great its acceleration is so small it cannot be detected. The acceleration of the stone, however, is very noticeable as it moves towards the ground.

The space round the earth is known as the **gravitational field** of the earth. The gravitational field is a region where a mass experiences a force although it has no direct contact with any other body. In the same way a magnetic field is a region where a magnetic pole experiences a force. An electric field is a region where an electric charge experiences a force. Magnetic and electric fields will be discussed in more detail later on. The idea of a field is useful in physics because it enables us to calculate the magnitude and direction of the force on a mass, pole or charge when it is placed at a particular point in the field.

The strength of the gravitational field is measured by the force exerted on a unit mass. At the surface of the earth the force on m kg is mg newtons, so the force on 1 kg is g newtons. The gravitational field strength at the surface of the earth is thus 10 N/kg. The gravitational field strength at the moon's surface is only about one sixth of the strength at the earth's surface so that weights on the moon are only one sixth of their value on earth.

Since the force of gravity obeys an inverse square law, the force gets less the greater distance we go from the surface of the earth, but although the force will get very small it will never be zero. At some point between the earth and the moon, for example, the gravitational attractions of the earth and the moon will be equal and opposite and the resultant will be zero. At this point an object would be **weightless** because there is no resultant force acting on it. The word weightless is often wrongly used, as, for example, when it is said that astronauts in a spacecraft are weightless because they have escaped from the gravitational field.

Weightlessness occurs when forces are balanced (as in the case of the body at a certain point between the earth and the moon) or when an object is falling freely. If a mass is hung from a spring balance fixed to the roof of a stationary lift, the weight will be recorded on the balance, figure 2.23a. If the lift were allowed to fall freely, as it could do if the cable broke, the balance would read zero, figure 2.23b. Both the lift and the mass have the same acceleration. Astronauts are trained for conditions of weightlessness in a diving aircraft which is really falling freely.

Projectiles

When considering the path of a projectile it is useful to resolve the velocity into two components in the same way as a force was resolved into components on p. 19.

Consider a small object projected horizontally from the edge of a table at the same instant as another object falls freely. The apparatus for carrying this out is illustrated in figure 2.24. Observation shows that both objects hit the floor at the same time though one falls vertically downwards and the other follows a curved path.

At the point A, figure 2.25, we can consider the velocity as having two components, a horizontal and a vertical component. Since both objects reach the

fig. 2.24 Apparatus for launching projectiles horizontally and vertically

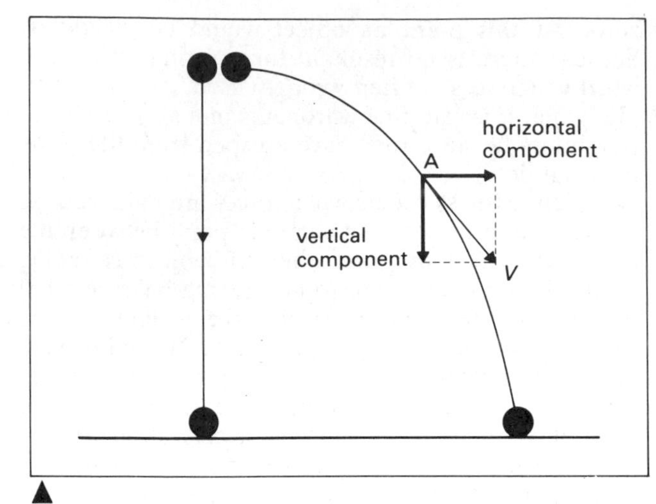

▲
fig. 2.25 Gravitational acceleration is unaffected by horizontal velocity (see also fig. 2.27)

floor in the same time the vertical motion is not affected by the horizontal motion. The horizontal velocity component remains constant, but the vertical velocity component experiences the normal acceleration due to gravity. The important point to realise is that the horizontal motion of a body following a curved path is not affected by the vertical motion.

When a body is projected horizontally from a point some height above the earth's surface it follows a curved path or trajectory. The curve is a **parabola**, figure 2.26. Neglecting air resistance there is no horizontal force acting on the body so the horizontal velocity remains constant. In the vertical direction, the body is falling freely and accelerating. It travels 5 m in the first second, 15 m in the next second, 25 m in the third second and so on. An object projected at an angle to the earth's surface also follows a parabolic path, figure 2.28.

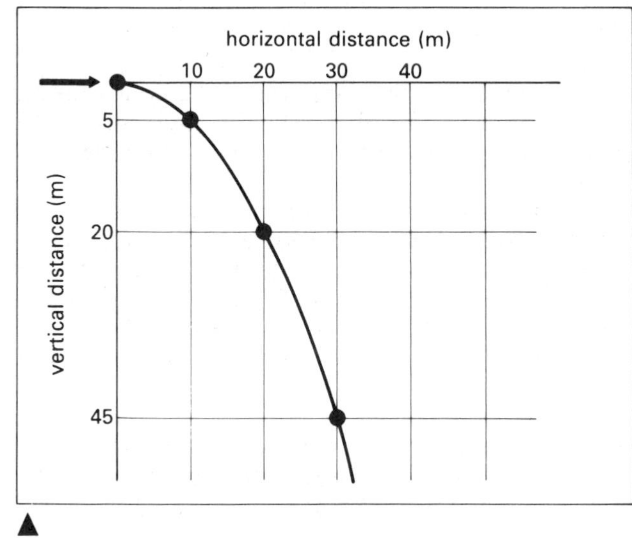

▲
fig. 2.26 A horizontally launched body follows a parabolic path

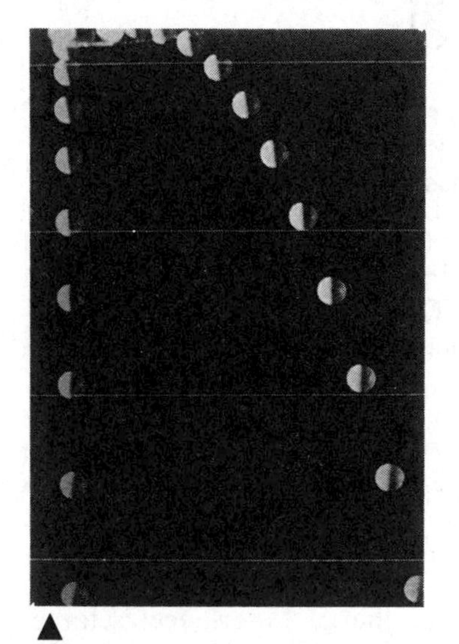

▲
fig. 2.27 Two balls set in motion at the same time, one falling freely, the other shot horizontally

fig. 2.28 Projectile launched from the ground follows a parabolic trajectory
▼

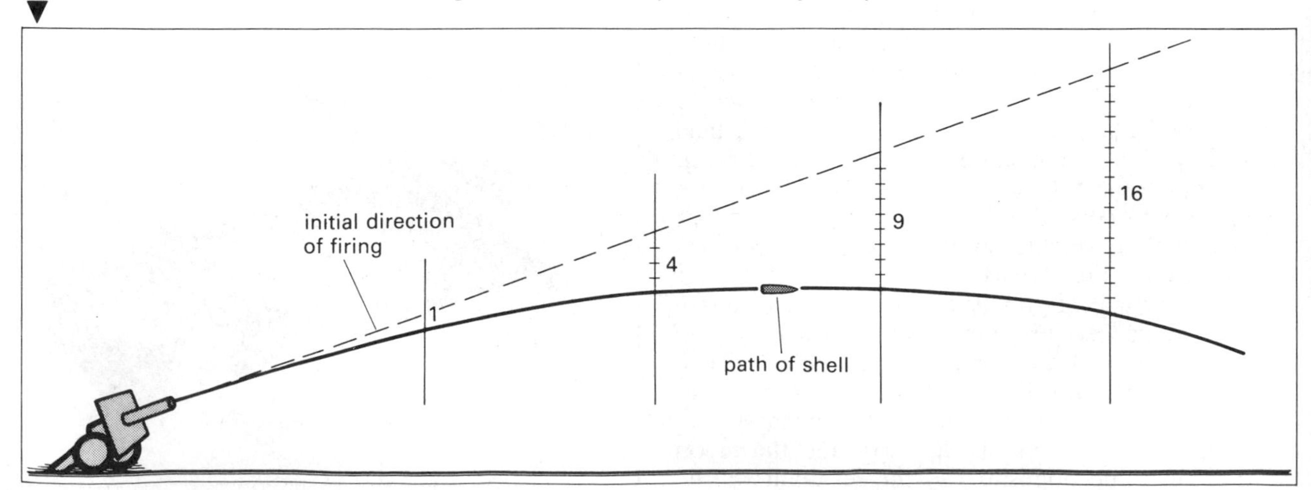

Satellites

It was a consideration of the behaviour of projectiles that led Sir Isaac Newton to the following problem. Suppose a body is projected horizontally from a point above the earth's surface and the velocity of projection is increased. At first the body will follow the path AB, figure 2.29. As the velocity is increased, the distance of B from the starting point will increase until, at a certain velocity, the body will follow the path C and go right round the earth, returning to the starting point A. A greater velocity still will produce path D, and at a high enough velocity it should be possible to follow the path E and escape from the earth altogether.

When a body follows a constant path or **orbit** around a planet or the sun it is called a **satellite** of the planet or the sun. The planets are satellites of the sun; the moon is a satellite of the earth. Artificial satellites can be shot into orbit using powerful rockets. If the satellite orbits outside the earth's atmosphere there is no air resistance and the orbit can remain constant, theoretically for ever!

Artificial satellites are used in radio communication over long distances. Radio waves are reflected from the various ionospheric layers in the atmosphere, but these layers are not very reliable reflectors. They do not reflect the very short radio waves used for television. A communication satellite is made to circle round the earth at just the right height and speed so that it takes exactly one day to complete one orbit. In this way it remains exactly above a fixed point on the earth and signals can be beamed to it. The signal is received, amplified and retransmitted in a narrow beam to a receiving station at another part of the earth's surface, figure 2.30.

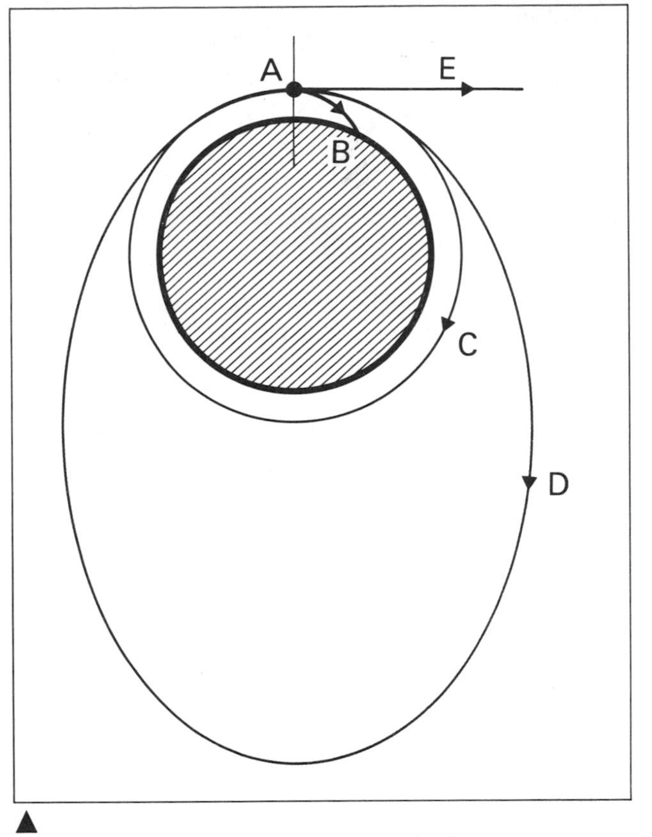

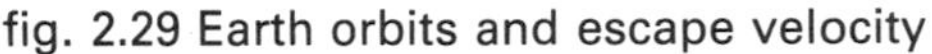

fig. 2.29 Earth orbits and escape velocity

fig. 2.30 Communications satellites orbit the earth at a height of about 35 000 km and relay telephone and telegraph messages and television programmes

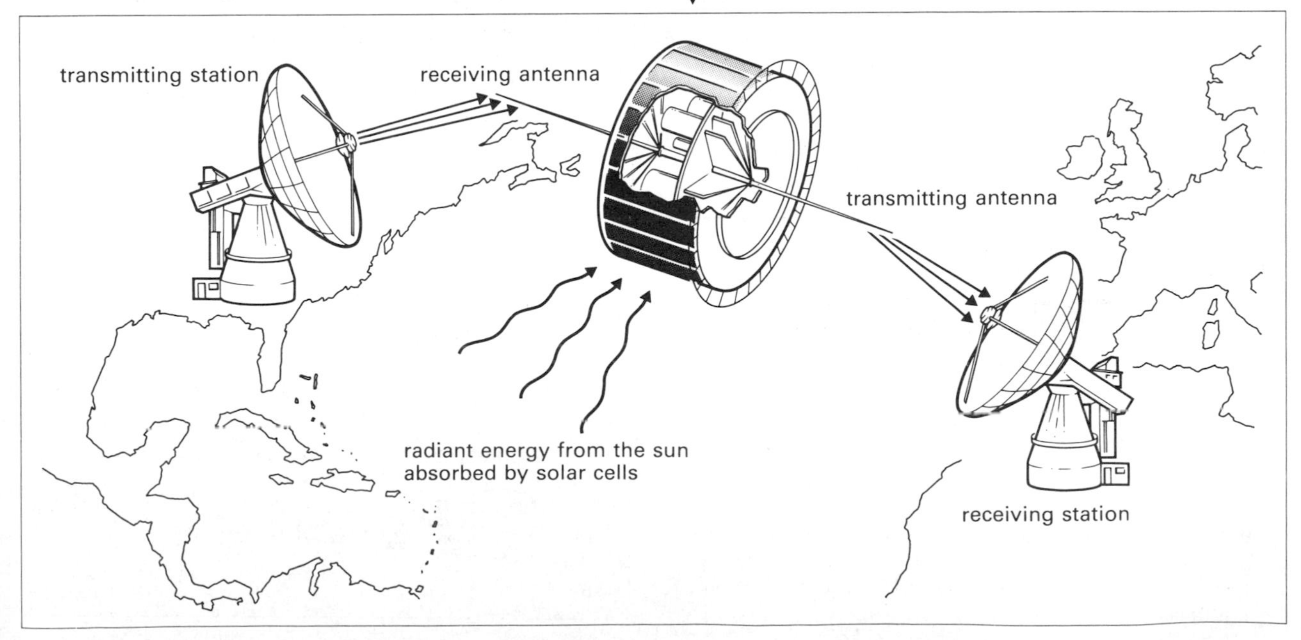

EXERCISE 2

For these questions the acceleration due to gravity can be taken as 10 m/s² and the earth's gravitational field strength as 10 N/kg.

In questions 1–4 select the most suitable answer.

1 Which of the objects **A** to **E** in figure 2.31 has the greatest resultant force acting upon it?

fig. 2.31

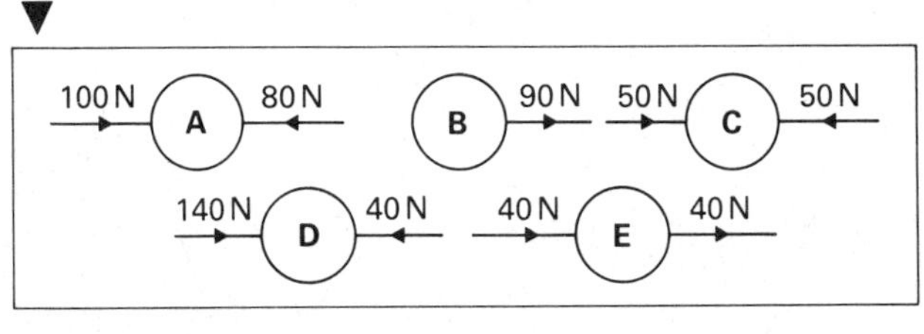

2 Which system gives the greatest acceleration?
A 65 N acting on 40 kg
B 10 N acting on 8 kg
C 18 N acting on 30 kg
D 80 N acting on 100 kg
E 90 N acting on 80 kg

3 An elastic spring doubles its unstretched length to 16 cm without exceeding its elastic limit when a force of 28 N is applied to it. When a force of 7 N is applied, the length of the spring is
A 2 cm
B 4 cm
C 7 cm
D 8 cm
E 10 cm

4 An object of mass m is acted on by a force F giving an acceleration a in the direction of the applied force, figure 2.32.

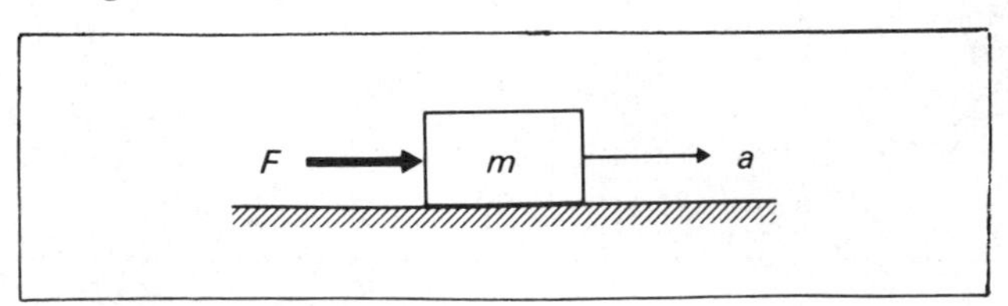

fig. 2.32

a What force is needed to give an acceleration $2a$ to the mass m?
b What force is needed to give an acceleration a to a mass $2m$?
c What force is needed to give an acceleration $3a$ to a mass $2m$?

5 The resultant of two forces 3 N and F is 5 N, figure 2.33. Using the scale diagram find the value of F.

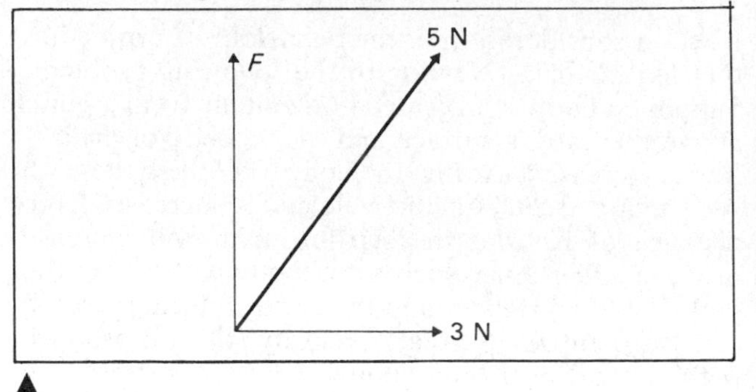

fig. 2.33

6 Figure 2.34a shows the experimental arrangements a student uses to test the elastic properties of a piece of rubber and the table shows the results of the experiment.
a The student concludes that the rubber under test obeys Hooke's law, at least for the loads he has used. (i) Do you agree with his conclusion? Explain your answer. (ii) What was the average extension of the rubber per newton of load?
b The same piece of rubber was then used to propel a model car of mass 0.1 kg, one end being fixed to the bench and the other end to the car which was pulled back, stretching the rubber as shown in figure 2.34b. The rubber was extended by 108 mm. (i) Calculate the value of the force exerted by the rubber on the car before it was released. (ii) Calculate the initial acceleration of the model car when it is released.

fig. 2.34

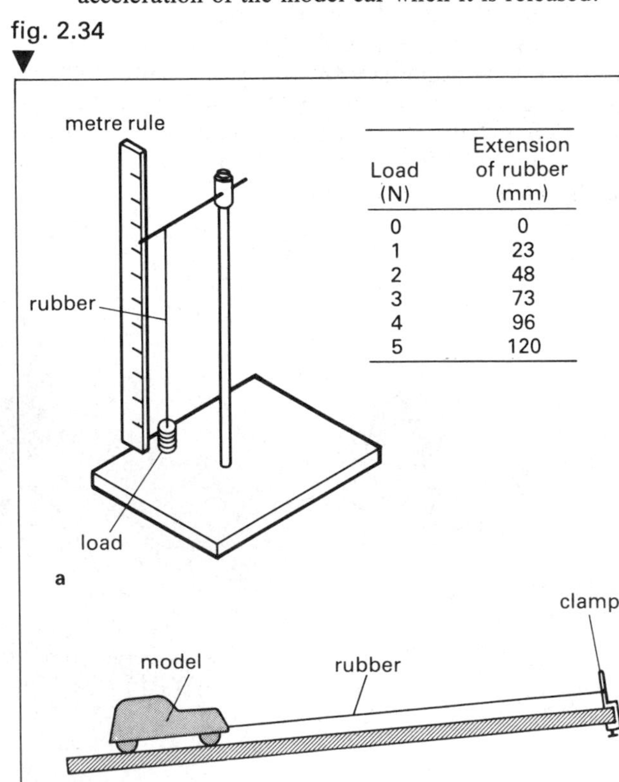

Load (N)	Extension of rubber (mm)
0	0
1	23
2	48
3	73
4	96
5	120

7 The apparatus shown in figure 2.35 is used to measure the extension of a spring for various forces applied to it. The results are given in the table.

Force (N)	Extension (cm)
0	0
2	4
4	8
6	12
8	16
10	20

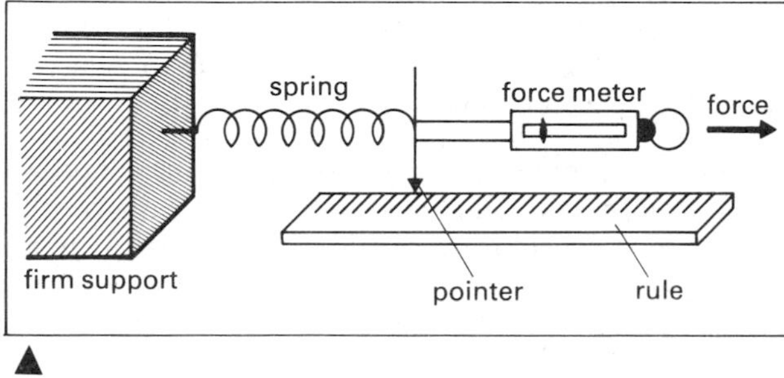

fig. 2.35

a (i) Describe briefly how you would use the apparatus to obtain the results. (ii) Plot a graph of force (vertical axis) against extension (horizontal axis), for the above results. (iii) From the graph what would be the extension of the spring if the load were 5 N? (iv) As the spring is extended to 20 cm, what is the average force applied?

b In figures A–D, all the springs are identical: a force of 1 N moves the bar 1 cm, as shown in figure A.

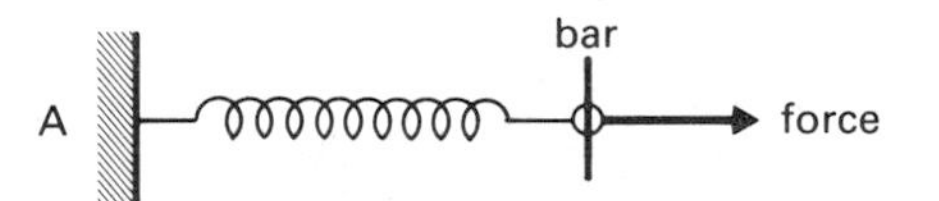

Calculate the force needed to move the bar by 1 cm in each of the following three cases:

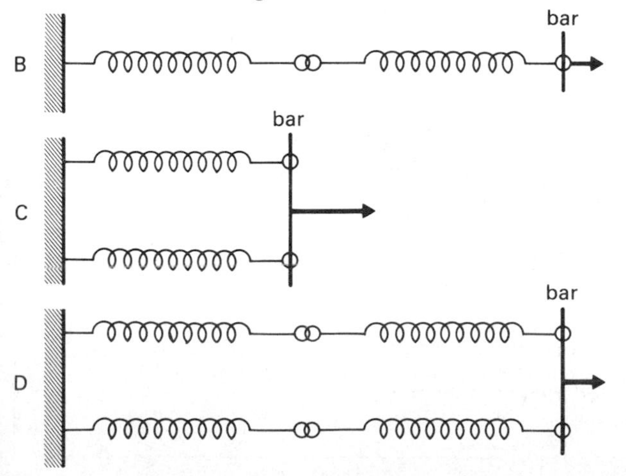

8 The figure shows a ball bearing falling through a column of oil. The distance fallen after certain times is given in the table.

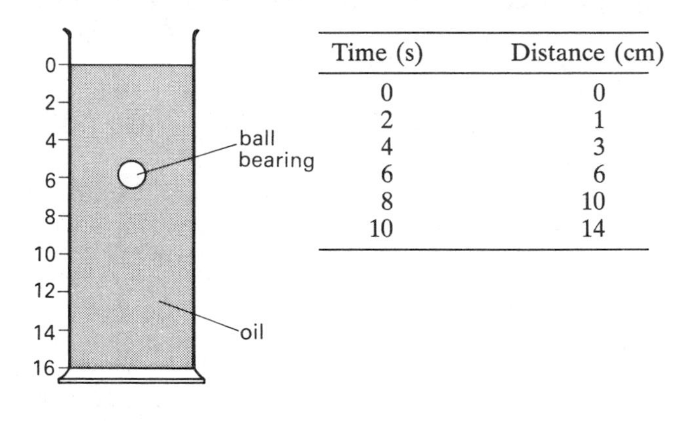

Time (s)	Distance (cm)
0	0
2	1
4	3
6	6
8	10
10	14

a What force is pulling downwards on the ball bearing?
b Name one of the forces acting upwards on the ball bearing.
c Which one of the following describes the initial motion of the ball bearing? (i) Uniform velocity of 0.5 cm/s, (ii) uniform velocity of 1.0 cm/s, (iii) uniform velocity of 2.0 cm/s, (iv) non-uniform velocity.
d When the ball bearing has a steady velocity, how will the downward force compare with the total upward force?
e State one factor which, if changed, would alter the velocity of the ball bearing.

9 **a** A space rocket in outer space fires its rocket motor. This motor exerts a constant force of 7.5×10^6 N. (i) If the mass of the rocket when the motor is fired is 5×10^5 kg, calculate the initial acceleration of the rocket. (ii) Fuel is used up at a rate of 3000 kg every second. What is the acceleration of the rocket 100 s after the motor is fired?

b A small model rocket motor is firmly attached to the top of a trolley as shown below.

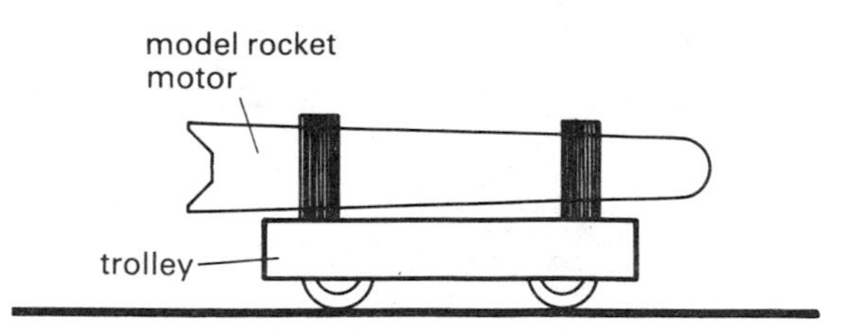

Describe how you would carry out an experiment using this apparatus to find the accelerating force produced by this rocket motor. In your description state (i) any other apparatus required, (ii) how you would carry out the experiment, (iii) what measurements you would make, (iv) how you would calculate the accelerating force.

3 NEWTON'S LAWS OF MOTION AND MOMENTUM

The whole of our treatment of motion and forces is based on Sir Isaac Newton's studies of moving bodies which he published in his *Mathematical Principles of Natural Philosophy* in 1687. The main ideas are summed up in Newton's three **laws of motion**. Probably Newton's greatest achievement was to realise that the laws could be applied, not only to objects moving close to the earth's surface, but also to the movements of the stars and planets. From this idea he developed his **law of gravitation**.

We have already met the first law as a definition of force and used the second law to measure force (p. 15) but the ideas are so important that we shall now look at them in more detail, including their experimental verification.

Newton's first law of motion

Every body remains in a state of rest or of uniform motion in a straight line unless it is made to change that state by an external force

We have already used the law to give a definition of force as that which changes or tends to change the state of rest or uniform motion of a body. It is an everyday observation that stationary objects do not suddenly begin to move of their own accord. The property of a body to remain at rest or, if it is moving, to continue with its motion unchanged is called **inertia** and Newton's first law is sometimes known as the law of inertia.

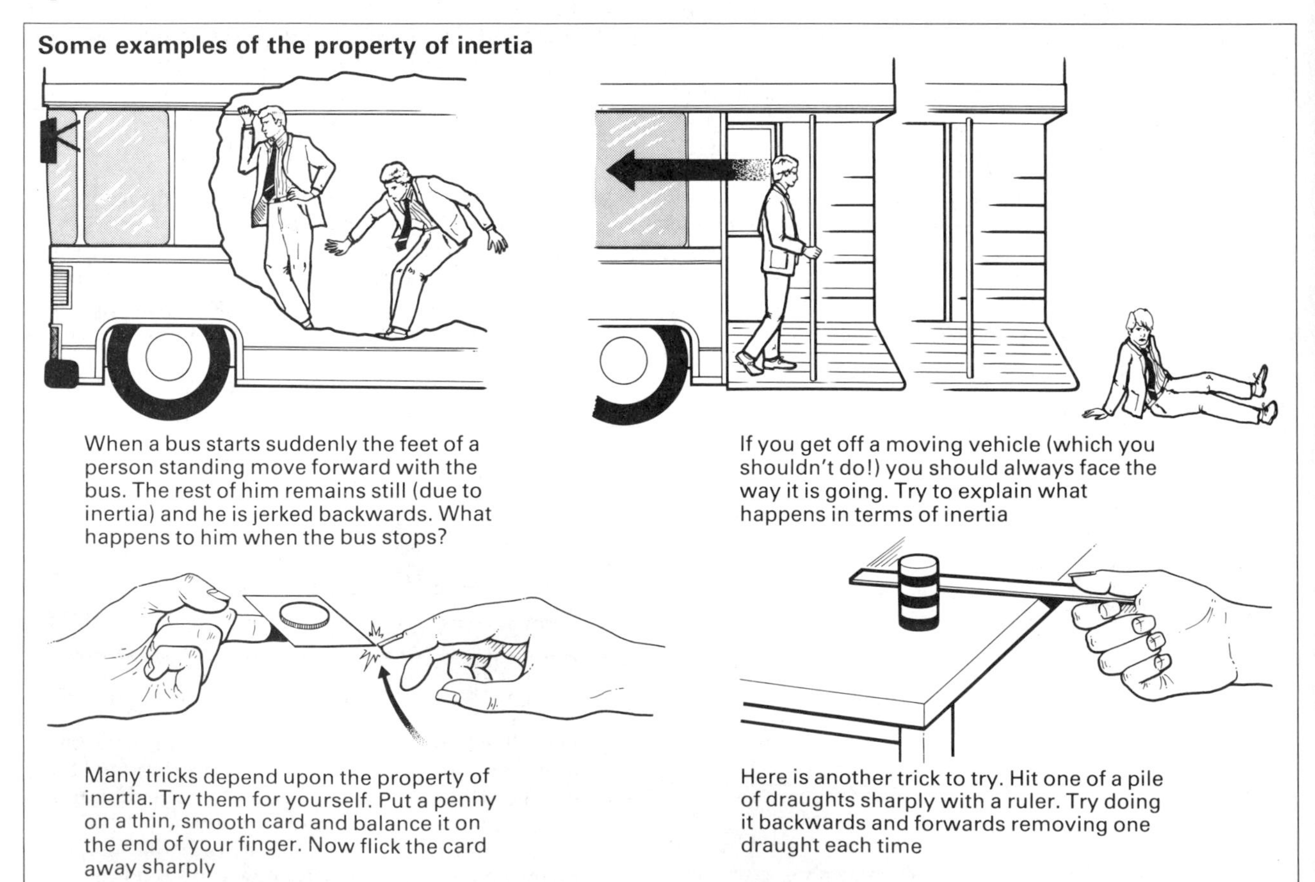

Some examples of the property of inertia

When a bus starts suddenly the feet of a person standing move forward with the bus. The rest of him remains still (due to inertia) and he is jerked backwards. What happens to him when the bus stops?

If you get off a moving vehicle (which you shouldn't do!) you should always face the way it is going. Try to explain what happens in terms of inertia

Many tricks depend upon the property of inertia. Try them for yourself. Put a penny on a thin, smooth card and balance it on the end of your finger. Now flick the card away sharply

Here is another trick to try. Hit one of a pile of draughts sharply with a ruler. Try doing it backwards and forwards removing one draught each time

fig. 3.1 The chair-o-plane: each rider follows a circular path determined by his or her mass and the speed of rotation of the machine

While it seems common sense that a body will remain at rest until an external force sets it in motion, it is not so easy to accept that a uniformly moving body would continue to move with constant velocity in a straight line if left to itself. This is because in practice we cannot eliminate all the forces which would retard the motion. Bodies moving through the air are retarded by air resistance. When a body slides over a surface there is the force of friction. If we make the resistance very small by sliding an ice-hockey puck over smooth ice then it will travel quite a long way in a straight line. From this we argue that if we made the resistance smaller still, the closer we would get to continual motion in a straight line and if we could remove all the resistance to the motion, Newton's first law would be verified. The law cannot be 'proved' by experiment.

Motion in a circle

When a body moves in a circle there must, according to Newton's first law, be some force acting on it which deflects it from its tendency to travel in a straight line. This force is directed towards the centre of the circle and is called the **centripetal force**.

If a particle is moving round a circle of radius r with a uniform speed v the time taken for one complete revolution is called the **period** or **periodic time.**

$$T = \frac{2\pi r}{v}$$

Although the speed is constant the velocity is continually changing because the direction of motion is changing.

In figure 3.2, at A the velocity is v_A while at B the velocity is v_B, both velocities being tangential to the circle. Since the velocity has changed there must be an acceleration. This acceleration is produced by the centripetal force and is directed inwards along the radius towards the centre of the circle. The magnitude of the inward acceleration is v^2/r and if the particle has a mass m the centripetal force is

$$F = \frac{mv^2}{r}$$

fig. 3.2 Motion in a circle

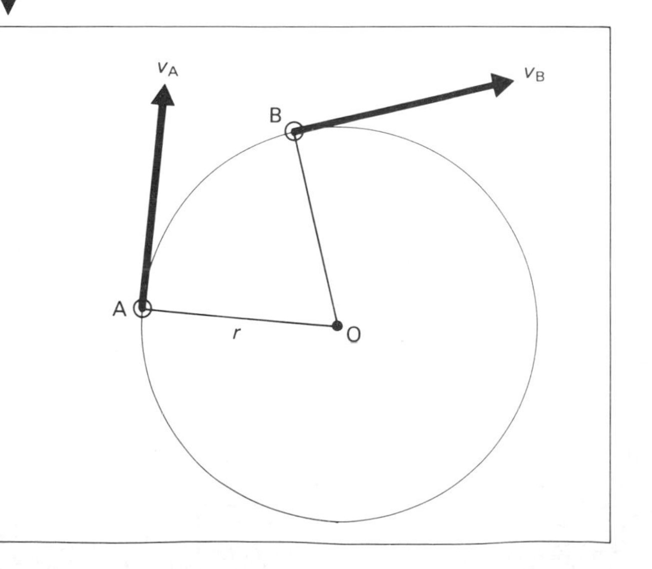

When a mass is tied to the end of a string and whirled round the centripetal force is supplied by the tension or pull in the string. When a train rounds a curve the inward force comes from the action of the rails on the rim of the wheels. When a car rounds a curve the inward acting force has to come from the friction between the tyres and the road. If this frictional force is not great enough the car skids off the road. The centripetal force to keep the planets in orbit round the sun is supplied by gravitational attraction.

Notice that the magnitude of the centripetal force depends on three factors:

- The greater the mass, the greater the centripetal force.
- The greater the speed, the greater the centripetal force.
- The smaller the radius of the curve, the greater the centripetal force.

If, when a mass is being whirled round on a piece of string, the string breaks, the mass flies off at a tangent. It does not move outwards in the direction shown by the dotted line, figure 3.3. There is no outward force acting on the mass. The force (supplied by the tension of the string before it broke) is radially inwards to produce the circular motion. There is often confusion and misunderstanding between centripetal and centrifugal forces. The centrifugal force is the pull felt by the hand holding the string and is equal and opposite to the inward force on the moving mass, figure 3.4.

Momentum

A force is necessary to set a body in motion because of the property of inertia. The more massive the body the greater will be the required force. Similarly, a force is necessary to stop a moving body. The more massive the body and the faster it is moving, the greater the force required to stop it. It is useful to have some quantity which combines the mass of a body and how quickly it is moving and this quantity is **momentum**.

The momentum of a body is defined as the product of its mass and its velocity.

momentum = mass × velocity

$$p = mv$$

Momentum is a **vector quantity** and its unit is **kg m/s**.

Since momentum is a product of two quantities, mass and velocity, a body of large mass moving slowly can have the same momentum as a small mass moving with high velocity. A slow-moving, heavy lorry could have the same momentum as a high-speed rifle bullet if the respective products of the masses and velocities are the same.

Newton called momentum the 'quantity of motion' possessed by a body.

Usually, we are concerned with the *change* in momentum brought about by some event, such as a collision or the application of a force.

Newton's second law of motion

The rate of change of momentum of a body is proportional to the applied force and takes place in the direction of that force

To understand exactly what this law means consider the case of a force F acting on a body of mass m for a time t and changing its velocity from u to v, figure 3.5.

fig. 3.3 Escape along a tangent

fig. 3.4 The centripetal force and the centrifugal force

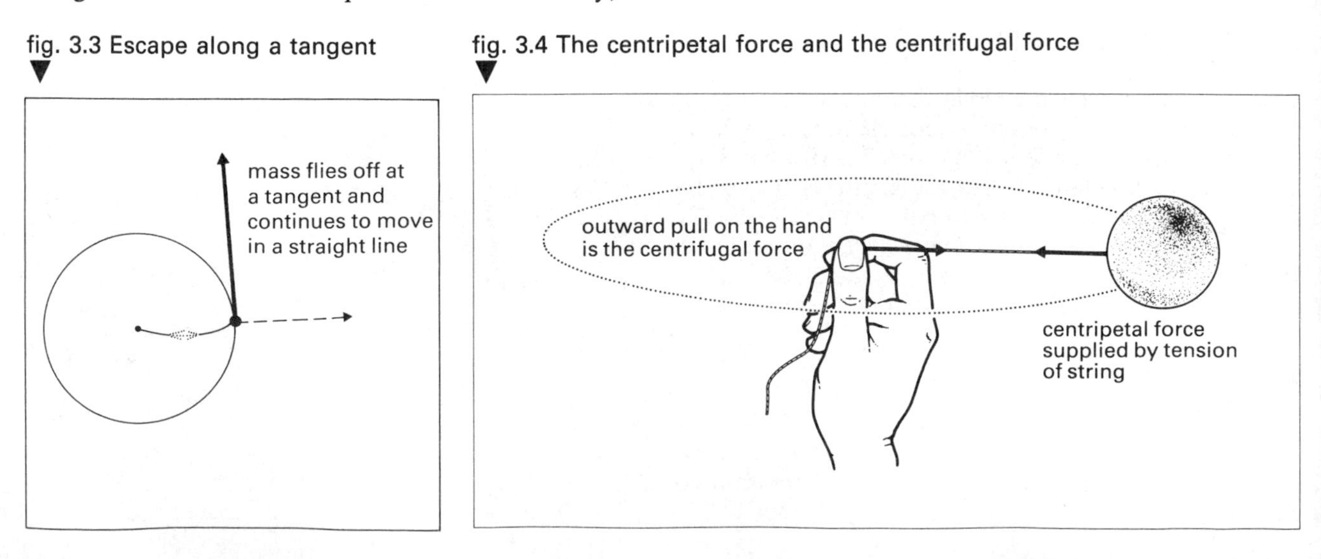

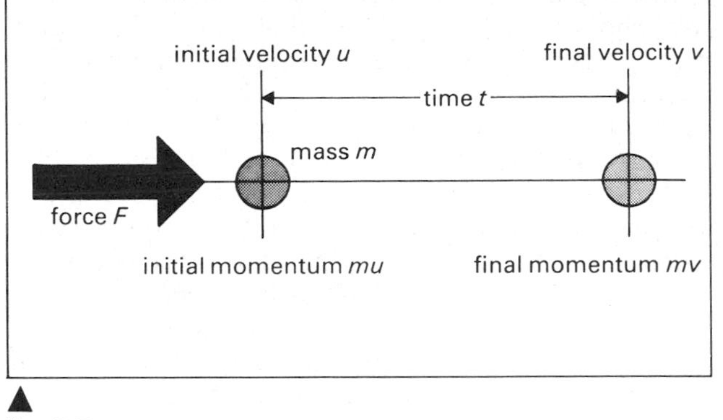

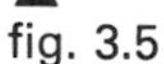
fig. 3.5

initial momentum $= mu$
final momentum $= mv$
change in momentum $= mv - mu$
since this change takes place in
time t, rate of change of momentum $= \dfrac{mv - mu}{t}$

According to Newton's second law this is proportional to the applied force F:

$$F \propto \frac{mv - mu}{t} \quad \text{or} \quad F \propto m\left(\frac{v - u}{t}\right)$$

but

$$\frac{v - u}{t} = \frac{\text{change in velocity}}{\text{time}}$$
$$= \text{acceleration}$$

therefore $F \propto ma$

Force is proportional to mass times acceleration, or

force = constant × ma

If we define our unit of force as that force producing unit acceleration in unit mass, we have $F = 1$ when $m = 1$ and $a = 1$ and so the constant = 1 and $F = ma$.

This is the justification for the relationship we have already used on p. 16 and the unit of force is the newton (N) which produces an acceleration of 1 m/s^2 in a mass of 1 kg.

Experimental verification of Newton's second law

Newton's second law cannot be 'proved' in the laboratory because it is an axiomatic statement on which Newton based his system of mechanics, but it can be tested and shown to work well in practice. We need a body which has no resultant forces acting on it, so that any change in motion we produce must be due to a known force which we apply to it. This is achieved experimentally by using a very freely running trolley, figure 3.6a, or a slider running on an air track which works on the same principle as a hovercraft, figure 3.6b.

The wheels of the trolley run on ball bearings and the trolley travels on a base board 2 metres long. Attached to the trolley is the tape of a ticker-tape timer so that the motion of the trolley can be investigated. Before an experiment is carried out the retarding effect of friction on the trolley must be eliminated. This is done by tilting the base board slightly until the trolley moves with constant velocity when started with a small push. The uniform velocity will be shown by the equal spacing of the dots on the tape. When the trolley moves with constant velocity the resultant force on it is zero (Newton's first law). If the mass of the trolley is changed a new friction compensation must be made because the friction depends on the force acting downward on the base board, i.e. the weight of the trolley.

fig. 3.6 Apparatus for the verification of Newton's second law

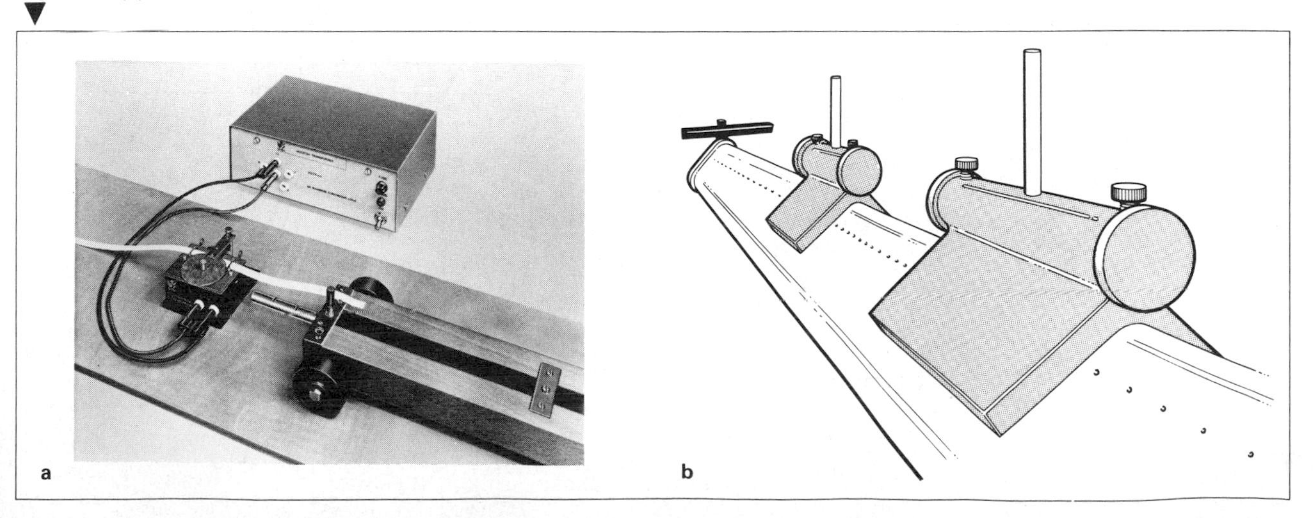

In order to investigate the validity of the expression

force = mass × acceleration

it is necessary to consider it in two stages.

If we keep the mass constant then the force applied should be directly proportional to the acceleration produced.

If we keep the force constant and change the mass the acceleration produced should vary inversely as the mass. The larger the mass the smaller the acceleration.

In preliminary experiments it is simplest to use any units of force and mass, since the relation $F = ma$ should hold for any *consistent* system of units. The mass can be changed by using a number of identical trolleys fixed one on top of the other. The force can be applied by an elastic cord stretched to a fixed length. One, two, three or more stretched cords would supply one, two, three or more units of force. You should link this idea with the work on Hooke's law and stretched springs in parallel on pp. 20-21.

Experiment to show that acceleration varies directly as the applied force when the mass is constant

A single trolley is used and the base board is friction compensated, figure 3.7. The trolley is pulled along by a single elastic cord with rings at each end. One ring is placed over the rod at the back of the trolley and the cord is stretched till the ring at the other end is level with the front of the trolley. This ring is kept in line with the front of the trolley as the trolley is pulled along. There is a constant stretching of the elastic and so a constant force is applied. A few trial runs should be done before carrying out the experiment. The timer is switched on and a run made. The experiment is repeated using two, three and four elastic cords. The acceleration produced in each case is worked out from the dots on the ticker-tape.

Assuming the timer to be connected to a 50 Hz alternating current supply, the dots are produced at intervals of $\frac{1}{50}$ or 0.02 s. The dots at the start of the tape are probably too crowded together for accurate measurement. Count two successive intervals of ten spaces and measure the distances d_1 and d_2 each of which are travelled in 0.2 s, figure 3.7.

$$\text{average velocity over the distance } d_1 = \frac{d_1}{0.2} \text{ cm/s}$$

$$\text{average velocity over the distance } d_2 = \frac{d_2}{0.2} \text{ cm/s}$$

$$\text{increase in velocity which has taken place in 0.2 s} = \frac{d_2}{0.2} - \frac{d_1}{0.2}$$

$$= \frac{d_2 - d_1}{0.2} \text{ cm/s}$$

$$\text{acceleration} = \frac{\text{increase in velocity}}{\text{time}}$$

$$= \frac{d_2 - d_1}{(0.2)^2} \text{ cm/s}^2$$

Two or three pairs of values of d_1 and d_2 are taken from each tape and an average value of $d_2 - d_1$ found.

fig. 3.7 Verifying that acceleration is proportional to applied force

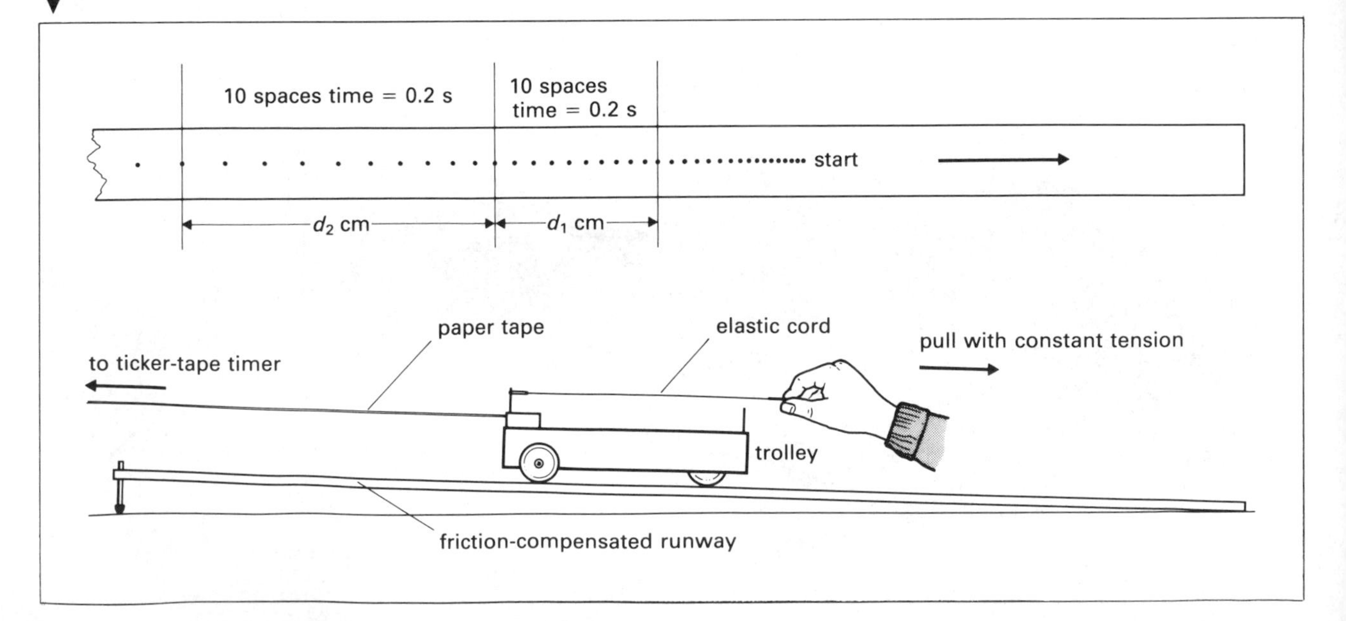

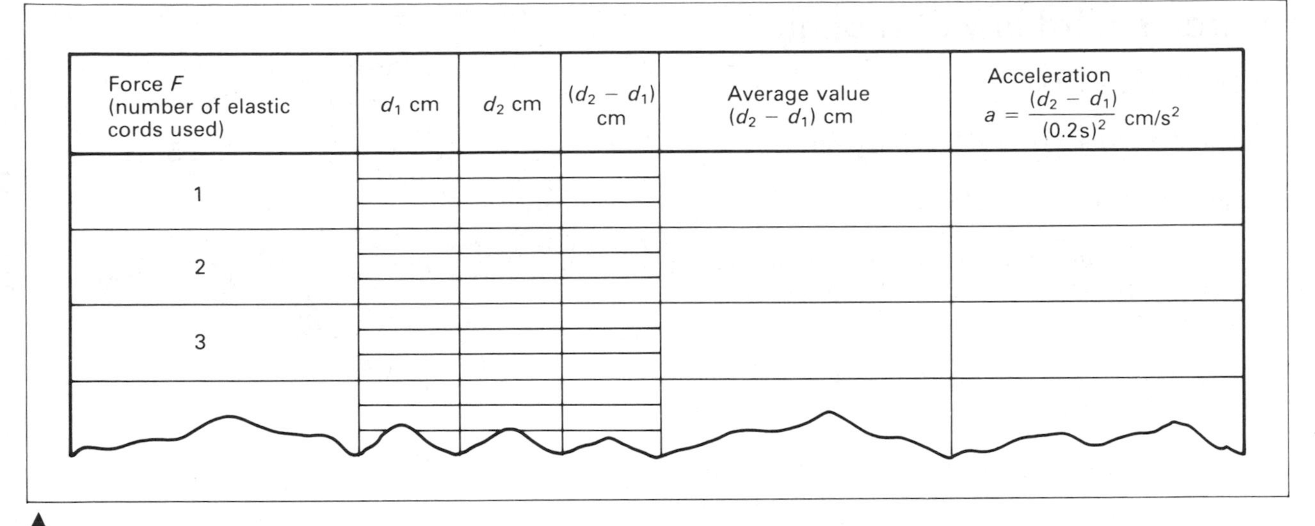

Force F (number of elastic cords used)	d_1 cm	d_2 cm	$(d_2 - d_1)$ cm	Average value $(d_2 - d_1)$ cm	Acceleration $a = \frac{(d_2 - d_1)}{(0.2\,s)^2}$ cm/s^2
1					
2					
3					

fig. 3.8

This will compensate to some extent for the fact that it is not too easy to draw the trolley along with a constant force. The results are collected in a table as shown in figure 3.8.

A graph of acceleration (a) is plotted against applied force (F). A straight line through the origin shows that acceleration is directly proportional to the applied force.

Experiment to show that acceleration varies inversely as the mass when the applied force is constant

The above experimental method is now repeated except that the mass is varied and the applied force is kept constant. The mass can be varied by using one, two, three or more trolleys, figure 3.9. For one type of trolley the mass can be varied by sliding in cylindrical weights. In this case the mass has to be found by weighing each time it is changed. Since larger masses have to be moved, it is probably best to use the constant force of two elastic cords throughout the experiment. As has already been stated, the adjustment for friction compensation has to be made each time the mass is altered. The results are collected in a table as shown in figure 3.10.

A graph of acceleration (a) is plotted against the reciprocal of the mass ($1/m$). A straight line through the origin shows the inverse variation relationship.

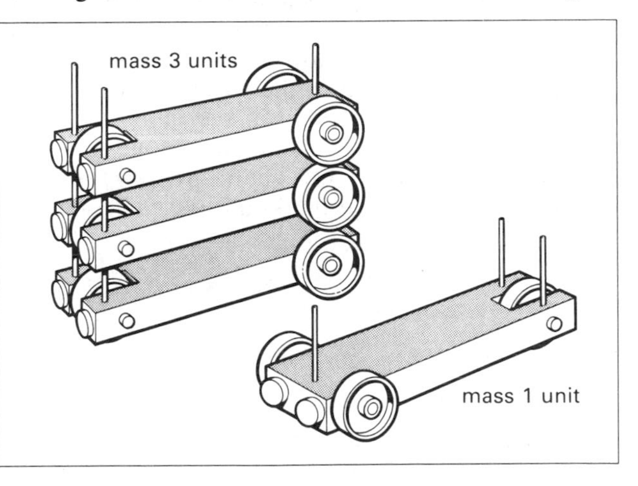

fig. 3.9 Stackable trolleys

fig. 3.10

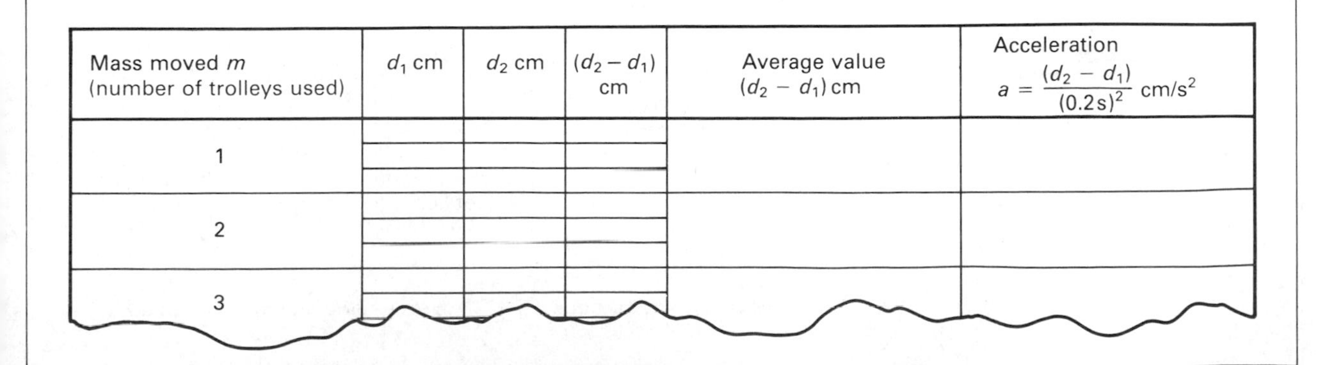

Mass moved m (number of trolleys used)	d_1 cm	d_2 cm	$(d_2 - d_1)$ cm	Average value $(d_2 - d_1)$ cm	Acceleration $a = \frac{(d_2 - d_1)}{(0.2\,s)^2}$ cm/s^2
1					
2					
3					

Newton's third law of motion

This law is often stated as

Action and reaction are equal and opposite

but a clearer statement would be

When one body exerts a force on a second body then the second body exerts an equal and opposite force on the first body

It is important to realise that the 'action' force and the 'reaction' force are acting on two different bodies.

On p. 16 the forces acting on a book resting on a table were discussed. It was pointed out that the book remained at rest because the forces acting on it were in equilibrium. The push of the table on the book was equal and opposite to the force of gravity on the book. These are not true action and reaction forces as they both act on the same body, i.e. the book. The true action and reaction forces described in Newton's third law are the weight of the book acting on the table and the reaction of the table on the book, i.e. the forces are acting on different bodies, figure 3.12.

This will become clearer if we consider the idea of a **free-body diagram**. In any discussion of forces it is necessary to choose a particular body, isolate it from its surroundings and then represent the forces acting on the body by force vectors. At the same time it is useful to think what causes the various forces. In the example of the book resting on the table, there are actually three bodies involved, the book, the table and the floor, figure 3.13. There are reactions between the table and the book and the table and the floor.

▲ fig. 3.11 The action of the boat on the man enables him to jump ashore; the reaction of the man on the boat causes it to move away from the landing stage

fig. 3.12 True action and reaction ▼

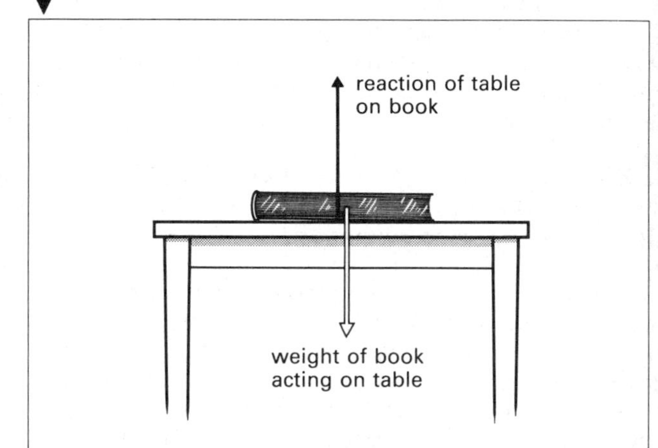

fig. 3.13 Free-body diagrams ▼

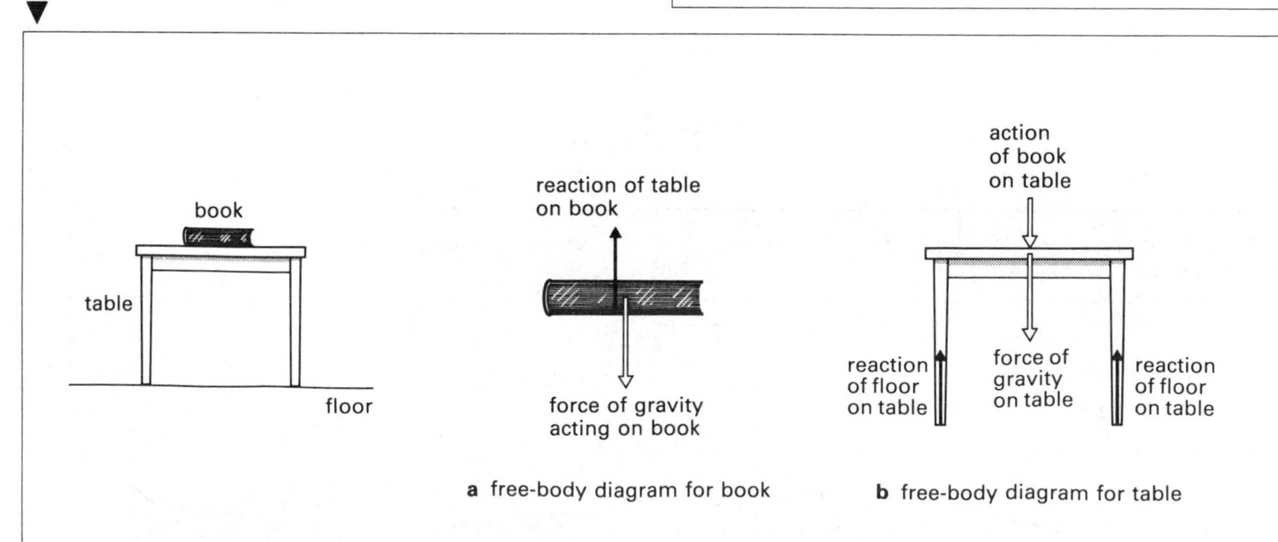

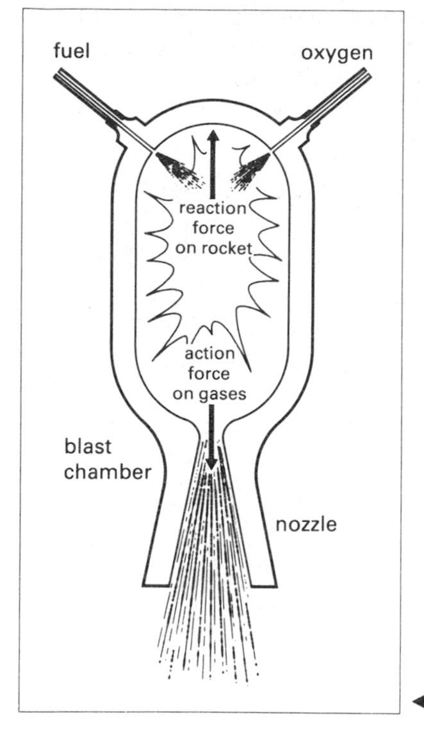

fig. 3.14 The rocket principle

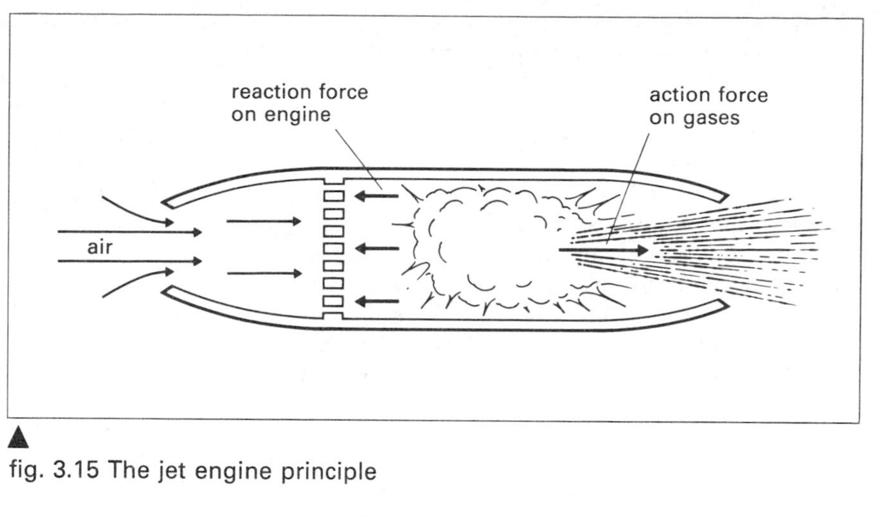

fig. 3.15 The jet engine principle

Rockets and jet engines

The small rockets used in firework displays and the large rockets for launching space satellites and space-craft work on the same principle, figure 3.14. The combustion of chemicals in the firework or the burning of a liquid fuel and oxygen in the satellite launcher produces a large volume of hot gas which can only escape at the rear of the rocket. Although the mass of gas escaping is quite small it has a very high velocity and so a very large momentum. An equal and opposite momentum is given to the rocket. The action of the rocket on the gases is equal and opposite to the reaction of the gases on the rocket so the rocket moves forward as the exhaust gases move backward. A recoiling gun is an example of the rocket principle. A rocket does not move by pushing against the air around it. It will work perfectly well in a vacuum. (Also, there will be no air resistance to reduce its motion.)

The jet engine works on the same principle as the rocket, figure 3.15. The difference is that while it carries its fuel supply with it the oxygen necessary for combustion is obtained from the atmosphere. Air enters the front of the engine, is compressed by a turbo-compressor and then passes into the combustion chamber into which the fuel (paraffin or kerosene) is injected. The combustion produces a high-speed jet of gases from the exhaust nozzle. The reaction of the gases on the engine is the thrust which drives it forward. Jet engines can only be used on aircraft which fly in the atmosphere. They would not work in space where there is no air supply.

The conservation of momentum

A very important principle follows from Newton's second and third laws.

Consider the example of a bullet fired from a gun. Equal and opposite forces act on the gun and the bullet (Newton's third law), figure 3.16. The forces are due to the expansion of gases produced during the explosion. The action of the gun on the bullet accelerates the bullet, the reaction of the bullet on the gun produces the recoil effect. Since these forces are equal in magnitude and last for the same time (i.e. the time for the bullet to travel down the barrel of the gun) the momenta given to gun and bullet must

fig. 3.16 An example of momentum conservation

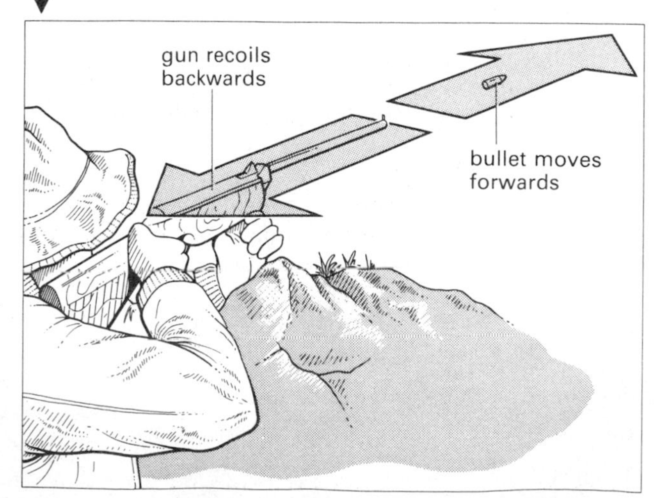

be the same. Newton's second law states that change in momentum is equal to the product of force and time. Since the momenta are the same

mass of gun × recoil velocity = mass of bullet × muzzle velocity

Momentum is a vector quantity so these momenta are in opposite directions and their vector sum is zero. The total momentum of the gun and bullet before the explosion was zero since there was no motion. So we have the conclusion that the momenta before and after the action are the same. This is generalised as the **conservation of momentum**:

When two (or more) bodies act on each other the total momentum before the action is the same as the total momentum after the action, provided no external forces act on the system

EXAMPLES ON MOMENTUM

Remember, momentum is conserved in actions where no external forces are applied. Also remember that

$$\text{momentum}\left[\frac{\text{kg m}}{\text{s}}\right] = \text{mass [kg]} \times \text{velocity}\left[\frac{\text{m}}{\text{s}}\right]$$

$$p = mv$$

$$\text{force [N]} = \frac{\text{change in momentum [kg m s}^{-1}]}{\text{time [s]}}$$

$$F = \frac{mv - mu}{t}$$

$$\text{force [N]} = \text{mass [kg]} \times \text{acceleration}\left[\frac{\text{m}}{\text{s}^2}\right]$$

$$F = ma$$

1 A bullet of mass 15 g travelling at 480 m/s strikes a target of mass 5 kg which is free to move. If the bullet embeds itself in the target, find their common velocity.

The collision is inelastic and momentum is conserved. Let the common velocity after impact be v m/s.

momentum before impact $= 0.015 \times 480$
$= 7.2$ kg m/s
momentum after impact $= (5 + 0.015)v$

momentum is conserved, therefore

$$5.015v = 7.2$$
$$v = 1.44 \text{ m/s}$$

2 An inflated balloon contains 2 g of air which escapes from the neck of the balloon at a speed of 4 m/s. Assuming that the balloon deflates at a steady rate in 2.5 s, what is the force exerted on the balloon?

change in momentum of the air $= mv$
$= 0.002 \times 4$
$= 0.008$ kg m/s

$$\text{force} = \frac{\text{change in momentum}}{\text{time}} = \frac{0.008}{2.5} = 0.0032 \text{ N}$$

This is the force exerted on the air. By Newton's third law an equal and opposite force is exerted on the balloon.

EXERCISE 3

In questions 1–4 select the most suitable answer.

1 Which of the following is *not* a vector quantity?
A acceleration
B displacement
C force
D momentum
E speed

2 Which of the following is a unit for momentum?
A kg m s^{-1}
B kg m s
C kg m^{-1} s
D kg m s^2
E kg m s^{-2}

3 A horse exerts a force on a cart. Which of the following forces is the reaction force?
A the force of gravity on the cart
B the pull of the cart on the horse
C the force of friction between the horse's hooves and the road
D the force of friction at the axles of the cart
E the force of friction between the wheels and the road

4 At a certain instant during a rocket's journey to Mars the booster rocket is fired, doubling the speed and halving the mass. If the original momentum of the rocket is P units, the new momentum is
A $P/4$ units
B $P/2$ units
C P units
D $2P$ units
E $4P$ units

5 State Newton's first law of motion and give three examples which illustrate the law.

6 a What is meant by inertia?
b Using the term inertia, explain why a driver of a car is advised to wear a seat belt.
c A tray is covered completely with a layer of small ball bearings. A 1 kg mass is on a small piece of hardboard on top of the ball bearings and a smaller mass is on another small square of hardboard. How could you use this apparatus to demonstrate the difference in inertia between two masses? State clearly your observations.

7 Define the terms *force* and *momentum*. How are they related? The speed of a train of mass 200 tonnes is reduced from 72 km/h to 48 km/h in 2 minutes. Find
a the change in its momentum,
b the average value of the retarding force.

8 Two students test the centripetal force equation $F = \frac{mv^2}{R}$ with the arrangement shown the in figure. The idea is to rotate the mass on the end of the thread in a horizontal circle keeping the radius R constant. This is done by keeping the paper flag attached to the thread stationary *just below* the tube. The lower weights provide the tension in the thread.

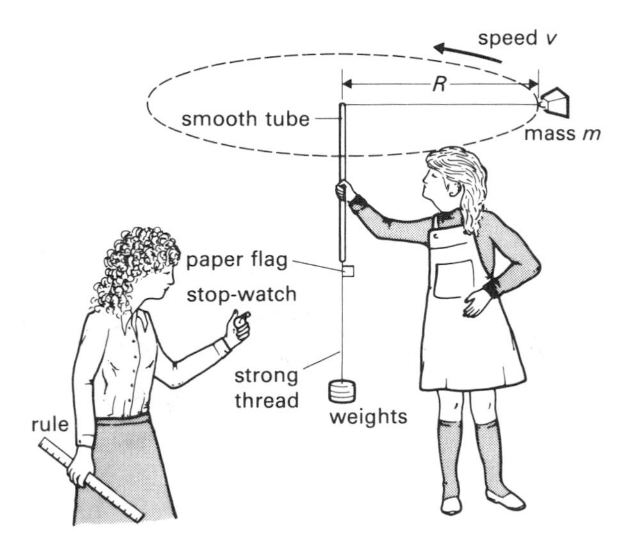

a In what direction does the centripetal force act?
b How is the size of the centripetal force found in this experiment?
c Describe how the radius R could be measured.
d Describe how the time t for one revolution would be measured.
e Why does the speed of the rotating mass equal $\frac{2\pi R}{t}$?
f If m = 0.01 kg, v= 2 metre/second and R = 0.5 m what should the centripetal force be?

9 It is required to cause a body of mass 500 g to accelerate uniformly from rest across a smooth horizontal surface so that it will cover a distance of 20 m in 4 seconds. Determine
a the acceleration which must be imparted to the body,
b the magnitude of the minimum force required to produce this acceleration and
c the momentum of the body at the end of the 4-second period.

10 If a net force acts on an object, the object accelerates.
a Give an everyday example of an accelerating object and say what provides the net force on it.
b Give an everyday example of an object at rest and explain what forces are in balance.
c Give an everyday example of an object moving at constant velocity and explain what forces are in balance.

11 An air rifle is used to fire a pellet into a block of Plasticine on a stationary toy train truck. From the measurements made, the speed of the fired pellet and other quantities can be calculated. The measurements are

mass of 100 pellets	= 110 g
mass of truck + Plasticine + one pellet	= 264 g
distance X	= 1 m
time for truck to travel distance X after pellet strikes truck	= 2 s

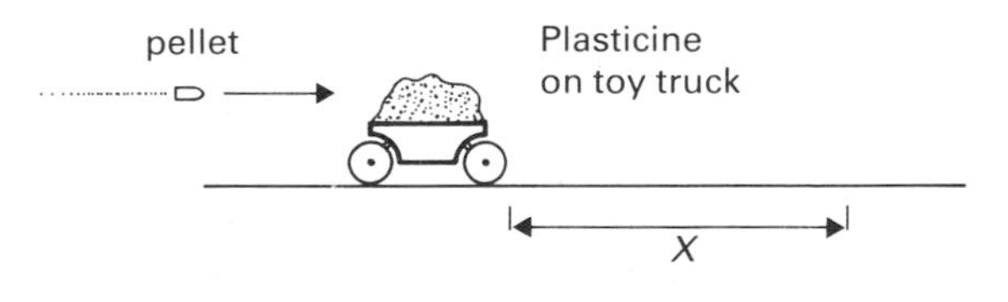

a Describe how the toy railway truck can be compensated for friction.
b What is the speed of the truck and contents after the pellet has hit it?
c What is the momentum of the truck and contents after the pellet has hit it? (State the units.)
d What is the mass of one pellet?
e Explain why measuring the mass of 100 pellets and then doing a calculation is better than measuring the mass of one pellet directly.
f Using the measurements taken in the experiment, the speed of the pellet before it hits the Plasticine works out at 120 metres/second. Show that momentum has been conserved in the collision.

12 When a man jumps ashore from a boat, the boat is simultaneously driven away from the bank. How is this explained? If the man and the boat are originally at rest and the man jumps from the boat with an initial horizontal velocity of 4 metres per second what is the velocity of the boat at that instant if the mass of the man is 75 kg and of the boat 200 kg.

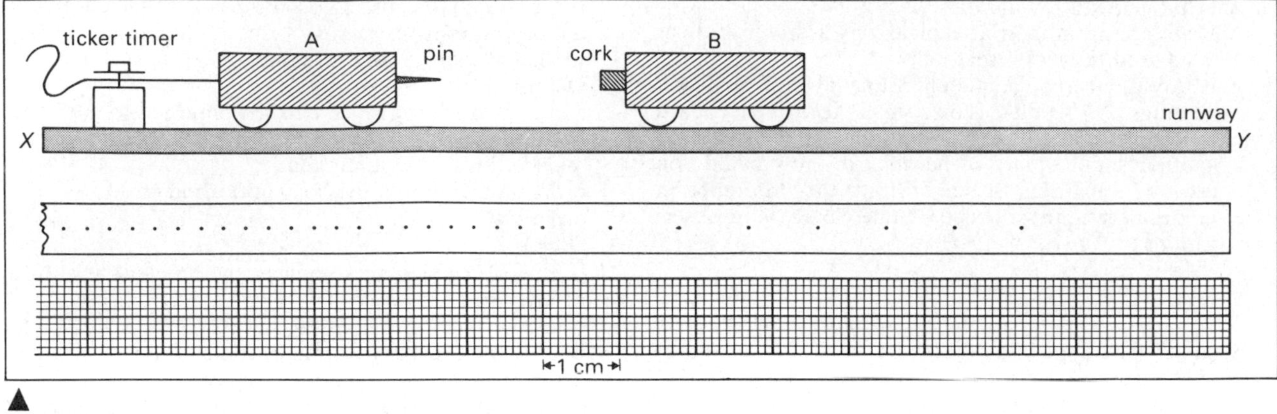

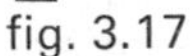
fig. 3.17

13 Figure 3.17 illustrates an experiment with trolleys and a ticker timer. Trolley A is given a slight push and collides with trolley B. The pin penetrates the cork so that both trolleys stick together and move as one. The tape obtained is shown, full size, alongside a millimetre grid in figure 3.17. The time interval between each dot is 0.02 s.

a Before carrying out the experiment, it is usual to make an adjustment which results in end X of the runway being slightly higher than end Y. (i) What is the purpose of this adjustment and why is it necessary? (ii) Has the adjustment been carried out in this case? Give a reason for your answer.

b Using the tape, determine (i) the average speed, in cm/s, of trolley A before the collision, (ii) the time which has elapsed before the collision takes place, and (iii) the average speed, in cm/s, of the two trolleys A and B after the collision.

c A textbook states 'When two bodies collide the total momentum is conserved'. Using the results given below, which were obtained in a further experiment, test the truth or otherwise of the above statement.

mass of trolley A	= 0.8 kg
mass of trolley B	= 2.4 kg
velocity of trolley A before collision	= 40 cm/s
velocity of trolley B before collision	= 0 cm/s
velocity of trolleys A and B after collision	= 10 cm/s

14 The diagram below shows a turbo jet engine.

a Write down the letters A to E and beside each put the name of the part to which the letter refers.

b Describe the working of this engine.

c What is the purpose of the turbine in this engine?

d Explain why the engine is suitable for travel at very high altitudes and yet not suitable for travel in space.

e A space rocket has no turbine and is much larger. Give reason.

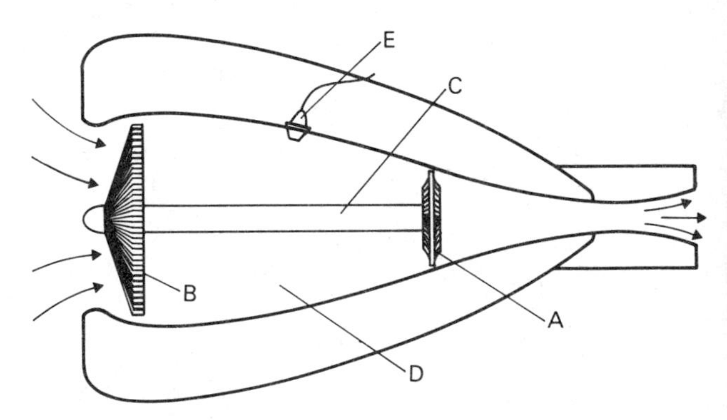

4 WORK, ENERGY AND POWER

The word 'work' is used in several different senses in everyday life. In science, work has a very precisely defined meaning. Mechanical work is done whenever anything is moved against a force or resistance. Notice that two factors are involved: there must be **movement**, and the movement must be against some **resistance**, figure 4.1.

It seems reasonable to argue that the bigger the force or resistance and the more movement you produce, the more work you will do. We measure the amount of work done by multiplying the size of the force by the distance moved.

work done = force × distance moved

The distance must always be measured in the direction in which the force is acting. In the case of lifting a weight against the force of gravity, the distance is measured vertically. In figure 4.2 the packing case is raised vertically by a distance of 6 metres and

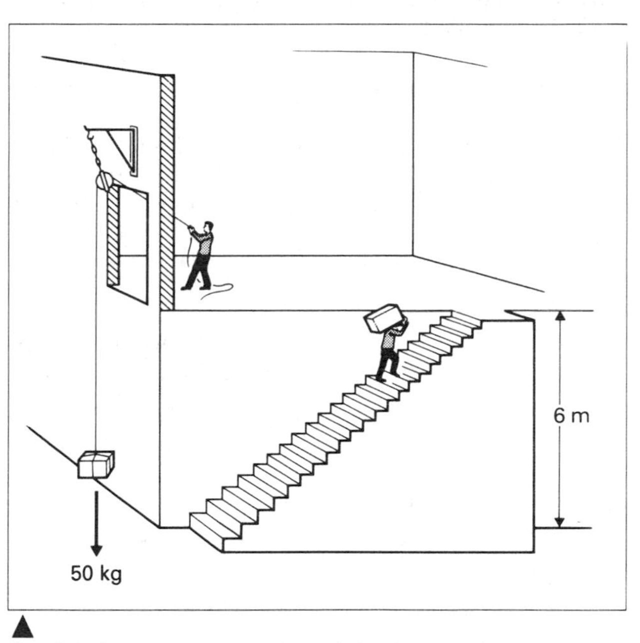

fig. 4.2 Same amount of work is done in both cases

fig. 4.1 Doing mechanical work

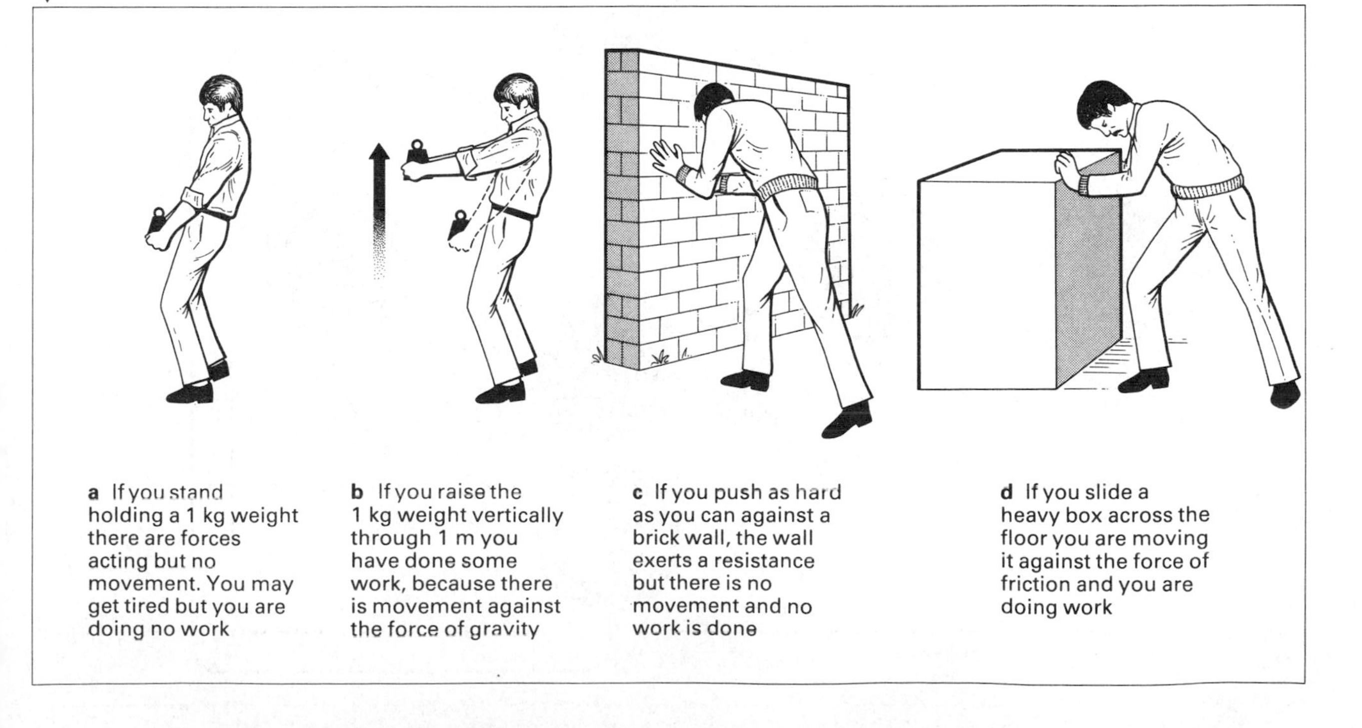

a If you stand holding a 1 kg weight there are forces acting but no movement. You may get tired but you are doing no work

b If you raise the 1 kg weight vertically through 1 m you have done some work, because there is movement against the force of gravity

c If you push as hard as you can against a brick wall, the wall exerts a resistance but there is no movement and no work is done

d If you slide a heavy box across the floor you are moving it against the force of friction and you are doing work

the work done is the same whether the case is hoisted up vertically or carried up the stairs. The actual distances travelled by the case may be different in the two ways of raising the case but the vertical distance is the same and so the work done is the same.

If the force is measured in newtons and the distance is measured in metres the work done is calculated in **joules (J)**.

One joule of work is done when a force of 1 newton moves through a distance of 1 metre measured in the direction of the force

joules = newtons × metres

In the example of the packing case the force downwards is 50 kg weight or 50 × 10 = 500 N. The distance moved is 6 m, so the work done is 500 × 6 = 3000 J.

EXAMPLES

In all examples for the calculation of work make sure that the force is in newtons and the distance travelled in the direction of the force is in metres.

Take 1 kg weight = 10 newtons.

1 Calculate the work done when
a a mass of 5 kg is lifted 50 cm,
b a mass of 300 g is lifted 150 cm.

a force = 5 × 10 = 50 N
distance = 0.5 m
work done = 50 × 0.5
= 25 J

b force = 0.3 × 10 = 3 N
distance = 1.5 m
work done = 3 × 1.5
= 4.5 J

2 A lift of mass 250 kg carries a man of mass 76 kg. How much work is done by the wire rope which draws the lift through a height of 15 m?

total mass = 250 + 76 = 326 kg
total force = 326 × 10 = 3260 N
distance travelled = 15 m
work done = 3260 × 15
= 48 900 J or 48.9 kJ

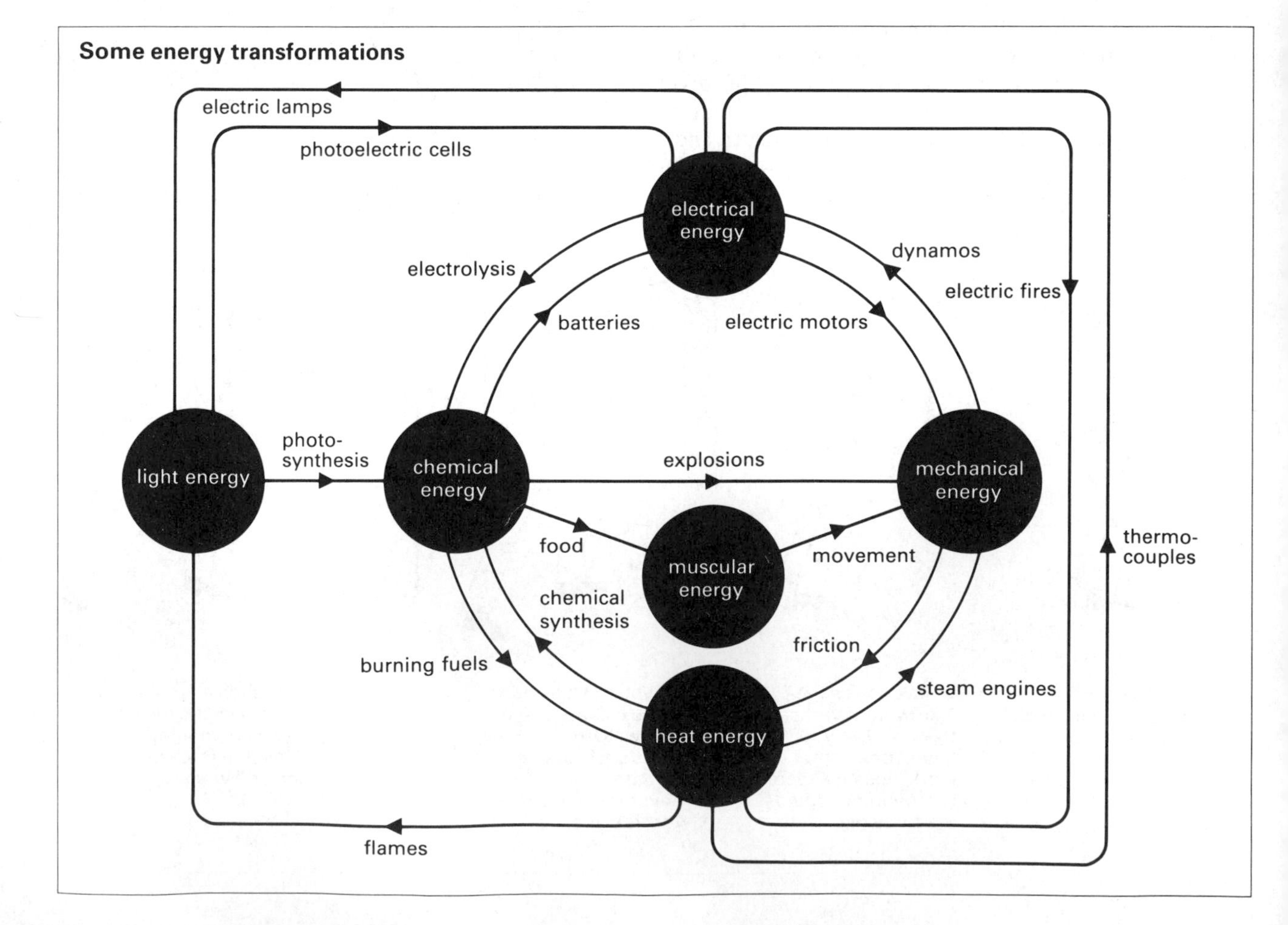

Energy

In order to do work we must have a source of energy. In the case of mechanical work, the source of energy produces the force which produces the movement. When we are lifting a weight we are using muscular energy, which in its turn comes from the stored chemical energy in the food we eat. Whenever work is done energy is needed.

Energy is the capacity for doing work

There are many different forms of energy. **Heat** is a form of energy. Heat can be transformed into **mechanical** energy in the steam engine, which can drive a generator to produce **electrical** energy. The electrical energy can be used to produce **light** energy and **sound** energy. **Chemical** energy is used in explosives, in the release of energy by the burning of fuels and by living things in the food they eat. Other forms of energy are **radiant** energy, **magnetic** energy and **nuclear** energy. Devices which convert one form of energy into another are called transducers.

One of the very important scientific ideas of the nineteenth century is that, although one form of energy can be converted into other forms, there is no loss of energy in the process. Energy cannot be created or destroyed; the total energy in the universe is constant. This principle is known as the **conservation of energy**.

Although no energy is actually lost in an energy conversion, nevertheless energy can be 'wasted'. It can be converted to a form in which it is not really useful to us. Generally, when one form of energy is converted into another form, some energy is wasted as heat. In the motor car engine, for example, we supply the energy in the form of chemical energy in the petrol. We want as much of this energy as possible to be converted into mechanical energy to drive the car along. Figure 4.3 shows that of the energy supplied over 70% is wasted as heat in the radiator and exhaust systems. More heat is produced by friction in various parts of the car. Only about 12% of the original energy is used to make the car move.

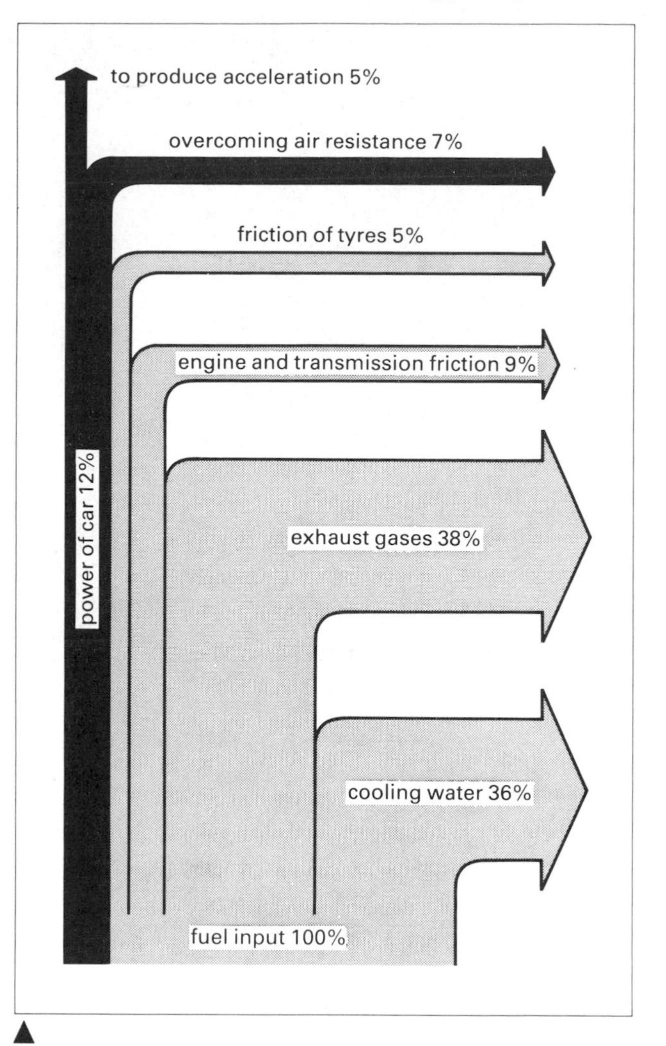

▲ fig. 4.3 Energy conversions in a motor car

Potential energy

When anything is lifted against the pull of gravity work has to be done. This work is stored in the lifted body. Because the body is higher than it was originally, it possesses more energy. When the heavy weight of a pile-driver is raised, work is done. The work done is stored in the weight as energy of position, figure 4.4. When the weight is released and falls, the energy of position is converted into energy of motion. This is the energy which drives the steel or wooden pile into the ground, doing work against friction and pushing the hard ground out of the way.

fig. 4.4 Energy of position ▼

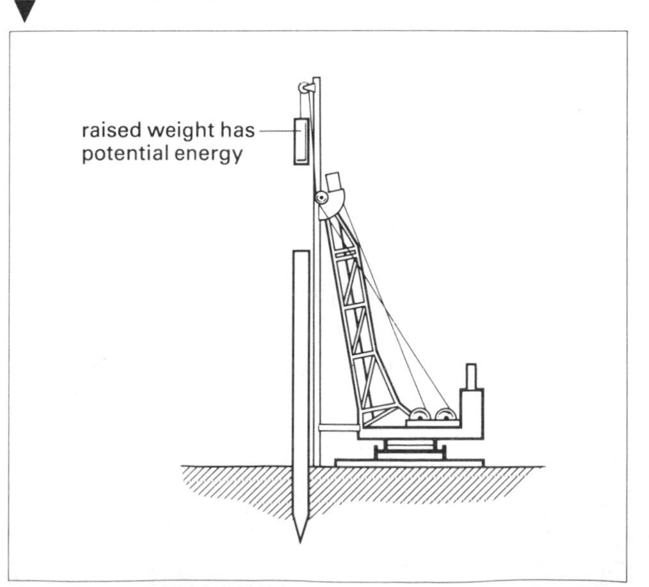

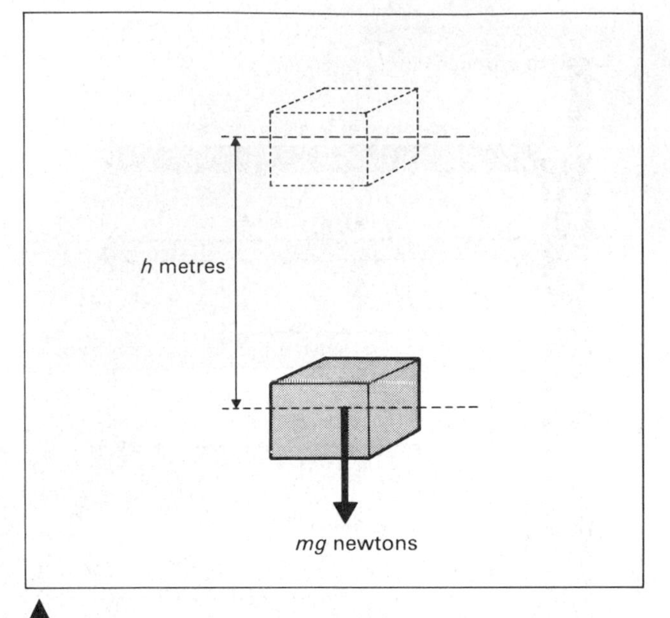

fig. 4.5 Raising a body increases its PE

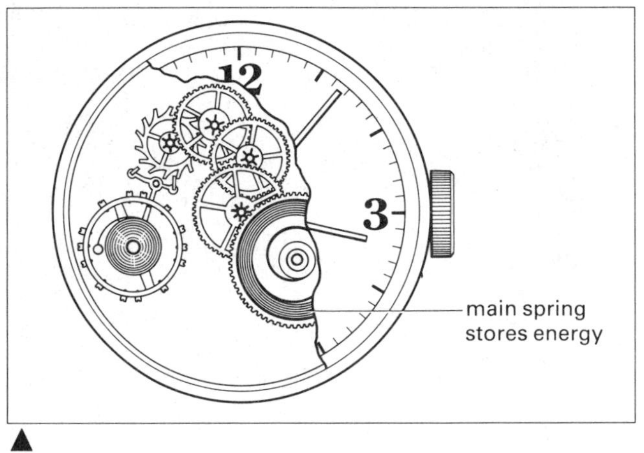

fig. 4.6 The wound-up spring of a clock or watch has stored potential energy which is slowly released as energy of motion in the moving hands

fig. 4.7 Conversion of PE into KE and vice versa

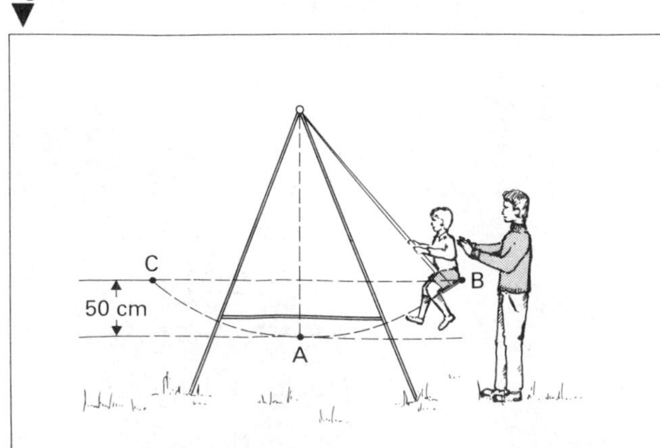

A body which possesses energy stored in it because of its position (or for any other reason) is said to have **potential energy**.

The potential energy of a body is the energy it possesses because of its position or state

The gravitational potential energy stored in a body when it is lifted to a new position can be easily found.

Figure 4.5 shows a body of mass m kilograms; the gravitational force on it is mg newtons. If the body is raised through h metres

work done = force × distance
= $mg \times h$
= mgh joules
work done = gain in gravitational potential energy

potential energy = mgh joules

Change in PE = weight × change in height

In figure 4.7 a boy weighing 40 kilograms is sitting on a swing. If he is pulled back so that he is raised vertically through 50 centimetres, the increase in his potential energy will be $mgh = 40 \times 10 \times 0.5 = 200$ J. When he is released he passes through A and (neglecting air resistance) would rise just as far to C at the other end of the swing. From B to A he is losing energy of position and gaining energy of motion. From A to C this energy of motion is used in raising him through 50 cm again. If there were no friction at all he would continue to swing backwards and forwards on his original store of 200 J.

Kinetic energy

In the example of the boy on the swing there was a constant interchange between energy of position (potential energy) and energy of motion. The energy a body possesses by virtue of its motion is called **kinetic energy**.

When a force acting on a body produces movement the force does work and the body possesses kinetic energy because of its motion.

Let a force F (N) move an object of mass m (kg) through a distance d (m), then

work done = force × distance
= $F \times d$
but force = mass × acceleration
so work done = $m \times a \times d$

The increase of velocity attained with an acceleration a acting through a distance d is given by

$$v^2 = 2ad \quad \text{or} \quad d = \frac{v^2}{2a}$$

$$\text{therefore work done} = m \times a \times \frac{v^2}{2a}$$
$$= \tfrac{1}{2} mv^2 \text{ J}$$

The gain in kinetic energy of the moving body = work done = $\frac{1}{2}mv^2$

kinetic energy = $\frac{1}{2}mv^2$ joules

There are many examples of the conversion of potential into kinetic energy and vice versa. The raised hammer in figure 4.11 possesses potential energy. As the hammer is brought down this turns to kinetic energy which we increase by making the hammer move faster. When the nail is driven into the wood the energy is converted into heat and strain energy in the fibres of the wood.

In figure 4.12 the car at A on the switchback has potential energy. This is converted to kinetic energy at B and then again to potential energy at C. The potential energy at C is less than that at A because work has been done in overcoming friction, so the heights of the hills must get gradually less as the ride proceeds.

▲
fig. 4.8 The stretched string of the bow possesses potential energy which will be converted into the kinetic energy of the moving arrow

fig. 4.9 These boys on their sledge possess potential energy. (Where did it come from?) When they slide down the hill it is converted into kinetic energy
▼

▲
fig. 4.10 The water at the top of a waterfall possesses both potential and kinetic energy. When it reaches the bottom it is all kinetic energy and could be used to drive a turbine

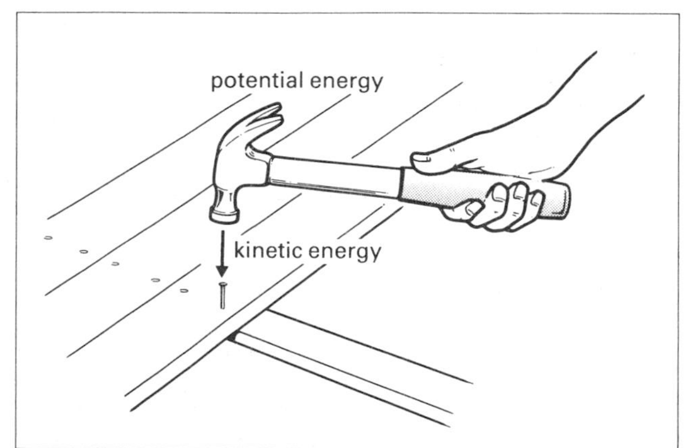

▲
fig. 4.11 Energy conversion in a hammerhead

fig. 4.12 Principle of the switchback
▼

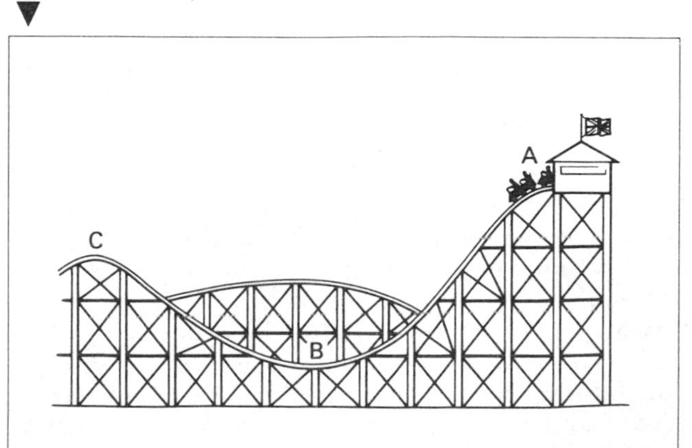

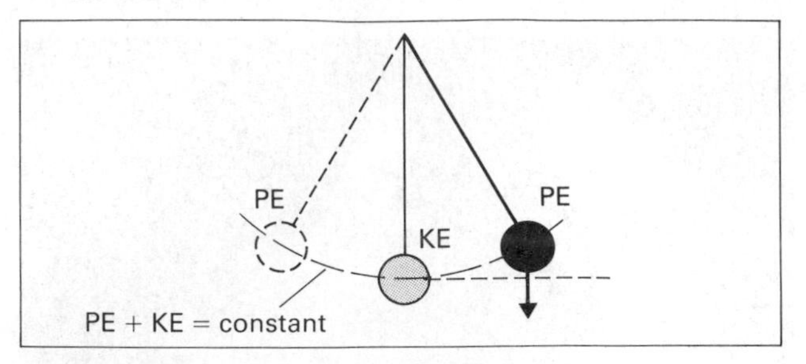

fig. 4.13

When a pendulum bob is drawn back to one side the bob rises above its rest position and gains potential energy, figure 4.13. When the bob is released this potential energy is converted into kinetic energy. At the lowest point, the bob is moving at its maximum speed and momentarily all the energy is kinetic. As the bob rises on the other side of the swing it loses kinetic energy and gains potential energy. If we neglect air resistance the sum of the PE and KE at any point is constant.

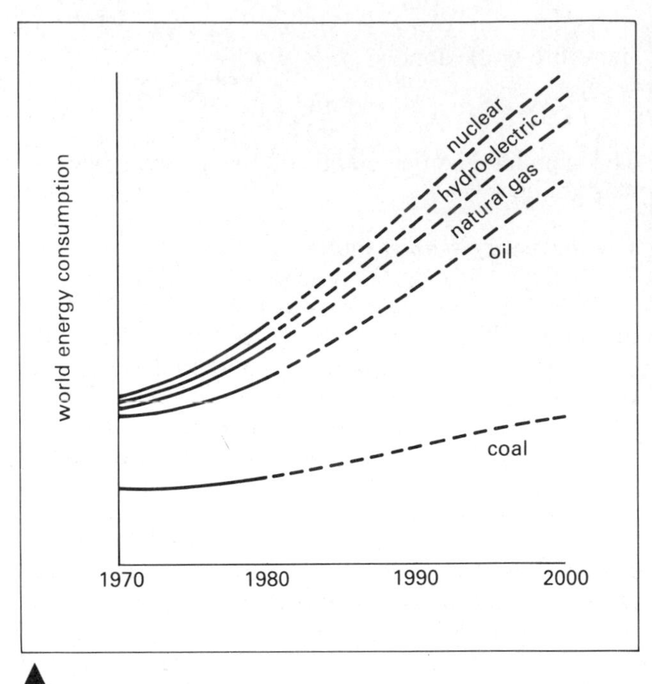

fig. 4.14 Trends in world energy consumption

Sources of energy

The countries of the world are using more and more energy. As countries become industrialised their rate of energy consumption increases rapidly, figure 4.14. It is important to list our sources of energy and if possible estimate how long they will last. Predicting the future is always difficult and often inaccurate. Nevertheless, we must try to do it as well as we can, because over-use of a particular source of energy today could have very serious effects in the future.

Muscular energy

Man and animals use their muscles to make things move. The energy comes from the chemical energy which is stored in food, and which is produced by solar energy in the process of photosynthesis carried out by green plants. Because the forces which can be exerted by even large numbers of men are relatively small, machines such as the pulley and the lever were invented to make work easier to do, figure 4.15.

fig. 4.15 Teams of many men were needed to raise the great stone pillars of ancient monuments

Wind power

Moving air can exert a very large force. This is obvious from the damage done by hurricanes. The energy of moving air is used in sailing ships and windmills. Wind power is at present being seriously investigated as a source of power for generating electricity and new types of wind engine are being tried out. In new types of windmill the sails revolve about a vertical axis so that they are turned steadily whatever the direction of the wind.

fig. 4.16 This wind turbine in South Wales supplies 200 kW of electricity to the national grid

▲
fig. 4.17 The turbine hall in a hydroelectric power station

Water power

The power of moving water was used to drive water wheels, and is now used in **hydroelectric** schemes where the kinetic energy of the falling water drives a turbine which in turn drives a generator. The energy which originally lifted the water to the high reservoirs came from the sun when evaporated water from the sea and lakes later fell as rain. When the demand for electricity is low the power station can use some of the power generated to pump water up to the high reservoir in reserve for when the demand increases.

Fossil fuels

The main modern sources of energy are the fuels **coal**, **oil** and **natural gas**. Coal is the fossilised remains of plants which flourished two hundred million years ago. Oil comes from the decomposition of marine organisms compressed under successive layers of sediment on the sea bed. Natural gas is mainly methane and is produced by the decay of material containing carbon. It is usually found associated with oil deposits. In all these cases the fuels contain chemical energy which can be released as heat on combustion, figure 4.18. The energy came originally from the sun. In the **steam engine** the heat is used to turn water into steam, and the pressure of the steam moves a piston sliding in a cylinder. In **petrol and oil engines** the fuel is burnt inside the cylinder and the hot expanding gases move the piston. These engines are known as **internal combustion engines**.

Solar energy

Fuels are storehouses of solar energy which is set free when the fuel burns. It should be possible to make use of the solar energy reaching the earth directly without waiting millions of years for the production of coal and oil. The problem is only just beginning to be solved and is discussed in more detail on pp. 126–127. In nature, solar energy is absorbed by green plants to bring about the conversion of carbon dioxide and water into sugars and later into starch and cellulose. Although only between 1% and 2% of the sun's energy is used in this way, it is at present the most efficient way of utilising solar radiation. The best thing to do may well be to grow large crops of green algae (a kind of pond weed). This can either be dried and burnt as a fuel or fermented to produce alcohol for use in internal combustion engines.

fig. 4.18 Fossil fuels
▼

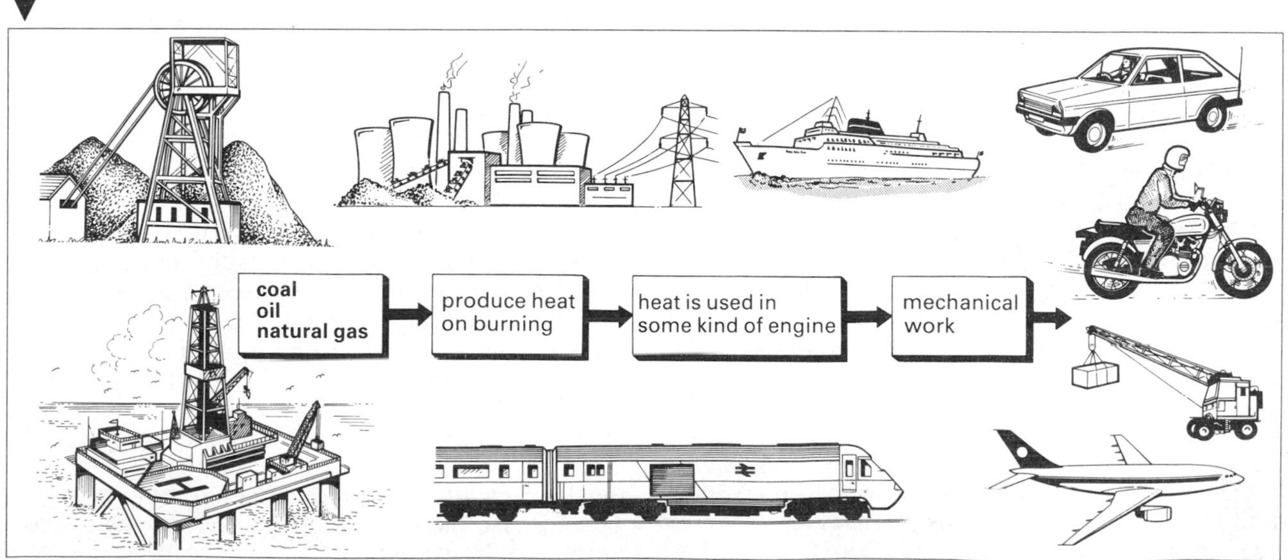

▲
fig. 4.19 Principle of harnessing tidal energy using a barrier across a river estuary

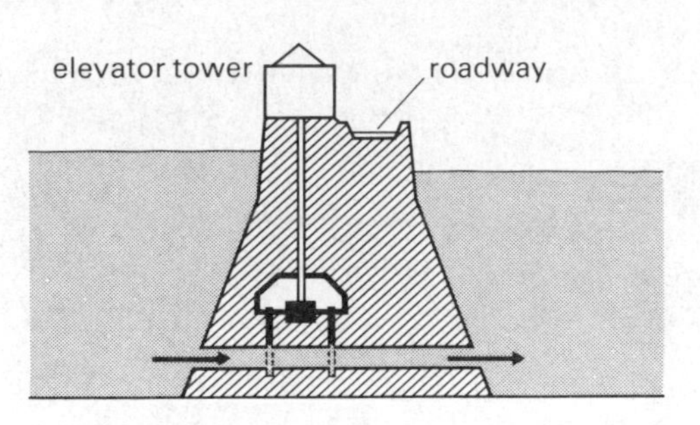

a High tide: water flows in through sluice gates

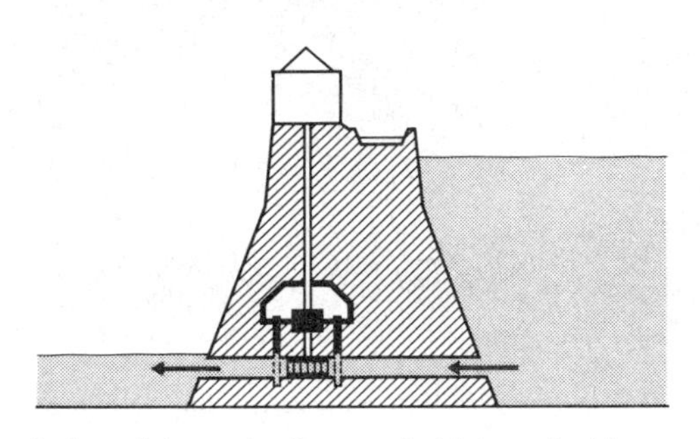

b Low tide: water flows out, driving a turbine

Geothermal energy

At many places on the earth, hot water and steam gush out of the surface as hot springs and geysers. The heat comes from the vast store of heat inside the earth. Holes can be bored and water pumped down one tube to be extracted as hot water up another tube. There are geothermal power stations in various parts of the world. At Larderello, in northern Italy, a geothermal power station produces sufficient electricity to run most of Italy's railway system.

Tidal and wave energy

The ebbing and flowing of the tides have been going on ever since the seas were formed. If only a part of the energy of this vast amount of moving water could be harnessed, we should have a very useful addition to our power supply. In theory the 'tide trap' is simple. A dam is built across an estuary. When the tide rises the sluice gates are opened and the tide flows to the landward side of the dam, figure 4.19a. When the tide begins to fall the gates are closed and the trapped water is allowed to flow out again, driving a turbine to generate electricity, figure 4.19b. The system is only worth considering when the difference between the tide levels is about 8 metres and the generators could only be worked for certain periods each day, but plans have been considered for building a tide trap across the Severn estuary.

The sea's surface, in addition to being affected by the rising and falling of the tides, is in constant motion in the form of waves. There is enough energy in the waves in the seas around Britain to generate half the present consumption of electricity. The problem is to extract this energy and convert it into a usable form. There are several ideas for doing this, one of which is shown in figure 4.20. The floats oscillate as the waves pass them and this oscillation operates an electric generator. The idea is attractive because winter, when the seas are roughest, is just the time when we need most electricity.

▲
fig. 4.20 Extraction of wave energy using a system of 'nodding' ducks or floats

Nuclear power

When the nuclei of large atoms such as uranium 235 are split by bombarding them with neutrons a very large amount of energy is released. The structure of atoms and details of the release of nuclear energy are given in Chapter 27. The atom-splitting process takes place in an atomic pile, and the heat produced is transferred to a boiler by a heat circulating system. Steam is produced to drive a turbine and generate electricity. Precautions have to be taken to prevent the leakage of dangerous radiation from the pile. There is also the very serious problem of the disposal of the radioactive waste.

Power

So far, in discussing work and energy, we have only considered how much work has been done or how much energy has been used. It is equally important to know how quickly a certain amount of work can be performed. When we speak of **power** we mean how quickly work is done.

Power is rate of doing work

$$\textbf{power (watts)} = \frac{\textbf{work done (joules)}}{\textbf{time taken (seconds)}}$$

When one joule of work is done in one second the rate of working is one **watt (W)**.

1000 watts = 1 kilowatt (kW)

The unit of power, the watt, is named after James Watt, the eighteenth-century engineer who wanted to sell his steam engines to mine owners. He wished to be able to say that his engines could do the work more quickly than horses could. His unit of power was the horse-power. One horse-power is equal to 746 watts.

EXAMPLE

A fork lift truck can raise a load of 315 kg to a height of 2 m in 20 seconds. What is the power developed?

$$\text{work done} = 315 \times 10 \times 2 = 6300 \text{ J}$$

$$\text{power} = \frac{\text{work done}}{\text{time taken}} = \frac{6300}{20} = 315 \text{ W}$$

Note: this is just under half a horse-power.

fig. 4.21 Fork-lift truck stacking empty containers

fig. 4.22 Measuring your own power

You can find your own power quite easily. Measure your weight, then get someone to time you with a stopwatch as you run up a flight of stairs as quickly as you can. Measure the vertical height you have raised yourself, figure 4.22. (What is the simplest way of doing this?)

$$\text{work done} = \text{weight (N)} \times \text{height (m)}$$

$$\text{work done per second} = \frac{\text{weight (N)} \times \text{height (m)}}{\text{time (s)}}$$

$$\text{power developed} = \frac{\text{weight (N)} \times \text{height (m)}}{\text{time (s)}}$$

You could not keep up this high rate of working for too long. An average man can work at about $\frac{1}{10}$ horse-power, or 75 watts.

EXERCISE 4

For these questions the acceleration due to gravity can be taken as 10 m/s^2 and the earth's gravitational field strength as 10 N/kg.

In questions 1–5 select the most suitable answer.

1 Which of the following is a scalar quantity?
A acceleration
B energy
C force
D momentum
E velocity

2 How much work is done when a mass weighing 400 N is lifted 3 metres?
A none
B 403 N m
C 1200 J
D 1200 W
E 133 J

3 A block of wood is pulled along a horizontal bench at a constant speed of 15 m/s by a force of 8 N. How much work is done against friction in 6 seconds?

A 720 J
B 120 J
C 48 J
D 20 J
E 3.2 J

4 An object is raised a few metres above the ground and then dropped. Neglecting the effect of air resistance, where in its fall will the gravitational potential energy and the kinetic energy be equal?

A at the instant it is released
B a quarter of the way down
C half the way down
D three-quarters of the way down
E all the way down

5 A mouse of mass 0.03 kg runs 2 m up a curtain in 4 s. Its power, in watts, is

A $0.03 \times 2 \times 4$
B $\dfrac{0.03 \times 2}{4}$
C $\dfrac{0.03 \times 4}{2}$
D $\dfrac{0.03 \times 10 \times 2}{4}$
E $\dfrac{0.03 \times 10 \times 4}{2}$

6 A lorry is loaded with 2000 kg of earth in 2 minutes by a mechanical digger. If each load has to be raised through a height of 3 metres, calculate

a the work done in loading the lorry,
b the power developed.

7 A wind pump raises 2500 N of water to the surface from 6 m below ground level in 300 seconds.

a Find the work done by the pump in that time.
b Calculate the power output of the pump.

8 Fig 4.23 shows a boy climbing some stairs. Each step rises 0.2 m and there are 10 steps. The boy weighs 500 N.

a How much work does he do in climbing the stairs?
b If he takes 10 s to climb the stairs, what power does he produce?

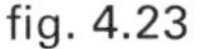
fig. 4.23

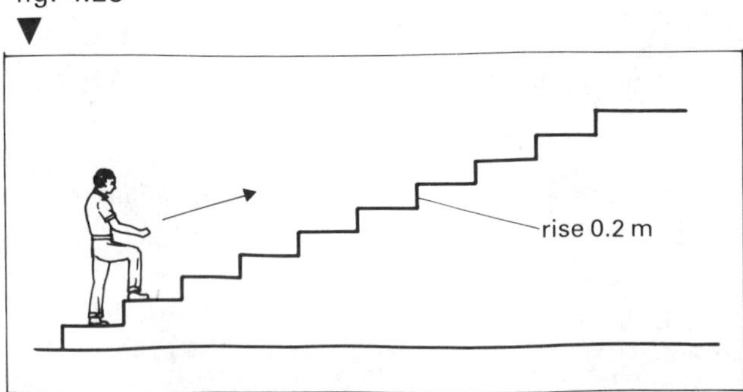

c How much work will he do in climbing the stairs if he takes two steps at a time?

9 a Joule, watt, kilowatt, kilowatt-hour. Which two of these units are units of energy, and which are units of power?

b Briefly describe how you would carry out an experiment to find your own useful power by running up a flight of stairs.

c Suggest a set of results for the experiment described in **b** and show how to calculate the answer for your own useful power. (Give the answer in any convenient unit of power.)

10 The diagram shows a ball A on three frictionless curtain runways.

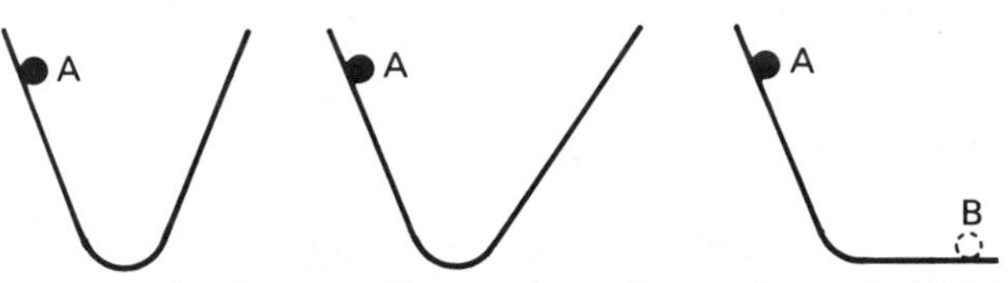

a On the first two figures draw the maximum heights the ball could reach on the other sides of the tracks.

b In the third figure what will be the energy of the ball in position B if it had 10 J of energy in position A?

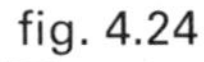
fig. 4.24

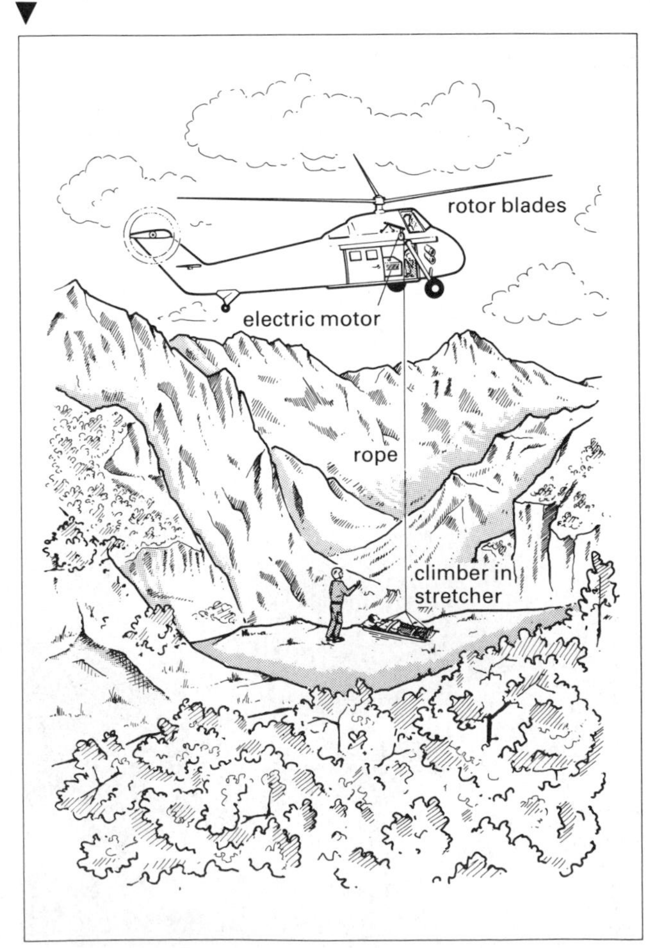

11 In a mountain rescue operation a helicopter hovers over an injured climber strapped to a stretcher. (See figure 4.24.)

a The total mass of the helicopter is 1555 kg. What upward force must be produced by the rotation of the rotor blades to keep the helicopter at a constant height? Give your reasoning.

b With the helicopter hovering at a constant height, the stretcher is hooked on to the rope which is then wound up by means of an electric motor in the helicopter. After a brief initial acceleration, the climber and stretcher are raised at a constant speed through a height of 30 m in 2 minutes. The total mass of the stretcher and climber is 100 kg.

When the climber and stretcher are being raised at constant speed what is (i) the tension on the rope? (ii) the upward force which must be produced by the rotating rotor blades so that the helicopter can remain at constant height? (iii) the output power of the electric motor?

c Give one reason why the *input* power to the electric motor is greater than the value calculated in **b** (iii).

d State how the tension in the lifting rope while the climber and stretcher are being raised at constant speed compares with the tension on the lifting rope during the period of initial acceleration. Give a reason for your answer.

12 Choose from the list below the types of energy which best complete *each* statement. You may choose any of the examples more than once. An example is done for you.

kinetic, potential, heat, chemical, nuclear, sound, electrical

Example: In a car accelerating from rest the *CHEMICAL* energy of the fuel is changed into the *KINETIC* energy of the car.

a In the element of an electric fire .... energy changes into .... energy.

b In a clockwork toy the .... energy of the spring changes into .... energy when it starts to move.

c In a microphone the .... energy produced by a musical instrument changes into .... energy.

d In a power station the .... energy of the uranium fuel changes into .... energy coming from the generator.

e As an accumulator is being charged .... energy is being changed into .... energy.

13 Three equal lengths of metal track are curved into the shapes shown in figure 4.25 and fixed vertically so that spheres can run on the tracks.

a Spheres are placed on the tracks at A and released. State, with reasons, whether each sphere will reach the end of its track.

b Describe the energy changes of the sphere running on track 1.

c Sketch a velocity–time graph for the sphere on track 1, explaining which part of the graph relates to each part of the track.

d A small light sphere and a heavier sphere are allowed to run separately on track 1. Compare the motions of the spheres, giving reasons for any differences or similarities.

fig. 4.25

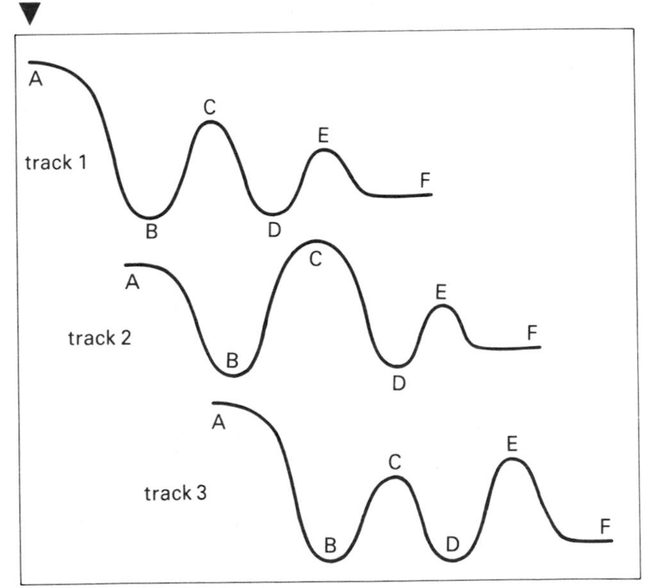

5 MACHINES

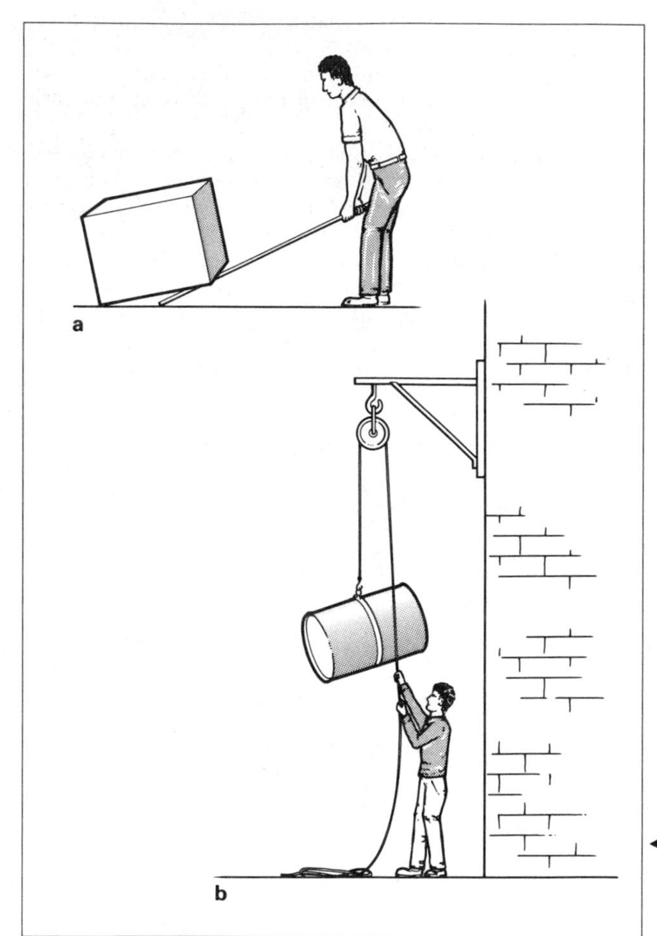

A machine is any device which enables us to move things more easily. For example, with a simple lever we can move a heavy object using a small force, figure 5.1a. With a single pulley we can raise a load through a distance greater than our own height, figure 5.1b.

The principle of all machines is the same, figure 5.2. We apply a force, the **effort** (E), to the machine, and something called the **load** (W) is moved by the machine. Ideally we want to move a large load using a small effort.

The ratio $\dfrac{\textbf{load}}{\textbf{effort}}$

is called the **mechanical advantage** of the machine.

The distances moved by the effort and the load are usually different.

The ratio $\dfrac{\textbf{distance moved by effort}}{\textbf{distance moved by load}}$

is the **velocity ratio** of the machine.

Machines make the transfer of energy more convenient. They can be regarded as force multipliers and sometimes as distance (or speed) multipliers.

fig. 5.1 Two types of machine

fig. 5.2 The principle of the machine

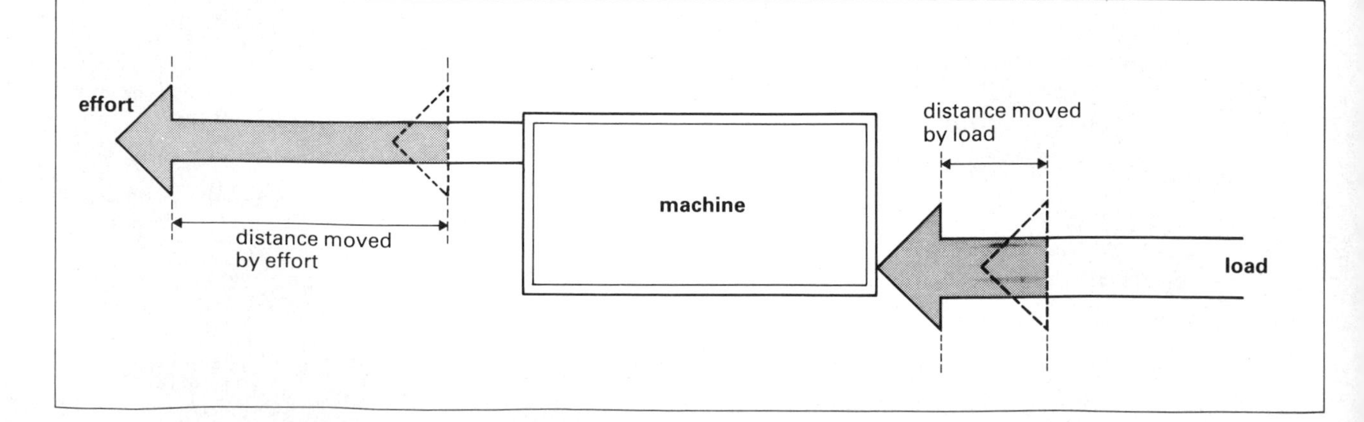

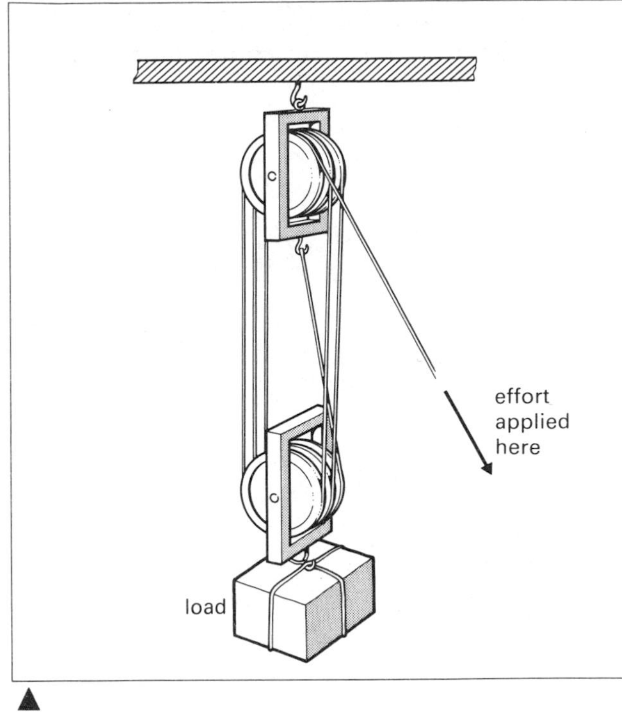

fig. 5.3 Block and tackle

fig. 5.4 Velocity ratio = 6

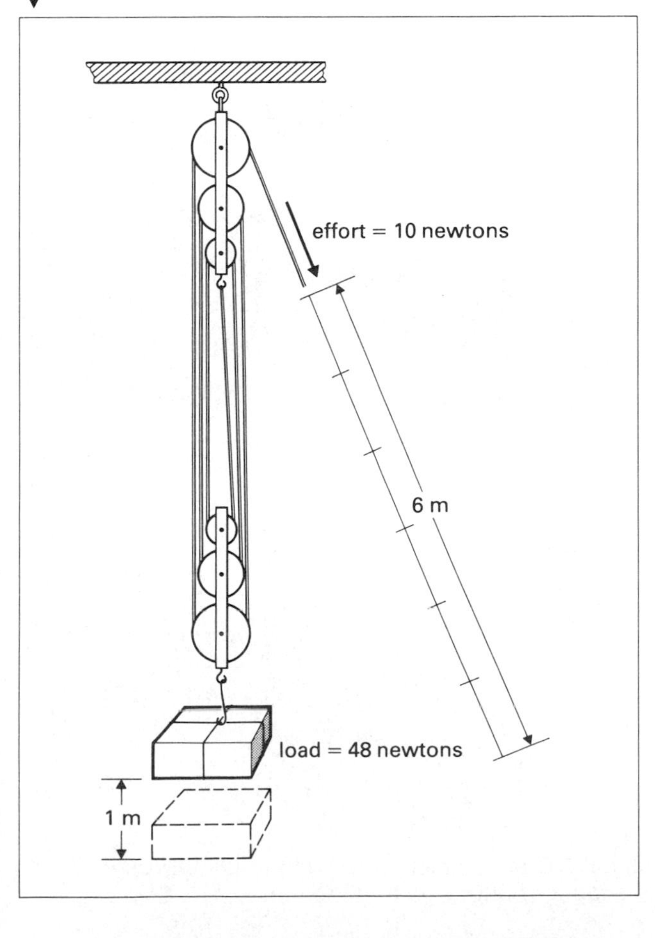

Pulley systems

A useful pulley system is the **block and tackle**, figure 5.3. In this machine a single rope passes round the pulleys which are arranged in two blocks or sheaves. Usually there is the same number of pulleys in each block. The mechanical advantage is theoretically equal to the number of ropes supporting the lower block (six in the diagram). In practice, however, it is less than this because of the friction at the pulley bearings and the weight of the lower pulley-block which also has to be lifted with the load.

The velocity ratio can always be worked out theoretically from a plan of the machine. In the block and tackle shown in the diagram, if the effort is pulled down 6 m the rope between the pulleys is shortened by 6 m. This shortening is divided among the six sections of the rope supporting the lower block. Each section shortens by 1 m and the load is lifted 1 m. So the velocity ratio is 6.

For a block and tackle with three pulleys in each block we know that the velocity ratio is 6. By experimenting we find that an effort of 10 N will lift a useful load of 48 N, figure 5.4. When the effort moves 6 m we do

$10 \times 6 = 60$ J of work

When the load is raised 1 m

$48 \times 1 = 48$ J of work is done

So we see that more work is put into the machine than we get out. The missing 12 J of work has been used in overcoming friction and lifting the lower pulley block. The fraction

$$\frac{\textbf{work got out}}{\textbf{work put in}}$$

tells us how **efficient** the machine is. It is usually calculated as a percentage.

$$\begin{aligned}\text{percentage efficiency} &= \frac{\text{work got out}}{\text{work put in}} \times 100 \\ &= \frac{48}{60} \times 100 \\ &= 80\% \text{ in this case}\end{aligned}$$

The velocity ratio can be worked out from the dimensions of the machine, but the mechanical advantage has to be determined experimentally for each machine because friction is a very variable quantity. (A well-oiled machine will have much less friction than a rusty one, for example.) When the velocity ratio and the mechanical advantage are known the efficiency can be worked out.

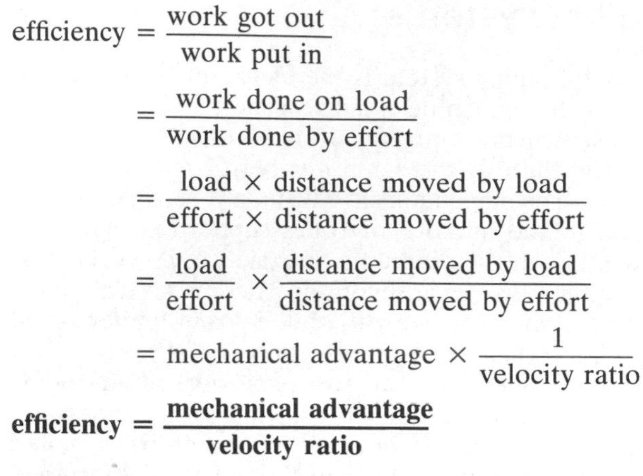

$$\text{efficiency} = \frac{\text{work got out}}{\text{work put in}}$$

$$= \frac{\text{work done on load}}{\text{work done by effort}}$$

$$= \frac{\text{load} \times \text{distance moved by load}}{\text{effort} \times \text{distance moved by effort}}$$

$$= \frac{\text{load}}{\text{effort}} \times \frac{\text{distance moved by load}}{\text{distance moved by effort}}$$

$$= \text{mechanical advantage} \times \frac{1}{\text{velocity ratio}}$$

$$\textbf{efficiency} = \frac{\textbf{mechanical advantage}}{\textbf{velocity ratio}}$$

The efficiency of a machine can never be greater than *one*, otherwise more work would be got out than would be put in. This would be a contradiction of the principle of the conservation of energy.

Experiment to investigate the variation of mechanical advantage and efficiency with the load

The velocity ratio of a machine is a constant depending on the particular design and dimensions of the machine, but as has already been stated the

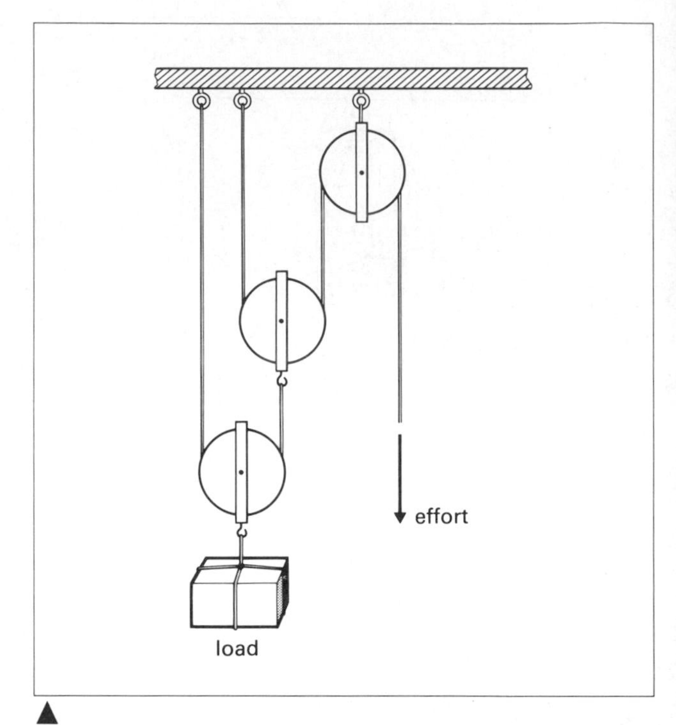

▲
fig. 5.5 This pulley system, which is different from the block and tackle, has two movable pulleys. The top pulley enables the effort to be made downwards. The velocity ratio is 4

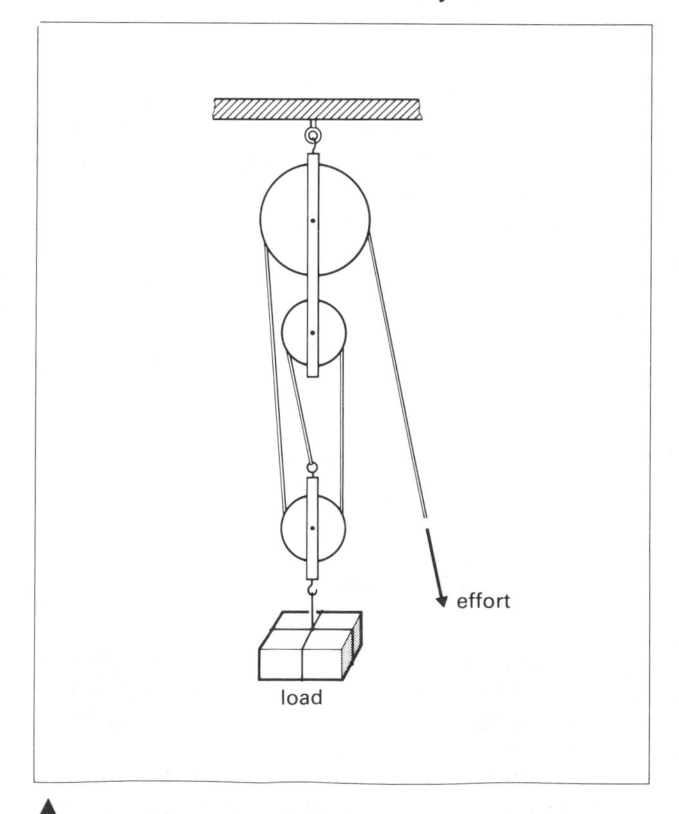

▲
fig. 5.6 With two pulleys in the upper block and one in the lower block the velocity ratio is 3

▲
fig. 5.7 Crane snatch block with multi-fall ropes capable of lifting up to 40 tonnes

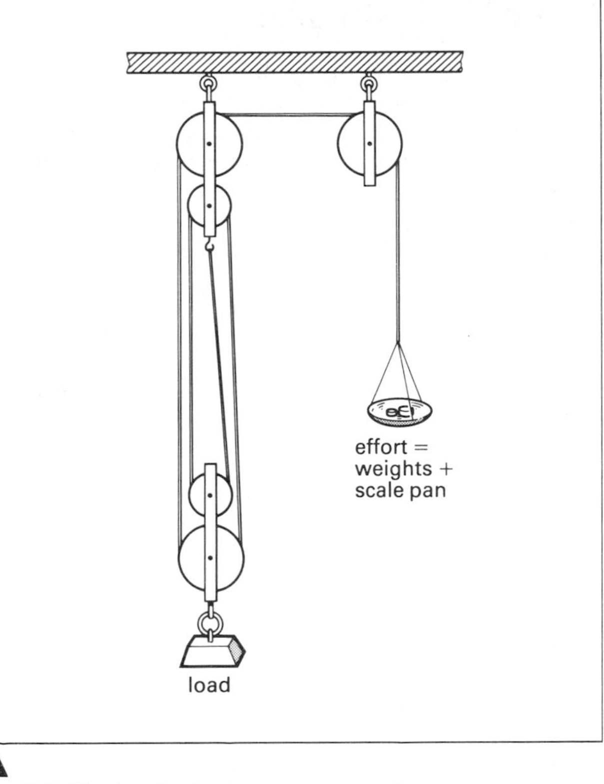

fig. 5.8 Mechanical advantage experiment

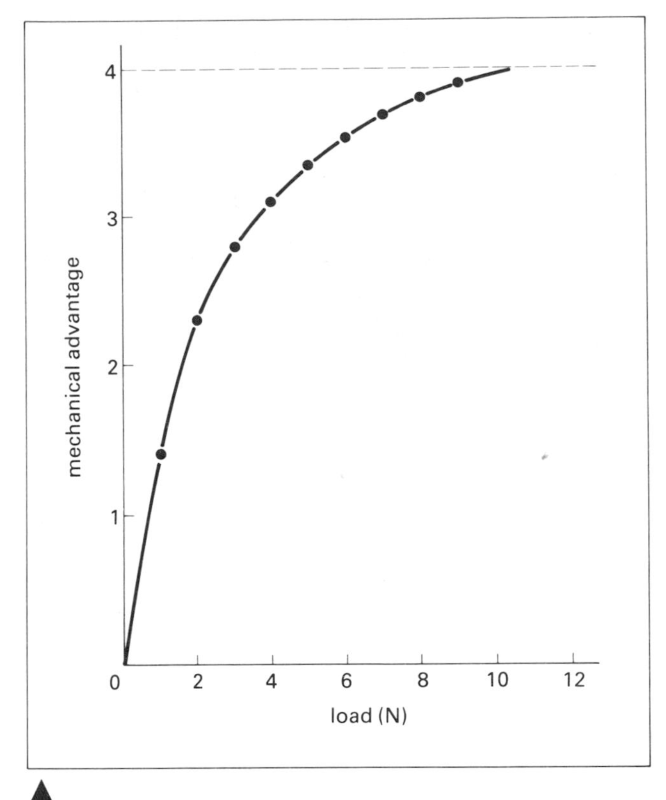

fig. 5.10 Mechanical advantage of a block and tackle

mechanical advantage varies and never in practice reaches its theoretical value.

The variation of the mechanical advantage of a machine with the load lifted is investigated as follows. The example shown in figure 5.8 is a block and tackle with two pulleys in each block but any simple machine can be used.

A load is hung on the lower block and weights are added to the scale pan until the load rises steadily when it is given a gentle start. The effort is the weight of the scale pan plus the added weights. The experiment is carried out for several different increasing loads. The velocity ratio is found by measuring a corresponding pair of effort and load distances. It can be assumed that the velocity ratio in this particular case is 4. The load and effort are measured in newtons remembering that the force in newtons is the mass in kilograms multiplied by g which is taken as 10 m/s^2. For each load the mechanical advantage and efficiency are worked out and tabulated as shown in figure 5.9. Graphs are drawn of mechanical advantage and efficiency against load.

Typical graphs are illustrated in figures 5.10 and 5.11. It will be seen that both mechanical advantage and efficiency increase as the load increases, and approach their limiting values: in this case 4 and 100%. The lower pulley block has to be lifted along with the load, but as the load increases the weight of the lower block is proportionately less compared with the load and so has less effect on the mechanical advantage. Friction also is proportionately less as the load increases.

fig. 5.9

Load (N)	Effort (N)	Mechanical advantage $= \frac{\text{Load}}{\text{Effort}}$	Efficiency $= \frac{\text{MA}}{\text{VR}} \times 100$

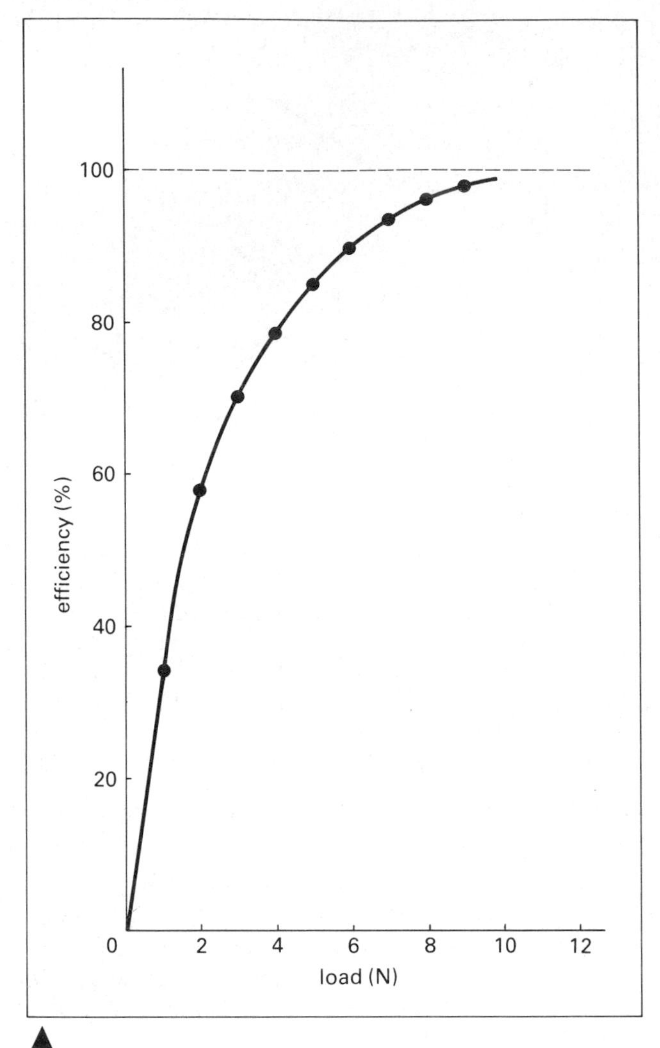

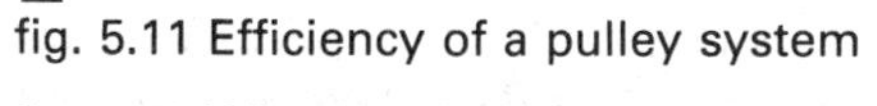

fig. 5.11 Efficiency of a pulley system

fig. 5.12 Wheel and axle

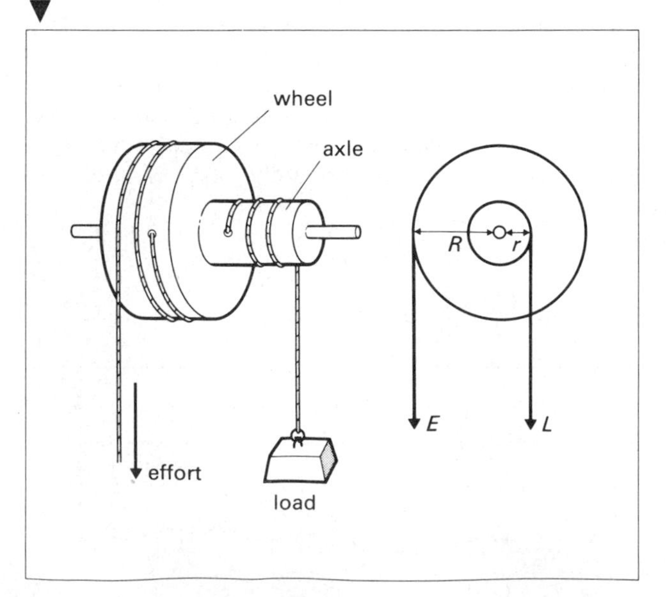

The wheel and axle

The wheel and axle can be thought of as two pulleys of different diameters fixed to the same axle, figure 5.12. The effort rope is wound round the larger pulley and the load rope round the smaller in the opposite direction. When the effort rope is pulled, the load rises.

When the wheel turns once the effort rope moves through a distance equal to the circumference, and the load moves through the circumference of the smaller pulley.

$$\text{velocity ratio} = \frac{\text{distance moved through by effort}}{\text{distance moved through by load}}$$

$$= \frac{2\pi R}{2\pi r}$$

$$= \frac{R}{r}$$

The inclined plane

The simplest and perhaps the most widely used machine is the inclined plane. If you want to load a packing case on to a lorry, you will find it easier to slide the case up a smooth plank rather than lift it straight up, figure 5.13a. The amount of work you do is more because there is work against gravity in raising the case, and also work against friction when the case slides along the plank. But you can move the case with a smaller effort. Early builders used the inclined plane to raise the heavy blocks of stone they used and now a builder uses it when he wheels his barrow up a plank, figure 5.13b.

fig. 5.13 Two uses of the inclined plane

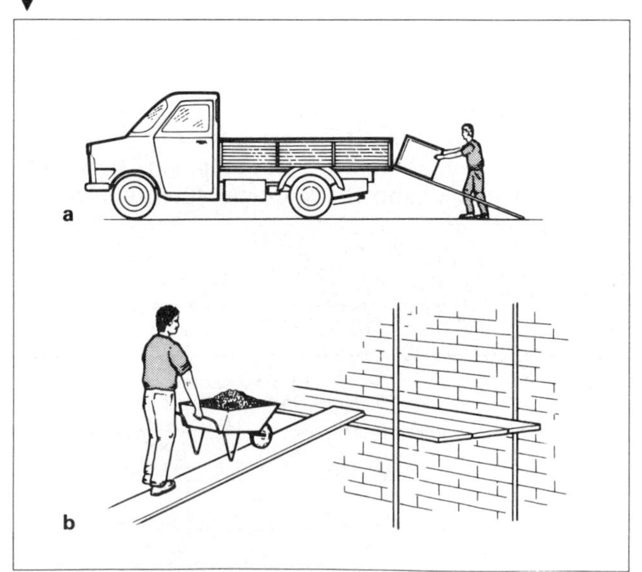

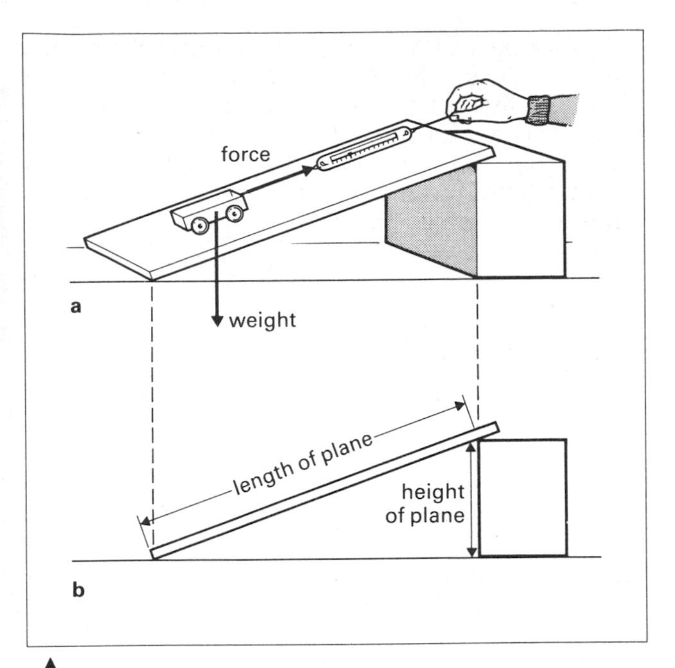

fig. 5.14

Looking at figure 5.14 and denoting mechanical advantage by MA and velocity ratio by VR, we have

$$\text{MA} = \frac{\text{weight of object}}{\text{force needed to pull it up plane}}$$

$$\text{VR} = \frac{\text{distance moved by effort along plane}}{\text{distance moved by load vertically}}$$

$$= \frac{\text{length of plane}}{\text{height of plane}}$$

The longer and more gradual the slope, the easier it is to move the object. It is easier to zig-zag up a steep hill than to climb directly up it.

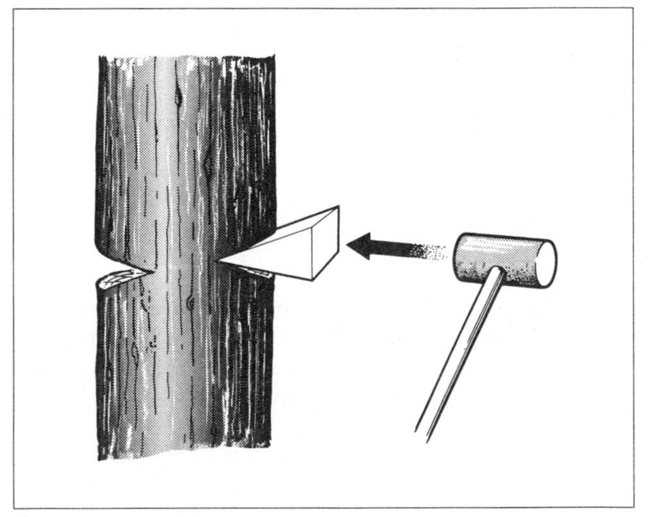

fig. 5.15 The wedge is a double inclined plane. It is useful when felling a tree

Gears

Gear wheels are often used in machines in order to change the speed of rotation (in effect velocity ratio) or the applied force (mechanical advantage) or to change the direction in which the power is transmitted. Different types of gear are shown in figure 5.18.

In figure 5.17 suppose the effort is applied to the larger wheel with 36 teeth. When this turns through one revolution the small wheel will turn through two (i.e. $\frac{36}{18}$) revolutions. The velocity ratio is thus $\frac{1}{2}$, and the mechanical advantage will have roughly the same value. The small wheel turns twice as quickly, but only exerts half the turning force. If the effort were applied to the smaller wheel, there would be a reduction in speed but an increase in force. In general terms:

$$\text{velocity ratio} = \frac{\text{number of teeth on load gear wheel}}{\text{number of teeth on effort gear wheel}}$$

fig. 5.16 The sharp edges of cutting tools like the chisel and axe are wedge-shaped

fig. 5.17

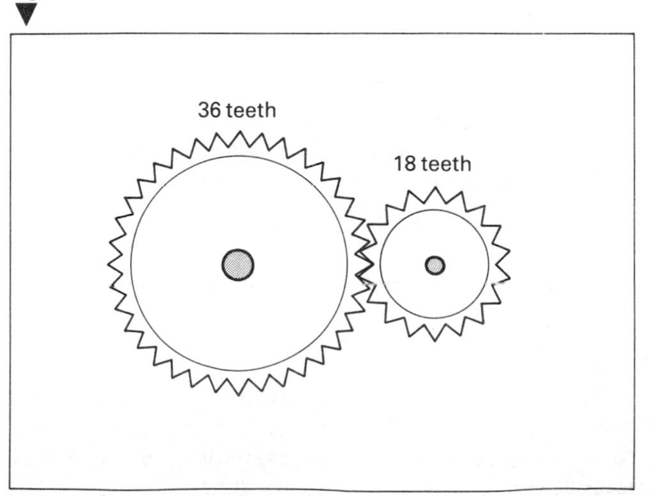

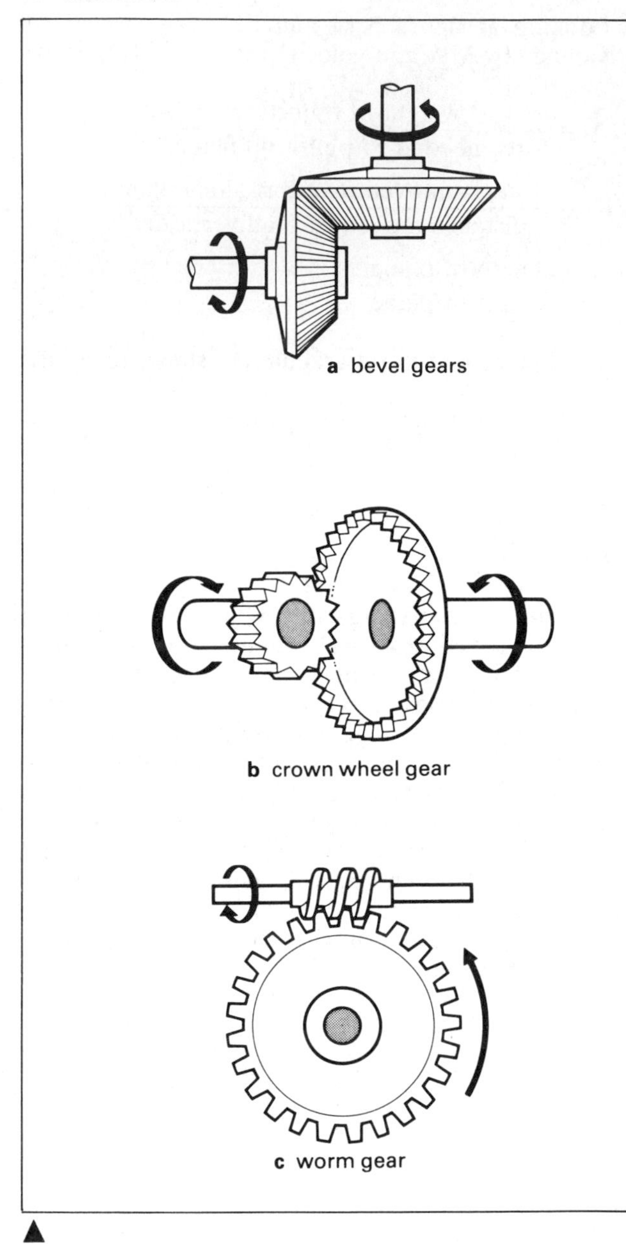

fig. 5.18 Three widely used types of gearing

Several gear wheels meshed together are called a **gear train** and the total gear ratio depends on the numbers of teeth in individual wheels.

In figure 5.19, for the first pair of wheels the velocity ratio is $\frac{48}{12} = 4$.

For the second pair of wheels the velocity ratio is $\frac{75}{15} = 5$.

The velocity ratio (or gear ratio) for the whole train is $4 \times 5 = 20$.

The driving wheel rotates twenty times to turn the driven wheel once. This is a reduction gearing of 20 to 1. There would be a corresponding twenty times increase in turning power or torque.

The gearbox of a car uses several gear trains, the ratios of which can be changed by moving the gear lever. The first (or low) gear is used to move a heavy load slowly as, for example, in starting the car from rest or going up a steep hill. The top (or high) gear is used when the car has gained speed. Typical gear ratios are

first gear	4 to 1
second gear	2.5 to 1
third gear	1.5 to 1
top gear	1 to 1

Figure 5.20 shows the gear trains in a simple (not synchromesh) three-speed gearbox. Wheel A is constantly meshed with wheel B. When the car is in neutral no wheels on the main shaft mesh with wheels on the layshaft. The main shaft can be moved by the gear lever so that wheel D meshes with wheel G as shown; the car is in first gear.

$$\text{gear ratio} = \frac{\text{product of numbers of teeth on driven (load) wheels}}{\text{product of numbers of teeth on driver (effort) wheels}}$$

$$= \frac{30 \times 35}{20 \times 15} = 3.5 \text{ to } 1$$

In second gear wheel F meshes with wheel C and the gear ratio works out at 1.5 to 1. In top gear the dog clutch is engaged and the drive is straight through. Figure 5.21 shows a four-speed gear box.

fig. 5.19 Gear train

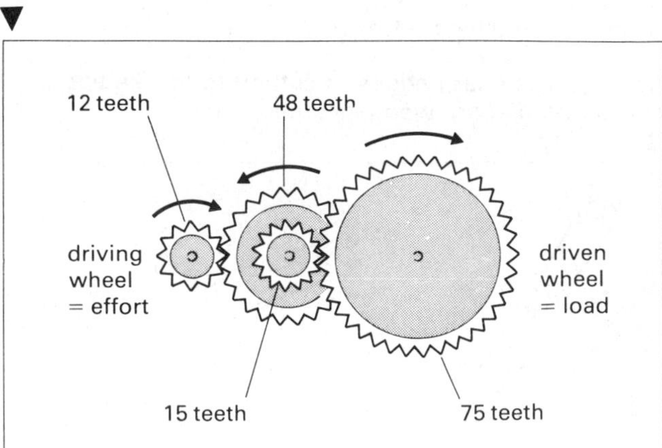

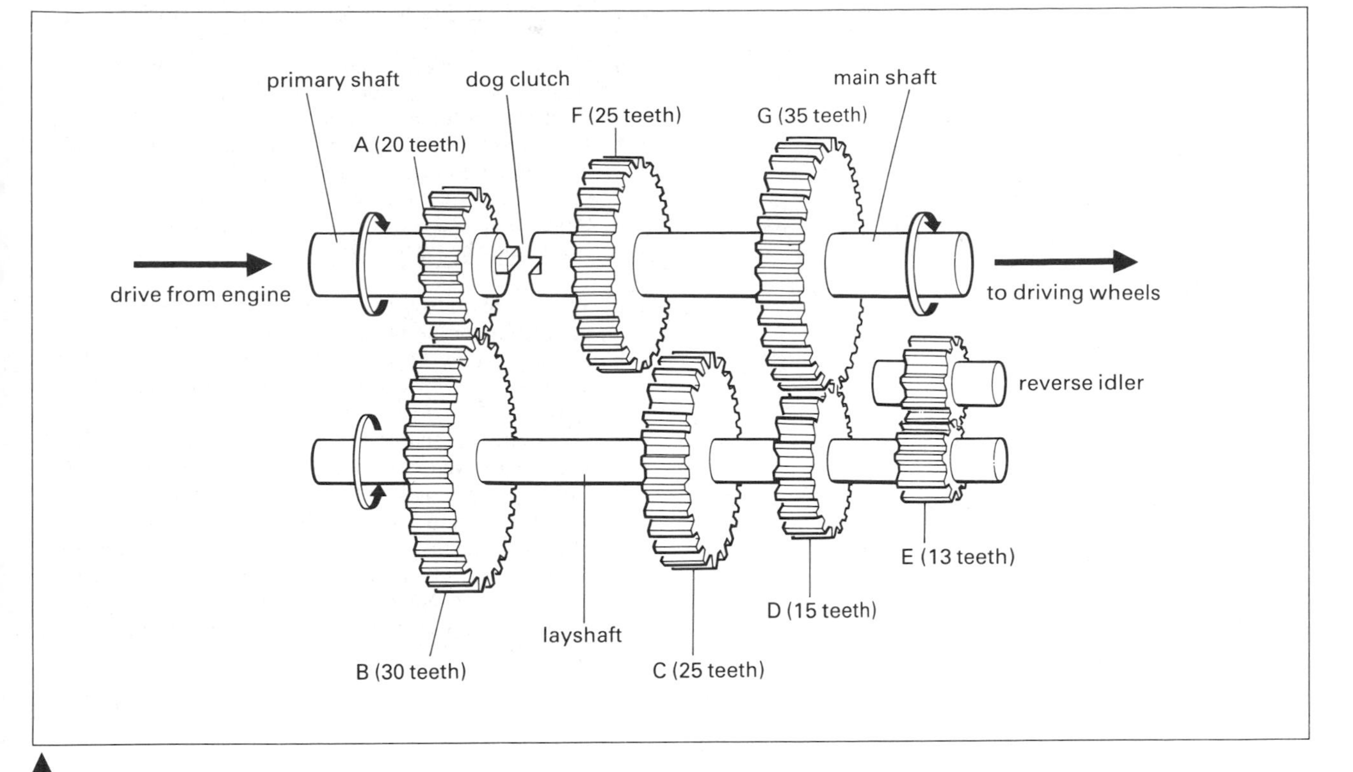

▲
fig. 5.20 Three-speed gear box

fig. 5.21 Four of the six gear positions in a four-speed car gearbox
▼

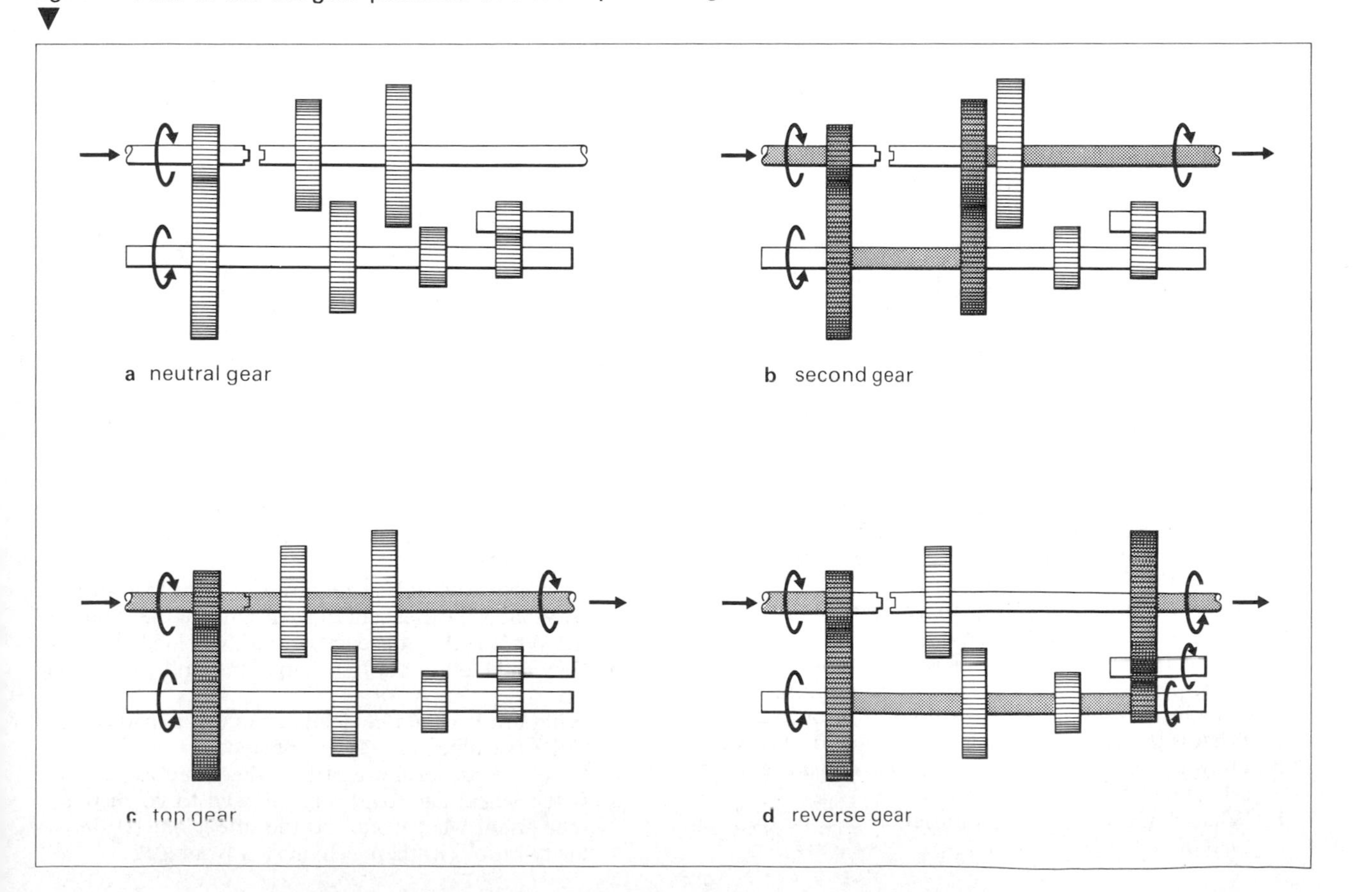

EXERCISE 5

In questions 1 and 2 select the most suitable answer.

1 The efficiency of a machine is equal to

A $\frac{\text{mechanical advantage}}{\text{velocity ratio}}$

B mechanical advantage × velocity ratio

C $\frac{\text{load}}{\text{effort}}$

D load × effort

E $\frac{\text{distance moved by effort}}{\text{distance moved by load}}$

2 An experimenter measures the mechanical advantage of a simple pulley system and obtains the value 5; its velocity ratio is 4.

A If this is possible, the efficiency of the system is 80%.

B This is impossible as the velocity ratio of a pulley system must be less than unity.

C This is impossible as it would imply that more energy is obtained from the machine than the work put into it.

D This would be possible only for a very inefficient machine.

E This would be possible only if the strings connecting the pulleys could increase in length.

3 The diagram shows a single pulley carrying a load of 10 N.

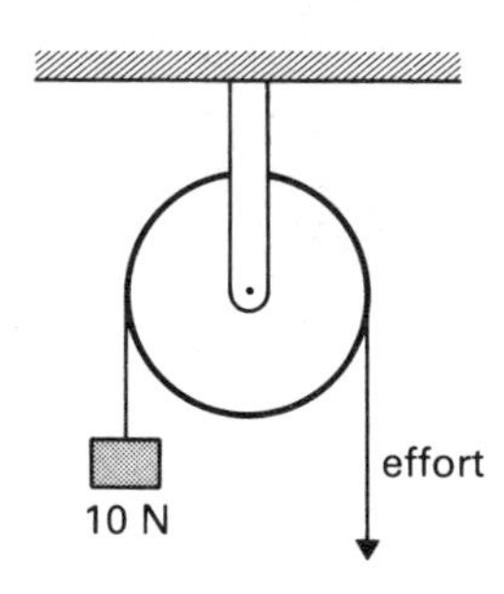

a What is the velocity ratio of the pulley?

b What is the mechanical advantage (i) If the pulley is 100% efficient? (ii) If the system is 50% efficient?

c Why is the pulley likely to be less than 100% efficient?

d What effort is needed if the machine is 50% efficient?

e If the load is lifted 2 m how much work is done on the load?

f If the pulley is only 50% efficient how much work is done by the operator to lift the load through 2 m?

g What is the advantage of using a pulley of this type?

h Draw a two-pulley system which you could use to lift a load of 500 N.

i What would be the minimum effort needed to lift a load of 500 N using this two-pulley system?

4 A block and tackle pulley system has a velocity ratio of 5.

a Draw a labelled diagram of the system marking clearly the load and the effort. Explain what is meant by the term 'block and tackle'.

b If the load is 600 N, calculate the effort which would be needed to raise it, assuming the system is 100% efficient. Explain why the effort needed would be larger than this value. If the load moves up 2 m how far would the effort move?

5 a What is meant by the term 'velocity ratio' when applied to a simple pulley system?

b Draw a pulley system with a velocity ratio of three, showing clearly the position of the load and the point at which the effort is applied.

c In a perfect machine the mechanical advantage is the same as the velocity ratio. Give two reasons why they would not be equal in the system you have described.

d A pulley system is used to lift a load of 500 N through a vertical height of 3 m. (i) How much work is done by the machine on the load? (ii) If the operator exerts a force of 200 N and needs to move the pulley string through 9 m how much work is put into the machine? (iii) What is the efficiency of this pulley system?

e A single pulley used to lift a load of 500 N needs an effort of 550 N but may still be of advantage to the user. Why?

6 The diagram shows a wheel and axle.

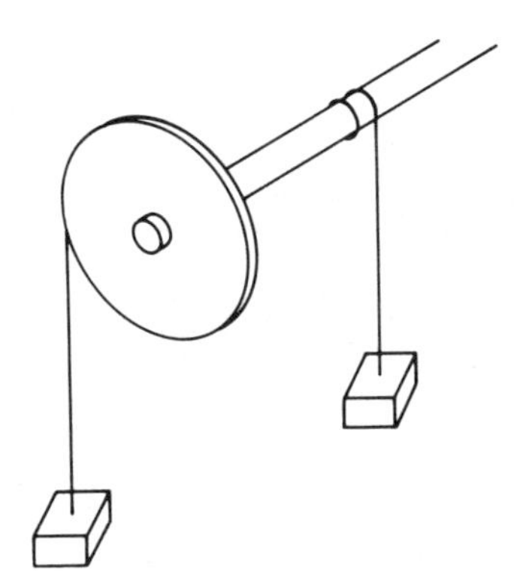

a Copy the diagram and mark on it (i) the load L, (ii) the effort E.

b If the radius of the large wheel is ten times the radius of the axle (i) how far will the effort move to raise the load 1 cm? (ii) What is the velocity ratio of the machine?

c If an effort of 10 N is required to lift a load of 80 N (i) What is the mechanical advantage of the machine? (ii) What is the efficiency of the machine?

d Why is the efficiency less than 100% in the wheel and axle?

e What effort would be needed to raise a load of 80 N if the machine were 100% efficient?

f Give one practical use of the wheel and axle.

g If the wheel and axle were allowed to go rusty and then used, what would be the effect on: (i) the velocity ratio? (ii) the mechanical advantage?

7 The diagram shows a rough rigid board arranged as an inclined plane in order to raise a rectangular load to a height of 1 metre above ground level. If the load has a mass of 120 kg determine, graphically or otherwise, the component of the weight of the load acting parallel to the incline, when the incline makes an angle of 20° with the horizontal.

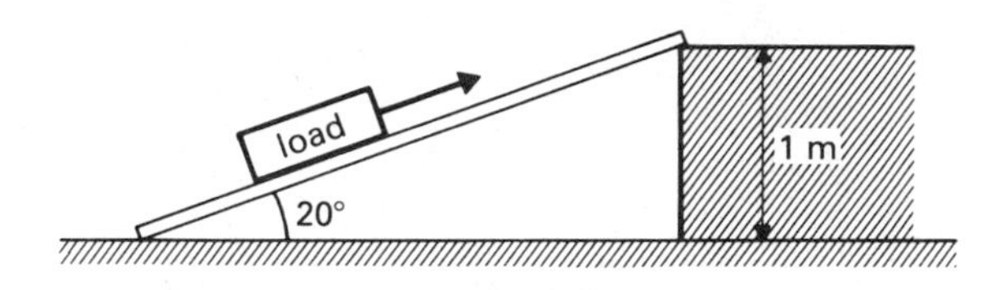

a Why may a smaller force than this, applied parallel to the plane, be sufficient to hold the load stationary but a greater load be needed to pull the load up the incline?

b Explain one advantage and one disadvantage of raising the load through the same vertical height using a longer board which reduces the angle that the incline makes with the horizontal.

c If a new incline is made of length 5 metres and a force of 400 newtons, parallel to the incline, raises the same load to a vertical height of 1 metre, calculate the work wasted.

d Discuss a benefit gained by having the load in the form of a cylinder that can be rolled up the incline.

8 A garden roller of weight 1500 N is pulled up a plank with a force of 300 N.

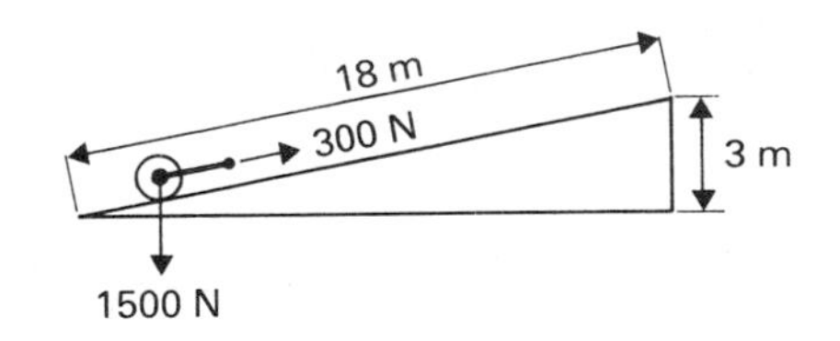

a What is the mechanical advantage of the machine?

b What is the velocity ratio of the machine?

c What is the efficiency of the machine?

d The roller is stopped on the inclined plane and because of the roughness of the surface it does not roll back. Draw a diagram to show the forces acting on the roller when it is in this condition.

6 LEVERS AND MOMENTS

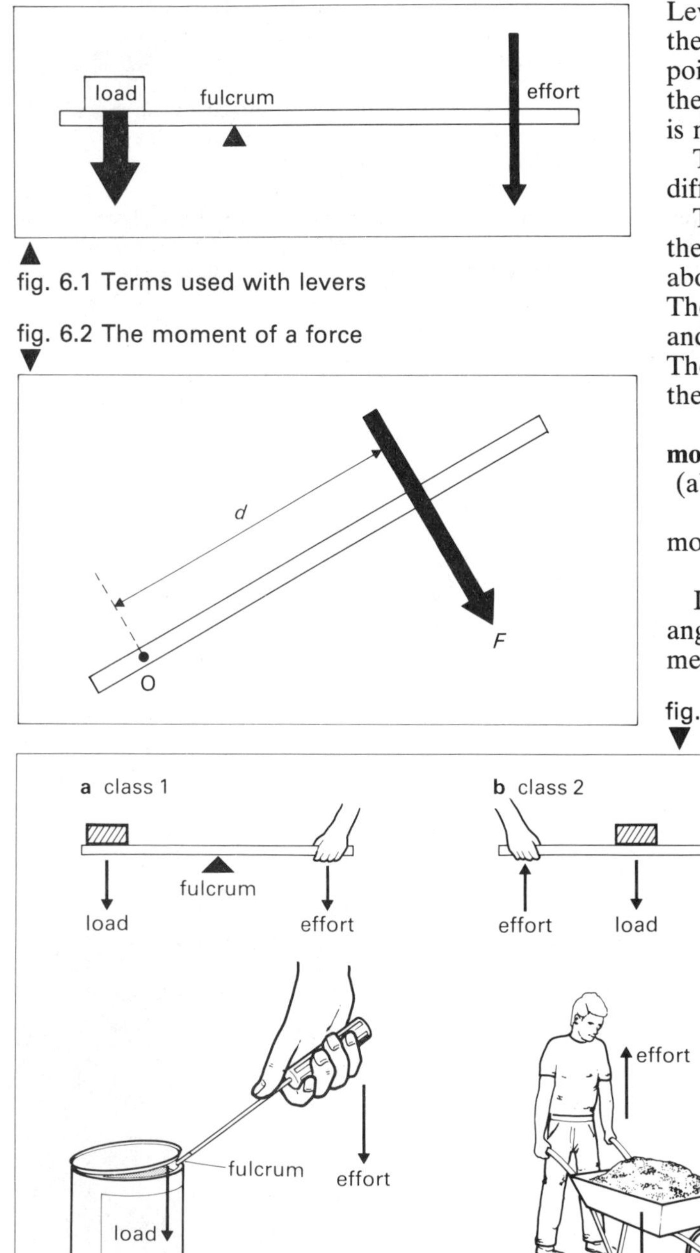

fig. 6.1 Terms used with levers

fig. 6.2 The moment of a force

Levers are very simple machines. All levers work in the same sort of way. The lever turns about some point called the **pivot** or **fulcrum**. We apply a force, the **effort** (E), usually near one end and the **load** (W) is moved by the other end, figure 6.1.

The effort, load and fulcrum can be arranged in different ways, figures 6.3a–c.

The force applied to a lever turns the lever about the pivot and we call the turning effect of a force about an axis the **moment** of the force about the axis. The turning effect depends on the size of the force and also on how far away from the pivot it is applied. The greater the distance from the pivot, the greater the turning effect. From figure 6.2,

moment of force = **force** × **perpendicular distance**
(about an axis) (from axis)

moment of F about O = $F \times d$ (**newton metres**)

In the majority of cases the force is acting at right angles to the lever and the 'perpendicular distance' is measured along the lever. If the force makes an angle

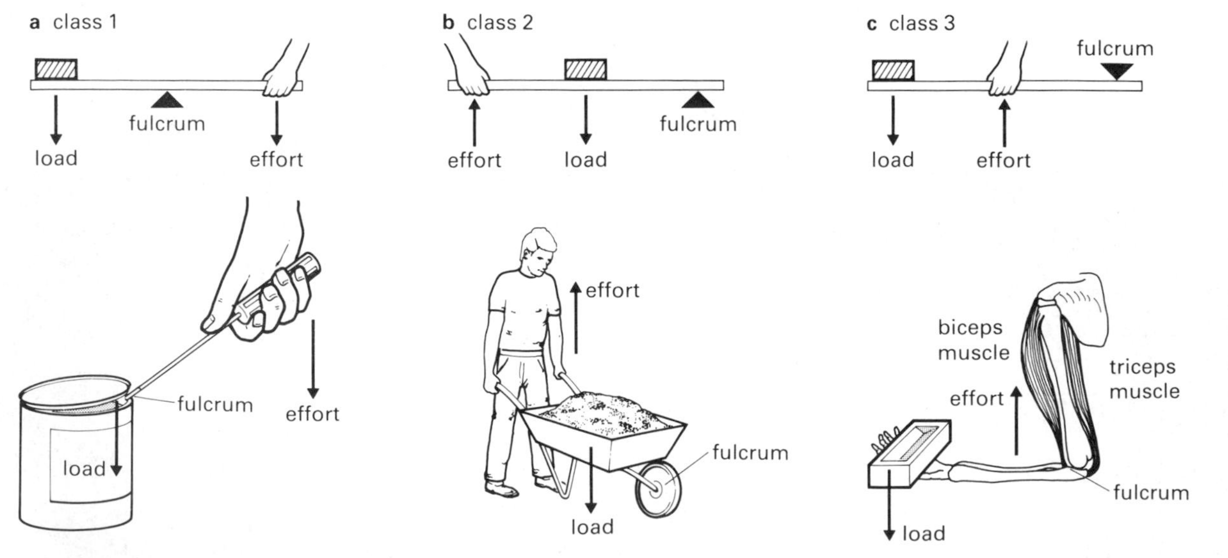

fig. 6.3 The three classes of lever

with the lever we have to find the perpendicular distance of the line of action of the force from the pivot, as shown in figure 6.4.

moment of F about O $= F \times x$

Levers can be used

- To produce large forces from smaller ones.
- To reduce the size of a larger force.
- To produce a large movement from a smaller one.
- To turn a large movement into a smaller one.

EXAMPLES

In each case, figures 6.5a and b and figure 6.6, work out where the pivot is and where the effort and load are applied.

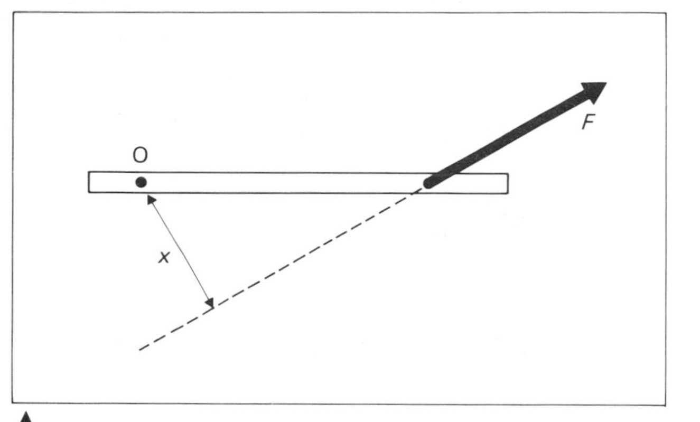

fig. 6.4 The moment of a force: general case

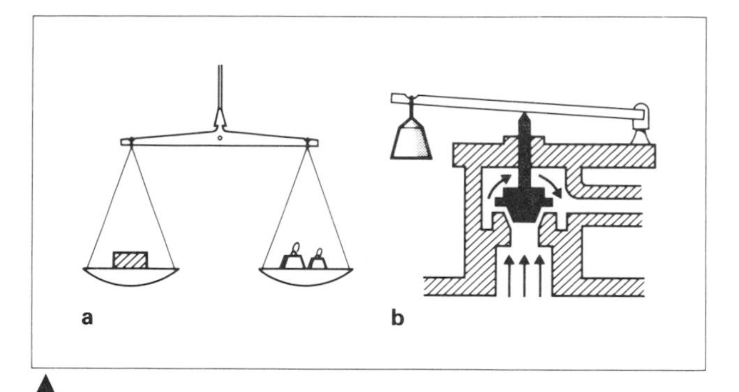

fig. 6.5 Two examples of single levers

The law of the lever

When two forces are applied to a body, as in the case of a lever, they will balance when their turning effects in opposite directions are equal.

From figure 6.7a, taking moments about O

$W \times a = E \times b$

mechanical advantage $= \dfrac{W}{E} = \dfrac{b}{a} = \dfrac{\textbf{length of effort arm}}{\textbf{length of load arm}}$

This relationship is often called the **law of the lever** and expresses the fact that if we want a large mechanical advantage we must have a long effort arm and a short load arm. But it is also obvious that the effort has to move through a greater distance than the load.

If several forces have turning effects on a body, then there will be equilibrium when the total turning effect in a clockwise direction is equal to the total anticlockwise turning effect. The sum of the clockwise moments is equal to the sum of the anticlockwise moments.

From figure 6.7b, taking moments about the pivot when the rod is balanced:

$W_1 \times a + W_2 \times b = W_3 \times c$

fig. 6.6 Some examples of double levers

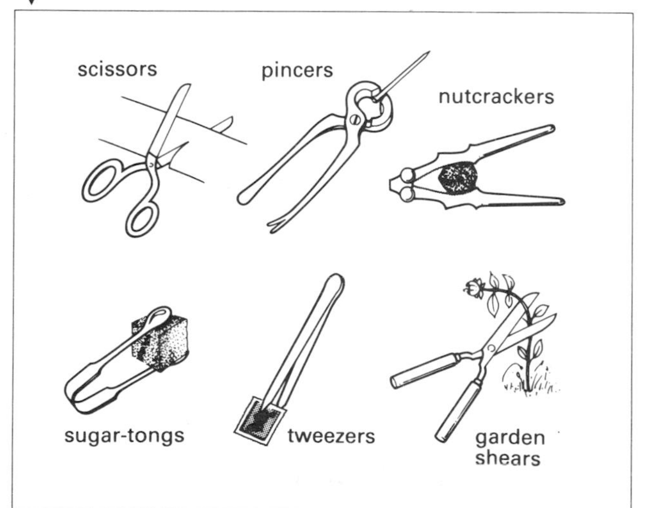

fig. 6.7

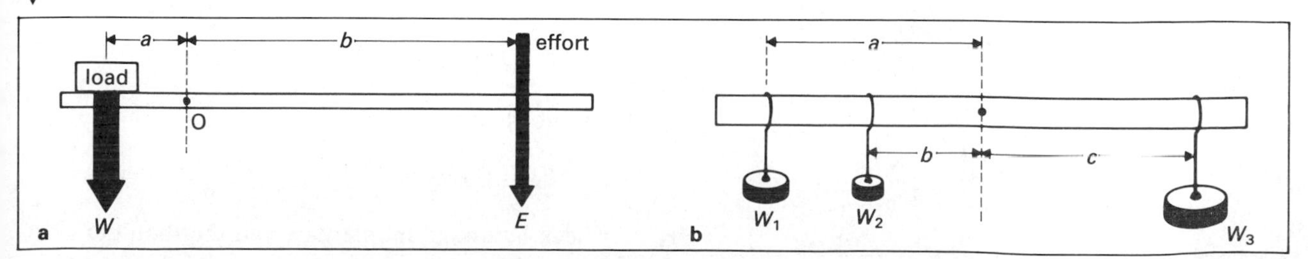

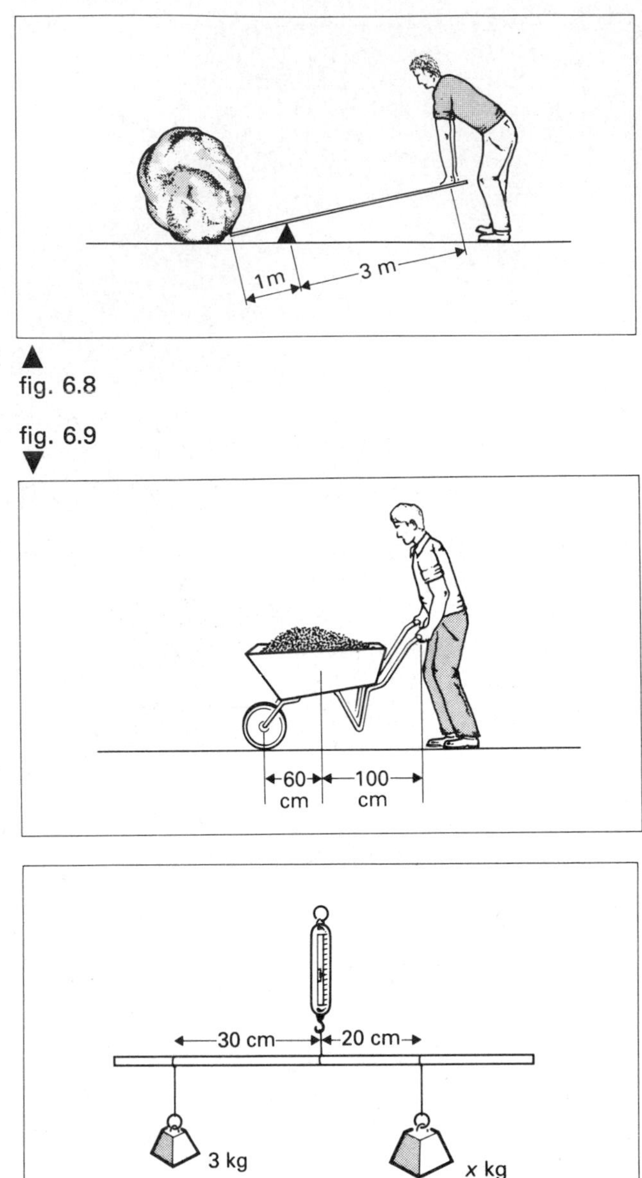

fig. 6.8

fig. 6.9

fig. 6.10

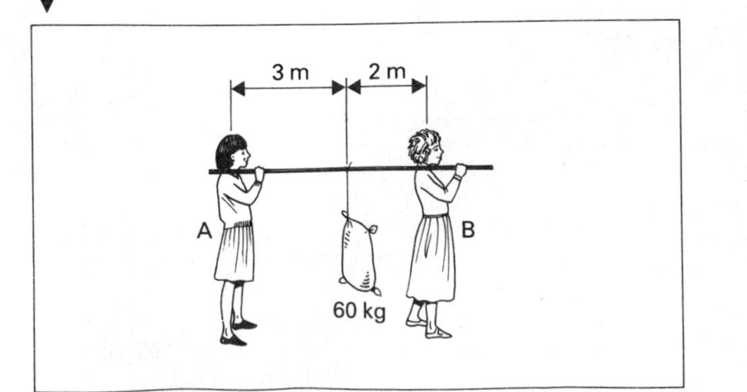

fig. 6.11

EXAMPLES

A moment is the product of a force and a distance and is strictly measured in newton metres. But provided we are consistent in our units we can use other force and distance units.

1 In figure 6.8 the man is pushing down with a force of 450 N. What is the force exerted on the boulder?

Let the force exerted on the boulder = F N
Taking moments about the fulcrum:

$$F \times 1 = 450 \times 3$$
$$F = 1350 \text{ N}$$

2 In figure 6.9 the load and the wheelbarrow weigh 50 kg. What force must the man exert to lift the barrow?

Let the force exerted by the man = P kg wt
Taking moments about the axle:

$$P \times 160 = 50 \times 60$$
$$P = \frac{50 \times 60}{160}$$
$$= 18.75 \text{ kg wt} = 187.5 \text{ N}$$

3 In figure 6.10 a wooden rod is suspended at its mid-point from a spring balance and remains horizontal. Find the unknown weight. If the rod itself weighs 0.5 kg, what is the spring balance reading?

Taking moments about the point of suspension:

$$x \times 20 = 3 \times 30$$
$$x = 4.5 \text{ kg}$$

total downward weight $= 3 + 4.5 + 0.5$
$= 8$ kg

The spring balance will read 8 kg.

4 In figure 6.11 two girls are holding a sack weighing 60 kg as shown. What force does each exert?

Let the forces exerted by the girls be F_A and F_B kg wt, respectively. Taking moments about the shoulder of girl B:

$$F_A \times 5 = 60 \times 2$$
$$F_A = 24 \text{ kg wt}$$

By subtraction

$$F_B = 36 \text{ kg wt}$$

Check by taking moments about the shoulder of A.

EXERCISE 6

In questions 1–3 select the most suitable answers.

1 What force, F, will be needed such that the uniform beam shown in the figure below, pivoted at its mid-point, will balance horizontally?
A 100 N
B 300 N
C 400 N
D 900 N
E 1200 N

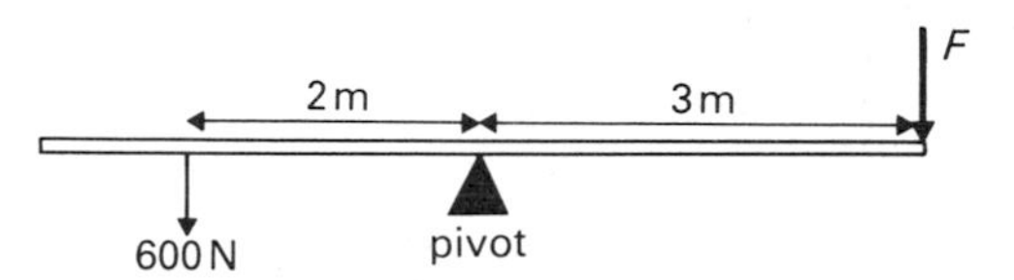

2 In the figure below, PQ is a uniform beam of length equal to 4 m, which is resting on supports at R, 1 m from P, and at S, 1 m from Q. If the beam has a weight equal to 150 N, what would be the minimum downward force, F, applied at Q which would lift the beam clear of the support at R?
A 150 N
B 100 N
C 75 N
D 50 N
E 37.5 N

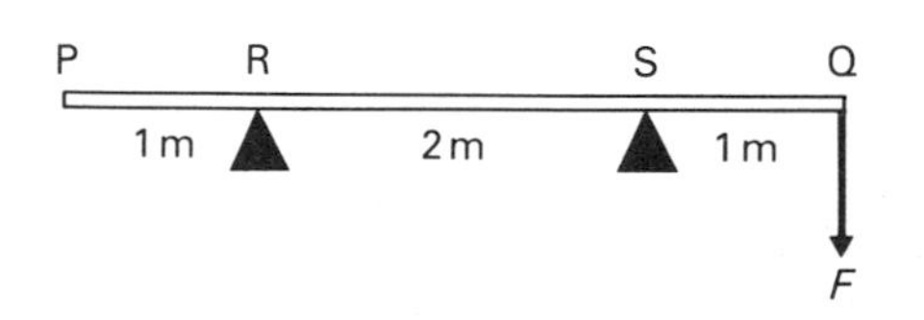

3 The shaft in an engine is subjected to two parallel but opposite forces of 500 N as shown in the figure below. Rotation of the shaft could best be prevented by applying
A a single force of 250 N
B a single force of 1000 N
C two forces of 500 N acting at right angles to one another
D two parallel but opposite forces of 500 N
E two parallel forces of 500 N acting in the same direction

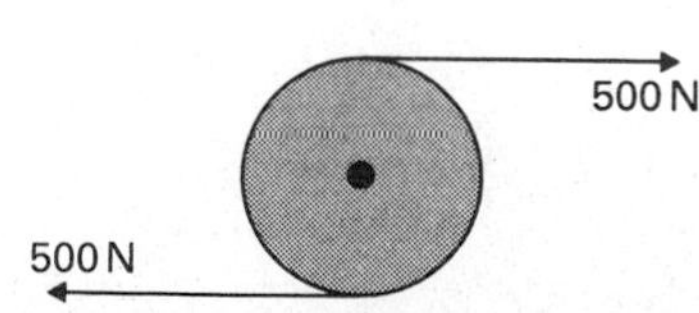

4 **a** State what is meant by (i) moment of a force, (ii) the principle of moments.
b The figure shows a long metal bar being used to lift a rectangular flagstone of uniform thickness. On the diagram mark (i) the fulcrum F, (ii) the best position to apply the effort E, (iii) the best direction to apply the effort.

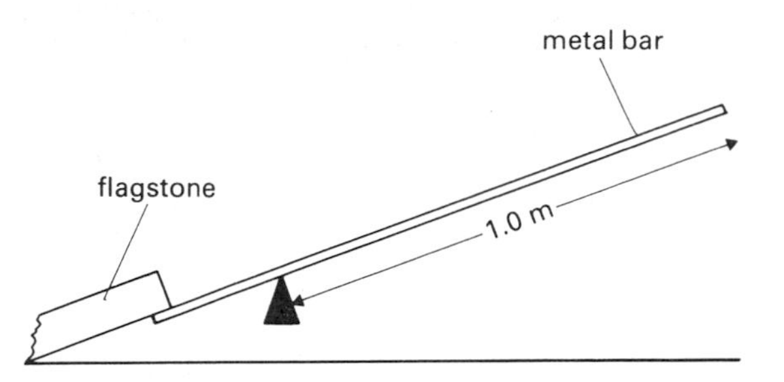

c What is the weight of the flagstone if its mass is 50 kg? (earth's gravity = 10 N/kg).
d If the bar is 1.1 m long, what is the minimum effort needed to lift the flagstone if the flagstone is horizontal?

5 In a region where the gravitational field strength is 10 N/kg, a 20 kg mass is supported from the middle of a light rod as shown below. What is the reading on the spring balance X?

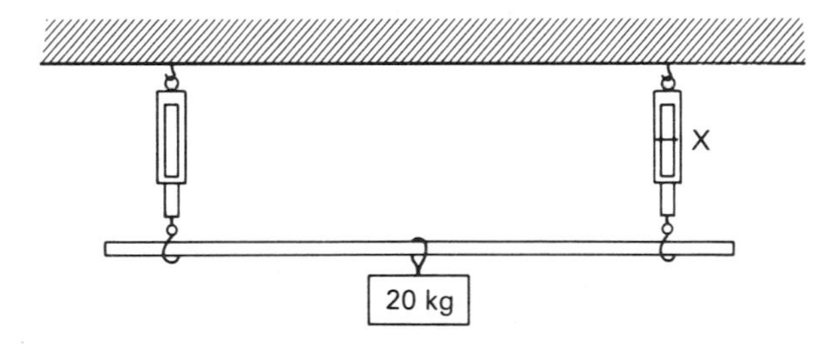

6 In the figure a crowbar is shown being used to lift a heavy stone.

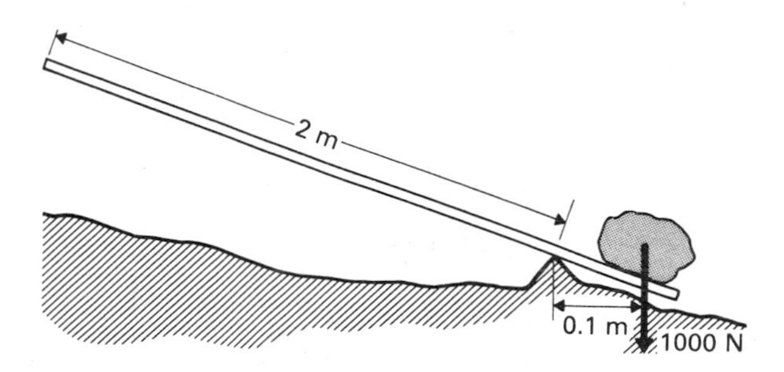

a Mark with an arrow on the diagram where you would push, and in what direction, to lift the stone with the *least* effort.
b Calculate the force which should be applied in order just to lift the stone, i.e. just to balance the weight of the stone.

7 Name a device which is a lever in which the load is applied between the fulcrum (pivot) and the effort. Draw a sketch of the device you have named and mark on it the load, the fulcrum (pivot) and the effort.

8 A boy, weighing 600 N, sits 6 m away from the pivot of a see-saw, as shown below. What force F, 9 m from the pivot, is needed to balance the see-saw?

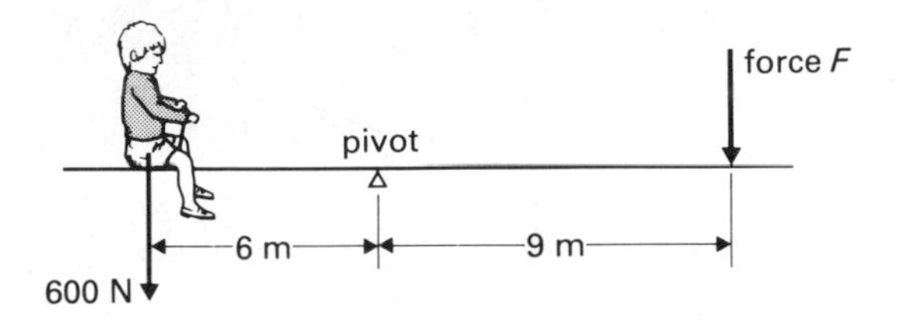

9 **a** State the principle of moments.

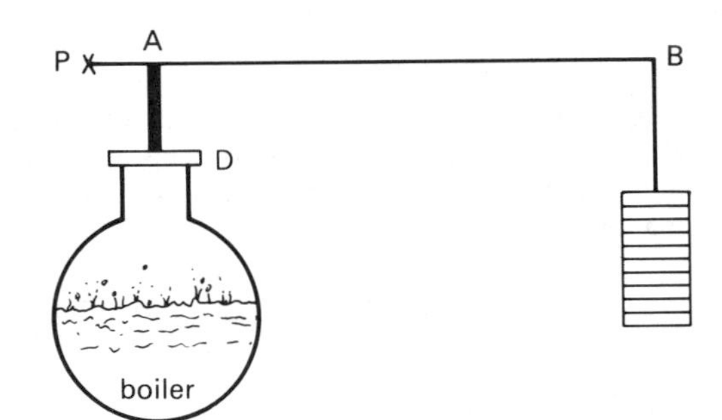

b The figure illustrates the safety valve for a boiler. A disc D is on a bar PB, 90 cm long. This bar is pivoted at P, 10 cm from A. Calculate the force at the disc D if a load of 50 N is applied at B.

10 Using the principle of moments, find the value of the force F acting at the point shown in the figure if the rod is to remain horizontal. The marks on the rod represent distances measured from the left-hand end in centimetres. (The weight of the rod can be neglected.)

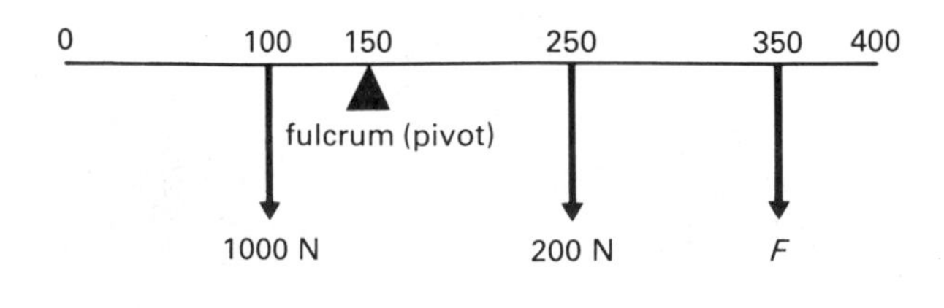

11 **a** Describe an experiment to show that for a body in equilibrium the sum of the clockwise moments about a point is equal to the sum of the anticlockwise moments about the same point.

b A painter stands on a uniform plank 4.0 m long and of mass 30 kg. The plank is suspended horizontally from vertical ropes attached 0.5 m from each end as shown in figure 6.12. The mass of the painter is 80 kg. Calculate the tensions in the ropes when the painter is 1.0 m from the centre of the platform. State briefly (no calculation required) how you would expect the tensions in the ropes to vary as the painter moves along the plank.

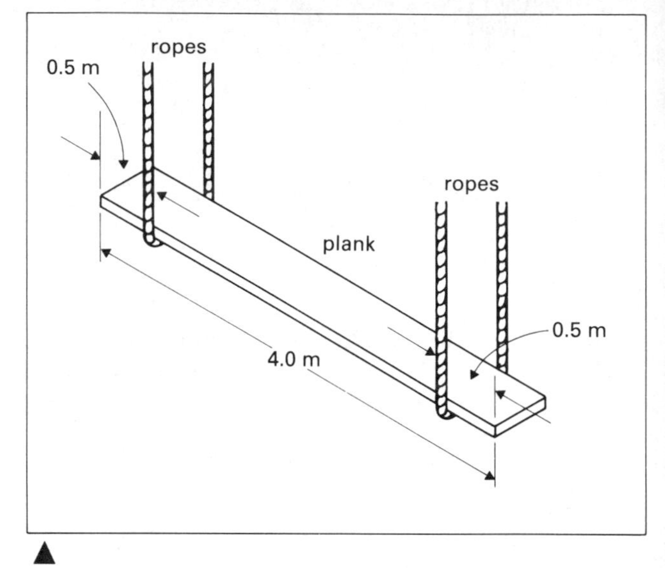

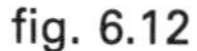
fig. 6.12

c Figure 6.13a shows an end view of the plank. Figures 6.13b and c show two other possible ways in which the plank may be supported by the ropes. State, giving your reasons, whether or not you would expect (i) the tensions to be the same in the vertical parts of the rope in each diagram, and (ii) the tensions to be the same in the sloping parts of the rope in figures 6.13b and c.

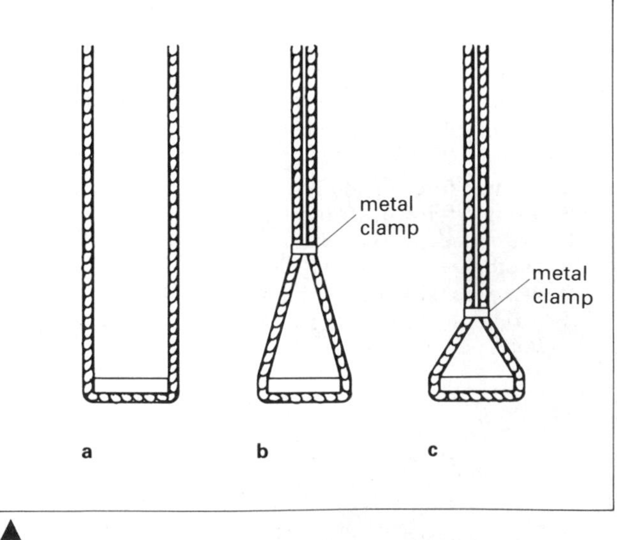

fig. 6.13

7 CENTRE OF GRAVITY AND STABILITY

The **centre of gravity** of a body is the point at which the whole weight of the body can be taken as acting, no matter in what position the body is placed. The centre of gravity of a body is in the same position as its **centre of mass**.

The centre of gravity may be regarded as the point of balance. It is not difficult to guess where the centre of gravity, or balancing point, of a flat, circular piece of metal is. The centres of gravity of regular and uniform (made all of the same material) objects are at their geometrical centres, figure 7.1.

The centre of gravity of a triangle can be found by drawing lines from the corners to the mid-points of the opposite sides, figure 7.2. These lines are called medians. The centre of gravity of the triangle is at the intersection of the medians, and is one-third of the way along any median from the side of the triangle: $DG = \frac{1}{3}DA$.

The centre of gravity of an irregularly shaped body must be found experimentally.

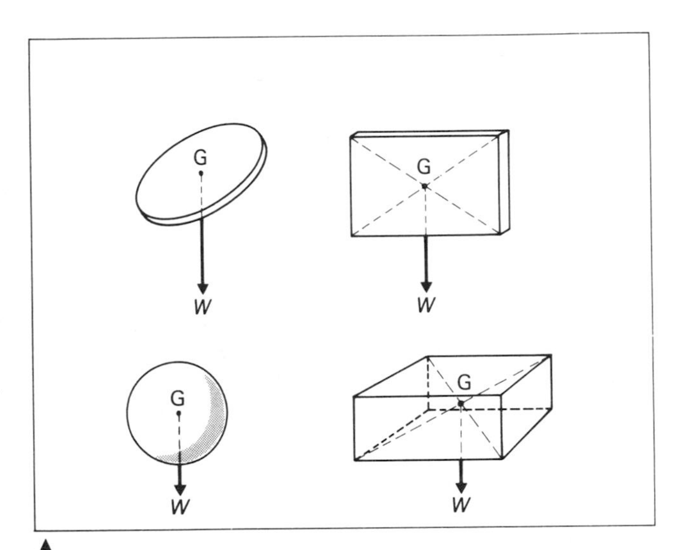

fig. 7.1 Cases of the CG at the geometrical centre

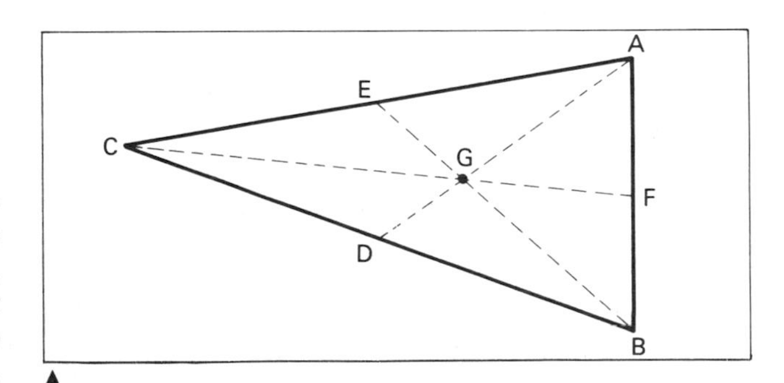

fig. 7.2 Centre of gravity of a triangle

When a body is suspended so that it can swing freely, it will come to rest with its centre of gravity vertically below the point of suspension

Hang an irregularly-shaped sheet of metal or cardboard from a nail and hang a plumb-line from the same nail, figure 7.3a. When the cardboard settles, mark the vertical line on it. Repeat from another point. The intersection of the vertical lines from two points of suspension will fix the centre of gravity, figure 7.3b. Hanging from a third point can be used as a check that the correct position has been found, figure 7.3c.

fig. 7.3 Locating the CG of an irregular shape

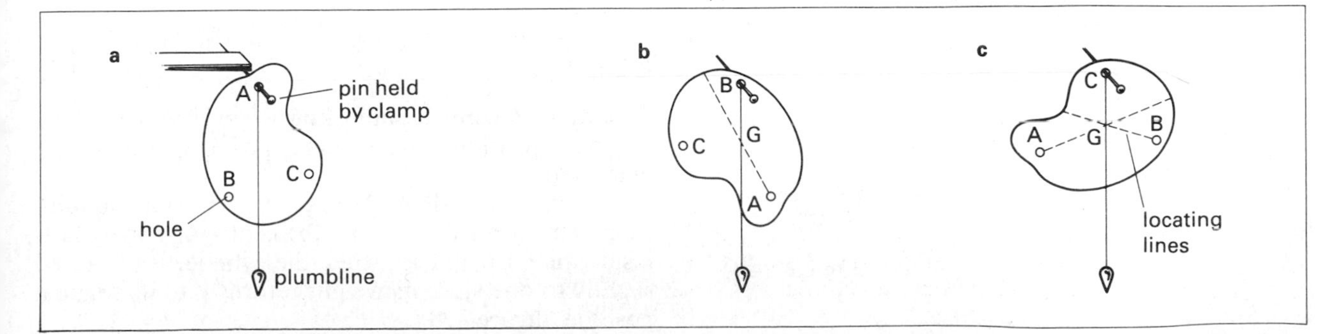

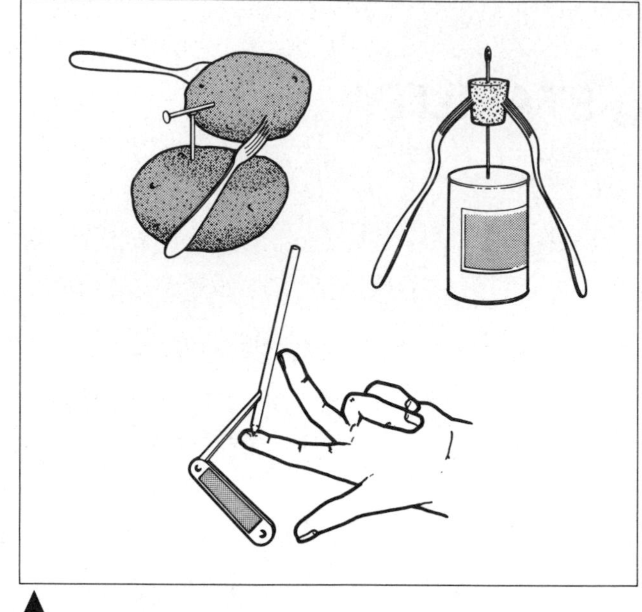

▲
fig. 7.4 These balancing tricks (and many similar ones) depend on the fact that the centre of gravity is vertically below the point of support

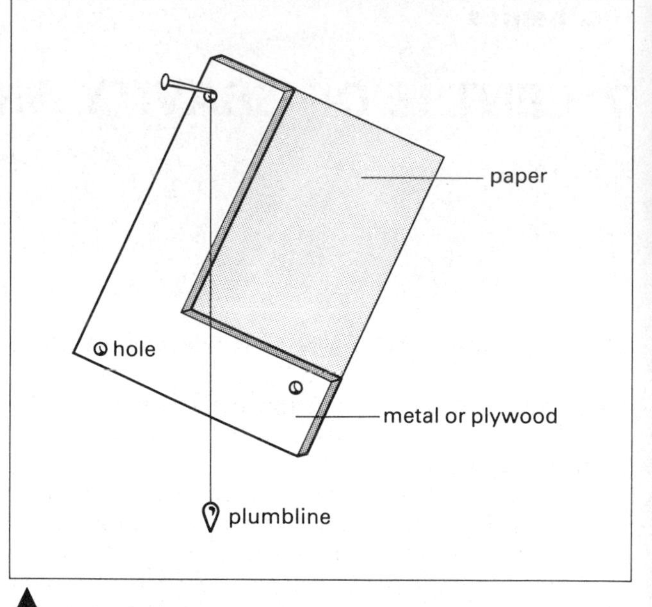

▲
fig. 7.5 Locating a centre of gravity which lies outside the actual material of a body

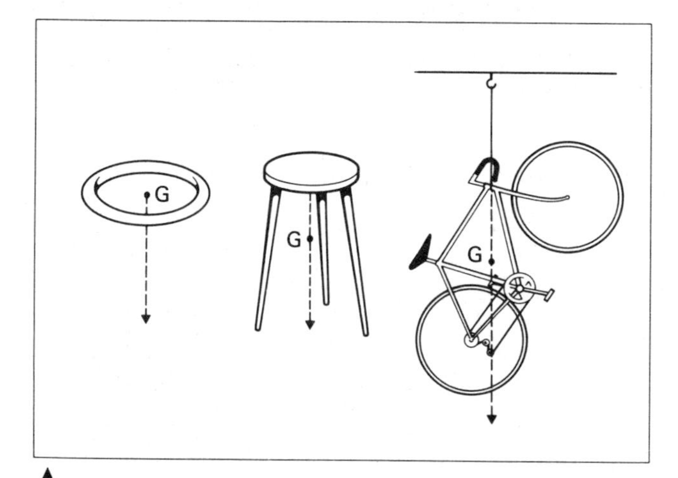

▲
fig. 7.6 Examples of the 'external' CG

fig. 7.7
▼

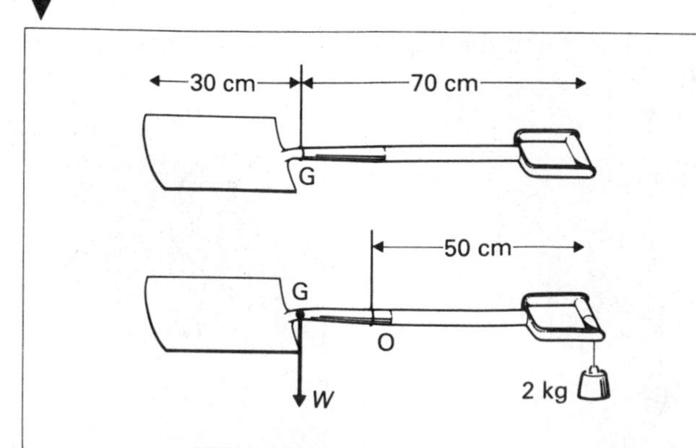

Cut a letter L out of metal or plywood and stick it on a sheet of thin paper to complete the rectangle, figure 7.5. Find the centre of gravity experimentally as discussed above.

This experiment shows that the centre of gravity of an object need not lie in the object itself, see figure 7.6.

EXAMPLE

A garden spade is 1 m long. It just balances in a string loop when the loop is 30 cm from the end of the blade. When a 2 kg weight is hung from the handle the balance point is at the centre of the spade. How much does the spade weigh? This illustrates a simple method of weighing without scales, figure 7.7.

Taking moments about O:

$$W \times 20 = 2 \times 50$$
$$W = 5 \text{ kg}$$

Stability

A body is in **stable equilibrium** when it returns to its original position after being given a small displacement.

A cylinder standing on its base is in stable equilibrium. The vertical line from the centre of gravity falls inside the base and so, when the cylinder is displaced slightly to one side, its weight returns it to its original position, figure 7.8.

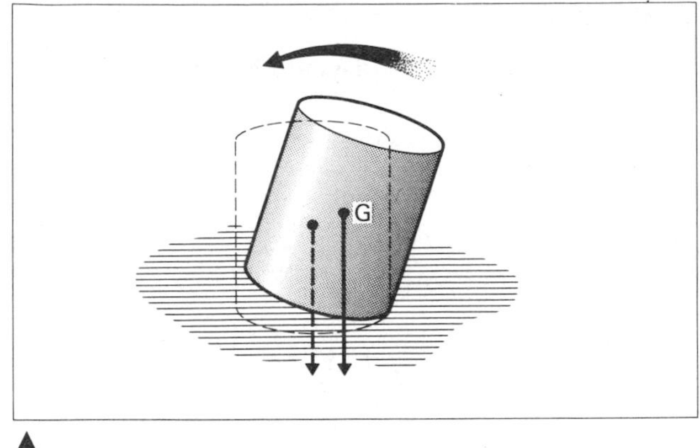

▲
fig. 7.8 Stable equilibrium

The lower the centre of gravity the further the body can be displaced and still return to its stable position when the displacing force is removed. When the vertical from the centre of gravity falls just outside the base, the body topples over. A vase with a good wide base is not easily knocked over, but if it is filled with tall flowers and leaves it becomes 'top-heavy' and is easily upset, figure 7.9.

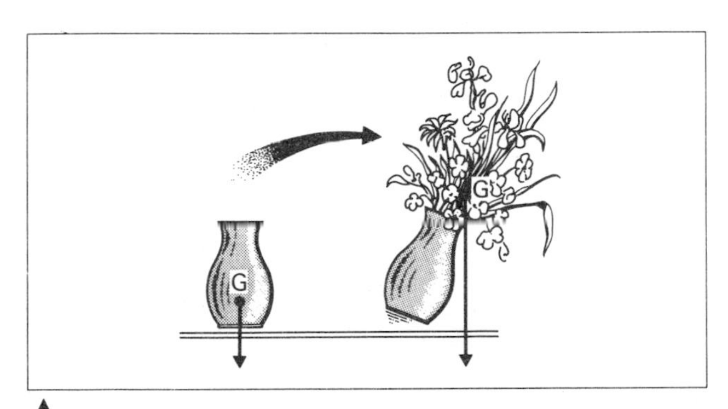

▲
fig. 7.9 A stable vase can be made unstable

fig. 7.10 Toys like this always stand upright. The centre of gravity is low due to the heavy lead base
▼

▲
fig. 7.11 A racing car has a low centre of gravity and a wide wheel-base to give it stability

fig. 7.12 This bus has been loaded with sandbags to represent the weight of the passengers. Although tipped over to this alarming angle, it does not topple because of its low centre of gravity due to the heavy engine and chassis
▼

A cylinder lying on its side is in **neutral equilibrium**. When it is given a displacement there is no tendency for it to return to its original position or to move further away from it. The vertical from the centre of gravity always passes through the point of contact with the surface, figure 7.13.

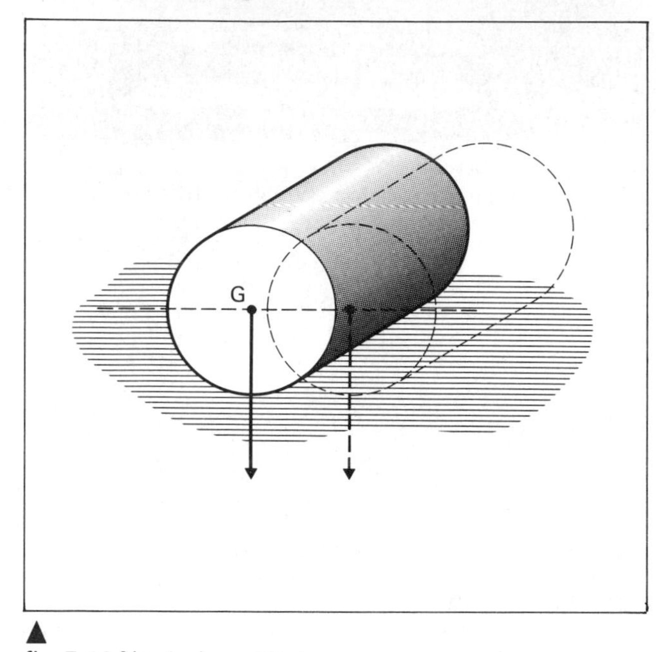

fig. 7.13 Neutral equilibrium

A bottle balanced on its neck is in **unstable equilibrium** because when it is given a small displacement it moves further away from its original position. The vertical from the centre of gravity soon falls outside the base and the bottle topples over, figure 7.14.

fig. 7.14 Unstable equilibrium

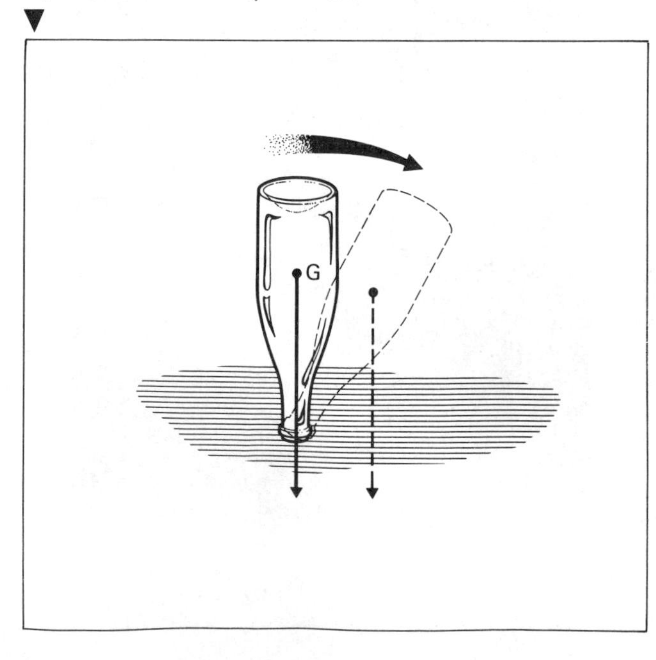

EXERCISE 7

In questions 1 and 2 select the most suitable answers.

1 Which of the bodies shown in figure 7.15 is in neutral equilibrium?

fig. 7.15

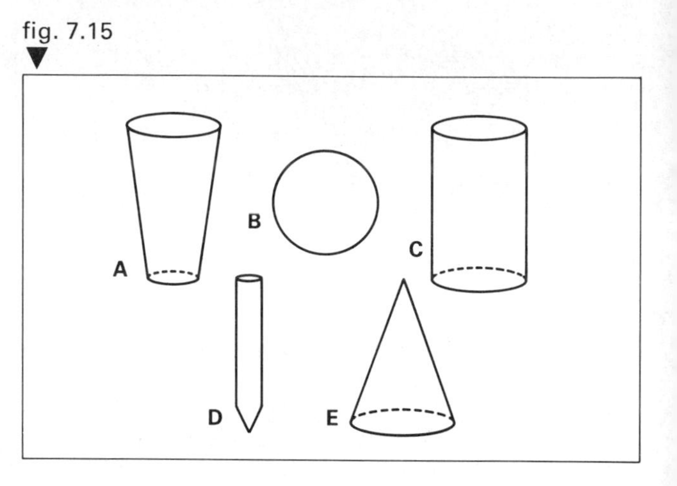

2 Figure 7.16 shows a shape cut from plywood of uniform thickness. Which of the points **A, B, C, D** or **E** is the most likely position of the centre of gravity?

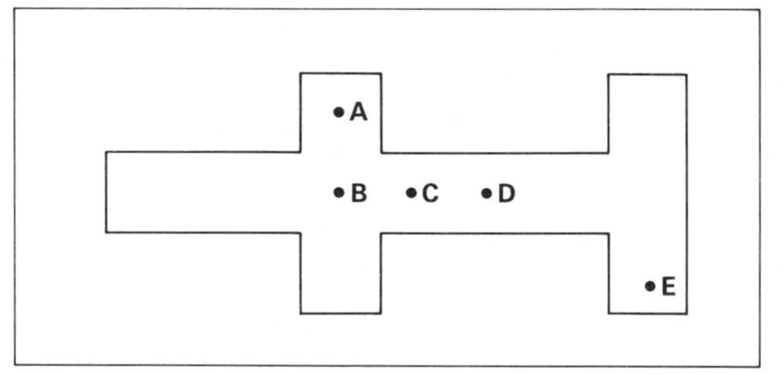

fig. 7.16

3 Figure 7.17 shows the front of a racing car.

a Mark with a cross the approximate position of the centre of gravity.

b What *two* features of the car give it great stability?

c Why are the tyres so large?

fig. 7.17

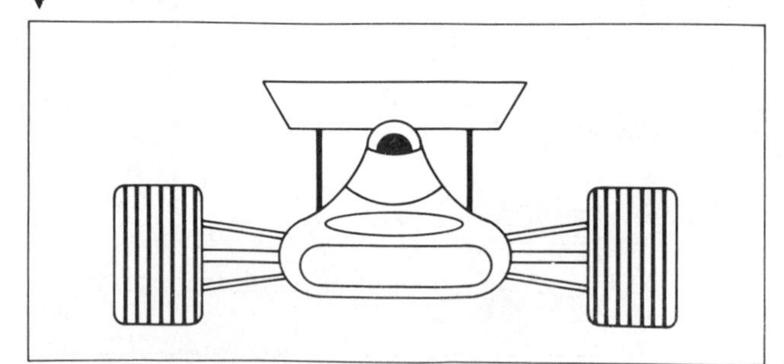

4 a What do you understand by the 'centre of gravity' of an object?

b Figure 7.18 shows a milk bottle in stable equilibrium. Draw two further diagrams to illustrate a milk bottle in (i) unstable and (ii) neutral equilibrium.

c Figure 7.19 shows the cross-sections of two symmetrically-designed table lamps, A and B. The stands in each case are solid.

(i) Mark with a cross in each diagram where you might expect the centre of gravity to be.

(ii) Which lamp is likely to be the more stable?

(iii) Give *two* reasons for your answer to part (ii).

d Usually on crowded buses passengers are asked to stand on the bottom deck, but not on the top deck. Give a reason why this is done.

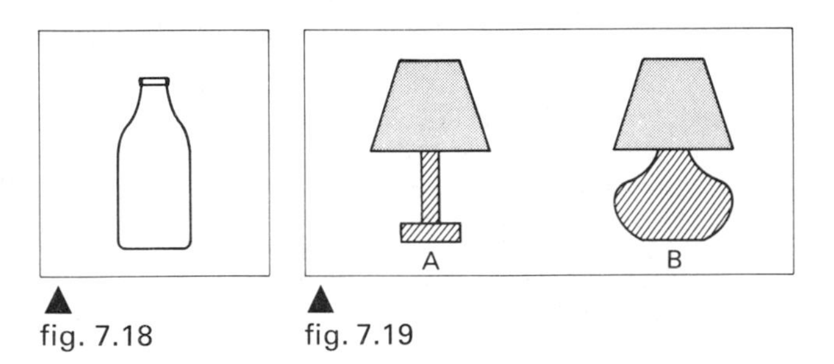

fig. 7.18 fig. 7.19

5 A non-uniform lever, with a weight of 5 N, rests on a pivot. When loads of 10 N and 30 N are hung from each end, as shown in the figure below, the lever balances horizontally.

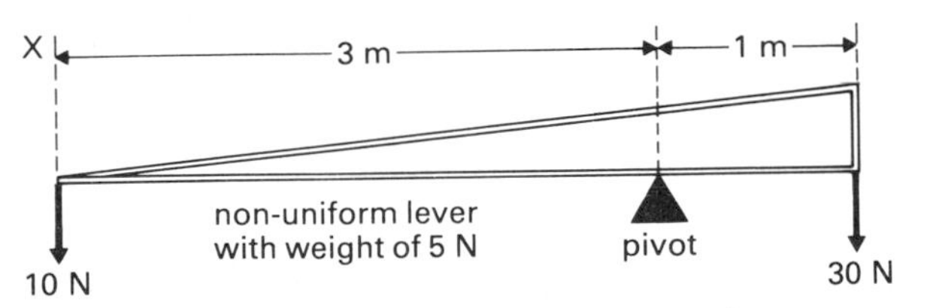

a What upward force does the pivot exert?

b How far from end X does the weight of the lever act?

c If the right-hand load were increased to 45 N, what total load would have to be hung from X to make the lever balance horizontally again?

6 Copy the side view of the aeroplane in figure 7.20 and mark in the position of the centre of gravity most likely to produce stable flight. Using large labelled arrows (e.g. ← **drag**) show the directions of the following forces: **lift, drag** (air resistance), **thrust**.

fig. 7.20

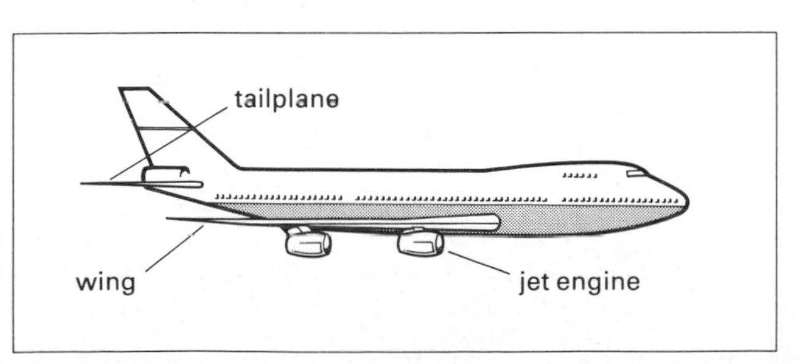

7 The rectangular top of a folding table is hinged to a wall and it is supported by one leg when in use. It has on it, in the position shown in figure 7.21, a container of water of total weight 40 N.

a Copy the diagram in sketch form and show the centre of gravity of the table top.

b Take moments about the hinge to calculate the upward force supplied by the table leg, given that the table top has a weight of 50 N. Show your working.

c What is the upward force supplied by the hinge?

fig. 7.21

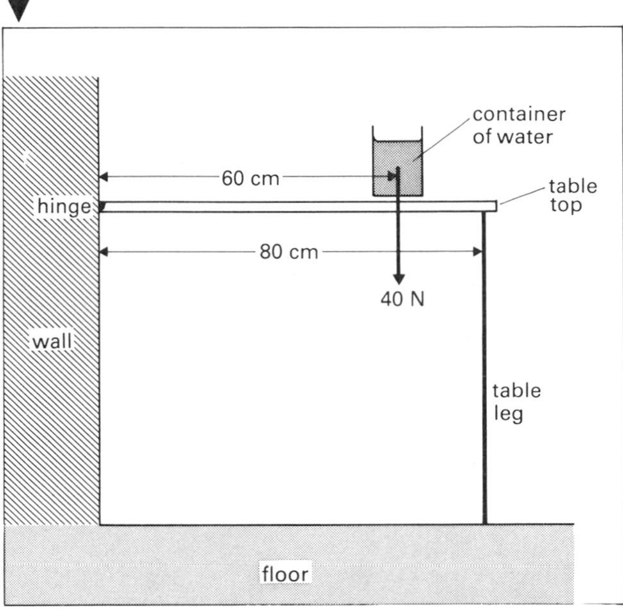

8 a The centre of mass of the thin sheet shown in figure 7.22 is found by allowing it to swing freely about a pivot through the small hole A until it comes to rest. A plumb-line suspended from the same pivot is allowed to come to rest and a line is drawn down it on the sheet. The operation is repeated from the small hole B. The centre of mass is where the lines cross. (i) Justify that the point of intersection of the lines is the centre of mass of the sheet. (ii) Suggest a better position for B, giving the reason for your choice.

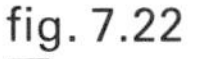

fig. 7.22

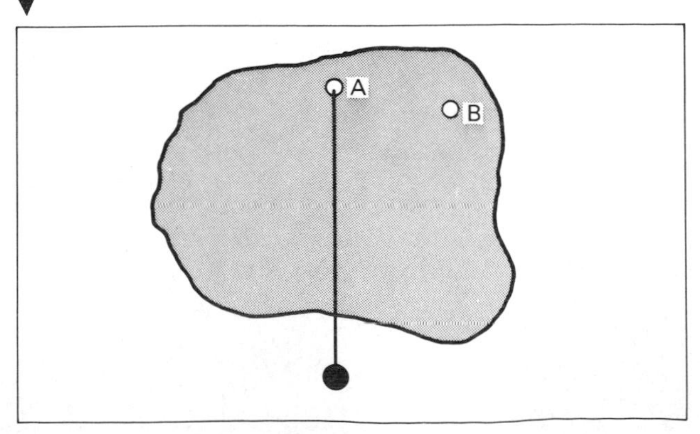

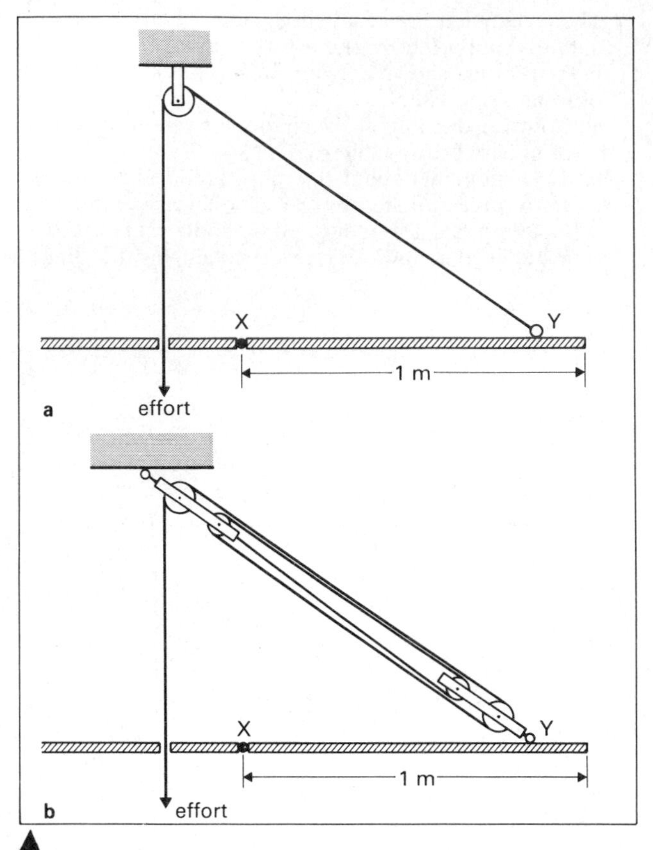

fig. 7.23

b The crew of a small boat going out into the open sea store heavy luggage below deck in the hold rather than lashing it on the deck. Explain, giving reasons, what may happen if the luggage is lashed on the deck.

c XY is a uniform trap door of mass 100 kg which is hinged at X. It could be opened by either of the systems shown in figure 7.23. (i) Explain what advantage the system shown in figure 7.23a would have over that shown in figure 7.23b and compare the efforts that would be required. (ii) Explain why Y is the best place on the trap door at which to attach the rope. (iii) How much work would be done in figure 7.23a in opening the trap door so that Y was vertically above X? (Neglect friction.) (iv) How much more work would be done in figure 7.23b if the bottom pulley block had a mass of 5 kg than if it had negligible mass? (Neglect friction.)

9 A double-decker bus standing on a sloping surface is tilted sideways. Draw a diagram to show the forces acting on it and explain how the design and use of the bus enable this tilt to be as large as possible without toppling.

An inclined plane has a surface rough enough to prevent slipping. Calculate, in each case, the maximum angle of inclination of this plane to the horizontal when:

a a uniform block 3 metres by 2 metres by 1 metre,

b a cylinder of length 1 metre and radius 1 metre

are placed in turn on the surface in the position of their greatest stability.

Name and explain briefly the type of equilibrium portrayed by a tightrope walker without a pole.

8 DENSITY

When we speak of a 'dense' crowd we mean a crowd in which people are tightly packed together. We can think of the density of a substance in a similar way. A dense substance is one in which a lot of matter is packed into a small space. If we are going to compare masses of materials we must compare the masses of equal volumes, figures 8.1a and b.

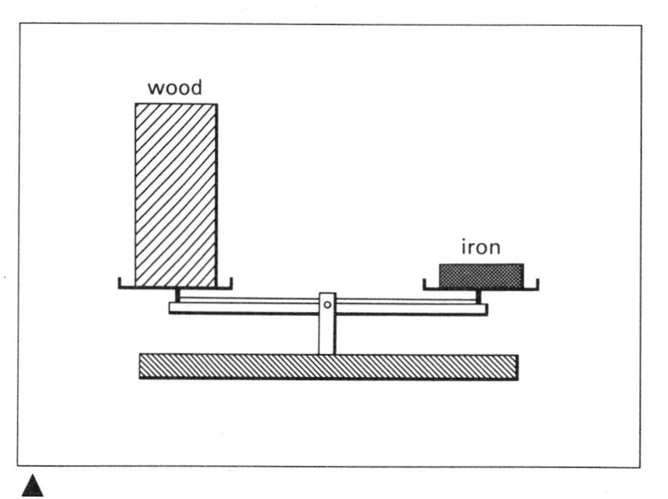

fig. 8.1a Equal weights different volumes

Density is the mass of unit volume of a substance

That is, the density of a substance is given by the expression

$$\textbf{density} = \frac{\textbf{mass} \text{ of object made of the substance}}{\textbf{volume} \text{ of object}}$$

$$\rho = \frac{m}{V}$$

Density is measured in **kilograms per cubic metre (kg/m^3)** or in **grams per cubic centimetre (g/cm^3)**.

The densities of some common substances are listed in table 8.1.

fig. 8.1b Equal volumes different weights

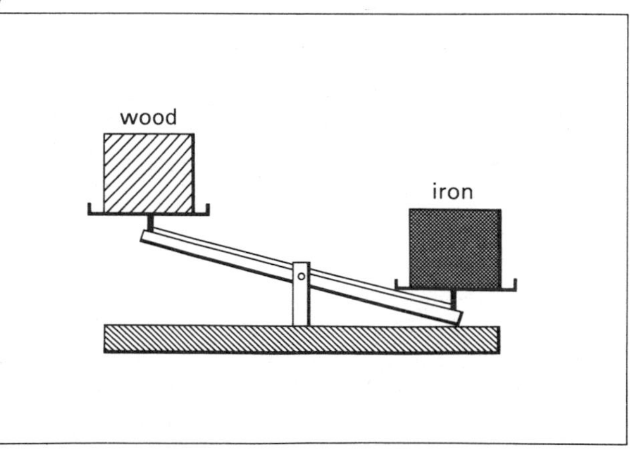

Density measurements

In order to determine the density of a material, we have to find the mass of an object made of the material and also the volume of the object. The mass can be measured directly with a balance very accu-

table 8.1 Densities of some common substances

Substance	Density (kg/m^3)	Substance	Density (kg/m^3)	Substance	Density (kg/m^3)
aluminium	2 700	balsa	200	air	1.29
copper	8 900	coal	1500	water	1000
gold	19 300	cork	250	ice	920
iron	7 900	deal	600	methylated spirit	830
lead	11 400	ebony	1200	paraffin	800
mercury	13 600	glass	2500	petrol	700
silver	10 500	oak	800	turpentine	870
uranium	18 700	rocks	2500	wax	900

rately, but the volume is more difficult to measure and the accuracy of the measurement is usually not so great.

The volume of a **regular solid**, such as a cube, rectangular solid or a cylinder, can be found by measurement of the appropriate dimensions and calculation of the volume using the appropriate formula.

The volume of an **irregular solid** can be found by measuring the volume of water displaced in a measuring cylinder directly or with the aid of a displacement vessel, figure 8.2. When reading a volume you should read the bottom of the curved meniscus and make sure that your eye is on the same horizontal level as the bottom of the meniscus, figure 8.3.

In order to find the density of a liquid, a known volume of the liquid is obtained using a measuring cylinder or more accurately a pipette or a burette, figures 8.4a–c. The known volume of liquid is placed in a weighed container and the mass of the liquid found.

It is sometimes convenient to compare the densities of materials with the density of some standard substance, e.g. water. The **relative density** of a substance is its density compared with the density of water.

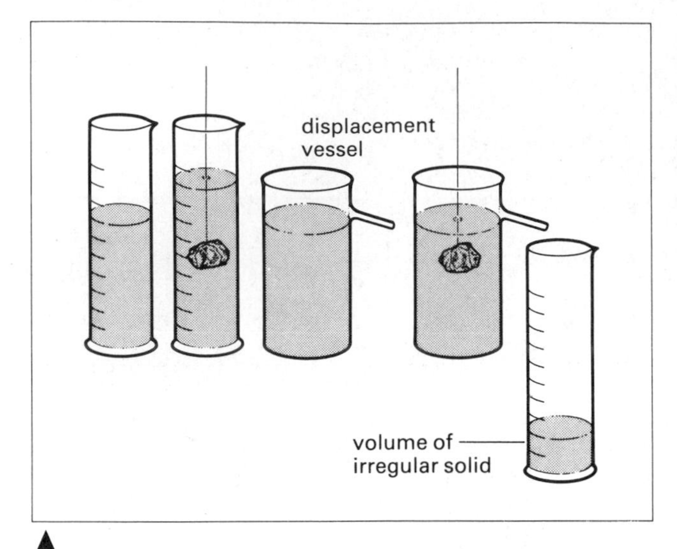

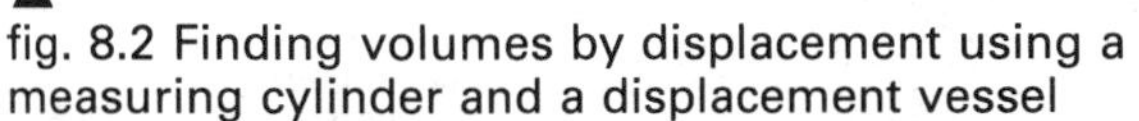

fig. 8.2 Finding volumes by displacement using a measuring cylinder and a displacement vessel

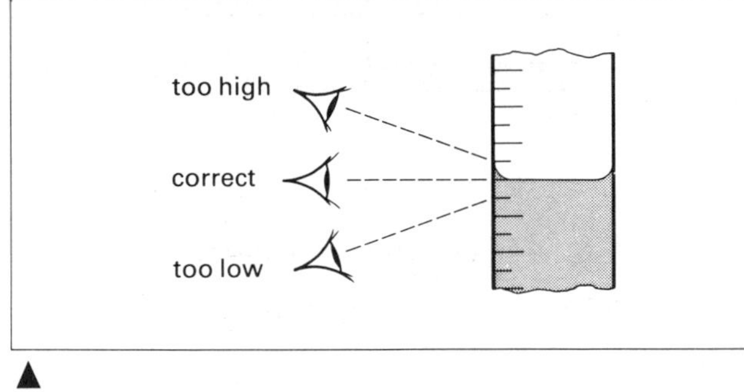

fig. 8.3 Correct way to read a volume

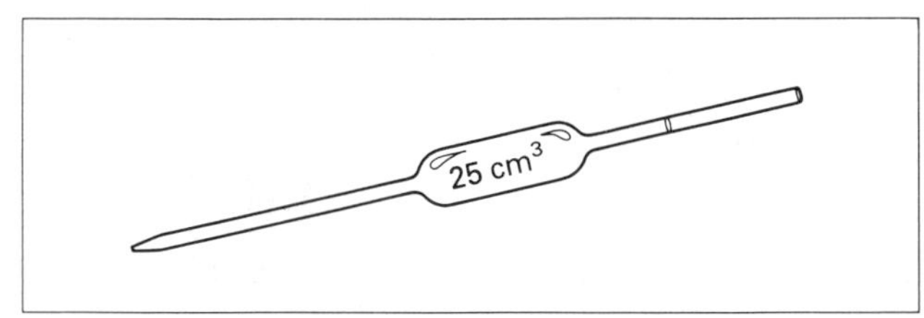

fig. 8.4a A pipette contains the volume marked on the bulb when it is filled to the mark on the stem

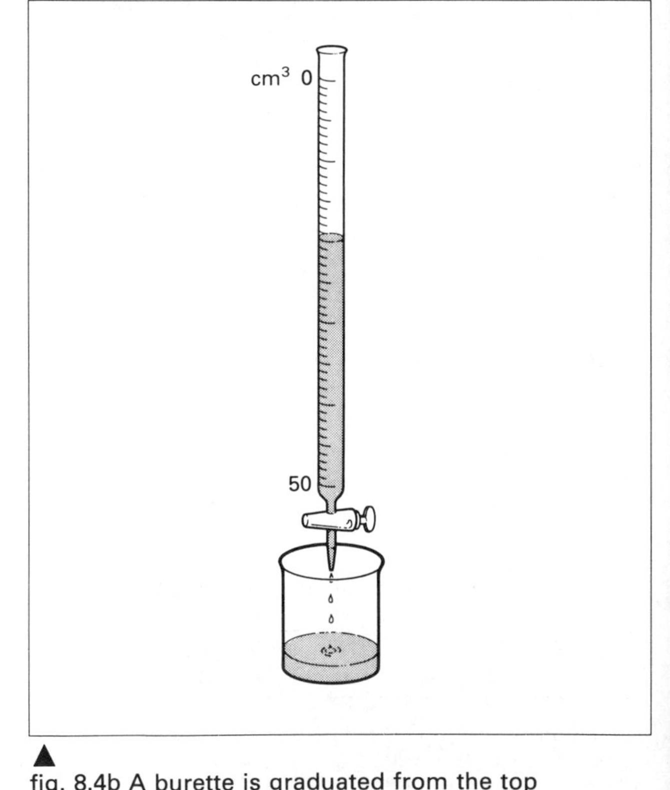

fig. 8.4b A burette is graduated from the top downwards and measures the volume of liquid run out from it

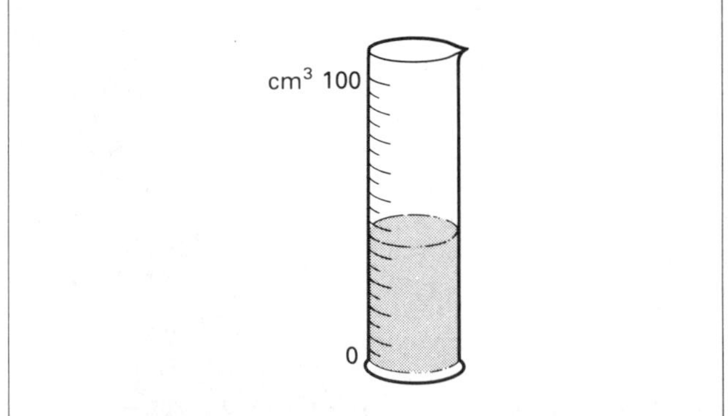

fig. 8.4c A measuring cylinder is graduated from the bottom upwards and measures the volume of the liquid poured into it

$$\text{relative density} = \frac{\text{density of substance}}{\text{density of water}}$$

An alternative definition is

$$\text{relative density} = \frac{\text{weight of substance}}{\text{weight of an equal volume of water}}$$

Since relative density is simply a ratio it has no units. Relative density used to be called **specific gravity**.

A useful piece of apparatus for getting a definite volume of liquid is the relative density bottle, figure 8.5. The bottle is full when the liquid rises just to the top of the fine hole in the glass stopper. Any excess liquid can be removed with filter paper. The bottle should be held by the neck only, otherwise the warmth of the hand will produce expansion of the liquid so that it overflows.

The relative density of a liquid can be found using the relative density bottle:

find the mass of the dry, empty bottle $= M_1$
the mass of the bottle full of water $= M_2$
the mass of the bottle full of liquid $= M_3$

$$\text{relative density of liquid} = \frac{M_3 - M_1}{M_2 - M_1}$$

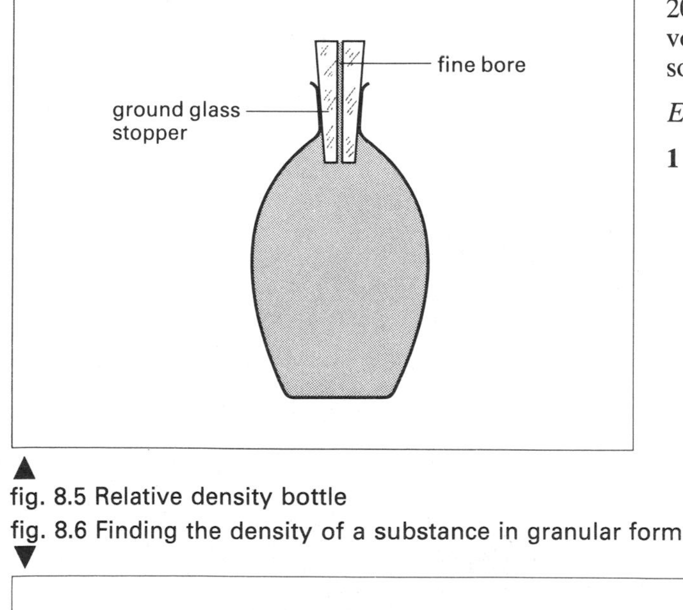

fig. 8.5 Relative density bottle

The relative density of a substance in powder form or in pieces small enough to go through the neck of the relative density bottle can be found as shown in figures 8.6a–c.

Weigh some lead shot on the pan of a balance. Then with the shot still on the balance pan find the weight together with a relative density bottle full of water. Now put the shot inside the bottle and see that the bottle is full of water right to the top of the stopper. Now the only difference between weighing (b) and (c) is the weight of the water displaced by the shot. This is equal to the weight of an equal volume of water so the relative density can be found.

The density of mixtures

When two substances of different densities are mixed together, the density of the mixture will lie somewhere between the densities of the constituents of the mixture, depending on the proportions in which they are mixed. The mass of the mixture will be the sum of the masses of the constituents, and the volume of the mixture the sum of the volumes, provided no volume change has taken place on mixing. Sometimes there is a volume change. For example, if 100 volumes of alcohol are mixed with 100 volumes of water, the volume of the mixture will be 190 volumes and not 200 volumes. Some contraction has taken place. Some volume change usually takes place on solution of a solid in a liquid.

EXAMPLES

1 An alloy is made by mixing 8.10 kg of aluminium of density 2700 kg/m^3 with 2.16 kg of zinc of density 7200 kg/m^3. What is the density of the alloy?

$$\text{density} = \frac{\text{mass}}{\text{volume}}$$

$$\text{volume} = \frac{\text{mass}}{\text{density}}$$

Therefore the volumes of the masses of aluminium and zinc can be calculated.

fig. 8.6 Finding the density of a substance in granular form

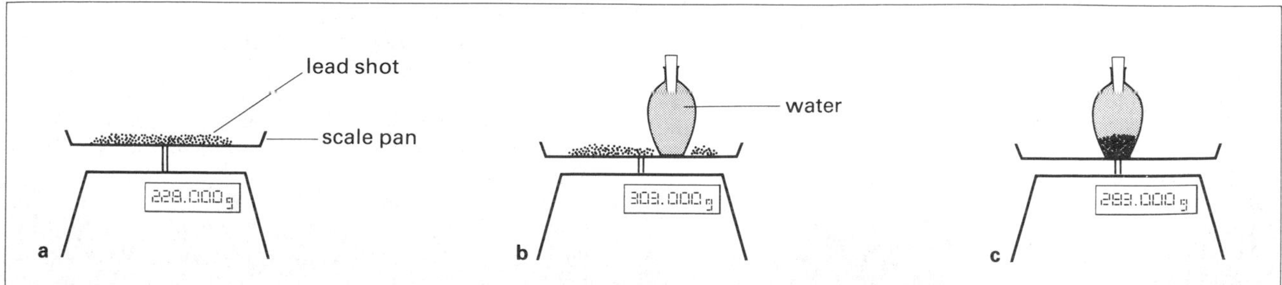

$$\text{volume of aluminium} = \frac{8.1}{2700} = 0.003\ \text{m}^3$$

$$\text{volume of zinc} = \frac{2.16}{7200} = 0.0003\ \text{m}^3$$

total mass = 8.1 + 2.16 = 10.26 kg
total volume = 0.003 + 0.0003 = 0.0033 m^3

$$\text{density of alloy} = \frac{\text{mass}}{\text{volume}} = \frac{10.26}{0.0033} = 3109\ \text{kg/m}^3$$

2 When 10 g of a powder (density 1.95 g/cm^3) is dissolved in 100 cm^3 of water the density of the resulting solution is 1.07 g/cm^3. What change in volume takes place on solution?

volume of water = 100 cm^3
mass of water = 100 g

$$\text{volume of powder} = \frac{10}{1.95} = 5.13\ \text{cm}^3$$

mass of powder = 10 g
total initial volume = 105.13 cm^3
total mass = 110 g

$$\text{final volume} = \frac{110}{1.07} = 102.8\ \text{cm}^3$$

change in volume = initial volume – final volume
= 105.13 – 102.8 = 2.33 cm^3

EXERCISE 8

In questions 1–4 select the most suitable answer.

1 A piece of metal has a mass of 140 g and a volume of 20 cm^3. Its density, in g/cm^3, is
A $\frac{1}{7}$
B 7
C 70
D 280
E 2800

2 A body has a density of 0.25 g/cm^3 (g cm^{-3}). If the mass is 120 g its volume is
A 0.002 cm^3
B 0.02 cm^3
C 30 cm^3
D 120.25 cm^3
E 480 cm^3

3 An empty relative density bottle has a mass 15.0 g. When completely filled with water its mass is 39.0 g. What will be its mass if completely filled with acid of relative density 1.20?
A 20.0 g
B 28.8 g
C 35.0 g
D 43.8 g
E 46.8 g

4 The volume of a large stone can be found by putting it into a displacement can which is full of water and collecting the water displaced in a measuring cylinder containing a known volume of water. The volume of the stone is
A the initial volume of water in the cylinder
B the final volume of water in the cylinder
C the difference between volumes of water in the cylinder
D the volume of water in the displacement can at the beginning
E the volume of water in the displacement can after the stone was put in

5 A glass stopper weighs 40 grams. It is placed in a measuring cylinder containing a liquid as shown in the figure. The cylinder gives the volume in millilitres.

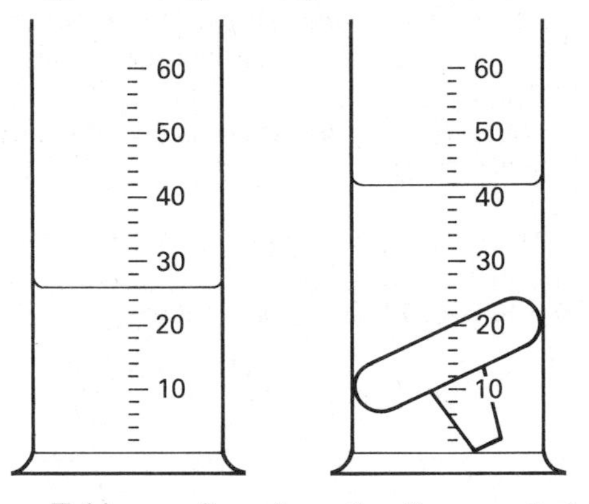

Taking readings from the diagrams find the
a volume of liquid,
b total volume of liquid and stopper,
c volume of the stopper,
d density of the stopper.

6 100 identical copper washers are put into an empty measuring cylinder and 50 cm^3 of water are added. The diagram shows the level of the water.

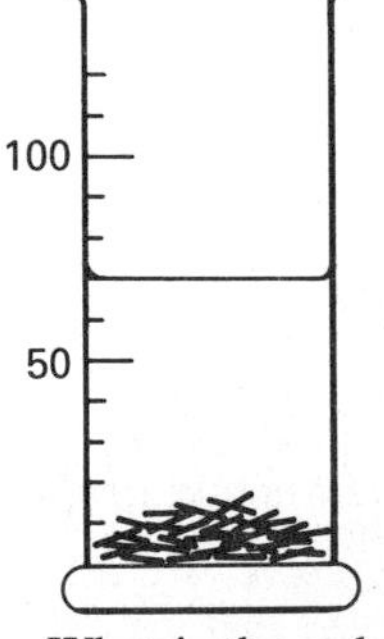

What is the volume of
a 100 copper washers?
b 1 copper washer?
c If all the copper washers together have a mass of 180 g, calculate the density of copper.

7 A light alloy used in aircraft is made by mixing 85% by mass of aluminium (density 2700 kg/m^3) with 15% by mass of magnesium (density 1740 kg/m^3). What is the density of this alloy?

9 PRESSURE

When two bodies are in contact under the action of a force pressing them together, the resultant behaviour depends not only on the size of the force, but also on the area of contact. For instance, if we try to walk over soft snow we shall sink in, but if we increase the area of contact by wearing snow-shoes or skis we can move over the surface. The downward force is the same in both cases, but the area of contact is different. The distribution of force over an area is called the **pressure**.

$$\textbf{pressure} = \frac{\textbf{force exerted}}{\textbf{area in contact}}$$

$$p = \frac{F}{A}$$

The unit of pressure is the **newton per square metre** (**N/m^2**). This unit of pressure is named the **pascal** (**Pa**).

The block of metal shown in figures 9.1a–c has a weight of 5000 N. Although the shape and weight of the block remain the same the pressure it exerts on the surface supporting it depends on which way up it is. The total force exerted is known as the **thrust** and this is the same in all three cases, namely 5000 N.

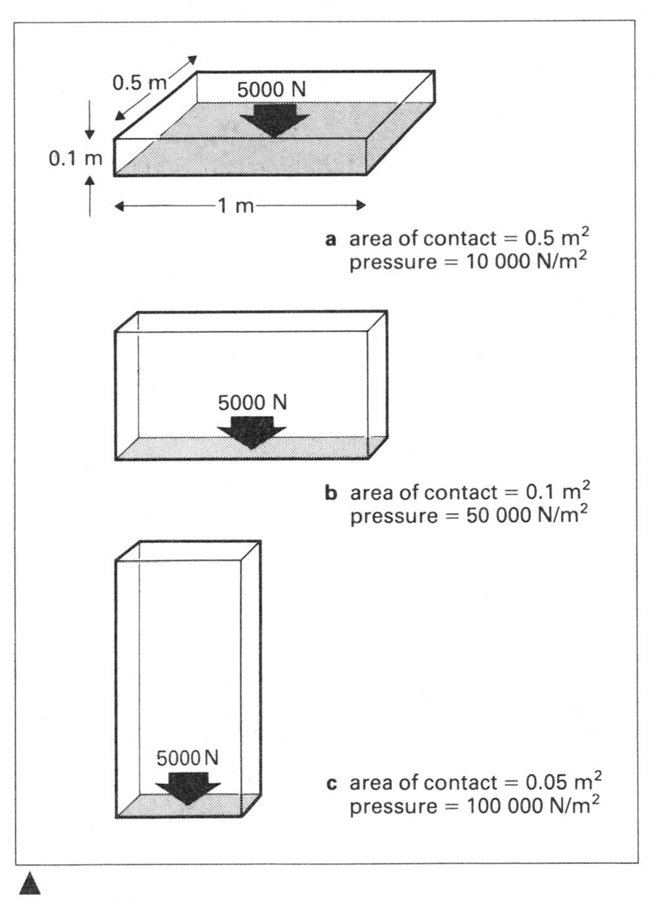

fig. 9.1 Different pressures from same thrust

Pressure in liquids

If we make a hole in the bottom or side of a tin full of water, the water flows out. The water must be exerting a pressure against the bottom and sides of the tin. If we push a piece of wood under water, it will rise to the surface as soon as it is released. The water must be exerting an upward force on the wood just as we have to exert a downward force to keep the wood under water. Liquids (and gases) exert pressure in all directions.

If we look at the 'spouting can' shown in figure 9.2 we see that the jet from the lower hole spurts out further than the jet from the hole at the top. This suggests that the pressure at the bottom of the can is greater than at the top. It is a fact that pressure in liquids increases with depth.

An instrument for detecting changes in pressure or for measuring pressure differences is a **manometer**. In

fig. 9.2 Pressure increases with depth

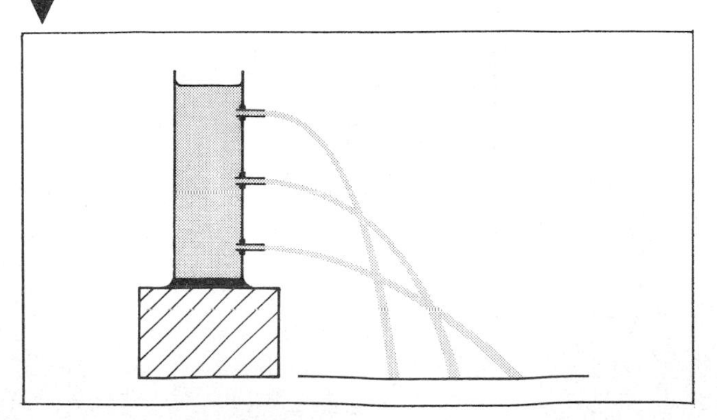

its simplest form it consists of a U-tube containing a liquid, figure 9.3. This could be oil, water or mercury, depending on the pressure to be measured. One side of the tube is connected to the pressure source to be investigated. With both sides at the same pressure, both liquid columns are at the same level. A difference in pressure is shown by a difference in the levels of the columns of liquid.

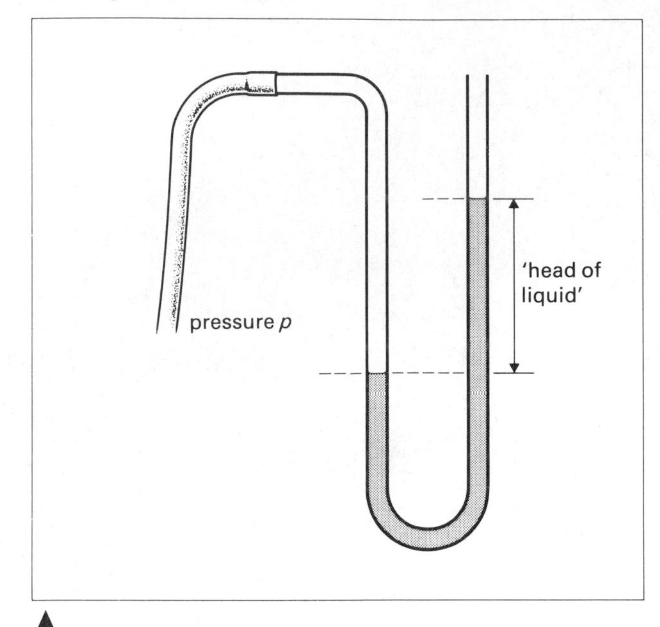

▲
fig. 9.3 Simple manometer

The manometer can be used to investigate the variation of pressure with depth in a liquid, figure 9.4.

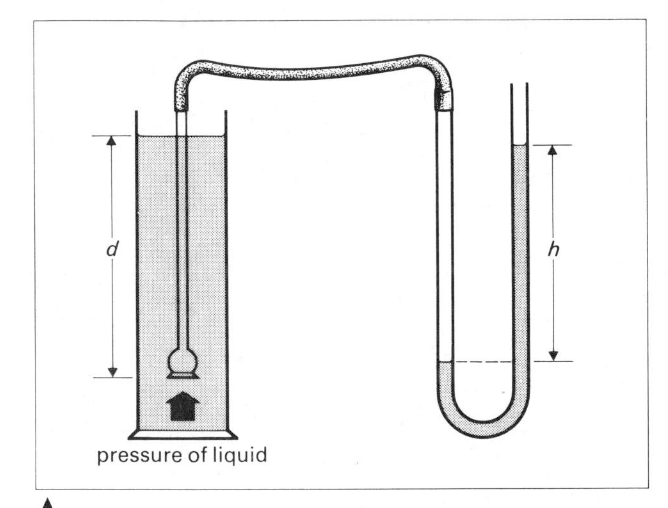

▲
fig. 9.4 Measuring change of pressure with depth

Position the end of the thistle funnel at various known depths below the surface of the liquid and for each position read the difference in levels of the manometer columns which measures the pressure at that particular point. Plot the pressure (h) against the depth (d).

The straight line graph produced shows that

Pressure in a liquid is proportional to depth

The pressure at a given depth in a liquid can be calculated.

Consider a column of liquid of height h and area of cross-section A, figure 9.5.

fig. 9.5 Pressure of a liquid column
▼

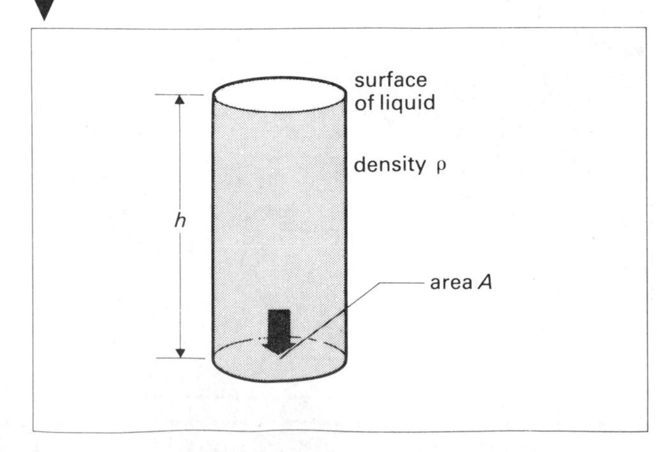

volume of liquid $= Ah$
mass of liquid $=$ volume × density
$= Ah\rho$
weight of liquid $= Ah\rho g$
$=$ downward force on the base

$$\text{pressure} = \frac{\text{force}}{\text{area}}$$

$$= \frac{Ah\rho g}{A}$$

$$= \boldsymbol{h\rho g}$$

From the result the pressure is directly proportional to the height (or depth) of the liquid column. It will be noticed that the result does not depend in any way on the area of cross-section. This can be demonstrated using the apparatus known as Pascal's vases, figure 9.6. The pressures at the bottom of each tube are the same so no movement of liquid takes place. The result is often summed up by saying 'liquids find their own level'.

The pressure is the same at all points at the same depth in a liquid at rest

A deep sea diver must wear a pressurised suit or use a diving bell because pressure increases with depth, figure 9.7. Even so, there are limits to the depth he can go. At high pressures, gases are forced to dissolve in the blood. The diver has to be decompressed slowly to allow these dissolved gases to escape slowly, otherwise the result can be fatal.

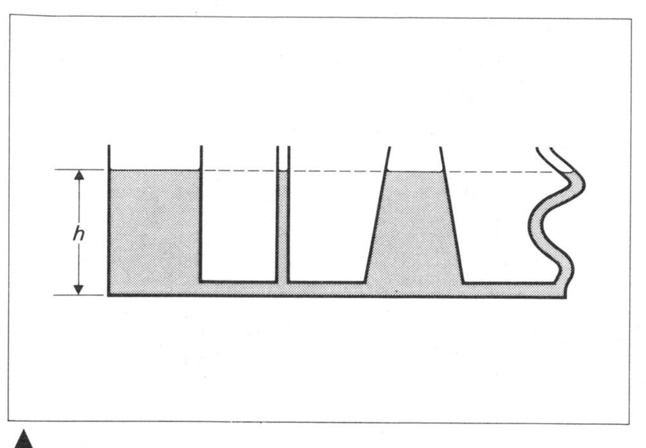

fig. 9.6 Pascal's vases

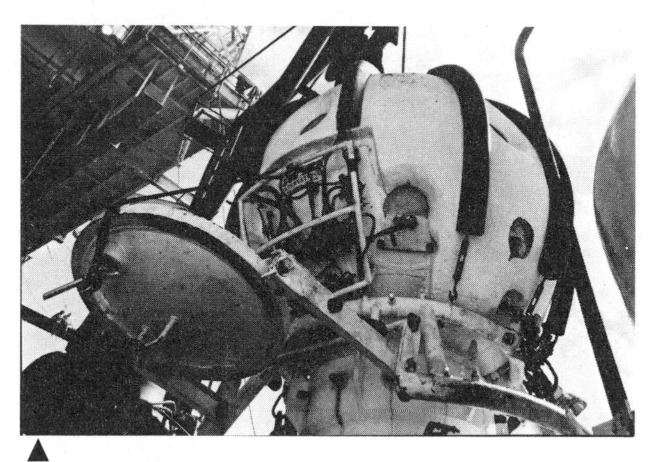

fig. 9.7 Deep-sea diving bell

Since pressure increases with depth, the force per unit area at the base of a dam is greater than at the top. For this reason dams are built much thicker at the base, figure 9.8.

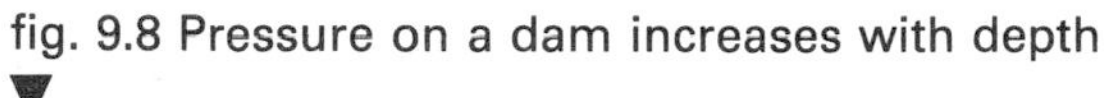
fig. 9.8 Pressure on a dam increases with depth

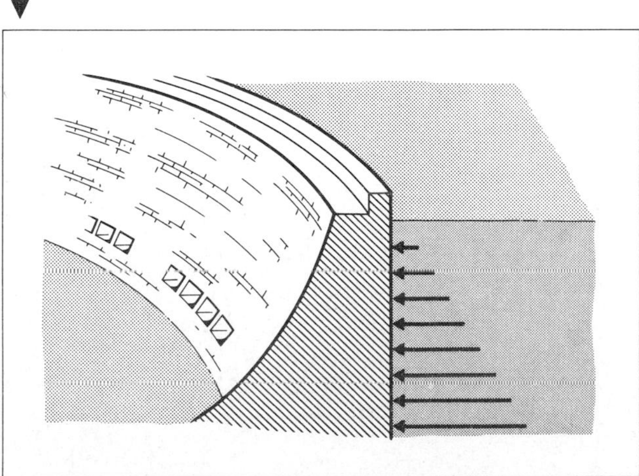

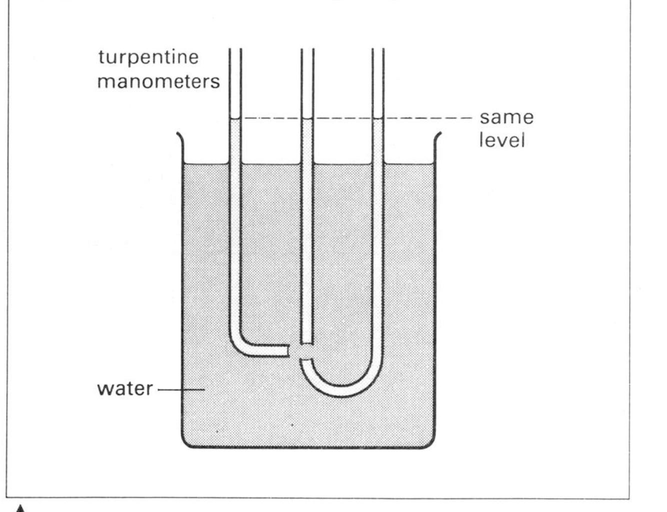

fig. 9.9 Pressure acts in all directions

Transmission of pressure

Fluids (liquids and gases) not only exert pressure equally in all directions, but also transmit pressure equally in all directions. An experiment by Robert Boyle illustrates the first point. Three tubes, bent as shown in figure 9.9, are filled with turpentine and lowered into a vessel of water. The tubes of turpentine are manometers measuring the pressure in different directions. The turpentine is maintained at the same level in all three tubes, showing that pressures upwards, sideways and downwards are the same.

Pressure in a fluid acts equally in all directions

When the piston of the syringe shown in figure 9.10 is pressed down, the water issues from all the holes perpendicularly to the surface of the sphere and equally in all directions.

fig. 9.10 Pressure is transmitted in all directions

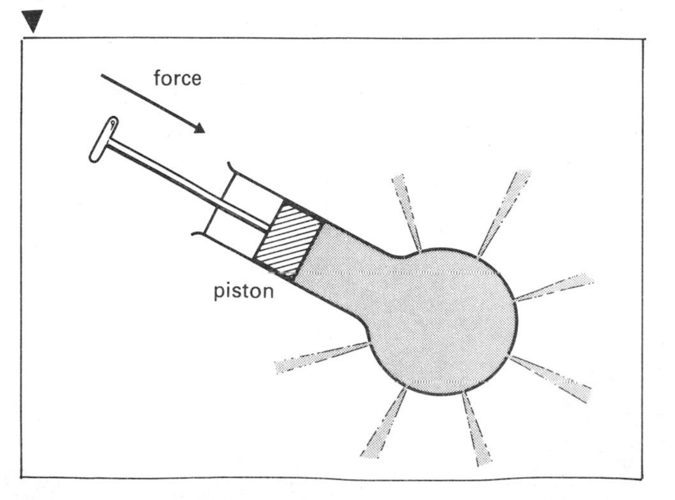

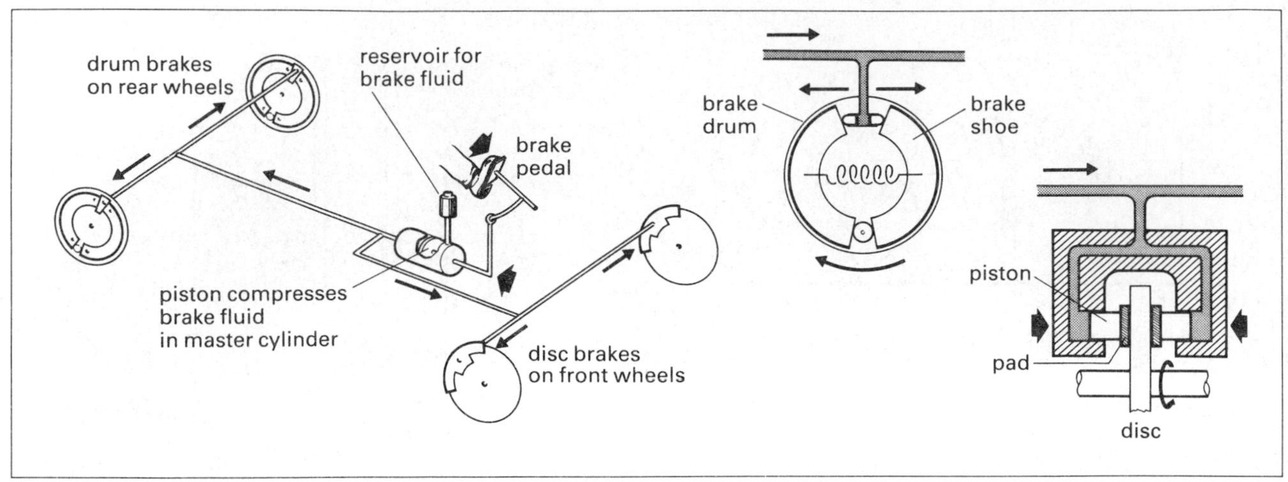

fig. 9.11 Hydraulic braking system of a car

Fluids transmit pressure equally in all directions

The transmission of pressure by liquids is used in many kinds of hydraulic machinery, figures 9.11 and 9.12. The principle of the **hydraulic press** is illustrated in figure 9.13. Suppose a force F_1 is applied to the small piston. The pressure is

$$\frac{F_1}{A_1} \quad \left(\text{pressure} = \frac{\text{force}}{\text{area}}\right)$$

This pressure is transmitted to the larger piston of area A_2. The total thrust on the larger piston is

$$F_2 = \frac{F_1}{A_1} \times A_2 \quad (\text{pressure} \times \text{area})$$

fig. 9.13 Principle of the hydraulic press or lift

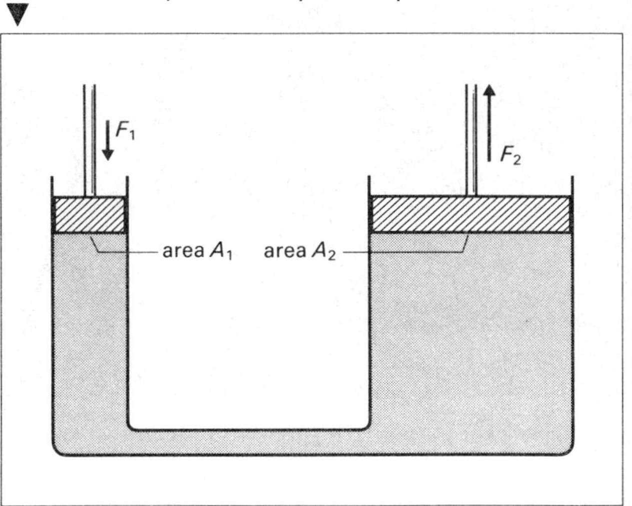

fig. 9.12 In a hydraulic car-jack the handle is pressed down and pushes the small piston forward. The pressure is applied to a larger piston which raises the lever and lifts the car

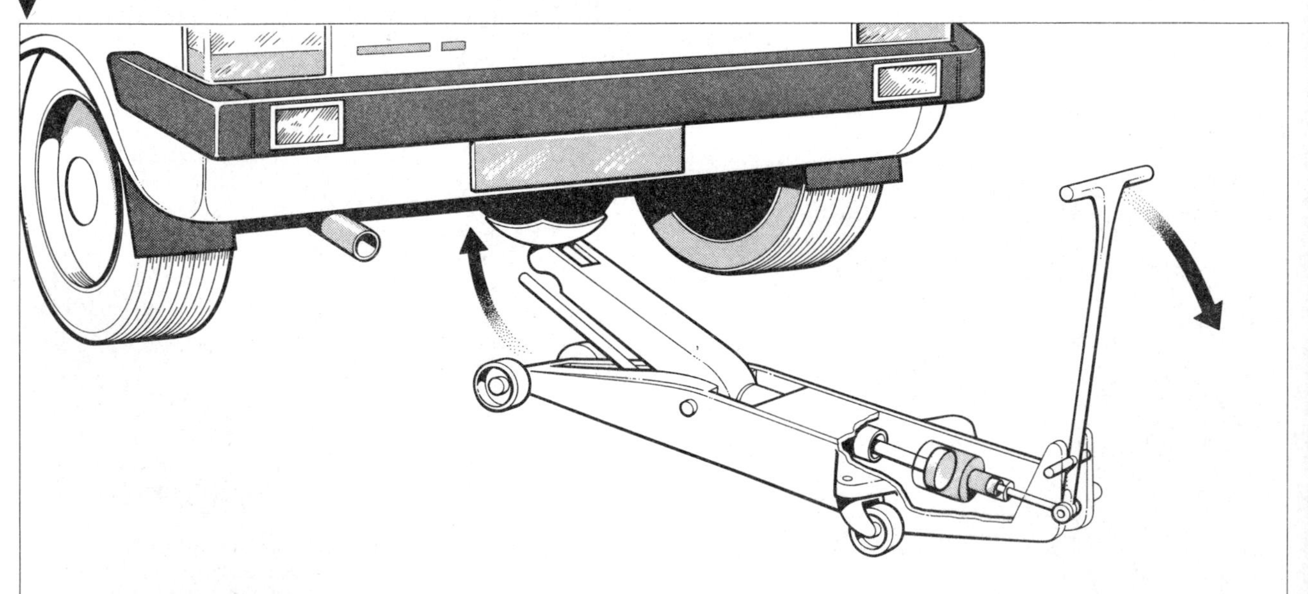

Acting as a machine this gives a mechanical advantage of

$$\frac{F_2}{F_1} = \frac{A_2}{A_1}$$

By making A_2 large and A_1 small, a large mechanical advantage can be obtained, although the distance moved by the larger piston will be correspondingly less than the distance moved by the small piston. The hydraulic press is very efficient because frictional losses are small.

The bourdon pressure gauge

The manometer measures pressure in terms of a head of liquid, but it is often useful to have a direct reading pointer instrument to measure pressure. The principle of the bourdon gauge, shown in figure 9.14, is that a curved metal tube of oval cross-section tends to straighten out when the pressure inside it is increased. As the tube straightens out the toothed quadrant rotates and turns the pointer round the scale. The scale has to be calibrated in the first place by using a series of known pressures measured by a manometer.

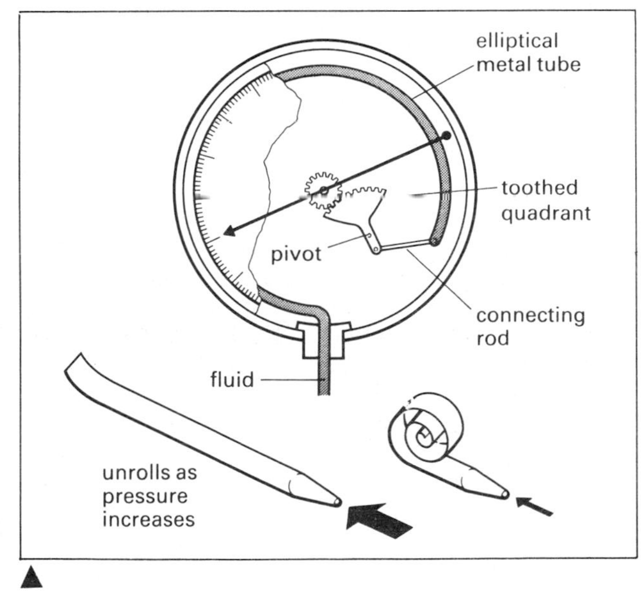

▲ fig. 9.14 Bourdon pressure gauge

fig. 9.15 ▼

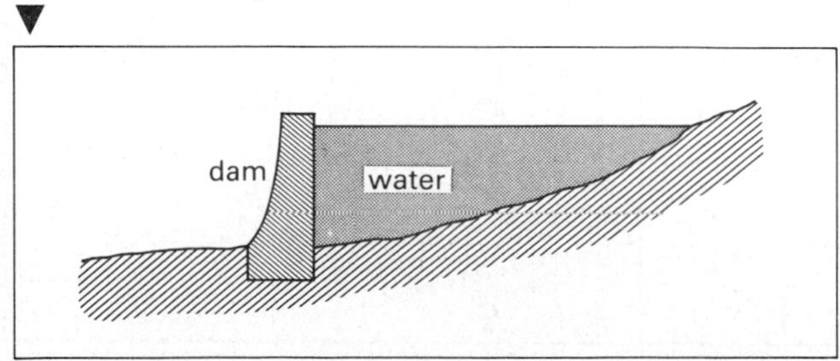

EXERCISE 9

In questions 1–3 select the most suitable answer.

1 Below the surface of a liquid the pressure acting in the liquid
- **A** decreases as the depth increases
- **B** increases as the depth increases
- **C** is independent of the depth
- **D** is the same for all liquids at a particular depth
- **E** depends only on the temperature of the liquid at a particular depth

2 A diving bell is to be used at great depths and so its walls are made very thick. The main reason for this is that
- **A** the water is much colder at great depths
- **B** water pressure increases with the depth of the water
- **C** the density of the water is much greater at great depths
- **D** ice forms in deep water and could crush the bell
- **E** a thin-walled vessel would float up to the surface

3 The hydrostatic pressure on the dam wall in figure 9.15 at the bottom of a deep reservoir depends upon the
- **A** depth of water
- **B** surface area of the water
- **C** length of reservoir
- **D** thickness of the dam wall
- **E** density of the material of the dam wall

4 a What average pressure is exerted on the ground by a man weighing 65 kg if the area of his boots in contact with the ground is 200 cm^2?

b If he is wearing ice skates with a blade area in contact with the ice of 5 cm^2, what pressure is exerted on the ice when he is skating on one foot?

5 a (i) Define the term pressure.
(ii) Give the SI unit of pressure.

b A motor vehicle weighing 10 000 N is supported on the road on four tyres. If the area of each tyre in contact with the road surface is $\frac{1}{100}$ m^2, calculate the pressure in each tyre.

6 a A force pump raises 60 m^3 of water to a height of 10 m. How much work is done? (density of water 1000 kg/m^3)

b If water is held in a rectangular container 5 m × 4 m × 3 m deep, what is the pressure in N/m^2 exerted by the water on the bottom of the container?

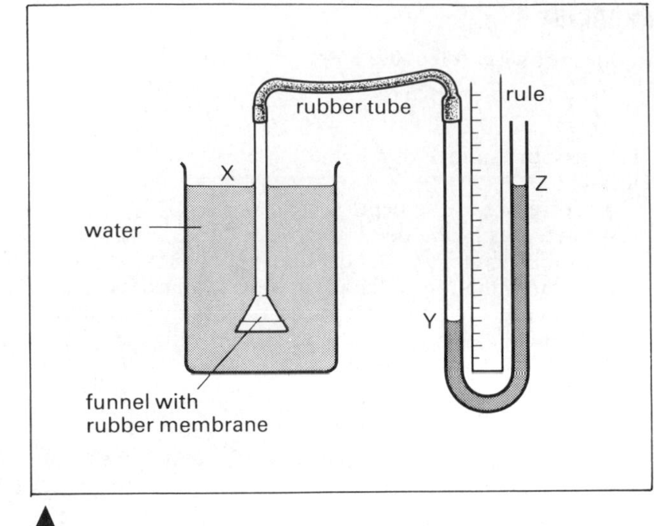

fig. 9.16

7 Look at figure 9.16.

a Why is the liquid in the U-tube likely to be water and not mercury?

b Why does the water not flow into the funnel?

c What is the pressure, relative to atmospheric pressure, exerted at (i) X? (ii) Y? (iii) Z?

d State what, if anything, would happen to the liquid levels in the U-tube if the funnel were (i) moved from side to side at the same horizontal level, or (ii) moved lower into the liquid.

e If a larger funnel were attached and held at the same horizontal level as in part **d** (i), what difference, if any, would be seen in the levels of the liquid in the U-tube?

f The weight of water in the container is 100 N and the base is square, measuring 0.2 m by 0.2 m. (i) What will be the pressure due to the water exerted on the base? (ii) What will be the depth of water in the container? (density of water = 1000 kg/m^3, gravitational force = 10 N/kg)

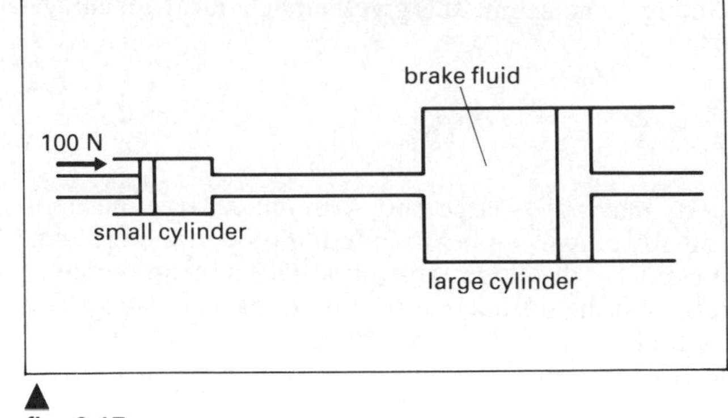

fig. 9.17

8 Figure 9.17 shows a simple mechanism used to operate the brakes of a motor car. The area (cross-sectional) of the small cylinder is 2 cm^2 and of the large cylinder 10 cm^2. The system is filled with brake fluid.

a If the brake pedal pushes against the piston in the small cylinder with a force of 100 N, what is the pressure exerted on the brake fluid?

b What is the force exerted by the brake fluid on the piston in the large cylinder?

c If the piston in the large cylinder moves 0.1 cm, how far would the piston in the small cylinder move?

d State the mechanical advantage of the system.

e What is the name given to this type of braking system?

9 a Figure 9.18 shows the hydraulic braking system in a car. Write down the numbers 1 to 6. Write down the name of each part labelled in the diagram beside the appropriate number.

b The brake pedal acts like a lever. Calculate the force applied by the foot in the situation shown in figure 9.19.

c What is the important physical property of brake fluid?

fig. 9.18

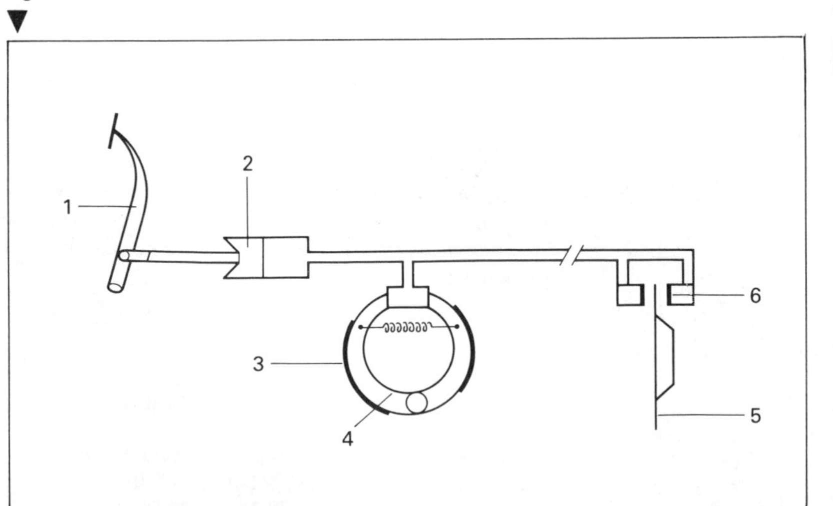

fig. 9.19

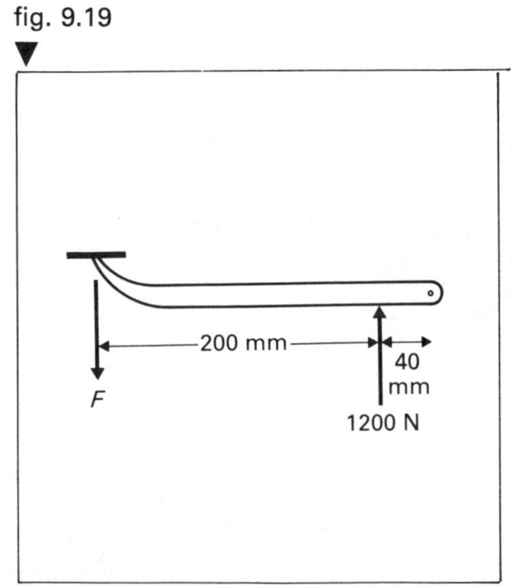

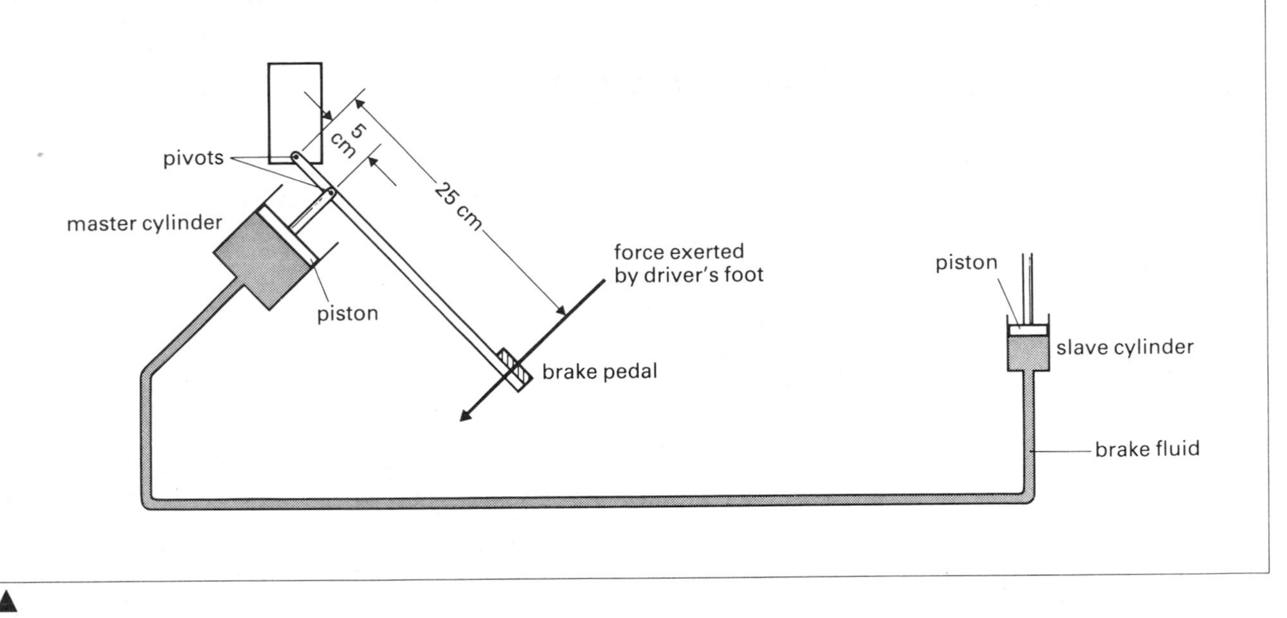

fig. 9.20

10 Figure 9.20 shows the brake pedal arrangement, the master cylinder, and the slave cylinder for a motor-car braking system. When the brake pedal is pressed down by the driver's foot the piston in the slave cylinder moves and operates the brake.

a (i) Which way will the piston in the slave cylinder move when the driver's foot is pressed down? (ii) Assuming that the driver's foot exerts a force of 100 N, calculate the force acting on the piston of the master cylinder. (iii) Explain why a force is transmitted to the slave cylinder piston.

b (i) If the master cylinder piston acts on a total surface area of 20 cm^2, calculate the pressure it exerts on the brake fluid. (ii) The slave cylinder piston has a surface area of 2 cm^2 in contact with the brake fluid. Calculate the value of the force acting on this piston.

c Explain why the brakes would not work properly if some air bubbles were accidentally left inside the connecting pipe.

d Give *two* other uses of hydraulic systems.

10 ATMOSPHERIC PRESSURE

Surrounding the earth is a layer of air, consisting of a mixture of gases, called the **atmosphere**. The earth retains its atmosphere because of the pull of gravity on the air molecules. The atmosphere becomes less dense with increasing height. Half of the total mass of the atmosphere occurs in the first 6 km.

The pressure of the atmosphere

The great pressure exerted by the atmosphere was first demonstrated in a spectacular way by Otto von Guericke (1602–86) at Magdeburg in Germany. Von Guericke had invented an air pump. With it he pumped the air from two hemispheres of thick copper fitted together with a leather ring soaked in oil. The sphere was just over 30 cm in diameter, and when there was a vacuum inside he discovered that he could not separate the two halves. In one experiment, performed in 1651, two teams of eight horses had to be used before the hemispheres could be separated. If the air was allowed to return the hemispheres could easily be pulled apart. The experiment can be repeated in the laboratory using smaller hemispheres or even two rubber suction discs pressed together.

Figure 10.2 shows another experiment which demonstrates the pressure of the air. This involves pumping the air out of a can. When the can is full of air the inside and outside pressures balance, but when the air is pumped out the can collapses due to the unbalanced atmospheric pressure outside.

The air pressure acts on our bodies in all directions, but pressures acting inside the body prevent us from being crushed.

fig. 10.1 Magdeburg hemispheres

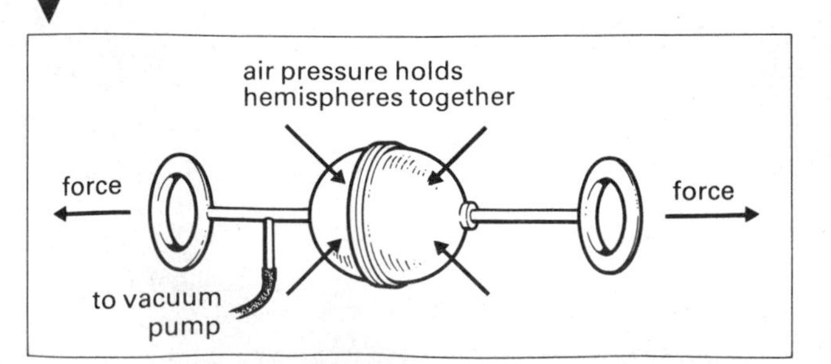

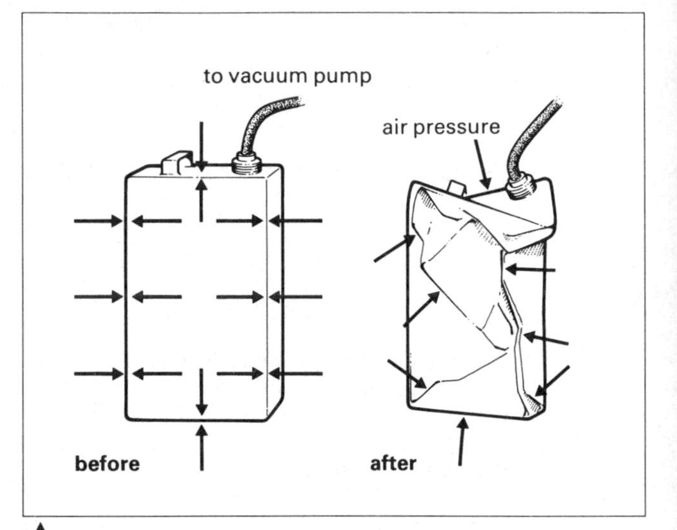

fig. 10.2 Collapsing can experiment

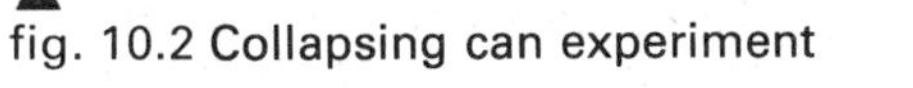

fig. 10.3 Measuring atmospheric pressure

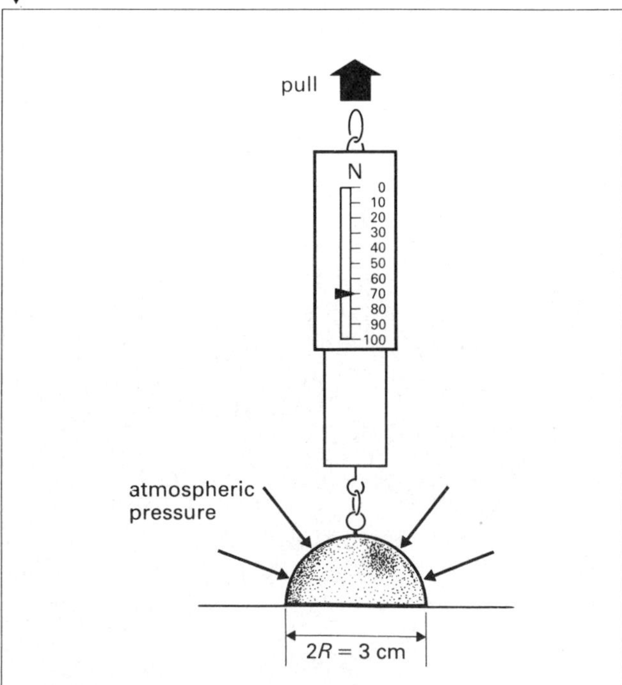

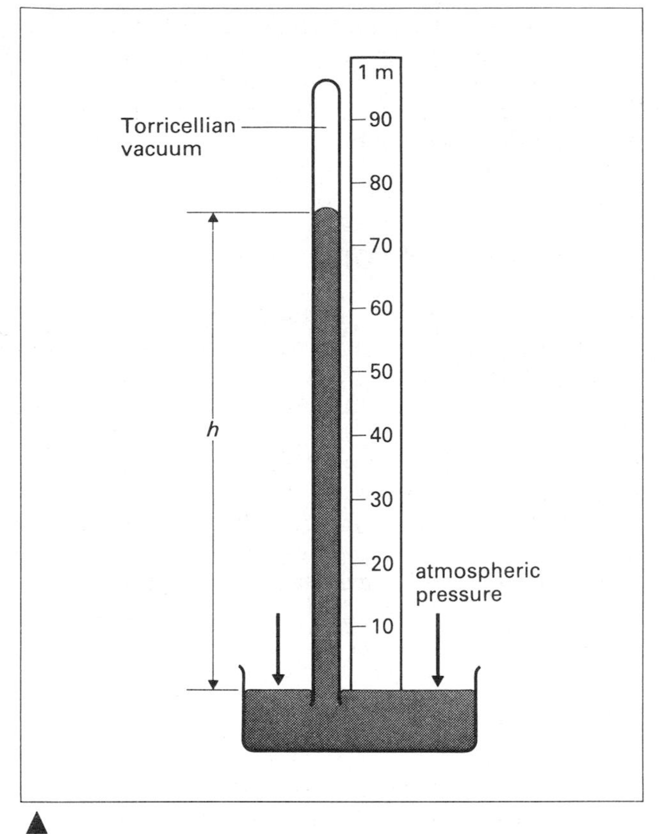

fig. 10.4 Simple barometer

Measurement of atmospheric pressure

A simple and direct (although not very accurate) method of measuring the pressure exerted by the atmosphere is to measure the force necessary to remove a suction disc from a smooth surface. This is shown in figure 10.3. The force is measured with a spring balance calibrated in newtons. As the disc is pulled away from the surface it becomes roughly hemispherical in shape and the diameter of the base should be measured with calipers just as it is leaving the surface. This enables the area over which the force is acting to be calculated, since the total thrust on a hemisphere is equal to the thrust over the circular end. The maximum pull registered on the spring balance is noted.

Typical results obtained:

radius (R) of disc when leaving surface = 1.5 cm
area (πR^2) = $3.14 \times (1.5)^2 = 7.06\ \text{cm}^2 \approx 7.0\ \text{cm}^2$
spring balance reading = 70 N

$$\text{pressure} = \frac{\text{force}}{\text{area}} = \frac{70}{7} = 10\ \text{N cm}^{-2}$$

$$= 100\ 000\ \text{N m}^{-2} \text{ or pascal}$$
$$= 100\ \text{kPa}$$

The standard atmospheric pressure is 101 290 Pa.

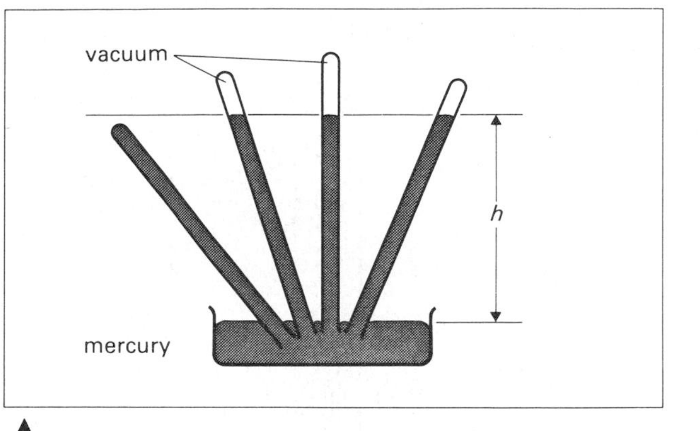

fig. 10.5 When the barometer tube is tilted, the vertical height of the mercury remains the same

The simple barometer

A barometer is an instrument for measuring the pressure of the air. The simple barometer shown in figure 10.4 consists of a tube about 80 cm long and closed at one end. The tube is completely filled with mercury (care being taken to remove all air bubbles), and then inverted in a dish of mercury. The mercury falls a little way in the tube, and then the column of mercury is supported by the pressure of the air on the free surface of the mercury in the dish. The vertical distance from the surface of the mercury in the dish to the top of the mercury meniscus in the tube is a measure of the atmospheric pressure. The average height of the column is 76 cm, since the air pressure varies from day to day. The space above the mercury column contains no air, although it does contain mercury vapour. It is called a Torricellian vacuum after the Italian, Torricelli, who first made a simple barometer in 1643.

Tilting the tube from side to side makes no difference to the vertical height of the mercury column since the pressure of a column of liquid depends only on vertical height.

Because the mercury barometer is commonly used to measure atmospheric pressure, the pressures are often given as the height of the mercury column supported. These heights can be converted to genuine pressure units using the expression for the pressure exerted by a column of liquid.

$$\text{pressure} = h\rho g$$

Taking

$h = 0.76$ m
$\rho = 13\ 600\ \text{kg/m}^3$
$g = 9.8$ N/kg

$$\text{pressure} = 0.76 \times 13\ 600 \times 9.8$$
$$= 101\ 290\ \text{Pa}$$

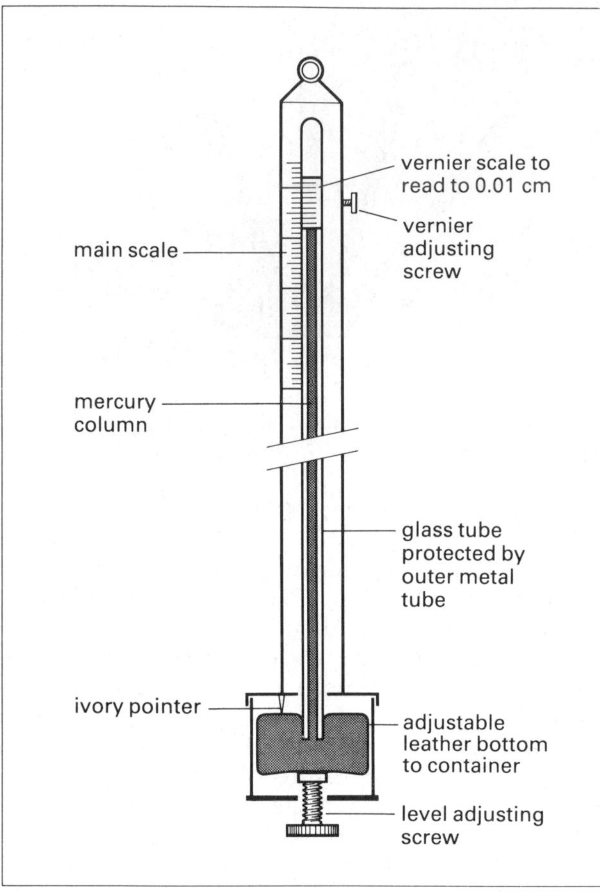

▲
fig. 10.6 Fortin barometer. The mercury is always set on the ivory pointer so that all readings start from the same place

Since the average atmospheric pressure is approximately 10^5 Pa, this figure is used by meteorologists and called a **bar**. The smaller unit, the millibar, is usually used on weather maps to label the isobars which are lines joining points where the atmospheric pressure is the same.

The standard Fortin barometer

When the atmospheric pressure changes, the height of the mercury in the tube changes, and so does the level of mercury in the dish. It is thus impossible to use a scale fixed to the tube to measure the height *h*. The Fortin barometer, figure 10.6, overcomes this difficulty by holding the mercury in a leather bag, the bottom of which can be screwed up and down. Before a reading is taken the mercury level is adjusted so that it coincides with the tip of an ivory pointer. This marks the zero of the main scale. An adjustable vernier scale at the top enables accurate readings to be taken.

The aneroid barometer

The mercury barometer is a rather clumsy instrument and is not easy to carry around. The **aneroid** barometer, figure 10.7, overcomes these difficulties. The word 'aneroid' means without liquid. The instrument consists essentially of a partially evacuated metal box which is prevented from collapsing by a metal spring. When the air pressure increases, the box is squeezed in. When the air pressure is low, the spring forces the

fig. 10.7 Aneroid barometer
▼

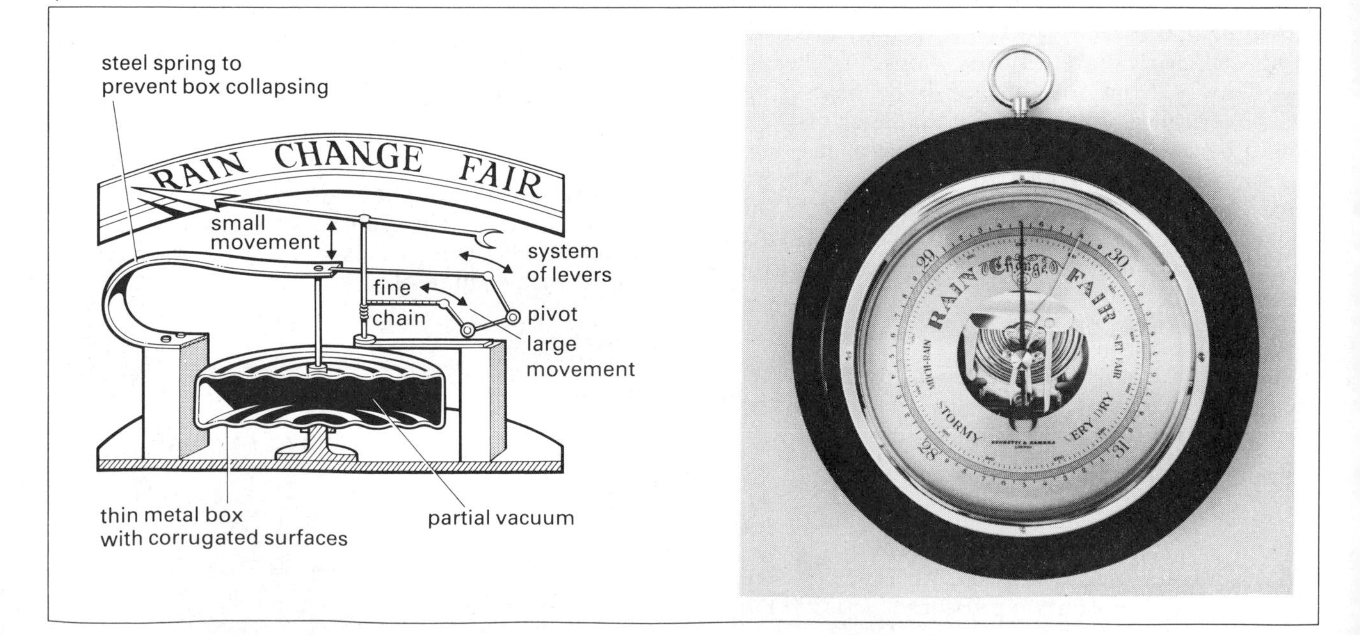

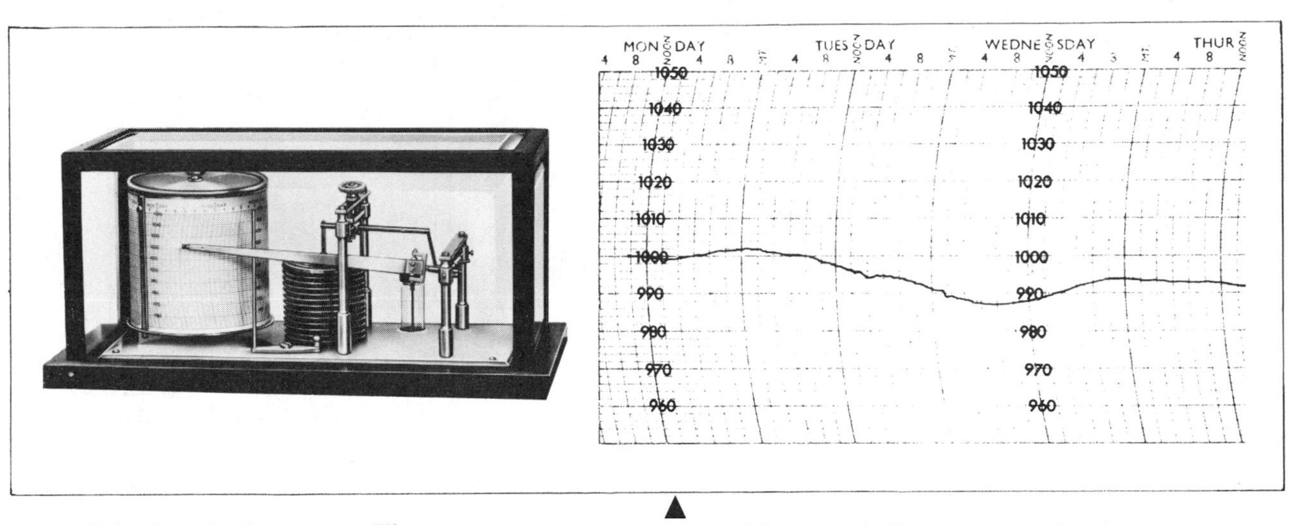

fig. 10.8 Barograph. Pressures on the chart are measured in millibars (mb). A pressure of one bar is equivalent to a barometric height of 75 cm of mercury

sides of the box further apart. These movements are magnified and transmitted, through a series of levers, to a pointer which moves over a scale. Sometimes the movement of the box works a lever with a pen on the end so that the pressure is automatically recorded on a chart. The instrument is then called a **barograph**, figure 10.8.

Since the atmospheric pressure changes with height above the earth's surface, getting smaller as the height increases, an aneroid barometer can be used as an **altimeter** or height measurer in aircraft.

fig. 10.9 The earth's atmosphere is sub-divided into several layers, each distinguished by certain properties. We live in the troposphere, which is a very thin band in comparison with the size of the earth and, indeed, with the other layers

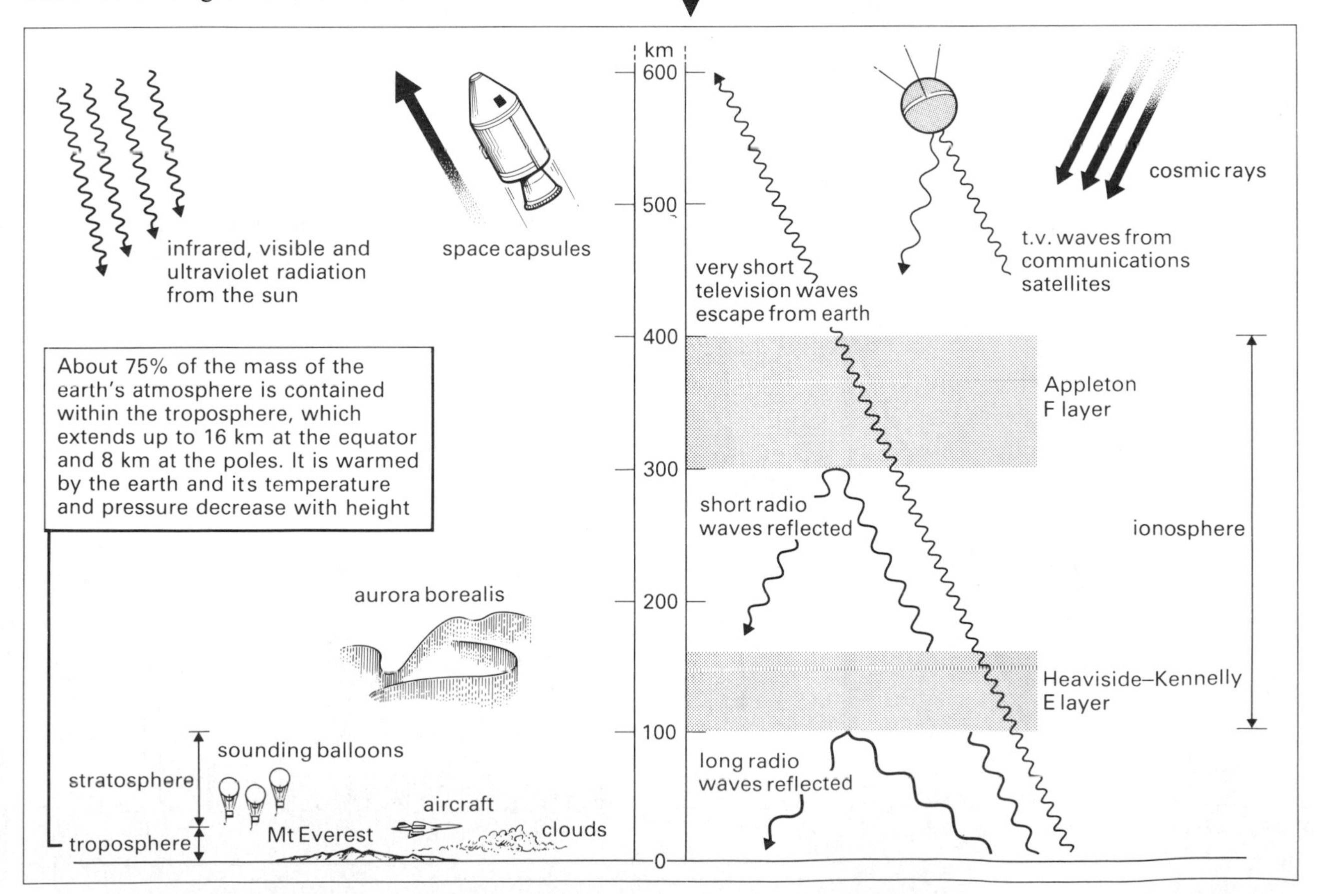

EXERCISE 10

In questions 1–4 select the most suitable answer.

1 Figure 10.10 shows a simple mercury barometer. The level of the mercury in the tube at S is much higher than that in the dish at R. One reason for this could be that
- **A** there is a vacuum above S which has drawn the mercury up to S
- **B** there is a vacuum above S so that the air pressure at R has forced the mercury up to S
- **C** the mercury rises to S due to capillary action
- **D** the temperature causes the mercury to rise to S
- **E** the air pressure in the tube above S is greater than the pressure outside so the mercury rises to S

2 An aneroid barometer may also be used to measure
- **A** the air temperature
- **B** the wind speed
- **C** the humidity of the air
- **D** the height above sea level
- **E** the direction of air currents

3 Figure 10.11 shows five different mercury barometers. In each diagram the enclosed space above the mercury is a vacuum. Which one is indicating the highest value of the pressure?

4 Assume that the atmospheric pressure at sea level is equal to the pressure exerted by 0.76 m of mercury and the acceleration of free fall does not vary with height. If the average density of air is 1.2 kg/m^3 and that of mercury is 13 600 kg/m^3, the atmospheric pressure at 3400 m above sea level is equal to the pressure exerted by
- **A** 1.060 m of mercury
- **B** 0.912 m of mercury
- **C** 0.760 m of mercury
- **D** 0.460 m of mercury
- **E** 0.300 m of mercury

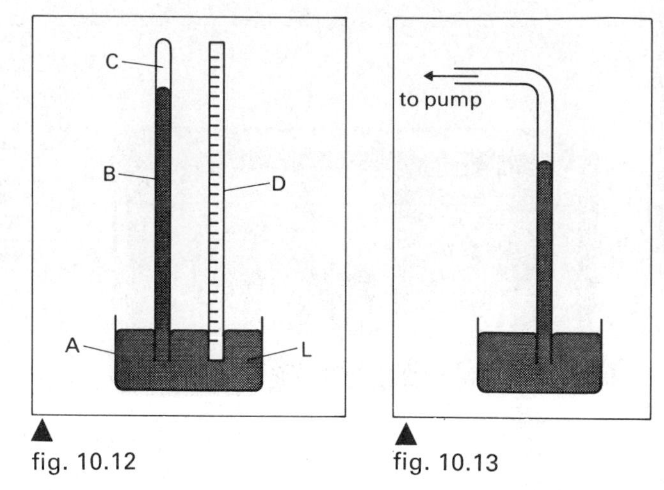

fig. 10.12

fig. 10.13

5 Figure 10.12 shows the outline of a simple barometer set up ready for use.
- **a** Name the parts labelled A, B, C and D. Name liquid L and explain why it is used rather than water. State the range of values that would be necessary on D.
- **b** Why is it necessary to measure air pressure regularly at weather stations? How are these values recorded on a weather map?

6 Figure 10.13 shows the lower end of a long vertical glass tube dipping into a dish of clean dry mercury which is open to the atmosphere whilst the other end of the tube is connected to a vacuum pump. When the pump is operated the mercury rises in the tube. Explain why this is so.

State the factors which will affect the maximum height to which the mercury will rise above the level in the dish when the pump is operated for some time. You may assume that there is sufficient mercury in the dish for the lower end of the tube to remain immersed throughout the experiment.

fig. 10.10

fig. 10.11

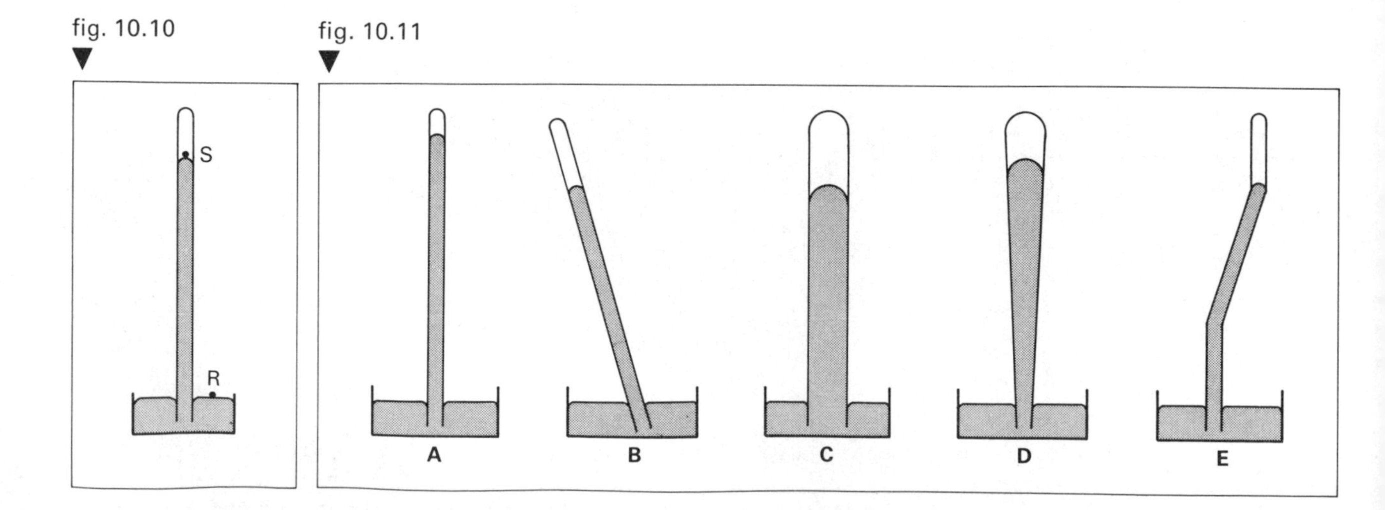

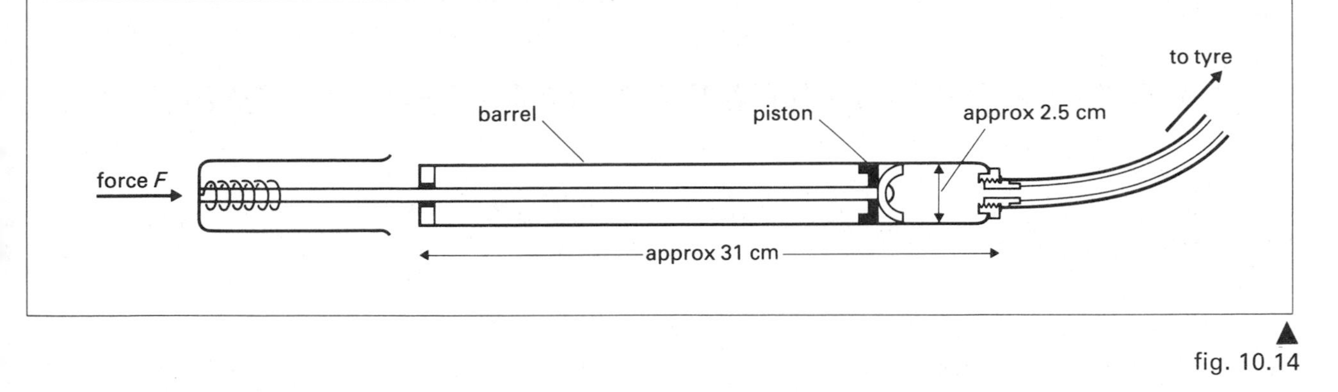

fig. 10.14

7 a The pressure of the air in the tyres of a four-wheeled car is 2×10^6 Pa in excess of atmospheric pressure. The area of contact of each tyre with the road is 0.008 m^2 when the car is empty. Calculate the weight of the car.

b Figure 10.14 represents a bicycle pump which is connected to a tyre. A force F is applied to the piston and as the piston moves work is done. Explain (i) what happens to the mechanical energy supplied to the piston; (ii) how the dimensions of the pump should be altered from those shown to make it more suitable for use with (1) tyres which are to be inflated to a higher pressure, (2) tyres which are to be inflated quickly but to a lower pressure.

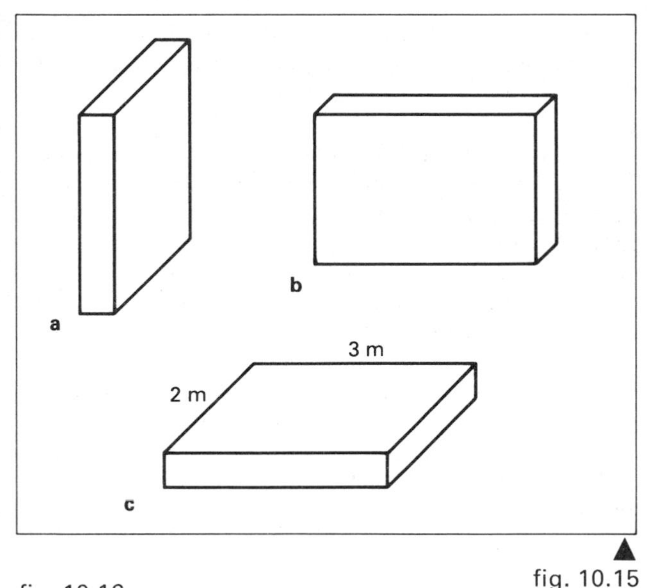

fig. 10.15

8 a The same concrete block is shown resting on the ground in thrcc positions in figure 10.15. In which position is it exerting the greatest pressure on the ground?

b If the dimensions of the largest face are 2 m width and 3 m length, what is the area of this face?

c What pressure does the block exert on the ground in figure 10.15c if the weight of the block is 36 000 N?

d Describe with the aid of a diagram an experiment which shows that the atmosphere exerts a pressure.

e Explain how atmospheric pressure makes it possible to use a straw to drink from a bottle.

f Figure 10.16 shows a glass tube inserted in a bottle containing some water so that air cannot enter the bottle once it is sealed. Someone attempts to drink the water and at first succeeds. What happens to the pressure of the air in the bottle? Why does the water then cease to flow?

g A submarine is at a depth of 40 m in fresh water of density 1000 kg/m^3. What is the pressure acting on the submarine due to the water? (Take $g = 10$ m/s^2)

fig. 10.16

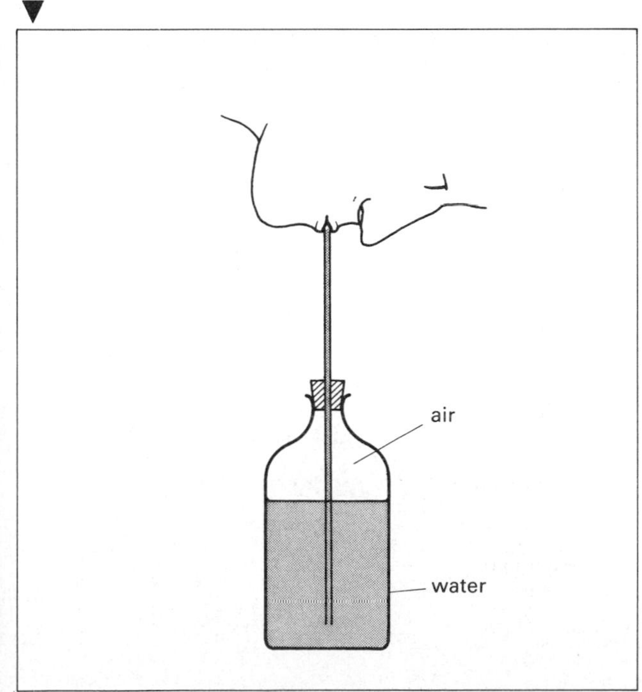

9 Describe, with the aid of a labelled diagram, how an aneroid barometer measures atmospheric pressure.

11 THE EFFECTS OF HEAT ON MATTER: EXPANSION

Heat is a form of energy. When heat energy is given to a body one or more of the following effects may be observed.

- The body gets hotter – there is an increase in temperature.
- There is an increase in size – expansion.
- There is a change of state from solid to liquid, or from liquid to vapour.
- The chemical composition changes.
- There is a change in electrical properties, e.g. resistance.
- There is a change of colour.

All these effects can be observed experimentally and some of them are used to measure the change in temperature which has taken place.

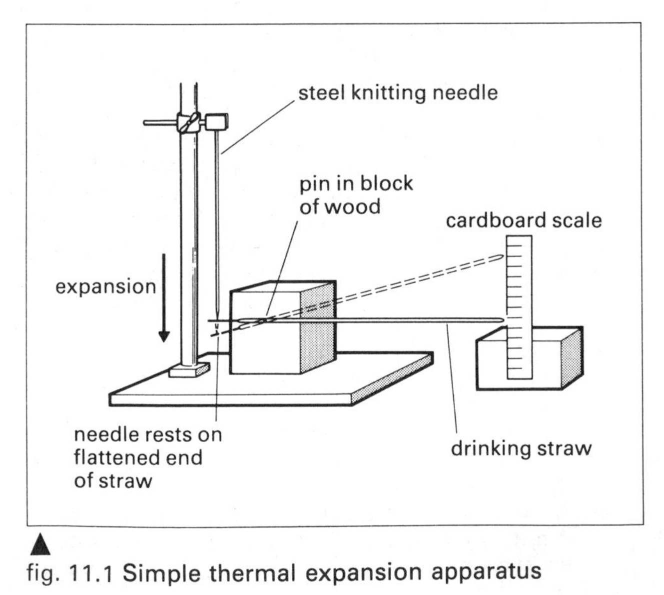

▲ fig. 11.1 Simple thermal expansion apparatus

fig. 11.2 Compound bar ▼

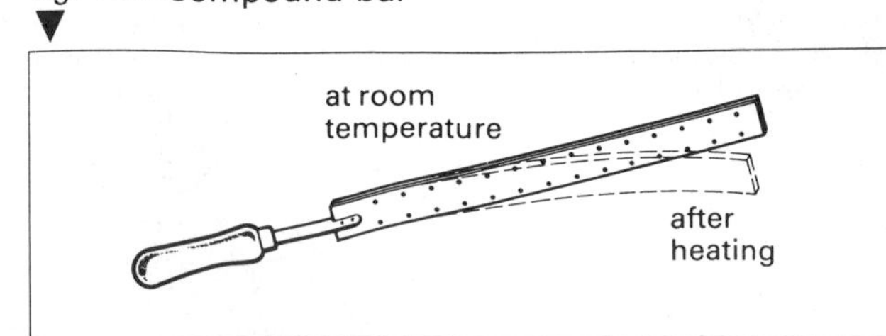

Expansion of solids

A variety of simple experiments will show that materials expand when heated and contract when cooled. Figures 11.1–11.3 illustrate some of these experiments.

The amount solids expand is very small. In figure 11.1 the straw lever magnifies the expansion of the steel knitting needle when it is heated and the movement is visible on the scale.

Figure 11.2 shows a compound bar made of strips of copper and iron riveted together. When heated the bar becomes curved, with the metal which expands most (copper) on the outside.

Figure 11.3 shows a steel bar which has been heated for some time so that it expands. Then a large nut is screwed up tightly, so that the cast iron bar is pulled tightly against the knife edges. The steel bar is cooled by pouring water on it and the cast iron bar snaps in

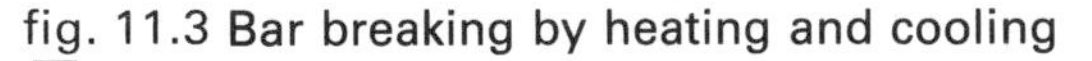
fig. 11.3 Bar breaking by heating and cooling ▼

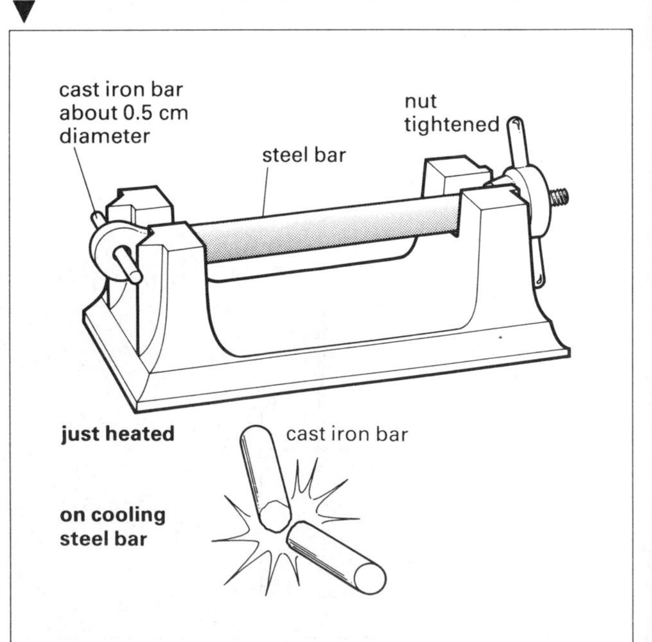

two, owing to the force exerted by the contracting steel. In order to snap a bar of this size, a pull of about a quarter of a tonne is necessary.

Expansion of liquids

When the coloured water in the flask shown in figure 11.4 is heated the rise of level in the tube shows the expansion.

The relative expansions of different liquids can be investigated with the apparatus in figure 11.5. For a true comparison equal volumes of the liquids must be heated by exactly the same amount.

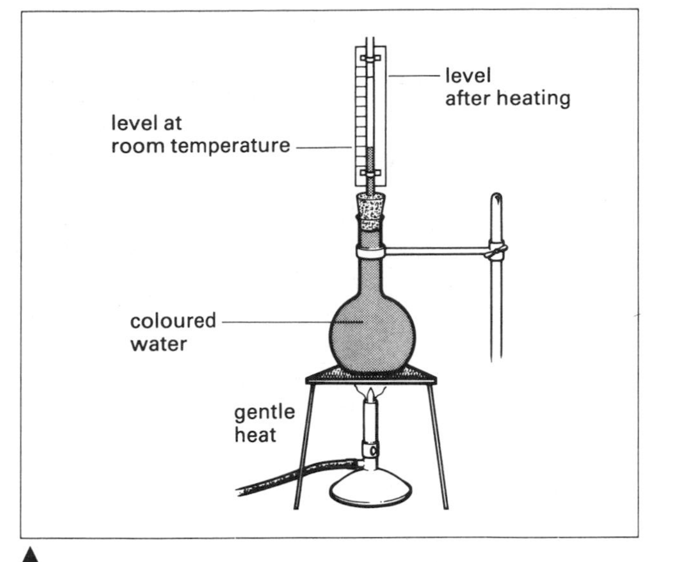

▲ fig. 11.4 Expansion of a liquid

fig. 11.5 Comparison of the expansion of liquids ▼

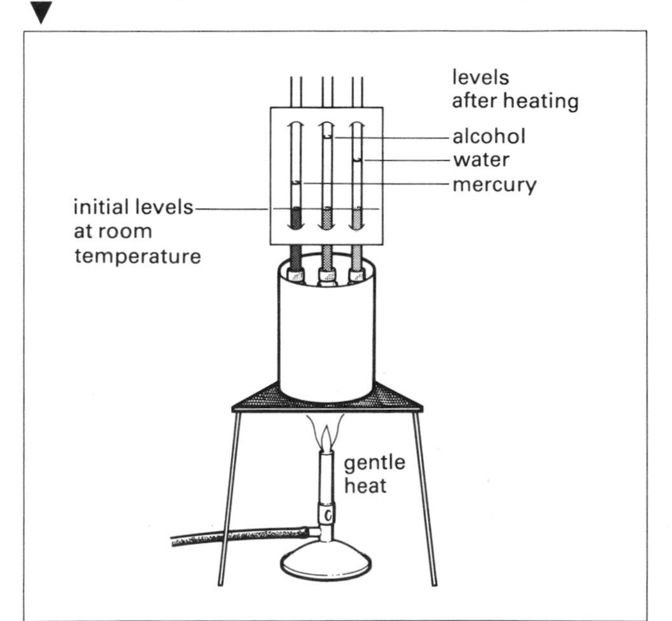

Expansion of gases

The expansion of air can be demonstrated very simply by the two ways shown in figure 11.6. Figure 11.6a shows air in a flask warmed with the hands. The expansion is shown by the rise in the level of the liquid in the tube. The same effect can be shown by expanded air bubbling out through water in a beaker, figure 11.6b.

Different solids and liquids expand by different amounts when heated through the same temperature range. All gases expand by the same amount, but we must remember that the volume of a gas can also be changed by changing the pressure. Solids expand less than liquids which, in turn, expand very much less than gases.

fig. 11.6 Demonstrating the expansion of air ▼

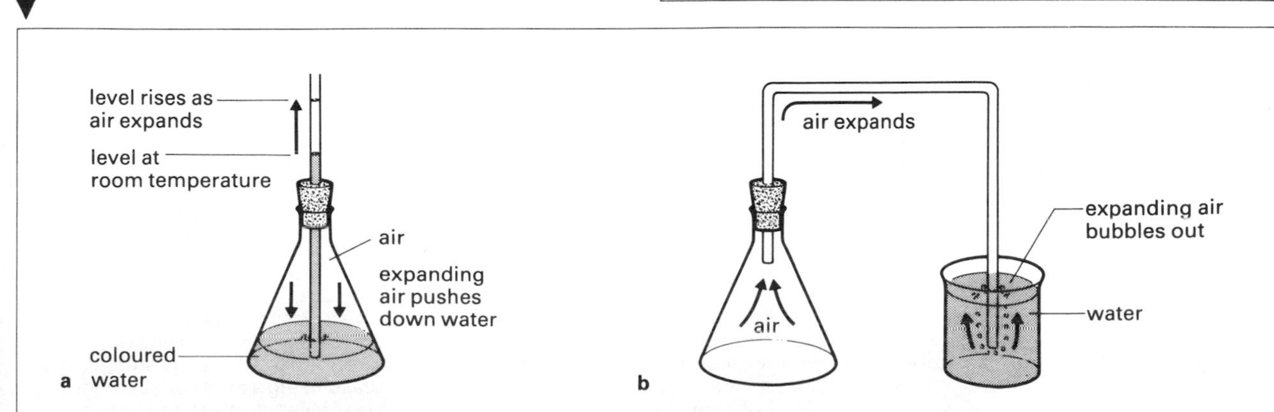

Some examples of expansion

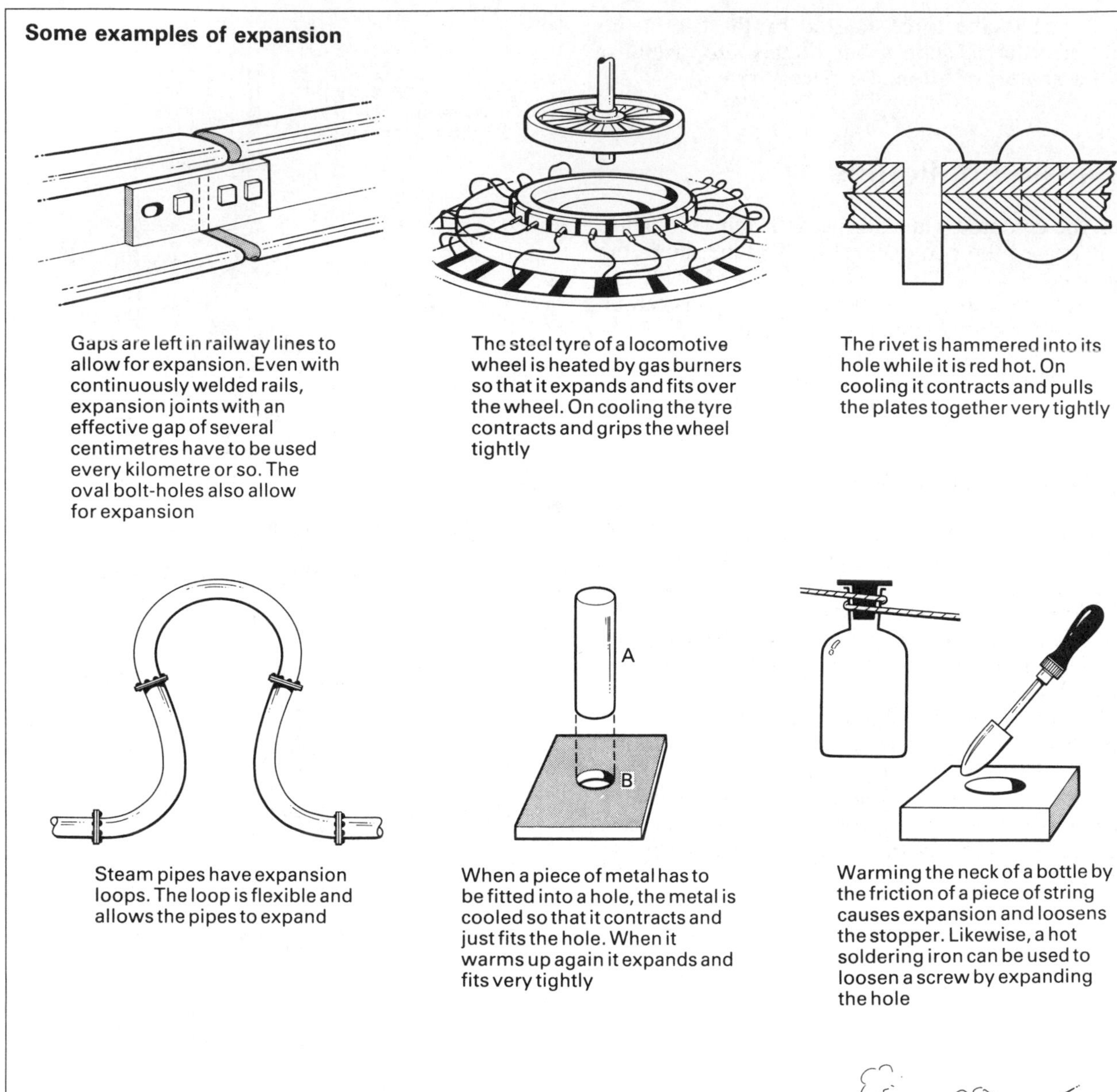

Gaps are left in railway lines to allow for expansion. Even with continuously welded rails, expansion joints with an effective gap of several centimetres have to be used every kilometre or so. The oval bolt-holes also allow for expansion

The steel tyre of a locomotive wheel is heated by gas burners so that it expands and fits over the wheel. On cooling the tyre contracts and grips the wheel tightly

The rivet is hammered into its hole while it is red hot. On cooling it contracts and pulls the plates together very tightly

Steam pipes have expansion loops. The loop is flexible and allows the pipes to expand

When a piece of metal has to be fitted into a hole, the metal is cooled so that it contracts and just fits the hole. When it warms up again it expands and fits very tightly

Warming the neck of a bottle by the friction of a piece of string causes expansion and loosens the stopper. Likewise, a hot soldering iron can be used to loosen a screw by expanding the hole

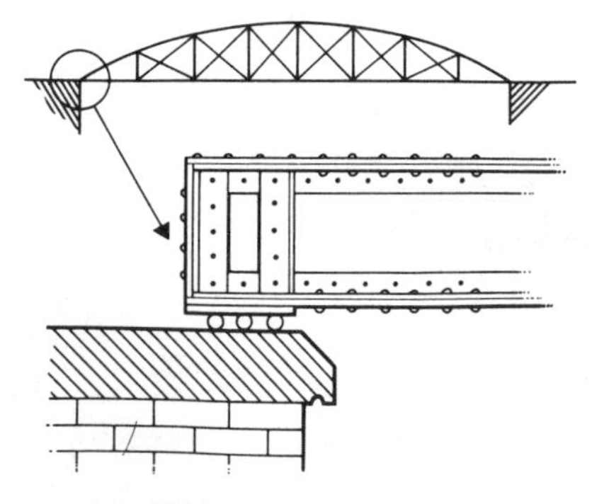

Iron bridges expand in hot weather and therefore sections are put on rollers to take up the movement and prevent buckling

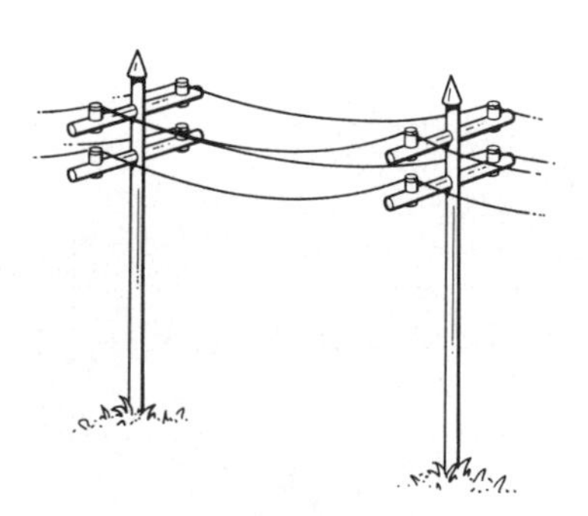

Telephone wires and other overhead cables are left slack in the summer so that when they contract in the winter they do not break

Thick glass cracks with boiling water because the inside expands before the outside, glass being a poor conductor of heat. Heat resistant glass has a very low expansivity

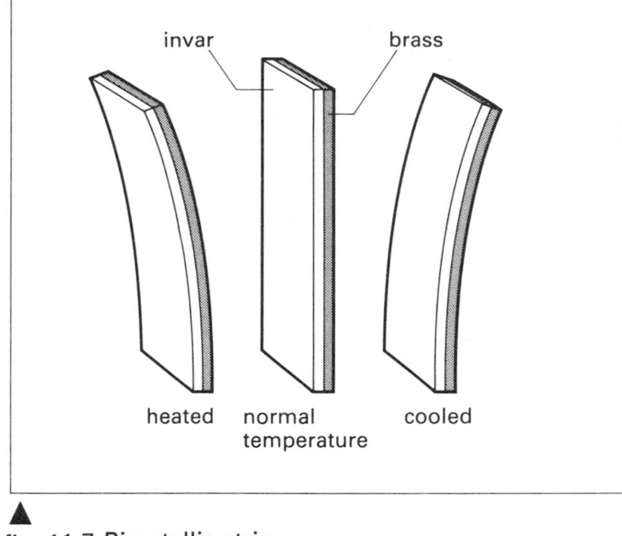

fig. 11.7 Bimetallic strip

Bimetallic strips

A compound bar, figure 11.2, made of two different metals riveted together, was used to demonstrate that different metals expand by different amounts when heated. The bar bent to one side with the most expansible metal on the outside of the curve. The greater the difference between the expansion of the two metals, the more the bar will curve. Bimetallic strips are made of two thin lengths of different metals welded together. The metals are usually brass and Invar, an alloy of nickel and steel which has a very small expansion. Bimetallic strips are very sensitive to temperature changes, bending one way when heated and the other way when cooled, figure 11.7.

A bimetallic strip is used as a **thermal switch**, figure 11.8. When the temperature increases the strip bends over to touch the contact screw and complete the electric circuit. It could be used as a fire alarm to ring a bell. Equally, the strip can be mounted the other way round so that a rise in temperature breaks the circuit.

Bimetallic strips are used in **thermostats**, which are devices for maintaining a constant temperature. There is a thermostat in an electric iron, figure 11.9. The bimetallic strip is used in thermometers (see p.99) and in the flashing indicator unit on cars.

Other types of thermostat depend on the difference in expansion of two separate metal components, figure 11.10, and on the expansion and contraction of metal bellows, figures 11.11 and 11.12.

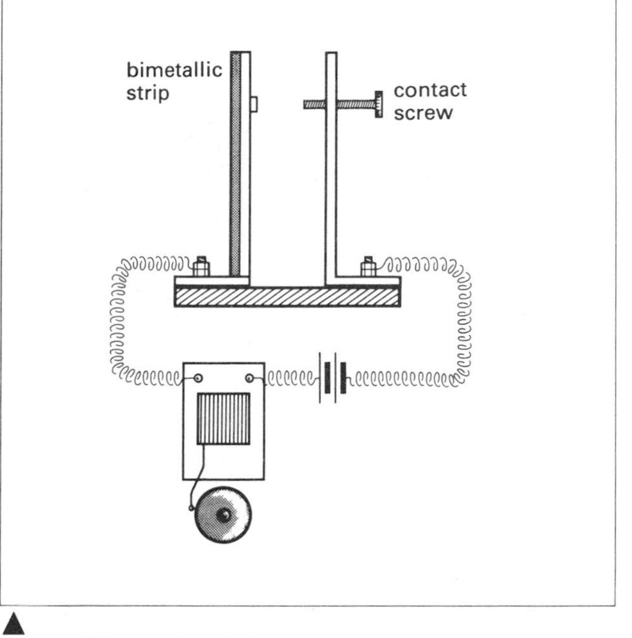

fig. 11.8 Thermal switch

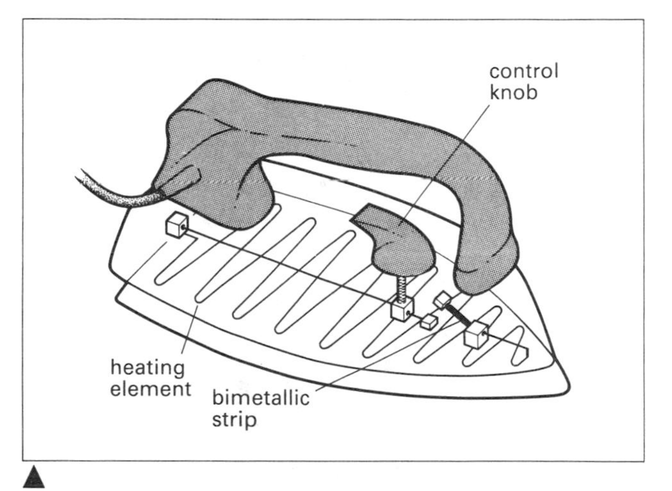

fig. 11.9 The thermostat in an electric iron uses a bimetallic strip which breaks the circuit when it expands and bends. The control knob alters the distance the strip must bend to cut off the electricity

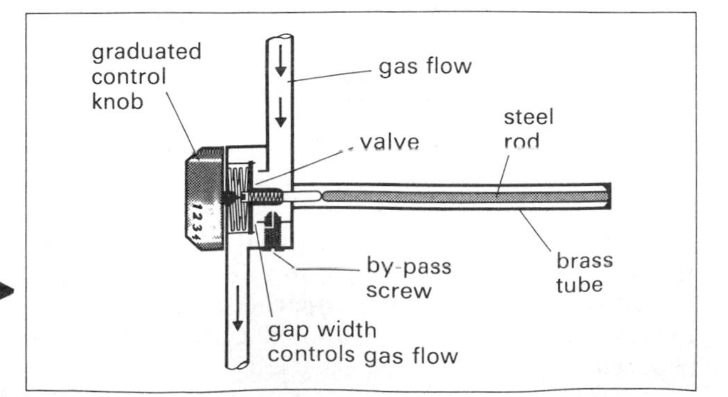

fig. 11.10 The Regulo thermostat on a gas cooker makes use of the expansion of a brass tube. As the tube gets hot and expands, it closes a valve and cuts down the gas supply

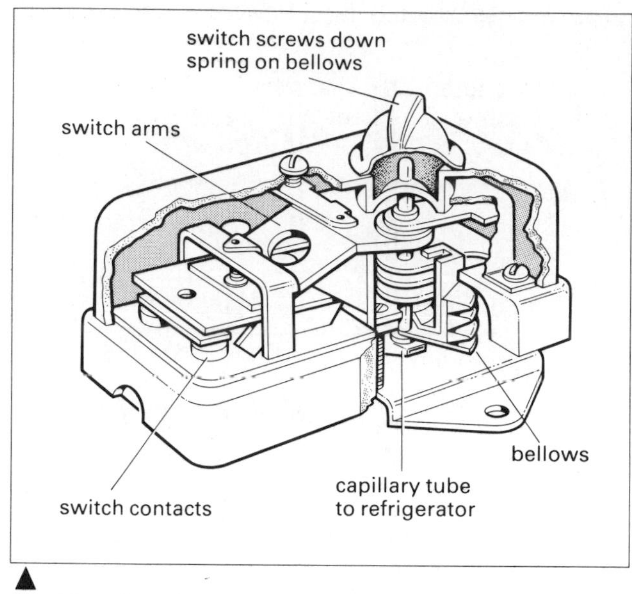

fig. 11.11 In one type of refrigerator thermostat, temperature changes inflate and deflate liquid-filled metal bellows which open and close a switch

fig. 11.12 Metal bellows filled with liquid are also used in some designs of car engine thermostat. When the engine is cold, the cooling water is not allowed to circulate and the engine warms up quickly

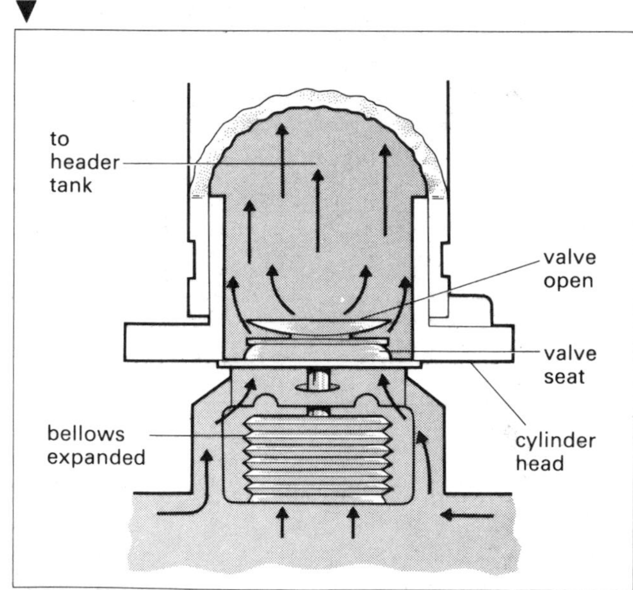

Real and apparent expansion

Solids expand in length (**linear expansion**), in area (**superficial expansion**) and in volume (**cubical expansion**).

Liquids and gases undergo cubical expansion. Since the liquid or gas must be held in a container we have to distinguish between the **apparent expansion** (relative to the container which also expands) and the **real expansion** of the liquid or gas itself.

Suppose we have a liquid in a glass vessel as shown in figure 11.13. At the lower temperature the level is at A. When the vessel and its contents are heated, let us imagine that the expansions take place in two stages. First, the vessel expands and the level drops to B. When the expansion of the vessel has taken place the liquid expands, and the level rises to C.

fig. 11.13 Apparent expansion of a liquid

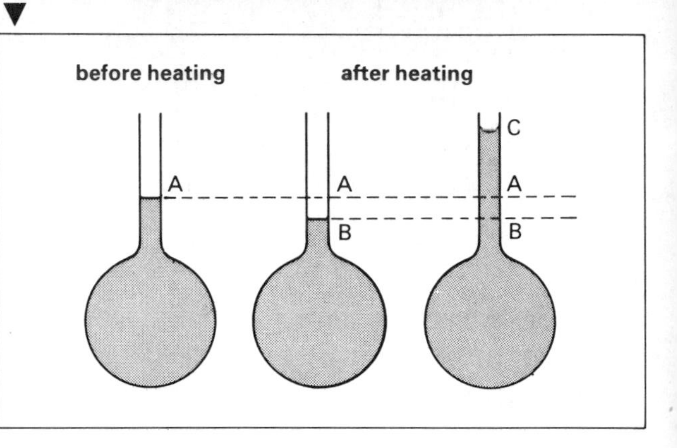

AB measures the expansion of the vessel
AC measures the apparent expansion of the liquid
BC measures the real expansion of the liquid

Since $\quad BC = AB + AC$

real expansion = apparent expansion + expansion of vessel

Calculating the expansion of a solid

Experiments show that the increase in length of a solid bar on heating depends on

- the original length of the bar,
- the increase in temperature,
- the material of which the bar is made.

The linear expansivity (α) of a material is equal to the increase in length of unit length when there is unit rise in temperature

$$\alpha = \frac{\textbf{increase in length}}{\textbf{original length} \times \textbf{temperature rise}}$$

The units of α are **per °C** or **per K**.

The expression can be written in the convenient form

increase in length = α × original length × temperature rise

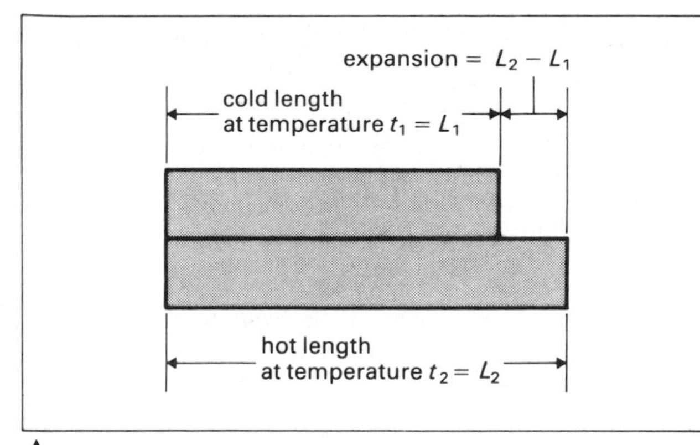

fig. 11.14 Linear expansivity

Using figure 11.14.

$$L_2 - L_1 = \alpha \times L_1 \times (t_2 - t_1)$$
$$L_2 = L_1 + \alpha L_1 (t_2 - t_1)$$

that is

$$L_2 = L_1 \{1 + \alpha (t_2 - t_1)\}$$

aluminium	0.000 012
brass	0.000 019
concrete	0.000 011
copper	0.000 018
glass	0.000 009
Invar	0.000 001
iron	0.000 012
platinum	0.000 010
Pyrex	0.000 003
silica	0.000 000 4
steel	0.000 011
zinc	0.000 026

table 11.1 Some common linear expansivities per °C

EXAMPLES

1 A bar of copper is 150 cm long at 0 °C. What would be its length at 90 °C? (Linear expansivity of copper = 0.000 018 per °C.)

increase in length = $\alpha L t$
= 0.000 018 × 150 × 90
= 0.243 cm
new length = 150.243 cm

2 A circular hole in a sheet of copper has a diameter of 20 cm. Through what temperature rise must the sheet be heated for the diameter of the hole to become 20.09 cm?

The hole expands as if it were a disc made of copper.

original length of the diameter = 20.00 cm
new length of the diameter = 20.09 cm
expansion = 0.09 cm

expansion = expansivity × original diameter × temperature rise

$$0.09 = 0.000\,018 \times 20 \times t$$
$$t = \frac{0.09}{0.000\,018 \times 20} = 250 \text{ °C}$$

EXERCISE 11

In questions 1–3 select the most suitable answer.

1 Figure 11.15a shows a metal tube at room temperature 20°C. Figure 11.15b shows the same tube through which steam at atmospheric pressure has been passed for a long time. Which of the following statements about the arrangement is most likely to be correct?
A The rod has expanded by 0.04 m for a 20 °C rise in temperature.
B The rod has expanded by 0.84 m for a 20 °C rise in temperature.
C The rod has expanded by 0.80 m for an 80 °C rise in temperature.
D The rod has expanded by 0.84 m for an 80 °C rise in temperature.
E The rod has expanded by 0.04 m for an 80 °C rise in temperature.

fig. 11.15

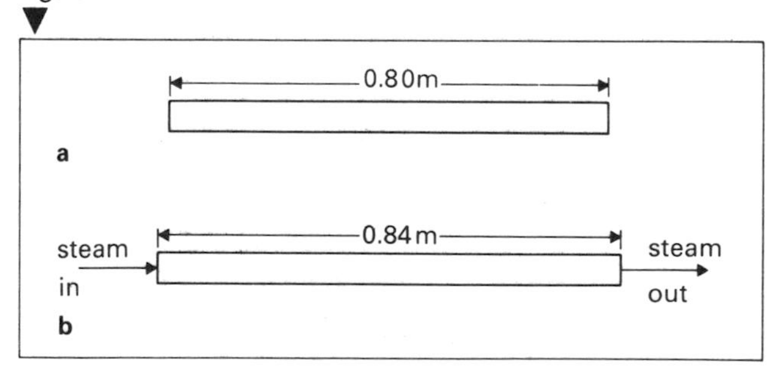

2 A metal rod expands by 1.00 mm when it is heated from 20 °C to 80 °C. Another rod, of the same material, twice as long as the first at 20 °C is heated from 20 °C to 50 °C. The second rod will expand by
A 0.25 mm
B 0.50 mm
C 1.00 mm
D 2.00 mm
E 4.00 mm

3 Which of the following statements about expansion is correct?
A Pyrex glass expands much less than ordinary glass under the same conditions.
B Liquids do not expand when heated.
C Gases do not expand when heated.
D Water expands uniformly from 0 °C upwards.
E Water expands when it melts.

4 Explain what is meant by
a apparent expansion of a liquid,
b real expansion of a liquid.
Describe an experiment which shows that liquids expand when heated.

5 By how much will a rod of iron, 2 m long, expand when its temperature is raised by 100 °C? (Linear expansivity of iron is 0.000 012 per °C.)

6 a A bimetal strip is two pieces of different metal bonded (riveted) together. Brass and iron are bonded together, as shown in figure 11.16. (i) Draw and label a diagram to show what happens when the strip is heated. (ii) Explain briefly why the change you have shown takes place.

7 a Describe an experiment that shows that a liquid expands on heating.
b The linear expansivity of copper is 0.000 017 per K. What does this statement mean?
c Draw a diagram of the apparatus you would use to measure the linear expansivity of copper.
d What measurement would you need to take and how would you use the results to calculate the linear expansivity of copper?

8 a Describe briefly, with a diagram, how you would demonstrate that a gas expands when it is heated.
b State *one* practical use of the expansion of a liquid.
c An overhead supply cable for an electric train is kept under tension by a spring, figure 11.18. Explain how the tension in the spring is altered by a rise in the temperature of the cables.

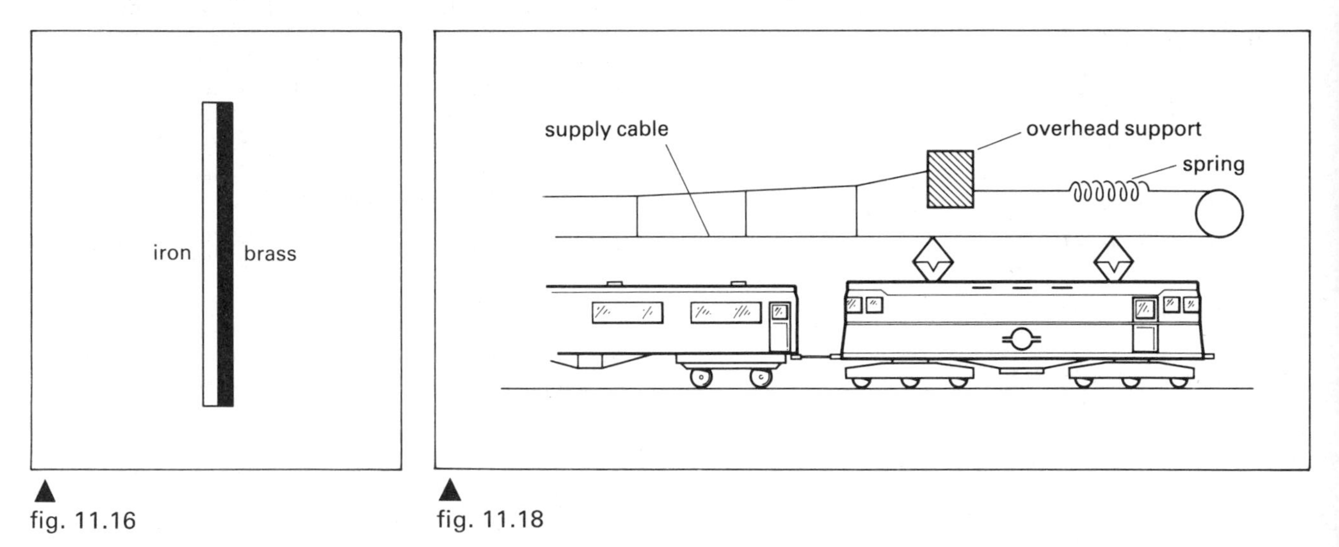

fig. 11.16

fig. 11.18

b One of the major uses of a bimetal strip is in a thermostat. Figure 11.17 shows a thermostat in an electric iron. (i) What are the parts labelled A, B and C? (ii) Describe how the thermostat will operate so that man-made fibres can be ironed at low temperatures and linen at high temperatures.

fig. 11.17

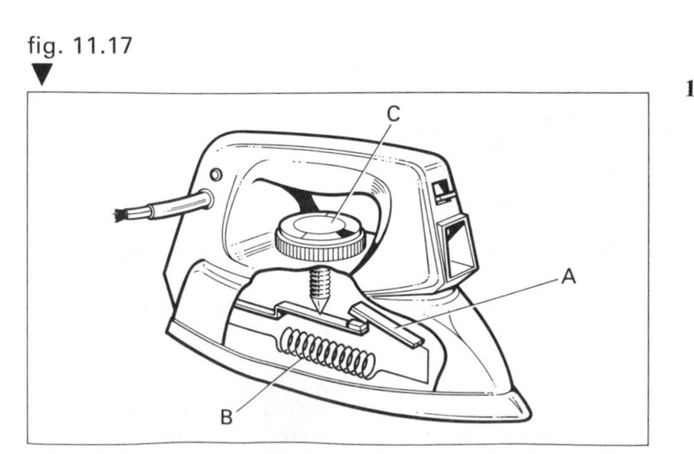

9 a Draw a labelled diagram of a thermostat and explain how it works.
b Draw sketches to show how the effects of thermal expansion are allowed for in each of the following: (i) long lengths of pipe, (ii) railway lines, (iii) telephone wires.
c A copper pipe is 5.0 m long when it is first installed at a temperature of 20 °C. Calculate its new length when carrying water at 100 °C. (Linear expansivity of copper to be taken as 0.000 02/°C.)

10 a Describe a laboratory experiment which shows clearly that a metal rod expands when heated with a bunsen. Draw a diagram of the apparatus used.
b Why is it necessary in part **a** to have a special means of making the expansion visible?
c Give *one* example, other than railway lines, where expansion is a nuisance and say how it is overcome.
d Explain what happens to the molecules in a rod when it expands on heating.
e A bimetallic strip makes use of expansion and can be used in a fire alarm. Redraw and complete figure 11.19 showing the electrical circuit used with such a strip and indicate which of the two metals must expand the most for the alarm to work.

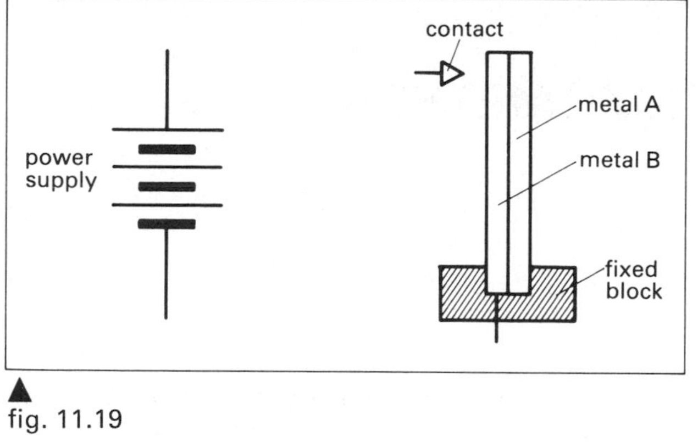

fig. 11.19

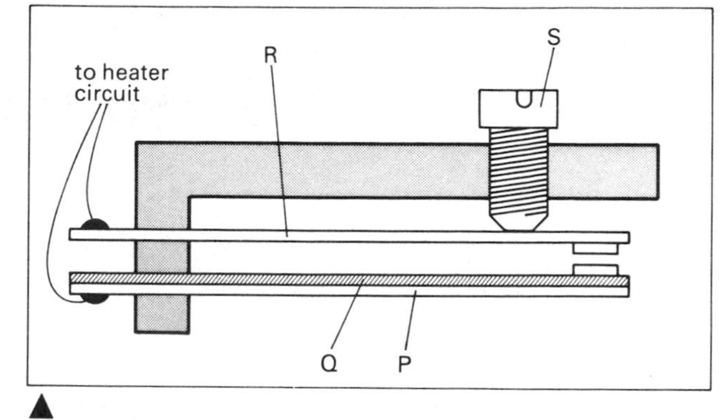

fig. 11.20

11 Figure 11.20 shows a type of thermostat used in an aquarium to maintain the temperature of the water between limits of 24.0 °C and 25.5 °C.

a Describe how the thermostat works.

b What adjustment would be made to the thermostat if it were required to maintain the temperature of the water between higher limits?

c Why it is preferable to position the heater near the bottom of the aquarium rather than near the top?

12 TEMPERATURE AND THERMOMETERS

Temperature is defined as the **degree of hotness** of a body. We can also think of temperature as the property which controls the direction of flow of heat. Heat flows from a high temperature to a lower temperature.

A **thermometer** is an instrument for measuring temperature. In order to calibrate a thermometer we choose two temperatures (called **fixed points**) and divide the interval between them into a number of equal spaces, called degrees. In this way we get a **temperature scale**.

The lower fixed point, the **ice point**, is the temperature of pure melting ice at standard atmospheric pressure.

The upper fixed point, the **steam point**, is the temperature of the steam from pure boiling water at standard atmospheric pressure.

On the **Celsius** (or centigrade temperature) scale, the interval between the fixed points of 0 °C and 100 °C is divided into one hundred degrees. On the Fahrenheit scale (which is still in use), the ice point is 32 °F and the steam point 212 °F.

The fundamental temperature scale is the **thermodynamic** or **kelvin** scale. The ideas behind the thermodynamic scale are outside the scope of elementary physics, although part of the reason for using this rather strange scale is explained in the section on the expansion of gases. The unit of temperature on this scale is the **kelvin (K)**.

Kelvin temperature = Celsius temperature + 273

Note: the accurate figure is really 273.15, but at this stage it is simpler to stick to the whole number 273.

fig. 12.1 The upper and lower fixed points

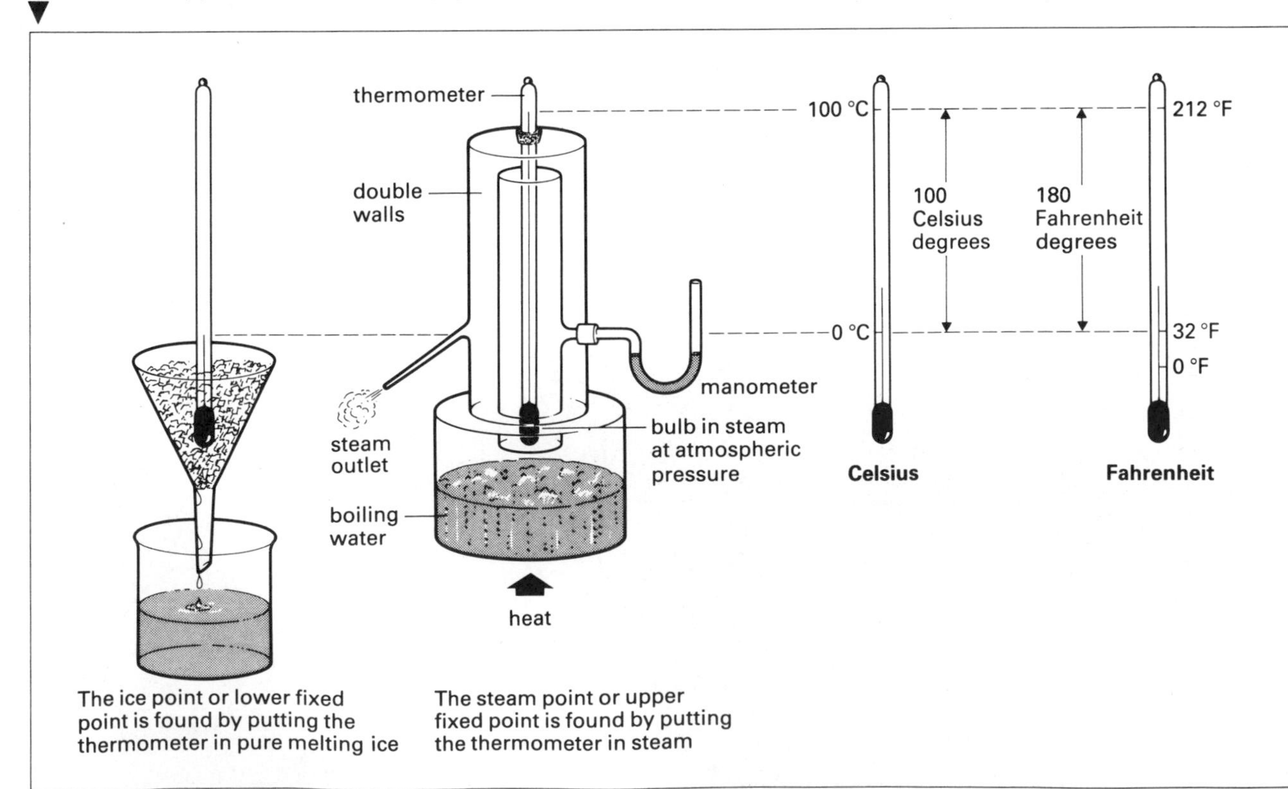

The ice point or lower fixed point is found by putting the thermometer in pure melting ice

The steam point or upper fixed point is found by putting the thermometer in steam

Thermometers

Any property which changes with temperature can be used in a thermometer to measure a temperature change. The most common property used is that of the expansion of a liquid in a glass tube. A **liquid-in-glass** thermometer, shown in figure 12.2, consists of a suitable liquid contained in a glass bulb at the end of a glass tube. The walls of the bulb should be thin to allow the heat to reach the liquid quickly, and the bore of the tube should be narrow so that a small expansion of the liquid produces a visible movement of the liquid in the tube.

The liquid commonly used is mercury or coloured alcohol. Mercury is used because (a) it has a low freezing point (−39 °C) and a high boiling point (357 °C), (b) the thread can be easily seen, (c) it expands regularly and gives readings which agree closely with other methods of measuring temperature. Alcohol freezes at −112 °C, so can be used for measuring low temperatures, but it boils at 78 °C and cannot be calibrated by using the steam point. Alcohol does not expand as uniformly as mercury.

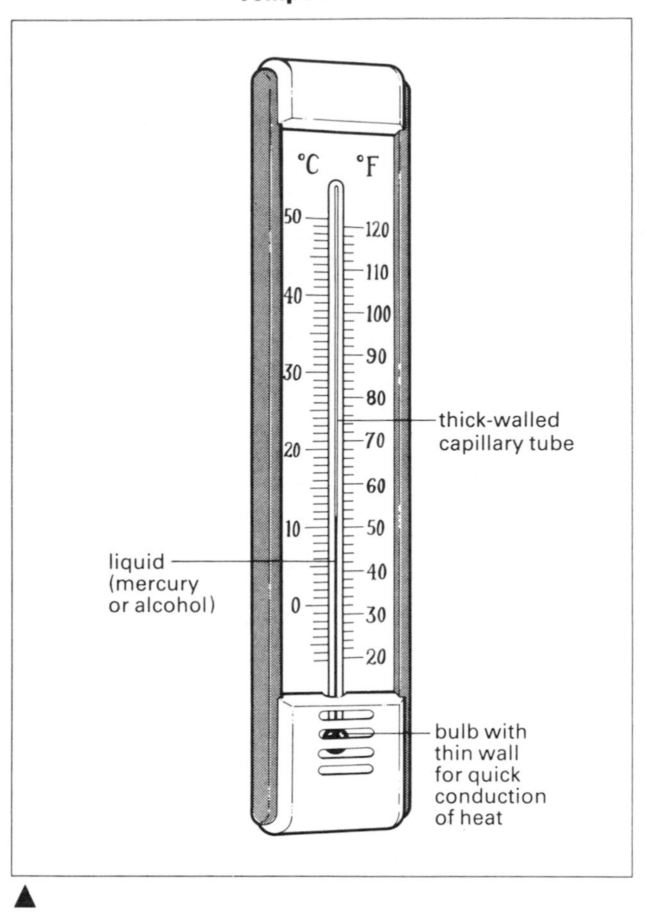

fig. 12.2 Liquid-in-glass thermometer

Clinical thermometer

The normal body temperature for humans is 36 to 37 °C (97 to 99 °F). When we are ill the body temperature often rises and an increase of 3 °C above normal can be dangerous. In order to take a temperature the doctor uses a **clinical thermometer**, figure 12.3.

The thermometer is placed in the mouth, under the tongue. As the mercury rises it can force its way past the kink in the tube. When the thermometer is removed from the patient and the mercury begins to contract the thread breaks at the kink and the reading can be taken. To set the thermometer ready for another reading it is given a sharp shake. In order to read the correct temperature, the thermometer must be left for a certain time (usually a minute or half a minute) in the patient's mouth. The time is marked on each thermometer.

fig. 12.3 Clinical thermometer

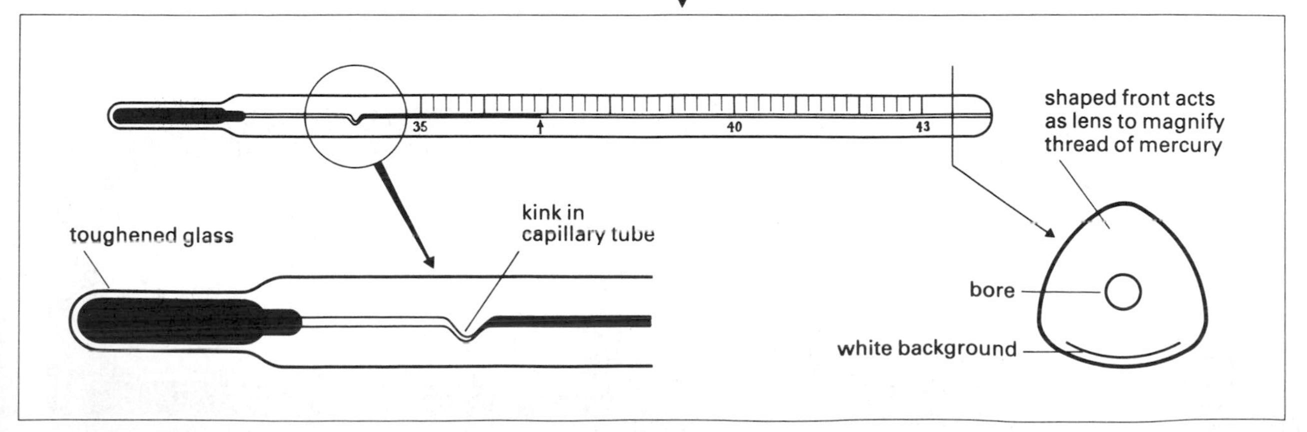

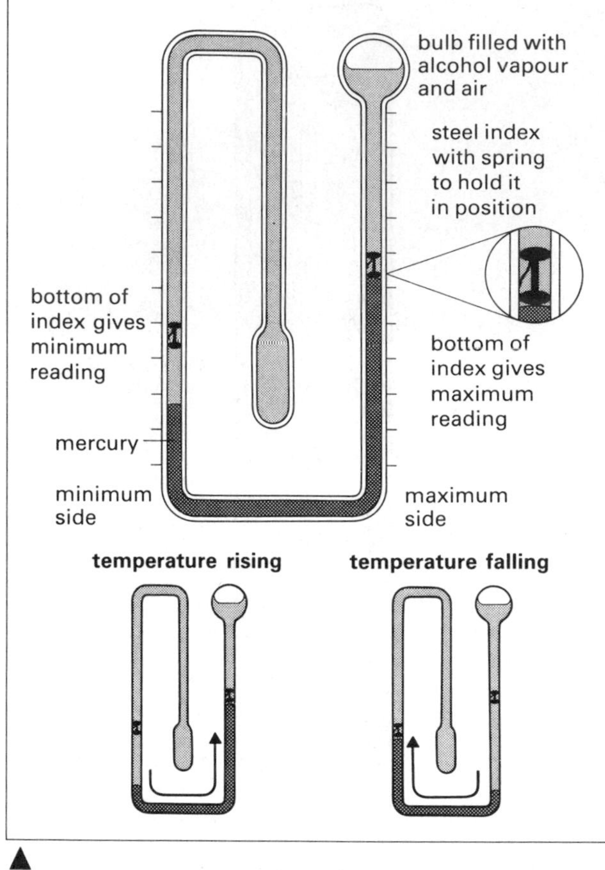

fig. 12.4 Maximum and minimum thermometer

Maximum and minimum thermometer

This thermometer records both the highest and lowest temperatures reached during the time it is set. Figure 12.4 shows the thermometer in detail. The main thermometer contains alcohol. The U-tube connecting the two sides of the thermometer is filled with mercury. The small bulb is filled with alcohol vapour and air. The pressure of this vapour keeps the mercury column pressed against the alcohol thermometer. The expansion or contraction of the alcohol causes movement of the mercury. This pushes tiny steel indices along the tubes. These indices remain in position when the mercury level changes. They are held against the walls of the tube by springs. The indices give the maximum and minimum temperatures reached. They can be re-set on top of the mercury using a small magnet.

Thermoelectric thermometer

When the junction between two different metals is heated an electromotive force is produced; heat energy is converted to electrical energy. This is the **thermoelectric effect**.

Usually two junctions are used, one of which is kept constant at the temperature of melting ice while the other junction is heated, figure 12.5. The voltage produced depends on the difference in temperature between the two junctions, and so the instrument can be used as a thermometer. It is called a thermopile. In the laboratory, copper and iron are convenient to use. Commercial instruments use couples of platinum and an alloy of platinum and rhodium. The electromotive force produced is of the order of a few millivolts. The meter can be calibrated to read temperature directly in degrees, figure 12.6.

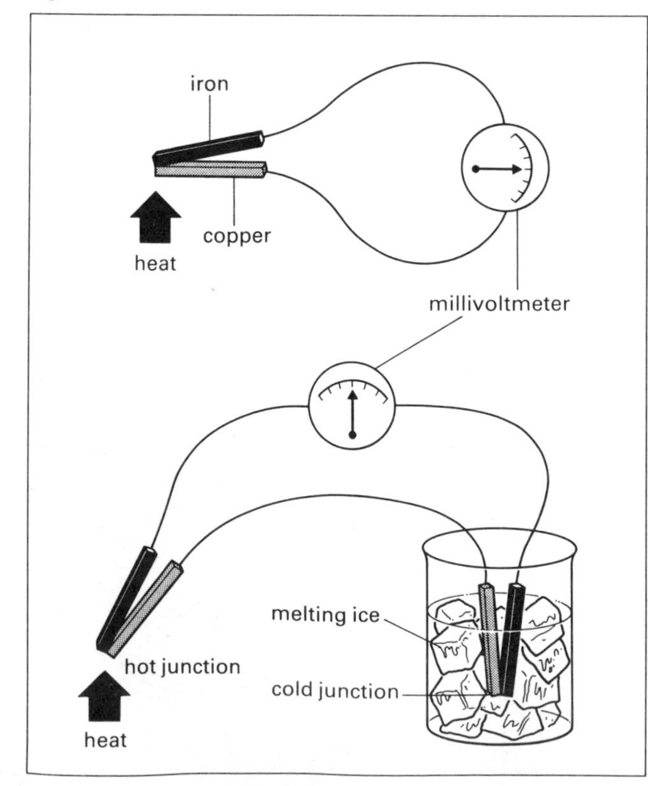

fig. 12.5 Thermoelectric effect

fig. 12.6 Thermoelectric thermometer

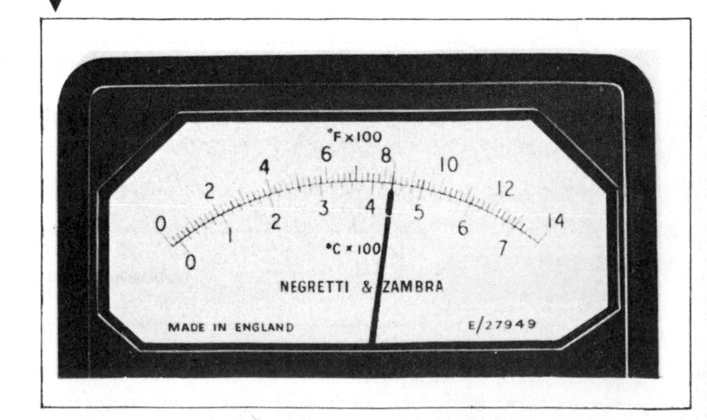

The advantages of this type of instrument are

- It can be used over a wide temperature range from –200 °C to 1500 °C.
- It can measure varying temperatures.
- The thermocouple is small and can measure the temperature over a small volume.
- The thermocouple can be at a distance from the recording instrument and, by suitable switching, temperatures at several different places can be measured on a central recorder.

Bimetallic thermometer

When a compound bar of two metals is heated it bends with the more expansible metal on the outside of the curve. The more it is heated the more it bends. The longer it is the more movement will take place at the end of the bar. In the bimetallic thermometer a long strip is used to give greater sensitivity. The strip is coiled into a spiral to make the instrument more compact, figure 12.7. One end is fixed, while the other is attached to a pointer which moves over a circular scale graduated in degrees. The instrument is robust, but does not respond quickly to rapidly changing temperatures.

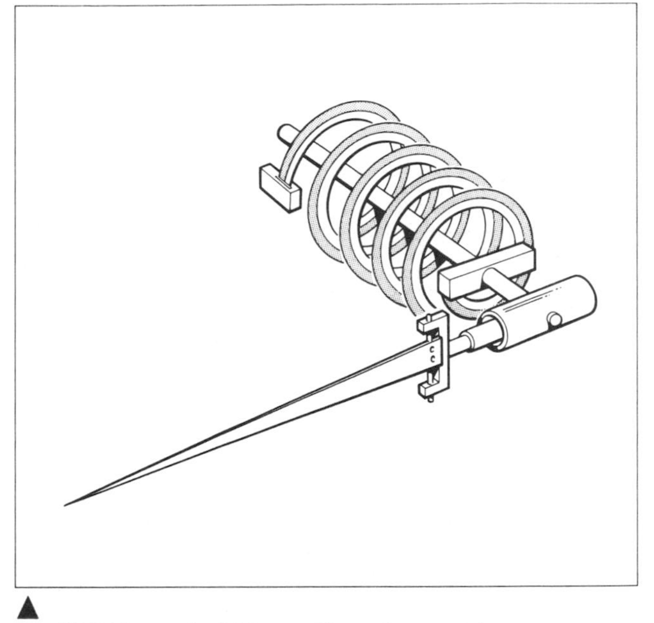

fig. 12.7 The coiled bimetallic strip combines temperature sensitivity with compactness

Other thermometers

The **constant volume gas thermometer** depends on the change in pressure of a gas heated at constant volume.

The **platinum resistance thermometer** depends on the change in electrical resistance of a heated element, see p. 276. The **optical pyrometer** depends on the nature of the radiation emitted by a hot body, see p. 123.

fig. 12.8 This thermometer (called a thermograph) uses a coiled bimetallic strip. The temperature is recorded on a revolving chart

EXERCISE 12

In questions 1–4 select the most suitable answer.

1 Mercury is commonly used in laboratory thermometers because
- **A** its silver colour absorbs light so that the readings may be seen easily
- **B** other liquids would not flow through the narrow capillary tube
- **C** it is a harmless liquid should the thermometer be broken
- **D** the length of the mercury thread increases uniformly with temperature rise
- **E** all other liquids combine chemically with the glass of the capillary tube

2 In order to make a mercury-in-glass thermometer respond more quickly to changes in temperature its construction should be changed by
- **A** making the capillary tube thinner
- **B** using thicker glass walls around the bulb
- **C** making the bulb much larger
- **D** using much thinner glass walls around the bulb
- **E** opening the top of the capillary tube to the air

3 The bulb of a clinical thermometer is made of thin glass so that the
- **A** mercury stays at its maximum reading
- **B** thermometer is more accurate
- **C** mercury is easier to see
- **D** mercury can absorb more heat
- **E** thermometer quickly registers the temperature

4 The clinical thermometer acts as one kind of maximum thermometer because it
- **A** contains mercury
- **B** has thick glass walls
- **C** is rather short in length
- **D** has a constriction at the bottom of its capillary tube
- **E** has a very thin capillary tube

5 Explain clearly the meaning of the terms **a** heat, **b** temperature.

6 a Describe how you would graduate an unmarked mercury-in-glass thermometer, no other thermometer being available.

b How can the sensitivity of a thermometer be increased?

7 Mercury and alcohol are two liquids commonly used in thermometers.

a Give *two* reasons why each is used.

b Under what conditions would it be better to use an alcohol thermometer than one containing mercury?

8 Describe, with a diagram, the action of a bimetallic strip, labelling the metals used.

Describe, and state a use for, a thermometer based on a bimetallic strip.

9 a What may a thermocouple be used for?

b Of what does a thermocouple consist?

10 Figure 12.9 shows a thermocouple to be used for measuring the temperature of a flame. Junction B was kept in a beaker of melting ice throughout the experiment. Junction A was placed in a beaker of hot oil and the temperature taken, together with the reading of the millivoltmeter. The results are given in the table. Finally junction A was taken out of the oil and placed in the flame. A reading of 11.5 mV was recorded.

a Draw a graph of meter reading against temperature of junction A (horizontal axis). Draw the best line, and extend it to 1200 °C. Your vertical axis will need a scale going up to 15 mV.

b Use your graph to estimate the temperature of the flame.

c What assumption is made in using the graph to produce the estimate?

d Explain why an ordinary laboratory (i.e. mercury-in-glass) thermometer could not be used to find the temperature of the flame.

fig. 12.9

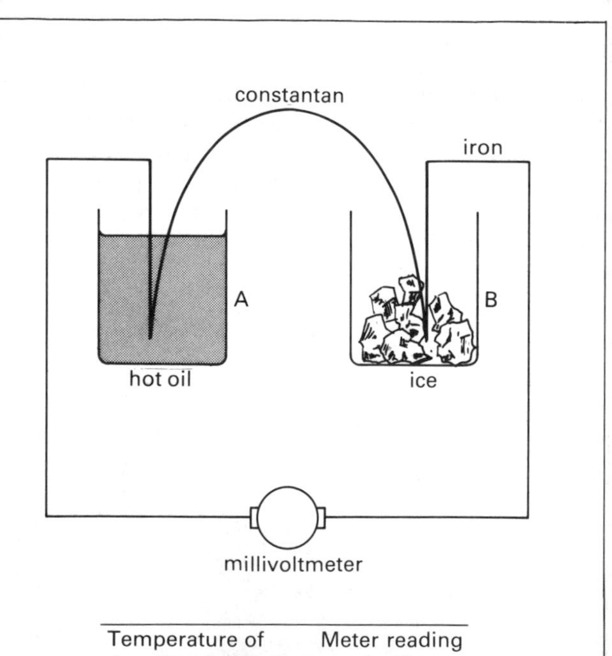

Temperature of junction A (°C)	Meter reading (mV)
90	1.2
140	1.8
190	2.4
245	3.2
300	3.7
340	4.4

13 HEAT CAPACITY

Since heat is a form of energy, quantities of heat are measured in joules. The effects of supplying the same quantity of heat to different bodies can be very different. Suppose a kilojoule of heat were supplied to a large tank of water and to an iron nail. The large quantity of water would only become a little warmer, but the iron nail would become red hot. The two bodies have different capacities for heat.

The heat capacity, C, of a body is equal to the quantity of heat required to raise the temperature of the body by one degree

$$C = \frac{\textbf{heat added to body}}{\textbf{rise in temperature produced}}$$

The unit of heat capacity is the **joule per °C** or **joule per kelvin (J/°C, J/K)**.

The heat capacity of a body depends on the quantity of matter (mass) the body contains, and also on the material of which the body is composed. Some bodies get very hot (like the iron nail), while others can absorb a large amount of heat without getting much warmer. In order to compare different substances we must compare equal masses.

table 13.1 Specific heat capacities (in J/kg °C) of some solids and liquids

▼

aluminium	900
brass	380
copper	400
glass	670
iron	500
lead	125
mercury	140
alcohol	2500
brine	3000
glycerine	2400
ice	2100
olive oil	2000
paraffin	2200
water	4200

The specific heat capacity, c, of a substance is the quantity of heat required to raise the temperature of unit mass of the substance through one degree

$$c = \frac{\textbf{heat added to the body}}{\textbf{rise in temperature produced} \times \textbf{mass of body}}$$

The unit of specific heat capacity is the **joule per kilogram °C**, or the **joule per kilogram kelvin (J/kg °C, J/kg K)**.

Notice that the term heat capacity refers to a particular body, while specific heat capacity refers to a particular substance:

to heat 1 kg through 1 °C requires c J of heat
to heat m kg through 1 °C requires mc J of heat
to heat m kg through t °C requires mct J of heat

In general

quantity of heat = mass × specific heat capacity × change in temperature

$$Q = mct$$

where m = mass in kg
c = specific heat capacity in J/kg °C
t = change in temperature in °C

It will be noticed from the table 113.1that water has the highest specific heat capacity, about twice as high as that of any other common substance. This means that it takes a lot of heat to warm water up, and so water is a good heat storer. Water is also a good coolant in industrial processes, because it will absorb a lot of heat.

EXAMPLES

1 How much heat is required to raise the temperature of

a 2.5 kg of water by 15 °C?
b 100 g of copper from 20 °C to 100 °C?

a quantity of heat = mct
= $2.5 \times 4200 \times 15$
= 157 500 J

b quantity of heat $= mct$
$= 0.1 \times 400 \times 80$
$= 3200$ J

2 What will be the resulting temperature if 2000 joules of heat are given to
a 200 g of water at 10 °C?
b 0.5 kg of aluminium at 20 °C?

$$\textbf{a}\ t = \frac{Q}{mc} = \frac{2000}{0.2 \times 4200}$$
$$= 2.38\ °\text{C}$$

The final temperature is 12.38 °C

b If the resulting temperature is θ, the temperature change is $\theta - 20$.

$$\theta - 20 = \frac{2000}{0.5 \times 900} = 4.44$$

$$\text{so}\quad \theta = 20 + 4.44$$
$$= 24.44\ °\text{C}$$

3 If 50 000 J of heat are given out when an iron ball of mass 0.5 kg cools from 250 °C to 50 °C, calculate the specific heat capacity of iron.

$$\text{specific heat capacity } c = \frac{Q}{mt}$$
$$= \frac{50\,000}{0.5 \times 200}$$
$$= 500\ \text{J/kg °C}$$

Measuring specific heat capacity

In order to measure the specific heat capacity of a material, a known quantity of heat must be given to a known mass of the material and the resulting rise in temperature measured. The quantities can then be substituted in the equation $Q = mct$, and c can be calculated.

In practice it is difficult to obtain an accurate value for the specific heat capacity because (a) some of the heat supplied is lost to the surroundings, although this can be minimised by efficient lagging of the apparatus, and (b) if the material is liquid it will have to be contained in a vessel which absorbs some of the heat supplied. Allowance can be made for this, but it makes the experiment more complicated.

The simplest way of supplying a known quantity of heat is by using an electrical heater connected to a joulemeter which measures the energy input, figure 13.1. A joulemeter works on the same principle as the domestic electricity meter which measures the amount of electrical energy used in the house and thus enables the cost of it to be calculated.

In an experiment to find the specific heat capacity of aluminium, the following measurements were made:

mass of aluminium = 0.5 kg
initial temperature = 15.4 °C
final temperature = 26.8 °C
initial joulemeter reading = 10 971 J
final joulemeter reading = 16 101 J
quantity of heat supplied = 16 101 – 10 971 J
= 5130 J
rise in temperature = 26.8 – 15.4 = 11.4 °C

fig. 13.1 Measuring the specific heat capacity

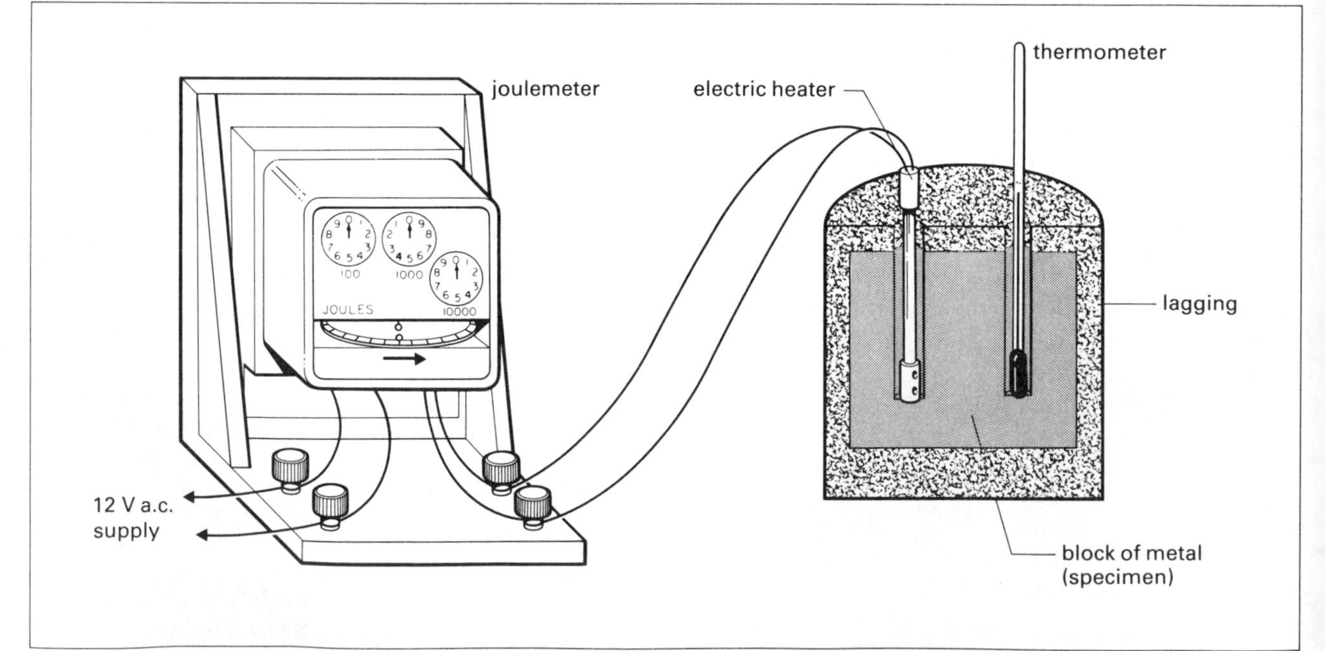

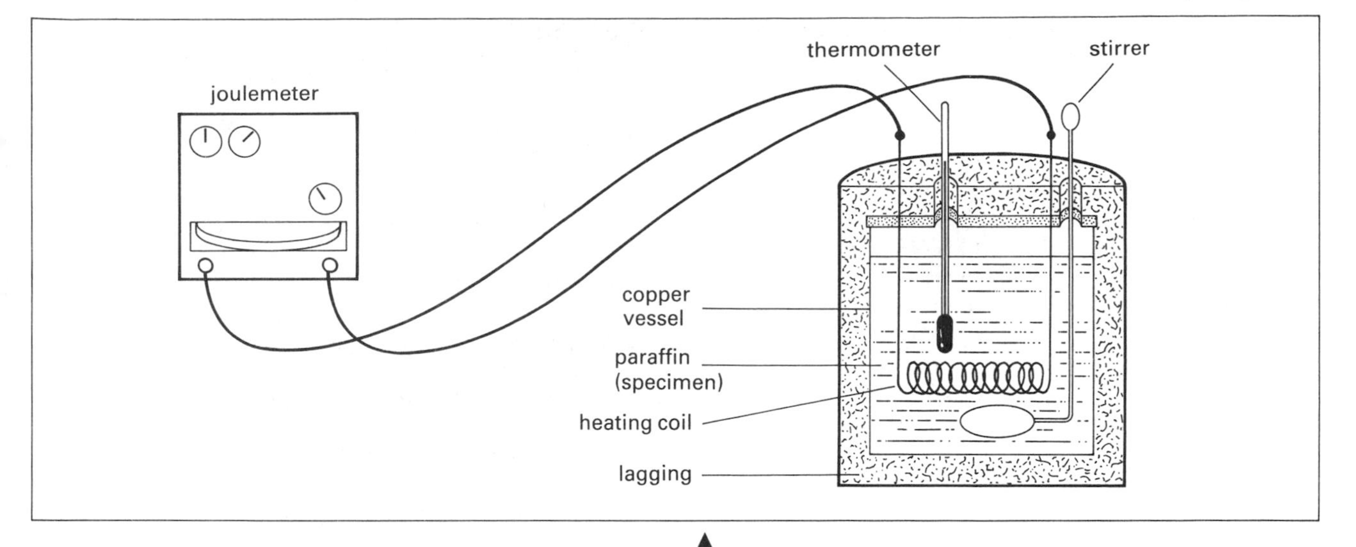

▲ fig. 13.2 Determining the specific heat capacity of a liquid (paraffin)

$$\text{specific heat capacity of aluminium} = \frac{\text{quantity of heat supplied}}{\text{mass} \times \text{rise in temperature}}$$

$$= \frac{5130}{0.5 \times 11.4}$$

$$= 900 \text{ J/kg °C}$$

In an experiment to determine the specific heat capacity of paraffin by electrical heating, 110 g of paraffin is placed in a copper vessel, figure 13.2. The vessel is well-lagged and contains a heating coil, thermometer and copper stirrer. The heating coil is connected to a joulemeter and low-voltage supply. The initial readings are taken and the paraffin is well stirred while heating takes place. The mass of the copper vessel and stirrer is 64 g.

initial temperature = 12 °C
final temperature = 20 °C
initial joulemeter reading = 6521 J
final joulemeter reading = 8662 J
total heat supplied = 2141 J

Assuming no heat is lost to the surroundings, some of this heat is used to warm up the copper vessel and the stirrer, and the remainder warms up the paraffin.

quantity of heat used to warm vessel and stirrer $= 0.064 \times 400 \times 8$
$= 204.8$ J

quantity of heat used to warm the paraffin $= 2141 - 204.8$
$= 1936.2$ J

$$\text{specific heat capacity of paraffin} = \frac{\text{quantity of heat}}{\text{mass} \times \text{rise in temperature}}$$

$$= \frac{1936.2}{0.11 \times 8}$$

$$= 2200 \text{ J/kg °C}$$

Heat produced by fuels and foodstuffs

When a fuel such as coal, gas or oil burns, chemical energy is converted into heat energy. When the digested products of foodstuffs are oxidised in living things during the process of respiration, energy is released. This supplies the living organism with mechanical energy to do work, energy for growth and repair of tissues and, in warm-blooded animals, heat energy to maintain the body temperature above the surroundings. Heating engineers and diet specialists need to know just how much energy is available from different fuels and foods.

The **specific calorific value** of a substance is the quantity of energy released when unit mass of the substance is completely burnt with excess of oxygen.

The unit of specific calorific value is the **joule per kilogram (J/kg)**. Sometimes the quantity is measured in joules per gram and the old heat units, calories, are unfortunately still in use. The small calorie is equal to 4.18 joules and the large calorie, generally used for food values, is a thousand times as large.

Specific calorific values are determined using a **fuel calorimeter**. The actual apparatus is complicated, but the principles are shown in figure 13.3. The specimen of fuel or food is contained in a crucible and is ignited by an electric heating coil. Oxygen under pressure enters the apparatus so that the specimen is completely burnt. The hot gases produced by the combustion are forced out of the bottom of the inner vessel and bubble up through the water in the surrounding container. From the rise in temperature and the mass of the surrounding water the quantity of heat produced can be calculated. The heat absorbed by the apparatus itself has to be taken into account.

table 13.2 Specific calorific values (in J/kg) of some fuels and foods

anthracite	35 000
coal	28 000
coal gas	28 000
coke	29 000
fuel oil	43 000
natural gas	56 000
paraffin	46 000
petrol	43 000
peat	23 000
wood	17 000
beef	15 000
butter	31 000
cheese	21 000
eggs	7 000
fats	38 000
margarine	33 000
milk	3 000
nuts	30 000
proteins	17 000
sugar	17 000

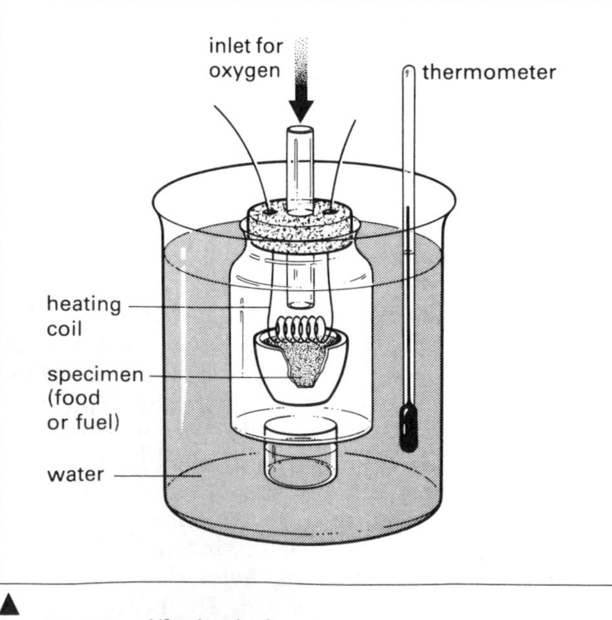

fig. 13.3 Food/fuel calorimeter

EXERCISE 13

In questions 1–4 select the most suitable answer.

1 In which of the following units is specific heat capacity measured?
A $J\ s^{-1}$
B $J\ K^{-1}$
C $J\ kg^{-1}\ K^{-1}$
D $J\ kg^{-1}$
E $J\ m^{-3}$

2 Which of the following represents a change in temperature?
A $\frac{\text{specific heat capacity} \times \text{mass}}{\text{energy}}$
B $\frac{\text{specific latent heat} \times \text{mass}}{\text{energy}}$
C $\frac{\text{energy}}{\text{mass} \times \text{specific heat capacity}}$
D $\frac{\text{energy}}{\text{mass} \times \text{specific heat capacity}}$
E $\frac{\text{specific latent heat}}{\text{mass}}$

3 A piece of iron mass m and specific heat capacity c, and a piece of aluminium of mass $2m$ and specific heat capacity $2c$ each received the same quantity of heat. The temperature of the aluminium rose by 8 K. By how much did the temperature of the iron rise?
A 2 K
B 4 K
C 8 K
D 16 K
E 32 K

4 2 kg of a substance of specific heat capacity 4 kJ/kg °C is heated from 20 °C to 50 °C. Assuming no heat loss, the heat required to do this is
A 15 kJ
B 36 kJ
C 160 kJ
D 240 kJ
E 400 kJ

5 A 45 W immersion heater is placed in a beaker containing 0.5 kg of water. The temperature of the water is observed to rise by 2 °C in 100 seconds.
a How much heat energy is given out by the heater in this time?
b How much heat energy is gained by the water?
c Give a reason why the heat given out by the heater is greater than the heat apparently gained by the water. (Specific heat capacity of water = 4200 J/kg °C)

6 6 g of a liquid fuel is required to raise the temperature of 2 kg of water through 30 °C. If a candle is used instead of the liquid fuel, 3 g of the candle burn away in raising the temperature of 2 kg of water through 5 °C. Assume specific heat capacity of water to be 4000 J/kg °C. By calculation, show which of the two fuels has the greater energy value. Calculate the energy value of the liquid fuel.

7 Describe an experiment you would perform to find the specific heat capacity of a liquid. State the precautions you would take to obtain an accurate result and show how you would calculate the result.

A saucepan of mass 0.75 kg containing 0.50 kg of water is placed on a gas burner. The initial temperature of the water is 20 °C. It takes 5 minutes before the water starts to boil. Find the rate at which heat is supplied to the water by the burner. (Specific heat capacity of water = 4000 J/kg K. Specific heat capacity of the material of the saucepan = 600 J/kg K.)

It is found to take less time to boil water and cook vegetables in a saucepan with a lid than in a similar saucepan without a lid. Explain why this is so.

8 An electric immersion heater raises the temperature of 1 kg of water from 20 °C to 35 °C in 3 minutes. How much heat energy is supplied to the water in 3 minutes? What is the power of the immersion heater? (Specific heat capacity of water = 4200 J/kg °C.)

9 a What is meant by specific heat capacity?

b Describe an experiment to determine the specific heat capacity of a solid.

c A copper block of mass 3 kg is heated in an oven. Its temperature rises 100 °C in two minutes. (Specific heat capacity of copper = 400 J/kg °C). (i) How much heat has the block absorbed? (ii) At what rate was the block absorbing heat? (iii) If only 50% of the heat generated by the oven was absorbed by the block, at what rate was the oven generating heat?

10 A heating engineer is asked to find out how much energy can be saved by using a lagged hot-water tank instead of an unlagged one. He has two similar tanks, one of which is lagged and the other unlagged. Each tank is fitted with a 5 kW immersion heater.

a (i) If each tank is filled with 90 kg of water at 10 °C, *estimate* the time needed to heat the water to 50 °C. (ii) State any assumption you have made in your calculation.

b Each tank has a thermostat attached to it which switches on the heater when the temperature of the water falls to 40 °C, and which switches off the heater when the temperature of the water rises to 50 °C. The temperature–time graphs, figures 13.4a and b show one cooling–heating cycle for each tank. (i) From the data given in the graphs, calculate the total time for which each heater would be switched on during a 24-hour period. (ii) Hence estimate the energy which would be saved during a 24-hour period by using a lagged tank instead of an unlagged one.

fig. 13.4

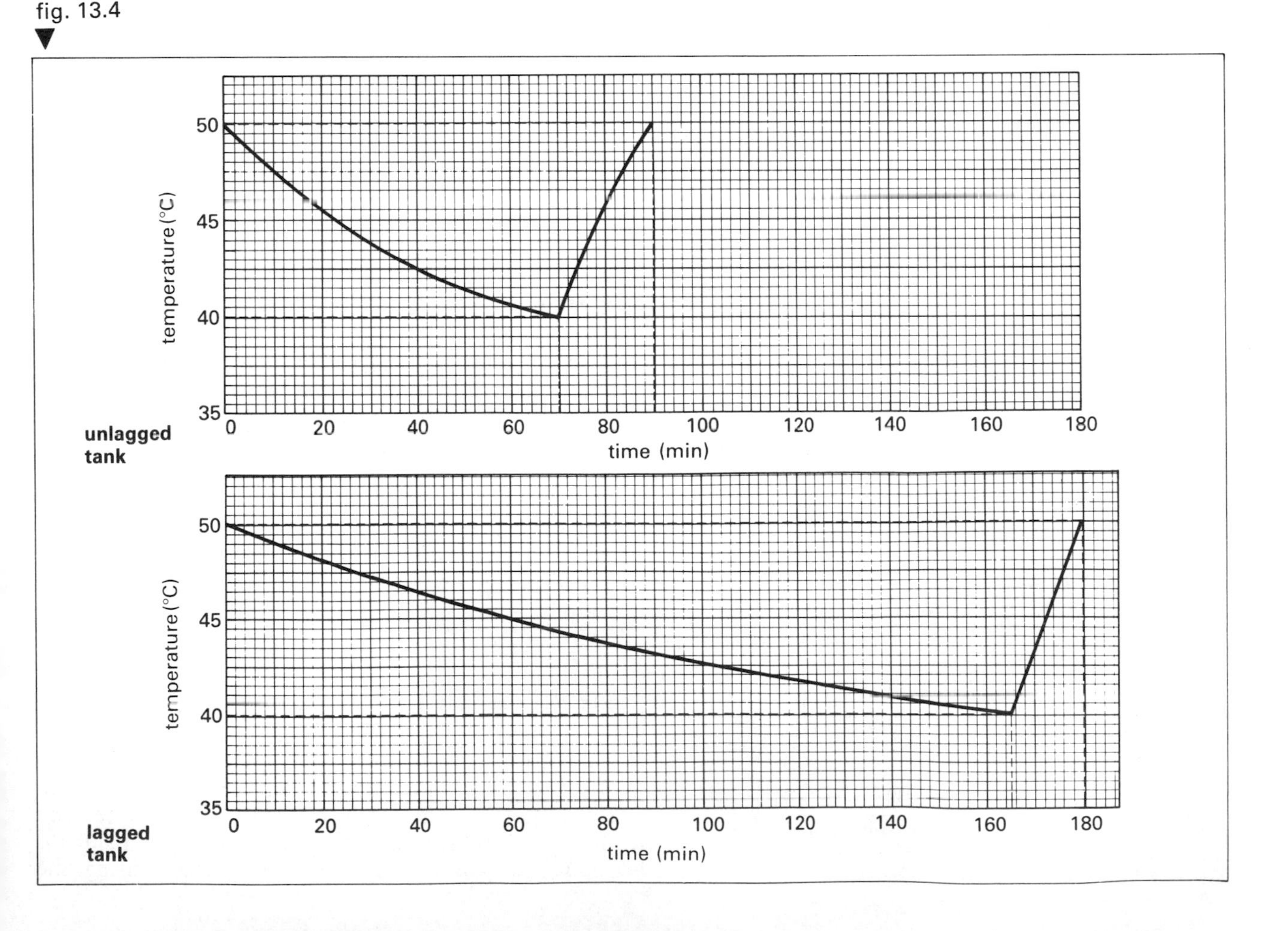

14 CHANGE OF STATE

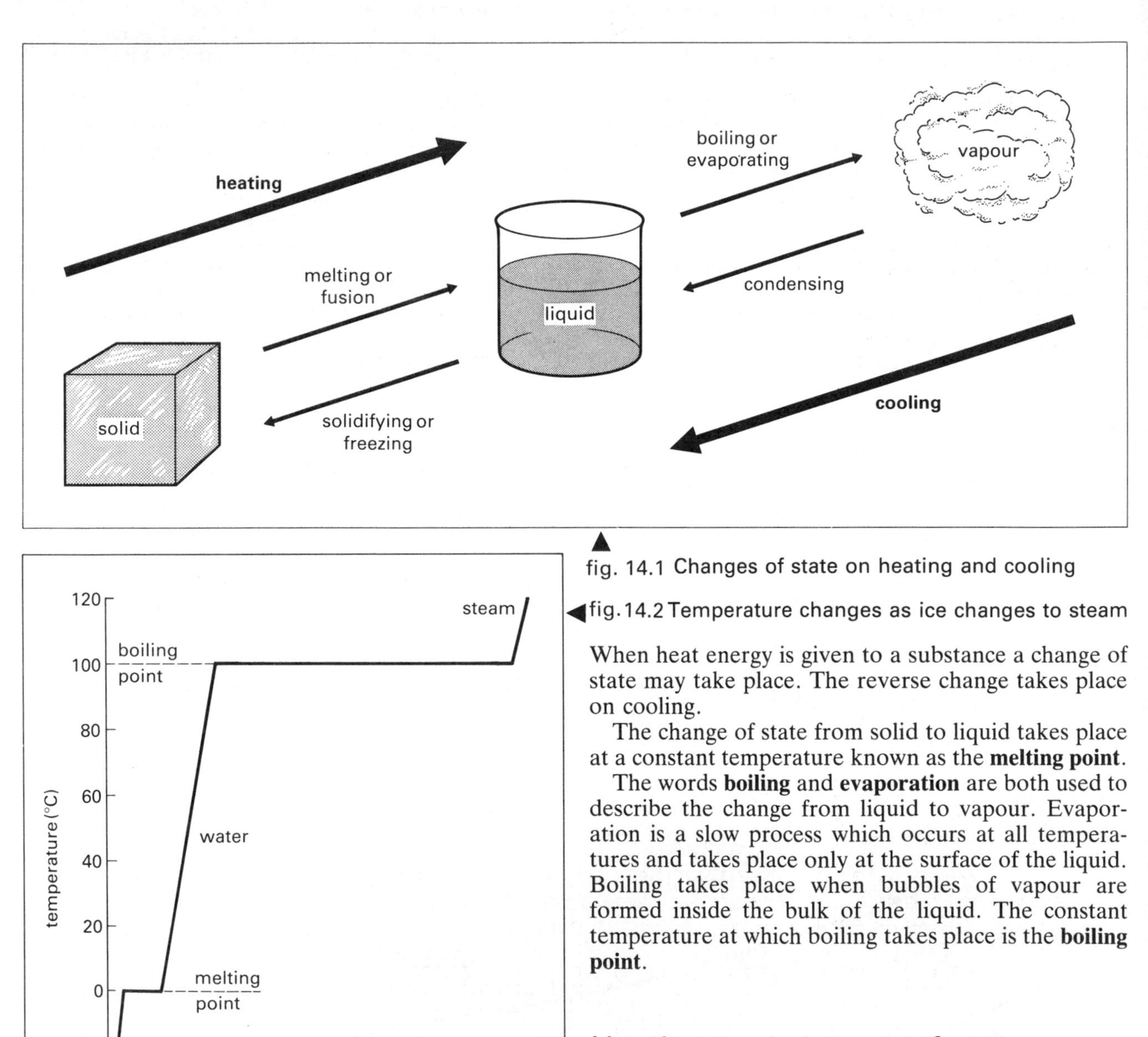

fig. 14.1 Changes of state on heating and cooling

fig. 14.2 Temperature changes as ice changes to steam

When heat energy is given to a substance a change of state may take place. The reverse change takes place on cooling.

The change of state from solid to liquid takes place at a constant temperature known as the **melting point**.

The words **boiling** and **evaporation** are both used to describe the change from liquid to vapour. Evaporation is a slow process which occurs at all temperatures and takes place only at the surface of the liquid. Boiling takes place when bubbles of vapour are formed inside the bulk of the liquid. The constant temperature at which boiling takes place is the **boiling point**.

Heating and change of state

Suppose a block of ice is supplied with heat at a constant rate, and the temperature is taken at constant time-intervals while the ice melts to water and then water boils into steam. A graph of the results is shown in figure 14.2. We notice that during the change of state from ice to water, and again from water to steam, the temperature remains constant, although in both cases heat is being supplied all the time at a constant rate. In order to change a solid into a liquid and a liquid into a vapour, heat energy is necessary. Since this heat energy does not show itself as a temperature change it is known as **latent heat**. The word 'latent' means hidden.

The specific latent heat of fusion is the heat required to change 1 kg of solid into liquid at the melting point without change of temperature

The specific latent heat of vaporisation is the heat required to change 1 kg of liquid into vapour at the boiling point without change of temperature

The unit of specific latent heat is the **joule per kilogram (J/kg)**, although the value may sometimes be given in joules per gram. The specific latent heat of fusion of water is 336 000 J/kg, and the specific latent heat of vaporisation of water is 2 260 000 J/kg. It will be noticed that it takes nearly seven times as much heat energy to change water into steam as it does to change ice into water. On the heating graph for water, the horizontal portion at the boiling point is nearly seven times longer than at the melting point.

Measurement of specific latent heat

In order to measure the specific latent heat of a substance at either the melting point or the boiling point, it is necessary to supply a known quantity of heat to the substance and measure the mass which has melted or evaporated. From these quantities the amount of heat to change the state of unit mass can be found.

fig. 14.3 Simple determination of the specific latent heat of vaporisation of water

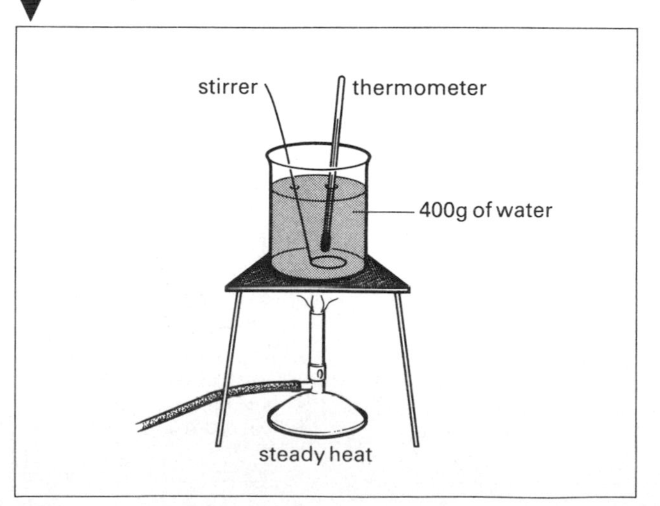

$$\textbf{specific latent heat} = \frac{\textbf{heat required to change the state of the substance}}{\textbf{mass of the substance}}$$

As with the experimental measurement of specific heat capacity, there are problems of accuracy arising from the loss of heat to the surroundings and heat absorbed by the containing vessel.

A simple determination of the **specific latent heat of vaporisation of water** can be made as shown in figure 14.3. Place a known mass of water (say 400 g) in a beaker or tin over a bunsen flame. Keep the flame steady and stir the water constantly, noting the temperature at minute intervals. Keep on after the water starts to boil and continue until some of the water has boiled away. Find the mass of water left at the end of the experiment.

mass of water at start = 400 g
mass of water at end = 326 g
water boiled away = 74 g

in 10 minutes temperature rose 50 °C (figure 14.4)
so in 1 minute temperature rose 5 °C

$$\begin{aligned}\text{heat energy supplied by bunsen} &= mct\\ &= 400 \times 4.2 \times 5\\ &= 8400 \text{ J/min}\end{aligned}$$

74 g of water boil away in 20 minutes

$$\begin{aligned}\text{heat energy to vaporise 1 g} &= \frac{20 \times 8400}{74}\\ &= 2270 \text{ J/g}\end{aligned}$$

fig. 14.4 Heating curve for water

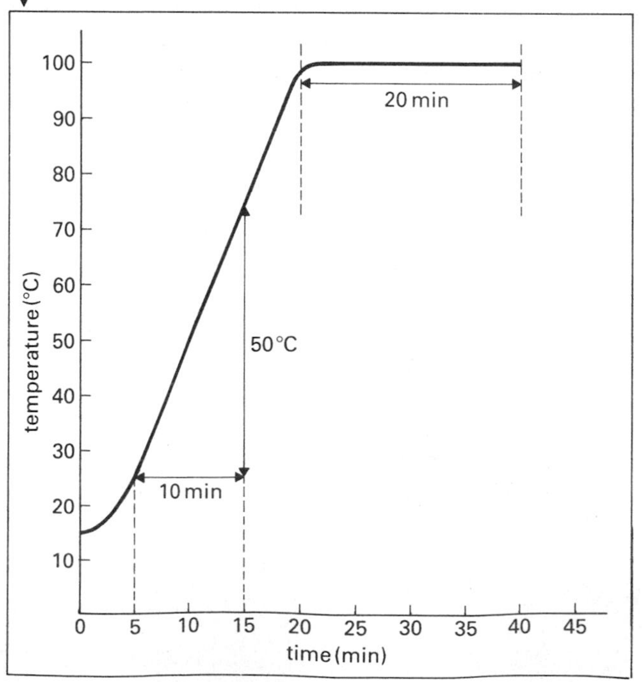

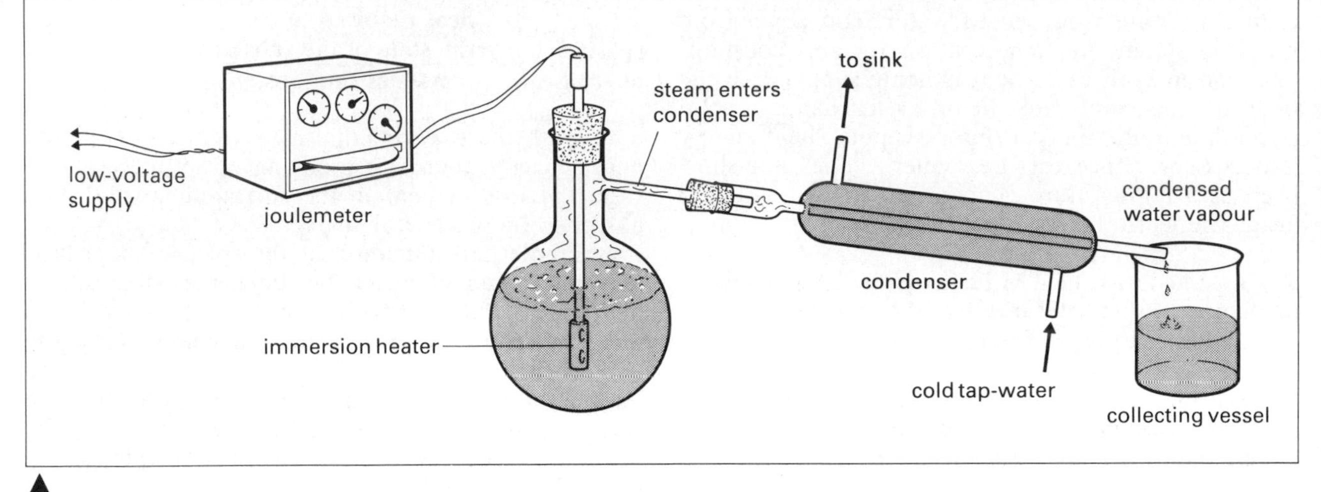

▲ fig. 14.5 Joulemeter apparatus for measuring the specific latent heat of vaporisation

The specific latent heat of vaporisation of a liquid can be found using electrical heating and measuring the energy input with a joulemeter, figure 14.5. The heater is switched on and the experiment left until all parts of the apparatus have reached a steady temperature and the condensed liquid is dripping into a collecting vessel at a constant rate. A weighed beaker is now placed under the outlet of the condenser for a known time, at the end of which the beaker and contents are again weighed to find the mass of liquid collected. The rate at which energy is being supplied can be found by reading the joulemeter over the same length of time. Notice that no temperatures have to be read because the liquid is always at its boiling point.

In an experiment to find the specific latent heat of vaporisation of water, the apparatus was kept going for ten minutes, during which time 12.4 g of water had condensed into the beaker and the energy used was 27 838 joules. This gives a value of 27 838/12.4 = 2245 J/g or 2 245 000 J/kg.

The **specific latent heat of fusion of ice** can be found using the method of mixtures. Ice is added to warm water in a metal vessel. The water and vessel cool, so losing heat which is used to melt the ice and raise the temperature of the water so formed to the final temperature of the mixture. Assuming no heat is lost to the surroundings:

heat given out = heat taken in

[heat lost by water in cooling + heat lost by vessel in cooling] = [heat used to melt ice + heat used to raise temperature of 'ice water']

Pieces of ice about the size of a lump of sugar are kept in water so that their temperature is 0 °C, figure 14.6a. The ice lumps are dried, figure 14.6b, and added one at a time to a known mass of water at a known temperature in a lagged vessel, figure 14.6c. If the ice is not dry, water at 0 °C will also be added

fig. 14.6 Measuring the specific latent heat of fusion of ice ▼

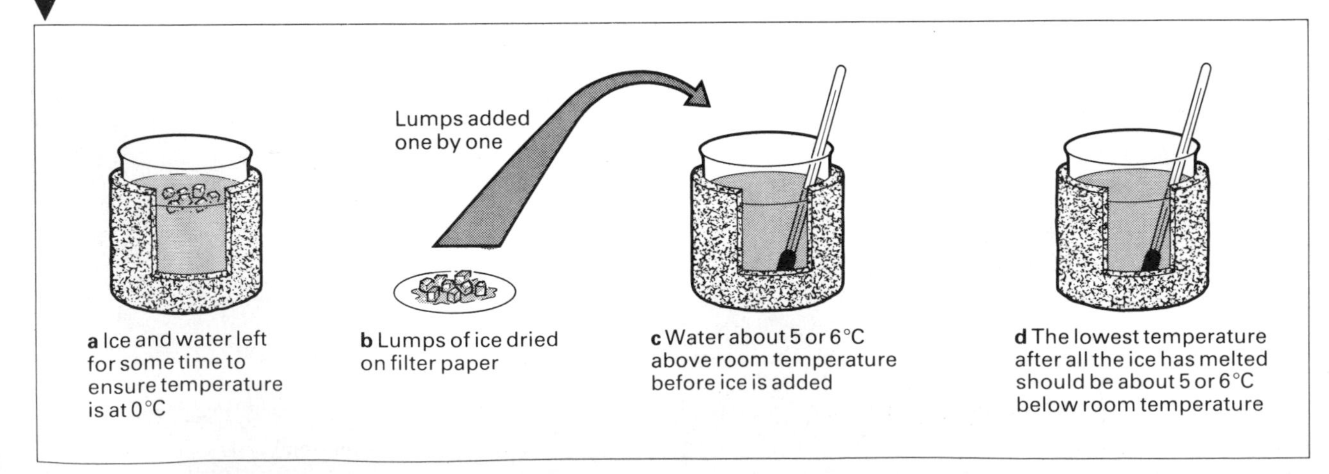

to the mixture, and this is not allowed for in the calculation. The water should be somewhat above room temperature. The water is well stirred and ice added until the final temperature is as far below room temperature as it was above room temperature at the start of the experiment, figure 14.6d. The idea of this is to try and balance out the heat lost to the surroundings in the first half of the experiment with that gained by the vessel and its contents when they are below room temperature in the second half of the experiment. Finally, the vessel and contents are weighed to find the mass of ice added.

mass of copper calorimeter and stirrer = 0.046 kg
mass of calorimeter, stirrer and water = 0.145 kg
mass of calorimeter, stirrer, water and ice = 0.169 kg

mass of water = 0.099 kg
mass of ice = 0.024 kg

initial temperature of water = 24 °C
final temperature of water = 5 °C
temperature of ice = 0 °C

heat lost by water in cooling $= 0.099(4200)(24 - 5)$ $= 7900$ J

heat lost by copper in cooling $= 0.046(400)(24 - 5)$ $= 349$ J

heat used by ice in melting $= 0.024 \times l_f$, where l_f is the specific latent heat of fusion

heat used to warm up ice water $= 0.024(4200)(5 - 0)$ $= 504$ J

heat taken in = heat given out

$$0.024 l_f + 504 = 7900 + 349$$
$$0.024 l_f = 7745$$
$$l_f = \frac{7745}{0.024}$$
$$= 322\ 708 \text{ J/kg}$$

Cooling curves

Some naphthalene crystals are melted in a test tube in a bath of hot water. A thermometer is put in the molten naphthalene and the tube removed from the water bath to cool. The temperature is recorded at minute intervals and a cooling curve is plotted, figure 14.7. The temperature remains steady at the melting point. As the naphthalene solidifies it gives out its latent heat of fusion, which just balances the heat lost to the surroundings so that the temperature remains steady. When all the naphthalene has solidified the temperature again falls.

Pure substances have a 'sharp' melting point. They change from solid to liquid at a single temperature. The value of the melting point can be taken as a measure of the purity of the substance. Mixtures usually have a range of temperatures over which they remain pasty in composition, as if part of the mixture had melted, but part had not. Paraffin wax contains several different compounds each with its own melting point. A cooling curve for paraffin wax would show a series of melting points, none of which would be sharply defined, figure 14.8.

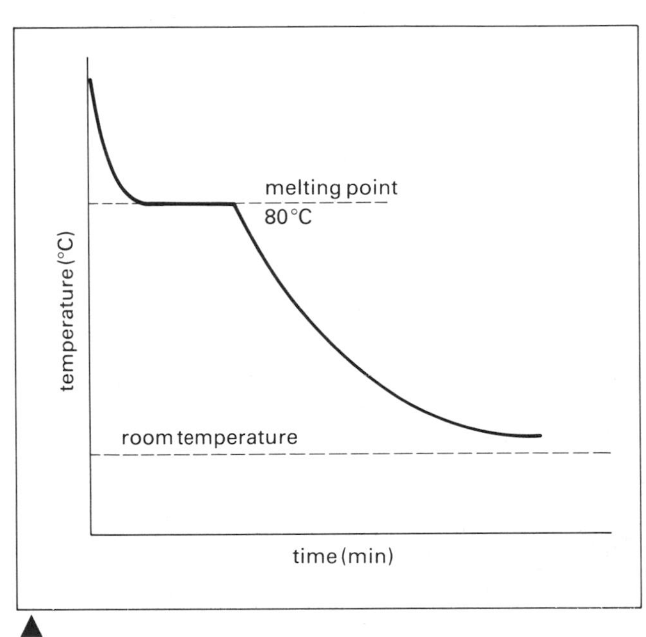

fig. 14.7 Cooling curve for naphthalene

fig. 14.8 Cooling curve for paraffin wax

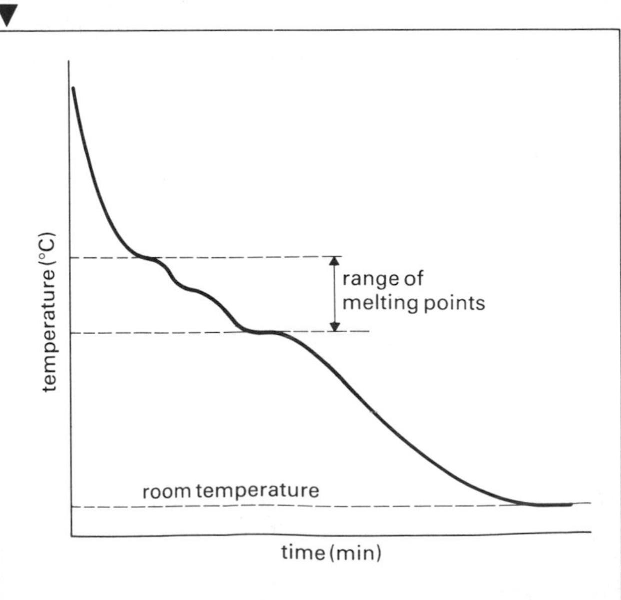

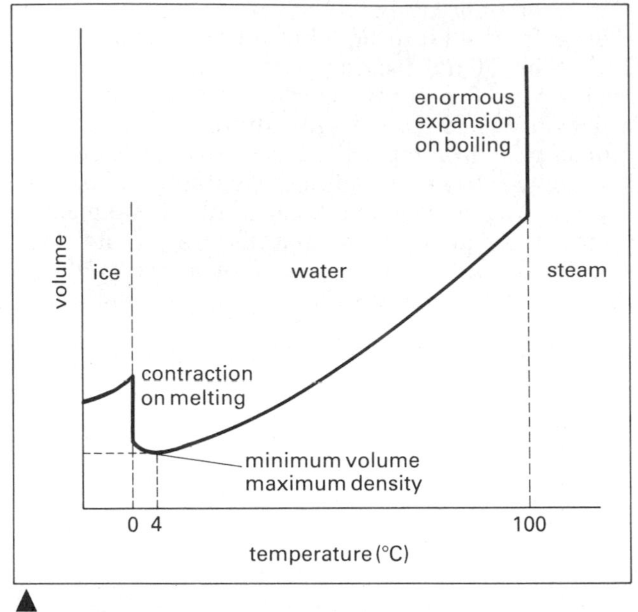

▲ fig. 14.9 Change in volume of water with temperature

▲ fig. 14.10 Milk, which is mainly water, expands on freezing

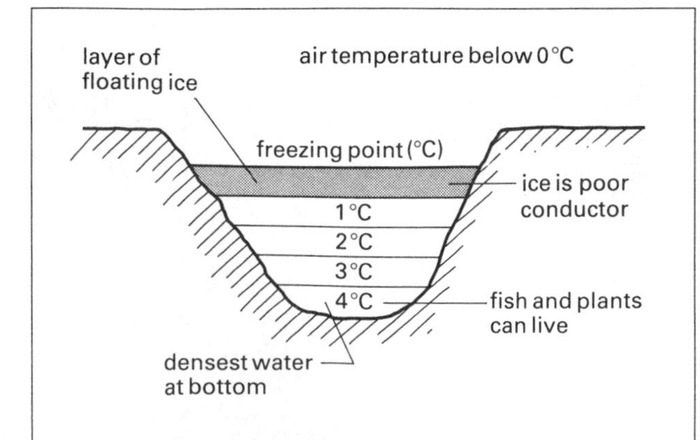

◀ fig. 14.11 Conditions in a frozen pond

Change of state and change of volume

Changes of state are usually accompanied by changes in volume and therefore by changes in density, since the mass remains the same. Most solids expand when they change to liquid and all liquids expand enormously when they boil and change to vapour. Water, for example, expands about 1650 times when it changes to steam. This large increase in volume is the cause of the great increase in pressure which, when controlled, drives a steam engine and, when uncontrolled, causes an explosion.

There are exceptions to the general rule. Cast iron and type metal (an alloy of antimony, lead and tin) expand when they solidify and therefore fill up all the corners of the mould and give good, sharp castings. Water behaves in a very different way.

The peculiar (or anomalous) behaviour of water

When water changes to ice there is an increase in volume. This is shown in figure 14.9. The expansion is demonstrated very nicely by the column of ice rising from the frozen bottle of milk, figure 14.10. The increase in volume — 100 units of volume expand to 109 units — means that the density of ice is less than that of water, so that ice floats on water. The expansion is also responsible for the bursting of water pipes during freezing, and for the cracking of rocks when pockets of water in crevices in the rock freeze during the winter.

Liquids expand when the temperature is increased, and contract when the temperature is decreased. Water behaves differently in that, as the temperature is raised from 0 °C to 4 °C, the volume actually decreases, and from 4 °C to 100 °C the volume increases. Water has a minimum volume and a maximum density at 4 °C. An important consequence of this anomalous behaviour of water is that ice forms first on the top of a pond and pond animals can live in the slightly warmer, denser water at the bottom of the pond, figure 14.11.

As the air temperature falls, the water at the surface of the pond cools, becomes denser and sinks to the bottom. This process continues until all the water is at 4 °C. When the surface water is chilled below 4 °C, it does not sink because it is less dense than the water beneath it. The surface cooling continues until freezing takes place. The thickness of the ice increases slowly, but only reaches the bottom of the pond in very severe winters. Otherwise, aquatic life would be killed off every winter.

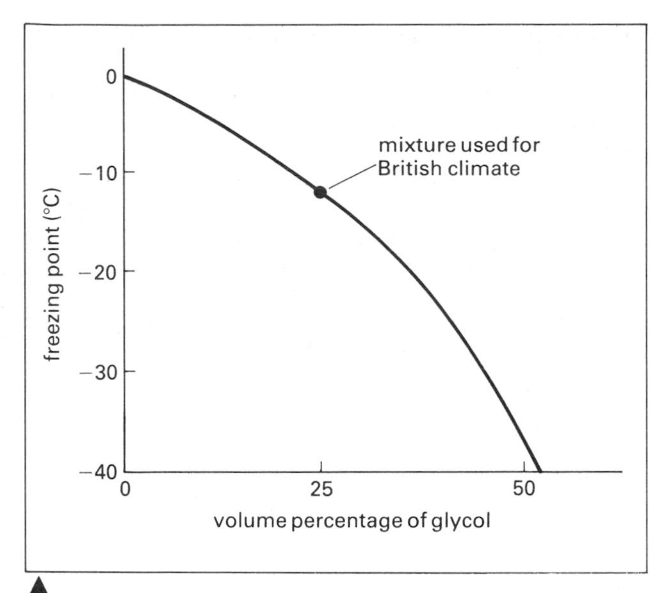

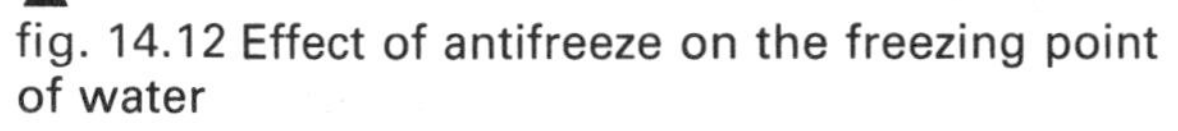
fig. 14.12 Effect of antifreeze on the freezing point of water

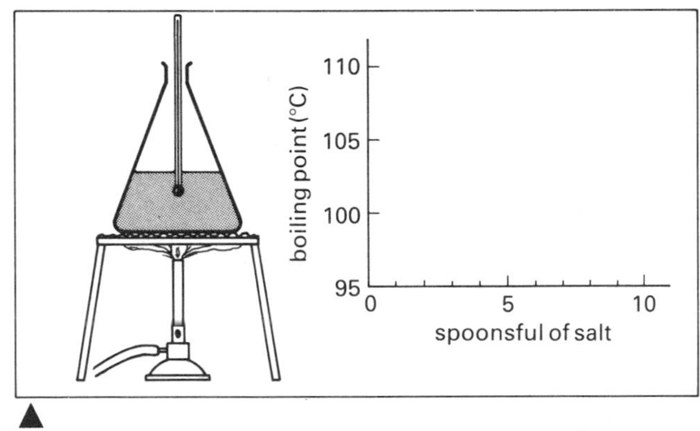

fig.14.13 Effect of salt on the boiling point of water

fig. 14.14 Effect of pressure on the boiling point of water

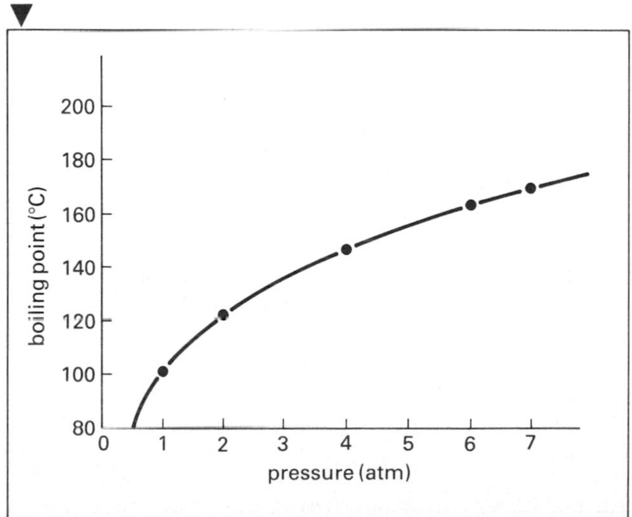

Factors affecting melting and boiling points

Dissolved material

The melting or freezing point of a liquid is changed by the presence of any substance dissolved in it. Dissolved material *lowers* the freezing point. Sea-water has a lower freezing point than freshwater, and the freezing point of a concentrated salt solution is lower still. The more solute is added, the lower is the freezing point. Salt scattered on ice or snow lowers the melting point of the ice to less than that of the air temperature, and the ice melts. Anti-freeze for car radiators is a solution of ethylene glycol in water. Figure 14.12 shows how the freezing point of the solution gets lower as the percentage of glycol is increased. The lowest temperature for freezing is about −40 °C for about a 50% mixture. For the British climate a 25% mixture is generally used.

The effect of dissolved materials is to *raise* the boiling point of the liquid. This can be investigated by the experiment shown in figure 14.13. Boil about 100 cm^3 of water in a flask and take the temperature of the boiling water. Make the volume up to the same point and add one teaspoonful of salt, noting the temperature at which the solution boils. Repeat the process several times and plot the boiling point against the amount of salt added. Notice what happens to the boiling point when the solution is saturated.

Pressure

The second factor which affects melting and boiling points is *pressure*, figure 14.14. The effect of increasing pressure is, in general, to lower the melting point and raise the boiling point.

The increase of boiling point with pressure is useful in steam engines, which work more efficiently with hotter steam, and in pressure cookers, which save both time and fuel because they work at a higher temperature.

An experiment which demonstrates the effect of pressure on melting point is shown in figure 14.15. A large piece of ice is supported on the ring of a retort stand. A copper wire is passed over the ice and a large weight (5–10 kg) hung on the wire loop. Slowly the wire passes through the ice, leaving a solid piece of ice behind. Since the wire is thin, the pressure below it is quite considerable. The freezing point of the ice underneath the wire is lowered below 0 °C, so the ice melts. The water flows above the wire, and since it is no longer under pressure, freezes at the normal freezing point. This melting and refreezing is called **regelation**. Another point to note is that the latent heat given out by the water above the wire when it refreezes is conducted by the copper to below

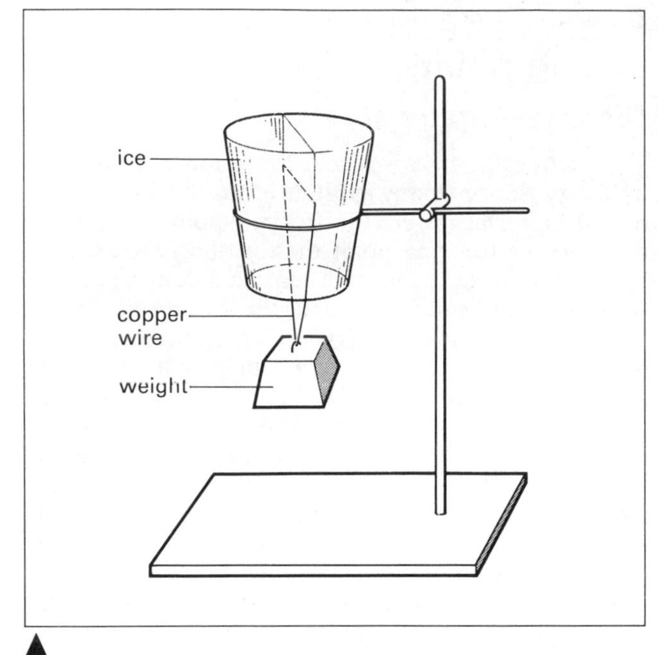

▲ fig. 14.15 Effect of pressure on the melting point of ice (regelation)

the wire where it helps to melt the ice. The experiment will not work with a piece of string because it is a poor conductor of heat. Two pieces of ice can be made to freeze together by pressing them hard against each other.

The pressure under the sharp blade of an ice skate melts the ice underneath it and so allows the skate to slip smoothly over the ice. Snowballs are made by melting the snow crystals under pressure and then allowing refreezing to take place, binding the snow into a hard ball. Making snowballs and skating are impossible in very cold weather.

The variation of boiling point with change of pressure can be determined with the apparatus shown in figure 14.16. As the pressure is slowly reduced simultaneous readings of temperature and pressure are taken and a graph of boiling temperature and pressure is drawn.

The pressure cooker

The pressure cooker, figure 14.17, consists of a strong vessel with a securely fitting lid. When the water boils, the pressure of the steam builds up inside the cooker and so the boiling point is raised. With all three weights on top of the valve the pressure is about twice atmospheric, and the boiling temperature is raised to about 120 °C. Foods cook much quicker at a higher temperature, so saving time and fuel.

Pressure cookers are used as sterilisers in hospitals, and in the canning of fruit and vegetables, because harmful bacteria are destroyed at 120 °C.

fig. 14.16 Determining the change of boiling point as pressure is reduced ▼

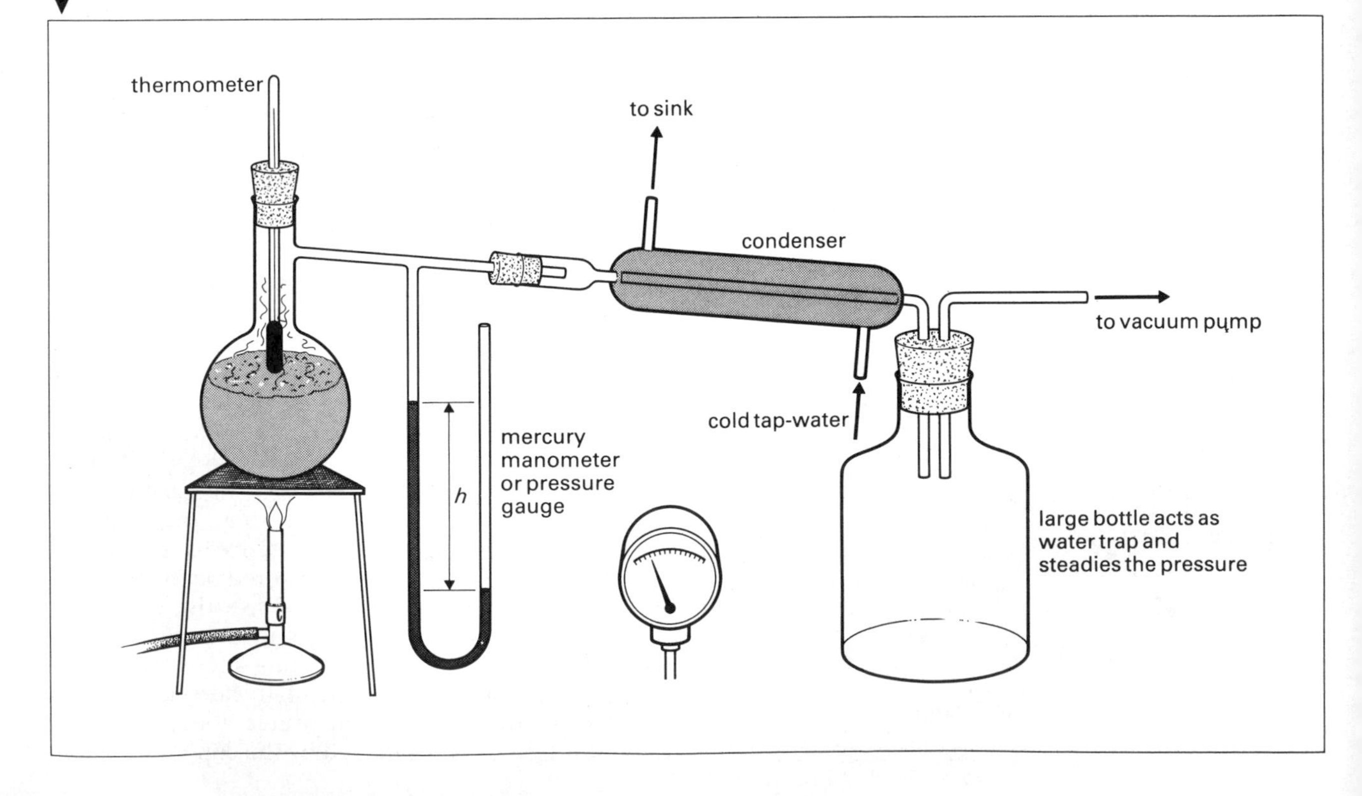

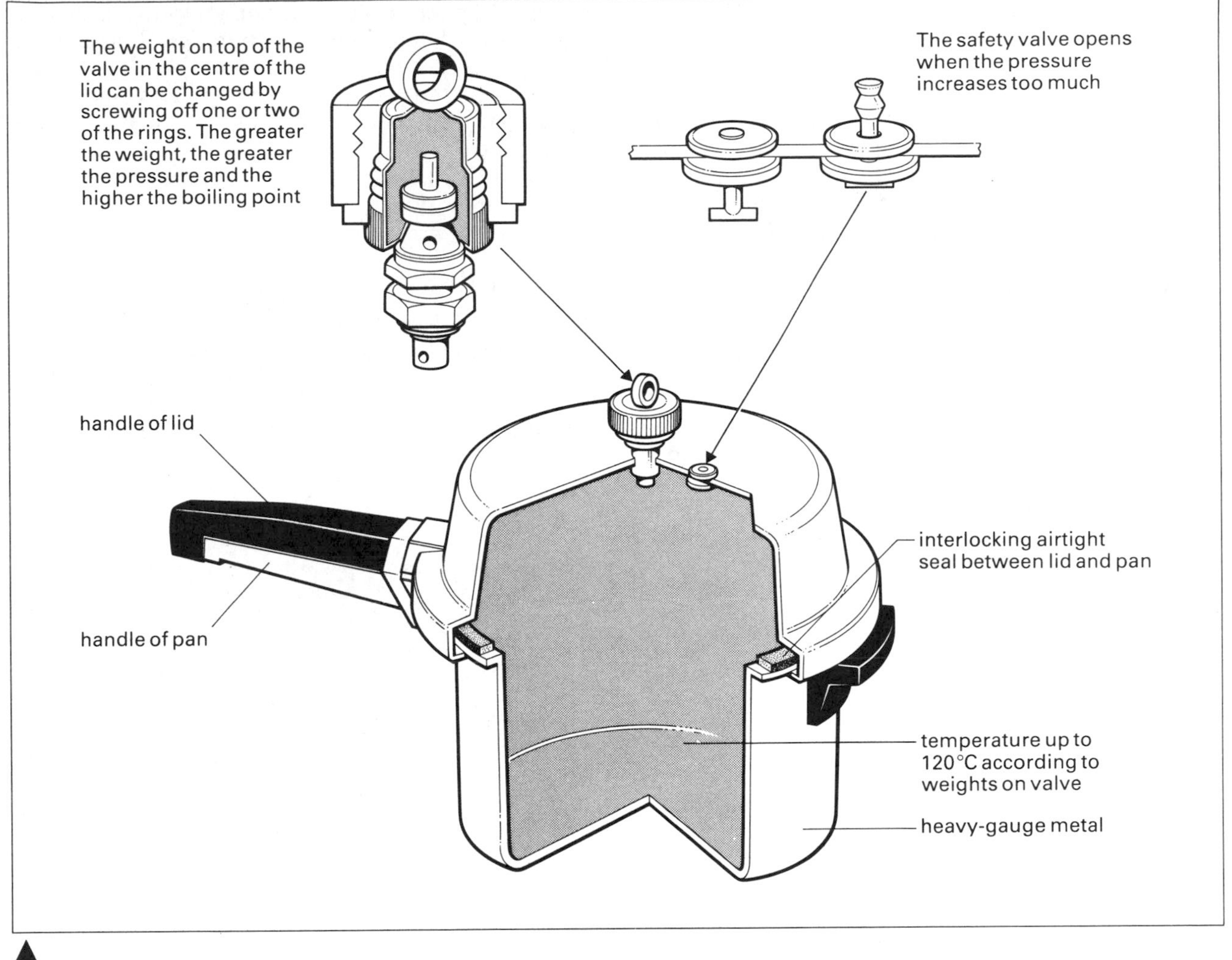

▲ fig. 14.17 Pressure cooker

fig. 14.18 Absorption and release of heat energy during changes of state ▼

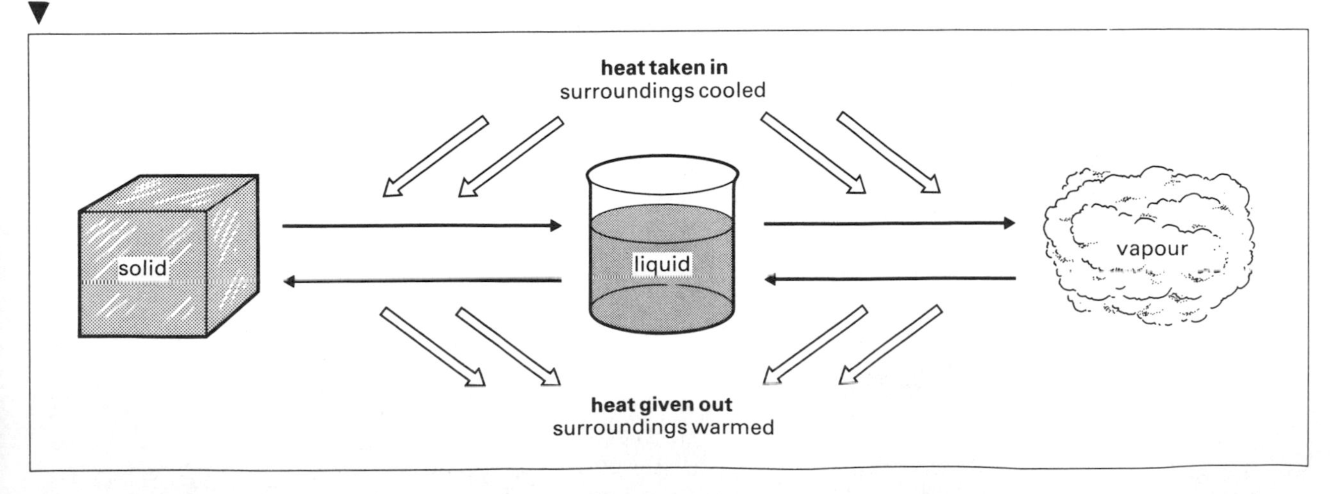

Cooling by evaporation

When a solid has to be changed into a liquid, or a liquid into a vapour, heat energy has to be supplied to bring about the change. This is illustrated in figure 14.18. When the reverse changes take place, heat is given out. When one gram of steam condenses, for example, 2260 J of heat are released. This is why a scald from steam is so much worse than a scald from boiling water. One gram of boiling water cooling down to body temperature would release only about 260 J.

A simple experiment to show the cooling effect of evaporation consists in evaporating a small quantity of ether in a beaker standing in a pool of water on a piece of wood, figure 14.19. When the ether evaporates rapidly as a result of air being blown through it, the necessary latent heat of evaporation is taken from the water, and it will soon be found that the beaker is frozen to the wood. The experiment should be carried out with good ventilation, for ether fumes are unpleasant.

Figures 14.20 to 14.22 show some examples of cooling by evaporation.

Refrigeration

In a refrigerator, figure 14.23, a liquid is constantly evaporating (absorbing heat). The vapour is then condensing (giving out heat) as it is pumped round a closed system of pipes. The refrigerator is really a machine for transferring heat from inside to outside the cabinet.

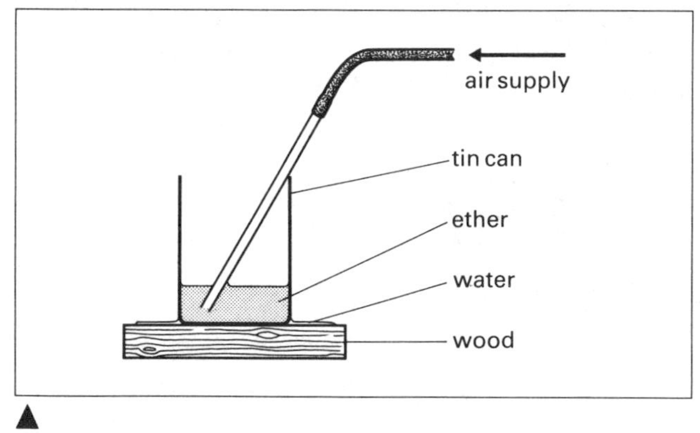

▲
fig. 14.19 Demonstration of the cooling effect of evaporation

▲
fig. 14.20 You feel cold after swimming, particularly if a breeze is blowing

fig. 14.21 Section of the skin. Heat generated in the muscles is carried by the blood to the body surface. The sweat glands give out sweat which evaporates and cools the skin
▼

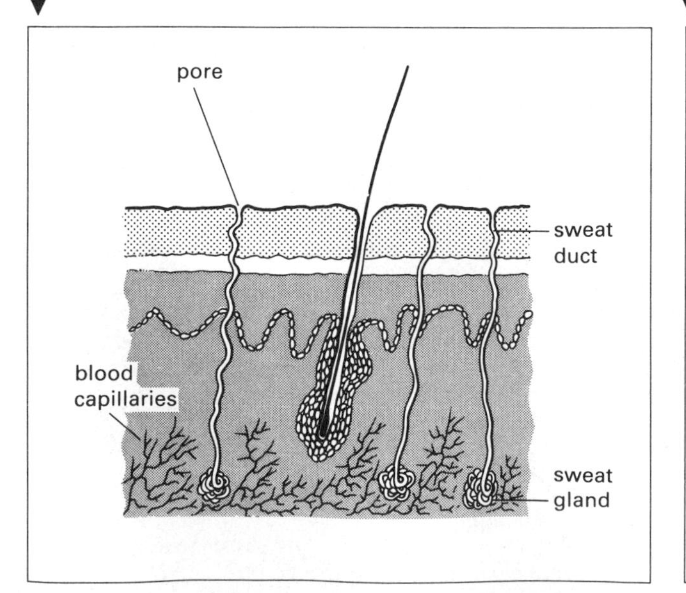

fig. 14.22 Wet and dry bulb hygrometer. The wet bulb shows a lower temperature than the dry one. Tables are supplied which enable you to calculate the **relative humidity** of the air
▼

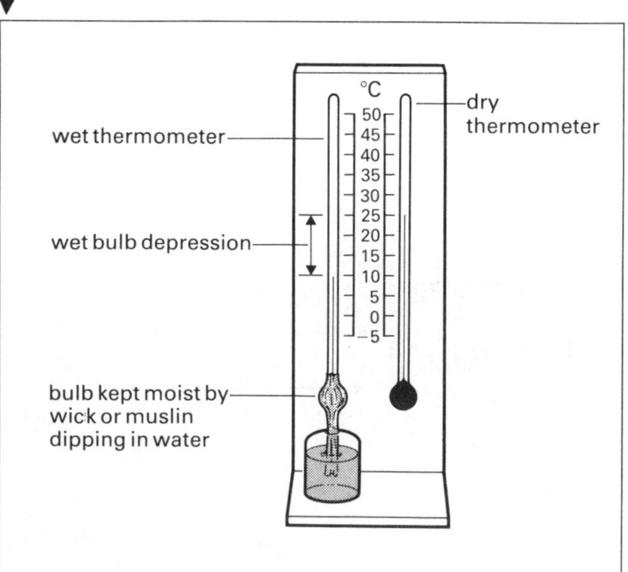

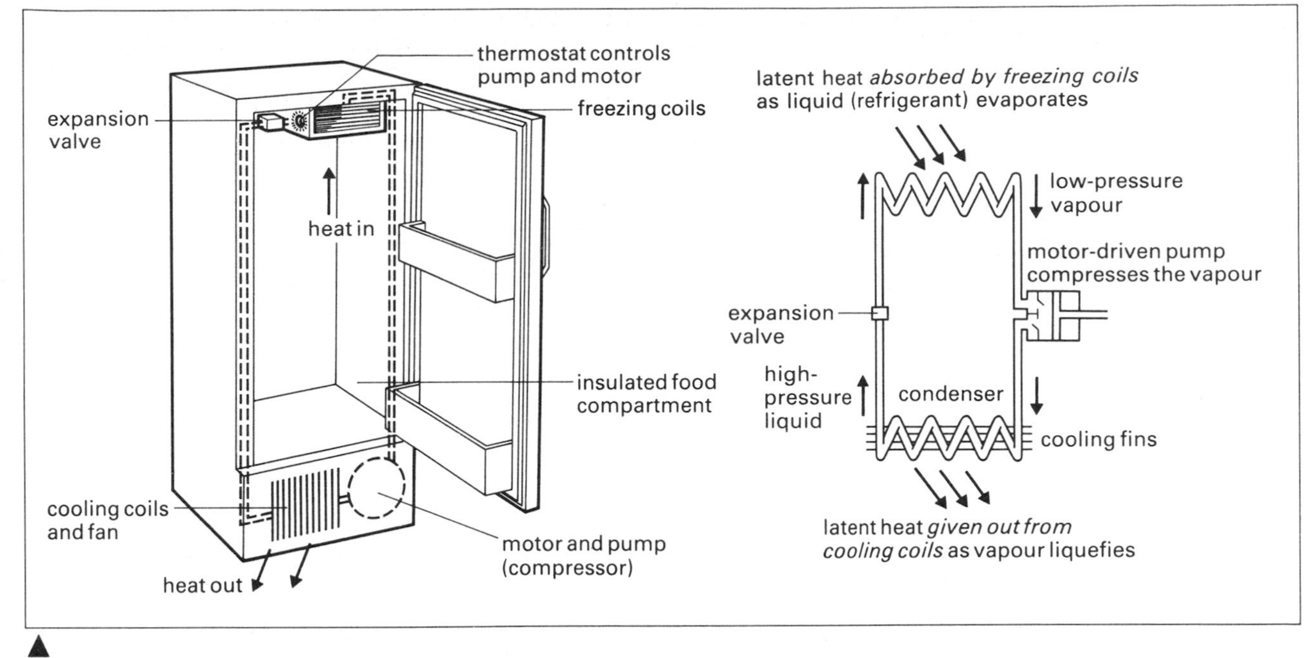

▲
fig. 14.23 Domestic electrical refrigerator

EXERCISE 14

In questions 1–6 select the most suitable answer.

1 Latent heat is
A residue heat left in a cooker when the current is switched off
B heat required to keep an even temperature
C heat required to make a metal expand
D heat required to change the state of a substance without a change in temperature
E heat capacity of a given volume of water

2 When a liquid evaporates it must
A become more dense
B change into a solid
C give out latent heat
D take in latent heat
E rise in temperature

3 An ice cube of mass 30 g melts and remains at 0 °C. The specific latent heat of ice is 336 J/g. The heat gained by the ice from the atmosphere is
A 112 J
B 306 J
C 366 J
D 5 040 J
E 10 080 J

4 Pure water at a pressure greater than one normal atmosphere
A boils at 100 °C
B freezes at 0 °C
C boils at a temperature above 100 °C
D freezes at a temperature above 0 °C
E boils at a temperature below 100 °C

5 Adding salt to pure water will
A have no effect on its boiling point
B have no effect on its freezing point
C raise its freezing point
D lower its boiling point
E lower its freezing point

6 The melting point of a substance is
A unaffected by changes of pressure
B the temperature at which vaporisation commences
C affected by changes of pressure
D the latent heat given out when a solid melts
E unaffected by the addition of impurities

7 a Using the axes shown in the figure, draw the graph you would expect to obtain when plotting temperature against time for a cooling liquid such as paraffin wax, which is solid at room temperature.

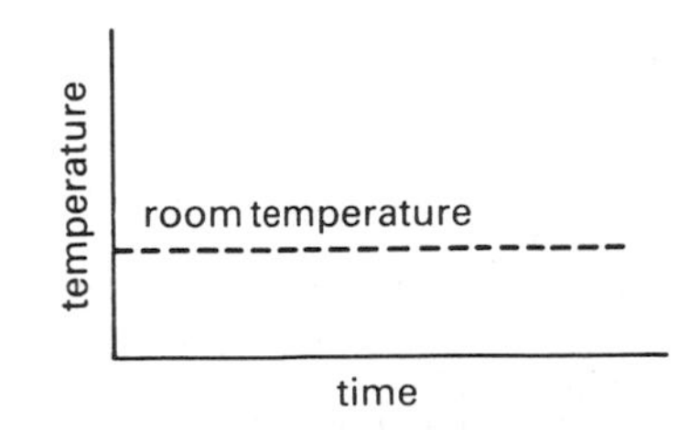

b Indicate on the graph which parts relate to (i) the liquid state, (ii) the solid state, (iii) the transition from liquid to solid.

c (i) Why is the graph the shape that it is in the part corresponding to **b** (iii) above? (ii) What important information about the cooling substance can be obtained from this part of the graph?

8 Calculate the heat given out by 200 g of a molten substance, at its melting point, as it solidifies without further change of temperature. Assume the specific latent heat of the substance to be 150 J/kg.

9 a What is 'latent heat of fusion'?
b Why would meat be kept longer by 1 kg of ice at 0 °C than by 1 kg of water at 0 °C?

10 a Why is it that 1 g of steam at 100 °C would cause a more serious scald than 1 g of water at 100 °C?
b Briefly, explain how your body loses heat when you perspire (sweat).

11 a Give *one* way in which evaporation and boiling differ, and *one* way in which they are similar.
b When a small quantity of methylated spirits (alcohol) at room temperature is poured on the back of the hand it feels very cold, whereas a small quantity of water at the same temperature feels much less cold. Why is this?

12 Brian sets up the apparatus shown in figure 14.24 to find the specific latent heat of ice. He takes large lumps of ice and places them in a filter funnel around the heater. He puts a beaker under the filter funnel and switches on the heater. A short time later he switches off the heater and removes the beaker. During the experiment he takes the following readings:

initial reading on joulemeter = 24 000 J
final reading on joulemeter = 34 000 J
mass of empty beaker = 60 g
mass of beaker + water = 110 g

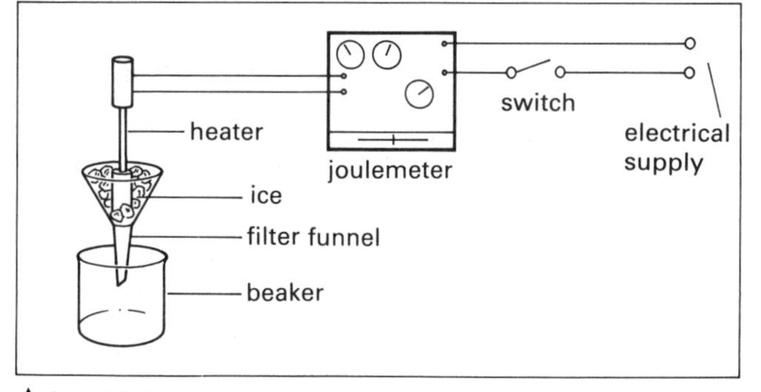

▲ fig. 14.24

a What value for the specific latent heat of ice do these results give?
b Jean tells Brian that his answer is inaccurate because some of the ice was melted by heat from the air. (i) Describe how Brian could find out how much ice was melted in this way. (ii) How would he use this result to get a more accurate value for the specific latent heat of ice?
c Explain why smaller lumps of ice would have improved the experiment.
d The value for the specific latent heat of ice listed in the data book is 3.3×10^5 J kg^{-1}. State in words what this means.

13 0.1 kg of water at 0 °C in a beaker was heated over a bunsen flame. Readings of the temperature were recorded at intervals until all water had been turned into steam. Some of the readings are given in the table:

Time (min)	0	1	2	3	4	5	9	13	17	21	25	29	32
Temperature (°C)	0	20	42	64		100							
						↑ water starts boiling after 5 minutes							↑ beaker dry after 32 minutes

a What is the temperature of the water between the 5th and 32nd minute?
b Draw a graph of temperature against time.
c Calculate the rate at which heat is supplied by using the equation:

$$\text{rate of supply of heat (J/min)} = \frac{\text{mass of water (kg)} \times 4000 \times \text{temperature rise of water (°C)}}{\text{time taken for water to boil (min)}}$$

d Give *two* reasons why this can only be an approximate answer.
e Calculate how much heat is supplied to the 0.1 kg of water to change it from water at 100 °C to steam.

14 Figure 14.25 shows the essential parts of a simple refrigerator circuit.
a Explain carefully what is happening in the pipes at A, stating clearly why this achieves the desired result.
b Why is A situated at the top of the refrigerator?
c Why are the fins metal, and what is their purpose?
d Name another machine in which metal fins are used to achieve the same purpose.
e The walls of the cabinet, B, are sometimes filled with crinkled aluminium foil. Why?
f A person who leaves open the refrigerator door to cool the room on a hot day in summer will not succeed. Give the reasons for this.
g Explain why a chest type deep freezer (lid at top) is thought to be more efficient than an upright type (door at side).

▼ fig. 14.25

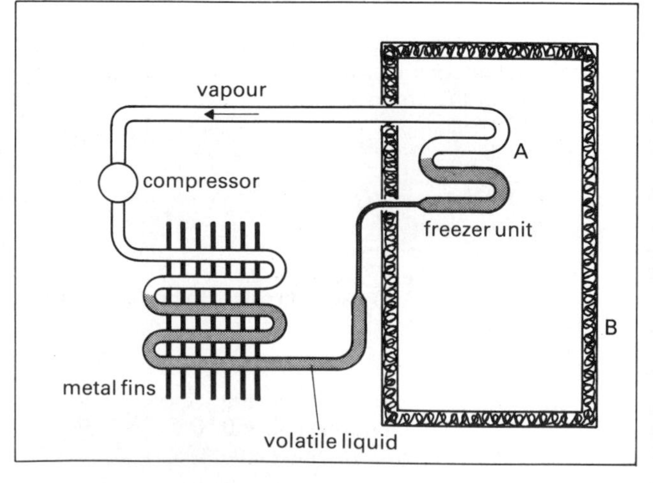

15 TRANSMISSION OF HEAT

Heat is transmitted from one place to another, from a higher to a lower temperature, in three different ways. These occur in different sets of circumstances and take place by different mechanisms.

- **Conduction** is the transfer of heat energy through a material, which is not at uniform temperature, from points of high temperature to points of low temperature, without movement of the material itself.

  Conduction is the method of heat transfer in solids. Liquids and gases are generally poor conductors.
- **Convection** is the transfer of heat energy by circulation of the material due to differences of temperature.

  Convection currents, by their very nature, can only occur in fluids.
- **Radiation** is the transfer of heat energy by means of electromagnetic waves without the need of a material medium.

  Radiant heat can travel through empty space and is the method by which heat energy reaches us from the sun.

Good and bad conductors of heat

Many simple experiments show that some materials conduct heat better than others. For instance, if copper and glass rods are held in a bunsen flame, the end of the copper rod will soon become too hot to hold, while the end of the glass rod remains quite cool, figure 15.1. Copper is obviously a better conductor of heat than glass.

The relative **thermal conductivities** of different materials can be found using the apparatus shown in figure 15.2. The ends of the rods are kept at the temperature of boiling water. The rods have small wire indicators on them and are coated with paraffin wax. The wax holds the indicators in place. As heat is conducted along the rod, each section of the rod may be raised to the melting temperature of the wax, so that the indicators slip down the rods. The indicator which moves the furthest shows the best conductor. The distances moved by the different indicators will put the rods in their order of conductivity, but the thermal conductivities are *not* proportional to the distances over which the wax melts.

fig. 15.1 Simple test for good and bad conductors

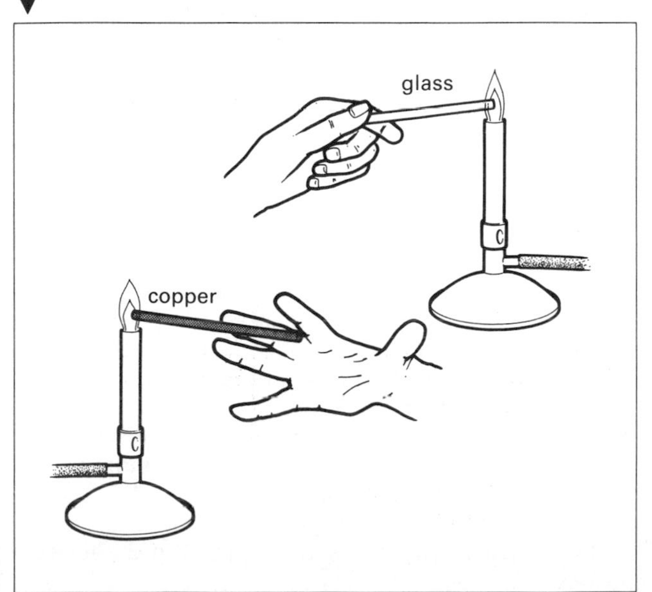

fig. 15.2 Comparison of thermal conductivities

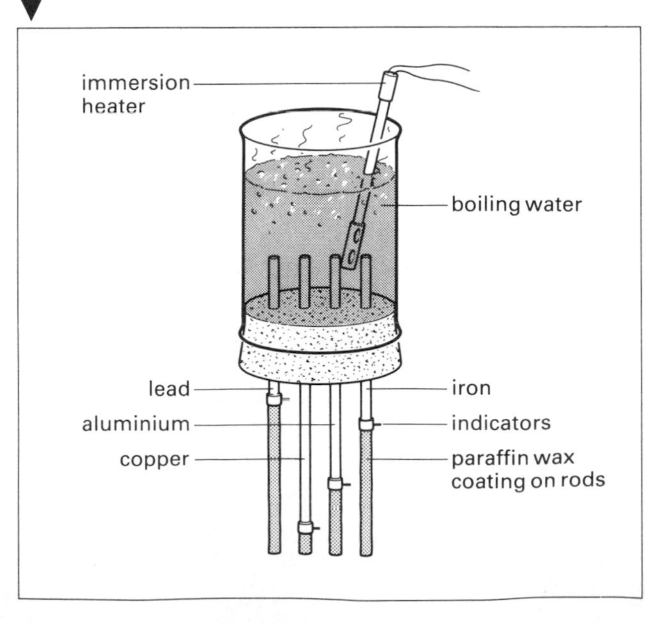

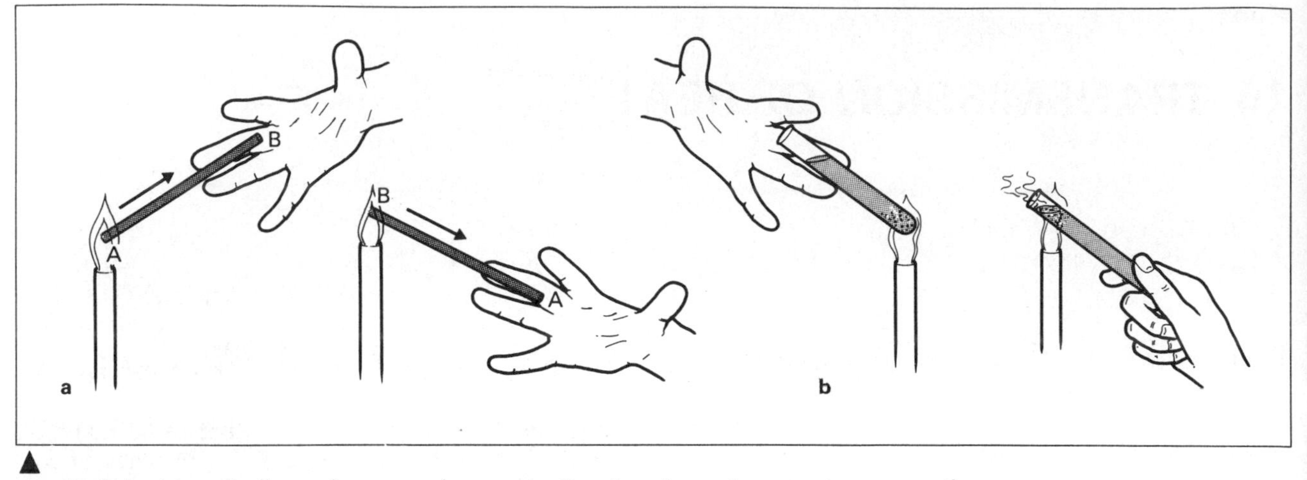

fig. 15.3 Heat travels through copper by conduction but through water by convection

Liquids and gases are generally poor conductors of heat. If we hold a copper rod at the top and heat it at the bottom, and then hold it at the bottom and heat it at the top, we find that the heat travels up and down the rod equally well, figure 15.3a. When we repeat the experiment with a test tube of water, figure 15.3b. we find that we can boil the water at the top of the tube while the bottom is still cool enough to hold. The heat travels up through the water by convection but is prevented from travelling downwards by the poor conductivity of the water.

Water can be shown to be a poor conductor of heat by the experiment shown in figure 15.4. Some lumps of ice are held at the bottom of a tube of water by a coil of wire. The water can be boiled at the top of the tube for quite a while without the ice melting.

Air is a bad conductor of heat (i.e. it is a heat insulator), but of course it moves easily by convection. If we want to use air as a heat insulator we must keep it still by trapping it in some way. This is one reason why we wear several layers of clothes to keep us warm. Loose materials like cotton wool and blankets are good heat insulators because a lot of air is trapped between the fibres. Birds' feathers and animals' fur act in the same way.

fig. 15.4 Demonstrating that water is a bad conductor of heat

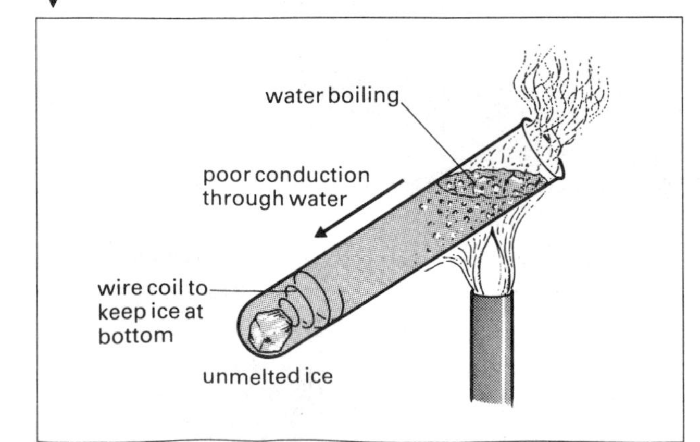

When we want heat to travel quickly and easily from place to place we obviously use a good conductor. When we want to stop heat escaping or travelling where we don't want it, we use a poor conductor — in other words, an insulator.

Good conductors of heat : all metals and alloys; silver and copper are very good conductors

Poor conductors of heat or heat insulators: water, asbestos, wood, china, plastics, glass, air; loose materials, such as cotton wool, because a lot of air is trapped between the fibres

The heat we supply to our houses by fires, radiators or central heating is constantly being lost to the cooler surroundings. The main routes of heat loss are shown in figure 15.9. Think about ways of cutting the heat losses to a minimum and so saving fuel.

Rate of heat conduction

The rate at which heat is conducted in any particular case depends on a number of factors. If we are considering, for example, the heat lost through a window the rate of loss of heat depends on

- the **area** of the window,
- the **temperature difference** between the two sides,
- the **conductivity** of the glass, the material of which the window is made,
- the **thickness** of the glass.
  The thinner the glass the more rapid the heat loss.

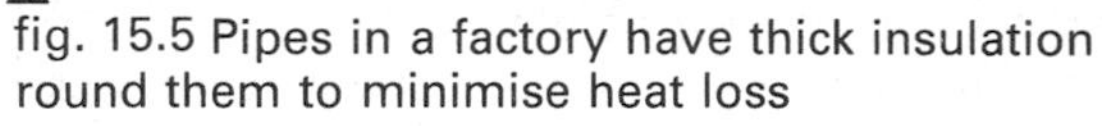

fig. 15.5 Pipes in a factory have thick insulation round them to minimise heat loss

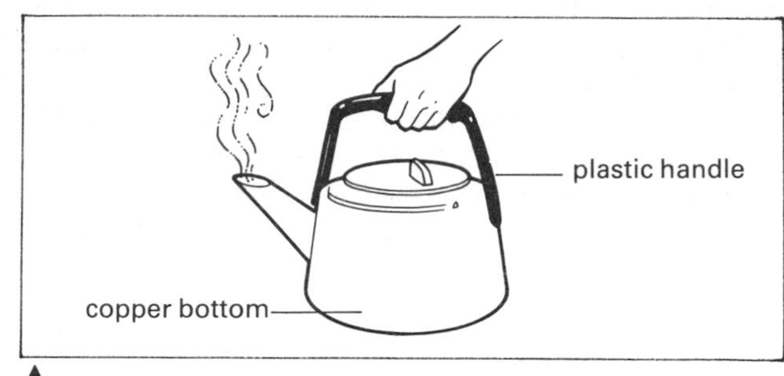

fig. 15.6 Kettles and pans often have copper bottoms (*good conductor*) and wooden or plastic handles (*insulator*)

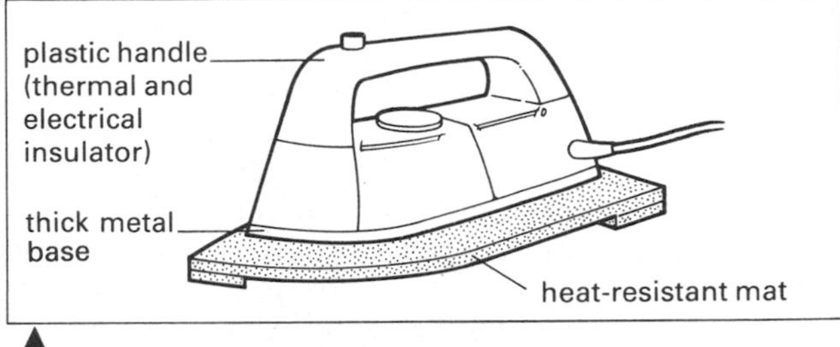

fig. 15.7 The electric iron has a plastic handle (*insulator*), thick metal base (*good conductor*) and stands on a heat-resistant mat (*insulator*)

fig. 15.8 A cavity is left between the outer and inner walls of a house. The air in the cavity prevents some loss of heat and thus the cavity wall is a better insulator than a solid wall of the same thickness. There is some improvement if the cavity is filled with insulating foam or slabs, as shown here

fig. 15.9 Heat losses in the home

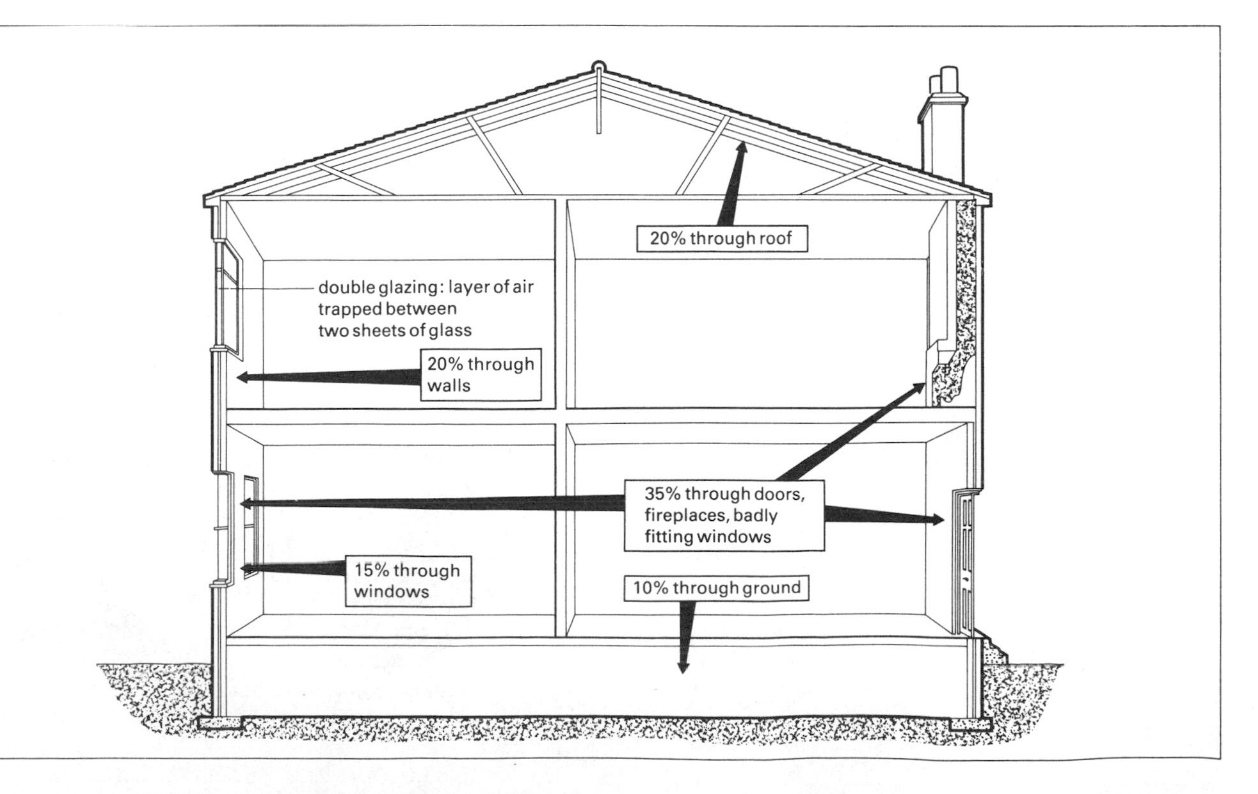

Convection currents

If part of a liquid is heated it expands and, since the mass remains the same, the density gets less. So a part of a liquid is less dense than its surroundings and will float up through the cooler liquid. This upward movement of hotter liquid (or gas) is a **convection current**. In the same way, if one part of a liquid is cooler than its surroundings it will sink.

At the bottom of a large glass beaker, in the centre, put one or two large pink crystals of potassium permanganate. (Other coloured substances would do, but this dissolves easily and colours the water deeply.) Very, very carefully pour down the side of the beaker some cold water so that the crystals are disturbed as little as possible. Place the beaker on a tripod (do not use a gauze) and heat with a small bunsen flame in the centre, figure 15.10.

The pink coloured water rises up the centre, fans out at the top and then descends at the sides of the beaker. Currents are formed in the water. The warm water at the bottom centre rises through the cooler surrounding water and cold water flows in from the sides and down from the top. So each part of the water comes in contact with the source of heat and gets warmed. The colour spreads through the whole of the water showing it is all warmed. These currents are **convection currents**.

Repeat the experiment described above with the crystals at one side, figure 15.11. Copy and draw in arrows to show the directions of the currents in the water.

Add some very fine aluminium powder or some aluminium paint to benzine in a test-tube. Aluminium paint is a suspension of fine particles of aluminium. Warm the bottom of the test-tube with your finger, figure 15.12. Say what happens and explain it.

Gases, like liquids, show convection currents. Warm air rises through the cooler air round it, and more air flows in to take its place.

Cut a spiral about 10 cm in diameter from fairly stiff paper. Allow it to hang loosely from a pin and hold it high over a bunsen flame, figure 15.13a. Explain what happens.

Make a paper propeller by cutting along the dotted lines and bending the blades downwards slightly. Balance it on a pin held in your hand, figure 15.13b. Now try it with the pin stuck in a piece of Plasticine on the bench. Why does it revolve in one case and not in the other?

fig. 15.10 Convection currents in liquids

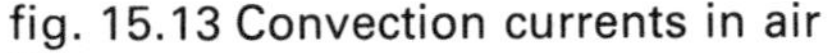

fig. 15.13 Convection currents in air

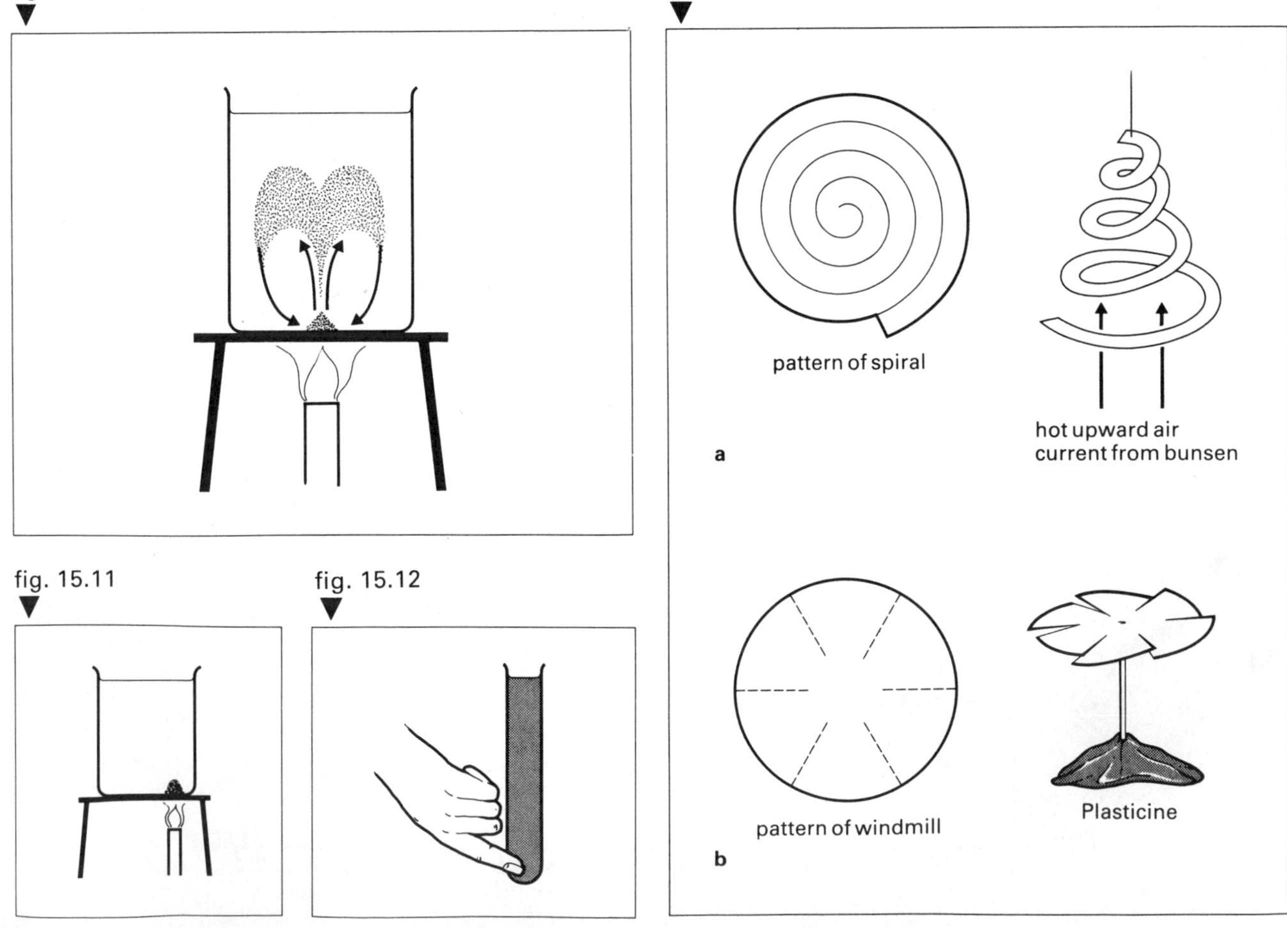

fig. 15.11

fig. 15.12

Examples of convection currents – 1

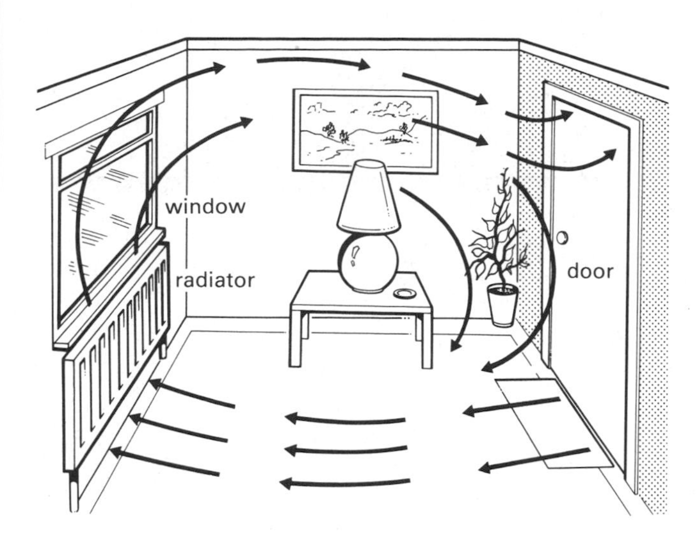

Central heating radiator: the air, warmed by conduction as it passes over the radiator, rises and sets up convection currents round the room

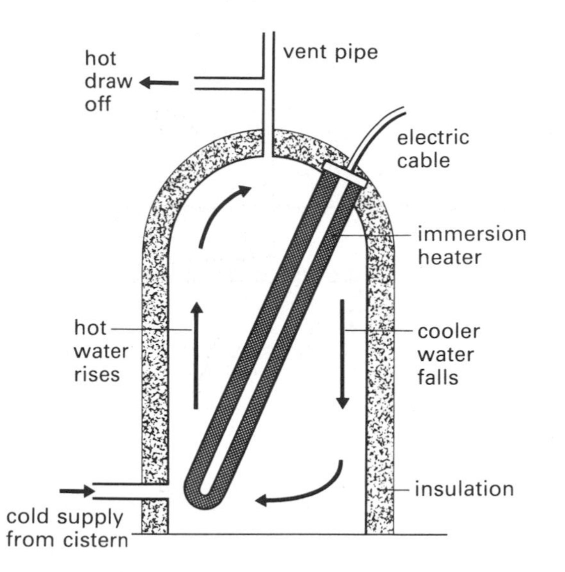

Electric immersion heater: the water is heated by conduction when in contact with the heater, rises and sets up convection currents in the cylinder, which heat all the water

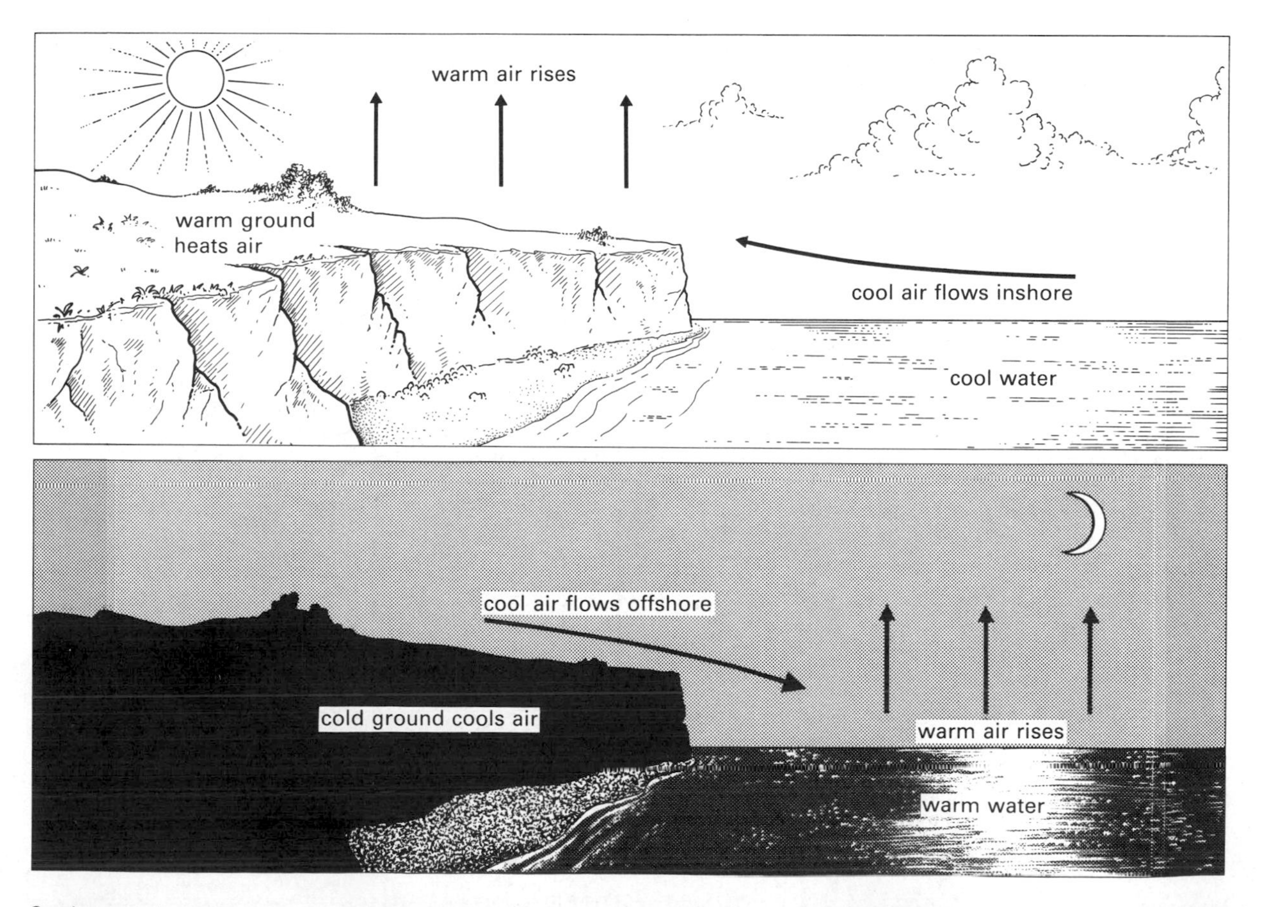

Sea breeze: during the day, the land heats up more quickly than the sea. As the warmer air rises above the land, it is replaced by cooler air from the sea

Land breeze: during the night, the land cools more quickly than the sea. As the warmer air rises above the sea, it is replaced by cooler air from the land

Examples of convection currents – 2

supply from mains
ballcock
cold water cistern
hot draw off
hot water cylinder
boiler without pump
a

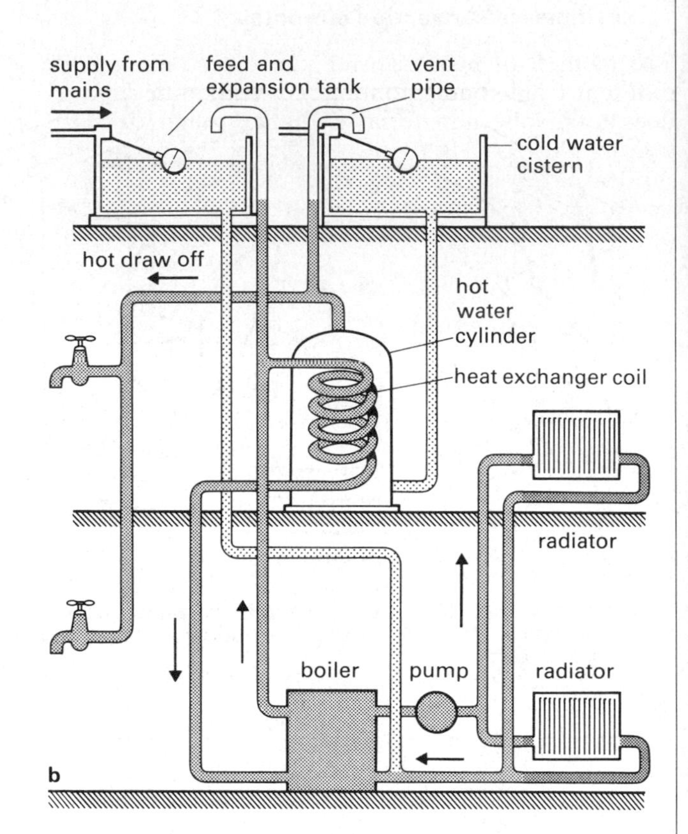

a Domestic hot water supply. In the simple 'gravity' system, hot water from the boiler rises and flows into the top of the cylinder, while the colder water flows from the bottom of the cylinder down to the boiler

b Combined hot water and central heating system. In the layout shown, part of the water heated in the boiler flows through the heat exchanger coil in the cylinder and heats the domestic hot water. The remainder of the hot water is pumped round the heating circuit.
In another typical layout both the domestic hot water and the central heating are pumped.

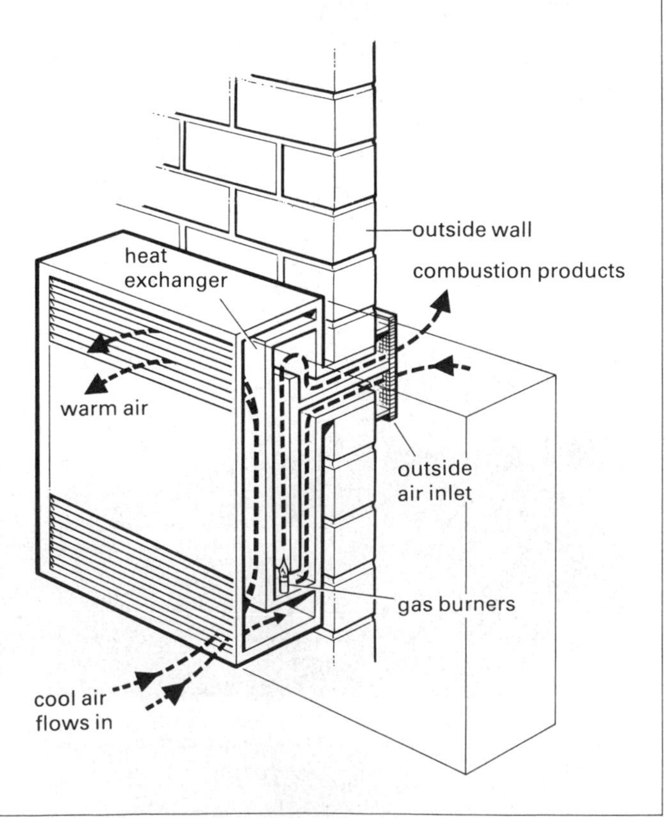

Balanced-flue convection heater: air from the room, warmed by conduction as it passes over the heat exchanger, rises and flows out into the room from the top of the heater. Air for combustion is drawn from outside the building

Radiation

The method of heat transfer known as radiation is different from conduction and convection because it does not involve a material medium through which to transfer the heat. Radiant heat or infrared radiation can travel through a vacuum in the form of electromagnetic waves and is therefore similar to light but of longer wavelength. (See the complete electromagnetic spectrum p. 197.) Radiant heat travels in straight lines at the same velocity as light, and can be reflected and refracted.

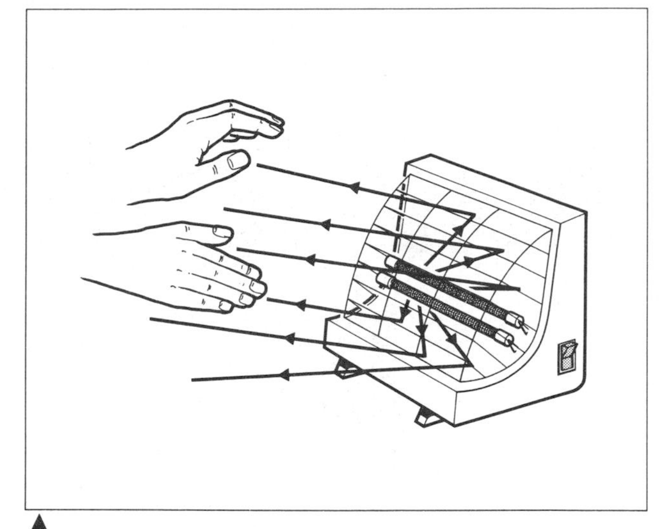

fig. 15.14 Radiant heat from the red-hot element of the radiator is reflected forward by the polished metal behind it

The radiation balance

All objects radiate heat (unless they are at absolute zero). At the same time, all objects are absorbing the radiant heat falling on them from their surroundings. If the rate of emission is the same as the rate of absorption the body is at a constant temperature. It is said to be in thermal equilibrium. Apparently nothing is happening, but this is really because two opposite processes are taking place at the same rate and balancing each other out. Such a state is called a dynamic equilibrium. The number of people in a building may be constant because people are continually entering and leaving at the same rate. If the rate of emission of radiation is greater than the rate of absorption, the object has a net loss of heat energy and therefore cools. If the rate of absorption is greater than the rate of emission, the temperature of the object rises. The rate at which an object radiates heat depends on

- the temperature,
- the surface area,
- the nature of the surface.

The rate of radiation is proportional to the fourth power of the absolute or Kelvin temperature. This means that, as the temperature rises, the rate of radiation increases very rapidly. If the temperature is doubled, the rate of radiation increases 2^4 or 16 times.

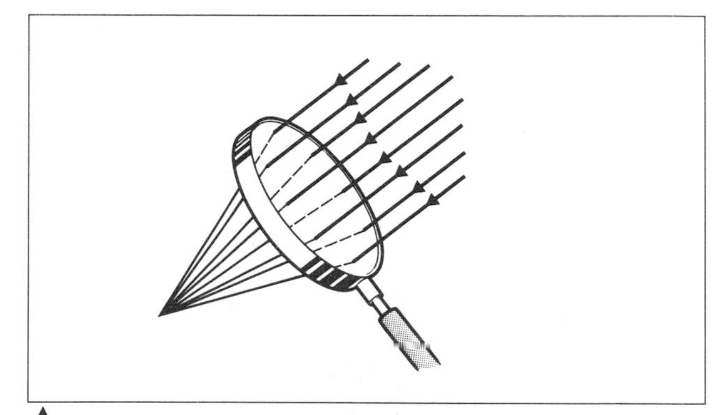

fig. 15.15 Light and heat radiated from the sun can be brought to a bright, hot focus by a lens

The radiating and absorbing properties of different surfaces

When we try to compare the radiating powers of different kinds of surface, we must make sure that the surfaces have the same area and are working under the same temperature conditions. Figure 15.16 shows two tins, one with a matt black surface and the other with a polished surface, each filled with nearly boiling water. The tins are left standing on a bench. The temperatures are taken at minute intervals and cooling graphs plotted. The black tin is found to cool more quickly than the polished one. We can conclude

fig. 15.16 Comparing emission powers of surfaces

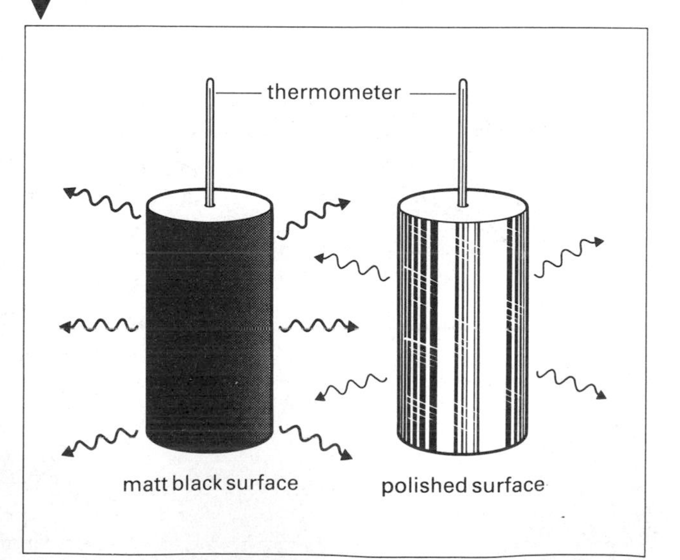

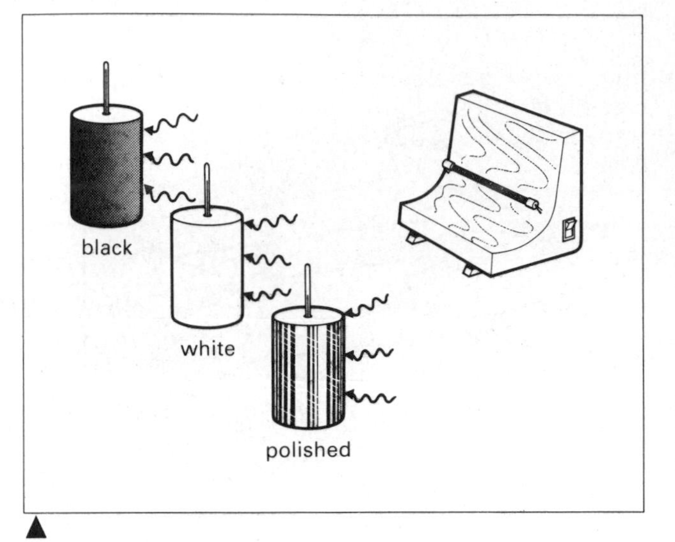

fig. 15.17 Comparing absorption powers of surfaces

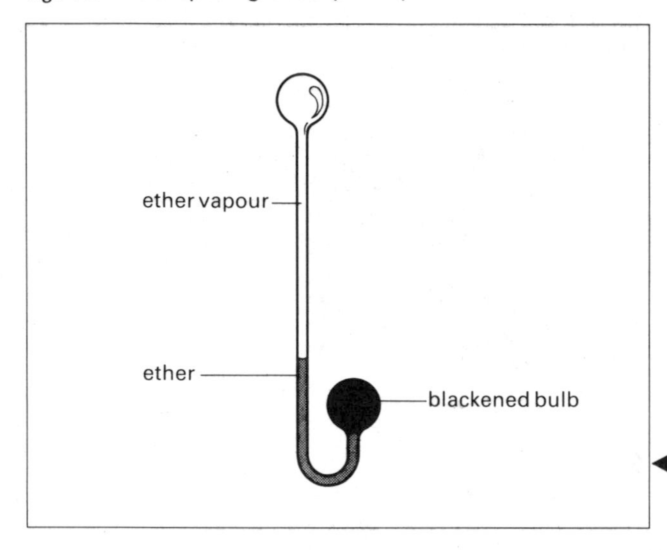

that a black surface is a better radiator than a polished one.

In order to find which surface absorbs heat radiation best, three tins filled with cold water are placed at equal distances from an electric radiator, figure 15.17. The temperature rises are noted in a given time. Or the temperatures of the tins could be noted at minute intervals and heating curves plotted. The black tin shows the greatest temperature rise and the polished tin the least.

Experiments such as these show that dull black surfaces absorb radiation better than shiny surfaces. Shiny surfaces are, of course, good reflectors and throw back the heat rather than absorb it. Black surfaces are also better radiators than polished ones. In fact all the observations on heat radiation can be summed up very neatly in the following way:

Good absorbers are good radiators
Poor absorbers are poor radiators

Detection of radiant heat

A thermometer with a blackened bulb will detect heat radiation falling on it. A more sensitive arrangement is the **ether thermoscope**, figure 15.18. The bottom bulb, which is blackened, is about half-full of coloured ether, and the space above the ether is full of ether vapour. When radiation falls on the bulb it is absorbed and warms the ether vapour which expands, forcing the liquid up the tube to the top bulb. A small quantity of radiation produces an easily visible effect.

fig. 15.18 Ether thermoscope

fig. 15.19 Thermopile and its use with Leslie's cube

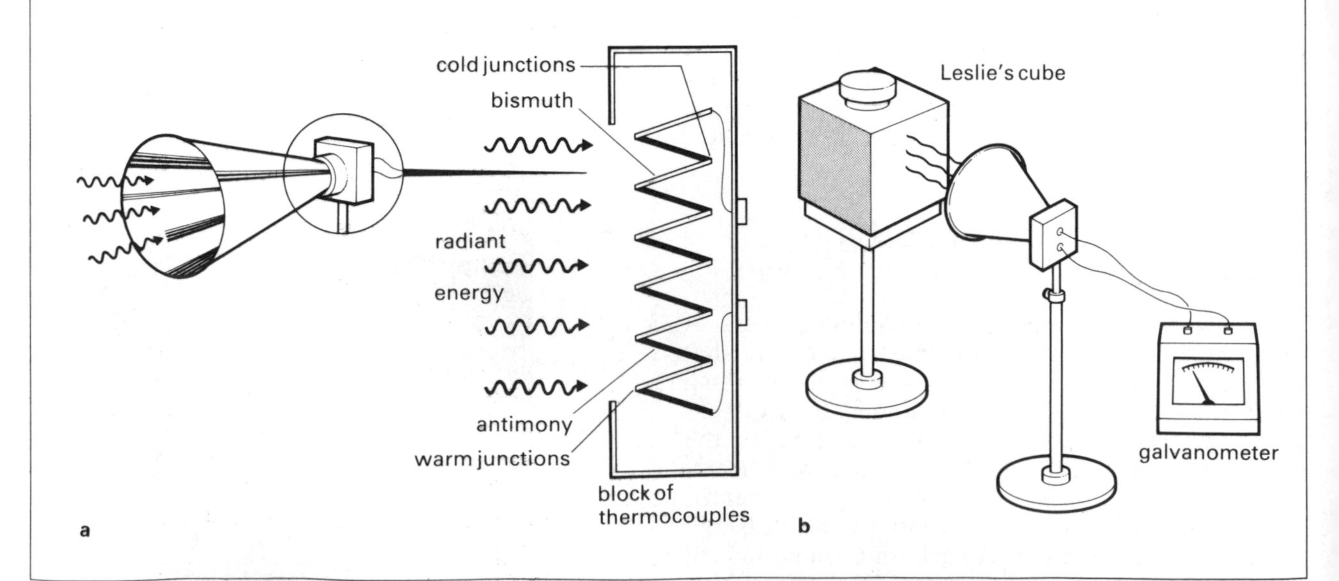

Another instrument used for the detection of radiation is the **thermopile**, which consists of a large number of thermocouples (see p. 98) connected in series so that their electromotive forces all add together, figure 15.19a. The thermopile is connected to a sensitive galvanometer.

The thermopile can be used with **Leslie's cube** to compare the radiating powers of different surfaces. Leslie's cube is a metal vessel with the four sides treated in different ways, e.g. dull black, shiny black, white, polished. The cube is filled with boiling water and the cone of the thermopile pointed, at a fixed distance away, at each surface in turn. The galvanometer reading gives a measure of the radiation emitted by each surface, figure 15.19b.

Wavelength and temperature

As the temperature of a body increases, the nature of the radiation it emits changes. The higher the temperature, the shorter the wavelength of the radiation emitted. This is shown by the colour changes of a metal as it is heated. At first it glows a dull red, then a bright red, then through orange until it is white hot, i.e. it is emitting white light. Solar radiation emitted at 6000 °C is of very short wavelength, figure 15.20.

In general, the shorter the wavelength the more penetrating the radiation is. Thus a sheet of glass will cut off the relatively long-wave heat radiation from a fire, but will allow the short-wave radiation from the sun to pass through. This has an important effect in the greenhouse.

fig. 15.20 Change of wavelength with temperature

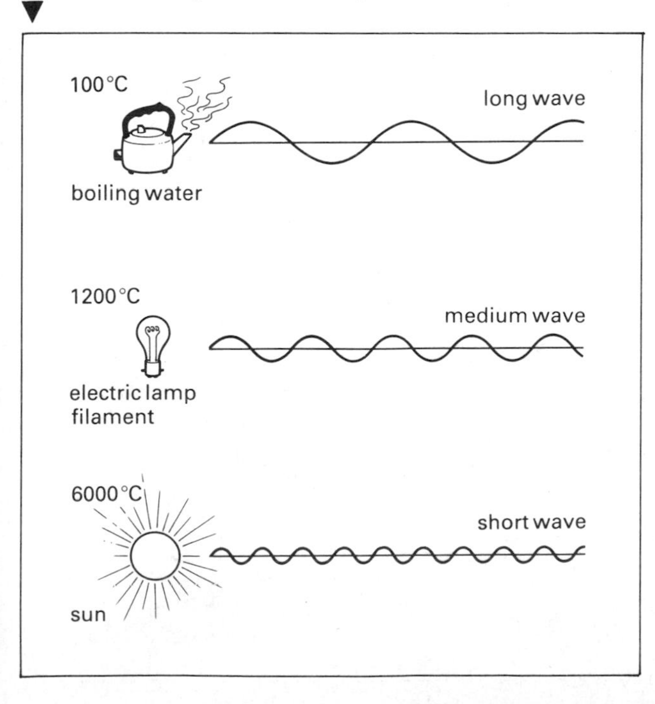

The greenhouse

The greenhouse, figure 15.21, acts as a heat trap because it allows the short-wave radiation from the sun through the glass. This radiation is absorbed by the contents of the greenhouse, which become warmer and emit their own radiation. But this is low-temperature, long-wave radiation which cannot penetrate the glass. Much of this radiation is re-radiated back into the greenhouse.

The atmosphere has a 'greenhouse' effect on the earth. The gases in the atmosphere allow most of the solar radiation to pass through, but absorb the terrestrial radiation and send it back to the earth.

This 'greenhouse' effect is increased by gases produced on earth, some naturally and some man-made. These gases include carbon dioxide from the burning of fuels – coal, oil and gas. The amount of carbon dioxide is also increasing because trees which use it for photosynthesis are being cut down. Other gases are oxides of nitrogen from fertilisers and methane from rubbish dumps. The propellant gases in aerosols (chlorofluorocarbons) and the gases produced by the breakdown of certain plastic containers are also increasing. The 'greenhouse' effect is necessary to keep the earth at a reasonable temperature, but if it increases the earth will get hotter and hotter, resulting in the melting of the arctic and antarctic ice caps and a resultant rise in the level of the oceans. There will also be a complete change in weather patterns.

fig. 15.21 A greenhouse acts as a heat trap

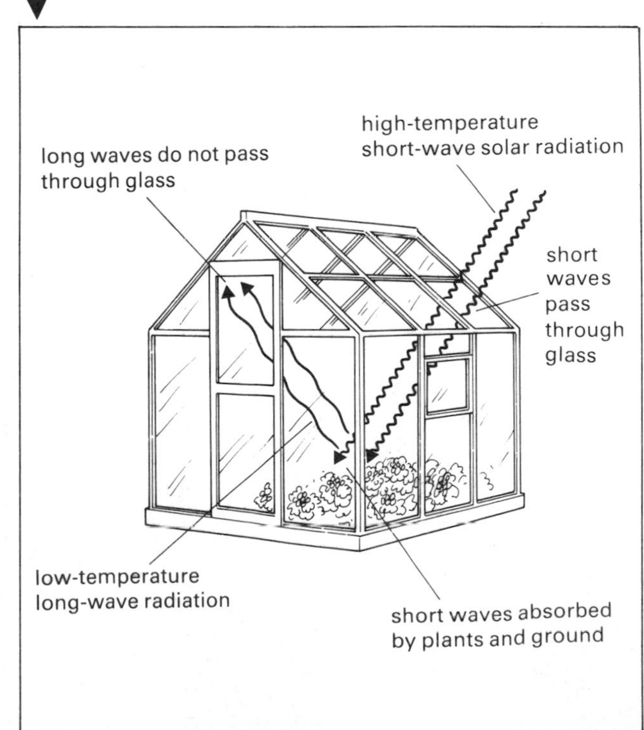

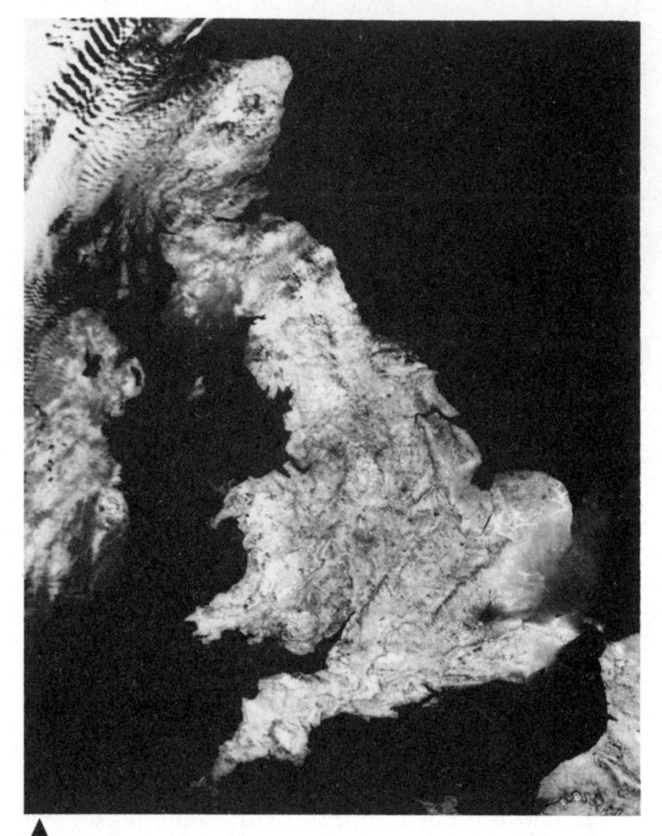

fig. 15.22 Photograph of the British Isles taken at night from a satellite, using the infrared radiation emitted by the ground

Infrared photography

It is approximately true to say that waves are scattered by particles which are about the same size as the wavelength. Water droplets forming cloud, mist and fog are larger than the wavelength of visible light, so visible light is scattered and we cannot see through clouds and fog. But infrared radiation has a wavelength larger than the water particles and so is not scattered so much. Using infrared sensitive films, we can take photographs through mist. Infrared photography can also be used at night, figure 15.22.

Solar radiation

The sun is an enormous radiator of energy. All our energy (with the exception of nuclear and tidal energy) comes to us or has come to us from the sun. The energy stored in fuels such as coal, oil and natural gas came from the sun as it was accumulated by plants and marine animals millions of years ago. The energy radiated by the sun in the form of heat and light is believed to come from the thermonuclear conversion of hydrogen into helium in the sun's interior. The surface temperature of the sun is about 6000 K and the internal temperature very much greater.

The amount of solar radiation reaching the earth is found by determining a quantity called the **solar constant**, which is defined as the quantity of energy per unit time falling on a surface of unit area placed at right angles to the sun's rays just outside the earth's atmosphere. The solar constant is measured at the earth's surface and allowance made for the passage of

fig. 15.23 Only about a half of the sun's radiation reaches the earth

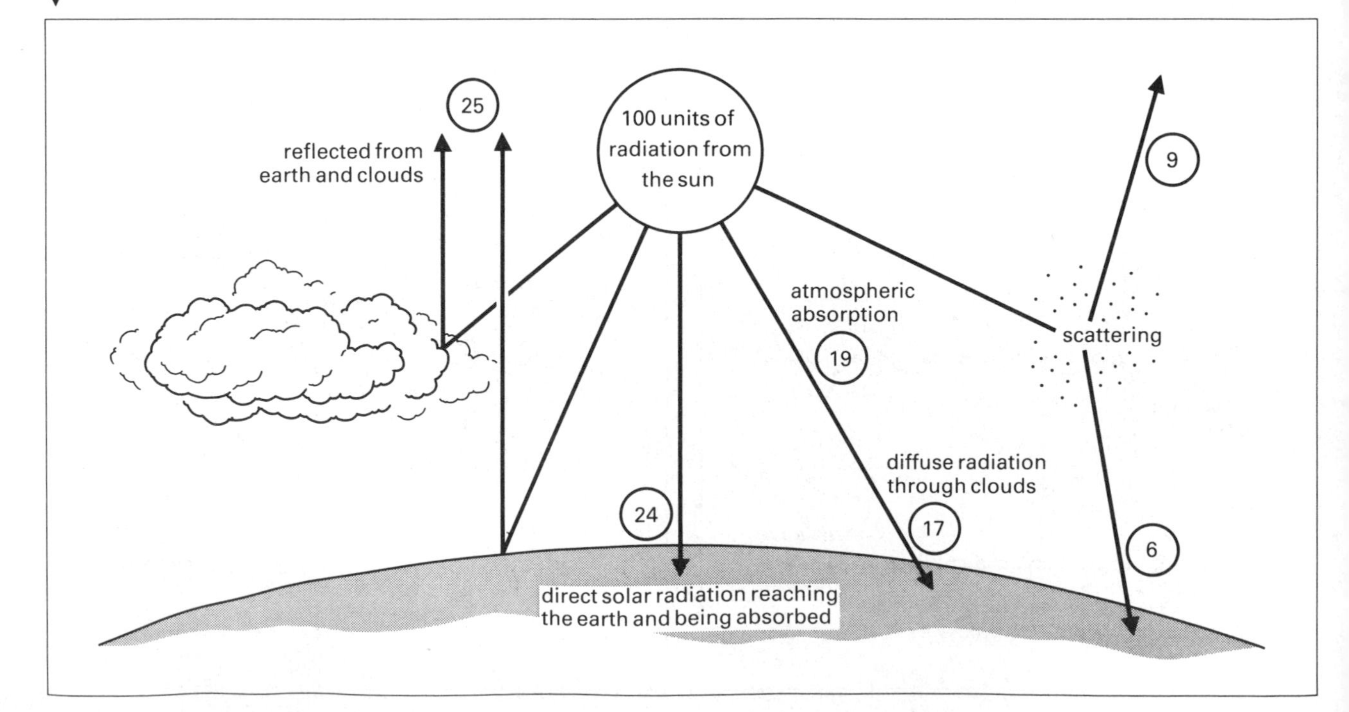

the radiation through the atmosphere and the clouds. In figure 15.23, the '100 units of radiation' represents the solar constant. It will be seen that about half (47%) of the radiation reaches the earth.

The solar constant is measured using an instrument called a **pyroheliometer**. One form of this consists of a thick copper disc mounted at one end of a tube, with a thin disc of equal diameter at the other end, figure 15.24. When the shadow of the copper disc just covers the lower disc, the surface of the copper disc is at right angles to the sun's rays. The disc is blackened so that it is a good absorber. The temperature of the disc is taken at the beginning and end of a known time interval. From the mass of the disc, the specific heat capacity and the rise in temperature, the quantity of heat absorbed can be calculated.

The value of the solar constant is 1.35×10^3 J per m^2 per second or 1.35×10^3 W/m^2. Assuming that about half this energy on an average reaches the earth, this gives us an energy source of about 650 W/m^2. The power reaching an area the size of a football pitch is 5 megawatts (5 million watts), so an enormous amount of energy is available, but at present it is difficult to make good use of it. We need to use large areas as collectors. The obvious places are deserts where there is plenty of space and sun, but the energy would have to be transmitted to where it could be put to practical use. Then, since the sun is not shining all the time, the energy would have to be stored. A theoretical problem is that good absorbers are also good radiators, and a good absorbing surface would lose heat as its temperature rose. The sun's energy is used in solar batteries (see p. 259), in solar heating panels for houses, figure 15.25, and in solar furnaces, figure 15.26.

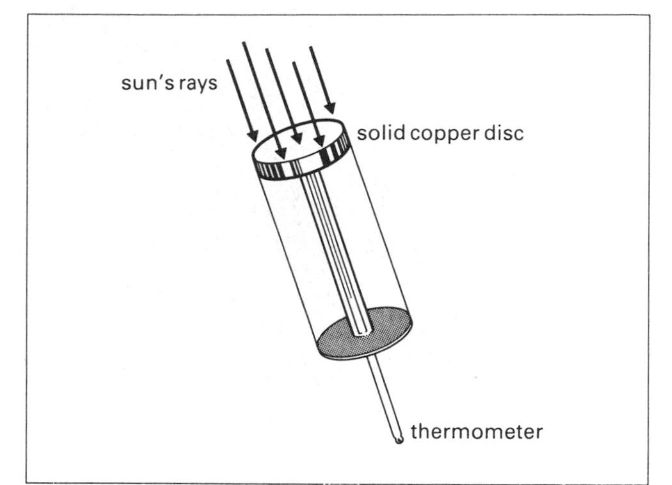

fig. 15.24 Pyroheliometer

fig. 15.25 Solar heating panels

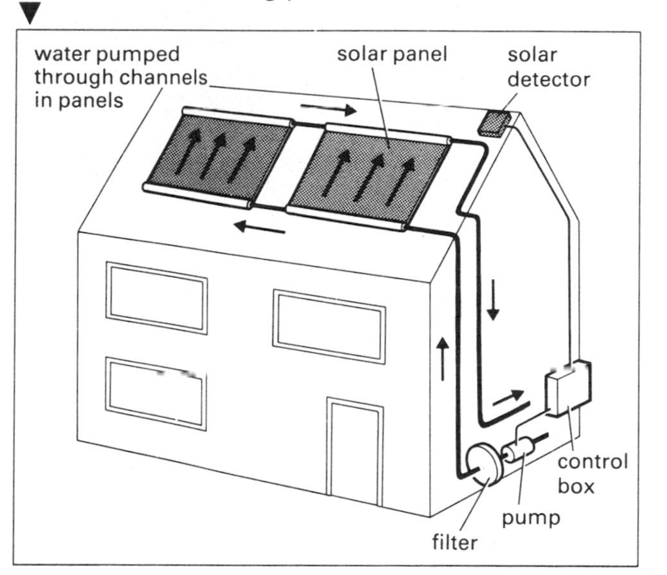

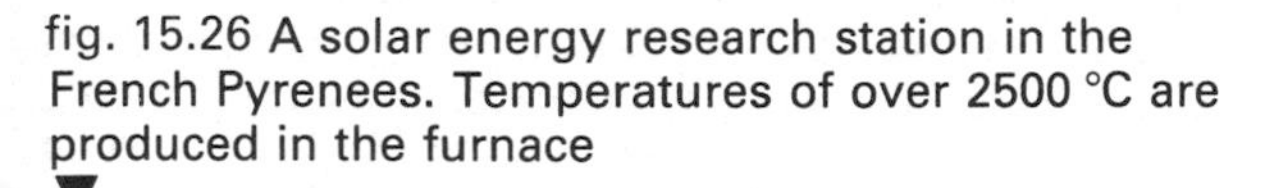
fig. 15.26 A solar energy research station in the French Pyrenees. Temperatures of over 2500 °C are produced in the furnace

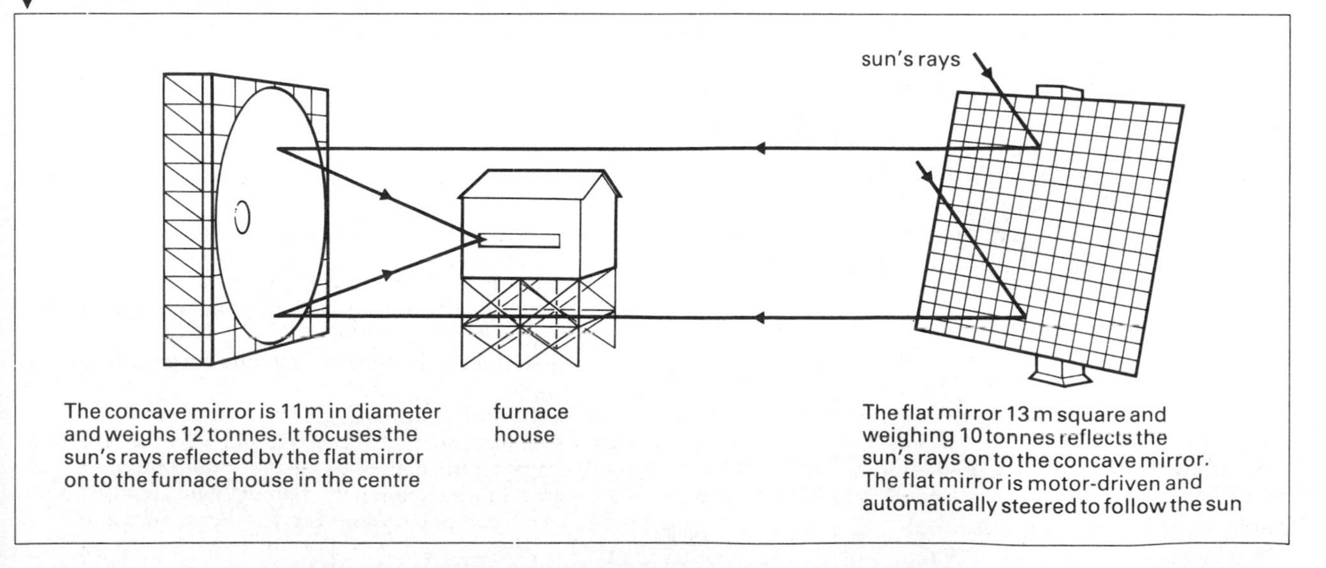

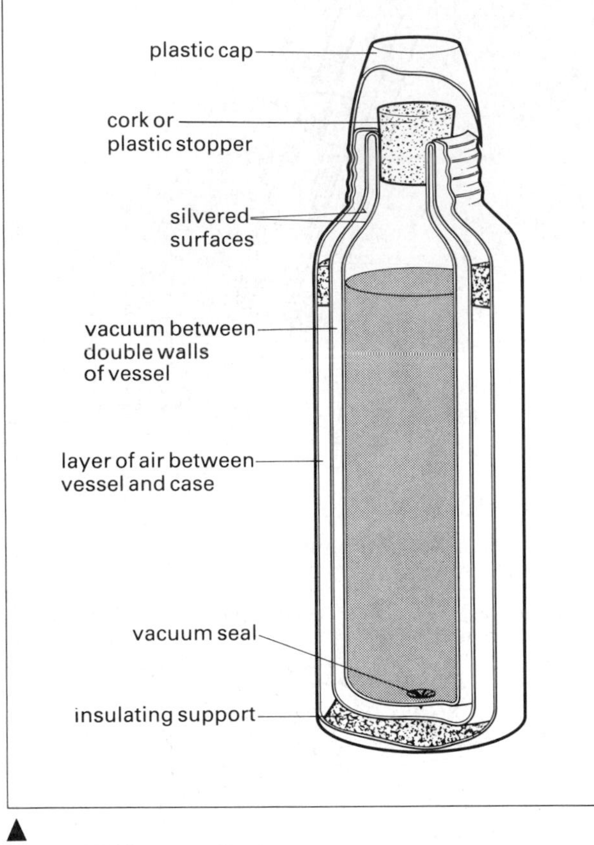

▲ fig. 15.27 Vacuum flask

The vacuum flask

The vacuum flask, figure 15.27, or Dewar flask, was designed about 1900 by Sir James Dewar to store the liquid gases he was producing. Since liquid air boils at about –200 °C, it was necessary to prevent heat reaching it. The vacuum flask is designed to keep materials at a constant temperature, either hotter or colder than their surroundings.

Conduction is reduced by using a glass vessel, a cork and a vacuum, all of which are poor conductors of heat.

Convection is prevented by the vacuum and by the layer of still air between the vessel and the case.

Radiation is reduced by silvering the inner walls of the double-walled glass vessel. Silvered surfaces are poor radiators.

Cooling by *evaporation* is prevented by the stopper, although this must not be used if a high pressure is liable to build up inside the flask.

EXERCISE 15

In questions 1–4 select the most suitable answer.

1 Of the following, the best example of heat transfer by conduction is
- **A** from an electric hotplate to the contents of a saucepan
- **B** from the sun to the earth
- **C** from the boiler to the hot cylinder of a domestic hot water system.
- **D** through the glass into a greenhouse
- **E** from an electric fire to a person sitting in a room

2 Convection takes place
- **A** only in solids
- **B** only in liquids
- **C** only in gases
- **D** in solids and liquids
- **E** in liquids and gases

3 Metal water pipes have been known to crack during severe frost because
- **A** metal is a good conductor of heat
- **B** water is a poor conductor of heat
- **C** metal contracts rapidly in these conditions and cracks
- **D** the volume of unit mass of water is greater than that of unit mass of ice at 0 °C
- **E** the volume of unit mass of ice is greater than that of unit mass of water at 0 °C

4 On a cold day, the metal part of a spade feels colder than the wooden handle. This is because
- **A** the wood is a better conductor of heat than the metal
- **B** the metal is at a lower temperature than the wood
- **C** the metal is a better conductor of heat than the wood
- **D** heat is conducted from the metal to the wood
- **E** the wood absorbs more heat from the air than the metal

5 Explain the following in terms of conduction, convection or radiation of heat.
- **a** A wooden spoon and not a metal one is used for jam making.
- **b** A woollen sweater keeps you warm.
- **c** Thatched houses are cool in summer and warm in winter.
- **d** Glider pilots find upward moving currents of air over towns.
- **e** Birds fluff out their feathers in cold weather.
- **f** The blackboard in a classroom stays at the same temperature as the rest of the room, although it absorbs more radiant heat than the light coloured walls.
- **g** Expanded polystyrene is a good heat insulator.

6 Two test tubes, one with its outside surface blackened, the other wrapped in aluminium foil, are filled with boiling water, stoppered and then left to cool, their temperatures being taken at regular intervals.

A sketch graph of their cooling curves is shown in figure 15.28. Room temperature is 15 °C.

a Which of the two curves shown, 1 or 2, shows the greatest initial rate of cooling?

b To which tube does the curve chosen in part **a** apply, the blackened one or the silvered one? Give a reason for your answer.

c Draw a third line *on the graph* to show what the cooling curve might look like for a blackened tube, *without a stopper*.

d What temperature would all three of the test tubes reach if left to cool for a long time?

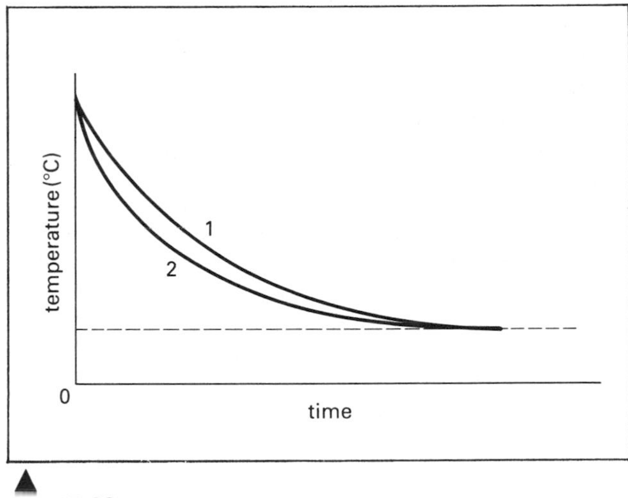

fig. 15.28

7 Figure 15.29 shows a vessel which is used to prevent hot liquids from going cold. It can also be used to keep cold liquids cool in hot weather.

a What is the name of this vessel?

b (i) What is the part labelled A usually made from? (ii) Why is this material chosen for this part?

c The walls of this vessel are made of thin glass but its inner surfaces, B, are specially treated. (i) What is special about these surfaces? (ii) Why are these surfaces treated in this way?

d (i) What is contained in the space C? (ii) What is the purpose of the space C?

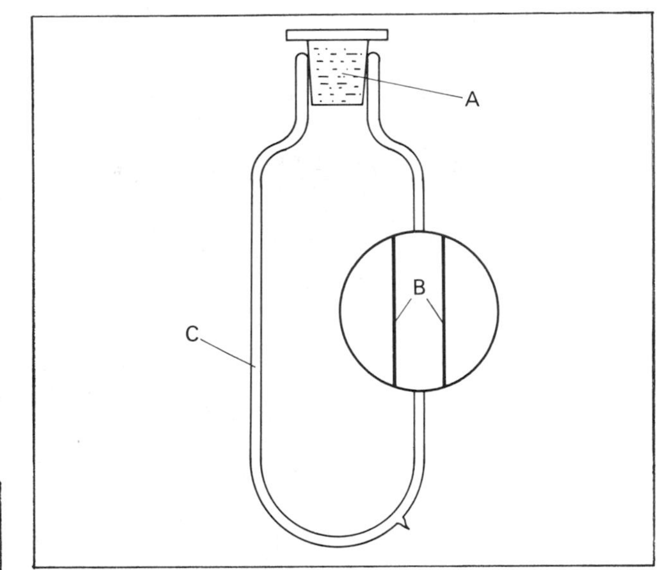

fig. 15.29

8 A student uses a small electrical heater to raise the temperature of a 1 kg block of aluminium. Every third minute he records that temperature with a thermometer inserted in a hole in the aluminium. The results are given in the table.

Time (min	Temperature (°C)	Time (min.)	Temperature (°C)
0	18.0	12	61.0
3	30.0	15	65.0
6	42.0	18	66.5
9	54.0	21	66.5

a (i) Draw a graph of temperature (vertical axis) against time for the block. (ii) Use your graph to find the likely temperature of the block 7.5 minutes after the start of the experiment.

b Using the same sheet of graph paper and the same set of axes, draw a sketch graph showing his likely results if he had used the same apparatus but had painted the aluminium black (label the sketch graph 'Second experiment').

c In another similar experiment using a 30 W heater the temperature of the 1 kg aluminium block rose from 18 °C to 23 °C in 150 seconds. (i) How much energy was supplied in that time? (ii) What was the rise in temperature of the block? (iii) Calculate the specific heat capacity of aluminium from this information.

16 ATOMS, MOLECULES AND THE KINETIC MOLECULAR THEORY

From careful observation of the behaviour of matter and from the results of many experiments, scientists believe that all materials are made up of a very large number of very small particles called **atoms**. In 1808 John Dalton published his ideas about atoms from observations made on the ways in which substances combine together chemically. This atomic theory remains one of the most important concepts in chemistry.

An atom is the smallest part of an element which can take part in a chemical change

A molecule is the smallest part of an element or compound which can exist by itself

In some cases, e.g. the gas neon, the molecule consists of a single atom. In other cases, e.g. oxygen, there are two atoms in the molecule. A molecule of carbon dioxide consists of an atom of carbon combined with two atoms of oxygen. For our present purpose it does not really matter whether we are dealing with atoms or molecules. The important thing to realise is that matter is made up of *particles*. For convenience we shall speak of these particles as molecules.

Later on we shall have to consider the fact that the atoms themselves are composed of still smaller sub-atomic particles such as protons, neutrons and electrons. A knowledge of the structure of atoms is important in the study of electricity and electronics.

fig. 16.1 How small are atoms?

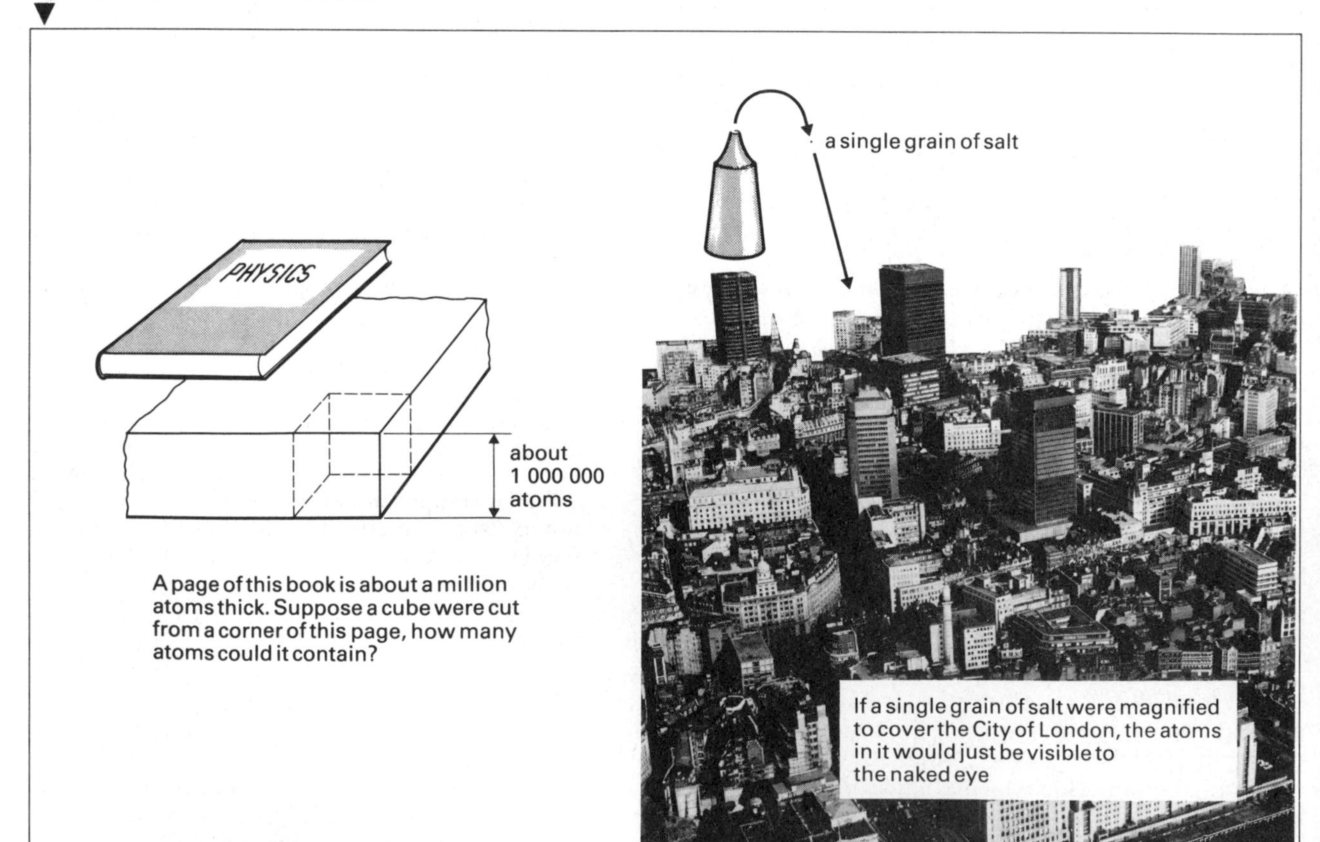

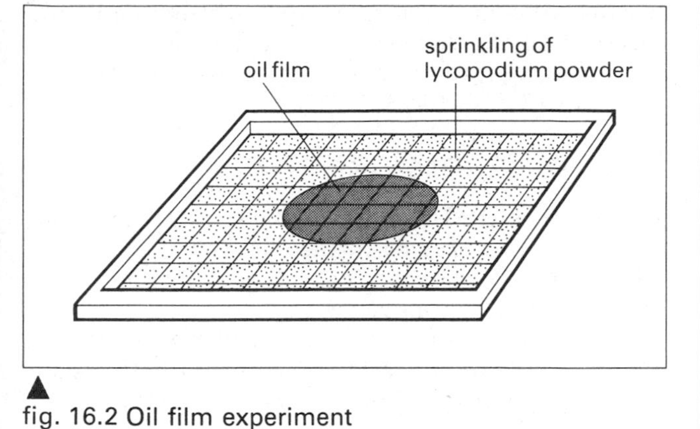

fig. 16.2 Oil film experiment

The size of molecules

Although it is not possible to see atoms and molecules (except with the electron microscope), it is nevertheless possible to make some estimate of their size.

If a single drop of oil is allowed to fall on a clean water surface, the oil spreads out in a very thin film. The size of the oil film can be seen clearly if the water surface has a fine powder (such as lycopodium) sprinkled on it, figure 16.2. The volume of the drop is the same as the volume of the film, which is equal to the area of film multiplied by the thickness of the film. So we can find the thickness of the film:

$$\text{thickness of film} = \frac{\text{volume of oil in the drop}}{\text{area of film}}$$

Now the film must be at least one molecule thick. We can say that the molecule size cannot be greater than the thickness.

The volume of the drop can be found most easily by letting fifty drops fall from a 1 cm^3 pipette and noting the total volume of liquid which has emerged. The pipette is graduated in hundredths of a cm^3, so that an accurate estimate of drop size can be made. The area of the film can be determined using a shallow tray with a glass bottom. A picture frame which has been made watertight can be used. Under the tray is a paper with a framework of centimetre squares so that the area can be found by counting squares. Sometimes the film forms a circular patch the diameter of which can be measured. Olive oil may be used, or a solution of oleic acid (1 part in 100) in alcohol. The alcohol dissolves in the water, leaving a film of oleic acid on the surface.

Measurements of this kind put the size of the molecule at about 10^{-9} m. Both olive oil and oleic acid are long, thin molecules which set themselves perpendicular to the water surface, so that it is really the length of the molecule which is measured. Since the molecules are about twenty atoms long, an estimate of atomic size can also be made.

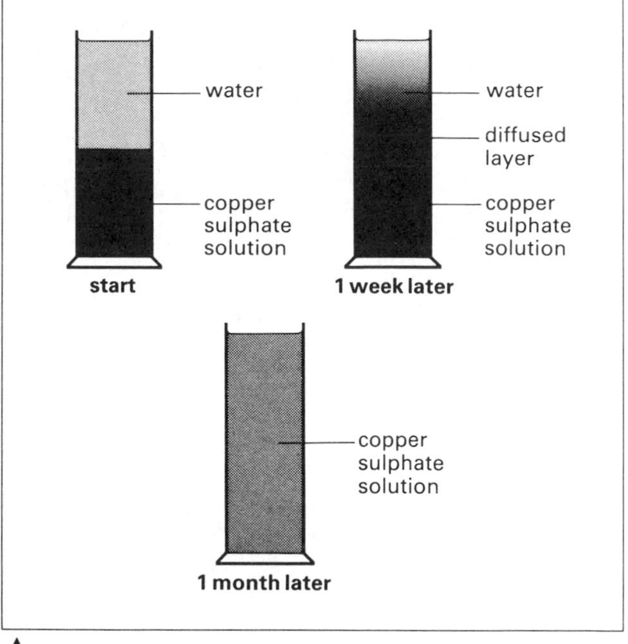

fig. 16.3 Diffusion of liquids

The movement of molecules

Once the idea of atoms and molecules had been well established at the beginning of the nineteenth century, scientists further believed that the molecules must be in motion. This idea was expressed in the **kinetic molecular theory**.

One phenomenon which supports the idea of moving molecules is **diffusion**. Diffusion is the spreading of a substance into a space which it has not previously occupied.

A cylinder is half-filled with a saturated solution of copper sulphate, figure 16.3. The top half of the cylinder is then carefully filled up with pure water so that the two layers do not mix. If the cylinder is left undisturbed for some time, the blue colour slowly spreads upwards, showing that movement of molecules is taking place.

Diffusion of gases is more obvious than that of liquids. If a bottle of strong perfume is opened, the scent soon becomes detectable all over the room. The scent molecules have spread rapidly through the air molecules.

The diffusion of gases can be dramatically demonstrated using bromine vapour, as shown in figure 16.4. Bromine is a brown liquid which evaporates readily to give an easily visible brown vapour. Bromine is a dangerous liquid producing burns on the skin, and the vapour is very irritating to nose and eyes, so it must be handled with extreme care by an experienced experimenter. Bromine can be bought in thin-walled glass bottles, and this is the safest way of handling it. Using the apparatus shown, the bottle of bromine is

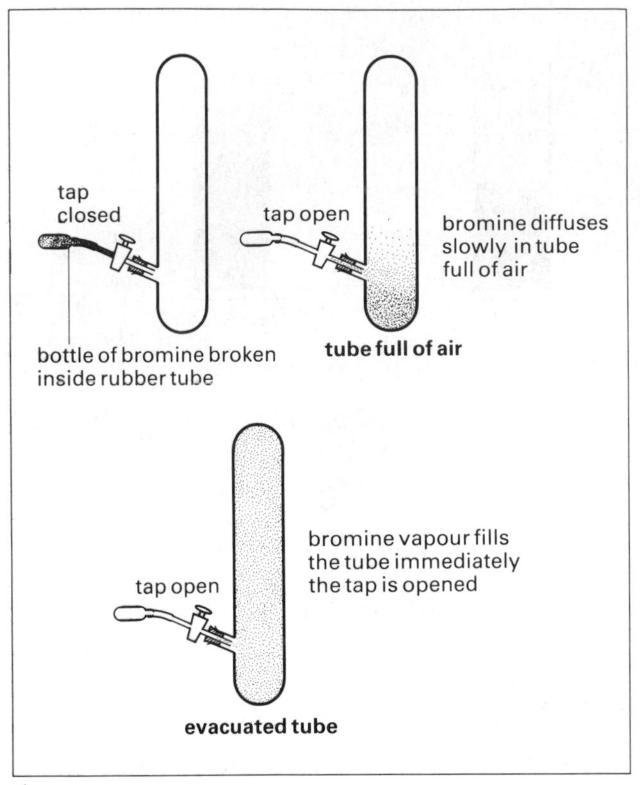

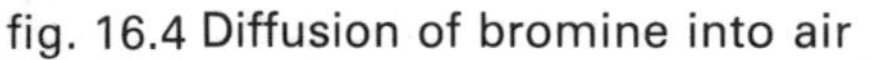

fig. 16.4 Diffusion of bromine into air

broken in the side tube and then the tap to the main tube is opened. The bromine vapour diffuses into the tube full of air and will have spread to the top of the tube in few hours. In the case of the evacuated tube the bromine vapour fills the tube instantaneously when the tap is opened, showing that in the absence of air molecules to impede their motion the bromine molecules move with great speed of the order of 500 m/s.

Gases will diffuse through the small holes in porous materials such as unglazed pottery. The air in the porous pot is diffusing out, and the carbon dioxide surrounding the pot is diffusing into the pot, figure 16.5. The movement of the water in the manometer shows that the pressure inside the pot is falling because the air molecules are diffusing outwards faster than the heavier carbon dioxide molecules are diffusing into the pot. The denser the gas, the more slowly it diffuses.

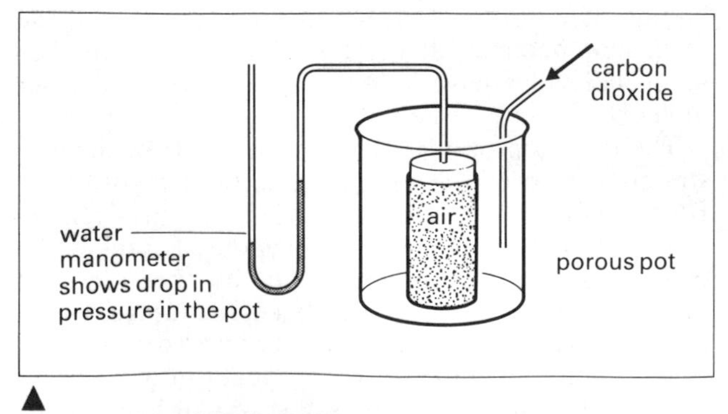

fig. 16.5 Gas diffusion through a porous material

Brownian motion

In 1827 Robert Brown, a botanist, was observing some pollen grains suspended in water under the microscope. He noticed that the grains were in constant and irregular movement.

We can repeat the experiment with a suspension of graphite particles in water, or smoke particles in air. The suspension of particles is placed on the microscope stage in a small glass cell and illuminated with a concentrated beam of light from the side, figure 16.6. The suspended particles scatter the light into the microscope and they appear as tiny bright specks of light which move in a continuous and random way. The explanation is that the movement results from the continuous bombardment of the particles by the moving liquid or air molecules which surround them. In order to have this effect on the particles, which are about a thousand times more massive than the molecules, the latter must be moving very quickly.

The observation of Brownian motion is direct experimental evidence that the molecules of liquids and gases are in constant random motion.

fig. 16.6 Observing Brownian motion

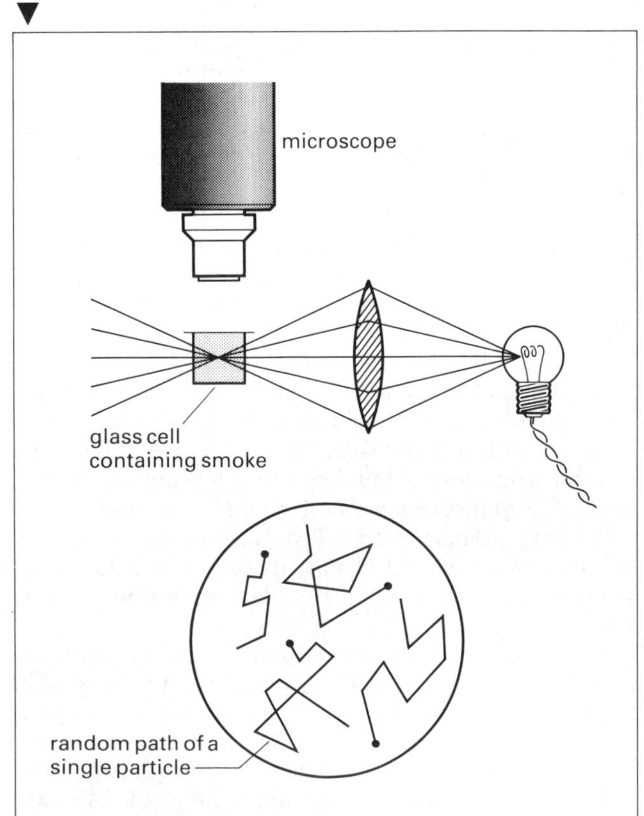

Solids, liquids and gases

The more obvious properties of solids, liquids and gases can be explained in terms of their molecular structure.

The fact that solids have a definite shape, figure 16.7a, suggests that the molecules are arranged in a regular pattern and are held pretty firmly in place. The forces of attraction which hold the molecules of solids and liquids together are known as **cohesive** forces. In solids the cohesive forces are very large. In order to overcome the cohesive forces between molecules of a bar of iron 1 cm^2 in cross-section, you would have to apply a pull of some 10 000 kilogram weight. These large cohesive forces give solids their hardness and firmness, figure 16.8a.

In a liquid the cohesive forces are less than in a solid, figure 16.8b. They are large enough to keep the molecules within a given volume, but are not sufficient to overcome the pull of gravity downwards. A liquid cannot 'stand up for itself'. In a gas the molecules are so far apart that there are no cohesive forces between them, figure 16.8c.

Molecules are always in motion. In a solid, the molecules vibrate about a given position. In a liquid, the molecules can move freely within the volume of the liquid. There are constant collisions between the molecules in liquids and gases, so that they are all the time changing their speeds and directions of travel. The average speed at which the molecules move or vibrate depends upon the temperature. When heat energy is supplied to matter, the average speed of the molecules increases and the distances through which the molecules move between collisions increase (except in the case of gases at constant volume). Thus, on being heated, substances will expand, as we know from observing simple experiments. When a solid is heated strongly (that is, supplied with a large amount of heat energy), the movements of the molecules may become so violent that the cohesive forces are partly overcome and the solid melts into a liquid. Further heating overcomes the cohesive forces completely, and the liquid changes to a vapour.

Evaporation

Evaporation is the change from the liquid to the vapour state. It takes place at all temperatures at the surface of the liquid. The molecules of a liquid are moving in random directions and with random speeds. Some are moving very slowly, some very quickly, with most of the molecules moving at about the average speed. Owing to the constant collisions of the molecules with each other and with the walls of the containing vessel, their speeds are continually changing. Some molecules moving upwards near the surface may have sufficient kinetic energy to escape completely from the liquid and so change to the vapour state, figure 16.9.

It is always the most energetic molecules which leave the liquid during evaporation, so the total energy of the liquid is reduced, i.e. the temperature drops. Evaporation causes cooling (see p. 114). Some escaping molecules may strike air molecules and be repelled back into the liquid.

Factors affecting the rate of evaporation

Any change which makes more molecules escape from the liquid surface in a given time will increase the rate of evaporation.

fig. 16.7 The three states of matter: solid, liquid and gas

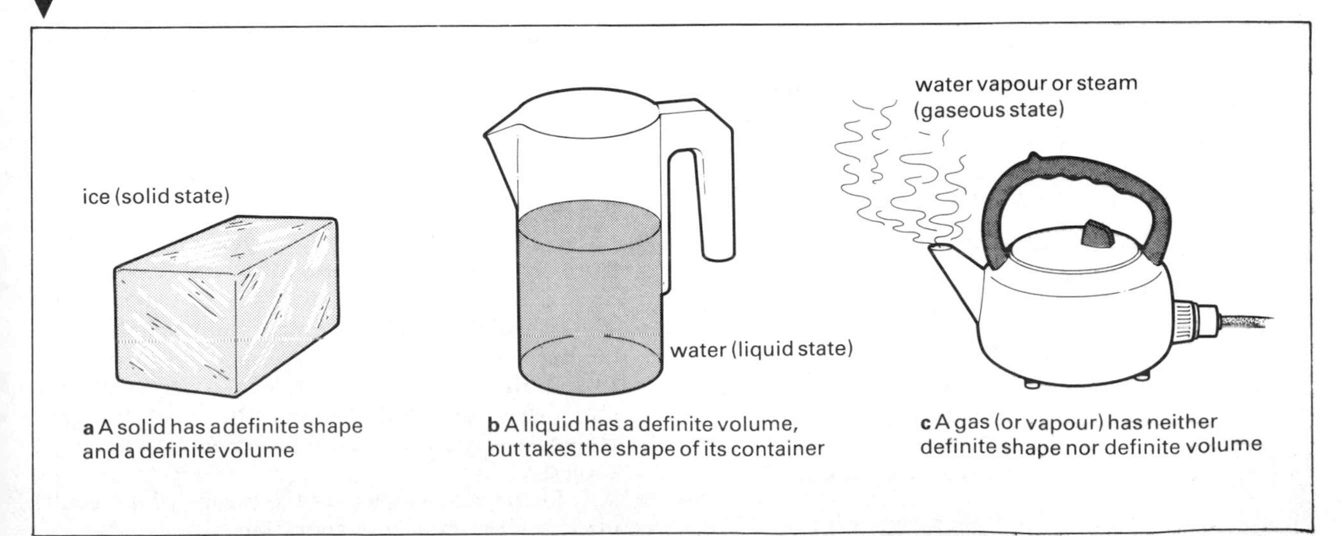

a A solid has a definite shape and a definite volume

b A liquid has a definite volume, but takes the shape of its container

c A gas (or vapour) has neither definite shape nor definite volume

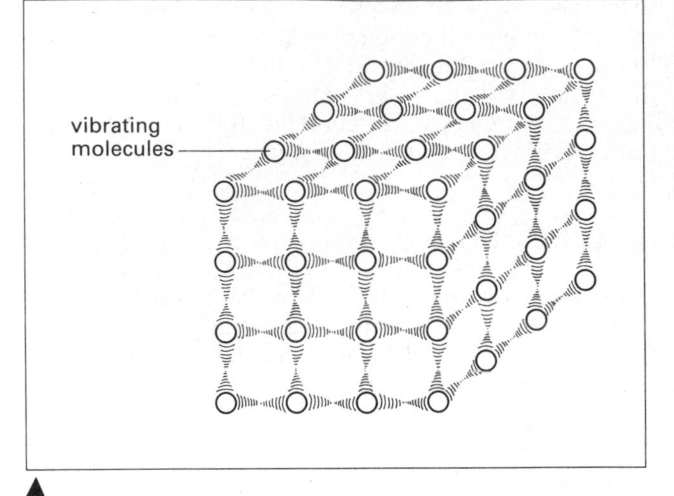

fig. 16.8a Molecular arrangement in a solid. The molecules can vibrate, but cannot move from their positions. It is difficult to compress or extend a solid

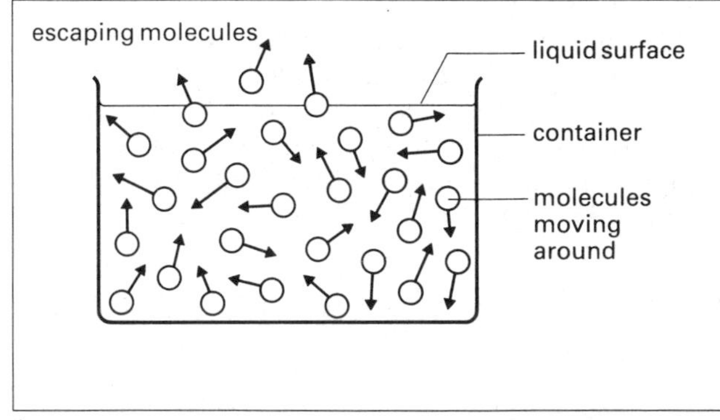

fig. 16.8b In a liquid the molecules are free to move about in all directions, but are still fairly closely packed. A few of the faster moving molecules can escape through the surface to form vapour

fig. 16.8c A gas has a very open arrangement of molecules which can move freely in all directions and 'fill' the vessel or space containing the gas

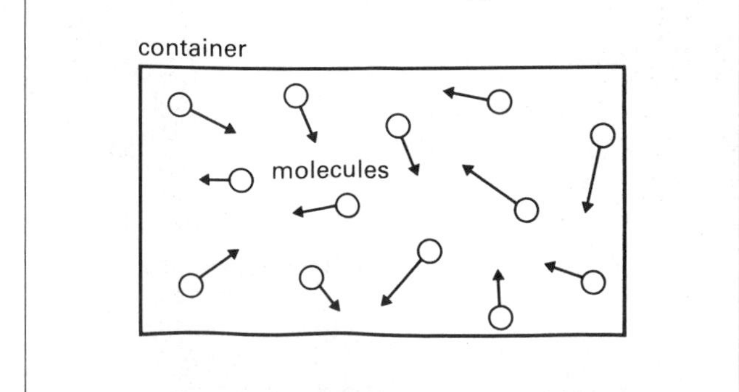

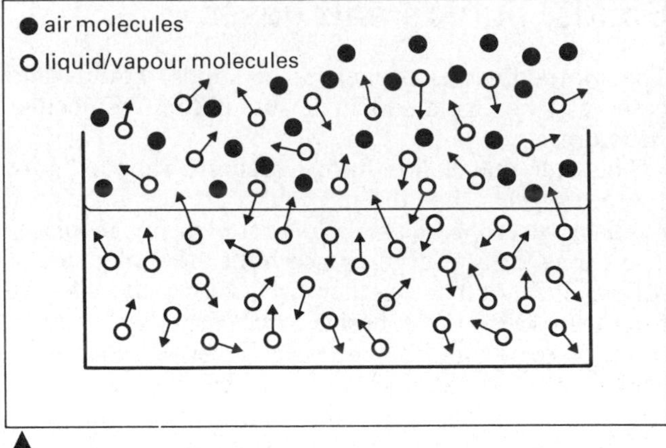

fig. 16.9 Evaporation

- The larger the surface area of the liquid, the easier it is for more molecules to escape. Large, shallow puddles dry up very quickly.
- The higher the temperature, the higher is the average kinetic energy of the molecules, so more molecules will have sufficient energy to escape.
- If a current of air passes over the surface of the liquid, molecules which escape will be swept away before they can return to the liquid, and the net number of molecules escaping is increased.
- Different liquids evaporate at different rates because their cohesive forces are different. Liquids with low cohesive forces between their molecules evaporate quickly.

Summary of kinetic theory ideas

- Solids, liquids and gases are made up of molecules.
- The cohesive forces are large in solids, much smaller in liquids, and negligible in gases.
- The molecules are in constant motion.
- Molecules in solids vibrate about fixed positions. In liquids they move randomly within a fixed volume. In gases they move freely in all directions.
- When heat energy is supplied, the kinetic energy of the molecules increases; since the mass of the molecule is constant, the speed must increase. This increase in energy usually produces expansion.
- For a change of state to take place, the cohesive forces have to be overcome; the energy needed to do this is the latent heat of fusion or evaporation.
- The collision of molecules with the walls of a containing vessel produces pressure.
- Evaporation is caused by molecules escaping from the surface of a liquid. Since the most energetic molecules escape, the liquid loses energy and so is cooled.
- Any factor which increases the escape of molecules increases the rate of evaporation.

EXERCISE 16

In questions 1–4 select the most suitable answer.

1 Which of the following describes particles in a solid?
A close together and stationary
B close together and vibrating
C close together and moving around at random
D far apart and stationary
E far apart and moving around at random

2 Which one of the following statements is correct?
A The molecules in a solid are all at rest.
B The molecules in a solid are freely moving in all directions.
C The molecules in a liquid are on average much further apart than those in a solid.
D The molecules in a liquid move with limited freedom.
E There are no attractive forces between molecules in a liquid.

3 Gas molecules enclosed in a sealed container exert a pressure by
A vibrating about a fixed position
B hitting and rebounding from the container wall
C hitting and rebounding from each other
D spinning about their centres
E repelling one another strongly

4 When pollen grains suspended in a liquid are observed under a microscope they are seen to be in continuous random motion. This is due to
A convection currents in the liquid
B small air bubbles in the liquid
C the acceleration of free fall
D the molecules of the liquid colliding with the pollen grains
E the pollen grains repelling each other

5 Figure 16.20 shows a carefully prepared tube. Layer S is colourless sugar solution, layer C blue copper sulphate solution and layer W is water. After a short while the boundaries of the copper sulphate become blurred.
a Explain what is happening.
b What is the name of the process?
c What will be the eventual appearance of the contents of the tube?
d Give an example where this process occurs in air.

fig. 16.20 fig. 16.21

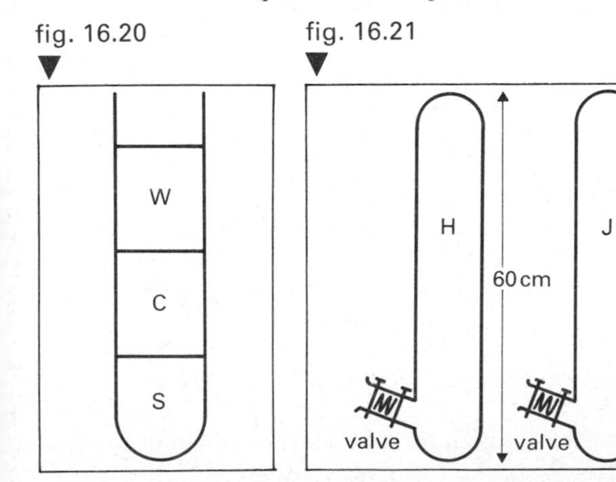

6 Figure 16.21 shows two identical glass tubes. Tube H contains air at normal pressure and tube J is evacuated. Both are at room temperature.
a When some bromine vapour (a brown coloured gas) is inserted through the valve on tube H several minutes pass before it can be seen at the top of the tube. (i) Explain how bromine gas, which is denser than air, reaches the top of the tube. [*Hint*: use your ideas on the kinetic theory of matter.] (ii) If the air in H was heated, and then the bromine inserted, would it reach the top more quickly or more slowly than before? Explain your answer.
b Explain what would happen if some bromine was inserted in tube J.
c (i) When a child's balloon is filled with hydrogen it becomes noticeably deflated within two hours, even though the neck is properly tied. Why is this? (ii) Explain why the same balloon would have stayed inflated longer if filled with air.

7 Figure 16.22 shows apparatus that can be used to demonstrate Brownian movement.
a What is inside the small container besides the smoke particles?
b Why is light shone into the small container as shown in the diagram?
c What is the purpose of the thin glass plate?
d Sketch what would be seen through the microscope.
e Sketch the path of a single smoke particle.
f Explain as clearly as you can (i) why the smoke particle moves, (ii) why it follows the path you have drawn.
g What scientific information does Brownian motion provide?
h Describe any other experiment you have seen or read about which supports the conclusion reached in part g.
i What happens to the molecules of a gas when the gas is (i) heated? (ii) compressed?

fig. 16.22

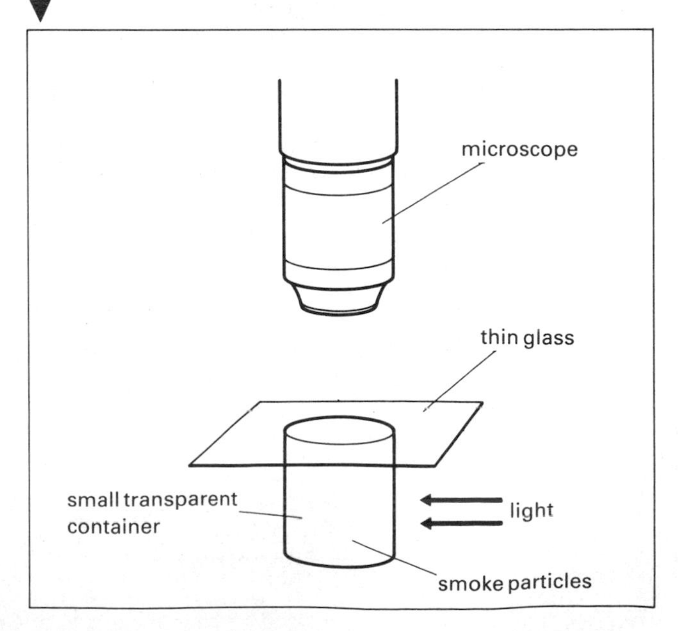

8 Figure 16.23 shows a beaker of water at 80 °C inside a bell jar containing air and connected to a vacuum pump.

a What happens to the number of molecules in the air in the bell jar when the vacuum pump is switched on?

b If the pump is left running what happens to the water in the beaker?

c Use your knowledge of the kinetic theory to explain your answer to **b**.

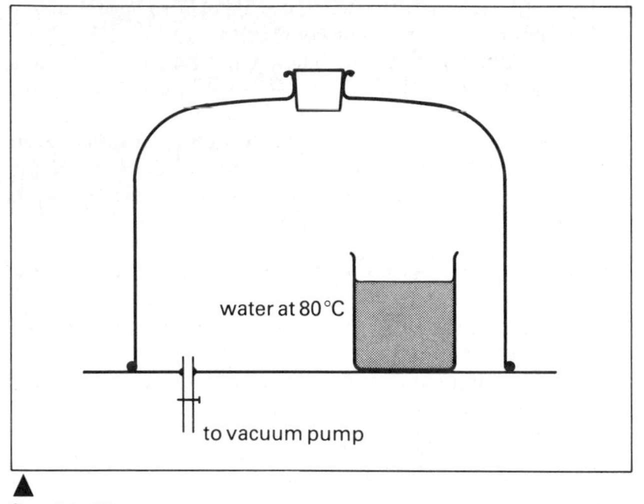

fig. 16.23

9 a Give *three* factors which affect the rate at which a liquid evaporates.

b How would you demonstrate *one* of these factors experimentally?

c Figure 16.24a shows air being blown down a tube into a beaker of a liquid, which evaporates readily. What happens to the reading on the thermometer?

d Why does this happen?

e What is the name given to the heat which a liquid requires to cause it to evaporate?

f Explain evaporation in terms of the movement of molecules.

g What happens to the molecules in a solid when it is heated?

h Figure 16.24b shows the rear window of a car. While the demister was heating the inside of the window and the outside was being sprayed with de-icing fluid (which evaporates rapidly), the window broke. Explain why this happened, using some of the principles you have explained in other parts of this question.

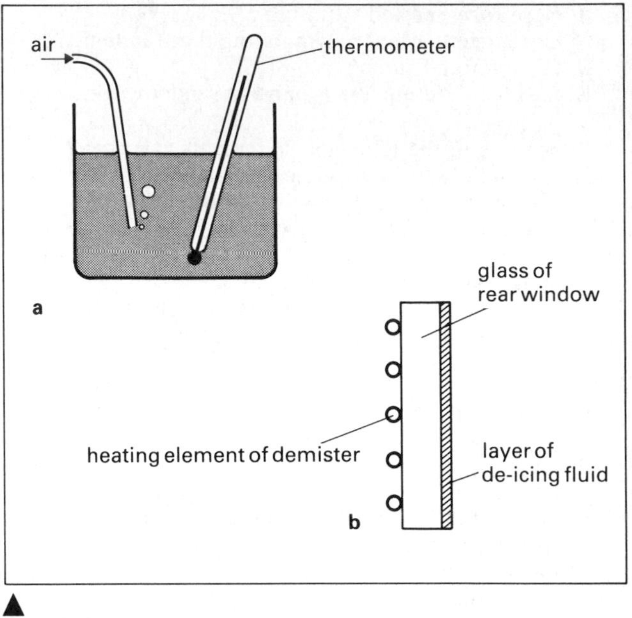

fig. 16.24

10 a What are the two basic assumptions of the kinetic theory of matter?

b How does the observation of a smoke cell help to confirm the kinetic theory of matter?

c How does the diffusion of gases also help to confirm the kinetic theory of matter? Give a particular example of diffusion in your answer.

d Air enclosed in a flask attached to a bourdon gauge shows a pressure reading due to the collision of gas molecules with the wall inside the apparatus. (i) What would happen to the pressure if the gas molecules lost some speed after each collision with the wall? Give a reason for your answer. (ii) What would happen to the air molecules inside the flask if they were heated and how would this affect the pressure?

e Explain in terms of the kinetic theory why the evaporation of a liquid causes cooling.

17 THE BEHAVIOUR OF GASES

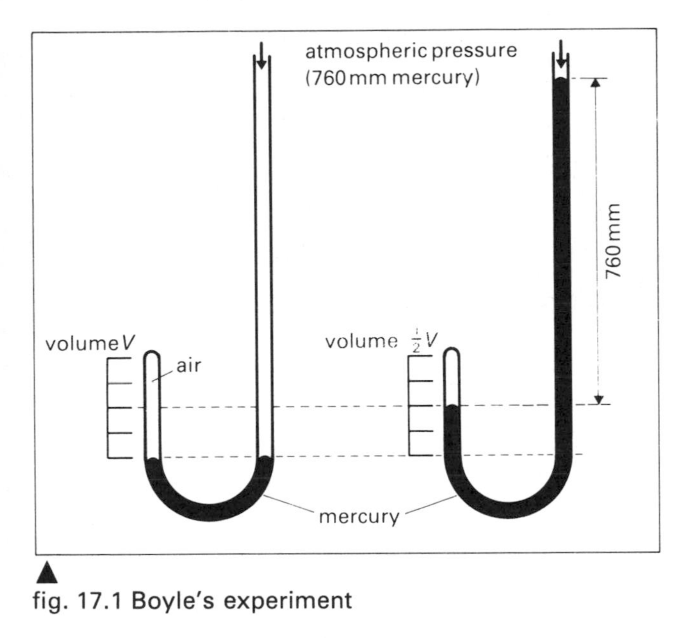

fig. 17.1 Boyle's experiment

fig. 17.2 Boyle's law apparatus

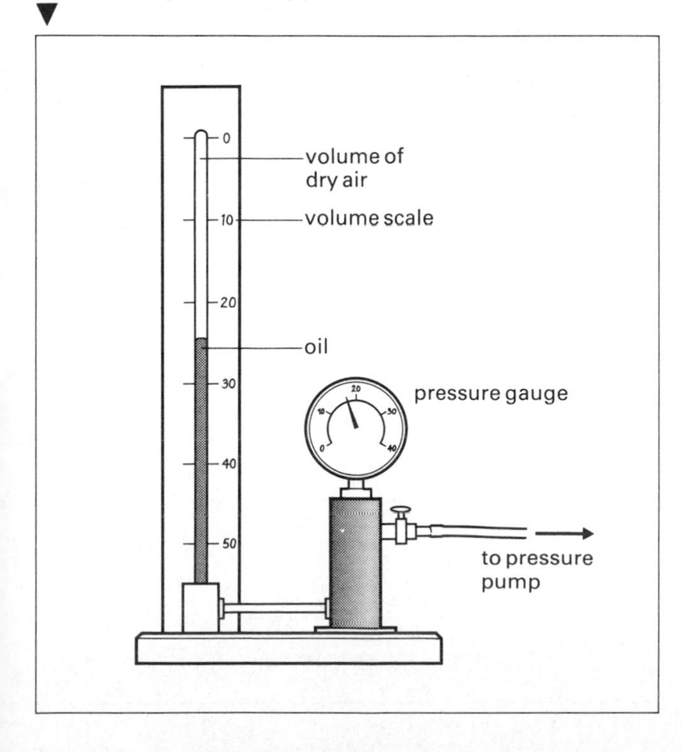

Gases, on the whole, behave in a simpler manner than solids or liquids. For example, the coefficient of expansion of all gases is (within certain limits) the same, while different solids and liquids have different coefficients. The kinetic molecular theory was first developed from a study of gases. It explains the similar behaviour of all gases by the facts that the gas molecules are quite far apart, and that there are no attractive forces between them, so that it does not really matter what kind of molecules they are.

We shall first study the behaviour of a reasonably large volume of gas, and then attempt to explain this behaviour in terms of molecular structure. Two problems arise in the investigation of gases. First, the gas has to be kept in some kind of container. Any change in volume of the container is very small compared with changes in volume of the gas, so that it can usually be ignored. A more serious problem arises from the fact that the volume of a fixed mass of gas (i.e. a fixed number of molecules) can be changed by altering either the pressure or the temperature or, indeed, both at the same time. In situations like this, we can find out how two of the variables are related while the third is kept constant. There are thus three cases to investigate:

- Variation of volume with pressure at constant temperature.
- Variation of volume with temperature at constant pressure.
- Variation of pressure with temperature at constant volume.

Variation of volume with pressure (temperature constant): Boyle's law

One of the most obvious properties of gases is that they are compressible. Solids and liquids are practically incompressible, even under pressures of hundreds of tonnes per square centimetre. If the hole at the end of the barrel of a bicycle pump is closed with the finger, you can easily push the piston halfway in. When you let go, the piston moves out again. Gases are springy.

table 17.1

Pressure p ($N/m^2 \times 10^5$)	Volume V (cm^3)	pV	$\frac{1}{p}$ ($1/Nm^{-2} \times 10^{-5}$)
2.5	20.0	50.0	0.40
2.2	22.7	49.9	0.45
2.0	24.8	49.6	0.50
1.8	27.2	49.0	0.55
1.6	31.1	49.8	0.63
1.5	33.0	49.5	0.67
1.3	37.7	49.0	0.77
1.1	45.0	49.5	0.91

Robert Boyle (1627–1691) investigated how the volume of a gas (air) altered when the pressure was changed. Boyle trapped some air in the short, closed limb of a J-tube and changed the pressure by pouring mercury into the long, open limb, figure 17.1. The pressure on the gas is equal to the atmospheric pressure plus the head of mercury in the long tube. A more modern apparatus measures the pressure directly, using a pressure gauge, figure 17.2.

The fixed mass of air under investigation in figure 17.2 is trapped in a strong glass tube by oil. The tube has a cross-section of 1 cm^2, so the volume can be read directly on the scale. Air can be pumped into the oil reservoir and the pressure read on the pressure gauge. A series of readings of pressure and volume are taken. These are given in table 17.1. The readings must be taken slowly, for when the air is compressed it warms up slightly and we are trying to work at constant temperature. After each change of pressure the column height is allowed to become steady.

It is clear from the results and the graph of p against V, figure 17.3, that, as the pressure increases, the volume decreases.

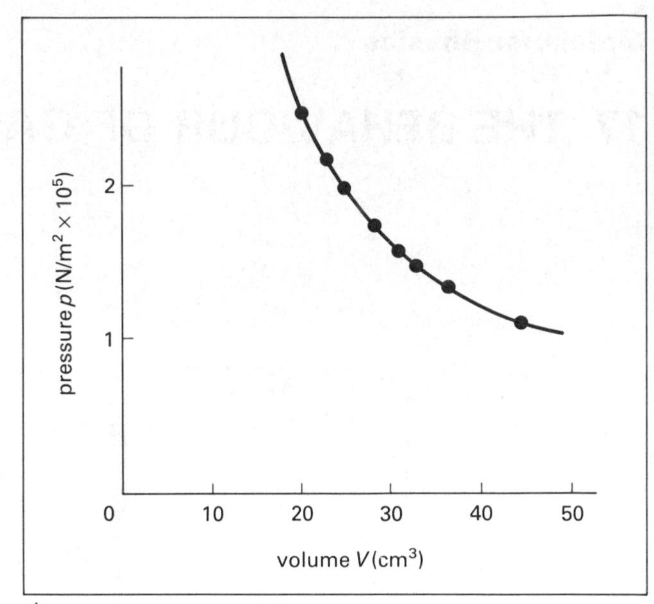

fig. 17.3 Graph of pressure against volume

The product pressure × volume is seen to be constant within the accuracy of the readings, so

pV = constant

This is one way of proving that the volume varies inversely as the pressure, which is further shown by the straight-line graph of V against $1/p$, figure 17.4. The results of experiments like this are summed up in **Boyle's law** which states

The volume of a fixed mass of gas varies inversely as the pressure if the temperature is constant

fig. 17.4 Graph of volume against 1/(pressure)

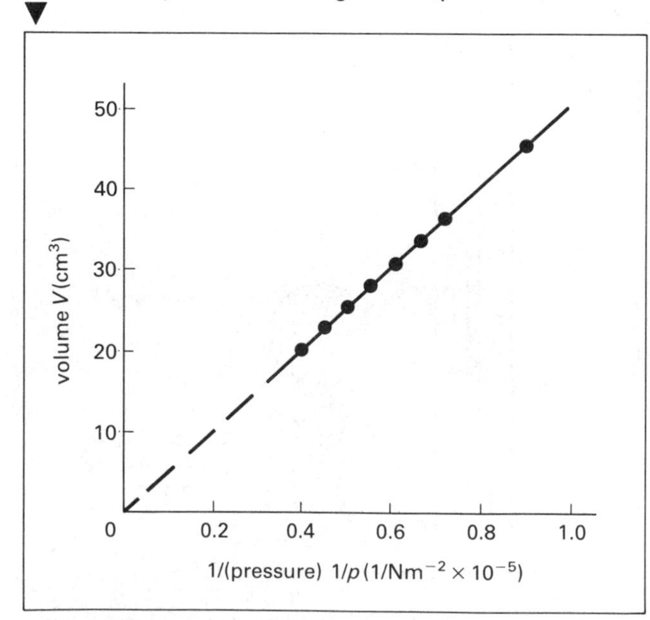

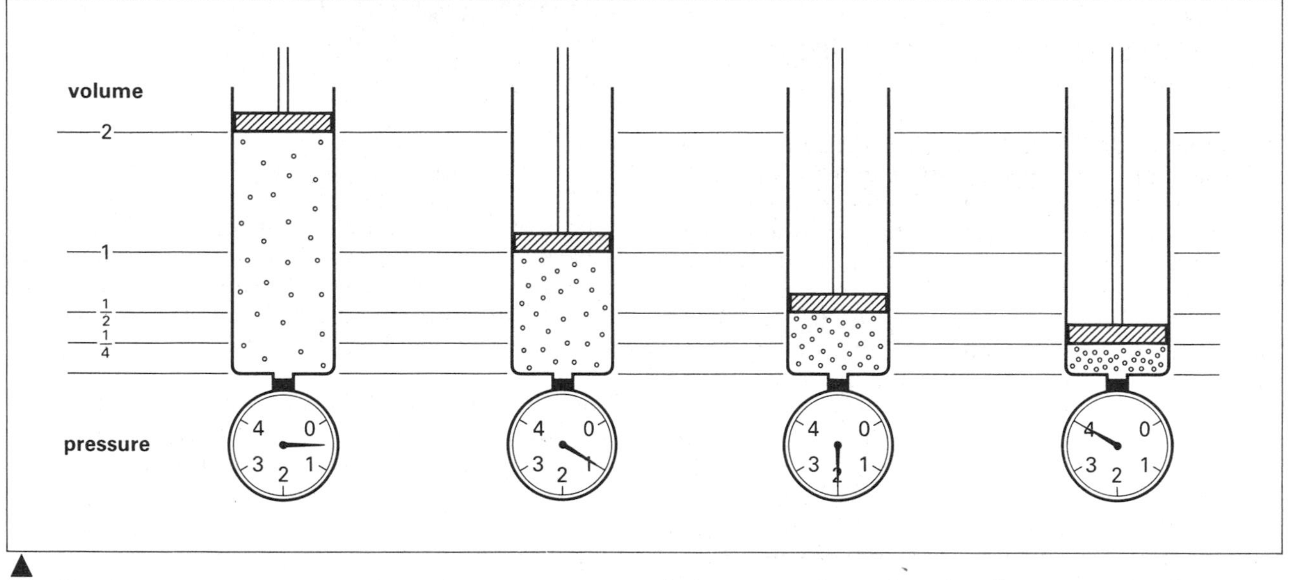

fig. 17.5 As the volume decreases, the pressure increases

The volume and pressure changes of a gas enclosed by a piston in a cylinder are shown in figure 17.5. In each case the number of molecules present is the same. The pressure of a gas results from the bombardment of the walls of the vessel by the rapidly-moving gas molecules. It is easily seen that, when the volume is smaller, the number of bombardments per second of the walls increases, because the molecules are more crowded and so the pressure increases.

The mathematical statement of Boyle's law is pV = constant. If the pressures on a fixed mass of gas at different times are p_1, p_2, p_3, . . . and the corresponding volumes are V_1, V_2, V_3, . . ., then

$$\mathbf{p_1V_1 = p_2V_2 = p_3V_3} \quad \text{and so on}$$

This is the most convenient formula to use when solving problems.

EXAMPLES

1 A mass of hydrogen has a volume of 1 m^3 at atmospheric pressure. What pressure will be needed to compress it into a cylinder of volume 0.0125 m^3?

p_1 = 1 atm $\qquad p_2$ = ?
V_1 = 1 m^3 $\qquad V_2$ = 0.0125 m^3

$$p_2V_2 = p_1V_1$$
$$0.0125p_2 = 1 \times 1$$
$$p_2 = 1/0.0125$$
$$= 80 \text{ atm}$$

This great pressure is the reason why the walls of gas cylinders are made of thick steel.

2 A narrow glass tube of uniform cross-section, open at one end and closed at the other, contains some air trapped by a thread of mercury 1 cm long, as shown in figure 17.6. When the tube is held vertically with the open end uppermost, the length of the air column is 37 cm. When the tube is inverted, the length of the air column is 38 cm. What is the value of the atmospheric pressure in cm of mercury?

Since the tube is of uniform cross-section, we can take the length of the air column as a measure of its volume. Let the atmospheric pressure be P cm of mercury.

initial pressure $p_1 = P + 1$
initial volume $V_1 = 37$

final pressure $p_2 = P - 1$
final volume $V_2 = 38$

$$p_1V_1 = p_2V_2$$
$$37(P + 1) = 38(P - 1)$$
$$37P + 37 = 38P - 38$$
$$P = 75 \text{ cm of mercury}$$

fig. 17.6

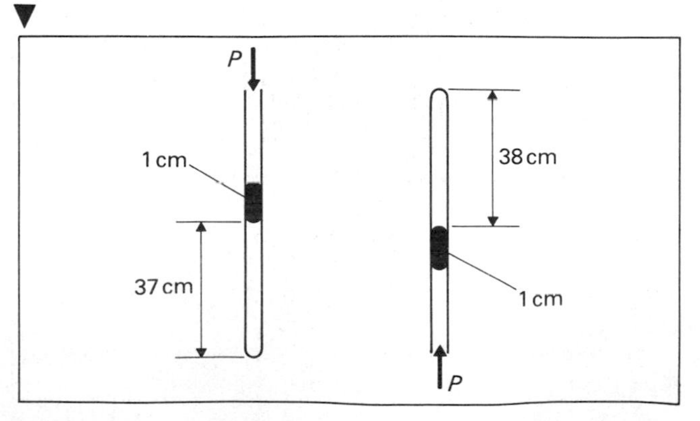

Variation of volume with temperature (pressure constant): Charles' law

A sample of dry air is trapped in a capillary tube by a pellet of mercury, figure 17.7. The tube is heated in a water bath using an immersion heater. The water must be well stirred to ensure that the whole length of the air column is at the same temperature, and the experiment should be carried out slowly to allow the heat to pass through the thick walls of the tube. Readings can be taken warming up and cooling down. Since the tube is of uniform cross-section and we only want to know how the volume changes with temperature, we can read the length of the air column and take this as a measure of the volume. Table 17.2 gives a series of typical readings. The pressure of the air is equal to the pressure of the atmosphere and the mercury index and we assume that this remains constant during the experiment.

When the volume (represented by the length of the column) is plotted against the temperature, the graph, figure 17.8, is a straight line. This shows that air expands uniformly with temperature, but it does not show that the volume is directly proportional to the temperature measured in degrees Celsius because the graph does not pass through the origin.

We can work out the coefficient of expansion with respect to the volume at 0 °C.

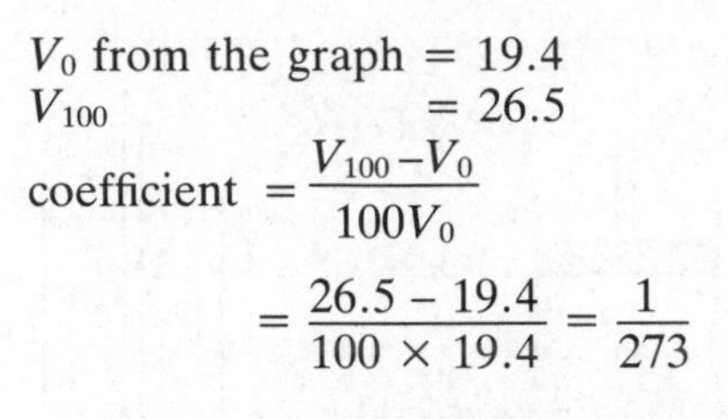

V_0 from the graph $= 19.4$

$V_{100} = 26.5$

$$\text{coefficient} = \frac{V_{100} - V_0}{100 V_0}$$

$$= \frac{26.5 - 19.4}{100 \times 19.4} = \frac{1}{273}$$

Experiments with different gases show that they all behave in the same way. They expand uniformly and have the same coefficient of expansion. The results are summed up in **Charles' law** (named after a nineteenth-century French scientist).

The volume of a fixed mass of gas increases by $\frac{1}{273}$ of its volume at 0 °C for every degree Celsius rise in temperature provided the pressure is constant

table 17.2

Temperature (°C)	Length of air column (cm)
16	20.4
28	21.1
32	21.7
40	22.1
56	23.3
66	24.0
80	25.0
100	26.5

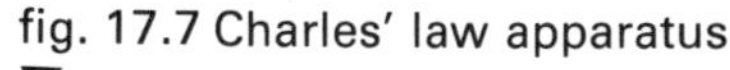

fig. 17.7 Charles' law apparatus

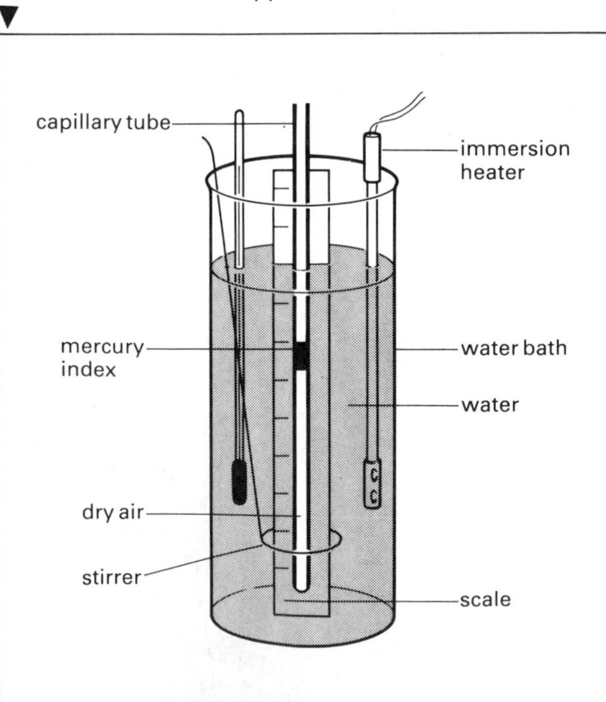

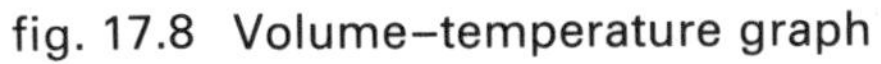

fig. 17.8 Volume–temperature graph

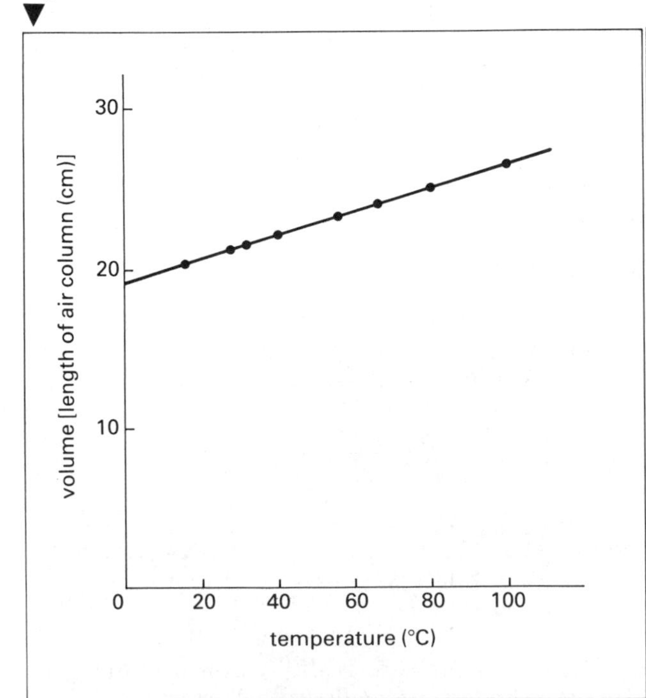

Variation of pressure with temperature (volume constant)

The final relationship to investigate is that between pressure and temperature at constant volume.

For this experiment the essential parts of the apparatus are a bulb to contain the air and some method of measuring the pressure. In one form of the apparatus, a pressure gauge is used, figure 17.9. In another, more traditional form the pressure is measured using a column of mercury, figure 17.10. In both cases the bulb is connected to the pressure measurer by a narrow capillary tube, so that as little air as possible is unheated. A series of readings of pressure and temperature are taken on heating and cooling, in each case allowing time for the air to attain the temperature of the water bath. See table 17.3.

When pressure is plotted against temperature a straight line results, figure 17.11, showing that pressure increases uniformly with temperature.

If we work out a **pressure coefficient**, which is the fractional increase of pressure for one degree rise in temperature between 0 °C and 100 °C, we have

p_0 from the graph $= 0.96 \times 10^5$ N/m^2
$p_{100} = 1.31$

$$\text{pressure coefficient} = \frac{p_{100} - p_0}{100 p_0} = \frac{1.31 - 0.96}{100 \times 0.96} \approx \frac{1}{273}$$

Just as with the relationship between volume and temperature, when the experiment is repeated with different gases the same result is obtained and this gives rise to the **pressure law.**

The pressure of a fixed mass of gas increases by $\frac{1}{273}$ of its pressure at 0 °C for every degree Celsius rise in temperature provided the volume is constant

table 17.3

Temperature (°C)	Pressure (N/m^2 × 10^5)
12	1.00
20	1.02
30	1.05
40	1.07
52	1.15
62	1.17
82	1.25
100	1.31

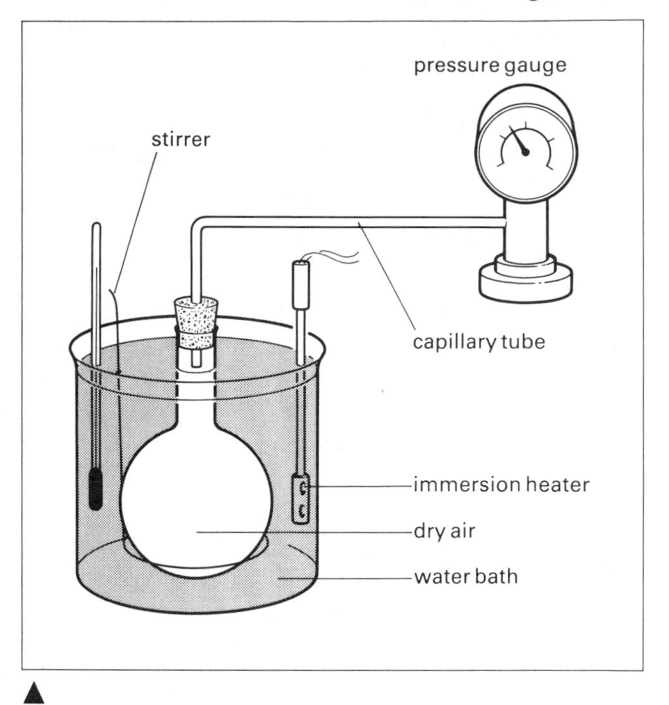

▲ fig. 17.9 Pressure–temperature apparatus

fig. 17.10 Investigating the variation of pressure with temperature at constant volume ▼

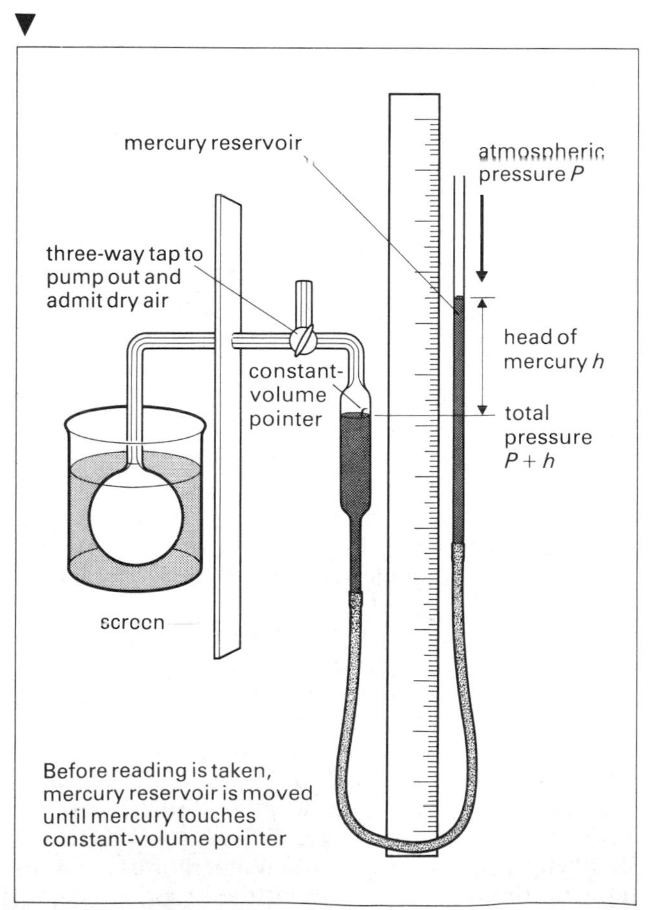

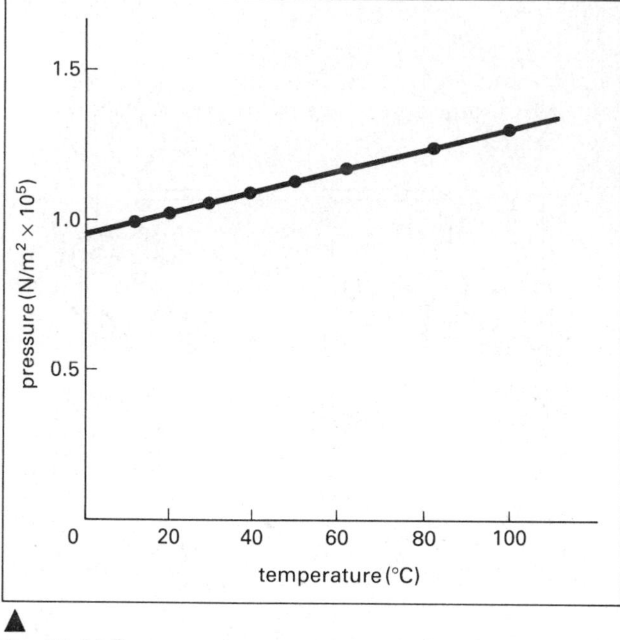

fig. 17.11 Pressure–temperature graph

Absolute zero and the absolute scale of temperature

The volume and pressure variations with temperature of gases behave in the same way. Both give straight-line graphs, figures 17.12 and 17.13, and both give coefficients of the same value. This is very good evidence that the molecules of all gases behave in the same way. If both graphs are extrapolated backwards they both cut the temperature axis at the same value, namely –273 °C. This means that, provided the gas goes on behaving in the same way when it is cooled below 0 °C, when it reaches –273 °C it would have zero volume and exert zero pressure. Since we cannot imagine a gas having a less than zero volume (i.e. occupying less than no space at all!), we conclude that –273 °C is the lowest possible temperature attainable and call it **absolute zero**.

In practice all gases would have liquefied and solidified before absolute zero is reached, but this does not alter the fact that, theoretically, there is a temperature below which it is impossible to go. At this temperature the molecules would have minimum energy.

Once we have established the idea of an absolute zero, it is an easy step to consider a temperature scale starting at this point – the **absolute** or **Kelvin scale** of temperature. Absolute zero is –273 Celsius, so that Kelvin temperatures are easily obtained by adding 273 to the Celsius temperature, figure 17.14. (Very accurate measurements now give absolute zero as –273.15 °C but we shall use the whole number for simplicity.) Kelvin or absolute temperatures are written with the symbol K, without the ° sign.

If we plot either pressure at constant volume or volume at constant pressure against absolute temperature, figure 17.15, we see that not only do we get a straight line but this straight line passes through the origin of the graph. This means that both pressure and volume vary directly as the absolute temperature. This simplifies both Charles' and the pressure laws, for we can write

***V* is directly proportional to *T* at constant pressure**
***p* is directly proportional to *T* at constant volume**

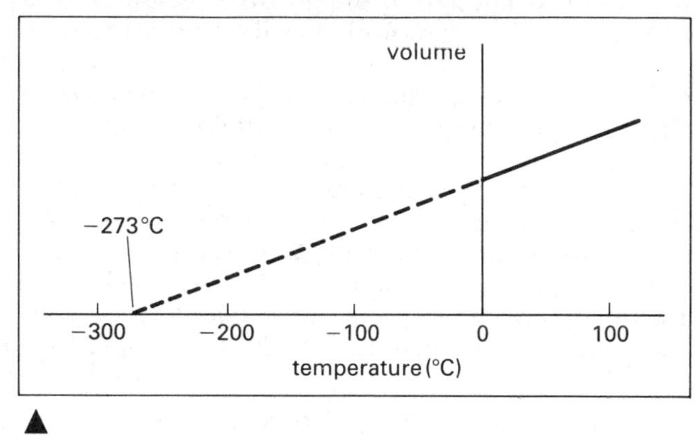

fig. 17.12

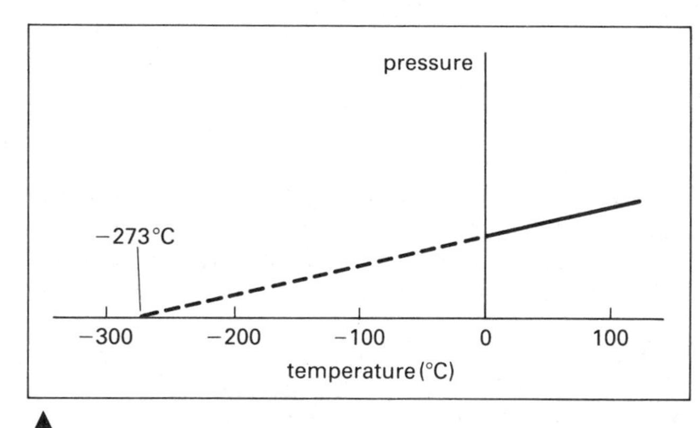

fig. 17.13

fig. 17.14

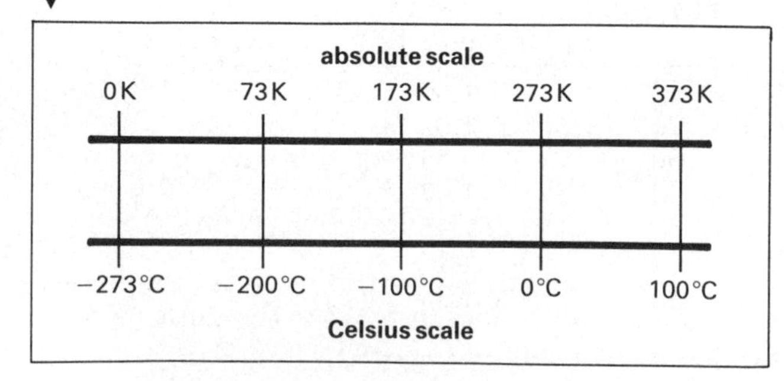

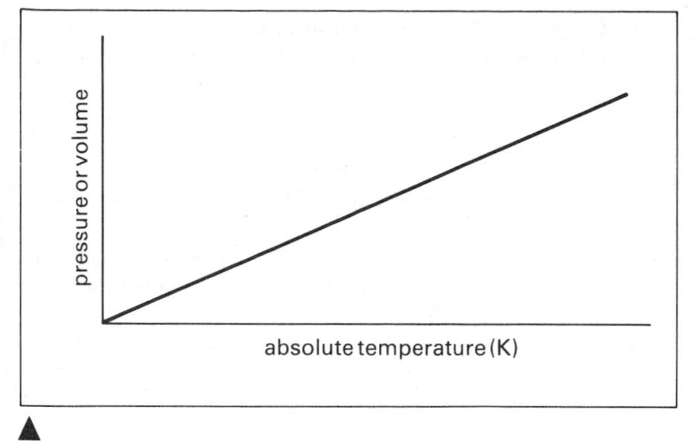

fig. 17.15

The gas equation

The three variables relating to a fixed mass of gas, volume (V), pressure (p) and Kelvin temperature (T), were dealt with two at a time, keeping the third constant, because this is the only way we can deal with them experimentally. What happens when all three change at the same time? The three separate relationships can be combined into a single expression which is called the **gas equation**.

Boyle's law: $p \propto \frac{1}{V}$ or pV = constant (for constant temperature)

Charles' law: $V \propto T$ or $\frac{V}{T}$ = constant (for constant pressure)

Pressure law: $p \propto T$ or $\frac{p}{T}$ = constant (for constant volume)

Combining these three into a single equation, we have

$$\frac{pV}{T} = \text{constant}$$

$$\frac{p_1V_1}{T_1} = \frac{p_2V_2}{T_2}$$

which is the gas equation for a fixed mass of gas.

Remember that all temperatures must be on the Kelvin scale.

The pressure and volume can be measured in any convenient units so long as they are the same on both sides of the equation.

EXAMPLE

A mass of gas has a volume of 400 cm³ at 40 °C and a pressure of 600 mm of mercury. Calculate the volume at 0 °C and a pressure of 760 mm of mercury.

$p_1 = 600$ mm $\quad p_2 = 760$ mm
$V_1 = 400$ cm^3 $\quad V_2 = ?$
$T_1 = 273 + 40 = 313$ K $\quad T_2 = 273 + 0 = 273$ K

$$\frac{p_2V_2}{T_2} = \frac{p_1V_1}{T_1}$$

$$\frac{760V_2}{273} = \frac{600 \times 400}{313}$$

$$V_2 = \frac{600 \times 400 \times 273}{313 \times 760} = 275.4 \text{ cm}^3$$

When comparing volumes of gases, they must always be under the same conditions of temperature and pressure. **Standard temperature and pressure (s.t.p.)** are 0 °C and 760 mm of mercury: 760 mm of mercury = 1 atmosphere = 1.01×10^5 N/m^2.

EXERCISE 17

In questions 1–5 select the most suitable answer.

1 For a fixed mass of gas, halving the pressure
A doubles the volume at constant temperature
B doubles the volume if the temperature doubles also
C halves the volume at constant temperature
D halves the volume if the temperature halves also
E does not affect the volume at constant temperature

2 When air is pumped into a car tyre at constant temperature, the pressure increases because
A more molecules are hitting the tyre
B the molecules are larger
C the molecules are moving faster
D the molecules have greater energy
E the molecules are closer together

3 When a fixed mass of gas is heated in a constant volume container, the pressure is
A increased because each molecule expands
B increased because the molecules strike the walls harder and more often
C increased because the hot gas will try to rise
D unchanged because the number of molecules remains constant
E unchanged because the volume remains constant

4 Which of the following temperatures on the absolute (Kelvin) scale corresponds to 100 °C?
A 0 K
B 100 K
C 273 K
D 373 K
E –273 K

5 Which of the following expressions represents the behaviour of a fixed mass of gas at a constant temperature? (p = pressure, V = volume, T = temperature measured on the Kelvin scale, k = constant)

A $pT = k$
B $V = kT$
C $p = kT$
D $p = kV$
E $pV = k$

6 a When you hold your finger over the exit hole of a bicycle pump and push in the handle you feel the pressure increase. If the speed of the molecules has not increased, why is there a greater pressure on the piston?
b The pressure in your bicycle tyre increases on a hot day even if you do not pump any more air in. Why is this?
c When a liquid evaporates it cools. This is particularly noticeable if you spill petrol, alcohol or ether on your hand. Explain carefully why this cooling occurs.

7 The apparatus shown in figure 17.16 is used to measure the volume (V) of a mass of air at different pressures (P). The table gives typical results

Pressure P (atmosphere)	Volume V (cm^3)	$P \times V$
1.0	30	
1.5	20	
2.0	15	
2.5	12	
3.0	10	
4.0	7.5	

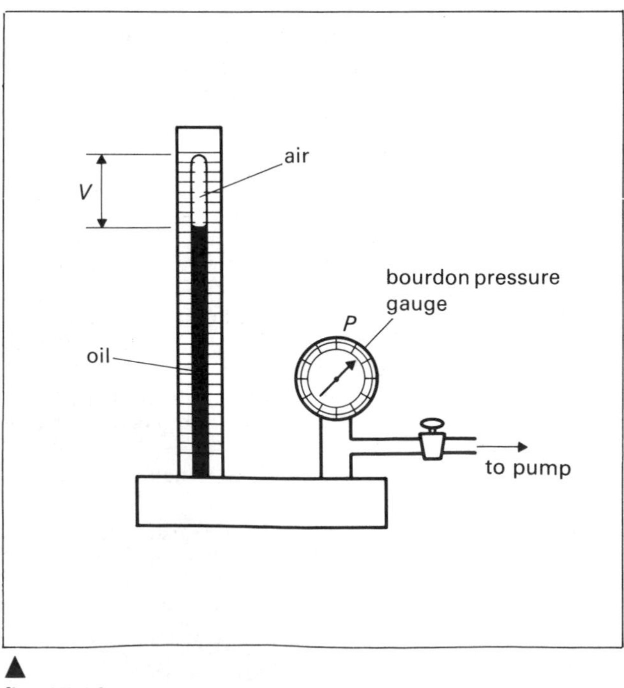

fig. 17.16

a Complete the third column of the table of results.
b From these results, what conclusion do you come to about the relationship between pressure and volume?
c What would be the volume of the air in the tube if it were compressed to a pressure of 5 atmospheres?
d Explain, in terms of the motion of molecules, why the pressure of a gas increases as it is compressed.

8 The kinetic theory is based on the idea of constantly moving gas molecules. Use this idea in answering *all* parts of this question.
a Describe the movement of molecules in a gas. Draw a diagram to show the type of path taken by a single gas molecule. Explain why changes in direction occur.
b Why do molecules in a gas continue to move? What would be the effect on the movement of (i) increasing the temperature, (ii) increasing the pressure?
c Why does a balloon increase in size when air is blown into it? What causes the pressure inside? Why is the air pressure inside greater than outside?
d With the aid of a labelled diagram, describe an experiment which is designed to show that air molecules are constantly moving.
e State *two* reasons why it is not possible to watch the movement of air molecules under a microscope.

9 a Describe how you would investigate the variation of volume with temperature of a fixed mass of dry air at constant pressure. Explain (i) how you ensure that the air is dry, and (ii) how the pressure is kept constant in your experiment. Draw a sketch graph showing the results you would expect to obtain.
b On aerosol cans there is a warning not to leave them in strong sunlight or to throw them on to a fire when empty. Give the physical reasons for this warning.
c Early in the morning the pressure in a car tyre was 2.00×10^5 Pa when the temperature was 7 °C. What would you expect the pressure in the tyre to be when the temperature has risen to 27 °C in the afternoon? (Assume that the volume of the tyre does not change.)

10 Figure 17.17 shows a collection of apparatus you could use to find how the pressure of air varies with its temperature.
a Draw a sketch showing the apparatus properly assembled for this experiment.
b Describe how you would take a set of readings like the ones given in the table.
c Plot a graph of the readings and find the pressure at which the temperature of the air is 40 °C.
d What difference would it make to the readings obtained if the flask held 500 cm^3 instead of 250 cm^3? (Assume that it can still be properly immersed in the water.)
e Given similar apparatus but without a thermometer and assuming pure ice water is at 0 °C and pure water boils at 100 °C, explain how you would use the apparatus to find (i) the boiling point of a saturated solution of salt, (ii) the temperature of a 'freezing mixture' of ice and salt.

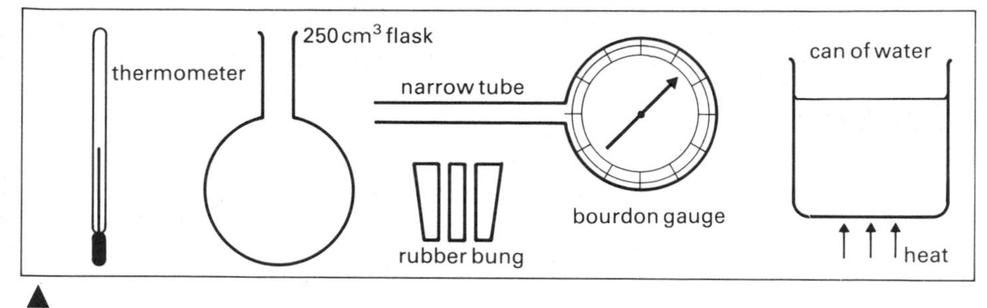

fig. 17.17

fig. 17.18

Thermometer reading (°C)	Gauge reading (N/m^2)
0	1.00×10^5
25	1.09×10^5
50	1.18×10^5
75	1.27×10^5
100	1.37×10^5

11 The table below gives the volume of a constant mass of dry air at different temperatures, the pressure remaining constant:

Temperature (°C)	15.0	40.0	65.0	80.0	90.0
Volume $(cm)^3$	7.0	7.6	8.2	8.6	8.8

a Draw a graph of volume against temperature.
b What is the volume of air at 0 °C?
c What is the gradient of the graph?
d Explain what information the gradient gives about the way in which the gas expands.

12 This is a question about the kinetic theory of gases.
a There is experimental evidence for saying that the molecules in a gas are in perpetual random motion. Name a suitable experiment and say briefly what is observed.
b How does the kinetic theory explain why a gas exerts a pressure?
c A sample of air is kept at a fixed temperature. It is compressed to half its original volume, causing the pressure to double. How does the kinetic theory explain this?
d What happens to the average speed of the molecules in a gas when the gas is heated to a higher temperature?
e What is the name given to the temperature at which the molecules of a gas would have no energy?
f Why does the temperature of a gas rise when it is suddenly compressed? (This happens, for example, when you push in the piston of a bicycle pump when your thumb is over the end.)
g Why don't all the molecules in the atmosphere fall to the ground (under gravity), forming a very dense layer a few metres deep?

13 a What is the relationship connecting the pressure, volume and temperature of a fixed mass of gas?
b A long uniform glass tube, sealed at its lower end, contains an air column of length 10 cm and a mercury thread of length 40 cm. It is supported vertically with the lower end in water at 27 °C, as shown in figure 17.18. (i) What is the pressure of the trapped air, if the atmospheric pressure is 75 cm of mercury. (ii) The water is slowly heated from 27 °C to 77 °C. What will be the pressure and length of the air column? (iii) Explain what differences there would be if the tube were only 50 cm long, all other measurements and temperatures remaining the same.

18 HOW LIGHT TRAVELS

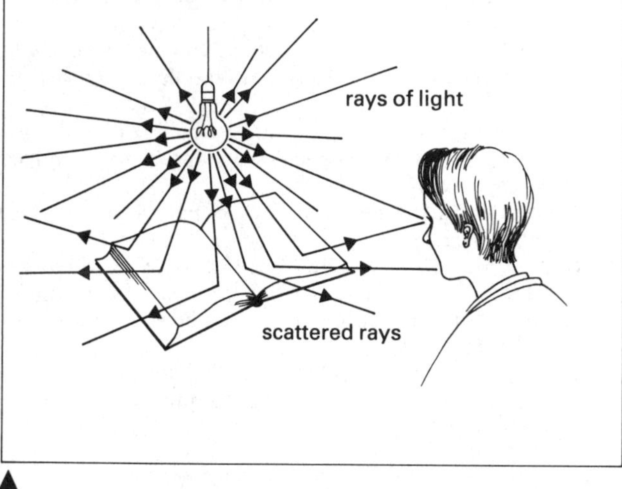

▲
fig. 18.1 We see the lamp by the light it **emits**; we see the book by the light it **scatters**

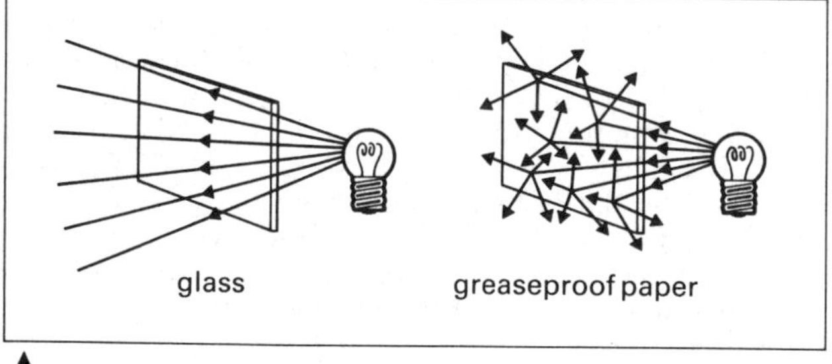

▲
fig. 18.2 Glass is **transparent**; greaseproof paper is **translucent**

fig. 18.3 White paper is opaque and scatters most of the light. Black paper is opaque and absorbs all the light
▼

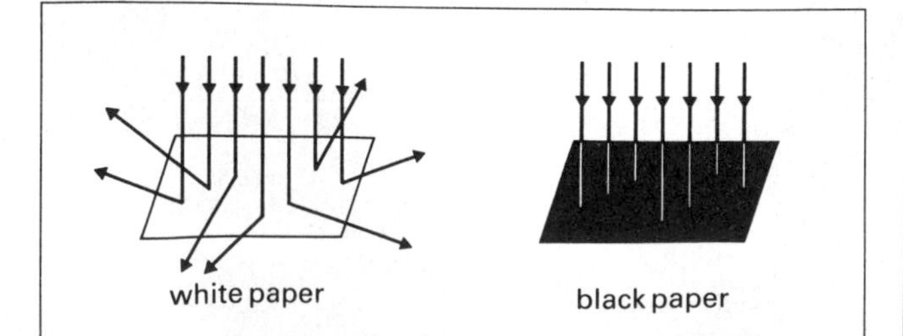

Light is a form of energy. Light energy stimulates the retina of the eye and produces the sensation of sight. We see an object when the light energy leaving it enters the eye. Some things – such as flames, the sun and the stars, the glowing wire in an electric lamp – give out light of their own. These are **luminous** objects. Objects which do not give out light of their own are **non-luminous**. Non-luminous bodies scatter the light falling on them in all directions so that we are able to see them.

Provided it remains in the same medium, light travels in straight lines. A straight line with an arrow on it showing the direction of travel is called a **ray**. A **beam** or **pencil** of light is a collection of rays, figure 18.4.

Shadows

Light will not pass through an opaque object and a shadow is formed. The nature of shadows and the positions in which they occur are evidence that light travels in straight lines. This is called **rectilinear propagation**, figures 18.5a and 18.5b.

fig. 18.4 Types of beam
▼

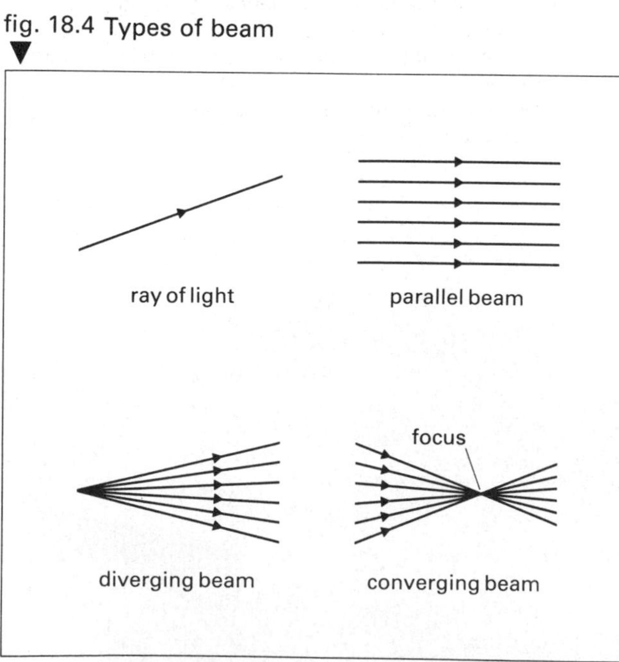

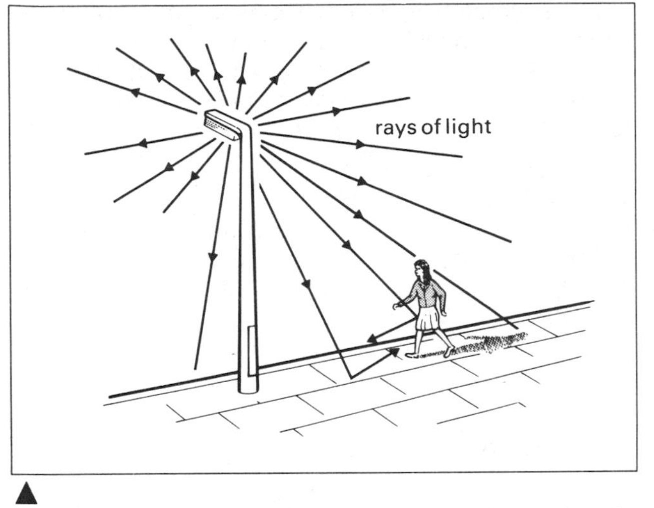

fig. 18.5a The end of the shadow, the girl's head and the lamp are in a straight line

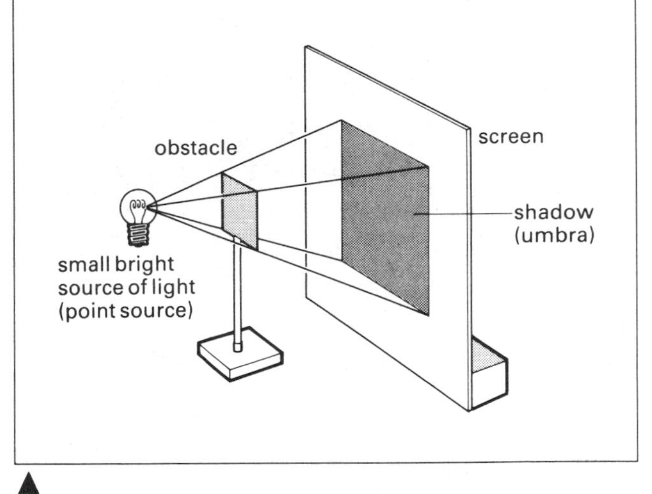

fig. 18.5b Straight lines drawn from the lamp through the corners of the obstacle tell us exactly where the shadow is going to be on the screen

When the source of light is not a point source the outline of the shadow becomes less clear, because light from part of the source reaches the edges of the shadow. This partial shadow or **penumbra** appears as various shades of grey. The outer edge is brighter than the inner edge because it is exposed to more of the source, figure 18.6. (The prefix *pen-* comes from the Latin word for almost, and *umbra* is Latin for shade: *pen-umbra* = almost shadow.)

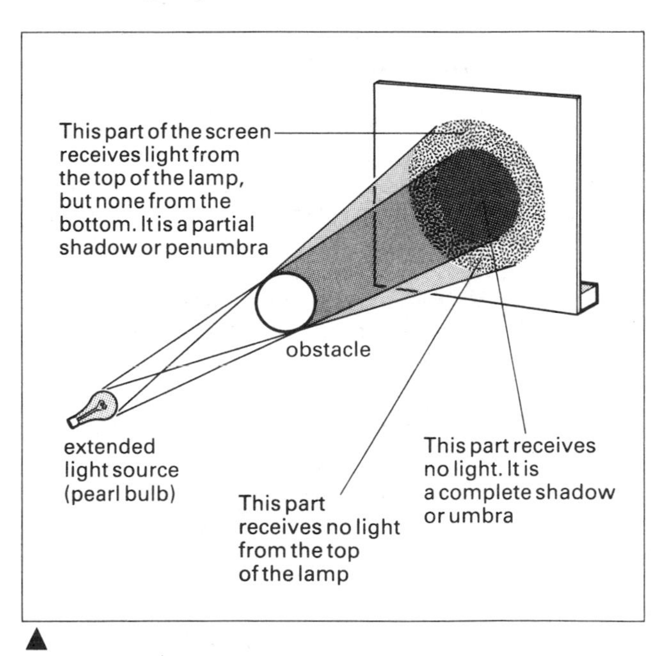

fig. 18.6 Shadow from an extended source

Eclipses

Eclipses are examples of large-scale shadow production. When the moon is between the earth and the sun the shadow of the moon may fall on the earth and for a short time the sun cannot be seen. This is an eclipse of the sun (**solar eclipse**). Since the moon travels round the earth once every lunar month (about 29 days) we might expect an eclipse of the sun once a month. But the orbits of the earth and the moon are not in the same plane. Total eclipses occur very rarely because the sun, moon and earth do not get in the same straight line very often. The next total

fig.18.7 Eclipse of the sun (not to scale)

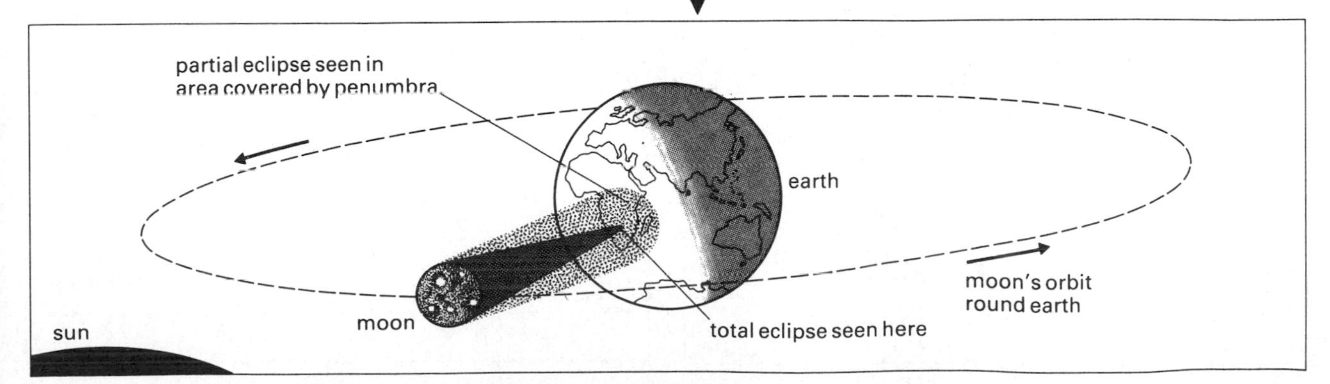

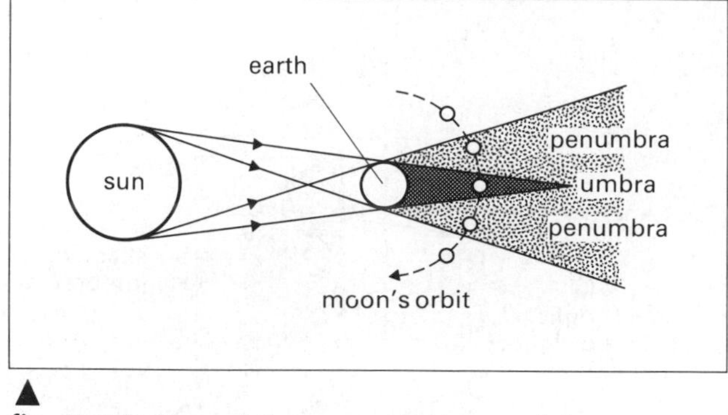

fig. 18.8 Eclipse of the moon (not to scale)

eclipse of the sun to be seen in Britain will be in 1999.

When the moon passes into the shadow of the earth there is an eclipse of the moon (**lunar eclipse**). This can only occur when the moon is on the opposite side of the earth from the sun, i.e. at a period when a full moon would occur, figure 18.8.

The pinhole camera

The **pinhole camera**, figure 18.9, is another piece of evidence for rectilinear propagation. The pinhole camera is a box with a pinhole in one side and a translucent screen at the opposite side. When the camera is pointed towards the window or a luminous object, an image is seen on the screen.

The image is a **real** image because it is really formed by rays of light on a screen. There are other images (for example see pp. 150, 156 and 175) where the rays of light do not pass through the image but only appear to come from it. These images are said to be **virtual** or imaginary images.

fig. 18.9 Pinhole camera

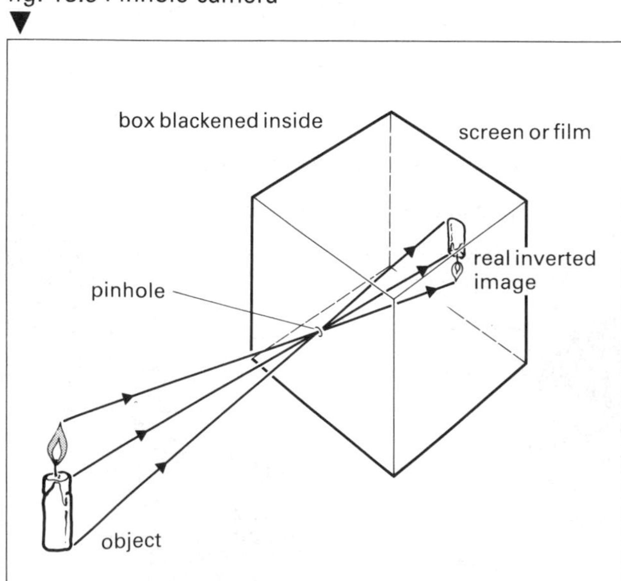

If the screen is replaced by a photographic film and a long exposure given, a photograph can be obtained. If straight lines (rays) are drawn from the top and bottom of the object passing through the pinhole they give the exact position of the image.

- The image is inverted.
- No focussing is required.
- A larger pinhole gives a brighter image, but it is less distinct.
- A longer box gives a larger image, but it is less bright.

EXERCISE 18

In questions 1–4 select the most suitable answer.

1 When a person walks in front of the projection beam in a cinema a black shadow is obtained on the screen. The shadow is black because
A the beam of light is diverging from the bulb to the screen
B white light has all the colours of the rainbow
C the projection bulb is very bright
D the source of light is very small
E the body stops some light reaching the screen

2 If a pinhole camera was made longer without altering the size of the pinhole, the image would be
A smaller and less bright
B larger and less bright
C smaller and brigher
D larger and brighter
E the same size and brighter

3 Figure 18.10 shows a vertical post drawn to scale. The ground behind the post is marked in metres. The sun shines so that its rays are at 45° to the horizontal. The length of the shadow of the post cast on the metre scale will be
A 1 m
B 2 m
C 3 m
D 4 m
E 5 m

fig. 18.10

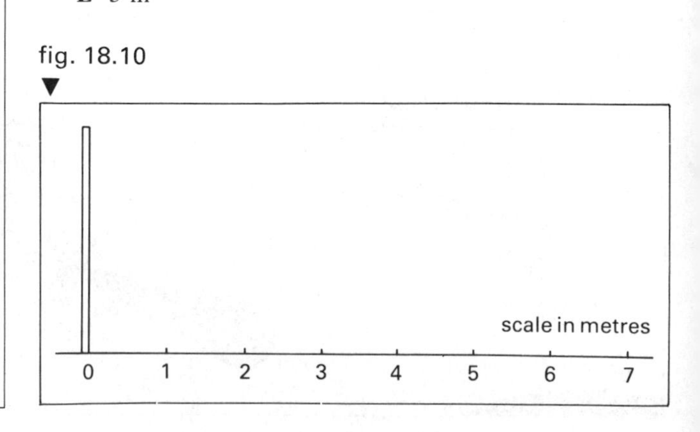

4 An eclipse of the moon occurs when
A the earth is between the moon and the sun
B the moon is between the earth and the sun
C the moon is at a point in its orbit when it is closest to the earth
D the shadow of the moon falls on the earth
E it is the winter solstice

5 Draw diagrams which illustrate the difference between the reflection of light from a matt surface and a polished surface. Label the surfaces.

6 Why do we believe that light travels in straight lines?

7 Figure 18.11 shows a wide source of light, a screen and two opaque objects.
Which of the points A, B, C or D receives no light at all?
Which of the four points receives most light?

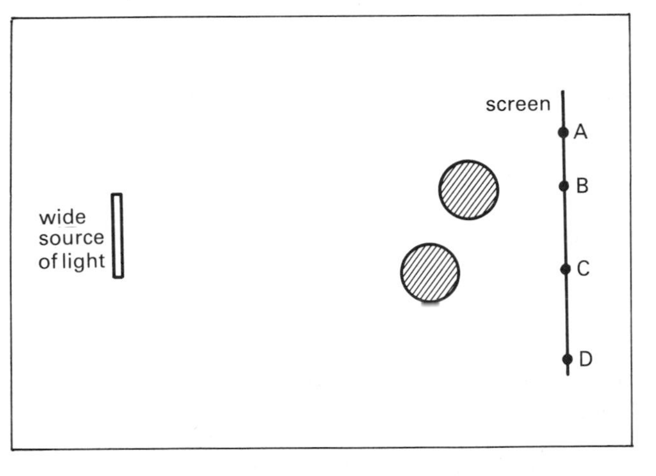

fig. 18.11

8 Draw diagrams to explain how it is that your shadow grows longer as you walk away from a street lamp.

9 Draw a diagram which illustrates an eclipse of the sun. Clearly mark and label
a A position where a *total* eclipse might be observed.
b A position where a *partial* eclipse might be observed.

10 Figure 18.12 shows a pinhole camera (not drawn to scale).
a How tall is the image on the screen?
b If the distance of the object from the front of the camera box were halved, what would happen to the size of the image on the screen (compared with its original size)?
c What happens to the clarity of the image if the pinhole is enlarged?
d What would be the effect on the image of making two pinholes in the camera?

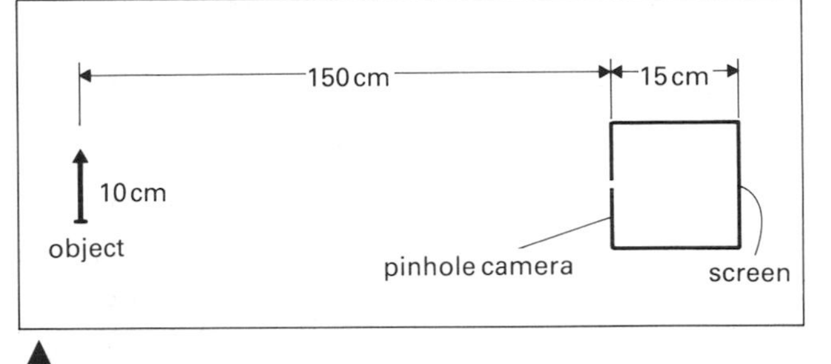

fig. 18.12

11 Figure 18.13 shows an opaque spherical object placed between a spherical light source and a screen.
a Draw a sufficient number of rays of light to define the regions of total shadow, partial shadow and no shadow at all.
b What important principle of light are you making use of when you draw your rays?
c Assuming there are small holes in the screen at the points P, Q and R, describe what you would see looking, from behind the screen, through each of these holes towards the object.
d If the screen were to be moved back to the dotted position the holes would now be at P′, Q′ and R′. What would you see now, looking through the holes?

Fig. 18.13

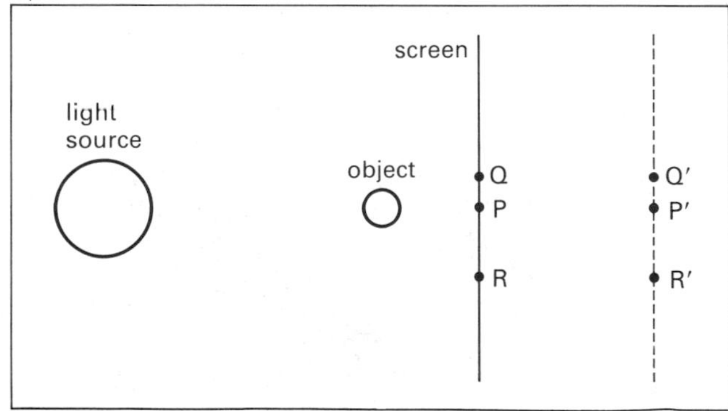

19 REFLECTION OF LIGHT

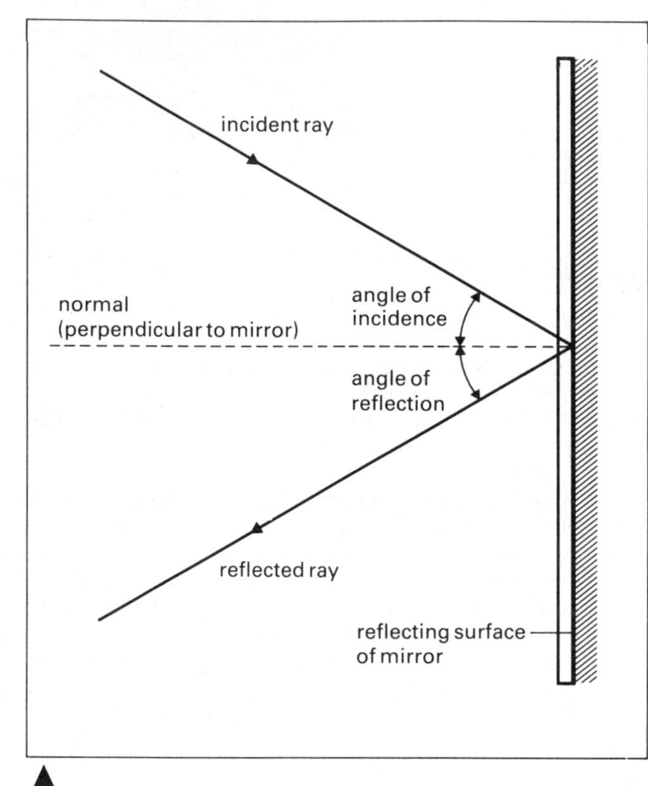

▲
fig. 19.1 Reflection from a plane mirror

When light falls on a smooth, highly polished surface it is **reflected** (= turned back). A piece of polished metal makes a good reflector. Glass mirrors have a thin layer of silvering deposited on the back of the glass, and this thin layer is protected with a coat of paint. A metal mirror reflects from the front surface, but a glass mirror reflects from the back surface, figure 19.1.

The laws of reflection

Law 1 The incident ray, the normal and the reflected ray all lie in the same plane, i.e. flat surface.

Law 2 The angle of reflection is equal to the angle of incidence.

Whatever kind of reflecting surface is used, the laws of reflection are always obeyed.

The image formed by a plane mirror

Probably the most common use of a mirror is to enable us to see ourselves. We say we are looking at our 'reflection' in a mirror. We should say that we are looking at our **image** in the mirror. The term image is used for any picture or reproduction of an object by means of rays of light. The image in a plane mirror is always

- The same size as the object and the same way up.
- As far behind the mirror as the object is in front and on the same perpendicular.
- Turned round sideways or **laterally inverted**.
- A virtual image, because the rays of light do not actually pass through it, but only appear to come from it.

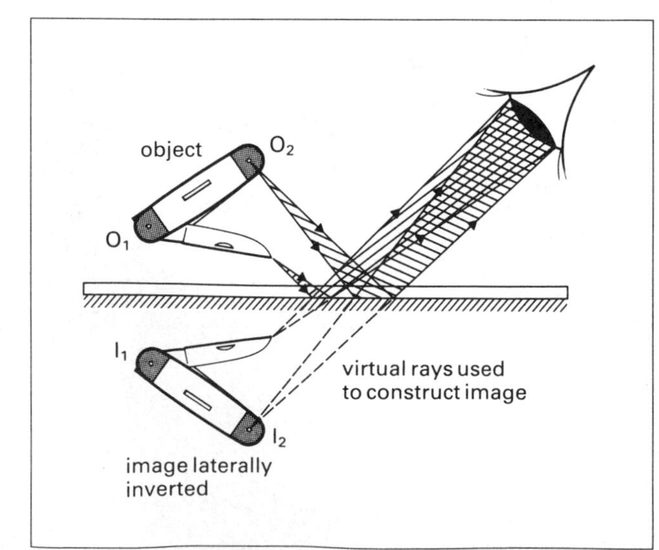

◀ fig. 19.2 How we see the image in a plane mirror. Rays from each point on the object are reflected by the mirror and on entering the eye appear to come from the corresponding point on the image

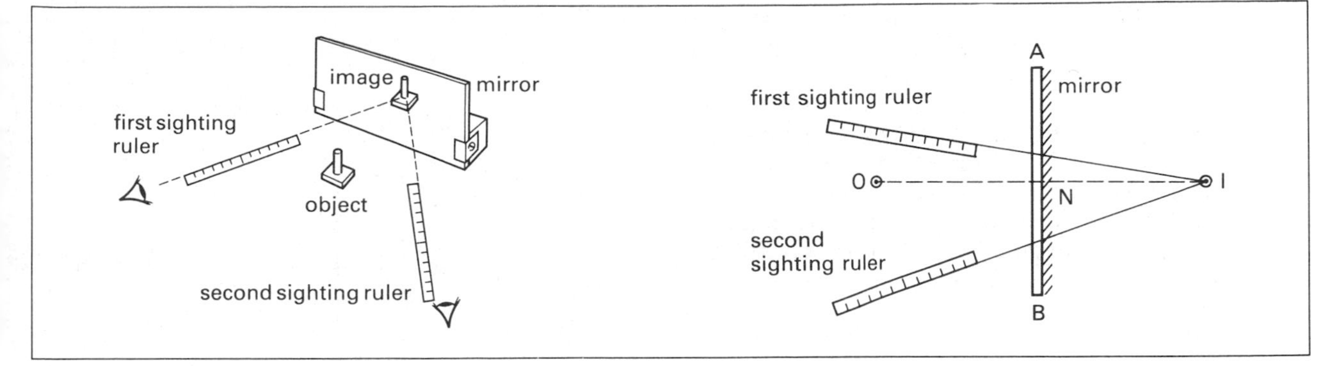

fig. 19.3 Locating the image by sighting

Locating the image in a plane mirror

The image formed by a plane mirror is a virtual or imaginary image, but the position of the image can be found exactly. There are two methods of doing this. The method shown in figure 19.3 is by *sighting*.

Stand an upright object in front of a mirror and move two rulers on the table top until, on looking along their edges, they point directly towards the image. You can use a third ruler to check the accuracy with which you have placed the first two. Mark the back of the mirror and the object position. Remove the mirror and extend the sighting lines back. The point where they cross gives the position of the image. Join O to I and measure the distances ON and NI and the angle ANO as shown in figure 20.3. Repeat the experiment with several positions of the object.

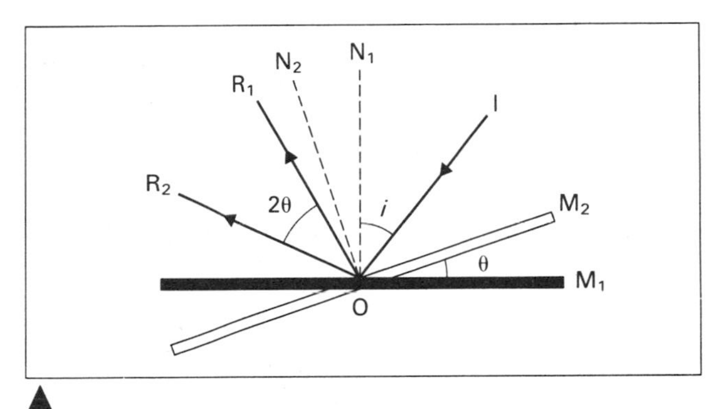

fig. 19.4 Reflection from a rotating mirror

fig. 19.5 Sextant

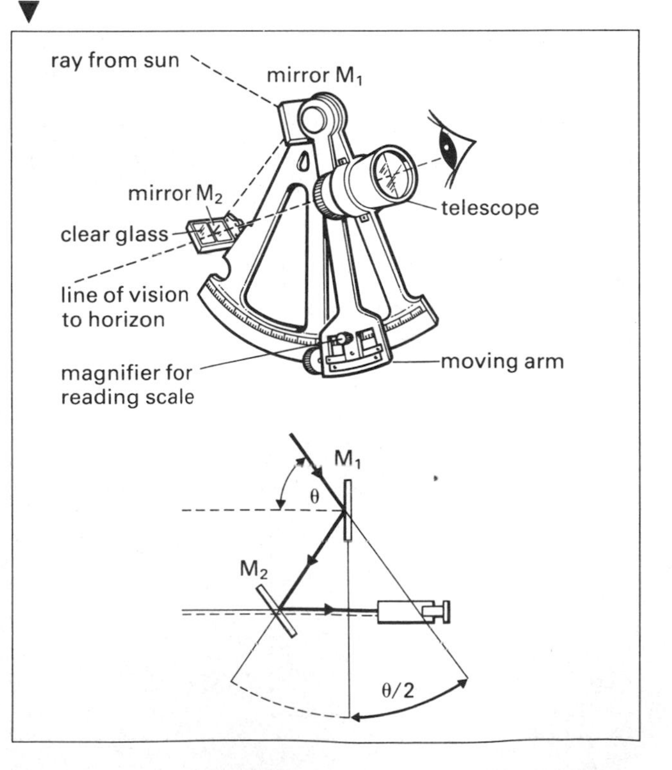

Reflection from a rotating mirror

When a ray of light is reflected from a mirror, the direction of the reflected ray depends on the position of the mirror. If the mirror is turned through an angle, the direction of the reflected ray is changed. For a fixed direction of the incident ray, the angle turned through by the reflected ray is twice the angle turned through by the mirror. Referring to figure 19.4, with the mirror in position M_1 the angle between the incident ray and the reflected ray is $2i$. The mirror is turned through an angle θ to position M_2. If the incident ray remains fixed the angle of incidence becomes $(i + \theta)$ and the angle between the incident and reflected rays is $2(i + \theta)$. The angle between R_1 and $R_2 = 2(i + \theta) - 2i = 2\theta$. The mirror need not rotate about the point of incidence; it can rotate about any point and the result is exactly the same.

The principle of the rotating mirror is used in mirror galvanometers where the reflected beam acts as a long, weightless pointer. The navigational instrument, the sextant, also uses a rotating mirror. The sextant is used to determine the angle of elevation of the sun or a star above the horizon. When the angle of elevation is known, the latitude can be obtained.

Light can enter the sextant's telescope via two routes: directly from the object and by reflection from the mirrors M_1 and M_2, figure 19.5. When the two images of the same object coincide, the mirrors are parallel and the scale reads zero. The instrument is set to zero on the horizon. Mirror M_1 is now rotated by moving the arm over the scale until the image of the sun or a star comes into line with the horizon seen directly through the telescope. The angle turned through by the mirror is half the angle of elevation. To save the trouble of multiplying by two, the degree divisions on the scale are half as large as they should be, i.e. 90° on the scale only takes up an arc of 45°. When the sun is viewed, filters are used to reduce the intensity and prevent damage to the eye.

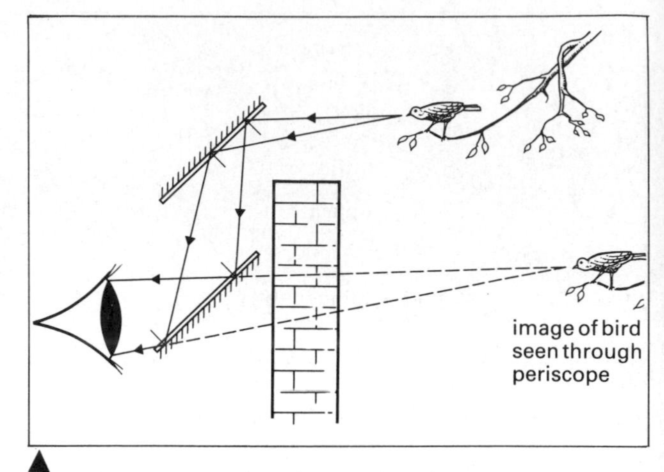

▲
fig. 19.7 Position of periscope image

Reflection from more than one mirror

One common use of two mirrors is in the **periscope** which enables one to see over obstacles. Two mirrors arranged at 45° in a tube enable an observer to see over an obstacle. Figure 19.6 is a simple diagram showing this. The accurately drawn ray diagram, figure 19.7, shows that the image does not appear in the same position as the object. It is moved downwards and further away by distances equal to the length of the periscope.

When light is reflected successively from two mirrors placed at an angle to each other, several images will be formed.

With two mirrors at right angles, figure 19.8, three images are formed. The image I_{12} is formed when light is reflected from M_1 and then from M_2.

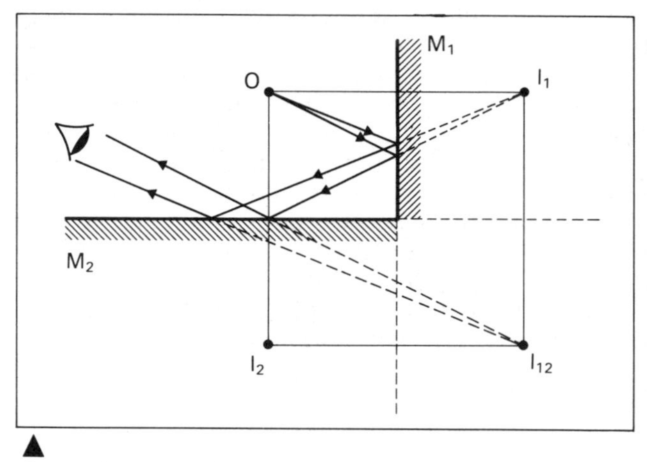

▲
fig. 19.8 Images formed in two mirrors at 90°

fig. 19.6 Principle of the periscope
▼

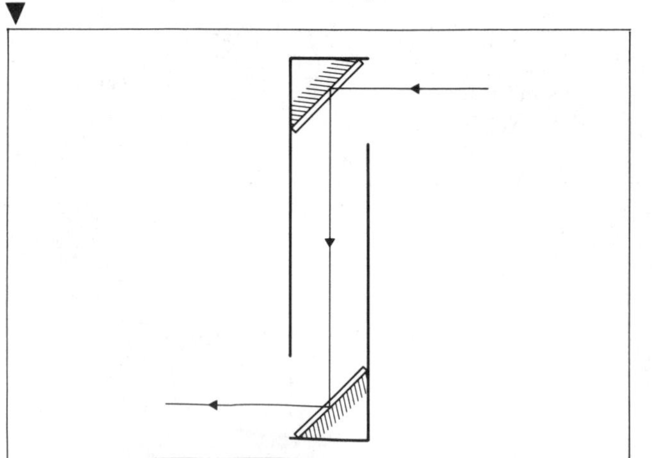

fig. 19.9 Images formed in two mirrors at 60°
▼

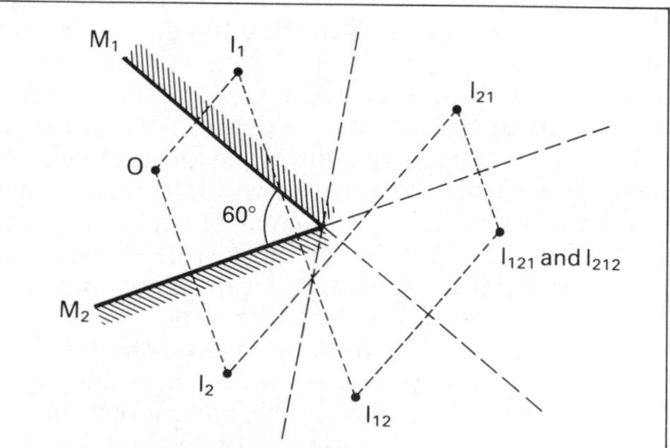

With two mirrors inclined at an angle of 60°, figure 19.9, five images are formed: I_{12} and I_{21} by two reflections and I_{121} (or I_{212}) by three reflections. The kaleidoscope uses reflections from two mirrors at an angle, and the attractiveness of the designs produced is due to their symmetrical appearance, figure 19.10.

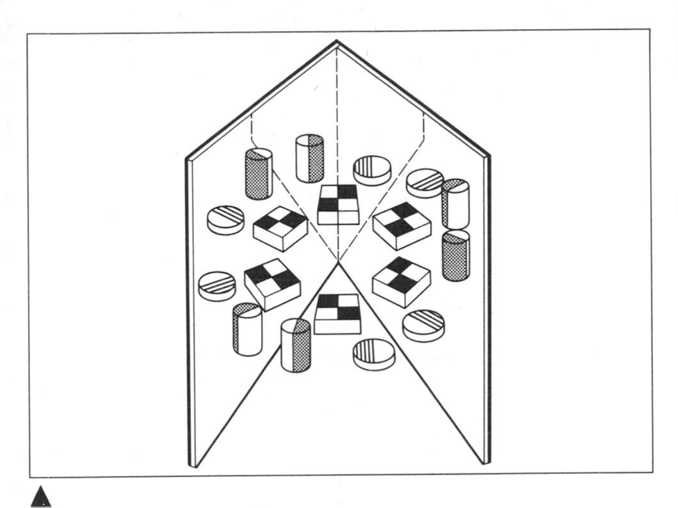

▲
fig. 19.10 The kaleidoscope contains two mirrors at an angle of 60°. Coloured paper, foil and wool are used as objects to form patterns

When a ray of light strikes one of two mirrors placed at right angles to each other and then the other mirror, it is reflected back parallel to its original path, figure 19.11. This is true whatever angle the ray makes with the first mirror, as can be easily shown using the geometrical properties of parallels. If we extend this idea to three dimensions, so that we have three mirrors mutually at right angles like the sides of a cube, it is still found that a ray is reflected back parallel to its original direction.

This principle is used in the moulded red reflectors on the rear of vehicles, in motorway reflector signs and 'cats' eyes' down the middle of the road, figure 19.12.

fig. 19.11 Reflection from two mirrors at 90°
▼

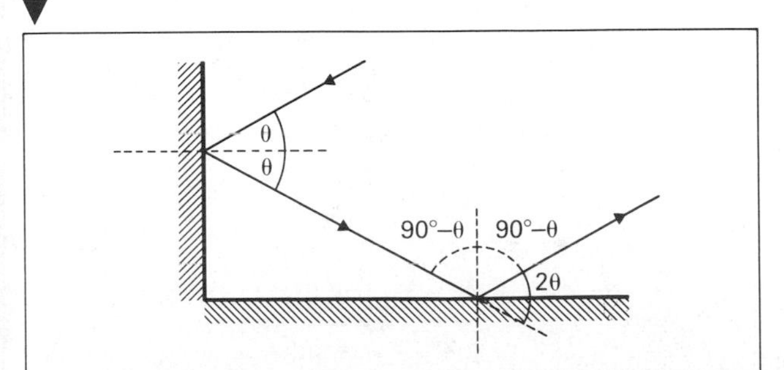

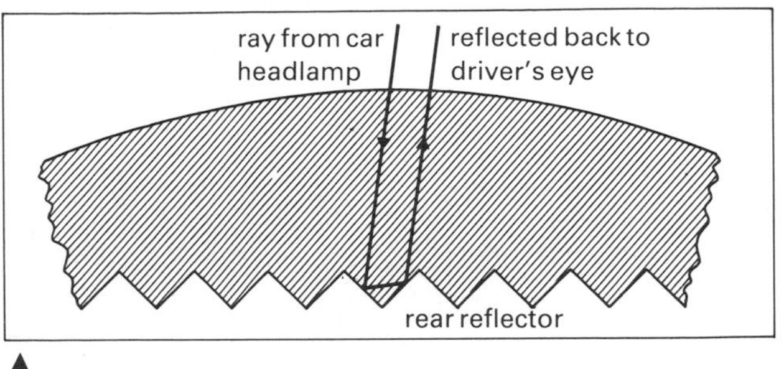

▲
fig. 19.12 Principle of the rear reflector

Curved mirrors

The simplest kinds of curved mirror are those which are parts of cylinders (*cylindrical mirrors*) or parts of spheres (*spherical mirrors*). If the polished surface bulges outwards towards you it is **convex**, and if it caves in away from you it is **concave**, figure 19.13.

From a beam of parallel rays the concave mirror produces a beam which converges to a **focus**, and then the rays spread out again. The convex mirror spreads out or diverges the beam. How much the rays converge or diverge depends on how curved the mirror is. Figure 19.14 shows the behaviour of a beam of

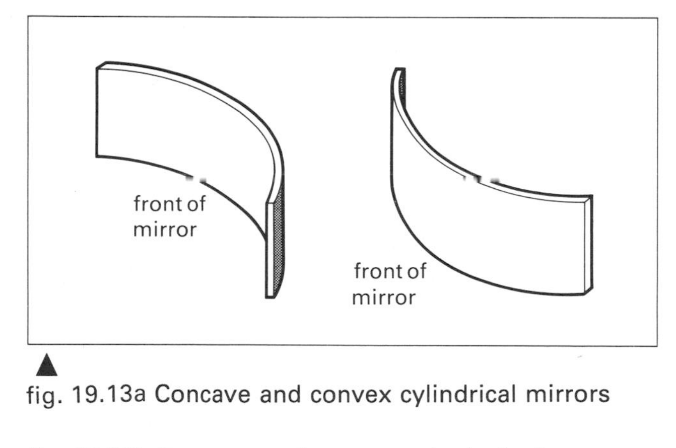

▲
fig. 19.13a Concave and convex cylindrical mirrors

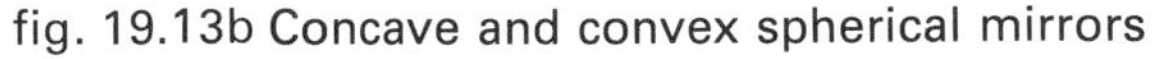

fig. 19.13b Concave and convex spherical mirrors
▼

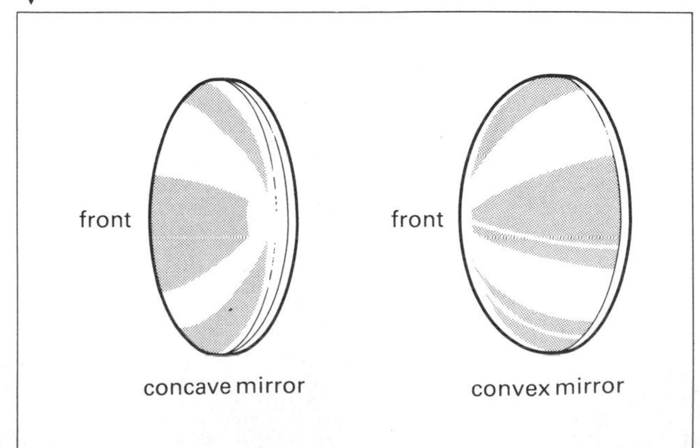

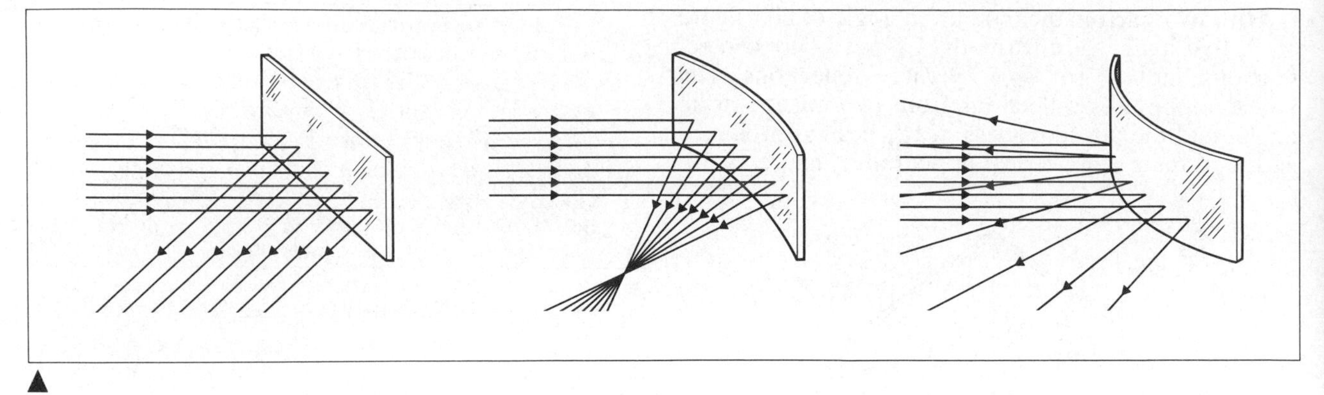

fig. 19.14 Reflection of a parallel beam from plane and curved mirrors

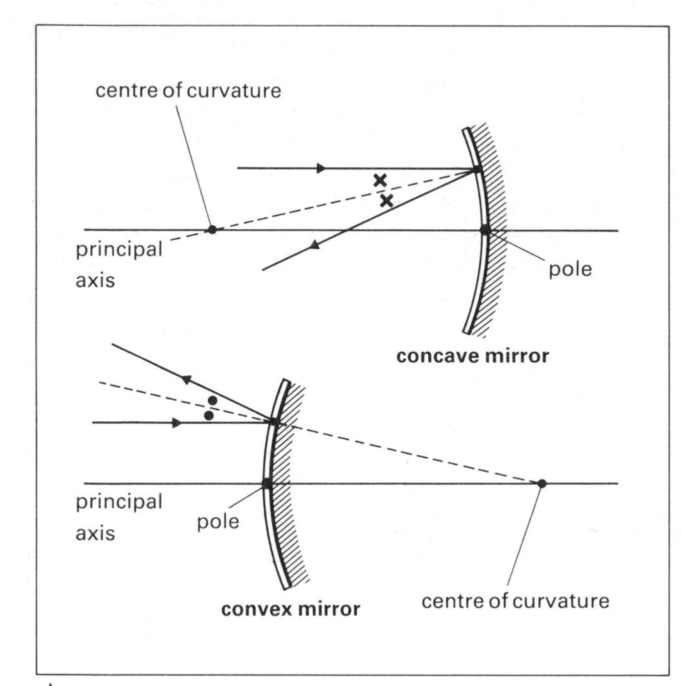

fig. 19.15 Terms used for curved mirrors

parallel rays falling on a plane, a concave and a convex mirror.

The laws of reflection hold good for curved mirrors just as they do for plane ones. If we think of a very small part of the mirror it will be practically flat. The radius of a circle is perpendicular to each small part of the circumference and can be taken as the normal. The radius through the middle point of the mirror is called the **principal axis** of the mirror. The centre of the reflecting surface is called the **pole**.

The distance of the centre of curvature from the pole of the mirror is the **radius of curvature**. When a beam of rays which are all parallel to the principal axis falls on a concave cylindrical or spherical mirror, the rays all converge to the **principal focus**, and the distance of the principal focus from the pole of the mirror is the **focal length**. For mirrors which are not very curved, the principal focus is half-way between the centre of curvature and the mirror.

The image in a plane mirror is always the same size and the same way up as the object, but the images in curved mirrors vary in size and nature with the kind of mirror used and the distance of the object from the mirror.

fig. 19.16a A concave mirror has a real focus

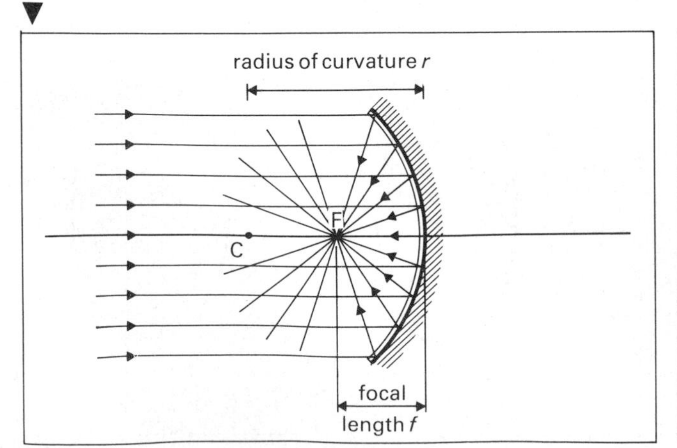

fig. 19.16b A convex mirror has an imaginary focus

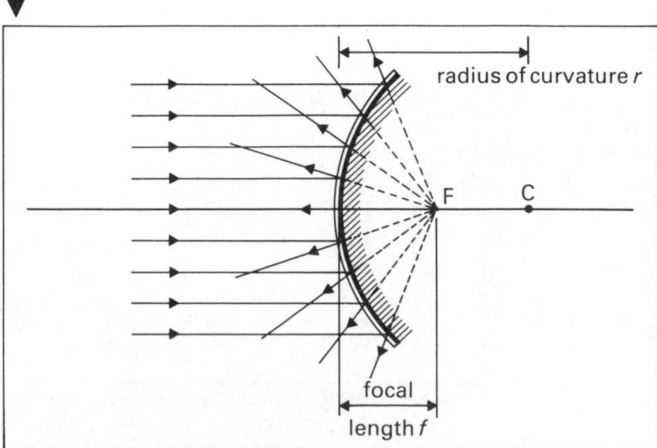

How to draw ray diagrams for curved mirrors

When an object is placed in front of a mirror, we want to know where the image will be and what kind of an image it is. Our general problem is that if light rays start from a point on an object, what happens to them after reflection by the mirror? We make the problem easier for ourselves by taking rays whose directions after reflection are known and can easily be drawn in a diagram, figure 19.17.

EXAMPLES

In figure 19.18, an object O is placed 60 cm from a concave mirror whose radius of curvature CP is 40 cm (focal length 20 cm). We can use a scale in which 1 cm represents 5 cm. Draw OB 2 cm high to represent the object. It is impossible on this scale to draw the mirror with its correct curvature. It is best to use a straight line for the mirror, only drawing the curved line to show what kind of a mirror you are using.

From B draw (i) a ray parallel to the principal axis and the reflected ray passing through F; (ii) a ray through F and the reflected ray parallel to the axis; (iii) a ray through C which is reflected straight back. It will be seen that all the reflected rays meet at the point M, which is the image of B. The complete image IM can now be drawn in. It is 30 cm from the mirror. Only two rays are really necessary to fix the position of M. In fact, all rays from B pass through M after reflection at the mirror. An eye placed as shown in the diagram sees a set of rays diverging from M. If we measure the image, we find it is 1 cm or half the length of the object.

The complete description of this image is **real**, because rays of light really pass through it; **inverted**, because it is the opposite way up to the object; **diminished**, because it is smaller than the object.

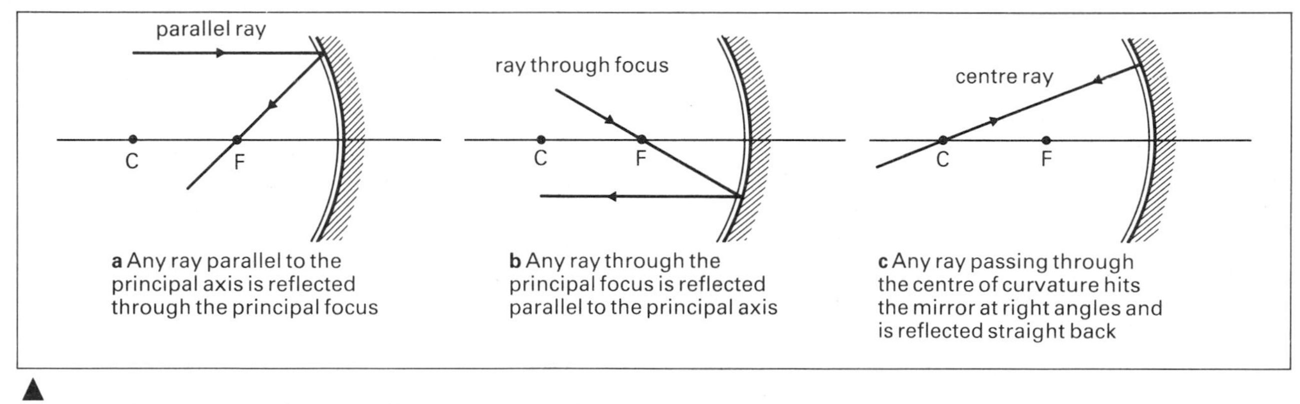

a Any ray parallel to the principal axis is reflected through the principal focus

b Any ray through the principal focus is reflected parallel to the principal axis

c Any ray passing through the centre of curvature hits the mirror at right angles and is reflected straight back

▲ fig. 19.17 The construction rays for a concave mirror

fig. 19.18 ▼

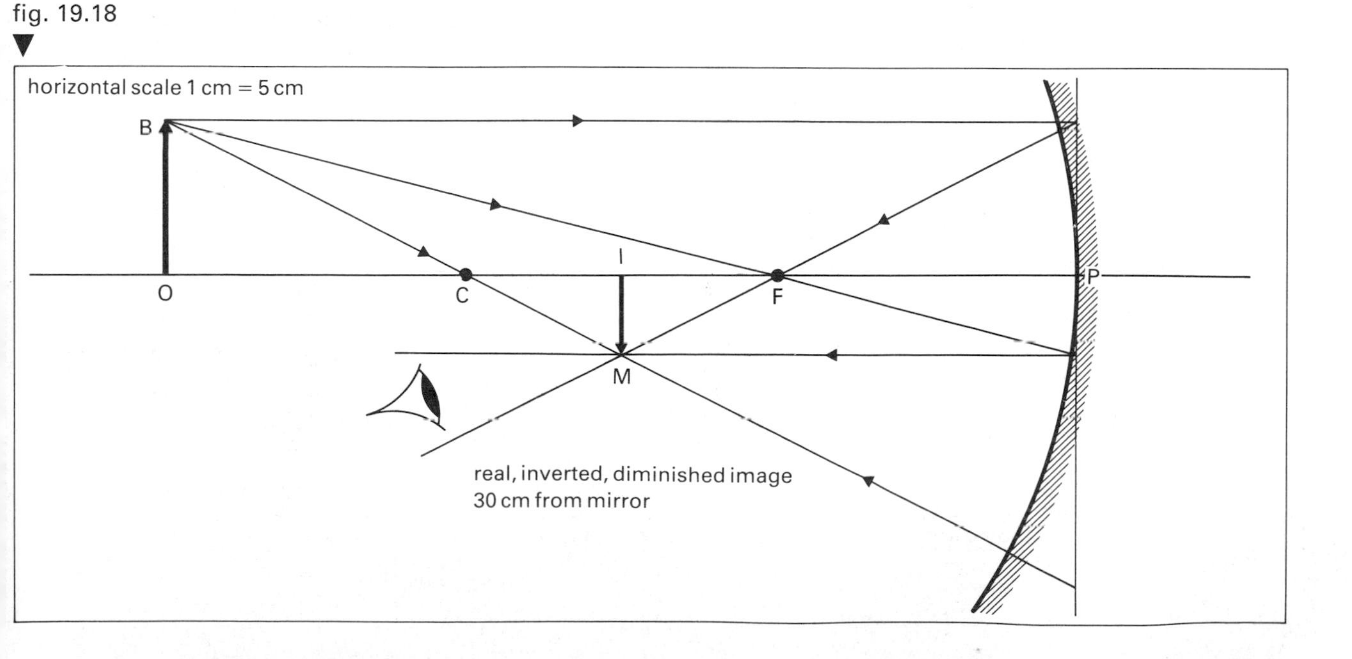

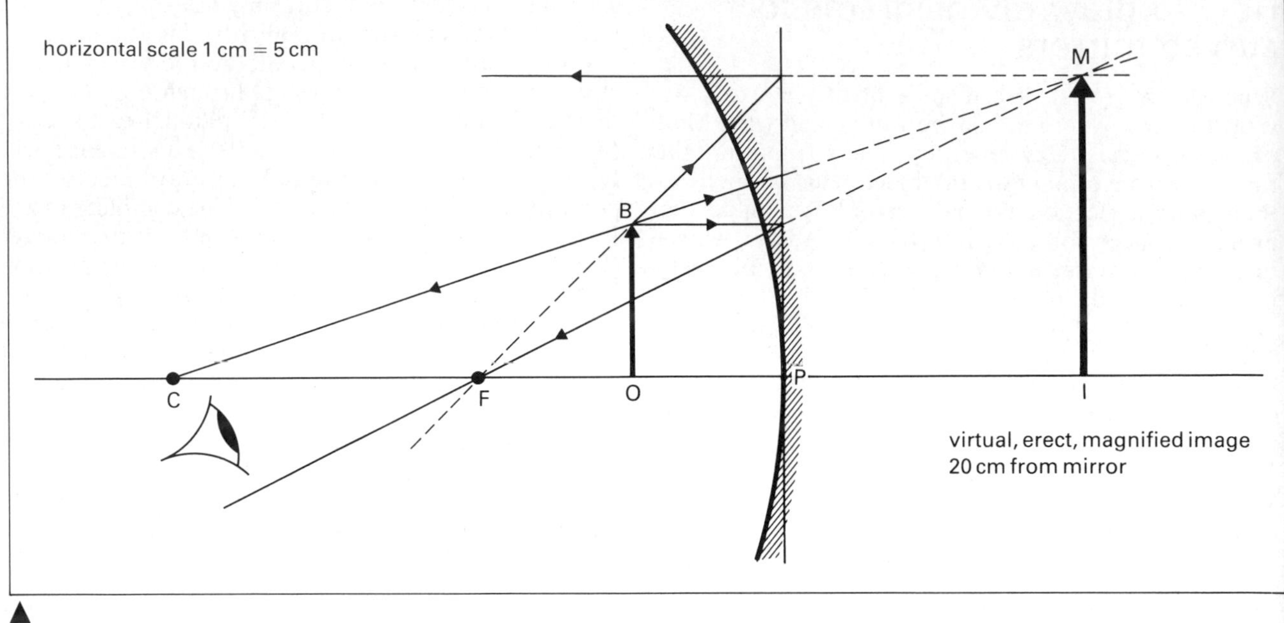

fig. 19.19

In figure 19.19, when the object O is 10 cm from the mirror (i.e. nearer to the mirror than the principal focus), we find on drawing the ray diagram that the reflected rays do not meet in front of the mirror. On drawing them back to the right of the mirror (use dotted lines because they are not real rays of light), we find that they meet at M, which is the image of B. The image is **virtual** or **imaginary**, because the light rays do not pass through it. We only imagine the reflected rays set out from M. It is **erect** because it is the same way up as the object; **magnified** because it is larger than the object. In this particular case, the image is 20 cm behind the mirror. This is the kind of image produced by a shaving mirror.

There are similar construction rays you can use with a convex mirror. These are shown in figure 19.20.

EXAMPLE

Figure 19.21 shows an object placed 20 cm from a convex mirror of radius of curvature 40 cm. Where is the image?

Repeat this diagram with the object at several different distances from the mirror.

What do you notice about all the images?

fig. 19.20 The construction rays for a convex mirror

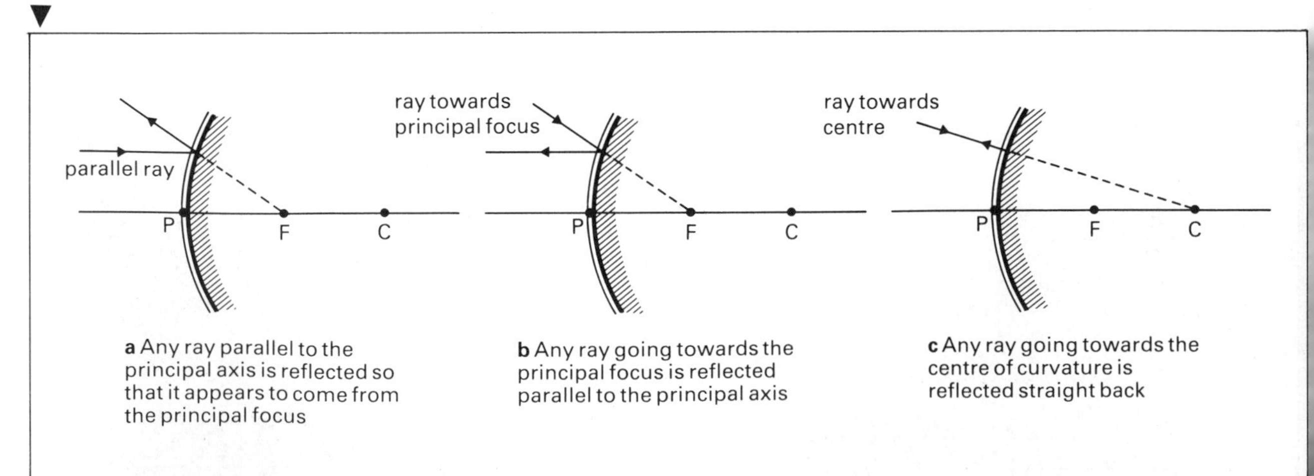

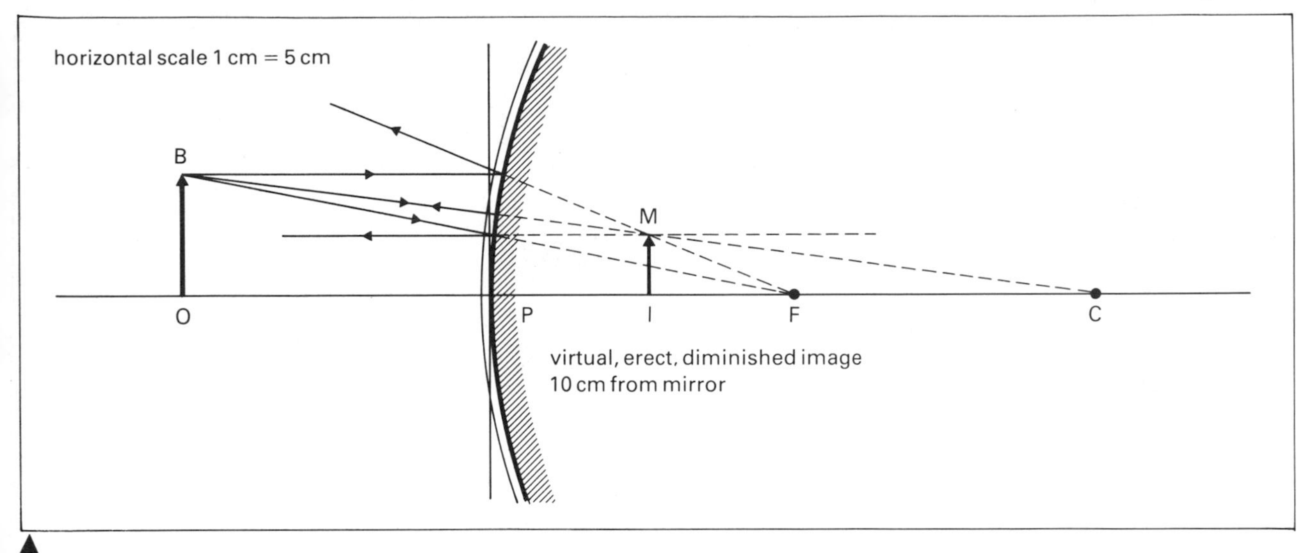

fig. 19.21

fig. 19.22 A *narrow* parallel beam of light can be obtained from a point source at the principal focus of a *spherical* mirror of *small* aperture. To obtain a *wide* parallel beam, a *parabolic* mirror must be used

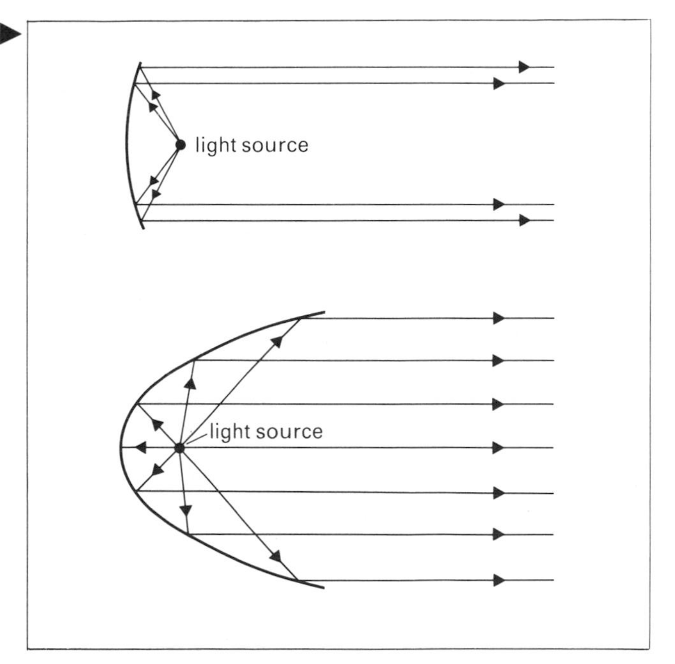

Summary of images formed by curved mirrors

The general positions and nature of the images formed by concave and convex mirrors are summarised in table 19.1.

table 19.1

Type of mirror	Position of object	Position of image	Nature of image
Concave	further away than C	between C and F	real, inverted, diminished
	at C	at C	real, inverted, same size
	between C and F	further away than C	real, inverted, magnified
	at F	at infinity	no image formed (reflected rays parallel)
	between F and P	behind the mirror	virtual, erect, magnified
Convex			images are always virtual, erect and diminished

Some uses of curved mirrors

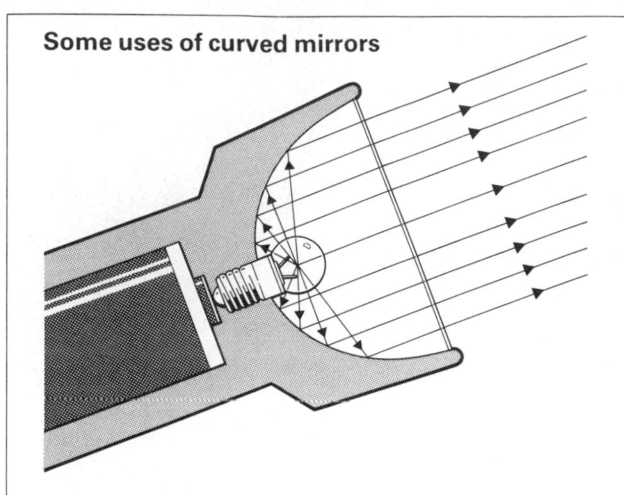

In a torch, the filament of the bulb is located at the focus of the concave reflector to give a parallel beam

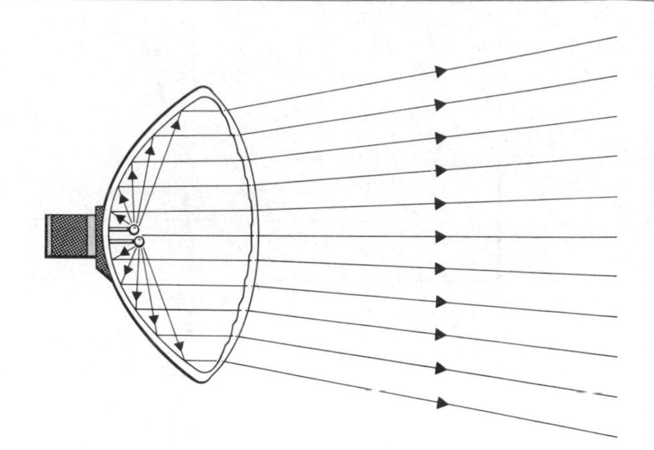

In the sealed-headlamp, the main beam filament is usually located at the focus of the concave reflector to give a parallel beam, which is then 'shaped' by the front lens composed of many prisms

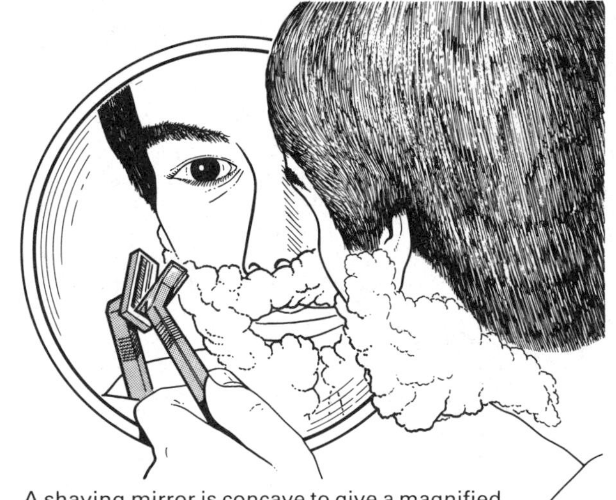

A shaving mirror is concave to give a magnified image of the face

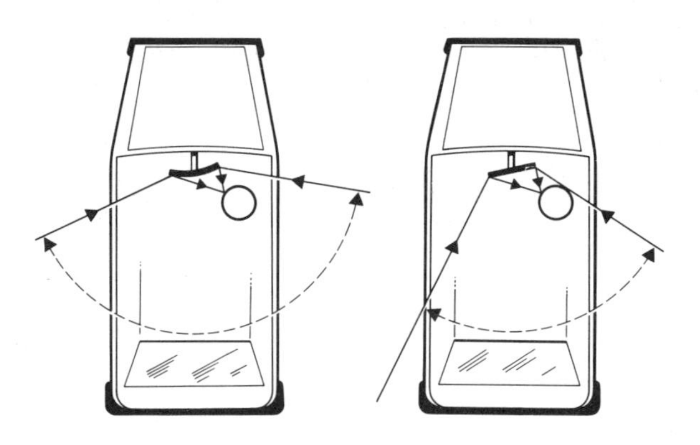

A convex driving mirror gives a wider field of view than a plane mirror of the same size, but the image is diminished

Curved parabolic reflectors are used in radio-telescopes. The reflector at Jodrell Bank, Cheshire, is 75 m in diameter and the receiving aerial is mounted at the principal focus. Curved reflectors are also used to beam radio signals to satellites.

EXERCISE 19

In questions 1–5 select the most suitable answer.

1 The image formed in a plane mirror is
 A real and the same size as the object
 B real and nearly the same size as the object
 C virtual and the same size as the object
 D virtual and nearly the same size as the object
 E virtual and half the size of the object

2 An optician's test card is fixed 80 cm behind the eyes of a patient, who looks into a plane mirror 300 cm in front of him, as shown in figure 19.23. The distance from his eyes to the image of the card is
 A 300 cm
 B 380 cm
 C 600 cm
 D 680 cm
 E 760 cm

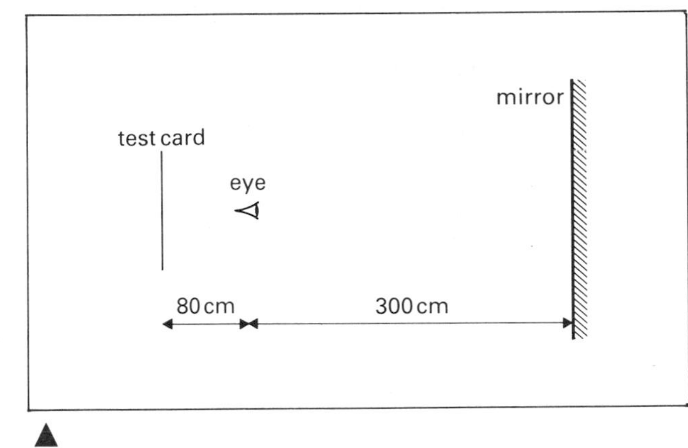

fig. 19.23

3 Two plane mirrors are supported with their surfaces at right angles and a small object is placed between them. How many reflected images of this object can be seen in the mirrors?
 A 1
 B 2
 C 3
 D 4
 E 5

4 A convex mirror is often used as an outside rear-view mirror on a car because
 A it has a wide field of view
 B it has a narrow field of view
 C the image is always magnified
 D the image is always real
 E the image appears the same size as the object

5 The image formed by a shaving or a make-up mirror is
 A real and erect
 B real and inverted
 C virtual and erect
 D virtual and inverted
 E virtual and the same size as the object

6 State the laws of reflection and indicate how you would verify them.

7 Complete the ray diagram in figure 19.24 in order to show how a plane mirror forms an image.

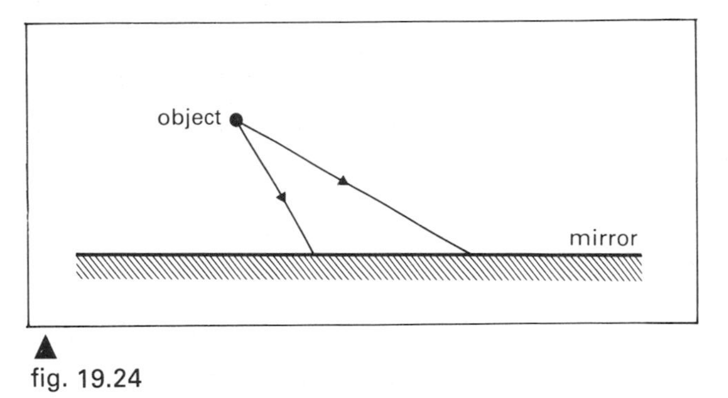

fig. 19.24

8 Figure 19.25 shows a left-hand glove in front of a plane mirror. Draw the image in the correct position. Draw the rays of light by which the eye sees the tip of the forefinger and the button of the glove.

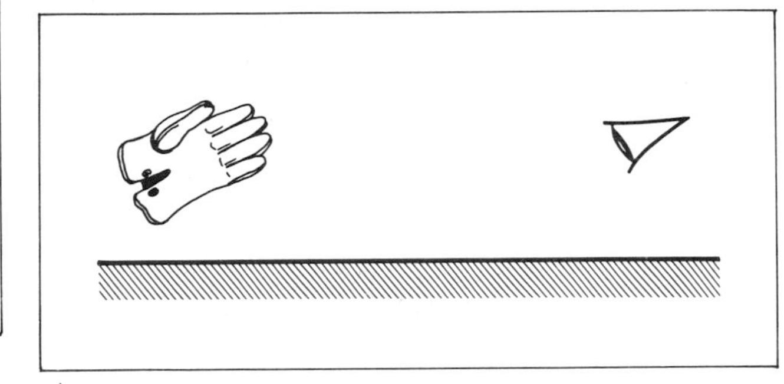

fig. 19.25

9 Draw a diagram of a simple periscope using mirrors, showing *two* parallel rays from a distant object through the system. State whether the final image is the same way up as the object or upside down.

10 Accurately copy and complete figure 19.26 to show the path of the ray of light. What common piece of motor vehicle equipment relies on the principle apparent in your answer?

fig. 19.26

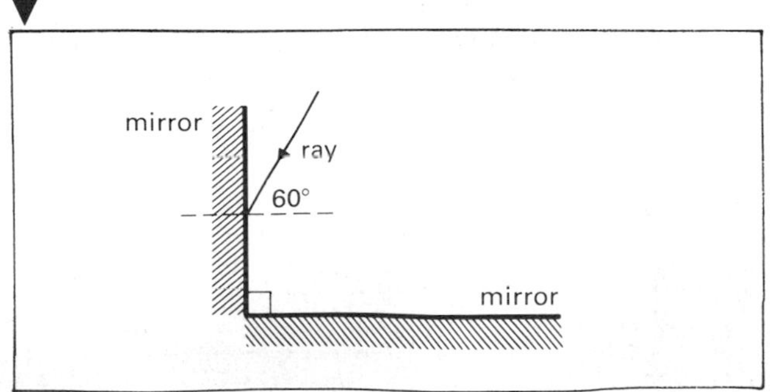

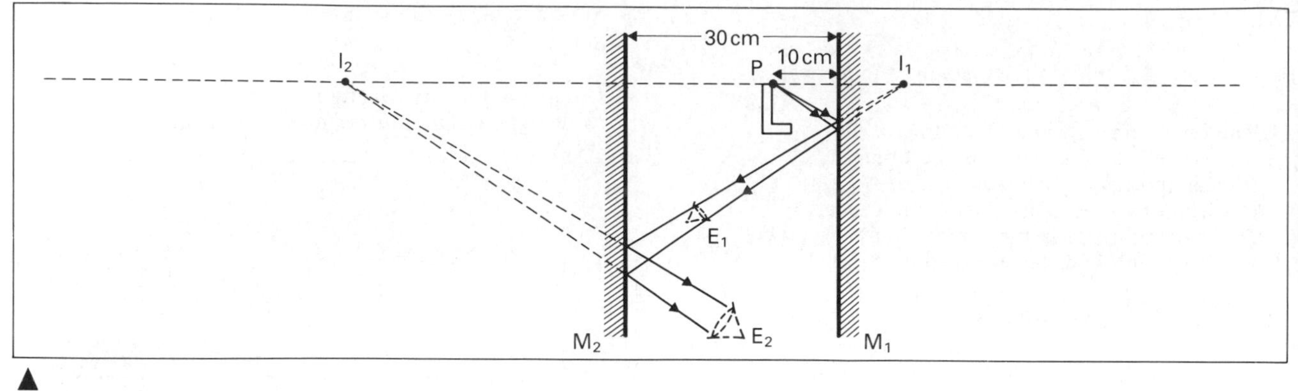

fig. 19.27

11 Figure 19.27 shows an object in the form of a letter L placed between two parallel plane mirrors M_1 and M_2 30 cm apart. The point P on the object is 10 cm from the mirror M_1. An eye positioned at E_1 will see the image I_1 produced by a single reflection and if moved to E_2 will see the image I_2 formed by a double reflection.

a (i) How far from mirror M_1 is the point I_1? (ii) How far from the mirror M_2 is the point I_2?

b Draw on the diagram the image of the letter L formed (i) at I_1, (ii) at I_2.

c When the eye is at E_2 another image can be seen in mirror M_2 formed by a single reflection. (i) Show the shape of this image formed in M_2. (ii) How far is this image behind M_2?

12 Show how the effect of a concave and a convex mirror on a parallel beam of light differs by completing the diagrams in figure 19.28.

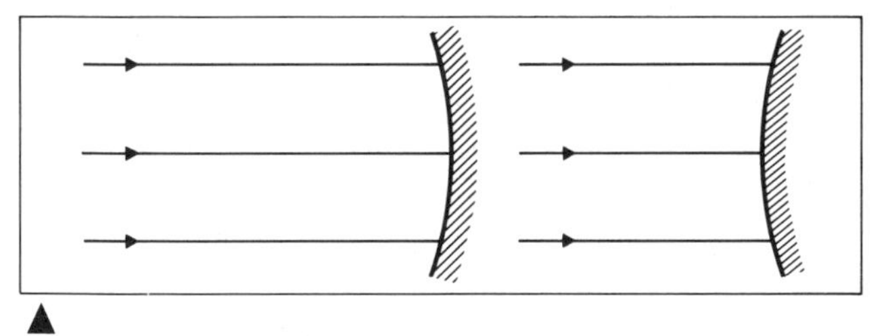
fig. 19.28

13 Figure 19.29 shows a concave mirror reflecting rays of light parallel to the principal axis. On the diagram label

a the pole of the mirror, P

b the principal focus, F

c the centre of curvature, C

d the focal length, *f*.

fig. 19.29

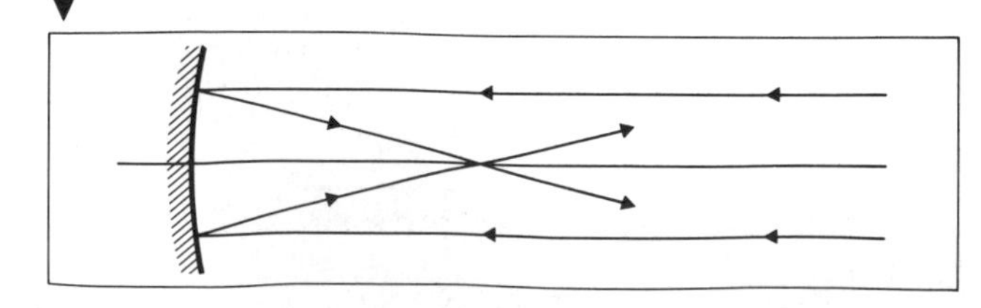

14 Three mirrors, one plane, one convex and one concave, are available for a series of experiments.

a Explain, with a diagram, how the plane mirror could be used to show the relationship between the angle of incidence and the angle of reflection.

b Describe carefully the appearance of the image of himself which the experimenter would see if he looked into each mirror in turn with the mirror held close to his face.

c Choose one of the mirrors which would be suitable for use as a small reflecting telescope. Draw a ray diagram to show how it would produce an image of a distant object.

d Choose a suitable mirror for use above the steps of a double-decker bus and state *one* reason for your choice.

e A dentist's mirror may produce a magnified virtual upright image of the back of a tooth. Choose a suitable mirror and draw a ray diagram to show how the required image is formed.

15 Draw diagrams to show (i) a convex mirror can be used to give a large field of view, (ii) a concave mirror can be used to give an enlarged upright image, (iii) a concave mirror and a small source of light can be used to produce an approximately parallel beam of light.

b An illuminated object and a concave mirror are used to produce a sharp image of the object on a screen. The corresponding magnifications (linear) and the image distances are given in the table.

Magnification m	0.25	1.5	2.5	3.5
Image distance v (cm)	20	40	56	72

Draw a graph plotting m along the vertical axis and v along the horizontal axis. Use the graph to find the image distance when $m = 1.0$.

What is the object distance when $m = 1.0$? What is the focal length of the mirror?

20 REFRACTION OF LIGHT

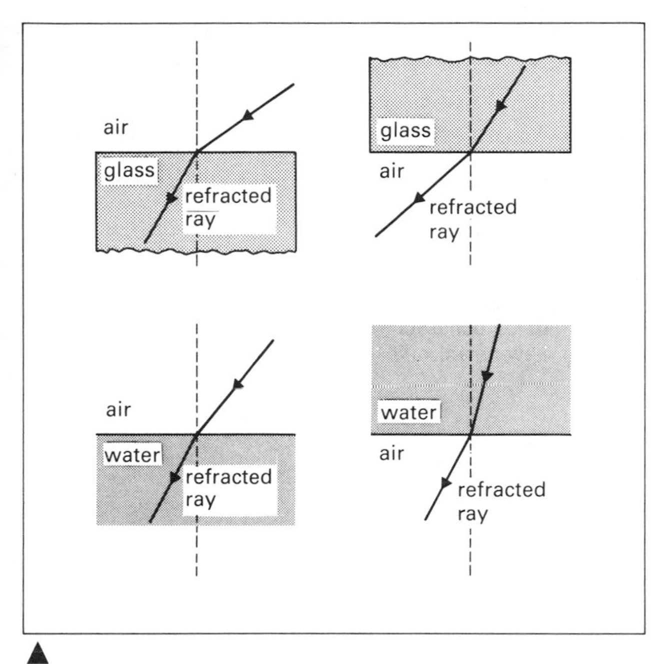

▲
fig. 20.1 A ray of light travelling from one medium into another is refracted

fig. 20.2 Terms used in refraction
▼

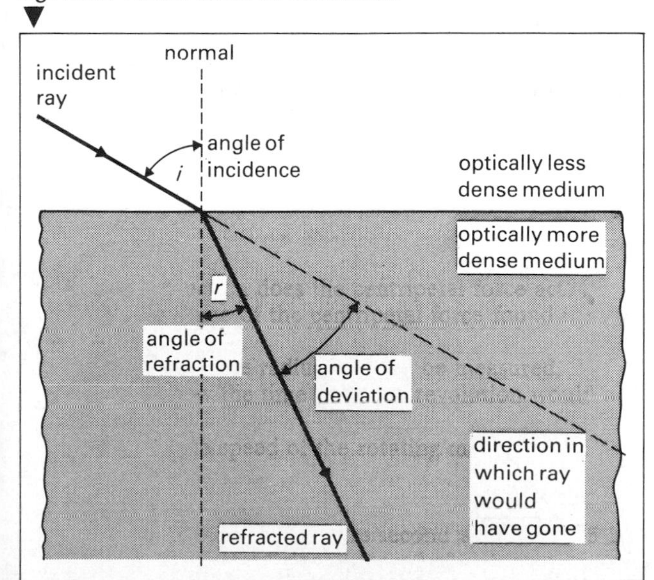

Light travels in straight lines so long as it remains in the same substance. When it passes from one transparent substance to another it is bent. **Refraction** is the change of direction of travel of light when it crosses the boundary between two media.

When light enters glass or water from air, it is bent *towards* the normal (the line at right angles to the surface). When it leaves glass or water for air, it is bent *away from* the normal. The bending is greater for the glass than for the water. Glass and water are said to be **optically denser** than air, and glass has a greater optical density than water.

When light passes from a less to a more optically dense substance, it is refracted towards the normal

When light passes from a more to a less optically dense substance, it is refracted away from the normal

The laws of refraction

Law 1 The incident ray, the normal and the refracted ray all lie in the same plane.

Law 2

The ratio $\dfrac{\textbf{sine of the angle of incidence } (i)}{\textbf{sine of the angle of refraction } (r)}$

is constant for light of a given colour (frequency) crossing the boundary of two given media.

This ratio is called the **refractive index** (n) for the boundary, that is

$$n = \frac{\sin i}{\sin r}$$

The refractive index of a transparent substance is measured when light enters the substance from a vacuum, but for all practical purposes we can take it as light entering the substance from air. The greater the value of the refractive index, the greater the bending of the light.

Refraction is caused by the fact that light travels at different speeds in different media and it can be shown that the refractive index is equal to the ratio of the speeds in the two media.

$$\text{refractive index of glass} = \frac{\text{speed of light in air}}{\text{speed of light in glass}}$$

This relationship is derived on p. 202

table 20.1 Some common refractive indices

alcohol	1.36
diamond	2.42
glass, crown	1.52
glass, flint	1.65
glycerine	1.47
ice	1.31
paraffin	1.44
Perspex	1.49
water	1.33

Images formed by refraction

The change in direction of light crossing a boundary between two substances affects the way in which we see objects. The man looking for a fish to spear, figure 20.3, sees an image of the fish. Light rays from the fish are bent away from the normal on leaving the water, and the direction in which they enter the eye shows that the fish will appear somewhere on the line AN. An accurate drawing using two rays of light fixes the image in the position shown. It is an imaginary image and if the man aims his spear at it he will only catch an imaginary fish!

All objects seen in water appear to shift their position, and the amount of shift depends upon their depth and the position from which they are viewed (see figures 20.4 and 20.5).

fig. 20.4 A stick appears bent when it enters water

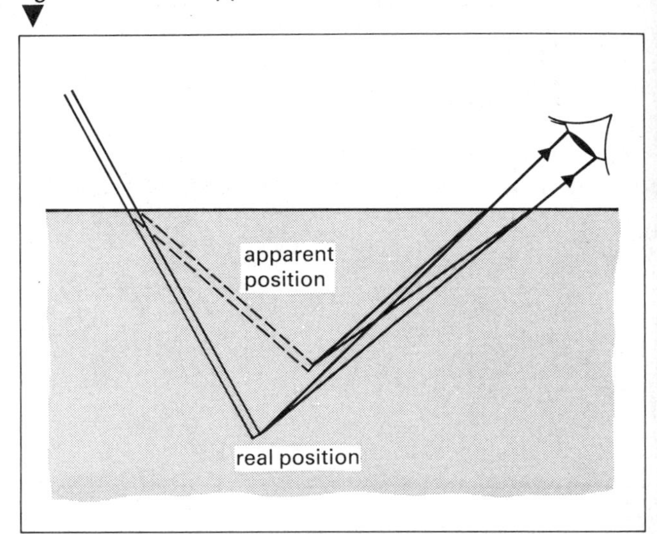

fig. 20.5 The problem of trying to spear a fish from a too oblique angle

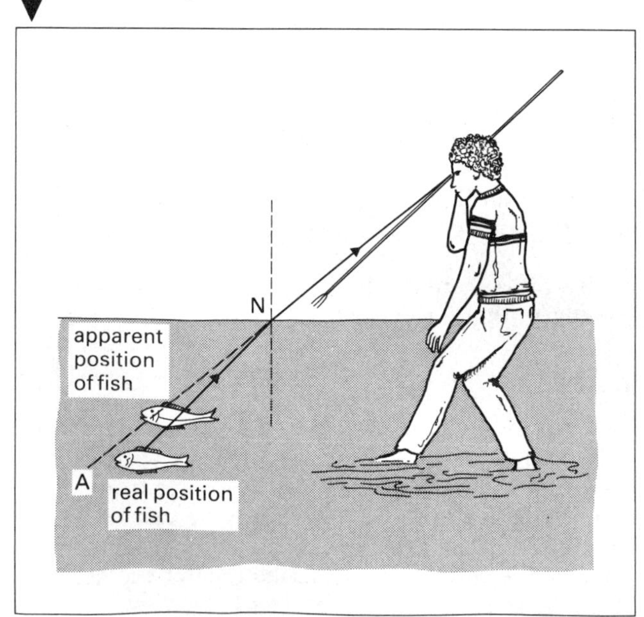

fig. 20.5 Place your eye so that you just cannot see a coin at the bottom of a dish. Then get someone to fill the dish with water

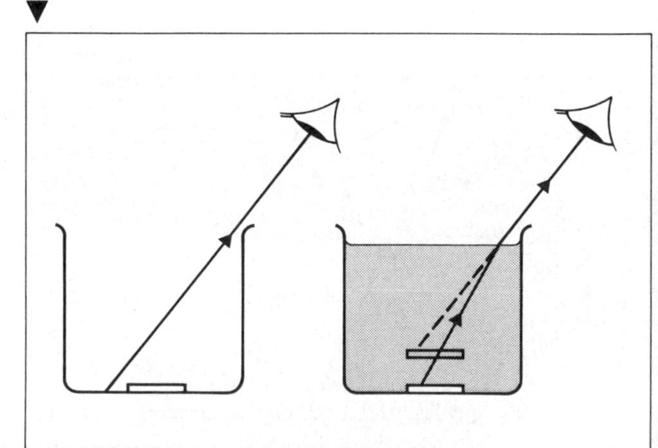

Real and apparent depth

The relationship between the real and apparent depth of a slab of transparent material can be found as follows.

In figure 20.6a, rays from O at the bottom of the slab are refracted at the surface and when they enter the eye appear to come from I, the image of O.

$$\text{refractive index } n = \frac{\sin \theta_1}{\sin \theta_2}$$

But $\theta_1 = \alpha$ and $\theta_2 = \beta$ (corresponding and alternate angles), therefore

$$\text{refractive index} = \frac{\sin \alpha}{\sin \beta}$$

$$\sin \alpha = \frac{NA}{AI} \quad \text{and} \quad \sin \beta = \frac{NA}{AO}$$

$$\text{Thus refractive index} = \frac{NA}{AI} \div \frac{NA}{AO} = \frac{AO}{AI}$$

Now if the beam of light is very narrow, A is very close to N and the ratio AO/AI will be very nearly equal to the ratio NO/NI, i.e. real depth to apparent depth. So we have shown that

$$\textbf{refractive index } n = \frac{\textbf{real depth}}{\textbf{apparent depth}}$$

In the case of water, $n = \frac{4}{3}$. The apparent depth is three-quarters of the real depth. For glass, $n = \frac{3}{2}$. The apparent depth is two-thirds of the real depth.

It is important to remember that this relationship only holds for normal viewing, that is when the rays are practically at right angles to the surface. If we view the bottom of the slab obliquely the apparent depth is very much smaller, figure 20.6b.

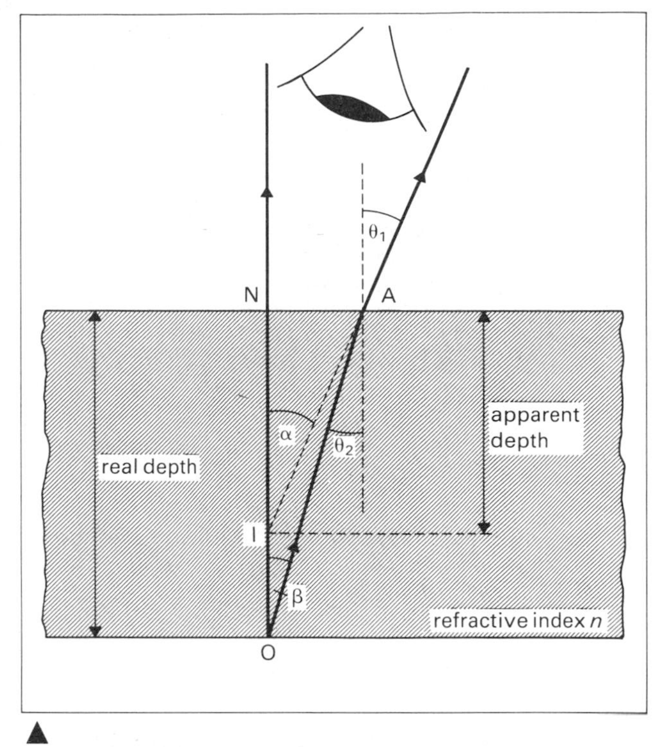

fig. 20.6a Apparent depth for vertical viewing

fig. 20.6b Reduced apparent depth in oblique viewing

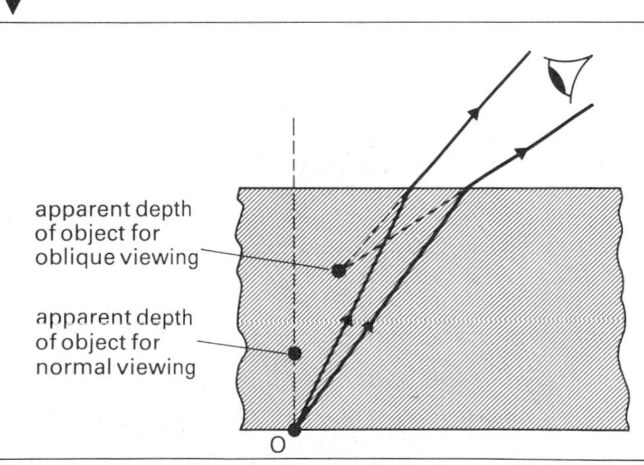

Determination of refractive index by real and apparent depth

The accurate measurement of real and apparent depth using a travelling microscope enables the refractive index to be measured. The method can be used for solids and liquids. The experiment is illustrated in figure 20.7.

The microscope is first focused on a mark (O) on a piece of paper. The scale reading is taken (y_1). A slab of material or a shallow dish of liquid is placed over the mark and the microscope focused on the image of the mark (I). The scale reading is taken (y_2). Finally, the microscope is raised still further till it is focused on the surface of the slab or liquid, which can be made clearer by sprinkling a little fine powder (e.g. lycopodium or French chalk) on the surface. The scale reading is taken (y_3).

$$\text{apparent thickness SI} = y_3 - y_2$$

$$\text{real thickness SO} = y_3 - y_1$$

$$\text{refractive index} \quad n = \frac{y_3 - y_1}{y_3 - y_2}$$

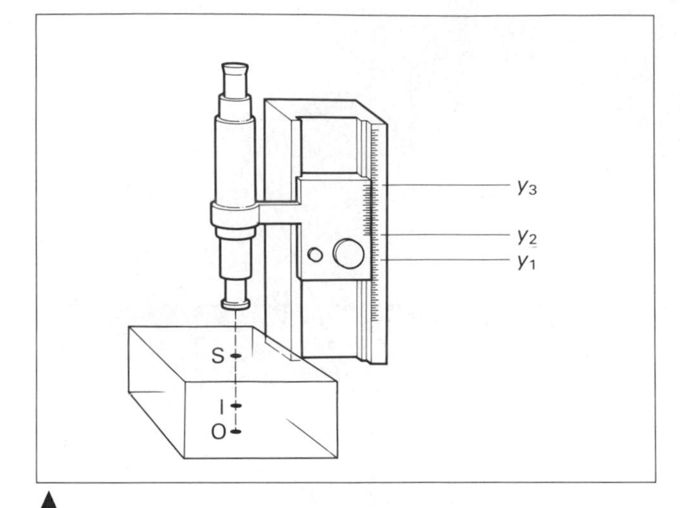

fig. 20.7 Refractive index by apparent depth method

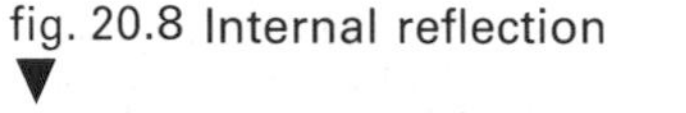
fig. 20.8 Internal reflection

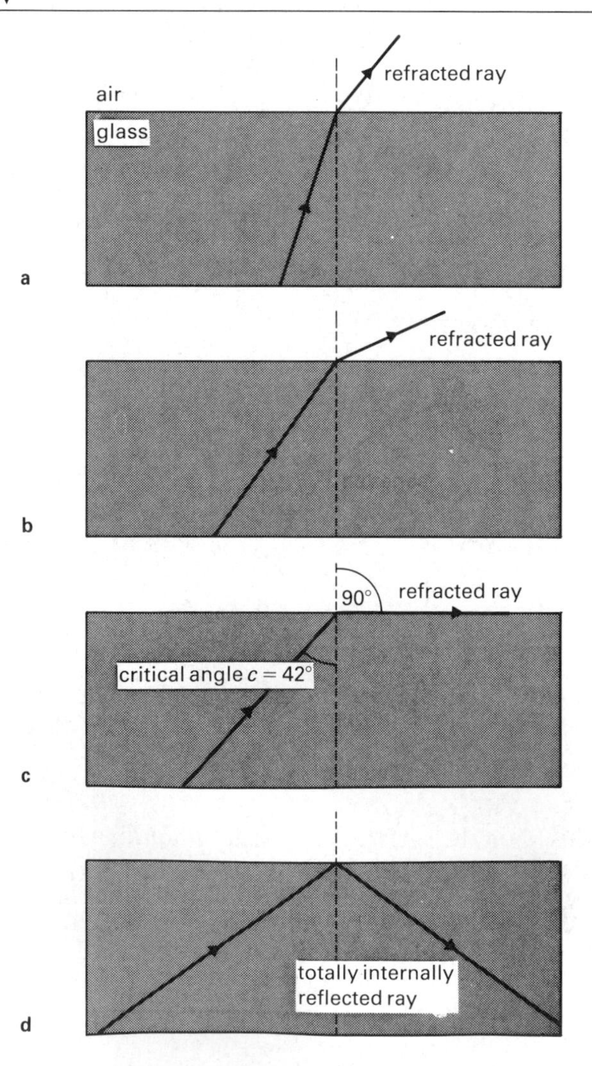

Total internal reflection and critical angle

Consider what happens to a ray of light leaving glass to enter air, figure 20.8. It is bent away from the normal (a). If the angle between the ray and the normal is increased it is bent more (b). As the angle with the normal is further increased, a stage is reached when the ray emerges just along the surface of the glass, and so there cannot be any more bending (c). You can easily find out, using ray apparatus, figure 20.9, that this occurs when the angle the ray in the glass makes with the normal is about 42° (the angle varies a little with the kind of glass used). It is an advantage to use a semicircular block of glass, because there are no corners to get in the way of the ray. If the angle is made greater than 42°, no light emerges and the ray is reflected back into the glass (d). This is known as **total internal reflection**, and the angle for which the light just gets out is called the **critical angle**. The size of this angle for a water/air surface is 49°.

We have treated reflection and refraction as if they took place quite separately, but when a ray of light strikes a glass surface some is always reflected and some passes through. It is because of the light reflected from the glass that we can see ourselves in the shop windows as we walk by. When light strikes a glass/air surface at an angle greater than 42°, reflection is complete, and that is why we use the name **total** internal reflection.

Glass/air surfaces are often used instead of ordinary mirrors, and they have these advantages:

- They do not need cleaning like a metal mirror which tarnishes.
- They are not as easily damaged as the silvered back of a glass mirror.
- They are stronger than thin glass mirrors.
- They do reflect all the light falling on them.
- Reflection is at a single surface, so only one clear image is formed.

fig. 20.9 Measuring the critical angle

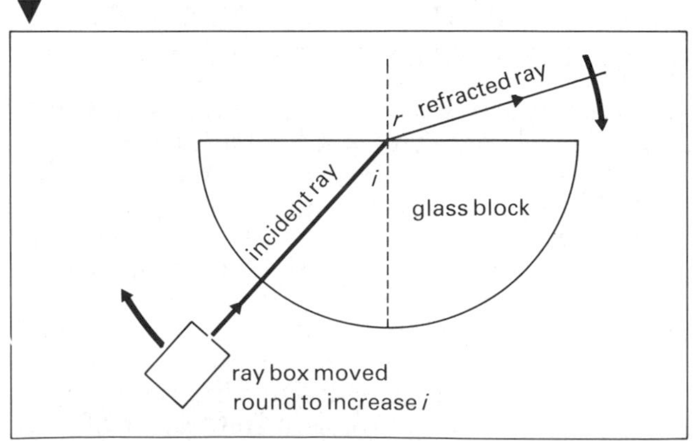

Examples of total internal reflection

Glass prisms are used in periscopes. Notice that a 90°, 45°, 45° prism must be used so that the angle which the light makes with the normal is 45°, and is greater than the critical angle, so that complete reflection takes place, figure 20.10. Other examples of the application of total internal reflection are shown in figures 20.11 to 20.13.

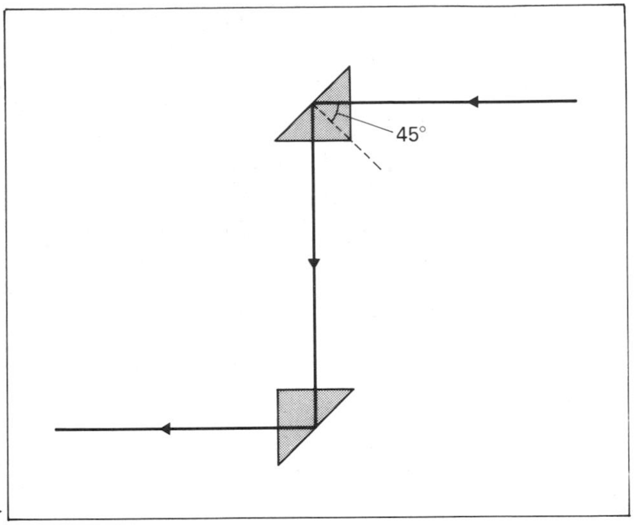

fig. 20.10 Total reflection by periscope prisms

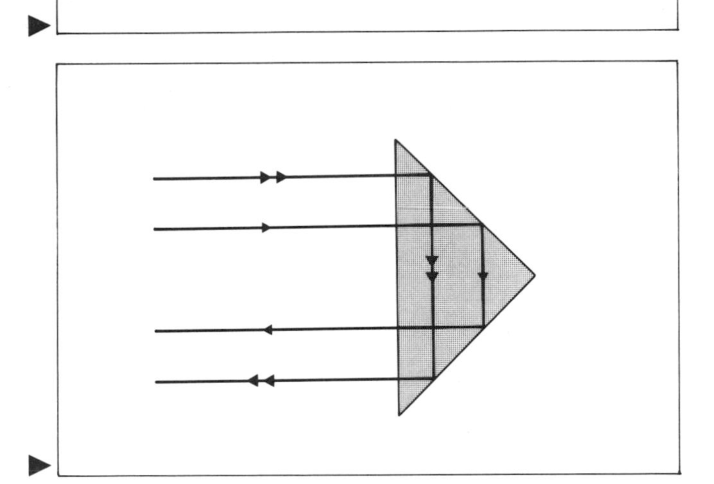

fig. 20.11 The rays are deviated through 180° and reversed top to bottom. When two such prisms are used, as in prismatic binoculars, the reversal is corrected

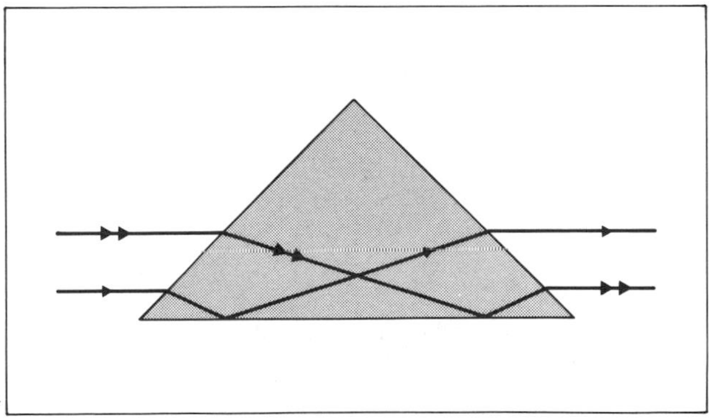

fig. 20.12 The direction of the beam is unchanged, but top and bottom are reversed. This prism is used to correct an inversion which has already occurred in an optical instrument

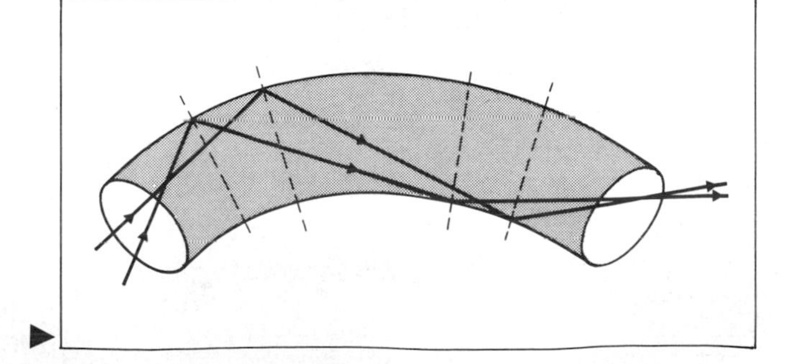

fig. 20.13 Light can be internally reflected in a rod of glass or Perspex. This is the principle of the 'light pipe' in fibre optics

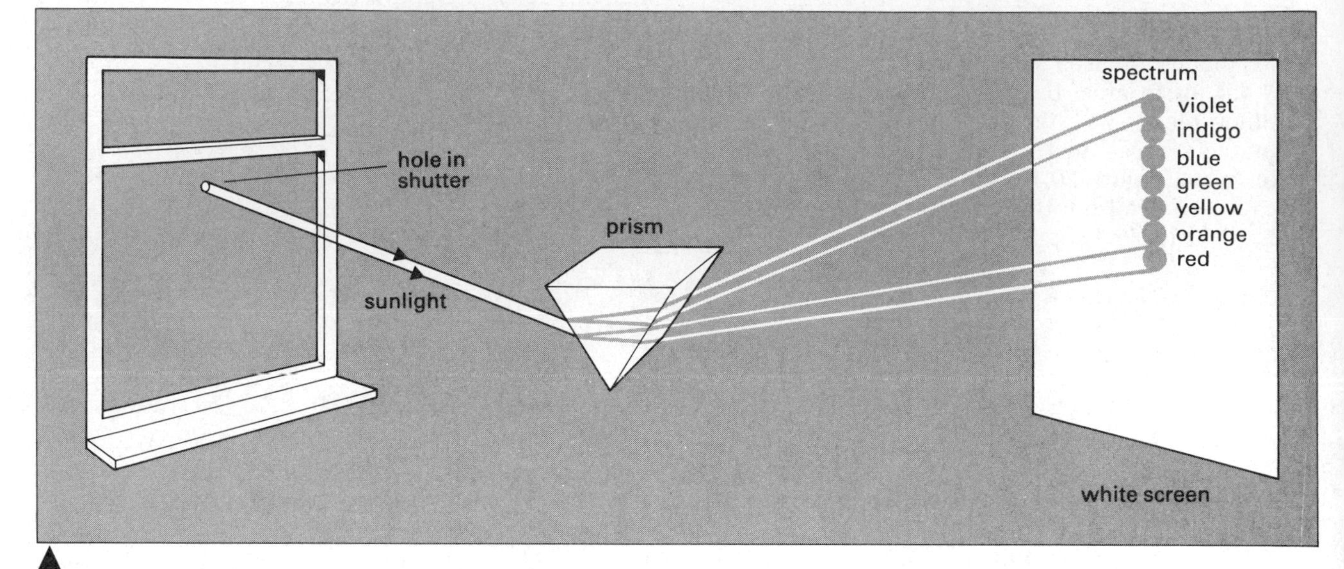

fig. 20.14 A reconstruction of Sir Isaac Newton's experiment

Formation of a spectrum

When a narrow beam of white light passes through a prism it is not only bent (refracted) as it enters and leaves the prism, but also spread out into the colours of the rainbow. These are usually given as red, orange, yellow, green, blue, indigo and violet, but many people cannot distinguish the indigo as a separate colour. This separation of white light into different colours is known as **dispersion** and the band of colours is called a **spectrum,** figure 20.14. The effect was first investigated experimentally by Sir Isaac Newton in the years following 1666.

Note that blue light is deviated more than red light, so the refractive index for blue light is greater than that for the red light, figure 20.15. White light is a mixture of all the colours of the spectrum. The prism separates the white light into its coloured components because of their differing refractive indices.

If a wide beam of light is used, there is overlapping between the emergent beams and the spectrum is not clear. In order to obtain a 'pure' spectrum it is necessary to use a lens to focus the spectrum clearly. Figure 20.16 shows the experimental arrangement. Without the prism in place, a sharp image of the single slit is focused on the screen in position (1). Now the prism is put in place and the screen moved to position (2). The prism is rotated slightly until the best result is obtained.

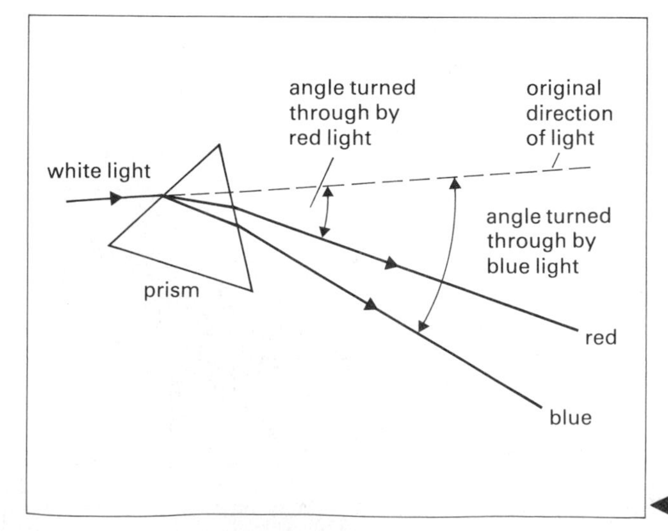

fig. 20.15 Dispersion of white light by a prism. Blue-violet light is bent the most

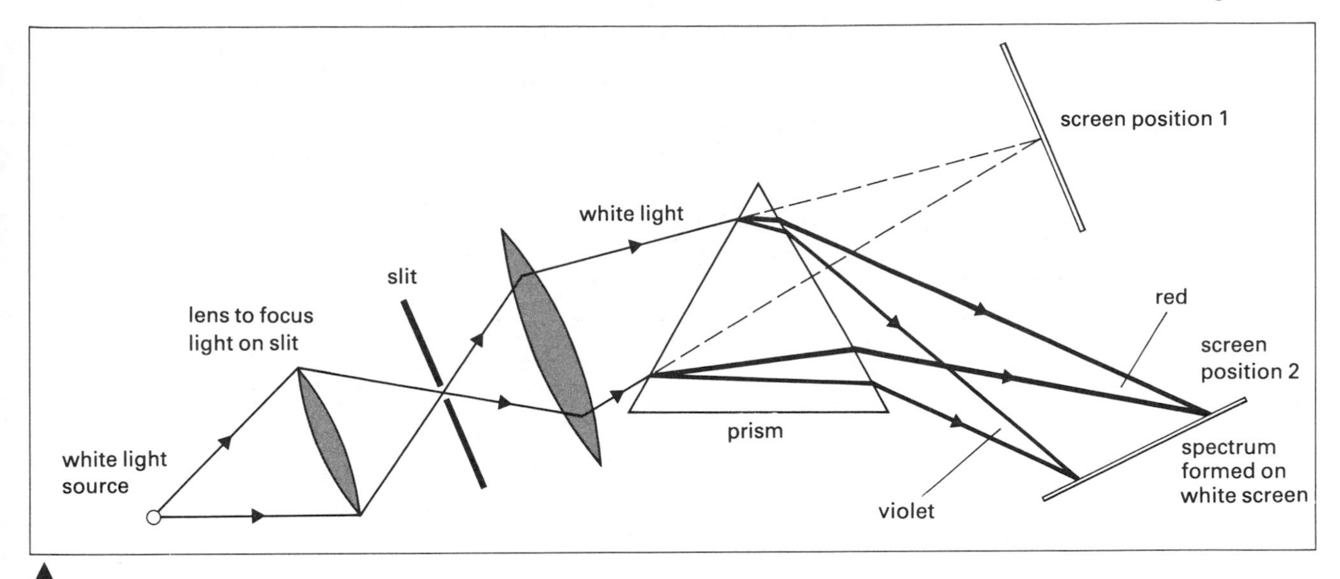

fig. 20.16 How to produce a pure spectrum

Recombining the colours to form white light

If Newton's idea that white light is really a mixture of coloured components is correct it should be possible to recombine the colours and reform white light. This can be done in two ways.

Newton first did it by using a second prism, placed the opposite way round from the first, so that the bendings took place in the opposite direction, figure 20.17. He now found he got a circular patch of white light on the screen because all the colours had overlapped and mixed on the screen, with the exception of a little red at one end and a little violet at the other.

The persistence of vision helps to show in another way that white light is really a mixture of the coloured lights of the spectrum. This is shown using Newton's colour disc, figure 20.18. Divide a cardboard disc into sections and colour them with the seven colours in roughly the proportions in which they occur in the spectrum. When the disc is spun quickly it appears white. Your eyes focus on one part of the disc, say the top, and a red image is formed on the retina. Next an orange image is formed and then a yellow, and so on until the disc has turned round once. If all these coloured images are formed within $\frac{1}{10}$ second they will mix on the retina of the eye and produce the sensation of white.

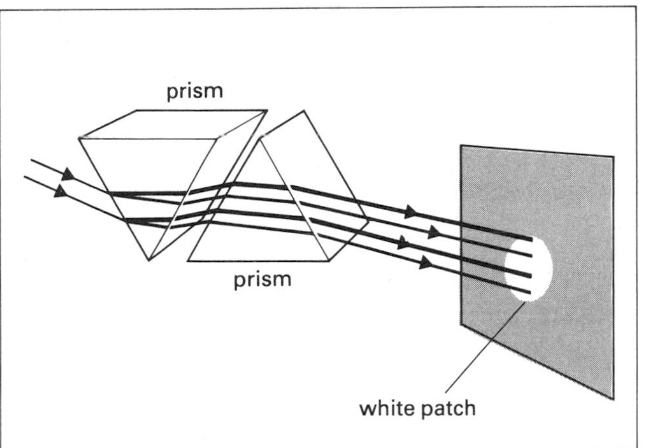

fig. 20.17 Recombination of dispersed light by a second prism

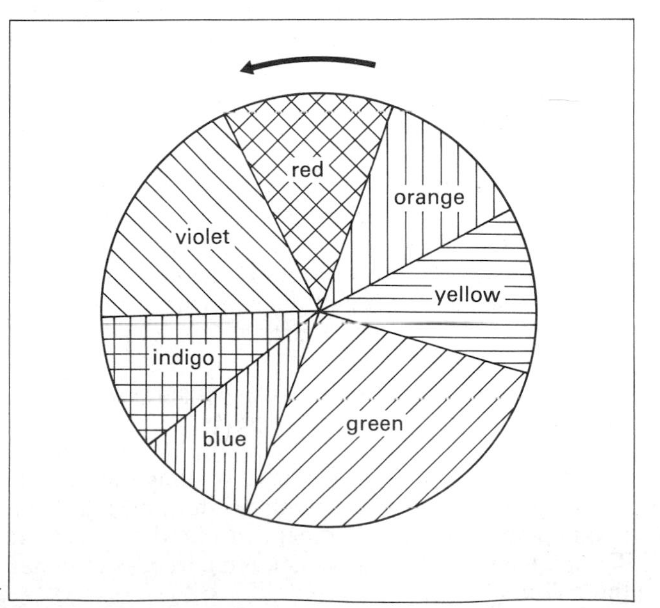

fig. 20.18 Newton's disc which, when spun, produces the sensation of white in the eye

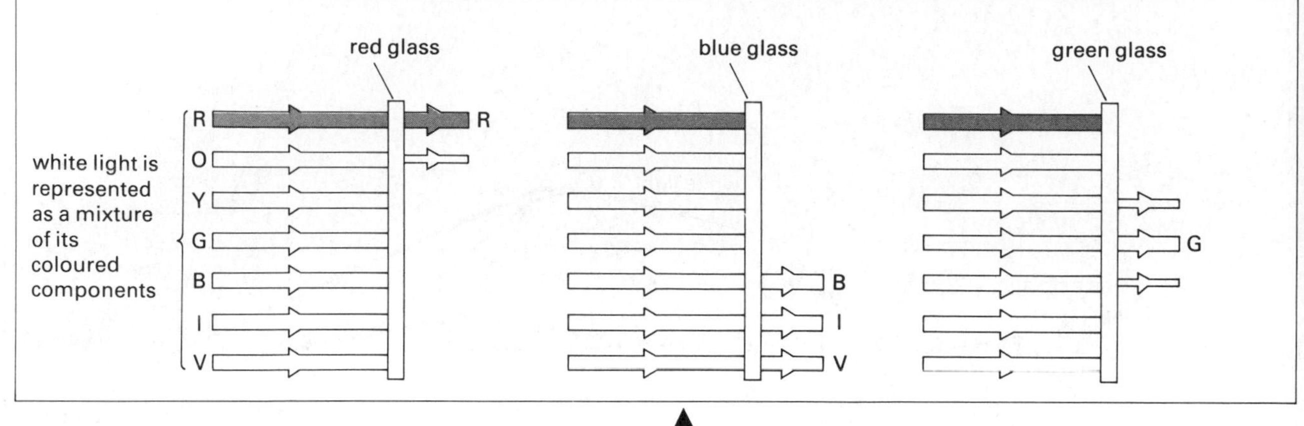

▲
fig. 20.19 Subtraction of colours

fig. 20.20 The appearance of objects in white light
▼

The colours of objects

When a piece of red glass or plastic is placed over the slit of the spectrum-producing apparatus the appearance of the spectrum changes, figure 20.19. The yellow, green and blue end of the spectrum disappears and only the red and probably some orange remain. Red glass allows only red light to pass through, and stops (absorbs) all the other colours of the white light. Objects look red when seen through a red glass because only the red part of the light scattered by them gets through the glass to our eyes. The red glass produces its coloured effect by subtracting part of the white light, and the coloured appearance results from what is left.

Transparent substances are coloured because they transmit parts of the spectrum and absorb the rest

Many coloured glasses and gelatines are not 'pure', and so transmit more than one colour of the spectrum. A yellow filter may let through orange and green; a green filter may allow yellow and blue to pass through and absorb red, orange, indigo and violet. You can only find out what a particular glass will do by testing it.

The colours of opaque objects are explained in a similar way. When white light falls on a red book, only red light is scattered and all other colours are absorbed. A violet flower looks violet because it scatters the blue-violet parts of the spectrum and absorbs the red and green parts, figure 20.20.

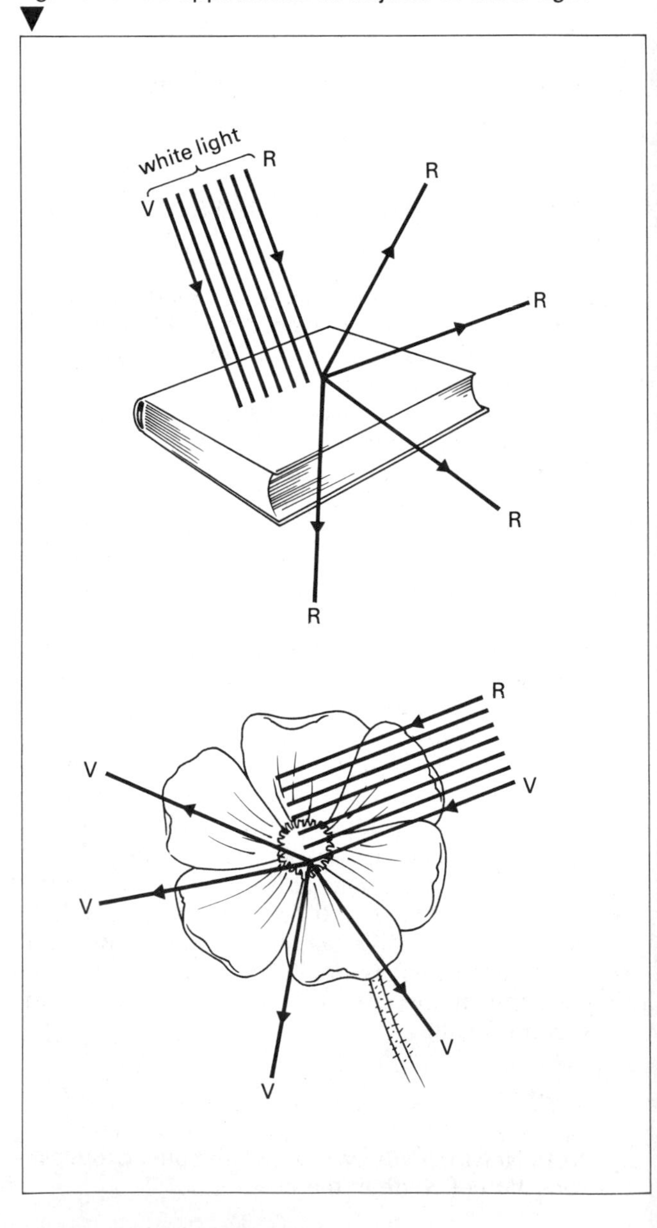

Coloured objects (and paints) absorb parts of the spectrum and scatter other parts

It is very difficult to make an absolutely pure paint, dye or pigment and most paints scatter not only their main colour but also those near to it in the spectrum. This explains the results we get when we mix together the colours in a paint-box.

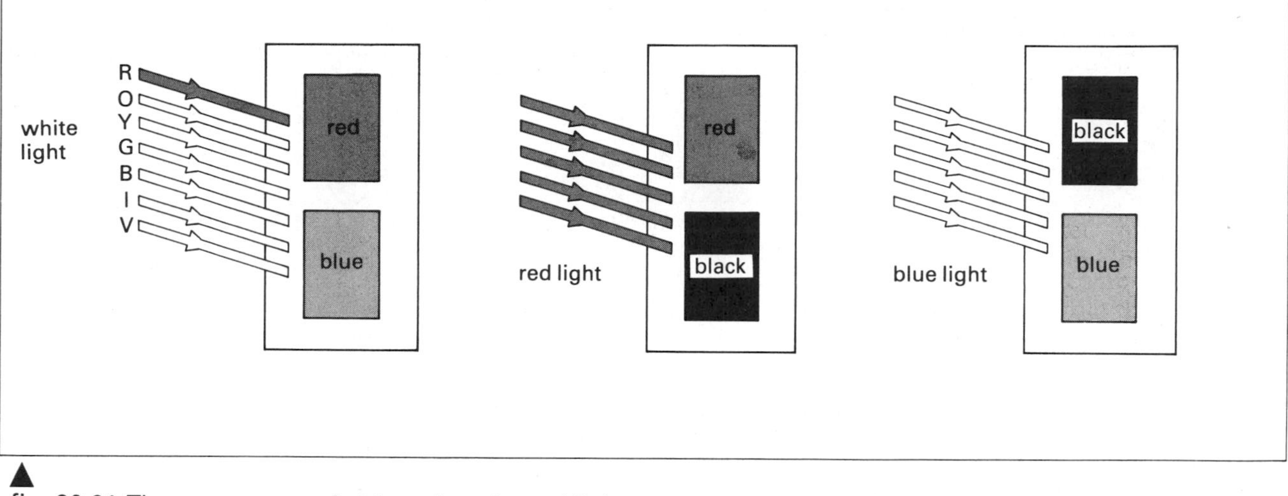

fig. 20.21 The appearance of objects in coloured light

The appearance of objects in different coloured lights

The colour of an object depends on the colour of the incident light falling on it. Suppose we have a red square and a violet square on a white background, figure 20.21. In white light we shall see them in their natural colours. If we shine a red light on them, the red square will still appear red because it scatters red light. The background will be red (although a lighter shade) because white scatters all colours. But the violet square will appear black because it absorbs red and there is no blue or violet light for it to scatter. Illuminated with blue light the red will appear black, because there is no red light to scatter.

It should be noted that in all the discussion on colour we have assumed we are dealing with perfectly pure colours. Pure filters and pigments are difficult to produce and in practice results may be somewhat different from those expected.

EXERCISE 20

In questions 1–6 select the most suitable answer.

1 A stick held in a pond appears to be bent at the surface because
- **A** the critical angle for water is 45°
- **B** total internal reflection has occurred
- **C** light travels faster in water than in air
- **D** rays of light from the eye bend into the water
- **E** rays of light from the stick are bent away from the normal at the surface

2 Which of the diagrams in figure 20.22 shows the passage of a ray of light through a glass block?

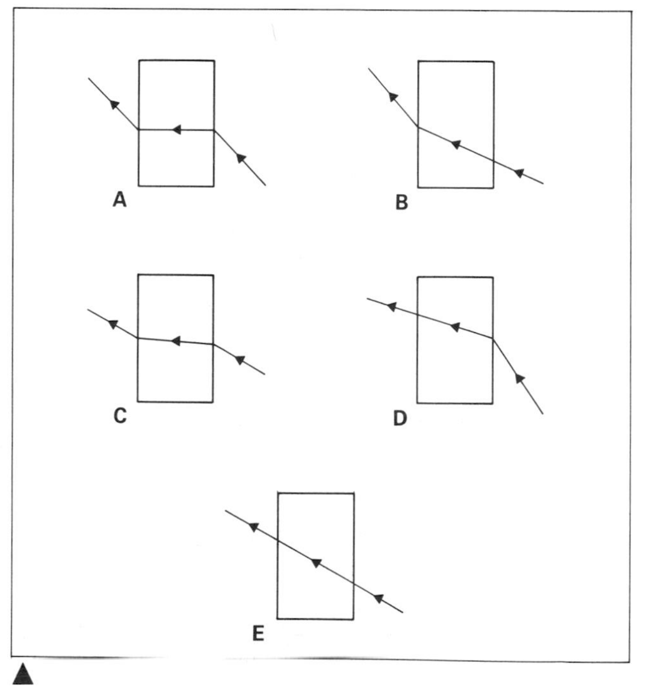

fig. 20.22

3 Figure 20.23 shows two rays of light passing through a rectangular block of glass. Which letter marks a point at which total internal reflection occurs?

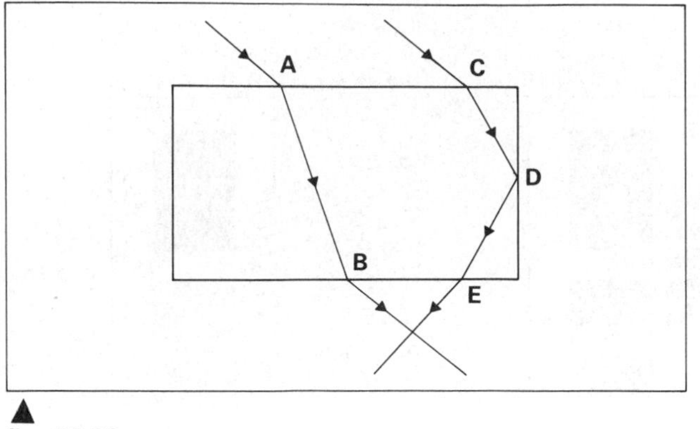

fig. 20.23

4 Figure 20.24 shows a ray of light incident on a triangular glass block. As the ray or rays pass through the prism, which of the following statements is correct?
A The blue ray is deviated more than the red ray.
B The red ray is deviated more than the blue ray.
C No dispersion of the ray occurs.
D Dispersion only occurs as the ray comes out of the prism.
E The ray passes through undeviated and undispersed.

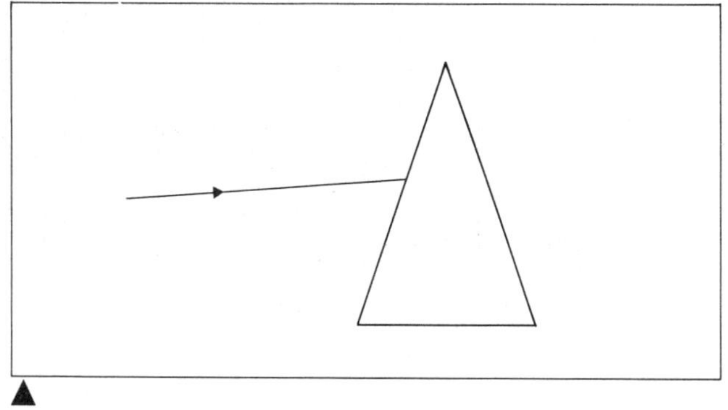

fig. 20.24

5 Which of the following statements is correct?
A Green light passes through a red filter.
B Green light shone on a green surface is all absorbed.
C Green light is split up into red and blue lights by a prism.
D Green light added equally to red light appears yellow.
E Green light shone on a red surface appears yellow.

6 In a dark room a red flag illuminated with blue light will appear
A red
B magenta
C blue
D black
E yellow

7 **a** Copy and complete figure 20.25 to show how the ray of light is refracted through the glass block, indicating clearly the angles of incidence and refraction and the direction of the emerging ray.
b What happens to the speed of light as it passes into the glass?
c If the refractive index of glass is $\frac{3}{2}$, how thick will a 0.1 m glass block appear to be to an observer looking at a coin under it?

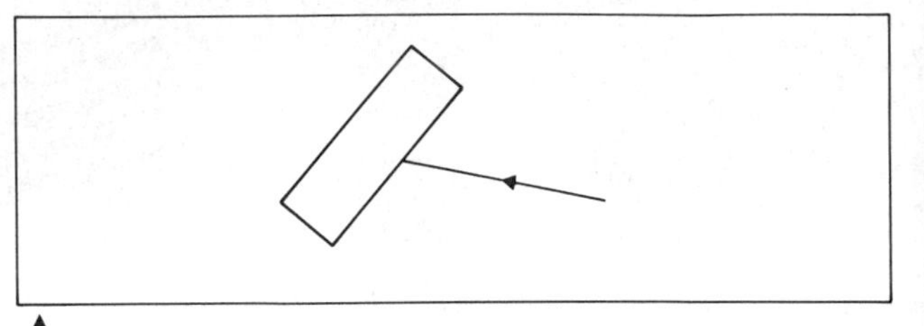

fig. 20.25

8 **a** What do you understand by refraction?
b Give *two* everyday examples in which refraction is observed.
c 'The refractive index of water is $\frac{4}{3}$.' What is meant by this statement?
d In an experiment to measure the refractive index of a glass block a boy set up the apparatus shown in figure 20.26. (i) Describe what the boy did in the experiment. (ii) What measurements did he take? (iii) If the results of his experiment were AB 12 cm, CD 4 cm, calculate the refractive index of glass.

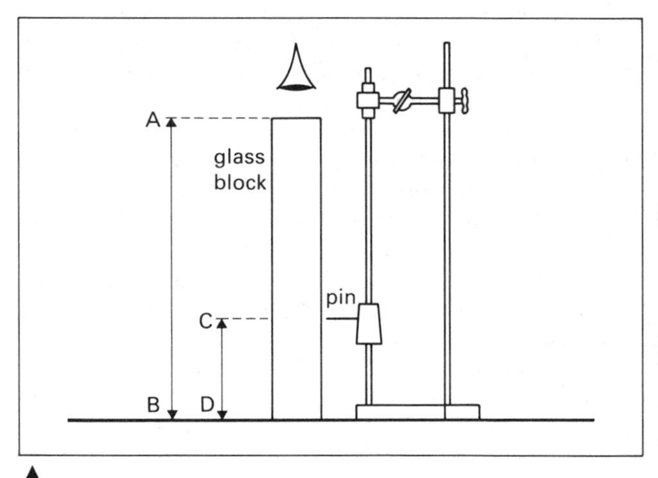

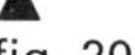
fig. 20.26

9 **a** (i) What is meant by critical angle? (ii) What happens when a ray of light strikes a surface at an angle greater than the critical angle?
b If the critical angle for glass to air is 42°, show by means of a diagram how a 45°, 45°, 90° prism can be used to turn a ray of light through 180°. Mark the values of the angles of incidence for the ray at each air/glass boundary.

10 Figure 20.27 shows three prisms A, B and C. A and C are solid glass and B is hollow with very thin glass walls. On each diagram draw how the ray Z would behave as it entered and left each of the three prisms.

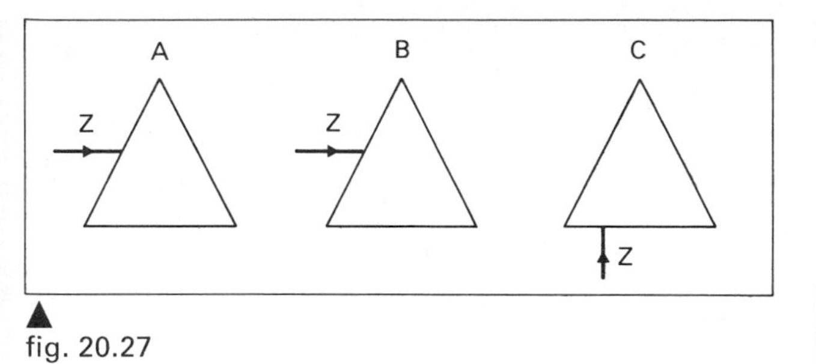

fig. 20.27

11 A ray of light is incident on the side AB of a glass block, of refractive index 1.5, at an angle of incidence of 60°, as shown in figure 20.28. The ray is reflected from BC and emerges after refraction from side CD.

a Draw a sketch showing the path of the ray through the block.

b Find the angle at which the ray is reflected from BC.

c Calculate the angle between the incident ray and the emergent ray.

fig. 20.28

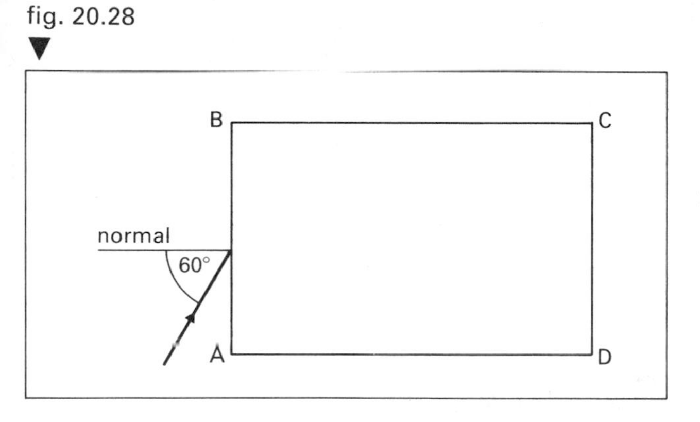

12 A beam of white light passes through a triangular glass prism on to a white screen as shown in figure 20.29.

a Copy figure 20.29 and write, in the correct position on the diagram, the names of any colours which would appear on the screen.

b What would be the effect of placing a red filter at position A-A?

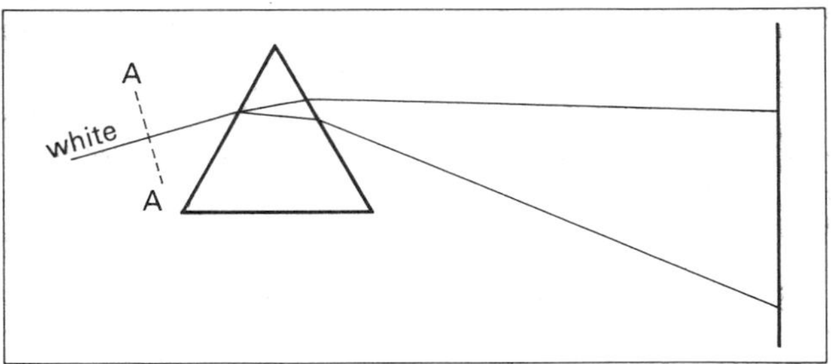

fig. 20.29

13 a Describe, with the aid of a labelled diagram, how to produce a pure spectrum from a point source of white light. Label the colours at the ends of the spectrum produced.

b Describe the appearance of the spectrum in **a** if it was produced on a red screen.

14 a With the aid of a sketch explain how you would construct a Newton's colour disc. State the effect seen when the disc is spun at an ever increasing speed. Briefly explain your answer.

b A surface is painted in red, yellow and blue stripes. Describe the appearance of the surface when the surface is strongly illuminated by (i) red light, (ii) yellow light.

21 LENSES

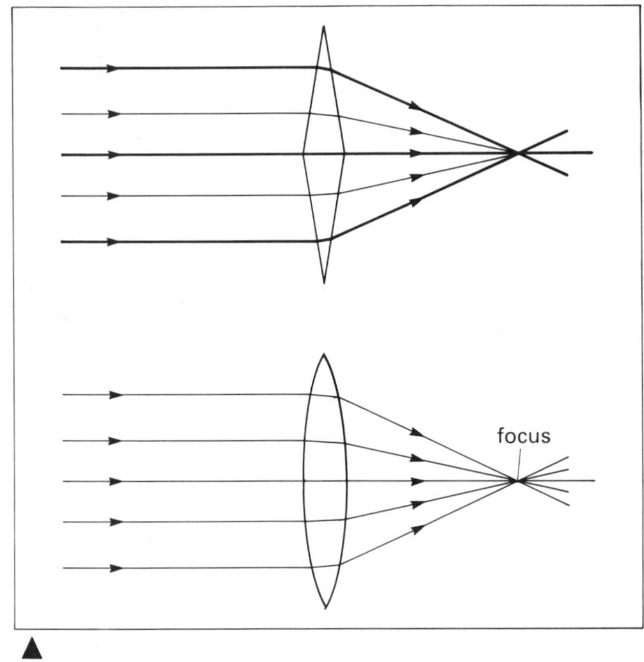

fig. 21.1 Converging lens. A converging lens produces a real focus

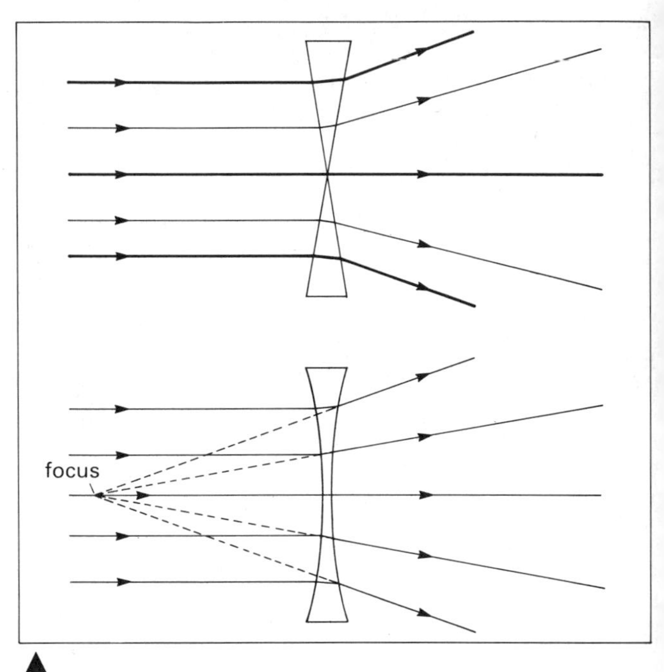

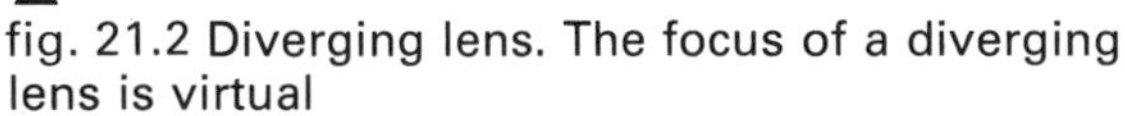

fig. 21.2 Diverging lens. The focus of a diverging lens is virtual

fig. 21.3 Various shapes of lenses

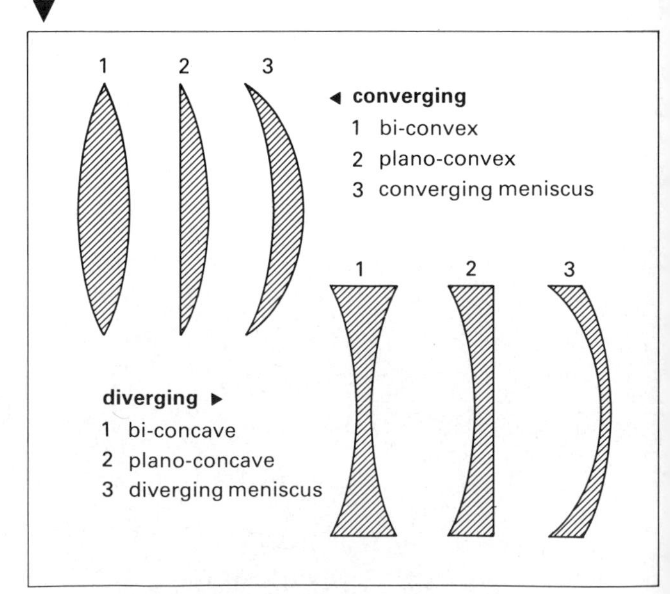

A lens is a piece of transparent material, usually glass, with one or both of its surfaces curved. Most often the surfaces are parts of spheres, but different kinds of curved lenses are made to do different kinds of jobs.

Two thin prisms placed base to base give us a clue as to how a lens works. Parallel rays falling on the lens are all brought to a focus at the same point. A lens which behaves in this way is a **converging lens**, figure 21.1.

If the prisms are placed point to point parallel rays are spread out and we have a **diverging lens**, figure 21.2.

Figure 21.3. shows some lenses as they appear in section. The three shown top left have one thing in common. They are all thicker in the middle than they are at the edges, that is, they are **convex**. These three lenses are converging lenses and would bring a parallel beam to a focus. The three lenses shown bottom right are thinner in the middle than they are at the

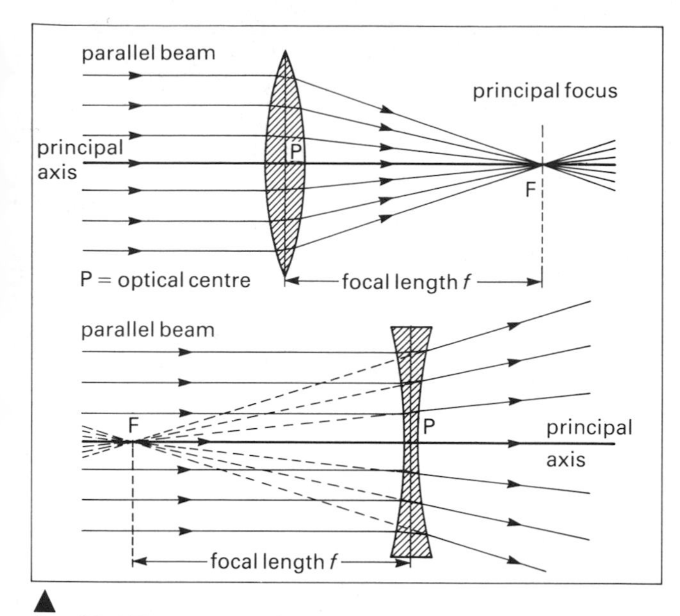

▲
fig. 21.4 Terms used for lenses

edges, that is, they are **concave**. They are diverging lenses.

The central point of the lens is the **pole** or **optical centre**, P, figure 21.4.

The line through the pole at right angles to the surfaces is the **principal axis** of the lens.

A beam of rays parallel to the principal axis of a convex lens is converged by the lens to the **principal focus**, F.

The distance of the principal focus from the optical centre of the lens is the **focal length**, f.

fig. 21.5 Weak and strong lenses
▼

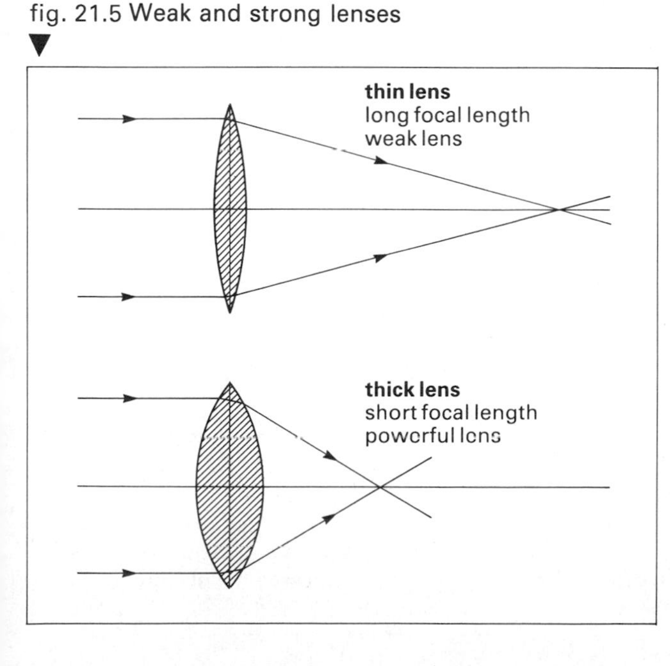

The power of a lens

It is useful to think of the power of a lens as its ability to bring parallel light to a focus quickly. Thus, a thin lens has a long focal length and is a weak lens; while a fat lens has a short focal length and is a powerful lens, figure 21.5. The shorter the focal length, the more powerful the lens. For this reason the power of a lens is measured by the reciprocal of the focal length. The unit of power is the **dioptre**.

The power of a lens in dioptres is equal to the reciprocal of the focal length in metres

Conventionally, converging lenses have a positive power and diverging lenses a negative power. Table 21.1 shows some instruments and their corresponding focal lengths and powers.

table 21.1 Some typical values of focal length and power

Instrument	Focal length (m)	Power (D)
spectacle lens	1	1
camera lens	0.1	10
magnifying glass	0.02	50
low-power microscope objective	0.016	62.5
high-power microscope objective	0.004	250

The idea of the power of a lens is particularly useful when dealing with combinations of thin lenses in contact with each other. The power of the combination can be found by adding together the powers of the components, figure 21.6. If a diverging lens is involved, its power (and focal length) must be taken as negative.

fig. 21.6 For this lens combination the power is $2 + 4 = 6$ D; focal length $f = \frac{1}{6}$ m
▼

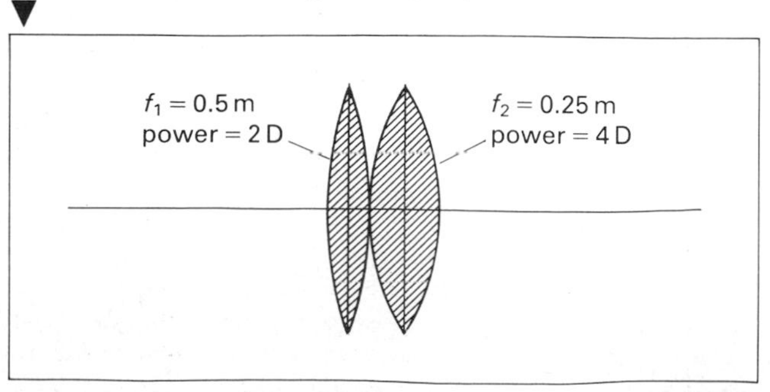

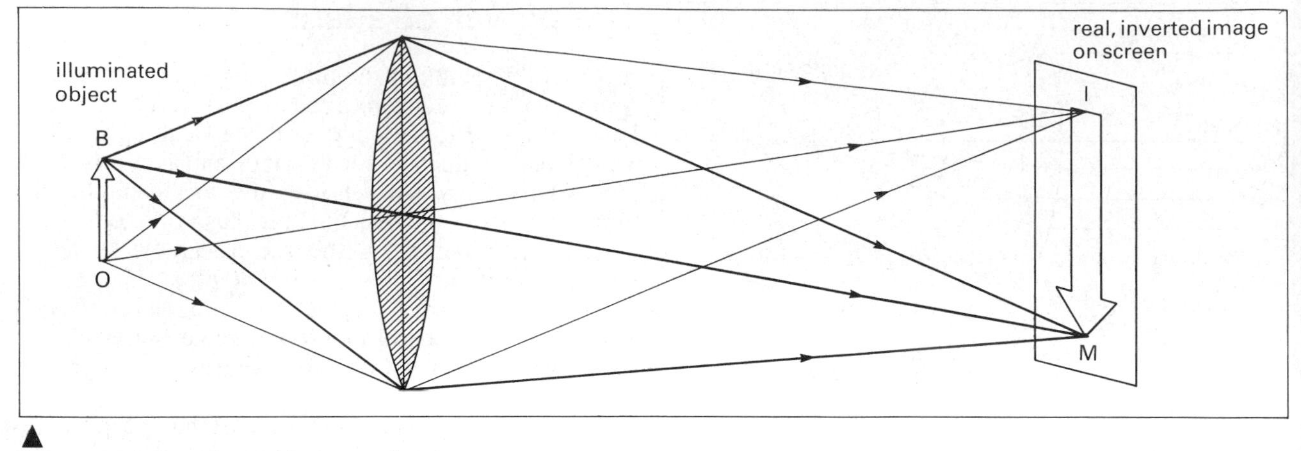

fig. 21.7 How an image is formed

How a lens produces an image

All the light rays falling on the lens from the bottom of the object are brought to a focus at I. Three of these rays are shown in figure 21.7. The lens turns the beam diverging from O into a converging beam with its focus at I. Similarly rays from the top of the object are brought to a focus at M. The light from every point on the object behaves in the same way and for each point on the object there is a corresponding point on the screen to which the rays are focused. In this way the complete image of the object is formed.

How to draw ray diagrams for lenses

Convex lenses

All rays from a point on the object, when they have passed through the lens, come to a focus to form a point on the image. Or, in some cases, they all appear to spread out from a point on an imaginary image. To draw ray diagrams we use three rays whose directions after they pass through the lens are known, figure 21.8.

EXAMPLES

An object is placed 15 cm from a convex lens of focal length 10 cm. Where is the image formed and what is its size?

Draw the three rays from B, the top of the object—one parallel to the axis, one through the centre of the lens and one through the focus. When the refracted rays are drawn in the proper directions, they all meet at M, the bottom of the image, figure 21.9. The image is twice the size of the object.

The position, nature and size of the image formed by a convex lens vary very much with the position of the object. The image may be real or virtual, inverted or erect, magnified or diminished. Consider the case of an object 5 cm from a convex lens of focal length 10 cm.

When we draw the three standard rays, it is clear that the rays emerging from the lens will never meet in a point to form a real image. They form a diverging beam. If we trace back their directions we see that they all appear to spread out from the point M, and we have an imaginary image, figure 21.10. If we are looking through the lens from the right-hand side we should see this enlarged image, and the lens would be actihg as a magnifying glass.

fig. 21.8 The construction rays for a converging lens

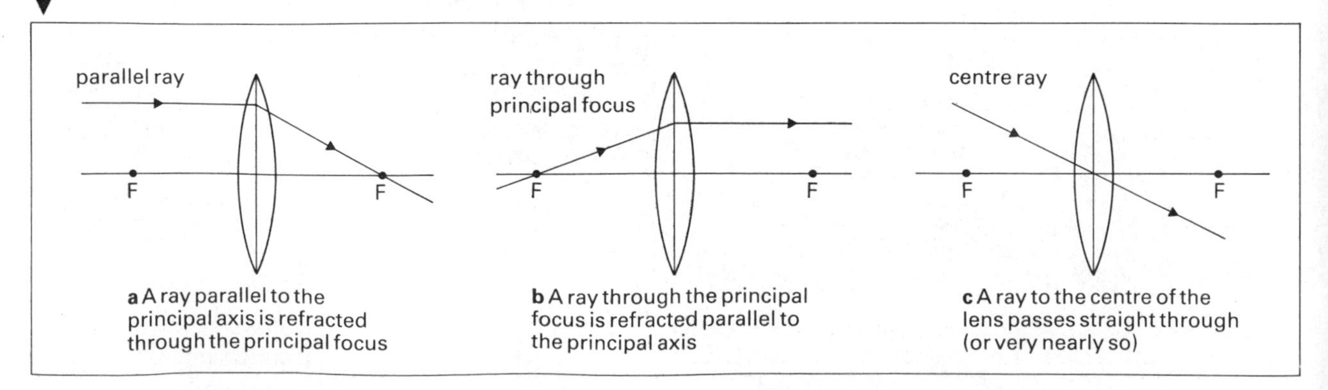

a A ray parallel to the principal axis is refracted through the principal focus

b A ray through the principal focus is refracted parallel to the principal axis

c A ray to the centre of the lens passes straight through (or very nearly so)

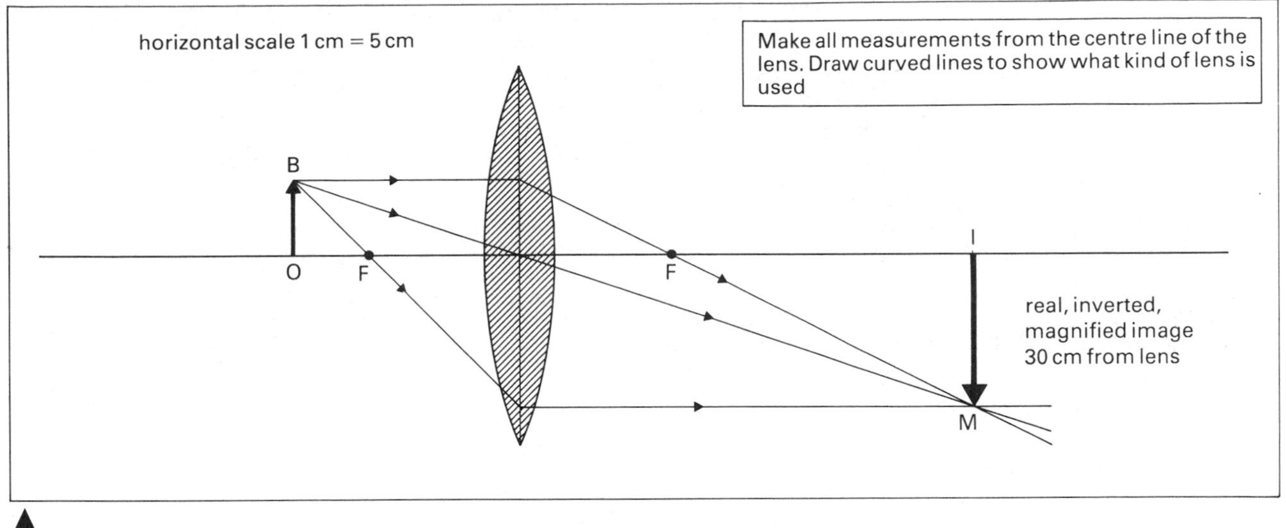

fig. 21.9

fig. 21.10

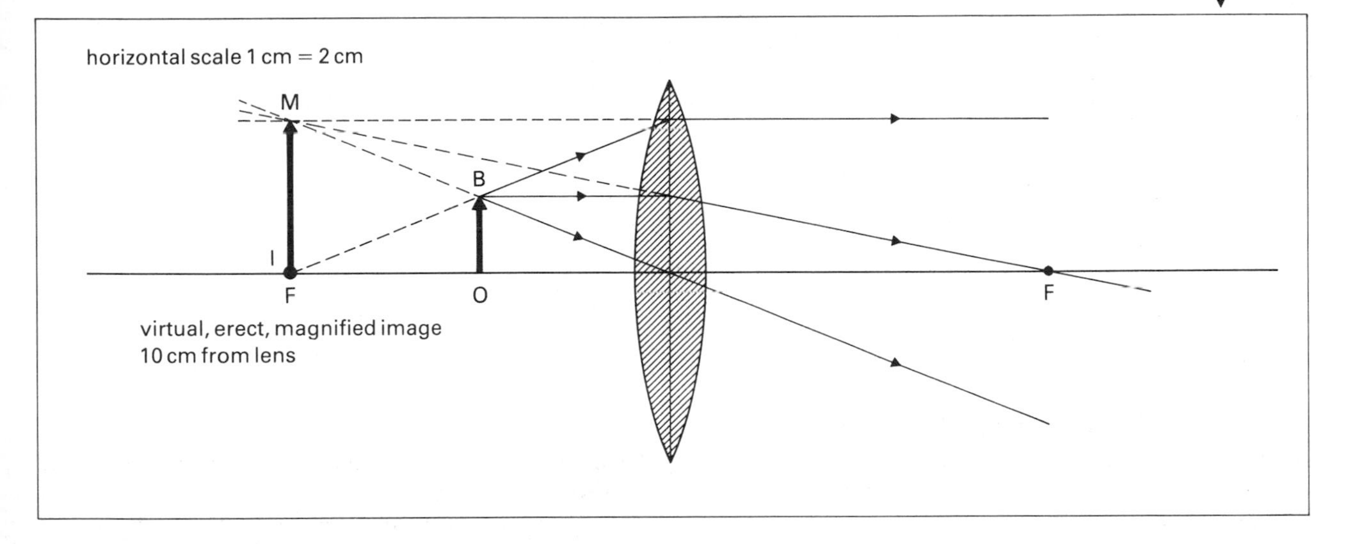

Summary of images formed by a convex lens

The general positions and nature of the images formed by converging lenses are summarised in table 21.2 below.

table 21.2 Image formation by converging (convex) lenses

Position of object	Position of image	Nature of image	Examples of use
between infinity and $2f$	between F and $2f$	real, inverted, diminished	telescope and camera
at $2f$	at $2f$	real, inverted, same size	erector lens in telescope
between $2f$ and F	further away than $2f$	real, inverted, magnified	projector
at F	light emerges parallel	no image formed	—
between F and lens	all positions possible	virtual, erect, magnified	magnifying glass and spectacles

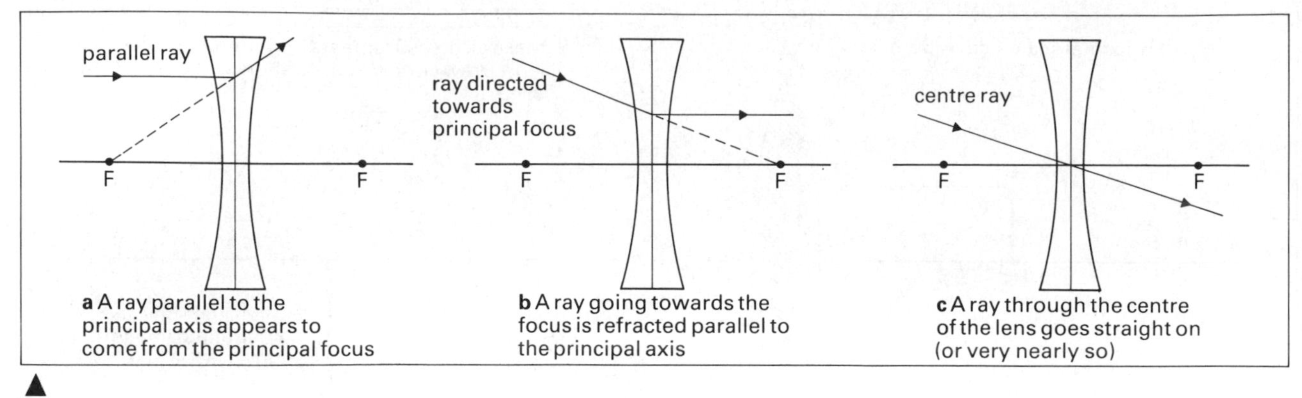

fig. 21.11 The construction rays for a diverging lens

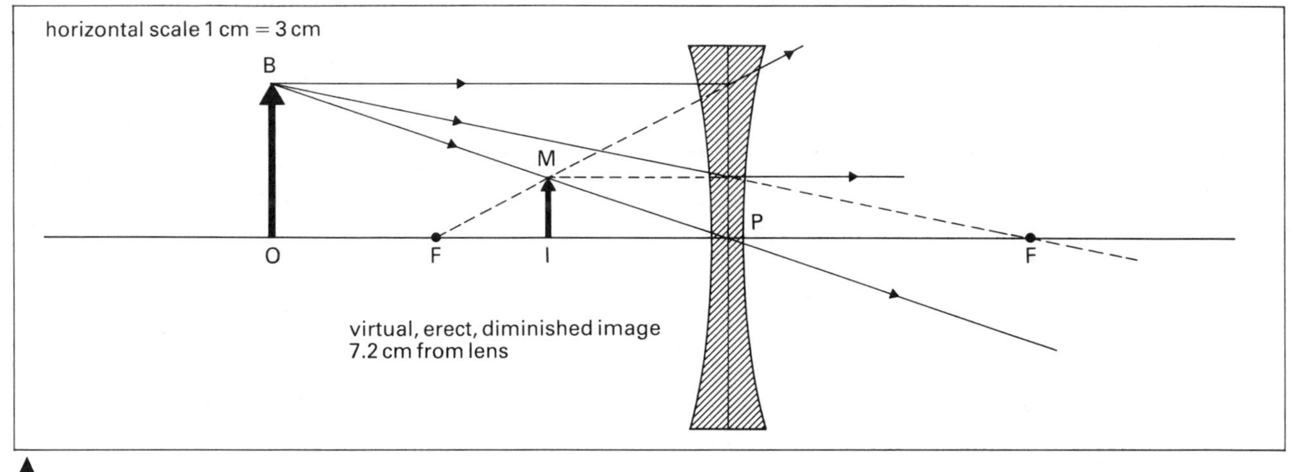

fig. 21.12

fig. 21.13 Obtaining a rough value for the focal length of a convex lens

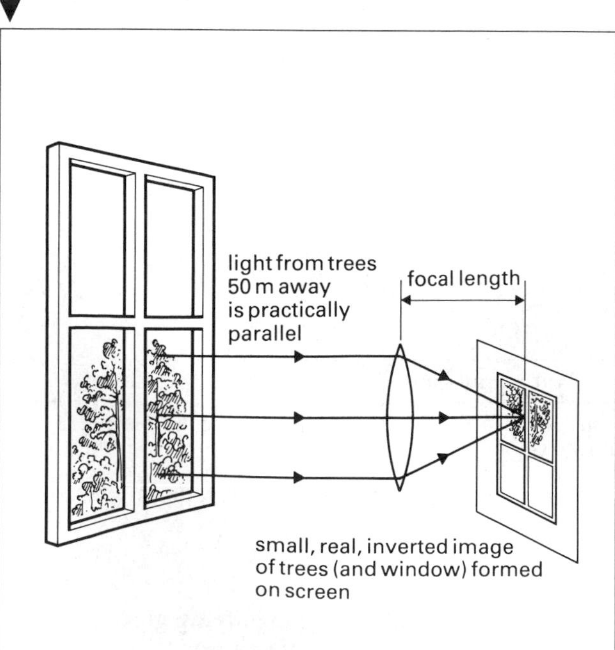

Concave lenses

For concave lenses the three main construction rays are shown in figure 21.11.

Concave lenses always produce diverging beams, so that we should not expect them to produce real images. Whatever the distance of the object from the lens, the image is always virtual, erect and diminished.

EXAMPLE

An object is placed 18 cm from a diverging lens of focal length 12 cm, figure 21.12. Where is the image formed?

Measuring the focal length of a converging lens

Using the parallel rays from a distant object, an image of the object can be focused on a cardboard screen, figure 21.13. The distance of the lens from the screen will give an approximate value of the focal length. Even if a more accurate method is going to be used, it is always worth while getting a rough value first.

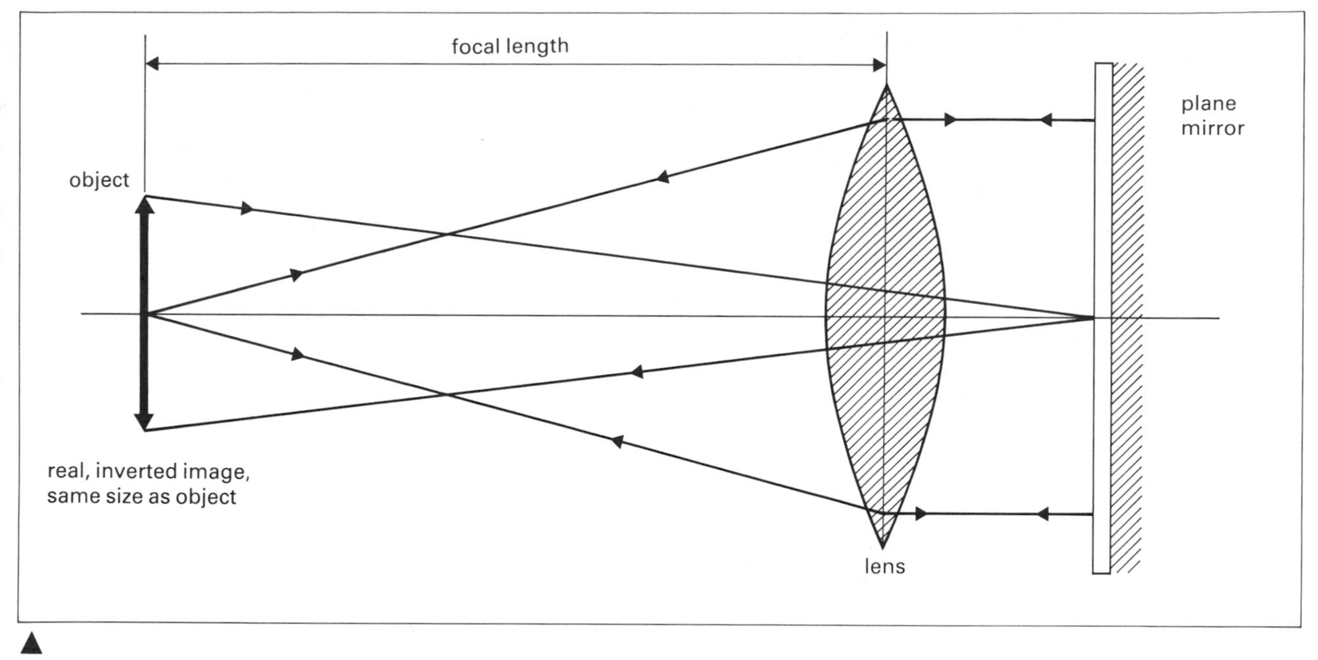

fig. 21.14

Light coming from the principal focus produces a parallel beam after passing through the lens. This is true for any point in the focal plane of the lens. If a plane mirror is placed at right angles to the beam, the light will be reflected back along its own path and, after again passing through the lens, will return to the principal focus to form an image, figure 21.14. If, using the mirror/lens combination, we can find a position where object and image coincide, we know they are at the principal focus, figure 21.15.

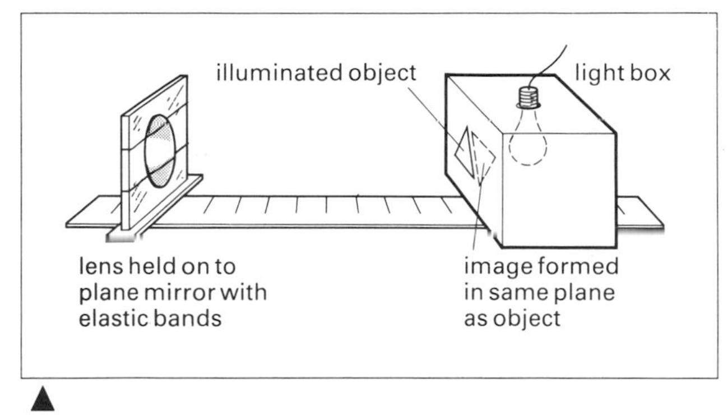

fig. 21.15 Apparatus for measuring the focal length of a converging lens

Magnification

The magnification produced by a lens is measured by comparing the size of the image with the size of the object. Generally we are concerned with the length of the image and object so

$$\textbf{magnification } (m) = \frac{\textbf{length of image}}{\textbf{length of object}}$$

In figure 21.16, triangles BPO and PIM are similar because they equiangular. Using the fact that the ratios of corresponding sides of similar triangles are equal, we have

$$m = \frac{IM}{OB} = \frac{PI}{PO}$$

$$\textbf{magnification} = \frac{\textbf{distance of image from lens}}{\textbf{distance of object from lens}}$$

fig. 21.16

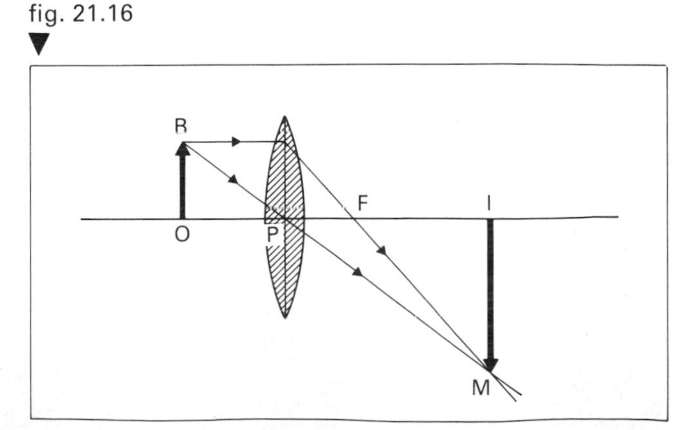

Some uses of lenses

See also Chapter 22 *The camera and the eye* and Chapter 23 *Optical instruments*

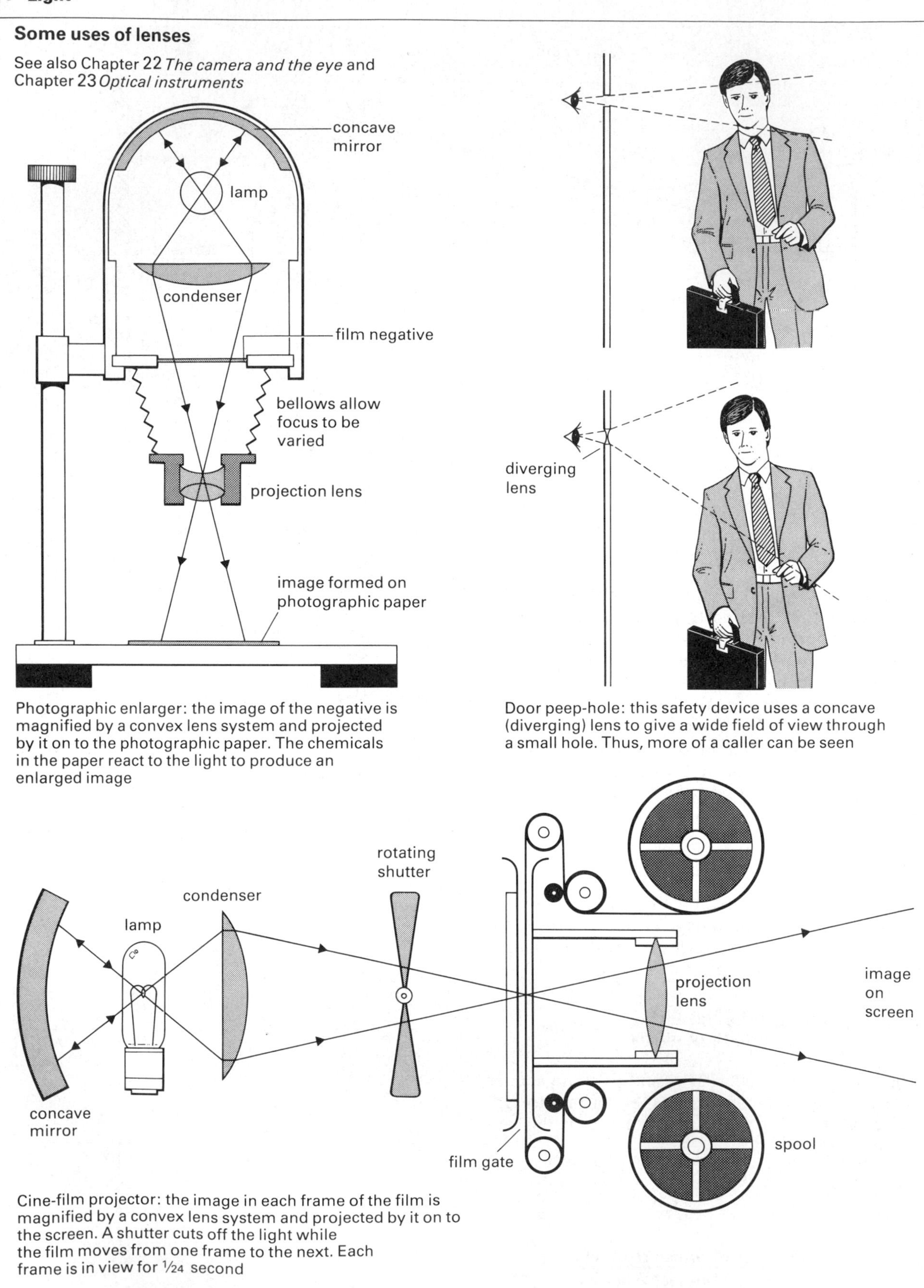

Photographic enlarger: the image of the negative is magnified by a convex lens system and projected by it on to the photographic paper. The chemicals in the paper react to the light to produce an enlarged image

Door peep-hole: this safety device uses a concave (diverging) lens to give a wide field of view through a small hole. Thus, more of a caller can be seen

Cine-film projector: the image in each frame of the film is magnified by a convex lens system and projected by it on to the screen. A shutter cuts off the light while the film moves from one frame to the next. Each frame is in view for $1/24$ second

EXERCISE 21

In questions 1–5 select the most suitable answer.

fig. 21.17

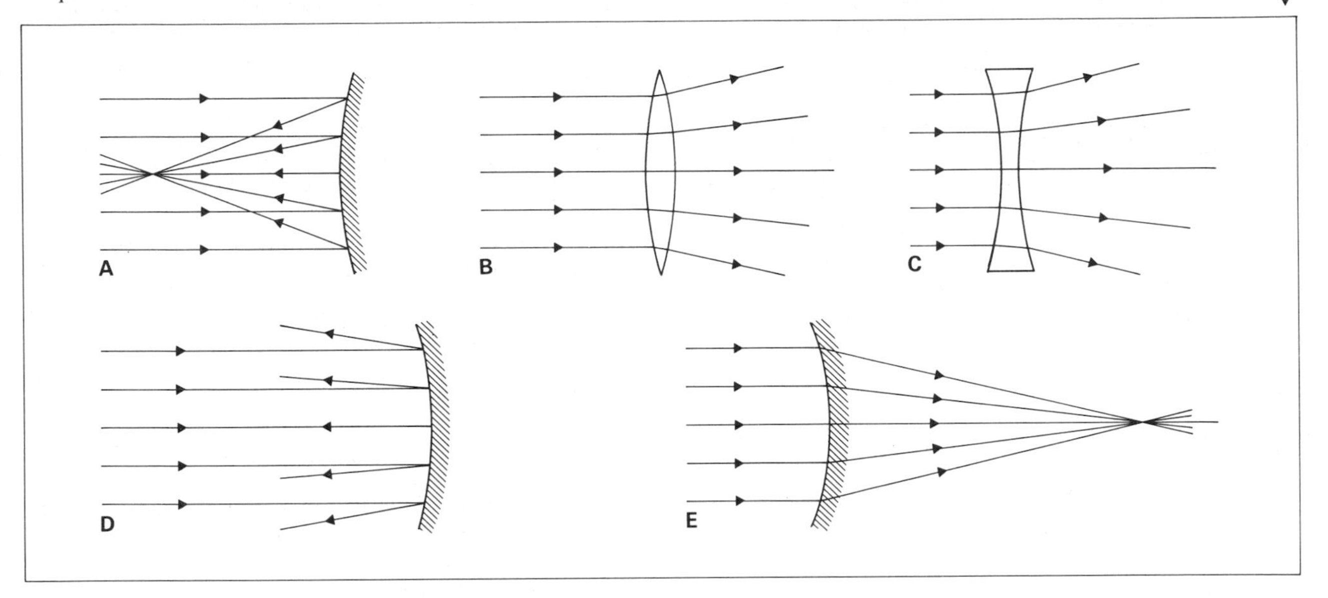

1 Which of the diagrams, **A** to **E**, in figure 21.17 shows rays of light correctly reflected or refracted by a lens or mirror?

2 An image of a distant coloured scene is projected, by means of a converging lens, onto a white screen. Which of the following statements concerning the image is *not* true?
A It is diminished.
B It is coloured.
C It is upright.
D It is real.
E It is formed close to a principal focus of the lens.

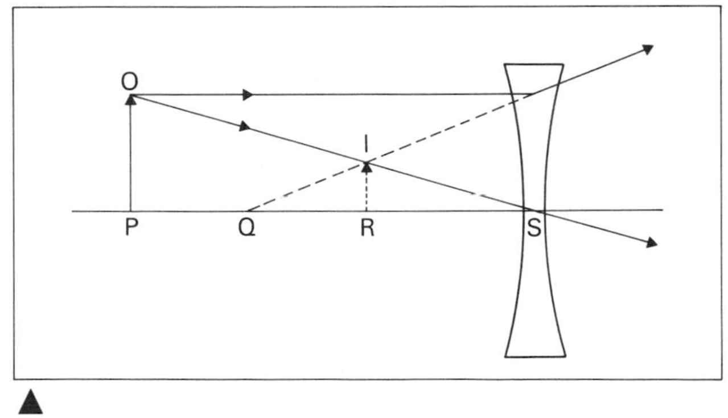

fig. 21.18

3 Figure 21.18 shows rays from the top of an object, O, passing through a diverging lens. The lens forms an image of O at I. Which of the following distances is the focal length of the lens?
A PQ
B PR
C PS
D RS
E QS

4 When a postage stamp is viewed through a converging lens which of the following could *not* be a description of the image formed?
A same size but upside down
B smaller and upside down
C magnified and upside down
D magnified and erect
E smaller and erect

5 An object is placed 12 cm in front of a convex lens of focal length 10 cm. The image formed will be
A virtual, upright and magnified
B real, upright and diminished
C real, upright and magnified
D real, inverted and magnified
E real, inverted and diminished

6 Figure 21.19 shows a concave (diverging) lens C, a convex (converging) lens D and a plane mirror E. On each diagram draw how the ray Y would behave as it continued its path.

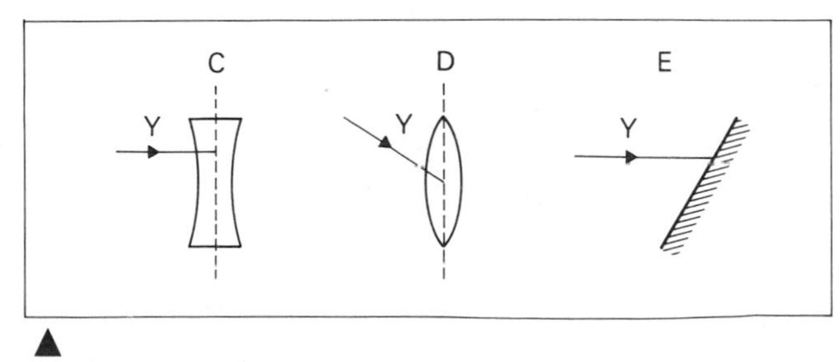

fig. 21.19

7 Explain how you would find the approximate focal length of a converging (convex) lens.

8 In figure 21.20 A, B, C and D are four open-ended boxes. Two rays of parallel light enter each box on the left-hand side and emerge as shown. Draw in each box the object which causes the rays to emerge as they are shown.

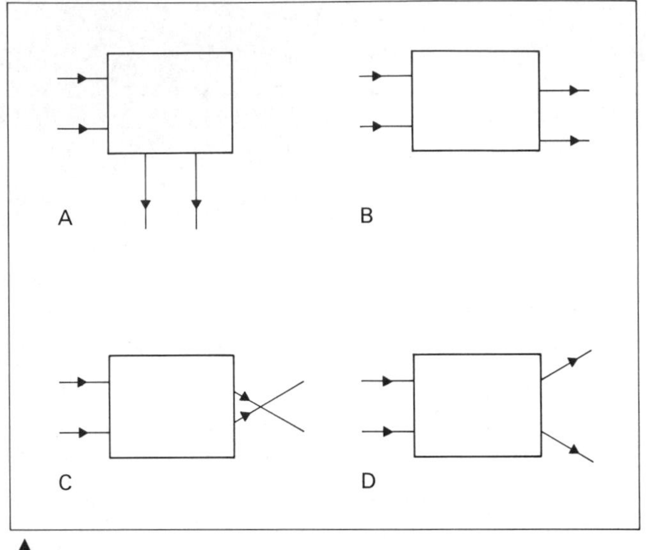

fig. 21.20

9 The ray diagram drawn to scale in figure 21.21 shows the way a slide projector throws an image on to a screen. Using the diagram (1 cm represents 5 cm)

a measure the height of the slide on the diagram
b calculate the true height of the slide
c measure the height of the image on the diagram
d calculate the true height of the image
e calculate the magnification
f draw on the diagram *one* other ray which leaves the tip of the arrow and passes through the lens to the screen.

fig. 21.21

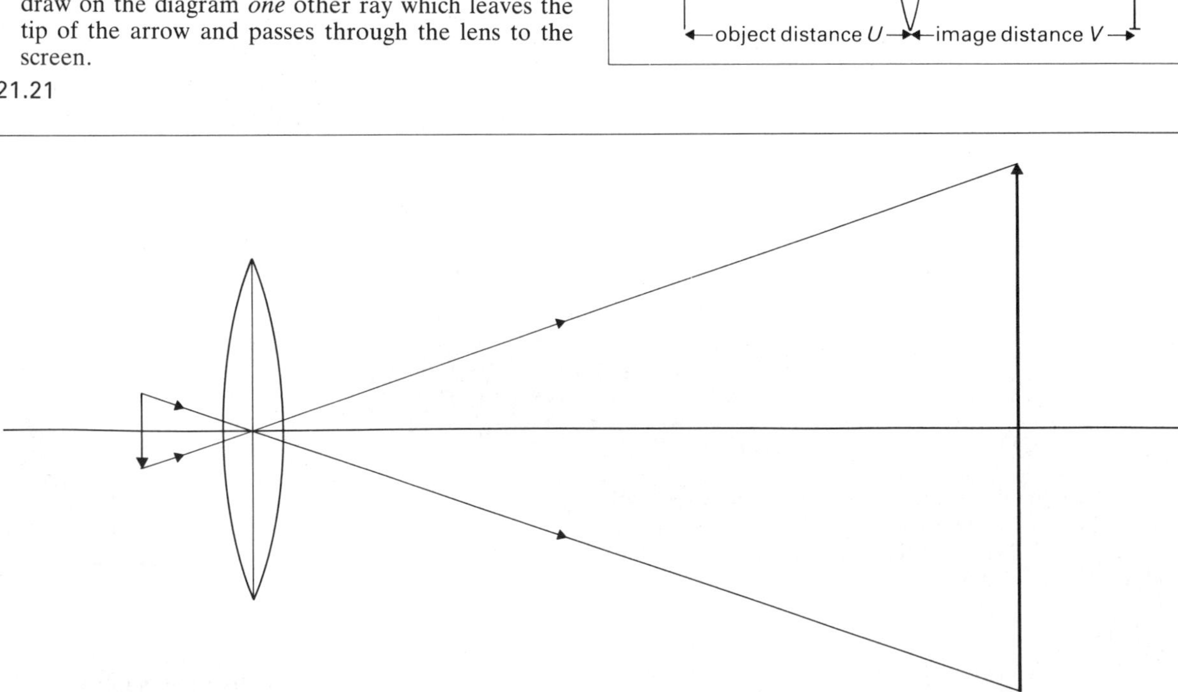

10 Figure 21.22 shows the relative positions of pieces of apparatus set up to show that a convex lens forms an image on a screen. The image distance V is found for each of 8 different object distances U, the image being clearly focused in each case. The results are listed in the table.

U	19	21	25	28	32	35	40	50	cm
V	71	52	37	32	28	26	24	21	cm

a Plot a graph of V (y-axis) against U (x-axis).
b Mark on the graph a point where $U = V$. From the graph find the value of U when $U = V$. Use this value to find the focal length of the lens.
c The image formed in this experiment is a real one. What does this mean?
d Draw a ray diagram to show how a convex lens may form a real image of an object.
e Where would the object crosswires be placed relative to the lens in order to produce an image which was not real? Draw a diagram showing the approximate positions of the object and image in this case, and indicate on the diagram the position of an eye which was viewing the image.

fig. 21.22

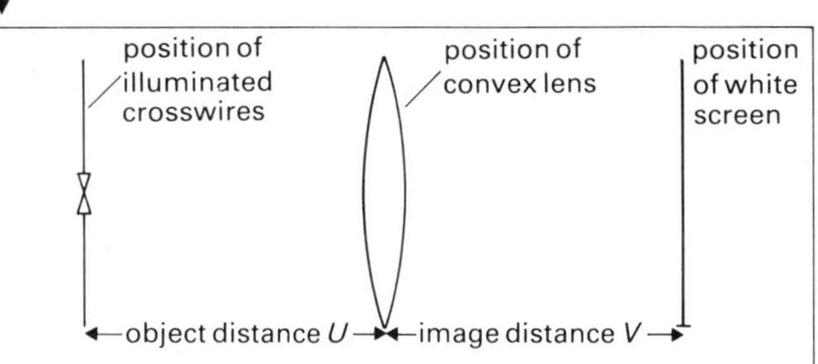

22 THE CAMERA AND THE EYE

The camera and the eye are alike in that they both use a converging lens system to produce a real, inverted, diminished image of objects in front of them, figures 22.1 and 22.2.

In the camera the screen is a light-sensitive film on which can be produced a permanent image. In the eye the light falls on light-sensitive cells and an impermanent 'image' is produced by the brain.

Although fundamentally the camera and the eye work in the same way, there are important differences between them. These will be discussed when each has been described in detail.

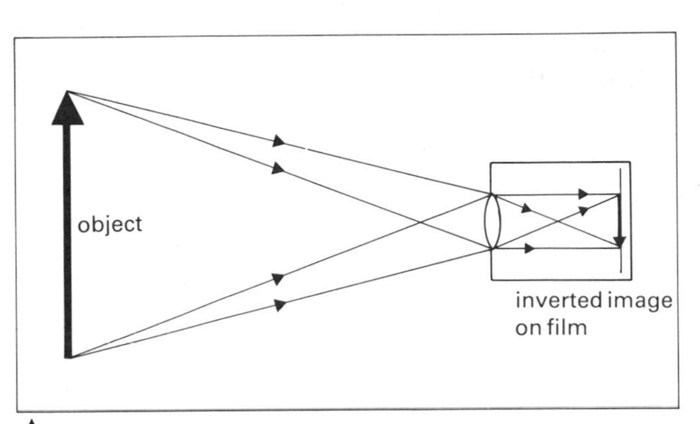

fig. 22.1 Converging lens system of the camera

The camera

The camera consists essentially of a light-tight box with a lens at the front and a sensitive film at the back, figure 22.3. The inside of the camera is blackened to prevent any light being reflected on to the film. In 'fixed focus' cameras the distance between the lens and the film is fixed. In other cameras the distance can be varied and objects at different distances can be focused sharply. It might be thought

fig. 22.2 Converging lens system of the eye

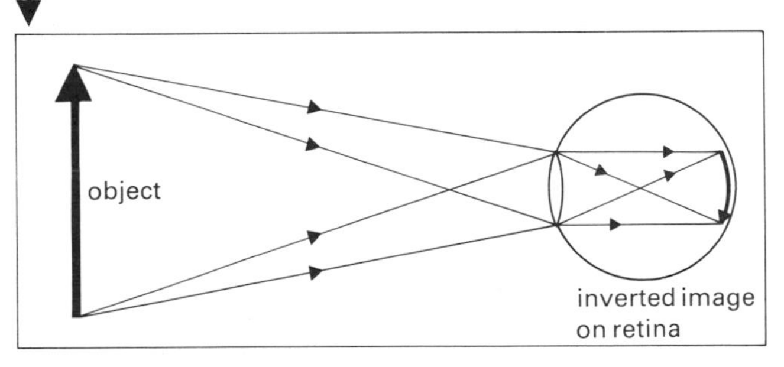

fig. 22.3 Image formation in a camera

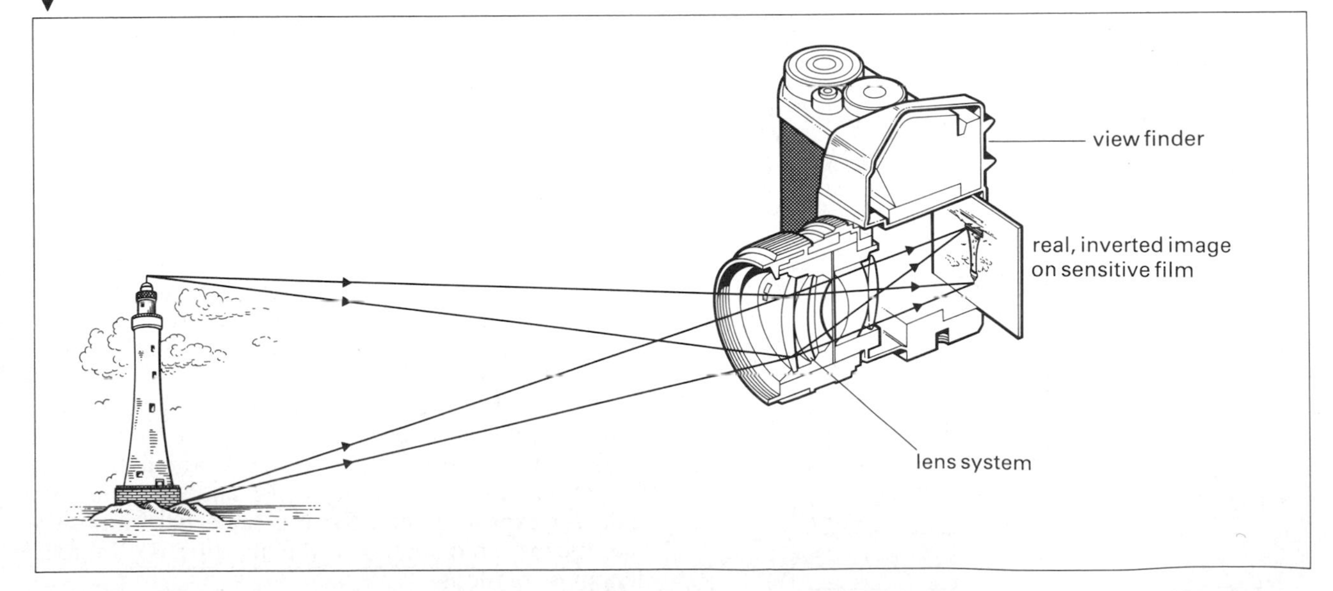

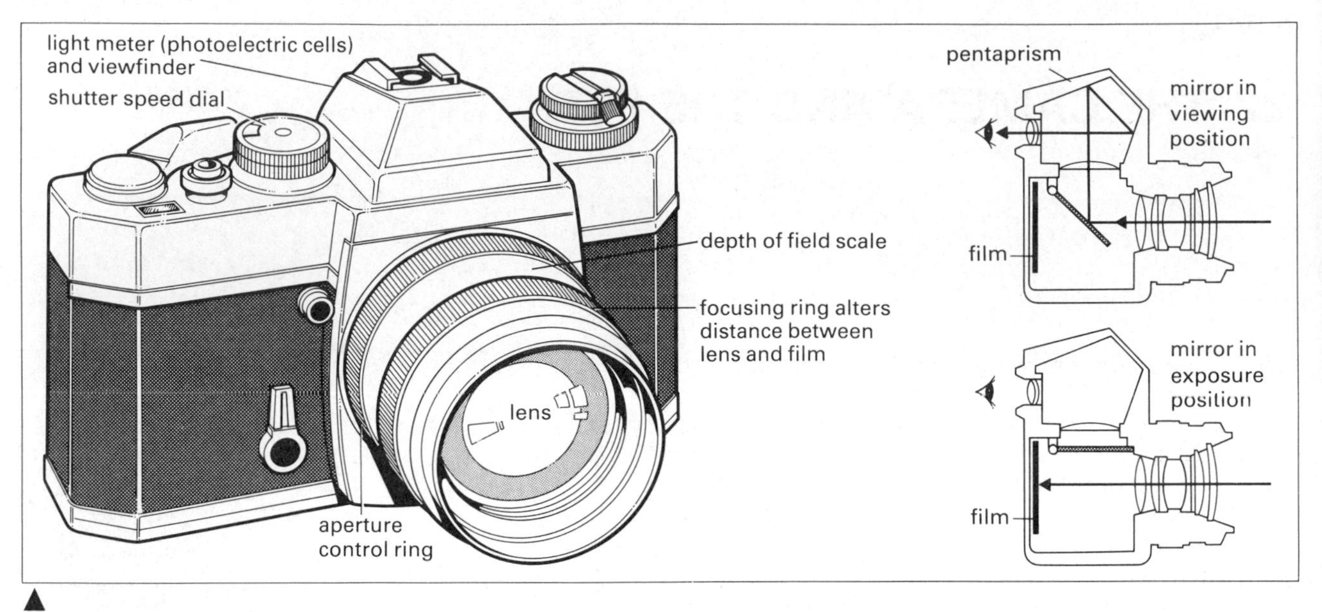

fig. 22.4 Controls on 35 mm single-lens reflex camera

fig. 22.5 The diaphragm can be adjusted to alter the size of the aperture. A large aperture is used on a dull day, and a small one on a bright day

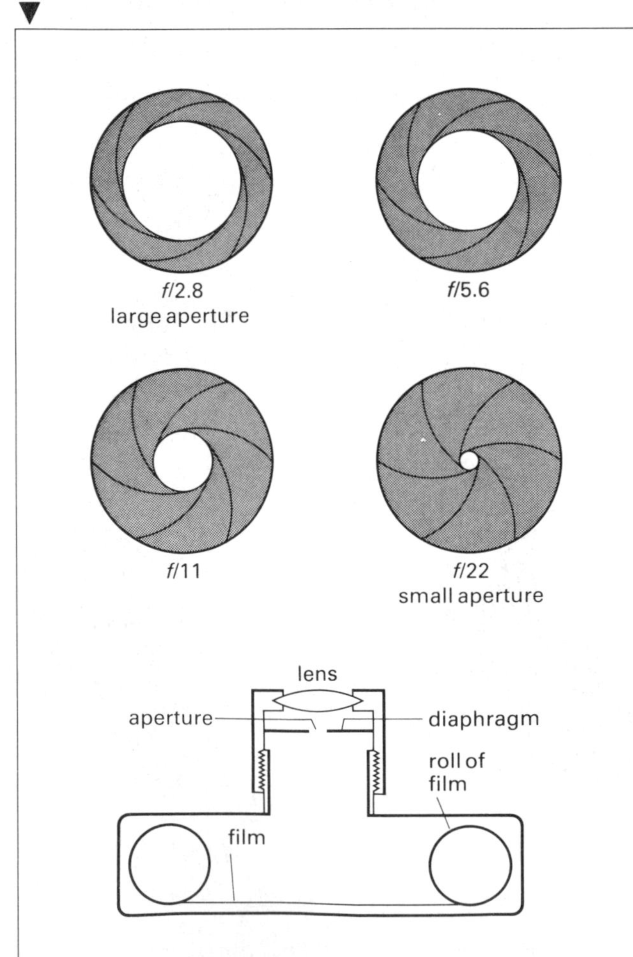

that a fixed focus camera could only focus objects at a fixed distance, but there is, in fact, a range over which an acceptably clear image can be formed. For example, if the lens has a focal length of 5 cm, an object at infinity will focus at 5 cm. An object 5 m from the lens will focus at 5.05 cm away. A reasonably sharp image of both these objects would be obtained and the camera would have a **depth of field** from 5 m to infinity.

In between the lens and the film is a spring-loaded shutter. The length of time the shutter is open controls the amount of light reaching the film, and so the amount of energy reaching the light-sensitive chemicals. The exposure time can be varied down to a very small fraction (e.g. $\frac{1}{500}$ th) of a second. Moving objects require a short exposure to produce a sharp image.

In the lens system itself is an **iris diaphragm**, figure 22.5. This is an arrangement of hinged and curved slides which enable the diameter of the opening through which the light enters to be changed while keeping the opening circular. The diameter of the opening is called the **aperture**.

Changing the size of the aperture obviously alters the amount of light reaching the film, but it also alters the depth of field. A large aperture allows a lot of light to enter, but has a small depth of field. A small aperture allows less light to enter (and so requires a longer exposure time), but has a greater depth of field. In practice, a compromise has to be made to take into account distance of object, light intensity and depth of field required.

The apertures are given as a series of ***f*-numbers**, e.g. *f*/4, *f*/5.6, *f*/8, *f*/11, *f*/16, *f*/22. The area of the aperture halves as one goes from one stop to the next, i.e. *f*/8 has twice the area of *f*/11 and so could have half the exposure time for the same light intensity. '*f*/8' means the diameter of the aperture is $\frac{1}{8}$ the focal length of the lens.

The eye

The human eye is very nearly spherical (diameter about 25 mm) and is mounted in a spherical socket so that it can be turned in any direction by three pairs of muscles. The front of the **sclerotic** is transparent and is called the **cornea**. The surface of the **retina** is made up of the nerve-endings of individual fibres of the optic nerve. The nerve-endings consist of **rods** and **cones**, and if light falls on either an impulse is sent to the brain and the person is conscious of the sensation of sight. The cones function in daylight and detect colours. The rods enable us to see in dim light and produce vision in shades of grey.

The **lens** is not rigid and its shape can be altered by the **ciliary muscles**. This ability to change the focal length of the lens is called **accommodation** and enables objects at different distances to be focused clearly. When a distant object is viewed, the ciliary muscles are relaxed. To view a near object the ciliary muscles are contracted, making the lens more powerful.

The **iris** gives the colour to a person's eyes and the diameter of the pupil can be altered. This has really very little effect on controlling the amount of light entering the eye and is probably more concerned with 'depth of field' like the aperture of a camera.

Most of the refraction of light takes place at the front surface of the cornea. The difference in refractive index between the lens (1.44) and the aqueous and vitreous humours (1.34) is very small, so that bending of the light rays at the lens surfaces is small. The lens acts like a fine adjustment device, and is concerned with the process of accommodation. For simplicity we usually consider the cornea and the lens to act as a single 'equivalent lens', figure 22.7.

The centre of the retina has more rods and cones and is the most sensitive part. This is known as the **yellow spot**. Where all the nerve fibres bunch together to leave the eye in the **optic nerve** there are no rods and cones and this is the **blind spot**.

Cover your right eye and hold the book so that the black spot in figure 22.8 is in front of your left eye. Look at the spot and bring the book slowly towards you till you find a position where the cross disappears. As shown in figure 22.8, this is when the image of the

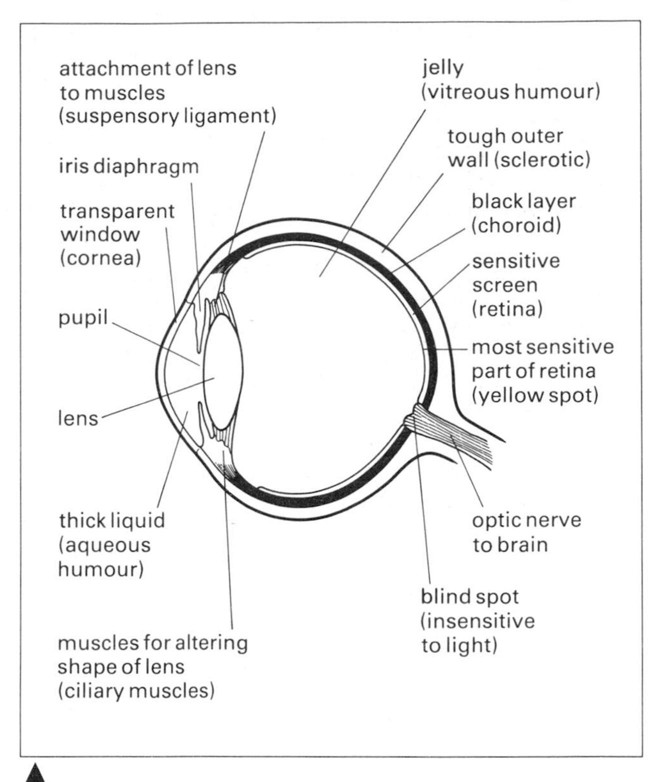

▲
fig. 22.6 The human eye (horizontal section through right eye)

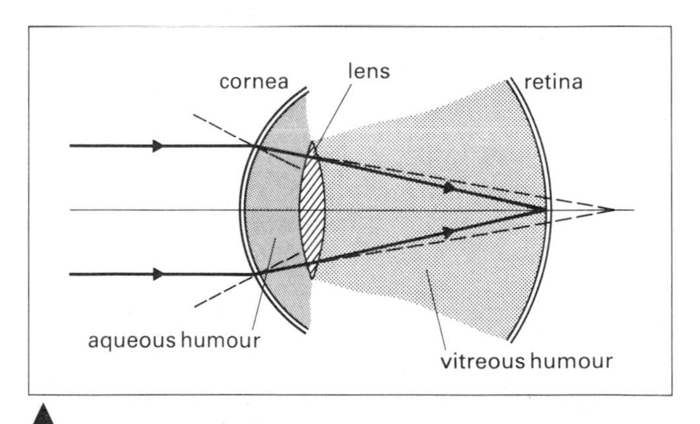

▲
fig. 22.7 The passage of light rays in the eye

fig. 22.8 Blind spot test
▼

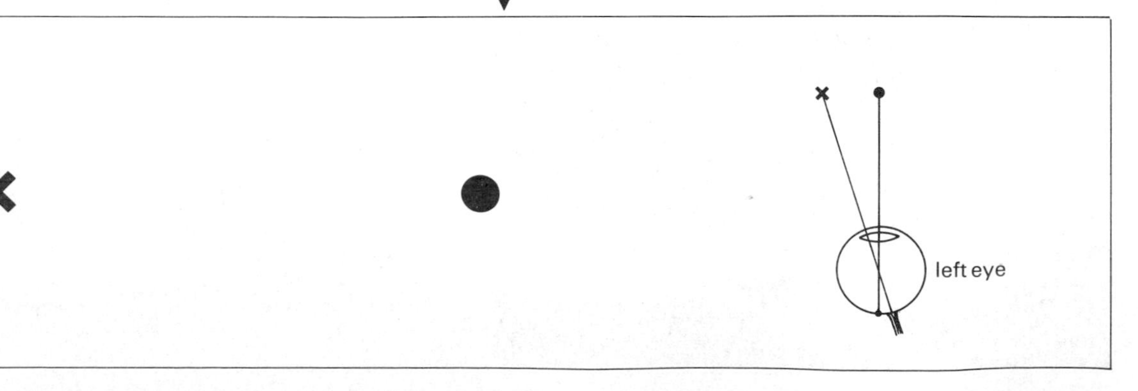

cross is just on the blind spot. If you try looking at the black cross with your right eye, you can make the spot disappear.

When an image is formed on the sensitive retina of the eye it remains there for about $\frac{1}{10}$ second. This is known as **persistence of vision**. Thus, if a light is flashed on and off (say) twenty times a second, the brain would receive the impression of steady illumination all the time. At the cinema a series of slightly different still pictures is shown on the screen at the rate of 24 every second, so that one picture blends with the next and we get the impression of movement. In fact, each picture is shown three times so that the 'flash rate' is 72 times per second and the resultant impression is free from flicker.

On the television screen, too, a rapidly moving spot of light builds up a picture in about $\frac{1}{25}$ second and, before this impression on the retina fades, a second slightly different picture appears and the two blend together.

Binocular vision – the advantages of having two eyes

One obvious advantage of having two eyes is that if one eye gets damaged we still have one left. Two eyes give us a wider angle of vision than only a single eye. Some animals have their eyes on the sides of their heads, and can almost see behind themselves.

Our two eyes give us slightly different views of an object. The right eye sees a little more of the right-hand side of the object, and the left eye sees a little more of the left-hand side. The brain combines both these views and we get an impression of the depth or solidness of the object. We see it in 'three dimensions'. A one-eyed view would be flat like a photograph. The effect of having two eyes can be imitated by a special camera with two lenses separated by just the distance between the eyes (about 6 cm). Two separate photographs of a scene are taken on separate pieces of film. These two slightly different views are placed in a 'stereoscope' so that the right eye sees only the right-hand view and the left eye the left-hand one. The brain combines the two images to give a 3-D (three-dimensional) effect.

A very interesting way of getting this 3-D effect makes use of colour photographs. It was also used in early stereoscopic films in the cinema. Two photographs are taken from different positions, whose separation is equal to the distance between a pair of eyes. The left-hand photograph is taken through a green filter and the right-hand through a red filter. They are then printed on the same piece of photographic paper. The photograph is now viewed through a special pair of spectacles having red glass for the right eye and green for the left, figure 22.9. Red glass will not let green light pass through, so the right eye sees only the right-hand picture and similarly the left eye sees only the slightly different left-hand picture. The combined effect is to give the impression of depth.

Our two eyes help us to judge distances more accurately. This is due partly to the fact that our eyes see two different images, and experience teaches us to judge distances using these two images. It is also due to fact that when we examine an object our eyes converge on it. The eyes converge more when looking at a near object than at a more distant one, figure 22.10. The angle of convergence is signalled to the brain. Our eyes are acting as a range-finder.

fig. 22.9 Viewing a 3D colour illustration with stereo glasses

fig. 22.10 Binocular vision. The angle of convergence gives a measure of the distance to the object

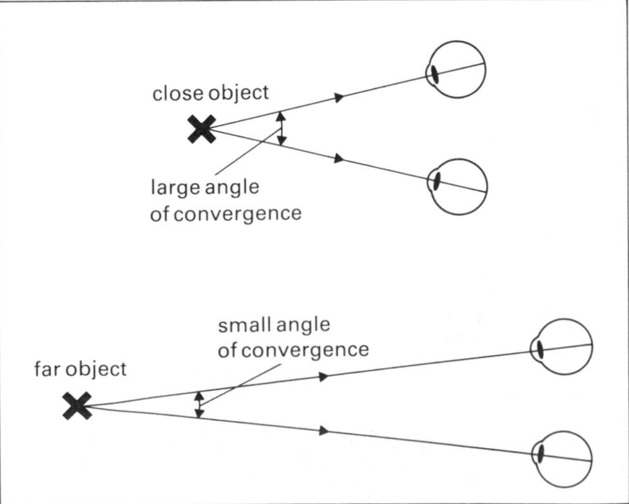

Defects of the eye and their correction

When the muscles of the eye are completely relaxed the eye is focused at the **far point**. For the normal eye this is at infinity, i.e. parallel light is entering the eye.

The closest point on which the eye can focus is the **near point**. For an average eye this is about 25 cm. The distance of the near point from the eye is the **least distance of distinct vision**, figure 22.11.

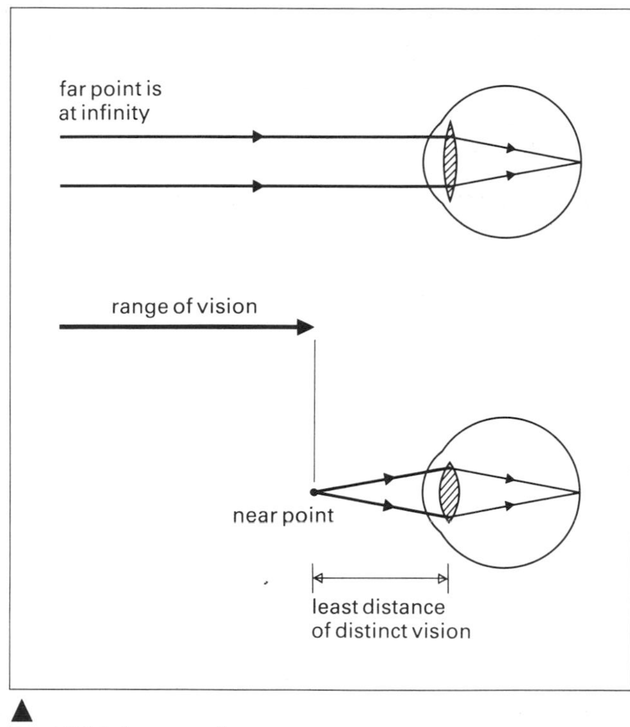

fig. 22.11 A normal eye

Lack of accommodation

As we get older the power of accommodation decreases. The eye lens becomes less flexible and it is more difficult for the ciliary muscles to make the lens thick enough to focus on close objects. The defect, which is caused by the natural process of ageing, is named **presbyopia**. It can be corrected by using a convex lens to assist the eye lens to become more powerful. Spectacles with convex lenses would be used for reading.

Short sight (myopia)

The short-sighted eye has too powerful a lens system, or too long an eyeball, so that the image of a distant object is formed in front of the retina. Both the far

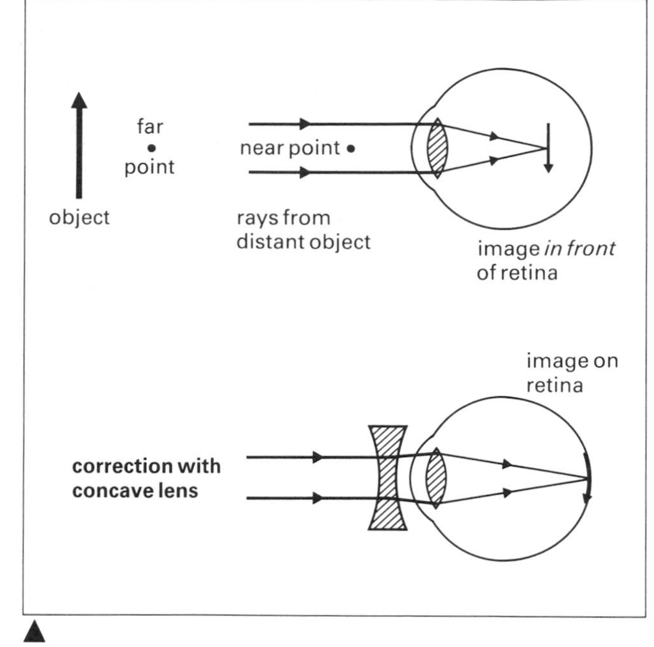

fig. 22.12 Short sight and its correction

point and the near point are closer to the eye and the range of vision is decreased, figure 22.12. Short sight can be corrected by using a concave lens of such a strength that, after passing through both the spectacle lens and the eye, parallel rays from a distant object are focused on the retina, figure 22.12.

Long sight (hypermetropia)

In the long-sighted eye the unaccomodated lens system is too weak or the eyeball too short, so that the image of a distant object is formed behind the retina when the ciliary muscles are relaxed. The near point is further from the cornea than for a normally sighted eye. Long sight can be corrected by using a convex lens of such a strength that, together with the eye lens, it forms an image of a distant object on the retina, figure 22.13.

Astigmatism

Astigmatism is caused by the curvature of the cornea (which does most of the refracting) being different in different directions. The result is that horizontal lines, for example, can be focused clearly while vertical lines are blurred. Astigmatism is detected by rotating a set of vertical and horizontal lines, as shown in

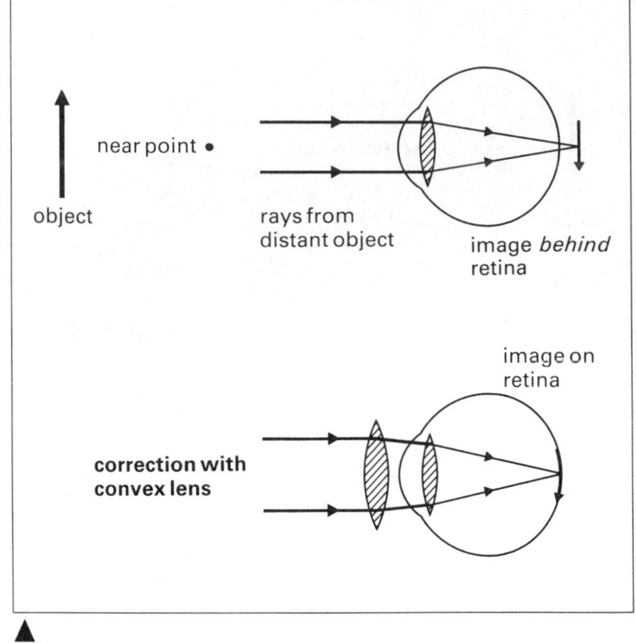

fig. 22.13 Long sight and its correction

figure 22.14. When one set of lines looks most distinct, the lines at right angles are least distinct. The defect can be cured by using cylindrical lenses with curvatures arranged so that the combined curvature of the lens and the cornea is the same for every direction.

fig. 22.14 Astigmatism test patterns

EXERCISE 22

In questions 1–5 select the most suitable answer.

1 A point of similarity between the eye and the camera is that they both
- **A** focus their image in the same way
- **B** have concave lenses
- **C** have shutters of variable speed
- **D** contain fluids
- **E** form an inverted image on their light sensitive surface

2 Which of the following statements about a lens camera is most likely to be correct?
- **A** The camera may be focused by moving the lens backwards or forwards relative to the film.
- **B** The camera may be focused by changing the shape of the lens.
- **C** The aperture is changed by moving the lens backwards or forwards relative to the film.
- **D** The aperture is changed by varying the shutter speed.
- **E** Changing the shutter speed automatically focuses the camera.

3 A part of the eye which does not change its shape when focusing objects at different distances but is responsible for a large proportion of the refraction of light rays from an object is
- **A** the retina
- **B** the pupil
- **C** the lens
- **D** the cornea
- **E** the iris

4 A long-sighted eye
- **A** can focus on nearby objects but not on distant objects
- **B** needs a diverging spectacle lens for correction
- **C** may have a lens with too short a focal length
- **D** may have an eyeball which is much deeper than normal
- **E** can form a clear image on the retina when viewing distant objects

5 A short-sighted eye
- **A** can focus on nearby objects but not on distant objects
- **B** needs a converging spectacle lens for correction
- **C** may have a lens with too long a focal length
- **D** may have an eyeball which is less deep than normal
- **E** can form a clear image on the retina when viewing distant objects

6 A 35 mm camera has a lens with a focal length of 50 mm.
- **a** What is the power of the lens?
- **b** Explain how the lens can be adjusted to produce a clear photograph of objects which are at distances ranging from 1 m to infinity.
- **c** A supplementary lens is fitted to the camera. This changes the focal length of the camera to 500 mm. What is the power of this combination?
- **d** A student takes a photograph of Buckingham Palace from the Palace gates, using the 50 mm lens. He then takes a second photograph of the Palace, using the 500 mm lens. What is the main difference between the photographs produced?

7 a Draw a labelled diagram to show the main parts of a single lens camera.

b Explain how the camera is adjusted to take photographs (i) of objects at different distances from the camera, (ii) under different light conditions. Quote examples in your answers.

8 a Figure 22.15 is a diagram of a *human eye*. Name the parts indicated on the diagram.

b When the eye views objects at different distances the lens changes shape. (i) What name do we give to the change of shape? (ii) Which of the lenses A or B in figure 22.16 is the correct shape for near vision?

c (i) Copy the diagram of the eye, figure 22.17, and on it show how light from a near object is not focused on the retina by an eye which is long-sighted. (ii) What type of lens is used to correct this defect?

d Explain why a person who has normal vision may need to wear spectacles as he grows older.

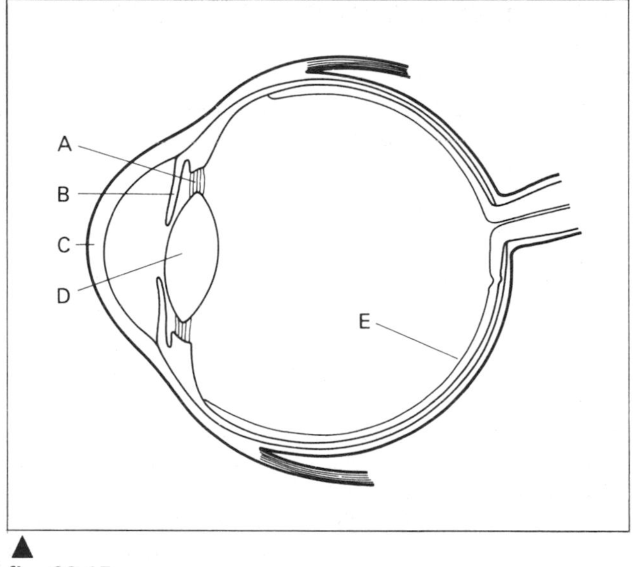

fig. 22.15

fig. 22.16

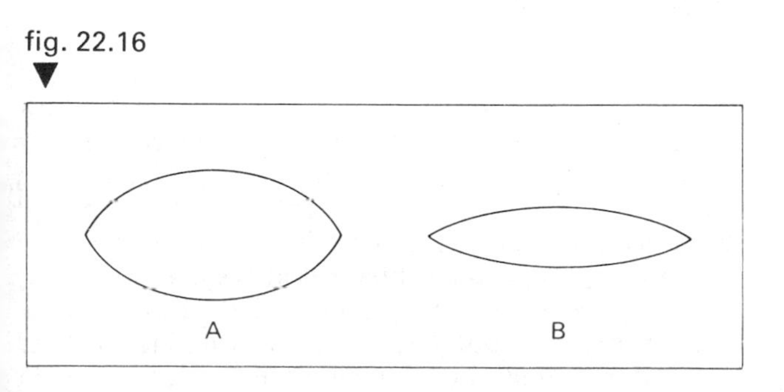

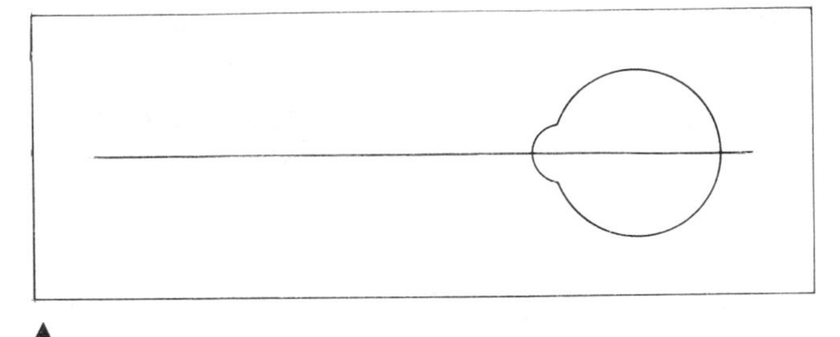

fig. 22.17

9 Figure 22.18 shows a simplified cross-section of the eye with rays coming from the near point and from a distant object to focus on the retina.

a Explain why the normal eye is able to adjust to form a clear image of near and distant objects in this way.

b Someone with short sight cannot focus distant objects on the retina. Draw diagrams to illustrate short sight and its correction using a suitable lens.

c Someone with long sight cannot focus a near object on the retina. Draw diagrams to illustrate long sight and to show how this defect is corrected by using an appropriate lens.

d What is meant by (i) loss of accommodation, and (ii) astigmatism? Explain how these may be corrected by suitable spectacles.

fig. 22.18

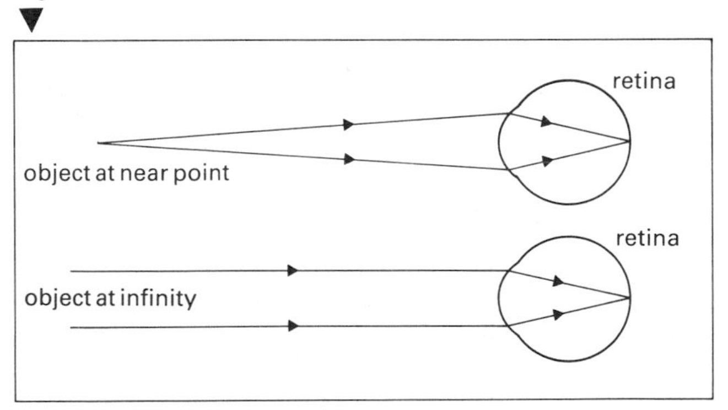

10 Which parts of the human eye play a part in converging light from a distant object?

What changes occur if the eye focuses on a near object? How does the human eye allow the correct amount of light to enter?

11 A movie film consists of a large number of still photographs which are rapidly shown through a projector. Why does an observer see a continuously moving picture and not a succession of individual pictures?

A cameraman takes a moving picture at 8 pictures per second. He then projects the picture with a projector that shows 16 pictures per second. What effect will this have on the picture projected?

23 OPTICAL INSTRUMENTS

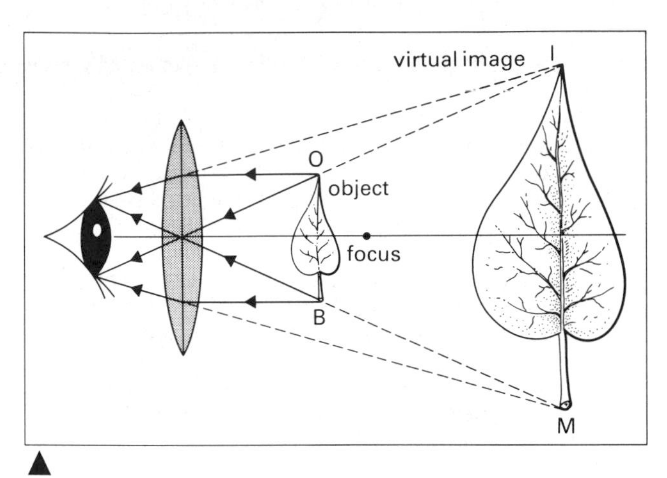

▲ fig. 23.1 Magnifying glass (simple microscope)

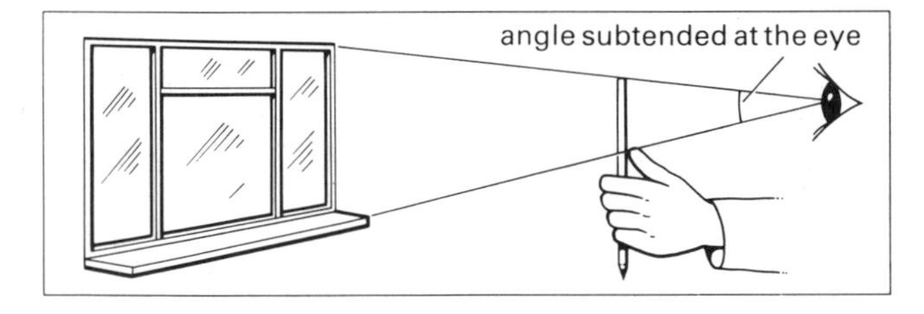

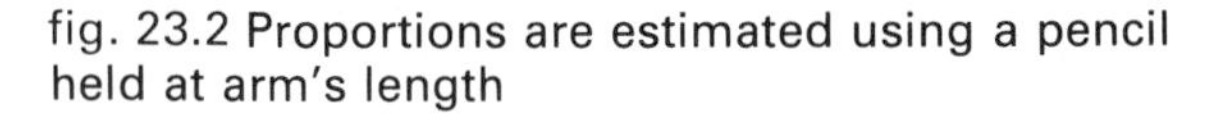
fig. 23.2 Proportions are estimated using a pencil held at arm's length

fig. 23.3 Aided and unaided vision ▼

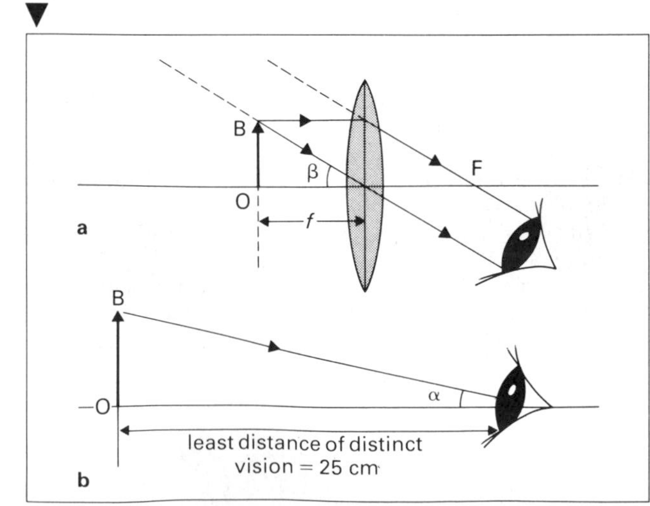

If our eyes have unusual jobs to do, it is possible to help them in their task by using some form of optical instrument. If we want to look at anything very small, or examine something closely, we can use a **magnifying glass** or a **microscope**. In order to see objects which are a long way off we use a **telescope**.

The magnifying glass or simple microscope

When an object is placed nearer to the lens than the principal focus a virtual, erect and magnified image is formed. This is the way in which a convex lens is used as a magnifying glass.

In figure 23.1, light rays which set out from O are refracted by the lens so that the beam of light entering the eye appears to have come from I. Similarly, the image of B is at M. The final image can be produced anywhere between the near point and infinity by altering the distance between the object and the lens. If an object is to be examined for some time it is as well to have the final image some distance away from the eye so that the rays from it are almost parallel and the eye muscles are relaxed.

Since the image is the same way up as the object, a lens can be used in this way as a reading glass, or a stamp magnifier, or by watchmakers to examine the small works of a watch.

Magnifying power

The apparent size of an object when we look at it depends on the actual size of the object and also how far away it is. A small object close to us can appear the same size as a large distant object. We judge size by the angle which rays of light from the top and bottom of the object make when they meet at the eye, i.e. by the **angle subtended at the eye**.

Artists make use of this fact when they estimate proportions by holding a pencil at arm's length and moving the thumb up or down until the upper part of the pencil seems to be the same size as the vertical feature being observed, figure 23.2. Both subtend the same angle at the eye. If we wish to produce magnification we have to increase this angle.

The **magnifying power** of an optical instrument is defined as follows:

$$\textbf{magnifying power} = \frac{\textbf{angle subtended at eye by image seen through instrument}}{\textbf{angle subtended at eye by object when viewed normally}}$$

Applying this idea to the magnifying glass, and considering the case when the eye is relaxed and the final image is at infinity, figure 23.3a:

angle subtended at eye by image seen through instrument $= \beta = \frac{OB}{f}$

angle subtended at eye by object when viewed normally $= \alpha = \frac{OB}{25}$

magnifying power $= \frac{\beta}{\alpha} = \frac{OB/f}{OB/25} = \frac{25}{f}$

The result shows that the smaller the value of the focal length, the greater the magnifying power of the lens. A lens with a focal length of 5 cm would have a magnifying power of 5 times.

The compound microscope

From the expression for magnifying power given in the last section it would appear that, by making the focal length of the lens small enough, a very high magnifying power could be obtained. But a small, short-focus lens would have very curved faces, and this would produce distortion of the image. Higher magnifications have to be produced using two lenses in the compound microscope, figure 23.4.

The objective is a short focus lens. The image is placed between one and two focal lengths from the lens and a real, inverted and magnified image (I_1M_1) is produced, figure 23.5. This image is arranged to fall just inside the focus of the eyepiece, which acts as a simple magnifying glass and produces further magnification. The final image (I_2M_2) is virtual and inverted with respect to the object. If when looking through a microscope you move the object to the right, the image moves to the left. The final image

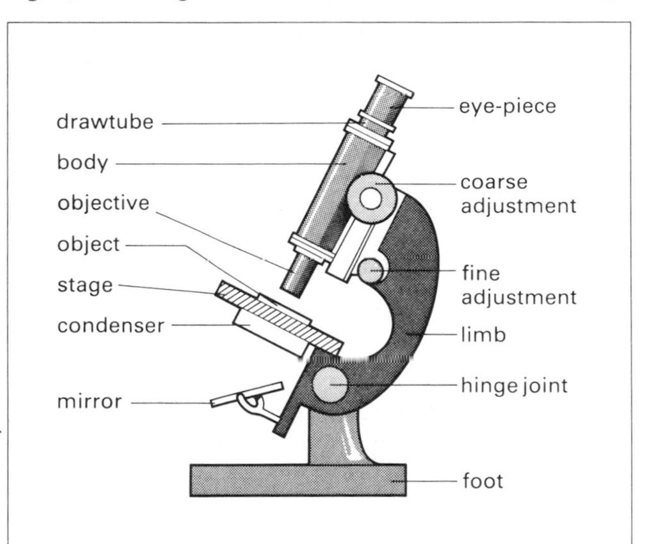

fig. 23.4 Compound microscope ▶

fig. 23.5 Formation of the final (virtual) image in a compound microscope ▼

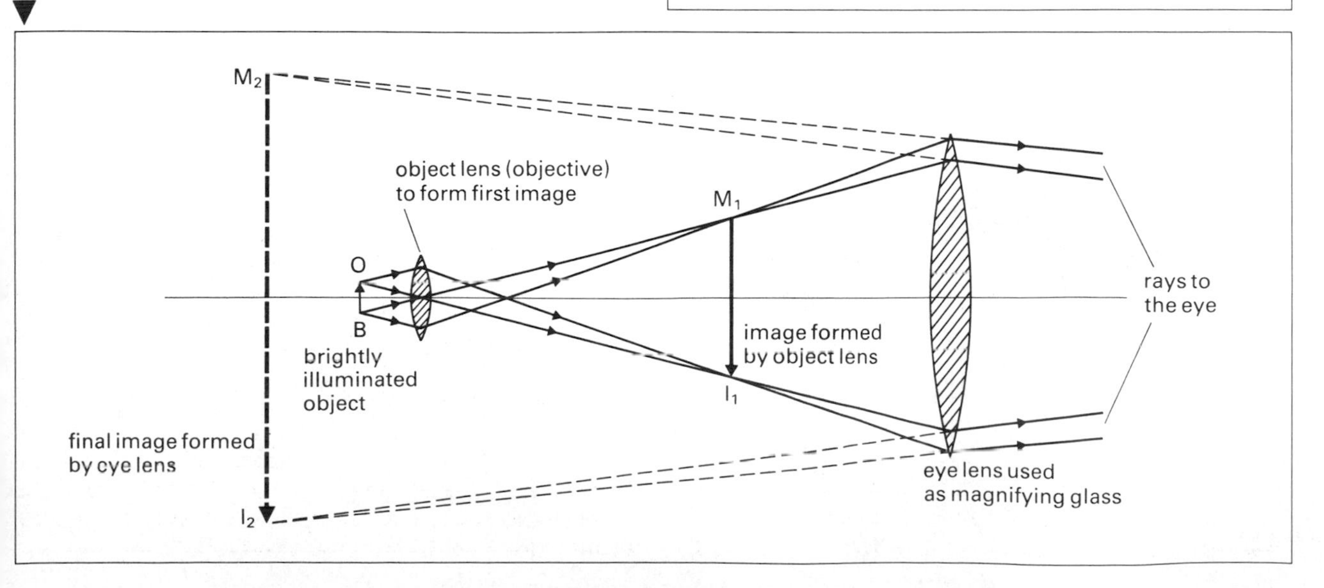

can be formed anywhere between the near point and infinity.

The magnification takes place in two stages, most magnification being produced by the objective. On most microscopes, objectives and eyepieces can be changed. Typical values are

low-power objective $f = 16$ mm, magnification 12 ×
eyepiece, magnification 10 ×

total magnification 120 ×

high-power objective $f = 4$ mm, magnification 50 ×
eyepiece, magnification 15 ×

total magnification 750 ×

The astronomical telescope

In the astronomical telescope, figure 23.7, a large diameter, long focal length objective lens forms an image (I_1M_1) of a distant object. The objective has a large area, so that it can collect as much light as possible and thus produce a bright image. It has a long focal length to produce large magnification. Since the object is distant, the image is formed at the principal focus of the objective. The intermediate image is viewed through the eyepiece (a shorter focal length lens) which acts as a magnifying glass.

The final virtual image can be formed anywhere from the near point to infinity, but is usually formed at infinity so that it can be viewed with a relaxed eye to avoid eyestrain. When the instrument is used in this way (in 'normal adjustment') the intermediate image is at the principal foci of both the objective and the eyepiece. The distance between the lenses is the sum of the focal lengths and the magnifying power can be shown to be

$$\frac{\textbf{focal length of objective}}{\textbf{focal length of eyepiece}}$$

The final image is inverted with respect to the object. This is no great disadvantage when viewing astronomical subjects, but makes the instrument unsuitable for looking at objects on the earth's surface.

fig. 23.6 The object lens of the telescope at the Yerkes Observatory in America has a diameter of 1 m and is the largest telescope lens ever made

fig. 23.7 Astronomical (refracting) telescope

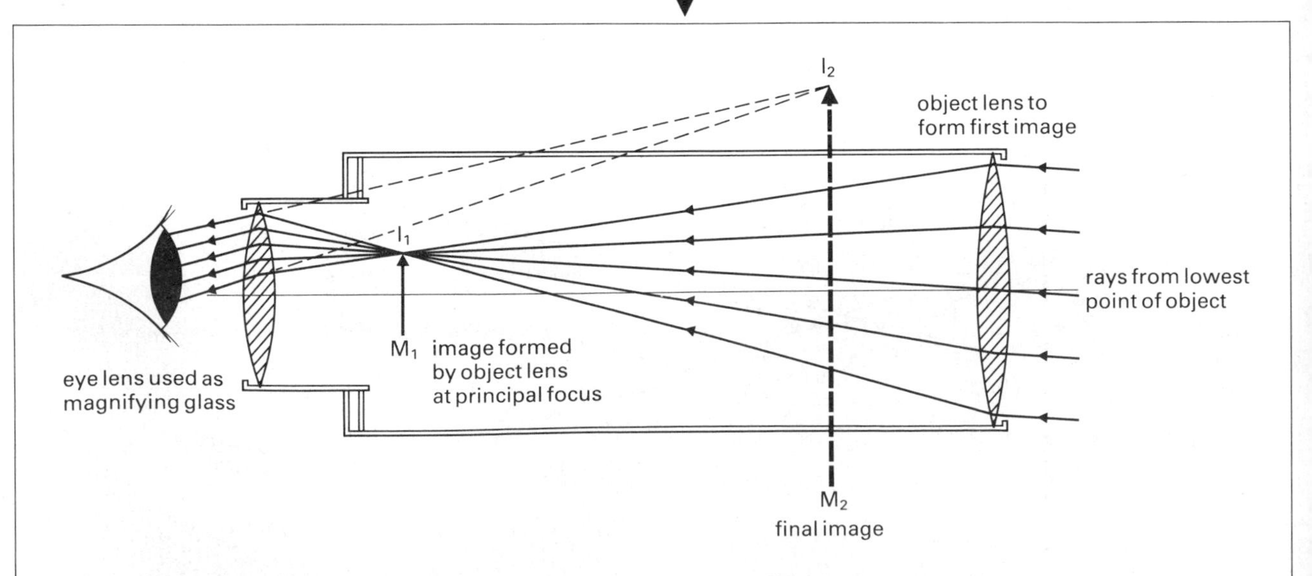

Prismatic binoculars

In order to reduce the length of the telescope the light rays are made to travel the length of the tube three times by using total reflecting prisms. The prisms also make the final image erect which is another advantage (see p. 165). One prism reverses the image from top to bottom, and the other from left to right, figure 23.8.

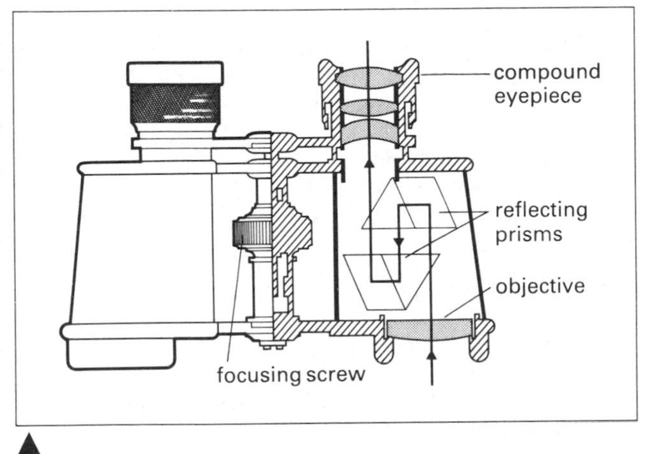

fig. 23.8 Prismatic binoculars

Terrestrial telescopes

Terrestrial telescopes for observing objects on the surface of the earth must obviously produce an erect image. Prismatic binoculars achieve this result by using two reversing prisms. Another method is to use an erecting lens placed between the objective and the eyepiece, figure 23.9. The Galilean telescope produces an erect image by using a concave eyepiece, figure 23.10. This system is used in opera glasses. Although the instrument is short and convenient to handle, both the field of view and the magnification are limited.

fig. 23.9 Terrestrial telescope with erecting lens

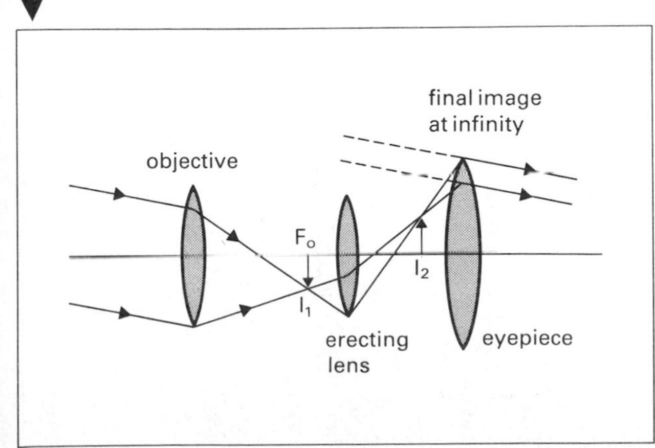

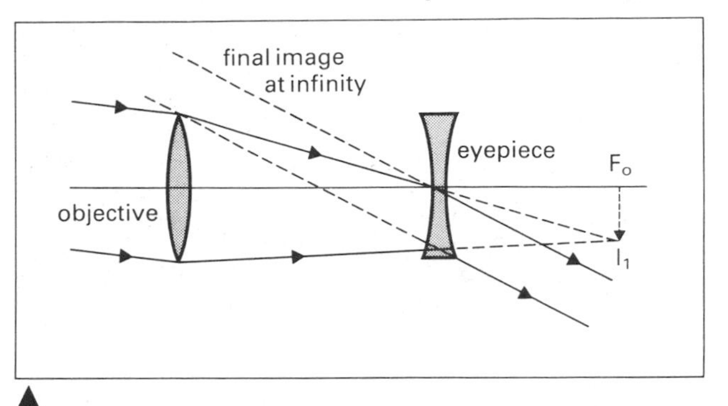

fig. 23.10 Galileo's telescope

The reflecting telescope

In order to observe very distant stars, a lens telescope would have to have a very large diameter objective so that it could collect sufficient light and produce the necessary magnification. But large lenses are difficult to make accurately. They are very heavy and have to be supported by the edges, thus the centre sags and the change of shape produces distortion. On the other hand, large mirrors can be supported from behind and so retain their shape. All large telescopes today are reflecting telescopes, figures 23.11 and 23.12.

Sir Isaac Newton constructed such a telescope in which he used a concave mirror to collect the light and then observed the image through an eyepiece.

fig. 23.11 The reflecting telescope at the Mount Pastukhov Observatory, in the Caucasus, is the world's largest optical telescope. Its mirror has a diameter of 6 m and weighs 40 tonnes. With this telescope, stars over 1 200 000 light years away can be photographed

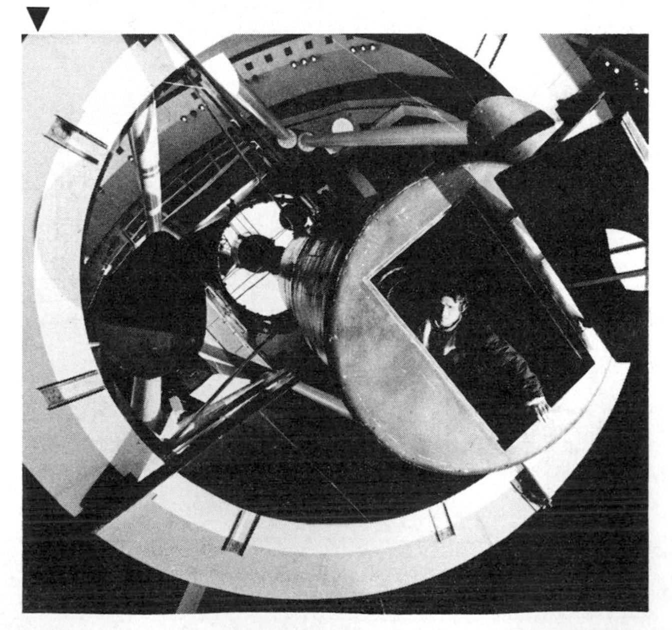

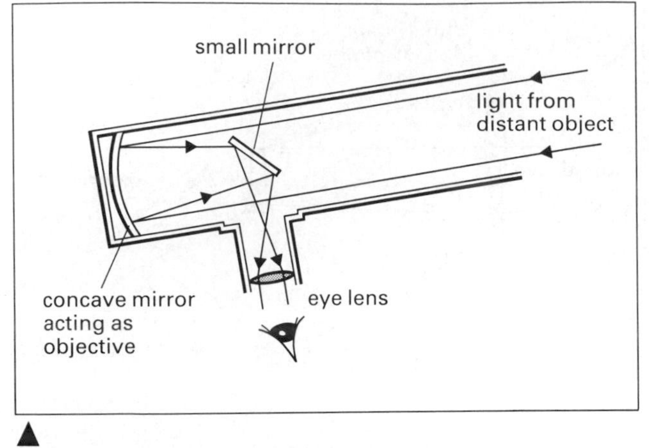

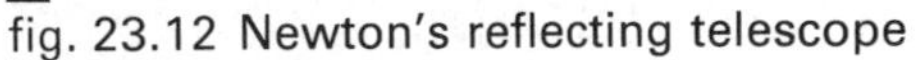
fig. 23.12 Newton's reflecting telescope

The projector

When an object is placed between one and two focal lengths from a convex lens the image is real and magnified. If the object is brightly illuminated this real image can be projected on to a screen, figure 23.13. Since inversion of the image takes place, the slide is put in the projector upside down and turned from right to left.

The lamp is placed at the centre of curvature of the concave mirror so that light going backwards is returned to the filament and the illumination increased. The condensing lens concentrates light which otherwise would not have passed through the slide. This increases the illumination and makes it more even. Focusing is carried out by altering the distance between the slide and the projection lens. The heat filter prevents infrared radiation from the lamp damaging the slide.

EXERCISE 23

In questions 1–4 select the most suitable answer.

1 When a convex lens is used as a magnifying glass the image is
A real and upright
B real and inverted
C virtual and upright
D virtual and inverted
E real and diminished

2 Which type of lens would be most suitable for the objective of a compound microscope?
A a long focus concave lens
B a long focus convex lens
C a short focus plano-concave lens
D a short focus convex lens
E a short focus concave lens

3 In an astronomical telescope, the best combination of lenses is
A a strong positive (converging) objective and a strong positive (converging) eyepiece
B a weak positive (converging) objective and a weak positive (converging) eyepiece
C a weak negative (diverging) objective and a strong positive (converging) eyepiece
D a strong positive (converging) objective and a weak positive (converging) eyepiece
E a weak positive (converging) objective and a strong positive (converging) eyepiece

4 The slide is put in a projector
A right way up and right way round
B upside down and right way round
C inverted and laterally inverted
D right way up and turned round
E turned through 90° and inverted

fig. 23.13 Slide projector

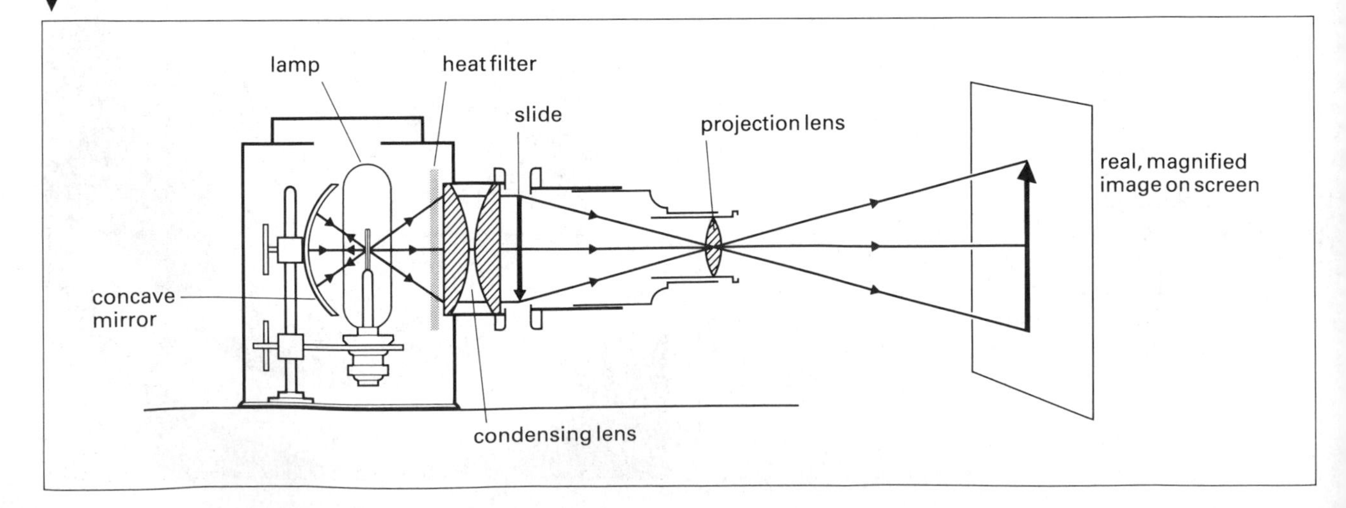

5 Convex lens   concave lens   convex mirror   concave mirror

Which of the above would be most suitable in the examples below?

a a magnifying glass
b a spectacle lens to correct for long sight
c the objective of an astronomical telescope
d the reflector of a car headlamp

6 a What type of lens is used as a magnifying glass?
b Suggest a suitable focal length in centimetres for a good magnifying glass.
c What advantage is gained from holding a magnifying glass close to the eye when using it?
d Draw a sketch showing how such a magnifying glass achieves its purpose. Label on your diagram object, image, eye, lens and focal points. Show the directions of any rays of light you use and draw any construction lines as dotted lines.

7 Draw a ray diagram to show how a magnified image is produced by a compound microscope. How can the instrument be changed to produce higher magnifications?

8 A student constructs a telescope using two converging lenses, A and B, mounted on a rule, figure 23.14.

Lens A has a diameter of 4 cm and focal length of 40 cm. Lens B has a diameter of 1 cm and focal length of 5 cm.

When looking at a distant object the lenses are separated by 45 cm.

a Which lens has the greater power?
b Why should the objective lens, A, have a large diameter?
c Indicate the approximate position of the image formed by the objective lens, A.
d How would you show experimentally that this is a 'real' image?

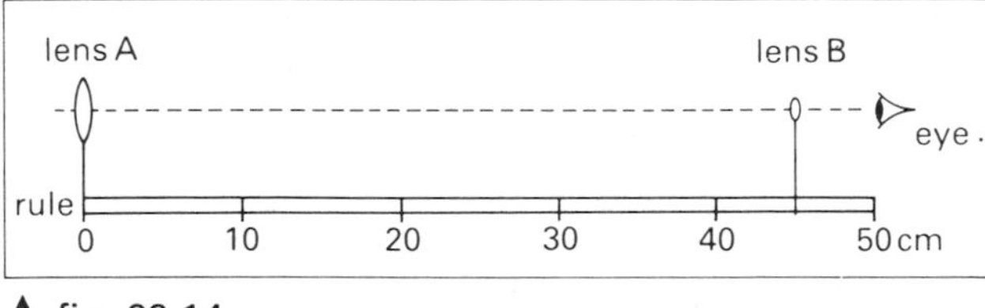

▲ fig. 23.14

9 Explain, in simple terms, why it would be difficult to use two astronomical telescopes as a pair of binoculars. Figure 23.15 shows how two 45° glass prisms may be positioned to produce a manageable-sized pair of binoculars. Complete the paths of the rays of light through the prisms.

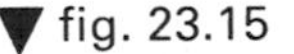

▼ fig. 23.15

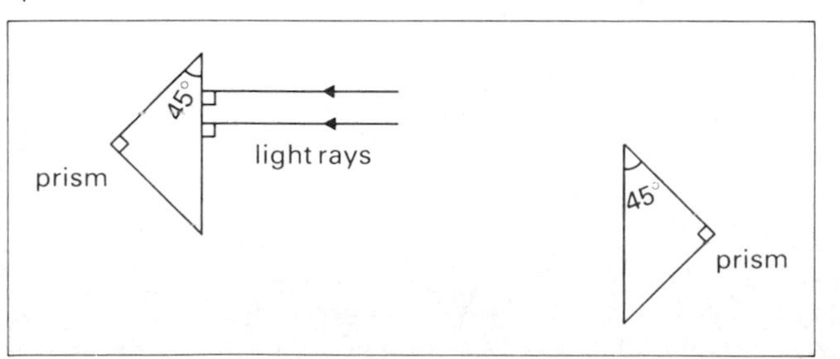

10 a Draw a labelled diagram of a slide projector, showing the position of the light source, condenser lens, slide, projection lens and concave mirror.
b (i) By scale drawing, or otherwise, find the position of the screen to give a focused image of a slide placed 12 cm from a projection lens of focal length 10 cm. (ii) Describe the nature of the image formed.

11 The three ray diagrams, figures 23.16a, b and c show the optical system for a *camera*, a *projector* and a *magnifying glass*, but *not* in that order.

a State which diagram is which.
b What type of lens is used in all of these instruments?
c Which *two* instruments give a magnified image?
d What is meant by a real image?
e Which *two* instruments give a real image?
f Why is the object shown inverted in figure 23.16a?
g How could a larger image be obtained in figure 23.16a still using the same lens?
h What would happen to the image in figure 23.16c if the lens was moved slightly to the right?
i Copy figure b on to your answer sheet, using the dimensions shown above, and then complete it by adding construction lines to show where the image is formed. State for the image (i) its height, (ii) whether it is erect or inverted, (iii) whether it is real or virtual, (iv) how far it is from the lens.
j Show, on the diagram you have drawn, a suitable position and direction of looking for an eye.

fig. 23.16 ▼

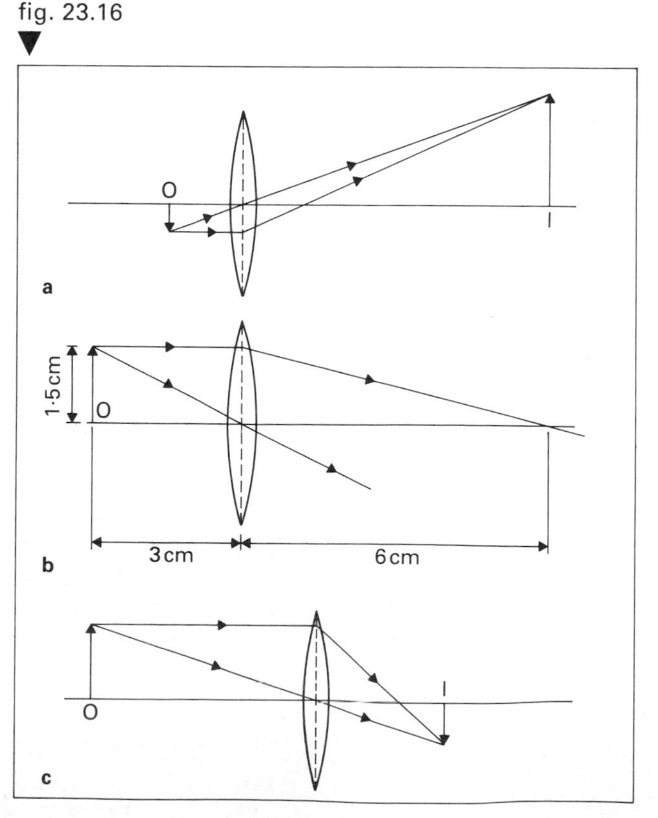

12 Figure 23.17 is a simple diagram of the components of a reflecting telescope.

a Copy the diagram and draw on it the path of two parallel rays of light from a distant object to the eye.

b Name the scientist who first designed a telescope of this type.

c In modern telescopes a parabolic shape is used instead of a spherical shape. What is the advantage of the parabolic shape?

d The largest telescopes are reflecting rather than refracting. Give *two* advantages of reflecting telescopes.

e Mirrors are silvered at the front rather than the back. What is the advantage of this?

fig. 23.17

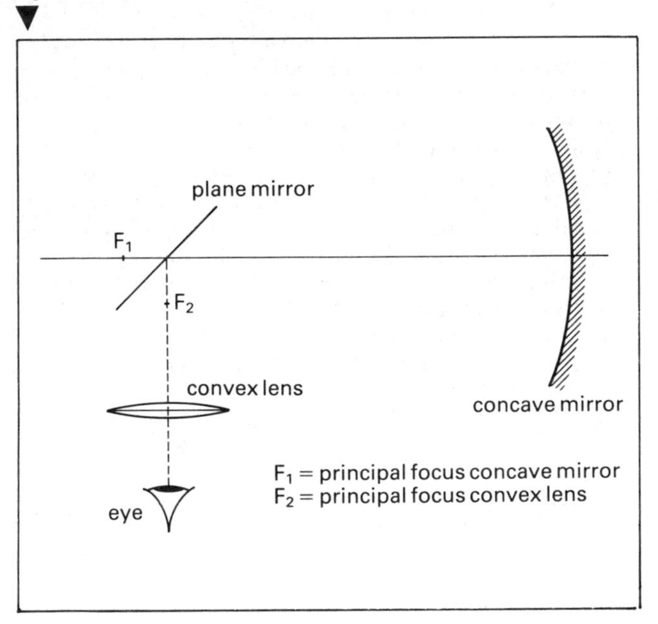

13 The following optical components are available:

Lenses	+1.00 D	−1.00 D	+20.00 D	−20.00 D	+33.33 D	−33.33 D
Mirrors	+1.00 D	−1.00 D	+20.00 D	−20.00 D	+33.33 D	−33.33 D

a (i) State which of the lenses listed above are convex (converging). (ii) State which of the mirrors listed above are convex.

b From the list above, to make a compound microscope, (i) select the most appropriate components; (ii) state the focal length, in mm, of each component; (iii) state which component would be used as the eyepiece.

c From the previous list, to make a refracting telescope, (i) select the most appropriate components; (ii) state the focal length, in mm, of each component; (iii) state which component would be used as the eyepiece.

d An object subtends an angle of 0.5° at the eye. The image formed by a magnifying glass subtends an angle of 2.5° at the eye. Calculate the magnifying power of the lens.

14 a What type of lens is used as the projection lens in a slide projector?

b State the nature of the image produced by such a projector, i.e. real or virtual, magnified or diminished, erect or inverted.

c Where is the slide placed relative to the projection lens?

d What other lenses are used in such a projector and what is their purpose?

e Sometimes such projectors contain a heat filter. What is the purpose of such a filter?

f Why does the lamp used in a projector have to be so powerful?

g Why is a mirror often mounted behind the projector bulb?

24 TYPES OF WAVE AND THEIR BEHAVIOUR

A rope or piece of rubber tubing about five metres long is fastened to the wall at one end. The free end of the rope is held in the hand and moved sharply up and down once. A 'hump' travels along the rope, figure 24.1a. Each bit of the rope is moved by the bit next to it and so progressively each bit of the rope moves up and down. A single movement transmitted along the rope is called a **pulse**.

A similar effect can be produced with a long 'slinky' spring lying on the bench, figure 24.1b. A side-to-side movement of one end produces a **transverse** pulse along the spring.

Using the same spring we can produce another kind of pulse which could not be produced with the rope. The end of the spring is moved forwards in the direction of the length of the spring so that the first few coils are compressed. The compression travels along the spring, figure 24.2. A pulse which travels in the same direction as the movement of the coils of the spring is a **longitudinal** pulse.

If, instead of a single movement of the end of the rope or spring, there is a continuous vibration then a **wave** travels along. The essential idea of **wave motion** is that a change (e.g. an up-and-down movement) occurring at one place is later repeated at other places. The change takes time to travel from place to place. There is a transfer of energy in the direction in which the wave is travelling, figure 24.3.

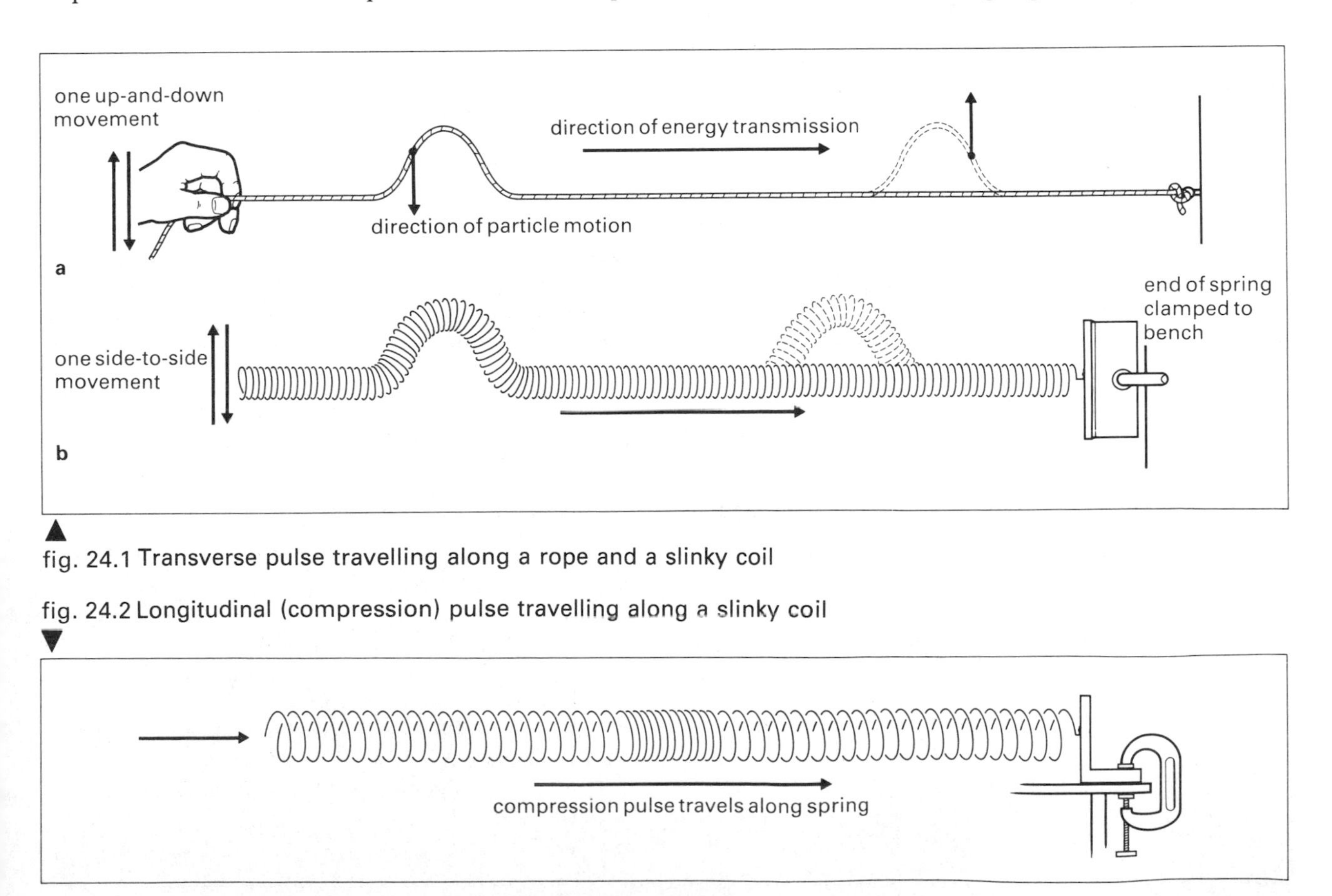

fig. 24.1 Transverse pulse travelling along a rope and a slinky coil

fig. 24.2 Longitudinal (compression) pulse travelling along a slinky coil

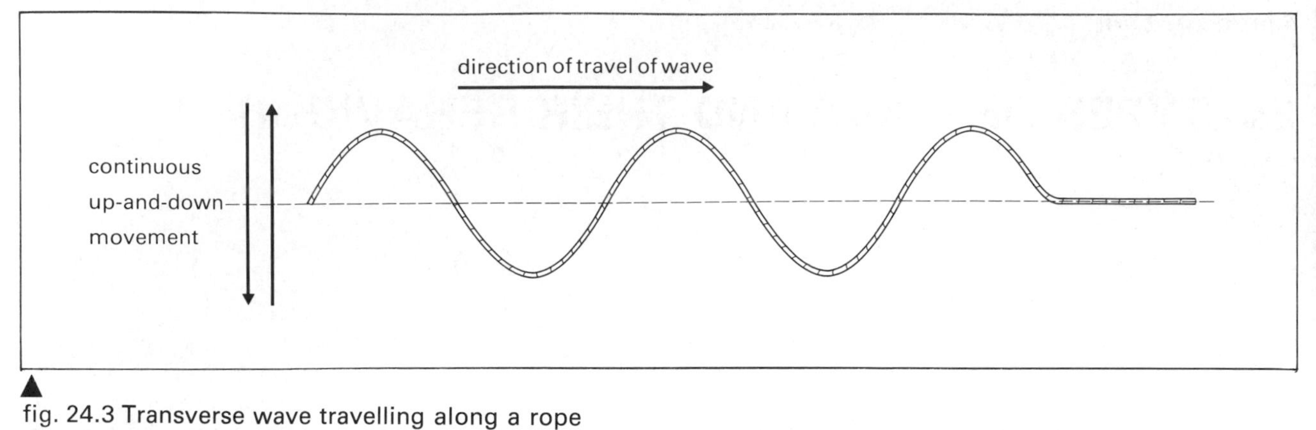

fig. 24.3 Transverse wave travelling along a rope

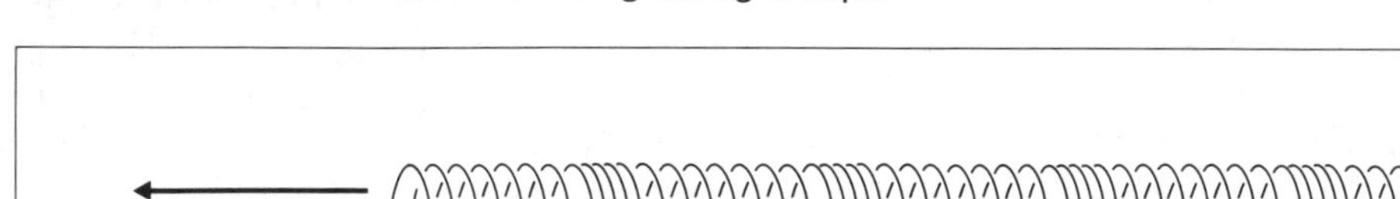
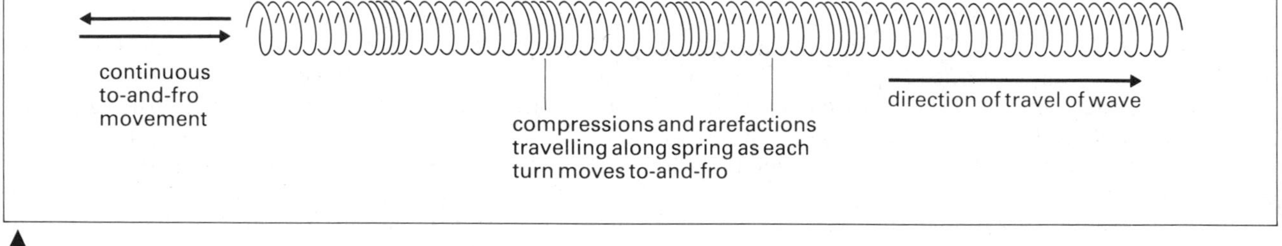

fig. 24.4 Longitudinal wave travelling along a spring

Transverse wave: the movement of the vibration is at right angles to the direction in which the wave is travelling

Longitudinal wave: the movement of the vibration is in the same line as that in which the wave is travelling

Both transverse and longitudinal waves are represented graphically by the same kind of curve. A **displacement–distance** graph, figure 24.5a, shows the position of *all* particles of the medium at the *same* instant of time. A **displacement–time** graph, figure 24.5b, shows the position of a *single* particle at *different* times.

The **wavelength** λ is the distance between successive crests or corresponding points of a wave.

The **amplitude** a is the maximum displacement from the mean position.

The **period** T is the time taken for one complete vibration.

The **frequency** f is the number of complete vibrations occurring in one second. The unit of frequency is the **hertz (Hz)** which is one vibration per second.

If the frequency is 100 Hz, i.e. 100 vibrations per second, the time for one vibration is obviously $\frac{1}{100}$ second:

$$\textbf{period} = \frac{\mathbf{1}}{\textbf{frequency}}$$

fig. 24.5 Graphs of wave motion: displacement against distance and against time

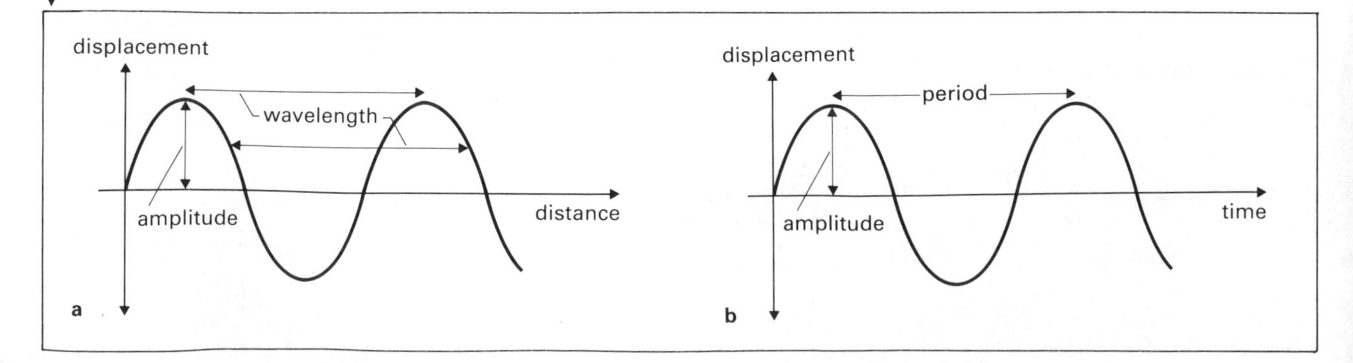

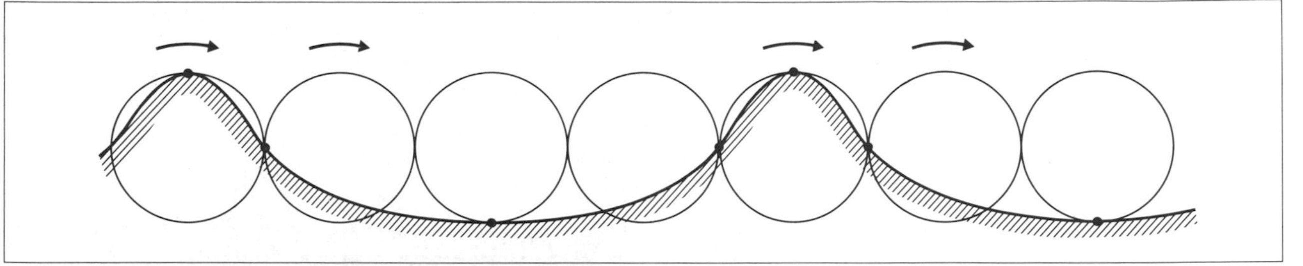

fig. 24.6 Ocean wave motion. The wave is travelling from left to right. Each water particle moves clockwise in a vertical circle. Each particle is shown 90° ahead of the one to the right

fig. 24.7 Circular ripples spreading outwards as a rain drop hits the surface of a pond

Different kinds of waves

In **mechanical waves** there is a change of position or a change of density which is transmitted through a material medium. In all cases there is a transfer of energy from one part of the medium to another, without any movement of the medium as a whole. Examples of mechanical waves are: ripples on a water surface, waves in stretched strings, sound waves in gases, liquids and solids, earthquake waves in the earth's crust. Ripples and waves in stretched strings are transverse waves. Sound waves and earthquake waves are longitudinal waves.

In ocean waves, figure 24.6, the movement of the water particles is circular, each particle being at a slightly different position in its circular path from the particles behind and in front. This circular movement is often shown when the crest of a wave curls forward as breakers approach the beach.

In **electromagnetic waves** there is a change in the intensity of an electromagnetic field. Electromagnetic waves are transverse waves and can travel through a vacuum. The properties, methods of production and detection of waves vary according to their wavelength (and frequency). The different types of electromagnetic wave are shown in figure 24.8, from which it can

fig. 24.8 The electromagnetic spectrum. Because the range of wavelengths (and frequencies) is so enormous, logarithmic scales are used. They increase in powers of ten. Radio and t.v. waves, for example, range from 10^{-1}m (10 cm) to 10^4 m (10 km). The visible wavelengths occupy a very small band compared with the other kinds of waves. (The inset showing the components of the visible spectrum is *not* to scale.)

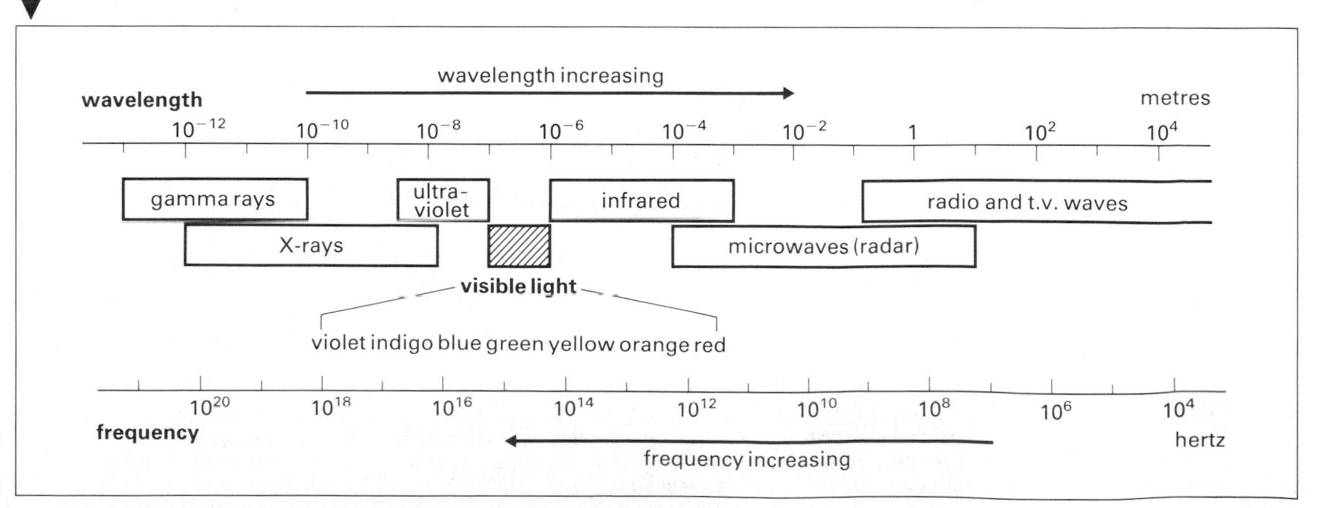

The microwave oven

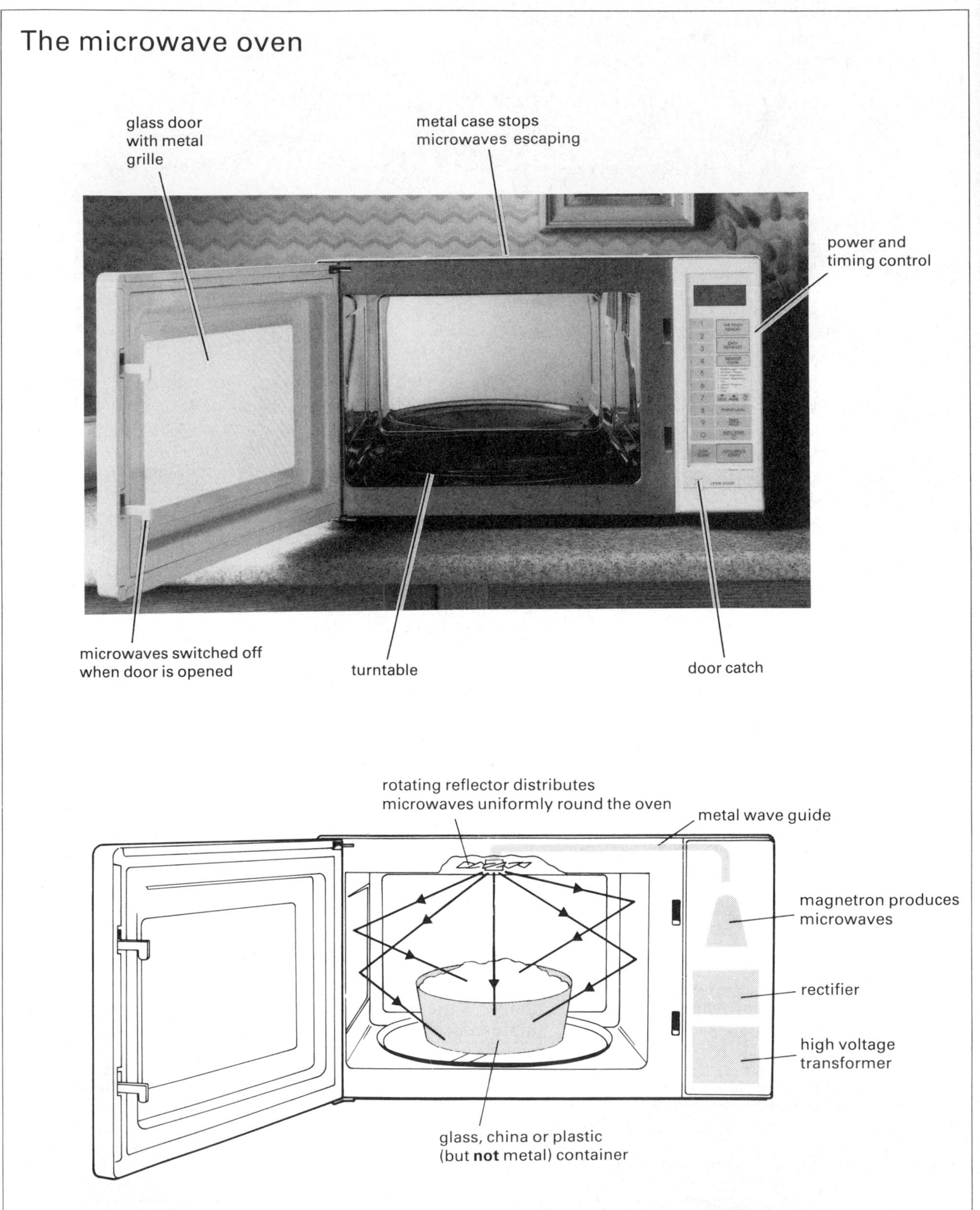

The microwaves have a frequency of about 2500 MHz which corresponds to the natural vibration frequency of water molecules (and certain groups of atoms in sugars and fats). The waves are absorbed by the water molecules which vibrate more rapidly producing heat within the food being cooked. The microwaves are not generated until the door is closed and the switch is pressed.

be seen that there is some overlap of the different types. This is because the same wavelengths of radiation can be produced in different ways. Some of the infrared waves you cook with in an electric grill will be of the same wavelength as some of the microwaves produced by an oscillator.

The relationship between wavelength and frequency

In figure 24.8 it is seen that, as the wavelength increases, the frequency decreases and vice versa.

Consider a wave source of frequency f Hz producing waves of wavelength λ metres, figure 24.9. In one second f waves each of length λ are produced so that the distance travelled is $f\lambda$. But the distance travelled in one second is the wave velocity c. Therefore

$$c = f\lambda$$

velocity = frequency × wavelength

This simple and useful formula applies to all types of wave motion.

EXAMPLES

The velocity of sound in air at room temperature is 340 m/s and the frequency of the middle C note is 256 Hz. What is the wavelength?

$$\text{wavelength} = \frac{\text{velocity}}{\text{frequency}} = \frac{340}{256} = 1.33 \text{ m}$$

All electromagnetic waves have a velocity of 3×10^8 m/s. So a radio station transmitting on a wavelength of 1500 m has a frequency given by

$$\text{frequency} = \frac{\text{velocity}}{\text{wavelength}} = \frac{3 \times 10^8}{1500} = 200\,000 \text{ Hz} \quad \text{or} \quad 200 \text{ kHz}$$

fig. 24.9

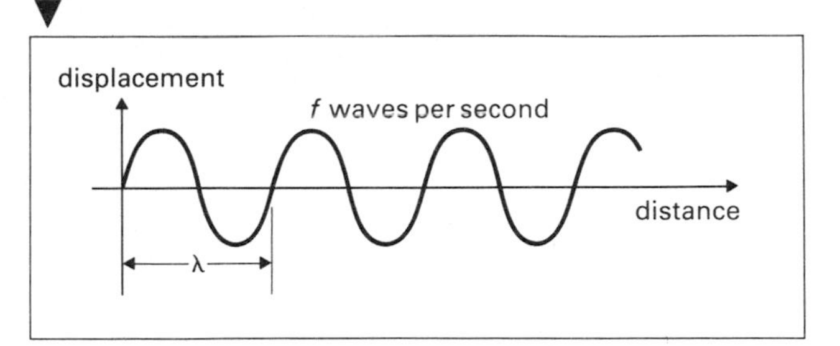

table 24.1 The electromagnetic spectrum

(wavelength decreasing ↓ from radiowaves to γ-rays)

Type of radiation	Production	Detection	Properties and uses
radiowaves	electrical circuits using inductors and capacitors to produce oscillations	receivers which can be adjusted to resonance with the signal's frequency	radio and t.v. communication, are reflected from ionised layers in the atmosphere
infrared	hot bodies, e.g. heated metals, furnaces, the sun	special photographic plates, thermopiles, skin	cooking, medical treatment, photography through haze and at night
visible light	incandescent bodies, electrical discharge tubes, some chemical reactions	eye, photographic plates, photoelectric cells	produce the sensation of sight, pass through transparent materials but not through opaque
ultraviolet	very hot bodies, mercury vapour lamps, the sun	photographic plates, photoelectric cells, fluorescence of certain materials	cause sunburn, kill certain bacteria and can be used for sterilisation, absorbed by glass
X-rays	electron bombardment of heavy metals such as tungsten	photographic plates, fluorescent screen	X-ray pictures in medicine, the shorter the wavelength the more penetrating, over-exposure is dangerous
γ-rays	emitted by the nuclei of radioactive atoms during disintegration	photographic plates, Geiger counters	very penetrating, can pass through 30 cm of lead or several metres of concrete, used in treatment of cancer, very dangerous

The ripple tank

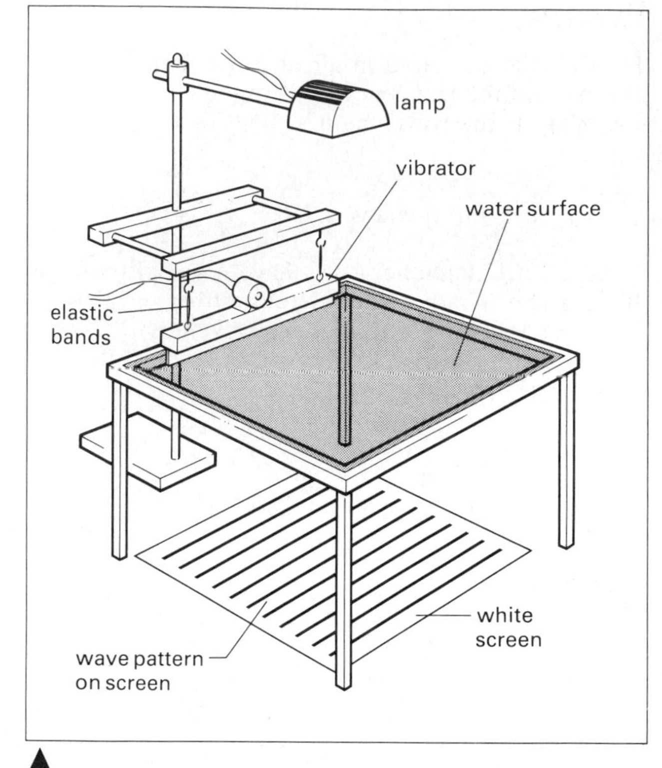

fig. 24.10 Ripple tank

The general properties of waves can best be understood by studying one particular type of wave and then extending this knowledge to explain the behaviour of other types of wave motion. Ripples on the surface of water are easily observed using a **ripple tank**, figure 24.10. A shallow tank of water has a glass base. The edges of the tank are lined with foam rubber or perforated metal to prevent the reflection of waves from the sides which would obscure the pattern being studied. A motor-driven vibrator at one side of the tank, figures 24.11 and 24.12, produces the ripples, and the frequency can be varied by varying the speed of the motor. The wave pattern is projected on to a white screen where the crests and troughs show as bright and dark bands.

The straight and circular bands representing the crests and troughs are called **wavefronts**. All the particles in a wavefront are performing exactly similar motions at the same time — they are said to be **in phase**. Motions which are not identical are **out of phase**.

Although waves are essentially moving phenomena it is sometimes necessary, in order to study them more clearly, to slow them down or stop them altogether. This can be done using a stroboscope. The flashing rate of the lamp may be mechanically or electronically controlled or the wave pattern may be observed through a simple hand stroboscope, which is a rotating disc with a number of holes or slits in it, figure 24.13. The picture is only seen when a hole is opposite the eye. When the frequency of the pictures seen is the same as (or a simple multiple of) the frequency of the wave source, the wave pattern appears stationary.

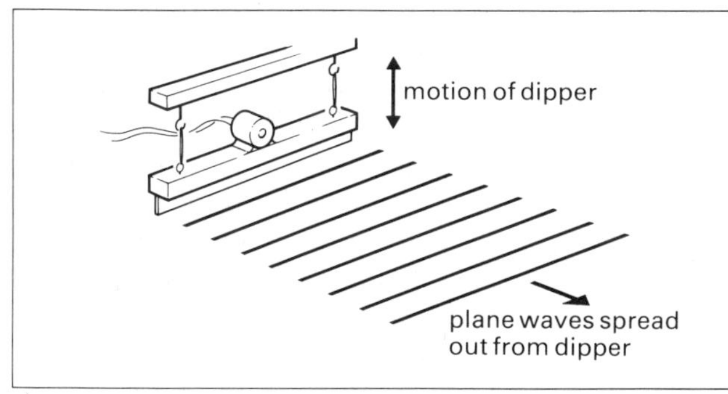

fig. 24.11 A vibrating wooden bar produces a set of **plane waves**

fig. 24.12 A single vibrating dipper produces a set of **circular waves**

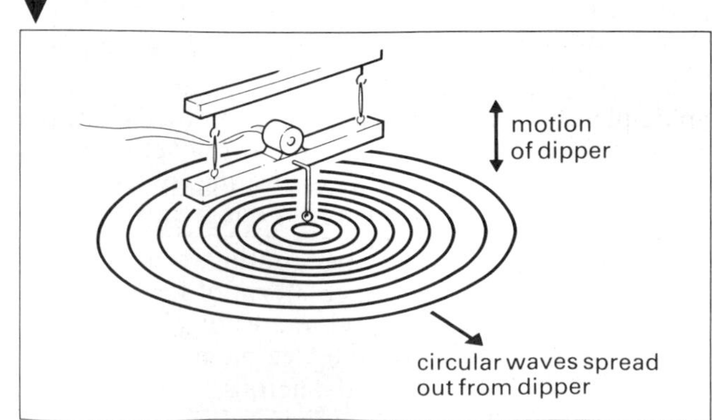

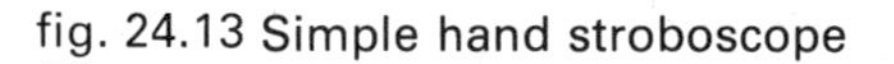
fig. 24.13 Simple hand stroboscope

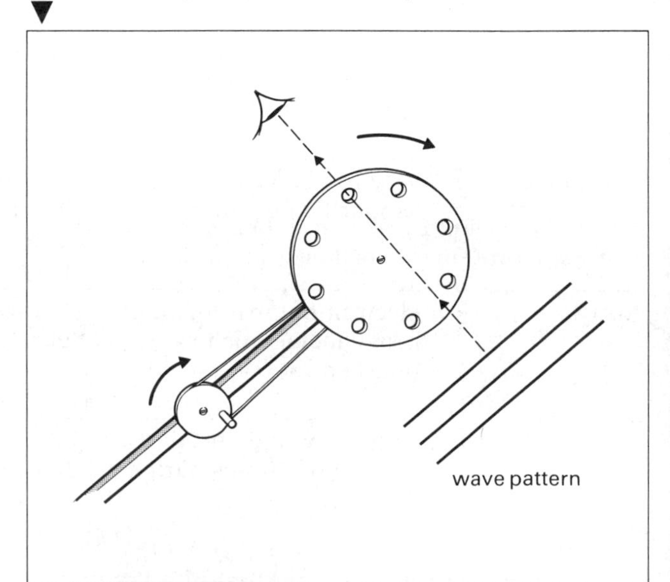

Properties of waves

Velocity and medium

Waves take time to travel from place to place; they have a finite velocity. Mechanical waves require a medium through which to travel. Electromagnetic waves can travel through empty space. The velocity depends on the nature of the waves and the nature of the medium through which they are moving:

velocity of waves in a stretched spring could be a few centimetres per second
velocity of sound waves in air at 0 °C is 331 m/s
velocity of sound waves in steel is 5000 m/s
velocity of light waves in a vacuum is 3×10^8 m/s
velocity of light waves in glass is 2×10^8 m/s

Reflection

When a metal barrier is placed at an angle to the path of plane waves in a ripple tank, the waves are reflected and the angle of incidence is equal to the angle of reflection, figure 24.14. Experiments with circular wavefronts and either a plane barrier or a curved reflector can also be carried out.

In figure 24.15 it will be seen that the waves are spreading out from a point source at O. After reflection, the wavefronts are circular and spread out as if they were generated at I. I is the image of O, and is as far behind the reflector as the source O is in front.

In figure 24.16 the reflected wavefronts converge to the point F which is the focus. After passing through F the wavefronts are convex and correspond to a diverging beam.

Refraction

The velocity of ripples on water depends on the depth of the water.

A plate of glass or Perspex covers half a ripple tank, and the water level is adjusted so that it just covers the plate. When a plane wave with its wavefronts parallel to the edge of the plate is produced and 'stopped' with the stroboscope, it will be seen that the wavelength (distance between two wavefronts) is shorter in the shallower water, figure 24.17a. Since all the wavefronts are 'stopped' by the stroboscope in both parts of the tank, the frequency must remain constant. Using $c = f\lambda$ and for constant frequency, a shorter wavelength means a lower velocity.

If the plane wavefronts meet the shallow water boundary at an angle, one end of the wavefront is slowed down while the other end in the deeper water continues to move at its original speed. The result is that the direction of the wavefront is changed or refracted, figure 24.17c. The connection between refraction and the change in velocity at the boundary

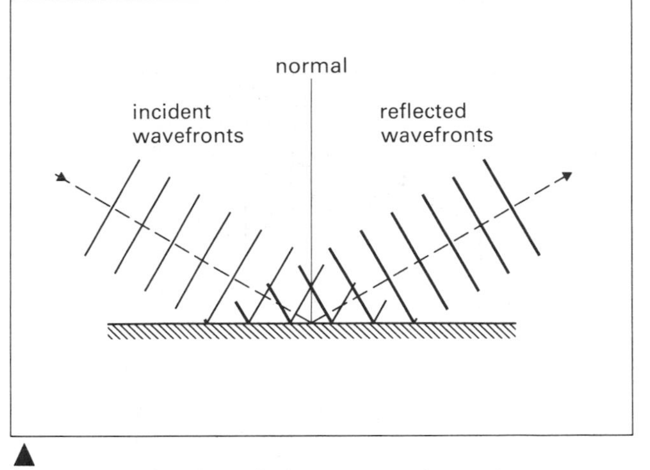

fig. 24.14 Reflection of plane waves by a plane reflector

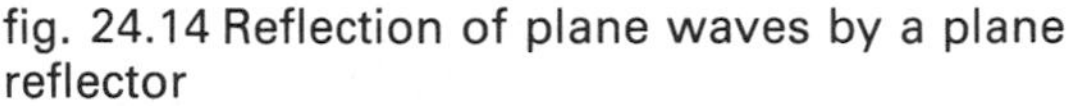

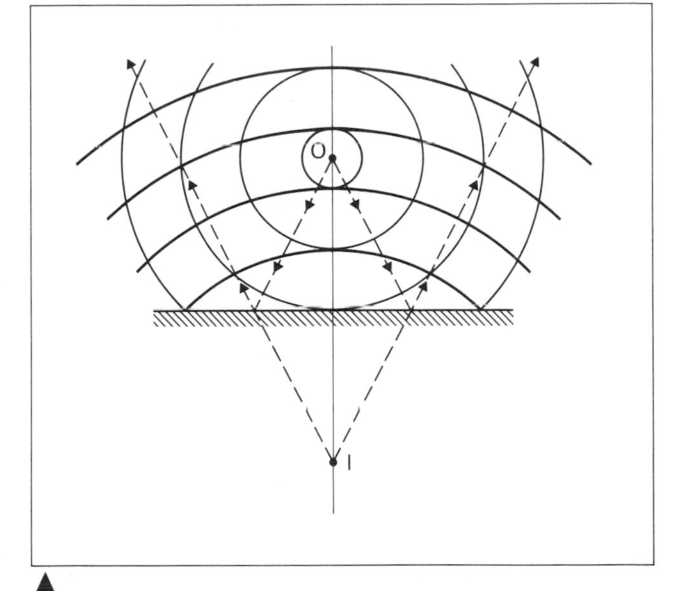

fig. 24.15 Reflection of circular waves by a plane reflector

fig. 24.16 Reflection of plane waves by a concave reflector

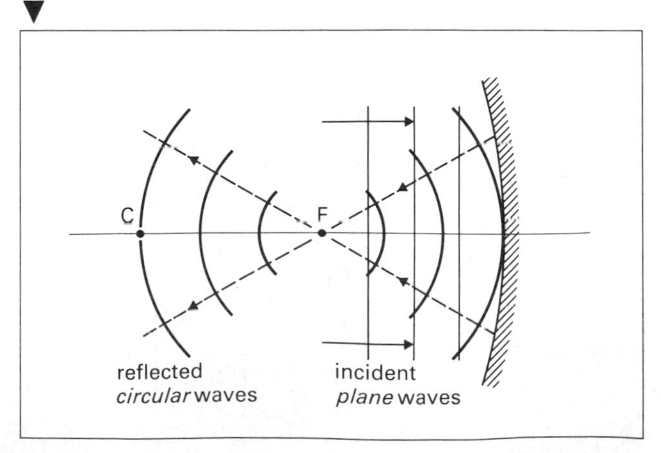

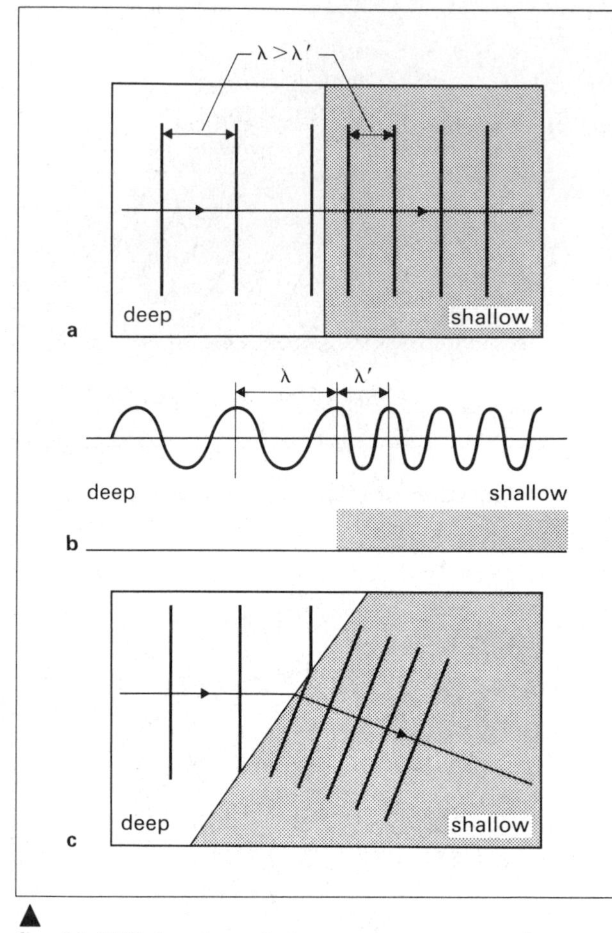

▲ fig. 24.17 Refraction of plane waves on meeting shallow water

fig. 24.18 Refraction of plane waves at a plane boundary ▼

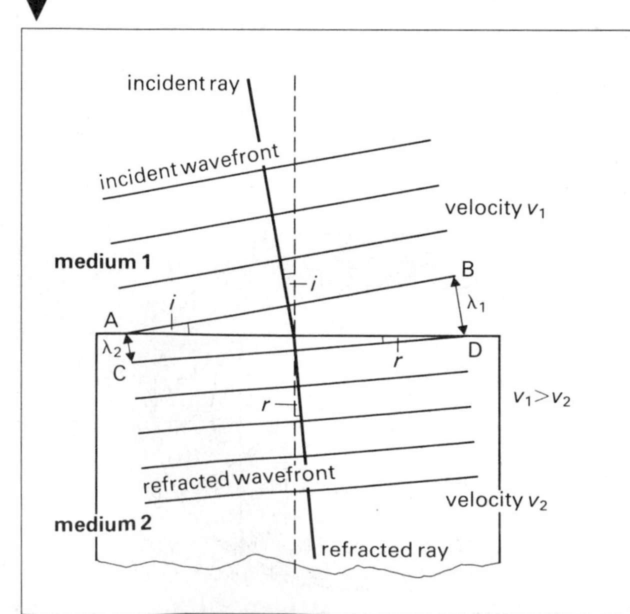

between two media is shown in figures 24.19–24.21.

Figure 24.18 shows a series of plane wavefronts crossing the interface between two media in which the waves travel with different velocities v_1 and v_2 Such a case would be a parallel beam of light passing from air to glass. When considering rays of light, the angles of incidence and refraction are measured between the incident and refracted rays and the normal. The geometry of the diagram shows that these angles are the same as the angles between the incident and refracted wavefronts and the surface separating the two media. By definition the refractive index is

$$n = \frac{\sin i}{\sin r}$$

In right-angled triangle ABD

$$\sin i = \frac{\lambda_1}{AD}$$

In right-angled triangle ADC

$$\sin r = \frac{\lambda_2}{AD}$$

$$\text{therefore refractive index } n = \frac{\lambda_1/AD}{\lambda_2/AD} = \frac{\lambda_1}{\lambda_2}$$

Since the frequency f of the waves is the same in both media and using the basic wave equation $v = f\lambda$, we can write

$$v_1 = f\lambda_1$$
$$v_2 = f\lambda_2$$

$$\text{from which } \frac{\lambda_1}{\lambda_2} = \frac{v_1}{v_2}$$

$$\text{therefore refractive index } n = \frac{v_1}{v_2}$$

The refractive index is the ratio of the velocities of the waves in the two media. This relationship is true for all types of wave motion.

Since light is refracted towards the normal when it enters an optically denser medium such as glass or water, it means that light travels more slowly in glass or water than it does in air, in accordance with the velocity ratio just derived.

$${}_{\text{air}}n_{\text{water}} = \frac{4}{3}$$

therefore

$$\begin{matrix}\text{velocity of light}\\ \text{in water}\end{matrix} = \frac{3}{4}\begin{pmatrix}\text{velocity of light}\\ \text{in air}\end{pmatrix}$$

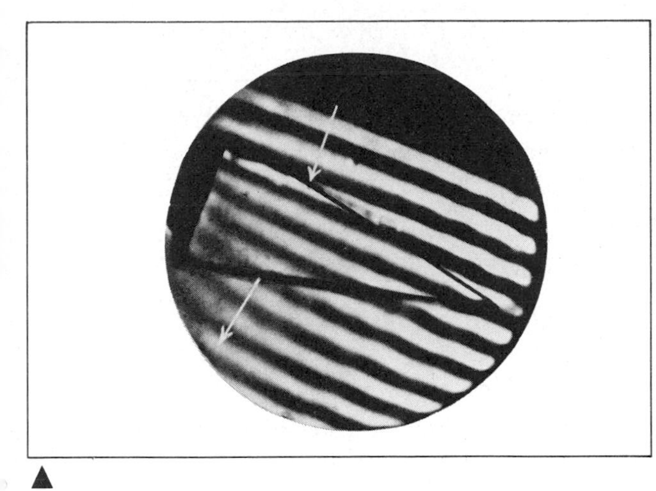

fig. 24.19 Refraction of plane waves by a prism

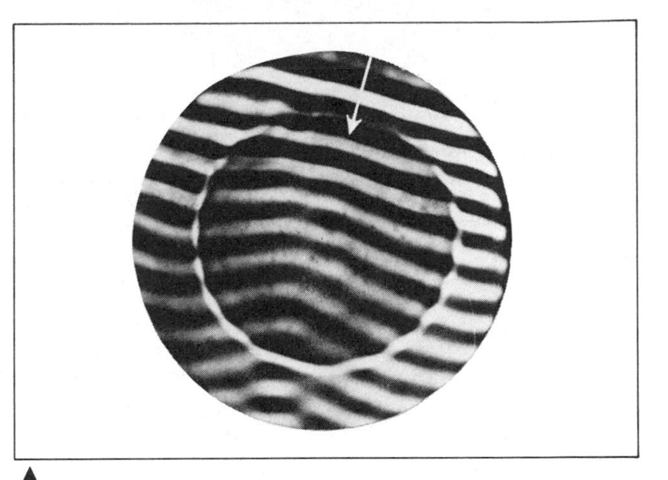

fig. 24.20 Refraction by a circular block. The slowing down of the wave in the block can be clearly seen

In the middle of the nineteenth century the French scientist Jean Foucault measured the velocity of light in water and found that it was about three-quarters of the velocity in air. This was a very important piece of experimental evidence in support of the idea that light is a type of wave motion.

Diffraction

When waves pass obstacles or through openings there is a spreading of the wavefronts into what we can call the 'geometrical shadow' region. This bending round corners is called **diffraction** and only occurs to any observable degree when the obstacle or aperture is of about the same size as the wavelength.

Since sound waves have wavelengths measured in metres, quite large obstacles and apertures will give diffraction effects. Because of the diffraction of sound waves, we can hear round corners although we cannot see round them.

The patterns produced by the diffraction of light round small obstacles are quite complicated, with light and dark fringes being formed, figure 24.22.

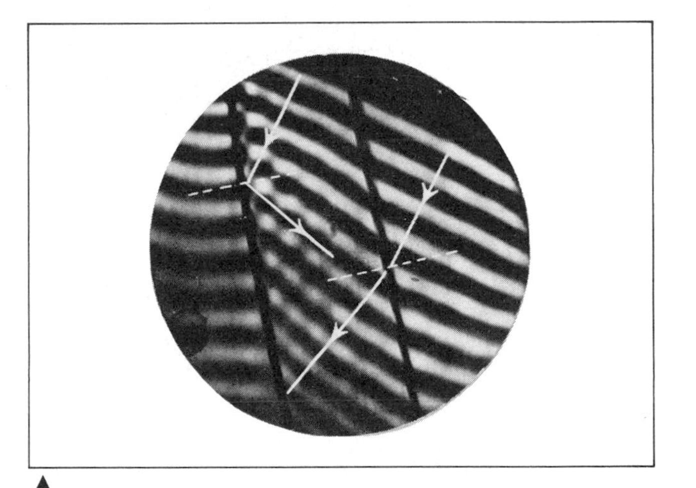

fig. 24.21 Refraction and internal reflection by a parallel-sided block

fig. 24.22 Diffraction of plane waves by a small aperture and an obstacle

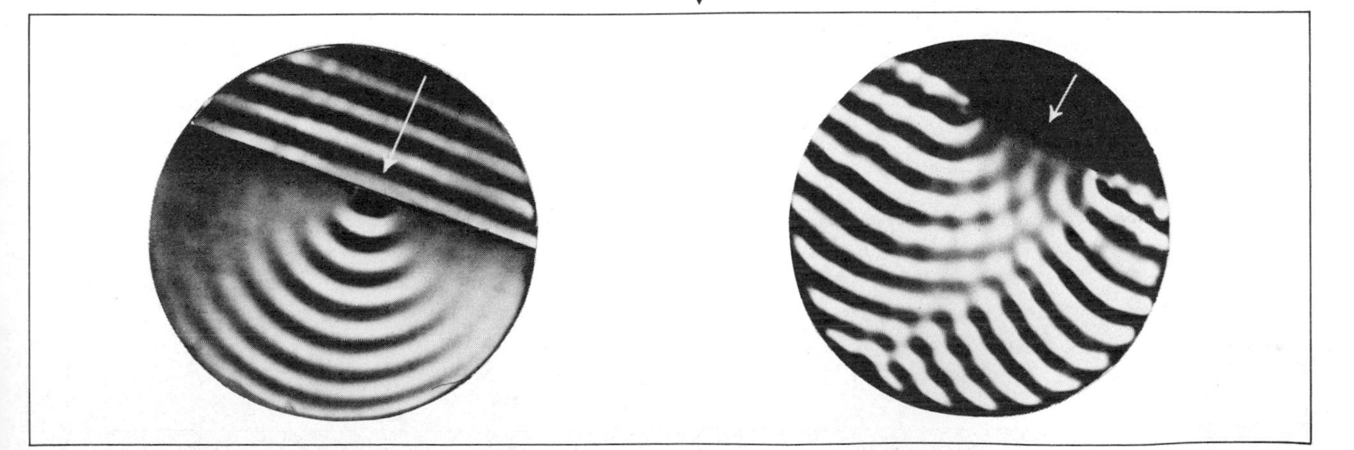

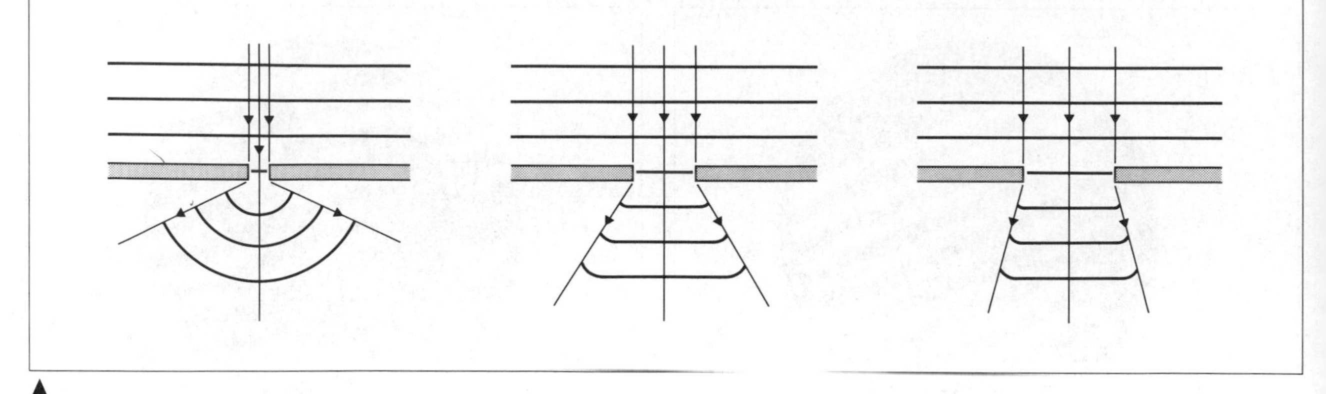

fig. 24.23 Diffraction of plane waves by apertures of different sizes

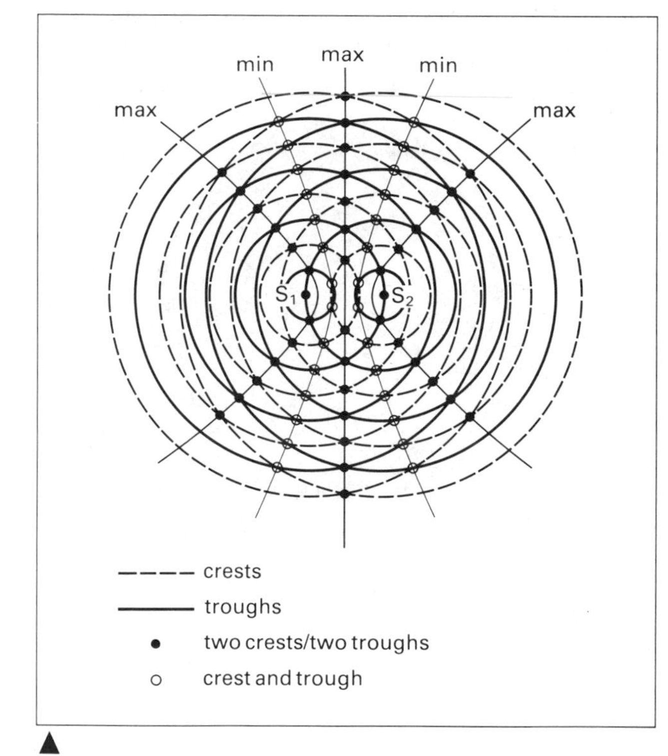

fig. 24.24 Interference pattern of two sets of circular wavefronts

Experiments with apertures of different widths, figure 24.23, and waves of different wavelengths give the following results.

- The smaller the aperture, the greater the amount of diffraction.
- The smaller the wavelength, the less the amount of diffraction.
- The wavelength, frequency and velocity of the waves remain unchanged during diffraction.

Superposition of waves and interference patterns

Any number of wave systems can travel through a medium (in the case of electromagnetic waves through space) at the same time. For simplicity, we will consider two sets of circular ripples in water produced by having two dippers fixed to the vibrating bar of the ripple tank. The dippers will have the same frequency and be in phase. Two sets of circular wavefronts spread out from S_1 and S_2, figure 24.24, and pass through each other. Where the wavefronts overlap or are **superposed**, an **interference pattern** is produced. When two crests or two troughs arrive at the same point at the same time, the displacement of the water surface will be twice as great. When a crest of one wave and a trough of the other wave arrive together, they cancel each other out and the displacement is zero. A whole series of maximum and

fig. 24.25a Constructive interference

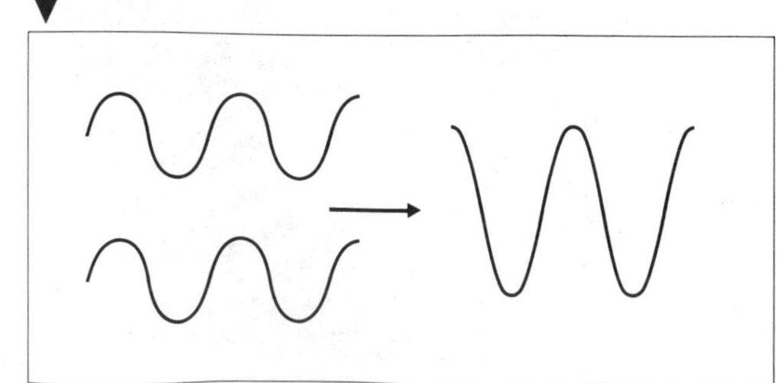

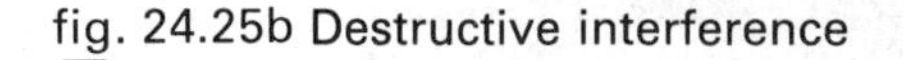

fig. 24.25b Destructive interference

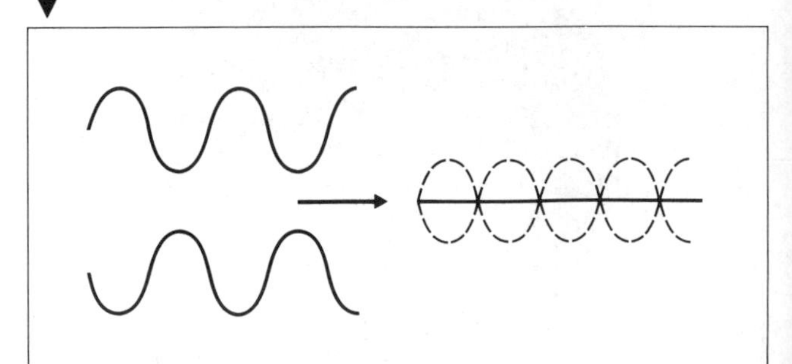

minimum (zero) displacements is produced, and these make up the interference pattern, figure 24.24.

When two waves meet and are in step (in phase), they reinforce each other and interfere **constructively**. The amplitude of the resultant wave is twice the amplitude of each component, figure 24.25a.

When two waves are half a wavelength out of step (out of phase), crest cancels out trough and the waves interfere **destructively**. The amplitude of the resultant is zero, figure 24.25b.

The lines of maximum (constructive interference) and minimum (destructive interference) disturbance are called **interference fringes**. The distance between the fringes is altered by changing the distance between the sources, and also by altering the wavelength. When the sources are closer together, or when the wavelength is longer, the interference fringes are more spread out.

An important point to realise is that the formation of an interference pattern can only be explained in terms of wave motion. The production of interference fringes is evidence for the presence of waves.

The wave nature of light

The behaviour of light during such processes as reflection and refraction has been known for a very long time, but the nature of light and the mechanism by which it travels from place to place were much more difficult to understand. In the seventeenth century the English scientist Sir Isaac Newton believed light to be a stream of particles, or corpuscles, emitted from the source in straight lines. The Dutch scientist Christian Huygens (1629–1695) believed light to be a form of wave motion. If light is a wave motion, then it should show all the properties of waves we have just been studying. That is

- travel with a finite velocity,
- obey the laws of reflection,
- obey the laws of refraction,
- be diffracted at small obstacles and openings,
- produce interference patterns by superposition.

Young's experiment

The first four properties of waves listed above had been observed in the case of light, but it was not until the beginning of the nineteenth century (1801) that Thomas Young showed that light could produce an interference pattern. It would seem fairly simple to carry out the fundamental interference experiment which was shown with the two vibrators in the ripple tank on p. 204. But, for complete destructive interference to take place, the two sources of wave-motion must be **coherent**. That is, they must have the same frequency, the same amplitude and a constant phase relationship. It is impossible to achieve these conditions with two separate sources of light. Young's idea was to use a single source, but to divide the light coming from it into two parts. The arrangement, shown in figure 24.26, is known as 'Young's slits'.

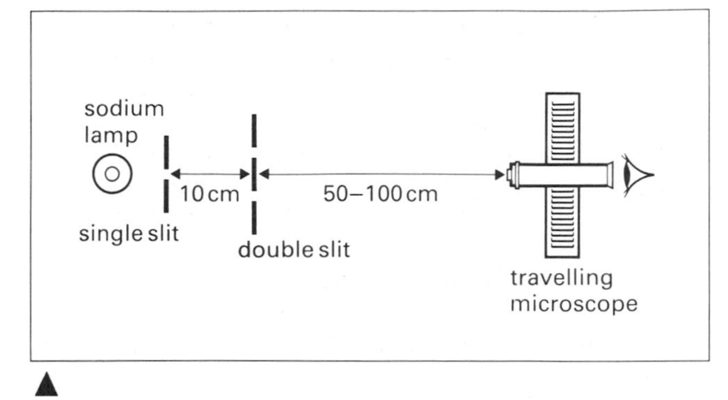

▲ fig. 24.26 Young's slits in plan view

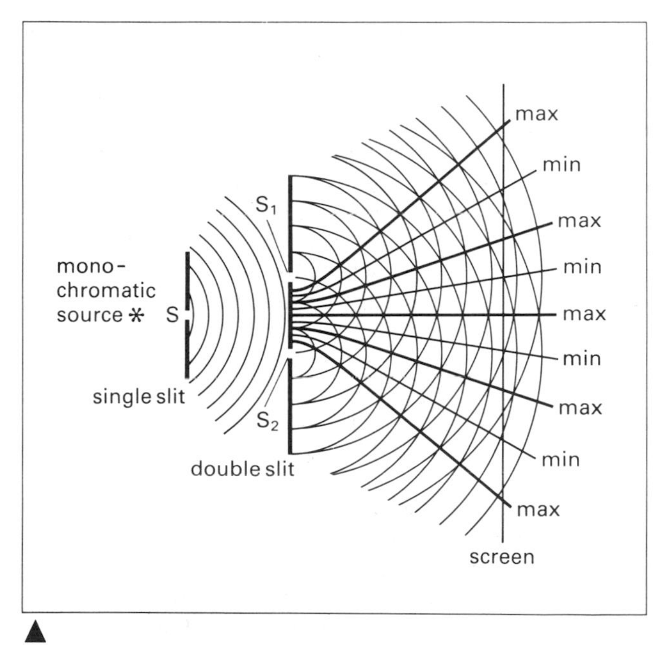

▲ fig. 24.27 Young's slits interference pattern

Young used a monochromatic (single colour) source of light with a single slit in front of it. The light from the single slit fell on two slits very close together. These two slits (S_1 and S_2), figure 24.27, acted as two new sources and, because they were both derived from the same single source, were coherent. A series of light and dark fringes — regions of constructive and destructive interference — were observed on the screen. Since the only explanation of interference fringes is in terms of waves, the experiment showed that light can be regarded as a wave motion.

Young's fringes can be observed in the laboratory, using the set-up shown in figure 24.26. The source can be a sodium lamp or a straight-filament low-voltage lamp with a colour filter. The double slits can be made by painting a glass slide with colloidal graphite and then ruling two parallel lines very close together with a fine needle and a steel ruler. The slits should be about 0.5 mm apart. It is very important that the lines of the double slit should be parallel to the single slit, and the double slit holder should be capable of rotation in a vertical plane in order to make this adjustment. Instead of a screen, the fringes can be observed with a travelling microscope so that measurements can be made.

Microwaves

All the properties of waves which were demonstrated with the ripple tank using water waves can equally well be demonstrated in the laboratory using very short radiowaves. It is important to do this because it shows that very different kinds of waves behave in essentially the same way. Very short radiowaves are used in **radar** (**ra**dio **d**etection **a**nd **r**anging). They have the advantages that they can be transmitted in very brief energy pulses and are not absorbed by the ionised layers in the earth's atmosphere. Radar pulses can be reflected from the moon.

The waves can be detected by a special receiver, and can be modulated so that, when amplified, the signal will work a loudspeaker, but we are not hearing the microwaves directly. When the meter reading is a maximum, or the sound from the speaker is at its loudest, we know the signal is being received.

Interference fringes can be produced using microwaves to perform Young's experiment, figure 24.29. The slits are between metal screens, and are about 1 cm wide and 6 cm apart. The receiver is moved along the scale AB, and readings of the meter giving the signal strength are taken at measured intervals. When the meter readings are plotted against distance along AB, very distinct maxima and minima are shown.

If the receiver is placed at a position of minimum signal strength where destructive interference is taking place and one of the slits is covered over with a strip of metal, the signal strength immediately rises, showing that the fringe system is really produced by the superposition of two wave systems.

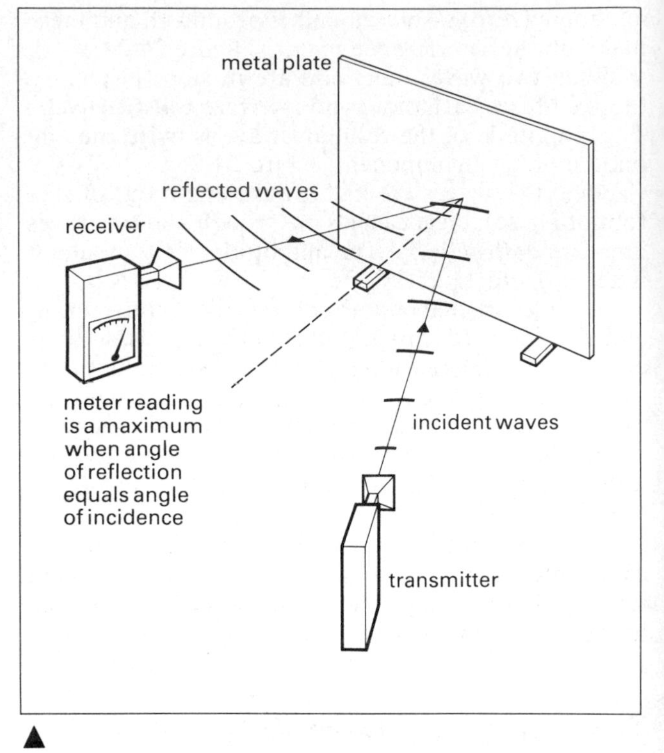

▲ fig. 24.28 Reflection of microwaves

fig. 24.29 Young's experiment with microwaves ▼

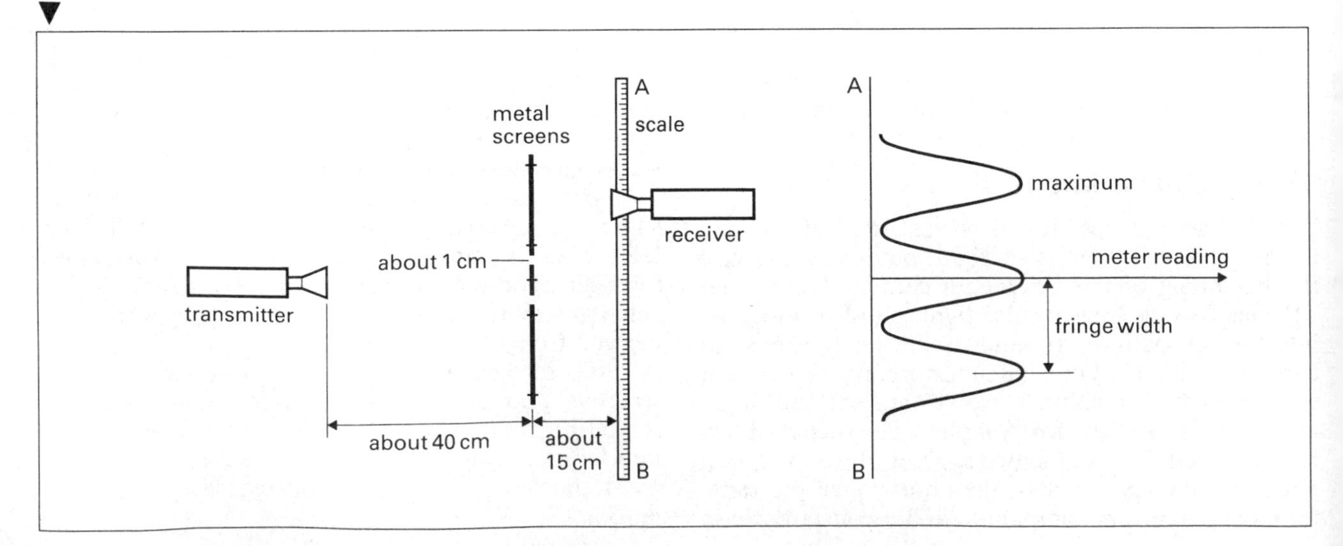

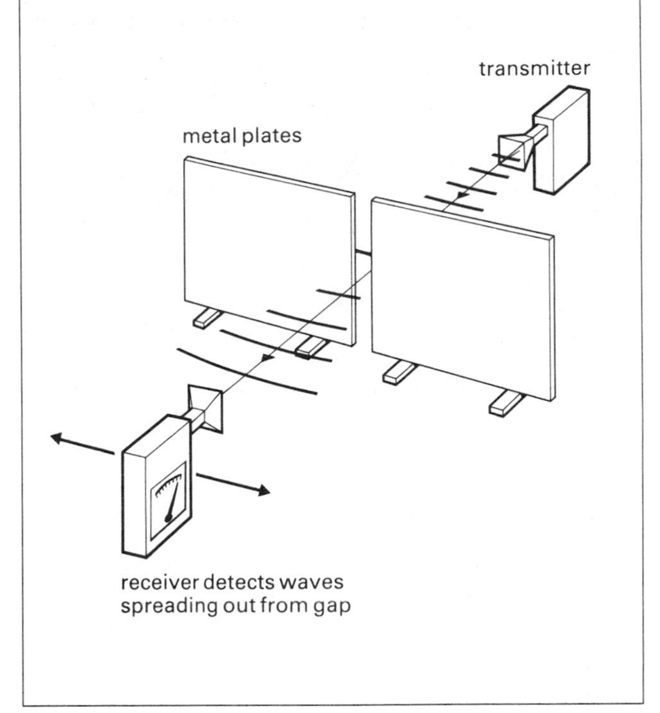

fig. 24.30 Diffraction of microwaves passing through a gap

EXERCISE 24

In questions 1–6 select the most suitable answer.

1 The basic difference between transverse and longitudinal mechanical waves is a difference in
A amplitude
B direction of vibration
C frequency
D medium through which they travel
E wavelength

2 A longitudinal wave could be
A a light wave
B a radio wave
C a wave on water
D a sound wave in open air
E a vibrating piano string

3 A transverse wave is drawn to scale in figure 24.31. What is the wavelength of this wave?
A 1 cm
B 2 cm
C 4 cm
D 8 cm
E 12 cm

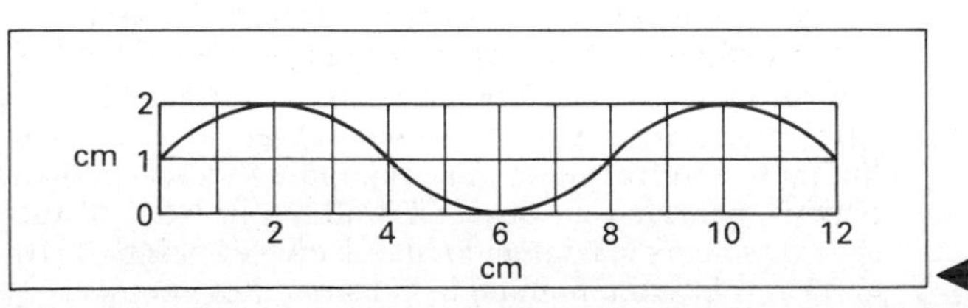

fig. 24.31

4 A note of 1600 Hz is sounded on a day when the speed of sound in air is 320 m/s. What is the wavelength in air of this note?
A 0.2 m
B 5.0 m
C 1280.0 m
D 1920.0 m
E 512.0 km

5 Visible light waves differ from ultraviolet waves in that visible light waves
A travel more slowly
B travel more quickly
C do not require a carrying medium
D vibrate transversely
E have longer wavelengths

6 Which of the following types of wave has the highest frequencies?
A X-rays
B infrared
C light
D ultraviolet
E radio

7 One end of a string is moved to and fro so that a transverse travelling wave is produced, as shown in the figure.

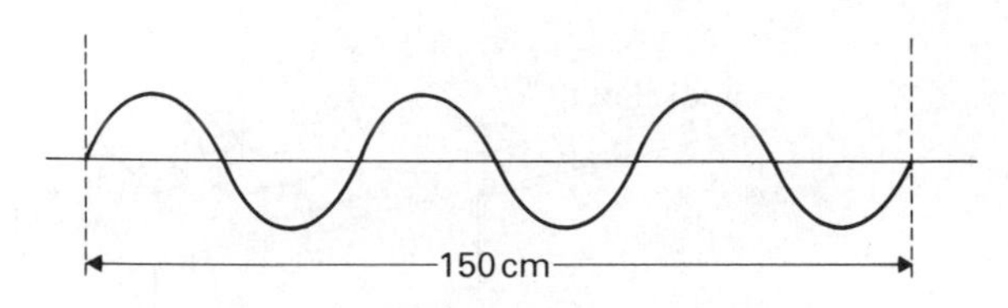

a What is the wavelength?
b If it moved to and fro 2 times per second, what is the frequency of the wave?
c What is the velocity of the wave?

8 The apparatus in figure 24.32 was used to show waves on water.
a What is meant by 'plane water waves'? Describe *one* way of producing these waves on the water surface.
b Why do the waves show up on the cardboard? Describe the pattern seen there.

fig. 24.32

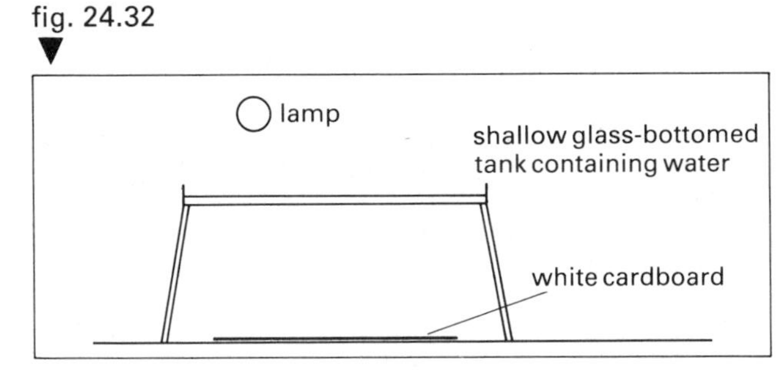

c What would be seen on the card if the tank was not quite level, so that the waves travel from shallower to deeper water? Explain your answer.
d Draw a labelled diagram showing the plane waves after passing them through a narrow opening between two metal barriers. (The opening is about the same as the wavelength of the waves.)
e Observations showed that the waves were produced at a rate of 9 complete waves every two seconds, and the distance between 10 successive crests was 30 cm. Calculate: (i) the frequency, (ii) the wavelength and (iii) the velocity of the waves.

9 The speed of red light in air is 3×10^8 m/s and the speed of red light in glass is 2×10^8 m/s. The wavelength of red light in air is 6×10^{-7} m (600 nm). Calculate, for red light
a the refractive index of glass,
b the wavelength of the light in the glass,
c the frequency of the light.

10 a Copy and complete the ray diagrams in figure 24.33 to show what happens to the light rays after hitting the mirror or prism.
b The diagrams in figure 24.34 show ripples in a ripple tank. Copy each one and in each case add three further ripples to show what happens when they are reflected or refracted.
c When light waves 'overlap', an interference pattern is formed, in which there are regions of extra brightness and regions of darkness. (i) How can light waves 'overlap' to give extra brightness? (ii) How can light waves 'overlap' to give darkness?

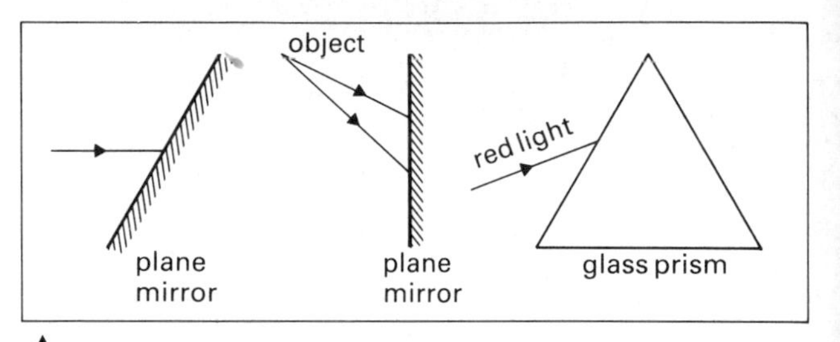

fig. 24.33

fig. 24.34

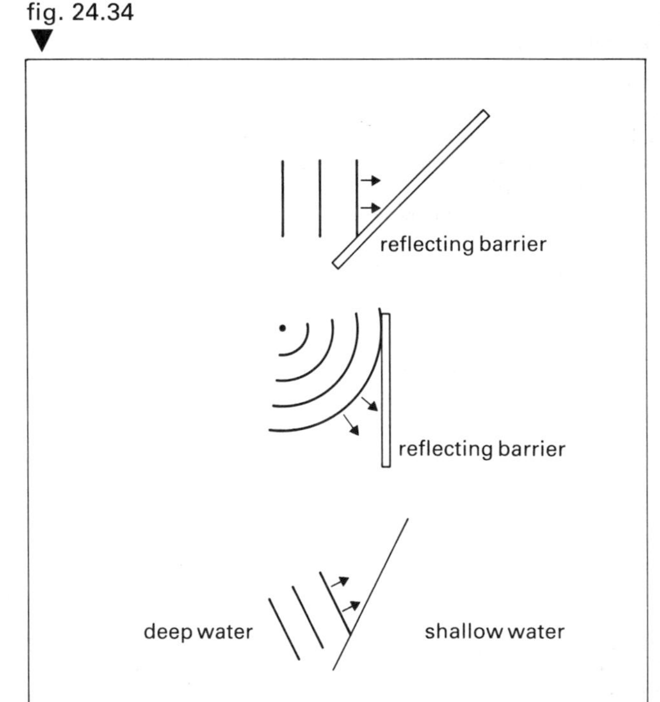

11 In order to demonstrate a 'Young's fringes' interference pattern with sound waves, two loudspeakers X and Y are set close to each other as shown in figure 24.35, each producing 'in step' sound waves of the same loudness and frequency. The equipment is set up on a playing field and an observer walking in front of the speakers hears maximum loudness only at points A, B, C, D and E and minimum in between these points.

a Why is it not satisfactory to perform this experiment inside a room?

b (i) Sound waves always arrive at C in step. Why is this? (ii) Sound waves always arrive at B in step. Why is this? (iii) Sound waves always arrive at A in step. Why is this?

c (i) If AX = 11 m and AY = 10 m, what is the wavelength of the sound used? (ii) If the frequency of the note emitted is 660 Hz, what is the velocity of sound in air? (Show your working.)

d (i) How would the spacing of the maxima, at the same distance from the speakers, be affected if the frequency of the sound was increased? (ii) Explain your answer.

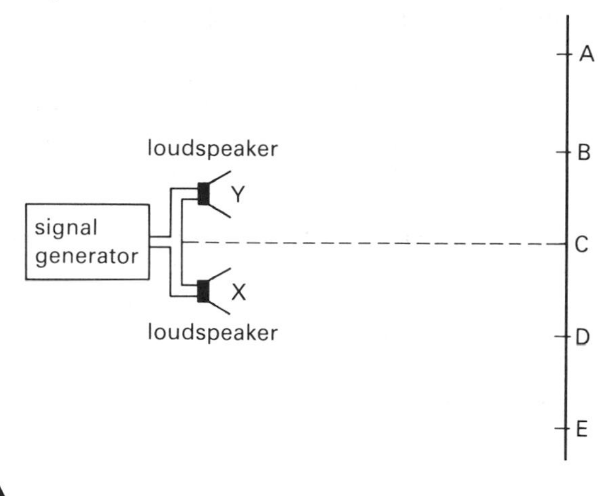

▲
fig. 24.35

12 The chart shows the different parts of the electromagnetic spectrum:

gamma rays	X-rays	ultra-violet	A	infra-red	radiowaves

a What radiation has not been named, labelled A on the chart?

b State *two* common properties of all the electromagnetic radiations.

c State a wavelength which would ensure that radiation of that wavelength would be a radiowave.

d What would be the effect on your body of receiving, separately (i) infrared radiation? (ii) ultraviolet radiation? (iii) gamma radiation?

e Explain, using a diagram, how X-rays can be produced.

f State how ultraviolet *or* infrared *or* gamma rays can be produced.

g What must happen to the electrons in a wire in order to produce radiowaves?

25 HOW SOUNDS ARE PRODUCED

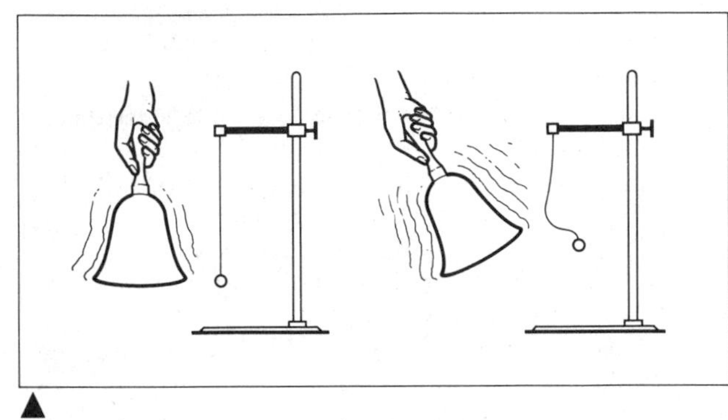

▲
fig. 25.1 Demonstrating the vibration of a bell

fig. 25.2 The human voice
▼

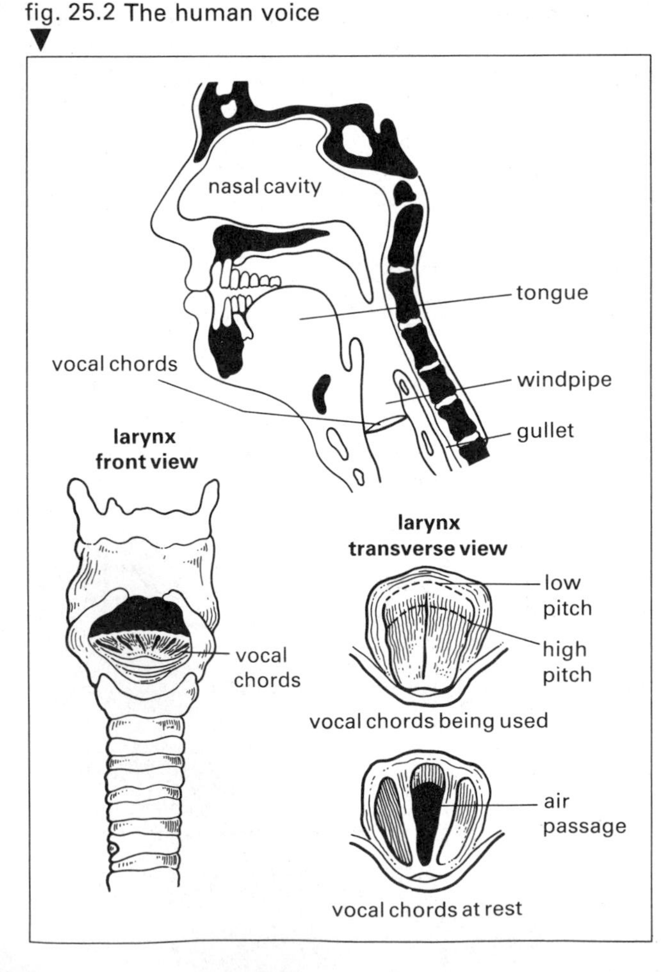

Sounds are always produced by movement of some kind. Sounds are produced by **vibrations** or rapid to-and-fro movements, and the energy of the sound comes from the energy of the vibrating source.

Often the vibrations are so small and so quick that our eyes cannot detect them (remember the persistence of vision). Figure 25.1 shows a hand bell being rung with a sharp blow of the clapper. If, while it is still sounding, it is allowed to touch a small piece of cork hanging on a thread, the cork is thrown violently to one side showing that the bell is really vibrating rapidly. If you touch a piano string while it is sounding you can feel the vibrations.

When you sing you can feel the vibrations by placing your fingers lightly on your throat. When we speak or sing the sounds are produced by vibrations of the vocal cords which are two thin, stretched skins made to vibrate when we breathe out, figure 25.2.

Three characteristics of sounds

Although all sounds are produced by vibration, the kind of sound we hear depends on several things. We can recognise the difference between high notes and low notes. We can also tell the difference between a soft sound and a loud one. If the same musical note is played on two different musical instruments, for instance a piano and a flute, it is quite easy to distinguish between them. These three characteristics of sounds are **pitch**, **loudness** and **quality**.

Pitch

The **pitch** of a sound depends on the **frequency of vibration** of the sound source. Low frequencies give low notes. High frequencies give high notes. The following experiments illustrate the fact that the faster the vibrations the higher is the pitch of the sound.

In figure 25.3, the cog wheels, rotated by an electric motor, have different numbers of teeth. When a piece of card is held against a rotating wheel a note is emitted. The greater the number of teeth and the faster the wheel is rotating, the higher the pitch. If the wheel has N teeth and is rotating at n revolutions per second, then the number of times a tooth hits the

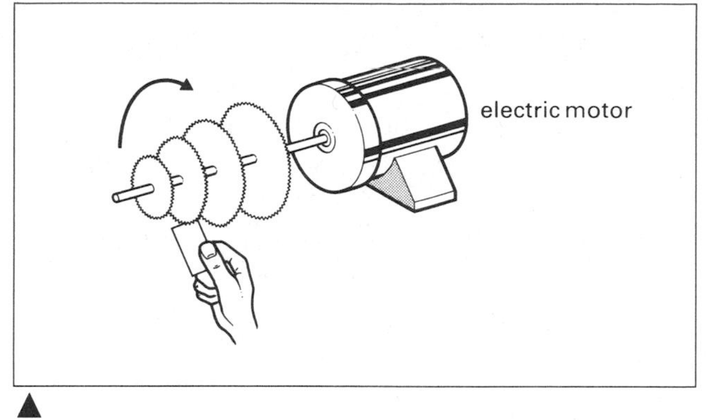

fig. 25.3 Savart's toothed wheels

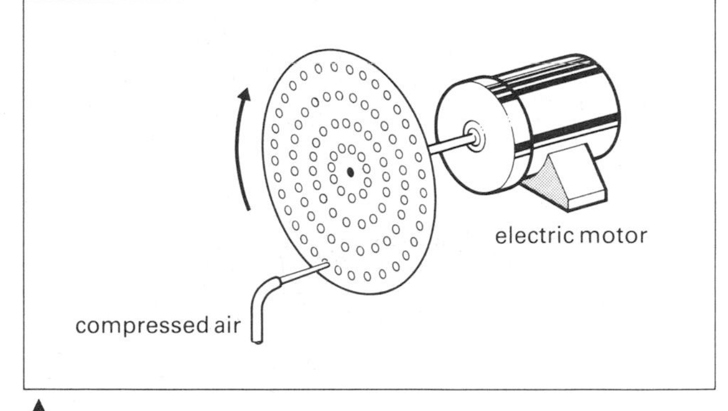

fig. 25.4 Simple disc siren

edge of the card in one second, i.e. the frequency of the sound, is nN hertz.

The siren works in a similar way, figure 25.4. When a jet of air plays on the rotating holes, little puffs of air get through each time a hole is opposite the jet. The more rapid the puffs, the higher the pitch. As with the toothed wheels, if there are N holes in one ring and the disc is rotating at n revolutions per second, the frequency of the sound is nN hertz.

When the vibrations occur regularly, we get a musical note. If the vibrations are irregular, we describe the sound produced as a noise. But the division of sounds into notes and noises is not a very exact one, and many people would call some kinds of music a noise! If the frequency of vibration is less than about 20 Hz, we hear the separate vibrations and they do not blend together to give a note. The human ear cannot detect very rapid vibrations — more than about 20 000 Hz — but birds and dogs hear high-pitched sounds which are inaudible to us.

Loudness

The second characteristic of a sound is **loudness**. This everyday word needs some explanation when used scientifically. A certain amount of energy is used in producing a sound, and it is this energy which produces the sensation of sound in the ear. The energy carried by a sound is measured by the **intensity** of the sound, the effect this energy has in stimulating the nerves in the ear, and the consequent sensation in the brain is the **loudness**. It will be realised that the loudness depends on the hearing organ of the person doing the listening. A sound of a certain intensity may be heard quite loudly by a person with normal hearing but be quite faint to a partially deaf person. This distinction should be borne in mind when we speak of the loudness of a sound.

The loudness of a sound depends on the **amplitude** of the vibration. The greater the amplitude, the louder the sound. More precisely, the loudness is

fig. 25.5 Apparatus to investigate pitch, loudness and quality of sound

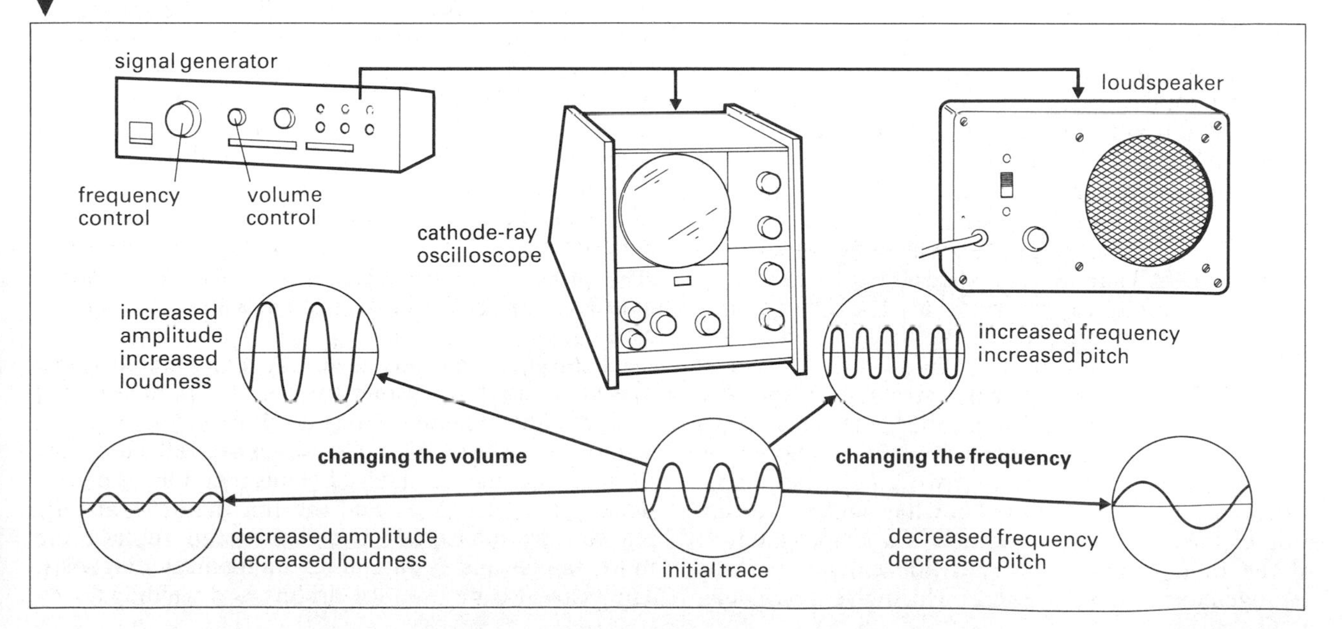

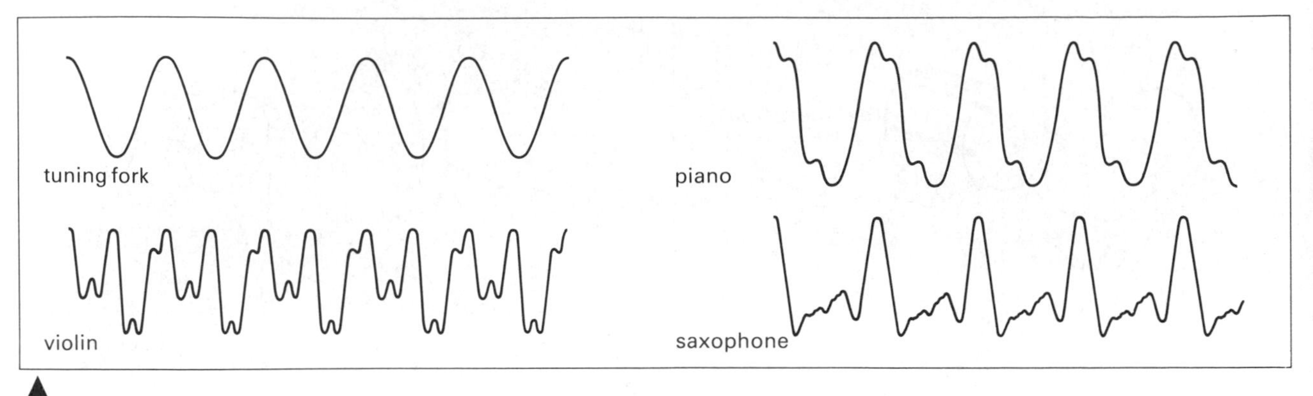

fig. 25.6 Waveforms for the same principal note from four different sources

proportional to (amplitude)^2. It also depends on the frequency of the sound, because the ear is not equally sensitive to sounds of the same intensity but different frequencies.

The dependence of pitch upon frequency, and of loudness upon amplitude, can be demonstrated with a signal generator and a cathode-ray oscilloscope, figure 25.5. The use of this experimental set-up also enables us to investigate what is meant by the quality of a sound.

In figure 25.5, the signal generator produces electrical oscillations which, when fed to the c.r.o., produce the **waveform** of the oscillations, and when fed to the loudspeaker produce an audible note. By changing the volume control on the generator the relation between volume and amplitude of the wave can be shown. By altering the frequency of the oscillations the relation between pitch and frequency can be demonstrated.

Quality

If the signal generator is disconnected and replaced by a microphone, the waveforms and sounds produced by different musical instruments can be investigated. When notes of the same frequency and intensity are played on different instruments, it is found that they each have a characteristic waveform.

The different waveforms shown in figure 25.6 indicate that each sound has a special **quality** of its own which enables us to recognise it. The quality of a sound really depends on the way it is produced. It depends on what is vibrating, and on the design of the instrument. In fact, each instrument produces a particular note plus a series of other notes at the same time, and we hear and recognise a combination of all the notes produced. The note we play is the **fundamental**, and the other notes are **harmonics**. All harmonics have a frequency which is a simple multiple of the fundamental. It is the different number and arrangement of the harmonics which give a particular instrument its own special quality. The note from a tuning fork is a simple note without any harmonics. The production of harmonics is discussed in detail in the section on vibration of strings.

Ultrasonics

The human ear can detect, as sound, vibrations of frequencies up to about 20 000 Hz. Vibrations of frequencies above this are said to be **ultrasonic** (beyond sound). Frequencies of up to several million per second can be produced. If the frequency is increased the wavelength is reduced, so ultrasonic vibrations are very short waves. As a result they are more penetrating, and can be beamed more accurately.

Ultrasonic waves are produced by vibrating crystals, such as quartz, and make use of an odd effect known as the **piezoelectric effect**. If a quartz crystal is compressed, electric charges are produced on the surfaces of the crystal, figure 25.7. If the crystal is stretched the charges are reversed. Conversely, if charges are applied to the crystal faces, it will expand or contract. If alternating charges are applied, the crystal vibrates. If the frequency of the applied alternating current is the same as the natural frequency of the crystal, we have resonance. The crystal is acting as a **transducer**, which is a device for converting one kind of vibration into another — in this case, electrical vibrations into mechanical vibrations. A vibrating quartz crystal is the basis of timekeeping in a quartz clock.

Ultrasonic vibrations have very many applications. The very rapid vibrations are used for removing dirt in ultrasonic cleaning baths. Various forms of echo-sounder use ultrasonics. The waves are reflected back from any change of material. Thus metal flaws can be detected, figure 25.8, and the thickness of fat on a pig can be measured. Ultrasonics can replace the more dangerous X-rays used in medical diagnosis. They can also be used for drilling and welding.

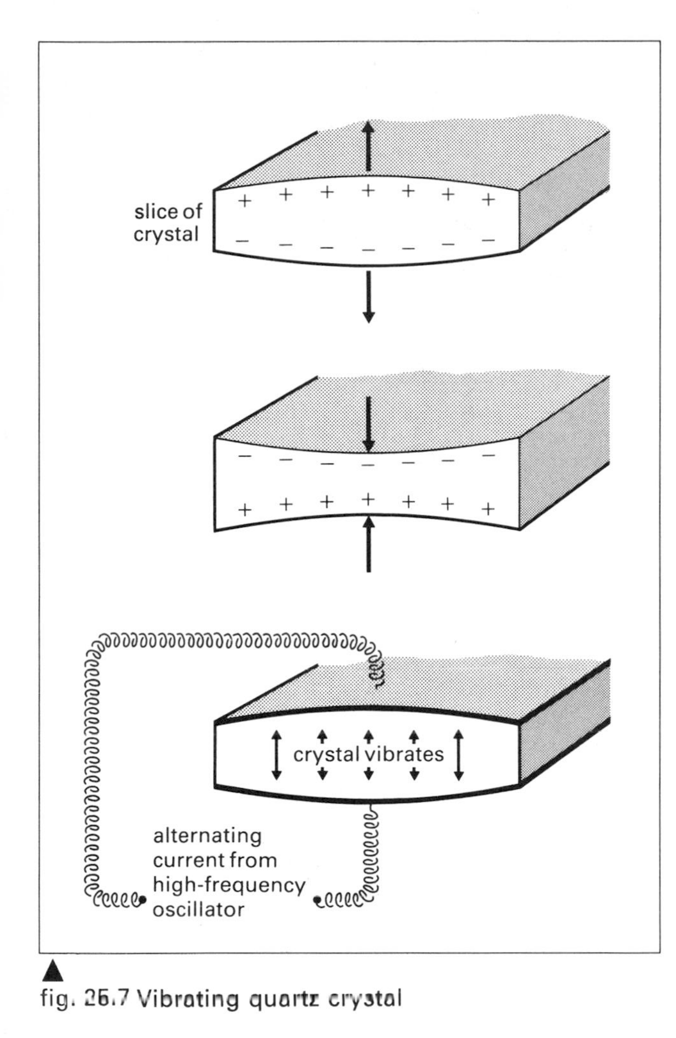

fig. 25.7 Vibrating quartz crystal

fig. 25.8 Flaw detection

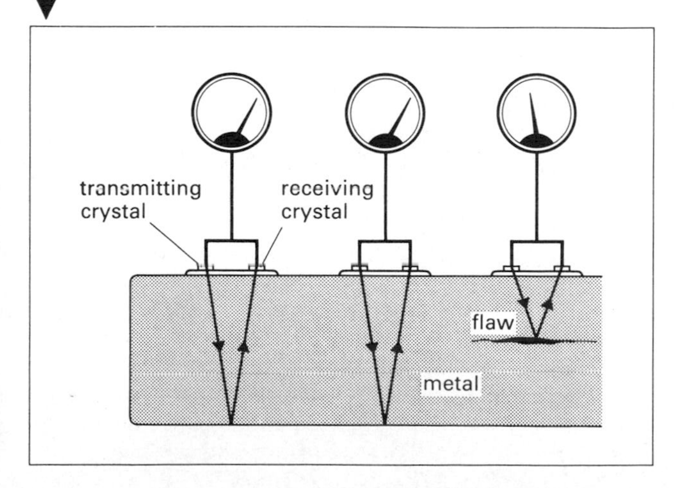

EXERCISE 25

In questions 1–5 select the most suitable answer.

1 All sound is caused by
- **A** electromagnetism
- **B** electricity
- **C** transverse waves
- **D** sine waves
- **E** vibrations

2 The physical factor that affects the pitch of a note is the
- **A** speed of sound in air
- **B** amplitude of the vibration
- **C** frequency of the vibration
- **D** density of the surroundings
- **E** number of overtones

3 The loudness of a sound is always reduced when there is a reduction in the
- **A** wavelength
- **B** amplitude
- **C** frequency
- **D** period
- **E** velocity

4 In a medium where the velocity of sound is 45 m/s a source of frequency 150 Hz will produce waves with a wavelength of
- **A** 0.3 m
- **B** 0.5 m
- **C** 3.0 m
- **D** 5.0 m
- **E** 10.0 m

5 A circular saw has 80 teeth. When cutting through wood it rotates at 2 turns per second. If the speed of sound is 320 m/s, what is the wavelength of the note produced by the saw?
- **A** $\frac{1}{8}$ m
- **B** $\frac{1}{2}$ m
- **C** 2 m
- **D** 8 m
- **E** 12 800 m

6 All sounds are caused by vibrations. What is vibrating in each of the following: **a** piano, **b** drum, **c** organ pipe, **d** tape recorder during playback?

7 A tuning fork is marked 480.
- **a** Explain what this marking means.
- **b** Calculate the wavelength of the note it will emit when sounding in air in which the velocity of sound is 340 m/s.

8 Describe simple experiments which demonstrate that, although sound is produced by vibration, it is possible to have vibration without the production of sound.

9 Explain the following.

a A siren can produce a note of fluctuating pitch.

b A trumpet and a violin playing the same note can nevertheless be readily distinguished by the ear.

c The hum of a gnat is higher than that of a bee.

10 a State the meaning of the *frequency* and the *wavelength* of a sound wave. How are these related?

b Figure 25.9a represents traces formed on an oscilloscope screen by two sounds. Compare the sounds which these waves represent.

c A length of thin wood is fixed, projecting from the edge of a bench as shown in figure 25.9b. The free end is made to vibrate. (i) Explain why no sound can be heard when the vibrating end is very long. (ii) Explain how the frequency of the note emitted varies with the length vibrating.

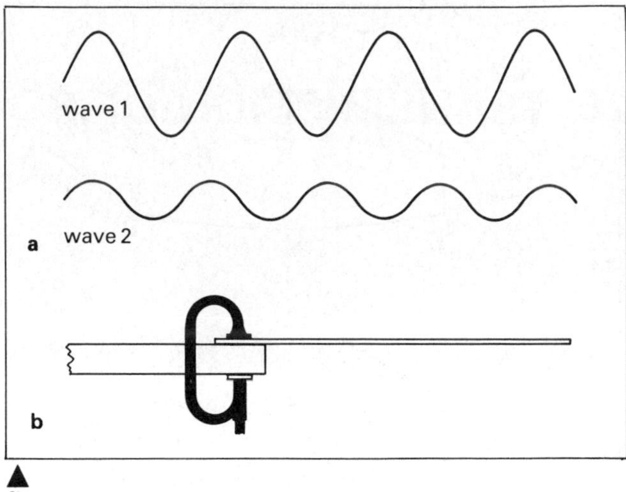

fig. 25.9

26 TRANSMISSION OF SOUND

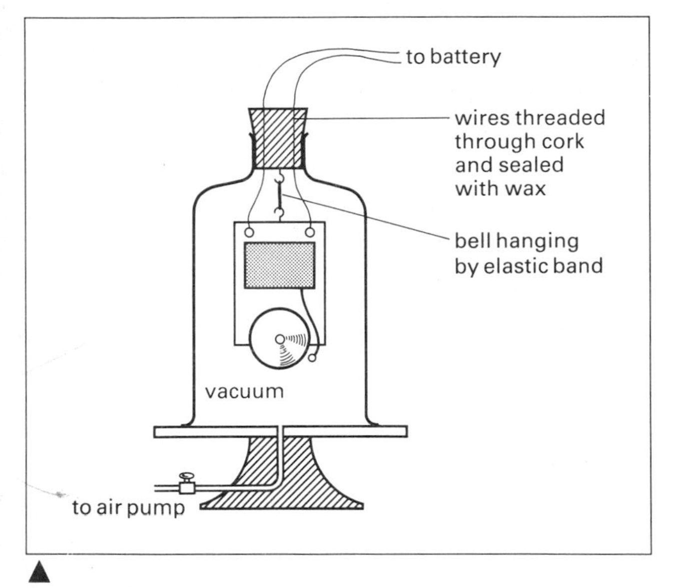

fig. 26.1 Sound cannot travel through a vacuum

fig. 26.2 Sound can travel through wood

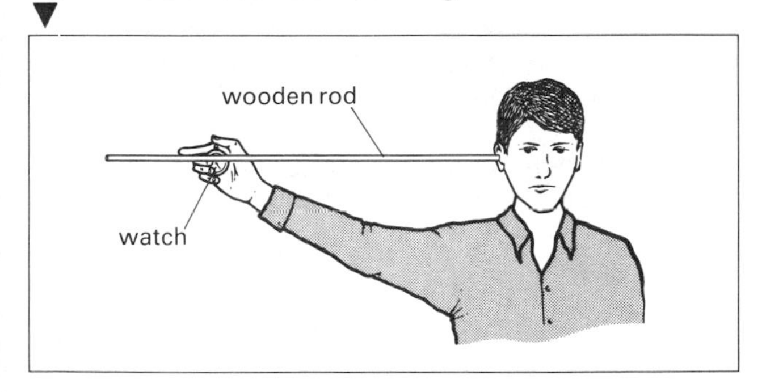

Sounds are caused by vibrations, which must have a material medium through which to travel. Sound will not pass through a vacuum.

An electric bell, figure 26.1, is suspended inside a bell jar by means of elastic bands. The wires go through the cork so that no part of the bell touches the glass. When the bell is connected up, we can both hear it ringing and see the hammer hitting the gong. The air pump is started, and as the air is drawn out the sound of the bell gets fainter and fainter. If a really good vacuum can be produced we cannot hear the bell at all, although it can still be seen to be working. When the air is let in to the bell jar the sound returns, so the conclusion is that sound will not travel through a vacuum.

Although sounds usually travel through the air, they will also travel through other substances. Hold a watch just far enough away from your ear so that you cannot hear it ticking. If you press the watch against a wooden rod held to the ear, figure 26.2, you will hear the ticking transmitted through the wood. Solids as well as gases allow sounds to travel through them. Sound also travels through water and other liquids.

When the railings, figure 26.3, are hit with a hammer, the boy with his ear to the railings hears two sounds, one very slightly after the other. One sound has travelled through the iron and one through the air. This suggests that sound travels at different speeds in different substances. Table 26.1 gives the velocity of sound in some common substances.

fig. 26.3 Two sounds are heard: one airborne, the other travelling through the railings

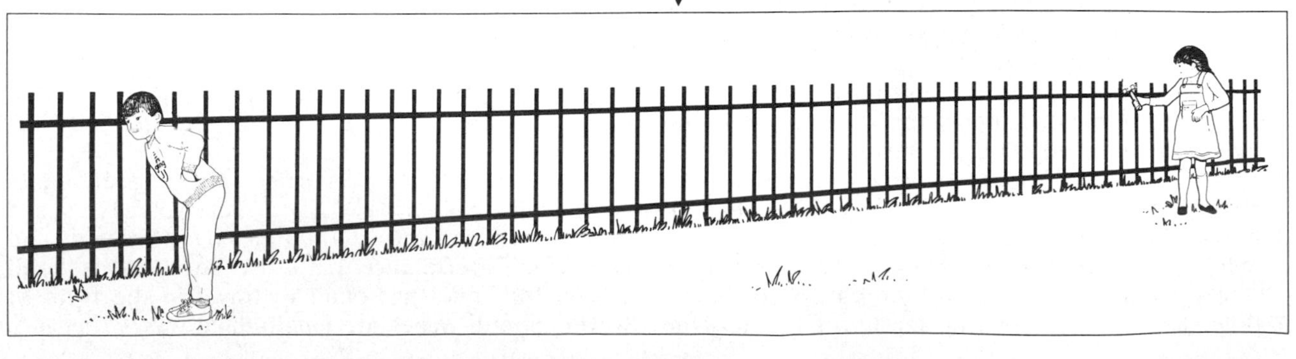

Measuring the velocity of sound in air

The velocity of sound in air can be measured by the large-scale experiment shown in figure 26.4. When the gun is fired the sound of the explosion is picked up by microphone (1) and the time is automatically recorded. The sound then travels a known distance, say 8 kilometres, and is picked up by microphone (2). The time is again recorded. Suppose the difference in time between the arrival of the sound at the two microphones is 24 seconds; then the velocity is given by 8000/24 = 333 metres per second. If a wind is blowing, the velocity should be found both with and against the wind and an average taken. The velocity of sound can also be measured in the laboratory.

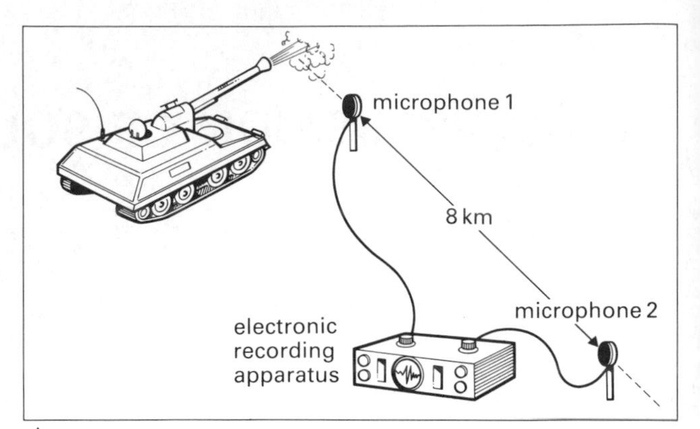

fig. 26.4 One way of determining the velocity of sound

table 26.1 Velocity of sound in various media

air at 0 °C	331 m/s
air at 20 °C	355 m/s
water at 20 °C	1457 m/s
iron	5000 m/s (approx)
wood	4000 m/s (approx)
rock	2500 m/s (approx)

If you are watching a man hammering in posts 350 metres away, figure 26.5, you will see the hammer hit the top of the post and one second later you will hear the sound. The time taken for the light to travel this distance is so very small (less than a thousand-millionth of a second) that we can ignore it. Another example of the difference in speed of light and sound is the fact that we hear the thunder after we see the lightning flash. If there is a 3 seconds interval between seeing the flash and hearing the thunder-clap, the storm is about a kilometre away.

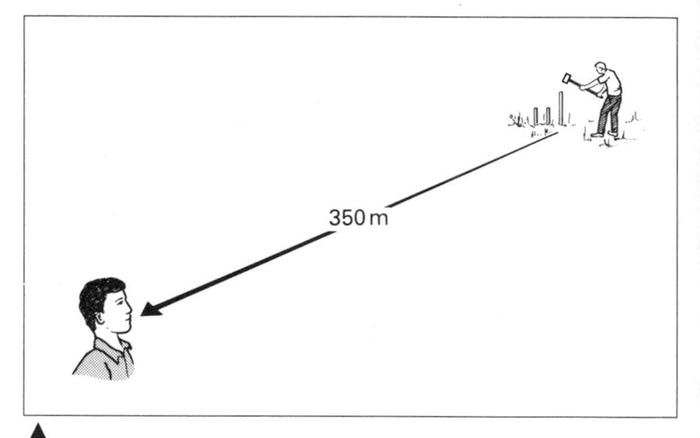

fig. 26.5 Sound takes 1 s, light takes no time

Factors affecting the velocity of sound

Sound waves require a material medium in which to travel and this medium must possess two properties. One property is **density** because this distinguishes material from non-material, such as a vacuum. The other property is **elasticity**, which is the ability to return to an original condition after being displaced. When a sound wave passes through air the air is compressed and must then expand again (and very quickly) to be ready for the next compression. In general terms, the velocity of waves increases if the elasticity increases and if the density decreases, but the relationship is not a simple one.

Because the velocity is changed if the density is changed, sound will travel at different speeds in different gases. The lower the density of the gas, the greater the velocity. Since the humidity of the air alters the density of air, sound travels at a different speed in dry air from moist air. Another factor which changes the density of a gas is the temperature. As the temperature increases, the density decreases and so the velocity of sound increases.

Velocity of sound in a gas is proportional to the square root of the absolute temperature

$$V \propto \sqrt{T}$$

The velocity of sound is independent of the frequency and loudness of the sound and also of the atmospheric pressure.

Sound travels in longitudinal waves

Sound shows all the properties of waves set out in Chapter 24, Sound travels at a finite velocity. It can be reflected and refracted. It can be diffracted, and can form interference patterns. So there is good reason for believing sound to travel in the form of waves. Sound waves are **longitudinal waves**, because

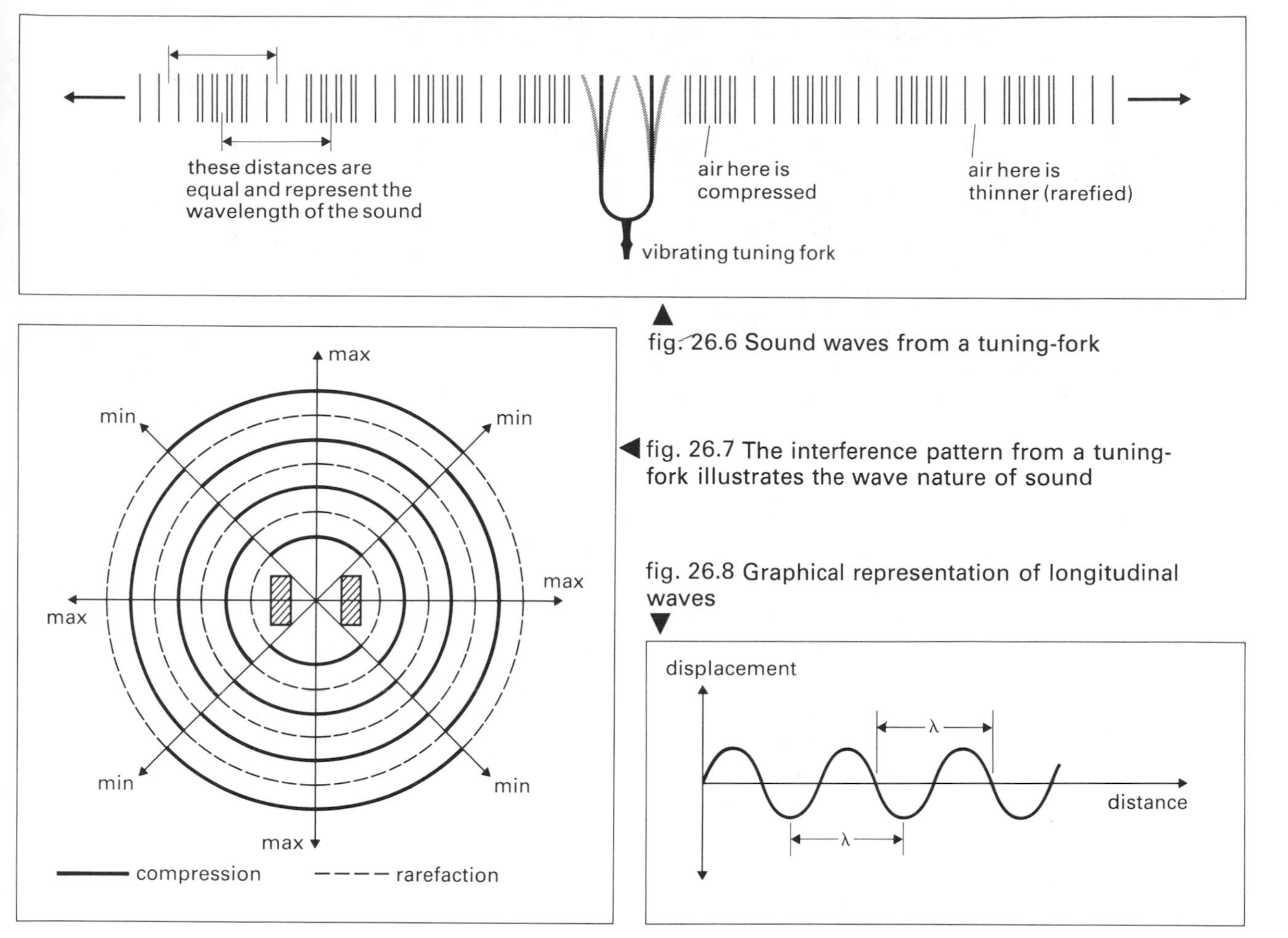

fig. 26.6 Sound waves from a tuning-fork

fig. 26.7 The interference pattern from a tuning-fork illustrates the wave nature of sound

fig. 26.8 Graphical representation of longitudinal waves

the movement of the medium is in the same line as the direction of travel.

When a tuning-fork is sounding, the prongs are moving inwards and outwards. When the prongs move outwards, the air next to them is pushed to one side and compressed. When the prongs move inwards, the air next to them can expand and becomes thinner or more rarefied. Thus a **compression** and a **rarefaction** are produced. As long as the prongs are vibrating, these compressions and rarefactions are produced at regular intervals and travel outwards from the fork through the air. Each particle of air moves regularly backwards and forwards in the same way as the fork does, figure 26.6.

When the prongs of the fork are moving inwards, rarefactions are produced outside the prongs, but a compression of the air will be produced between the prongs. When the prongs move outwards, the compressions are outside the prongs and a rarefaction is produced between them. Thus two longitudinal waves are produced, travelling outwards from the fork in directions at right angles. These two waves are just out of step by half a wavelength, and where they overlap destructive interference takes place. There will be four directions in which the wave systems cancel out (marked 'min' in figure 26.7). In effect, a set of interference fringes is formed, and since wave motion is the only explanation of interference this is a simple way of demonstrating the wave nature of sound. If a tuning-fork is struck and held close to the ear and then rotated slowly through one complete turn there will be four positions of maximum sound and four positions in which no sound is heard.

Although sound waves are longitudinal waves it is convenient to represent them diagramatically by the usual type of wave diagram, figure 26.8, because there is no simple way of drawing a longitudinal wave. The crests can be regarded as compressions and the troughs as rarefactions.

Reflection of sound

The reflection of sound can be demonstrated by using two tubes of metal about 1 m long, figure 26.9. (Cardboard tubes will work but not quite so well.) A clock is used as the source of sound, and the ticking is prevented from reaching the observer's ear directly by the drawing board. Without the reflecting board the sound is inaudible. The reflecting board is put in place and turned until the sound of the ticking is heard most clearly. The angles ABD and DBC will be about equal. The angle of reflection is equal to the angle of incidence. This can be tested in another way by replacing the clock with a lamp, the reflecting screen with a mirror, and looking down the second tube.

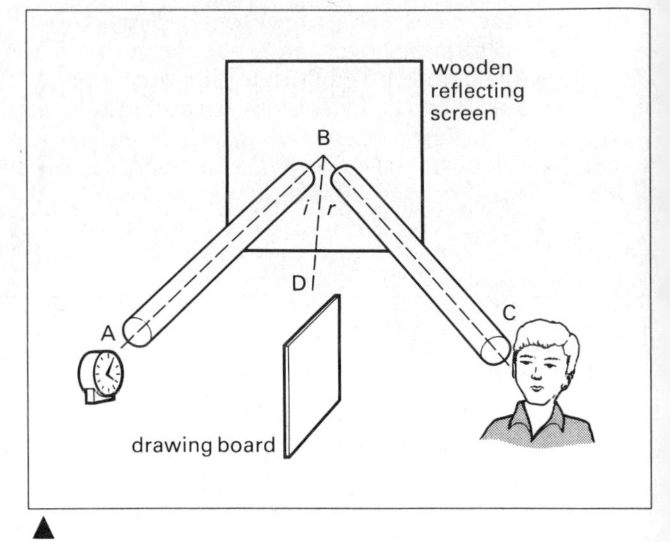

fig. 26.9 Demonstrating that sound obeys the laws of reflection

Sounds are reflected by any large surface and the reflection is called an **echo**. If a shot is fired 700 metres from a cliff face, the sound will be heard again after four seconds, figure 26.10. If the time can be determined accurately, then the distance from the cliff can be worked out if the speed of sound is known. If both time and distance can be measured, this can be a simple way of finding the speed of sound.

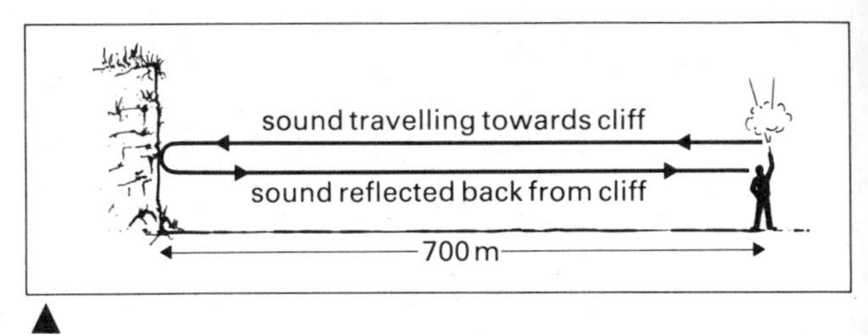

fig. 26.10 An echo

EXAMPLE

An explosion in a quarry takes place at a distance of 70 m from an observer. An echo from a tall cliff 50 m beyond the source of the explosion is heard by the observer 0.5 seconds after he sees the explosion, figure 26.11. Calculate the velocity of sound in air.

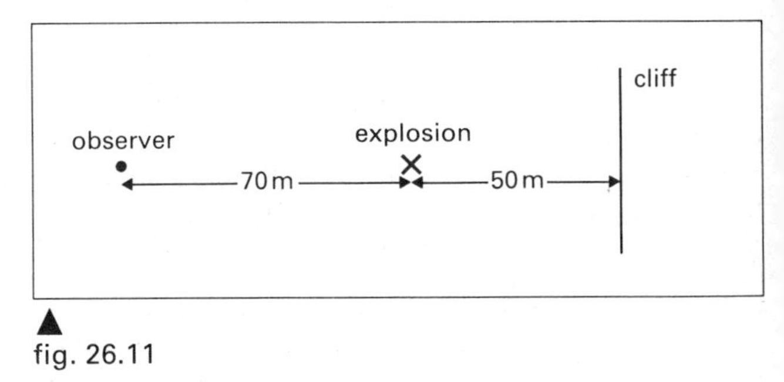

fig. 26.11

When solving problems like this, the important thing is to work out the total distance travelled by the sound and the time taken. The sound of the explosion travels to the cliff (50 m) and then back to the observer (50 m + 70 m).

total distance travelled = 170 m
time taken = 0.5 s

$$\text{velocity} = \frac{\text{distance travelled}}{\text{time taken}}$$

$$= \frac{170}{0.5} = 340 \text{ m/s}$$

Echoes in a large concert hall are a nuisance. The sound from a speaker or performer on the stage may reach a listener in the gallery by several routes, figure 26.12. It will travel directly to the listener, and also may be reflected from various surfaces such as the

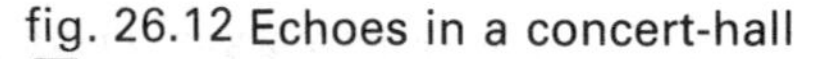

fig. 26.12 Echoes in a concert-hall

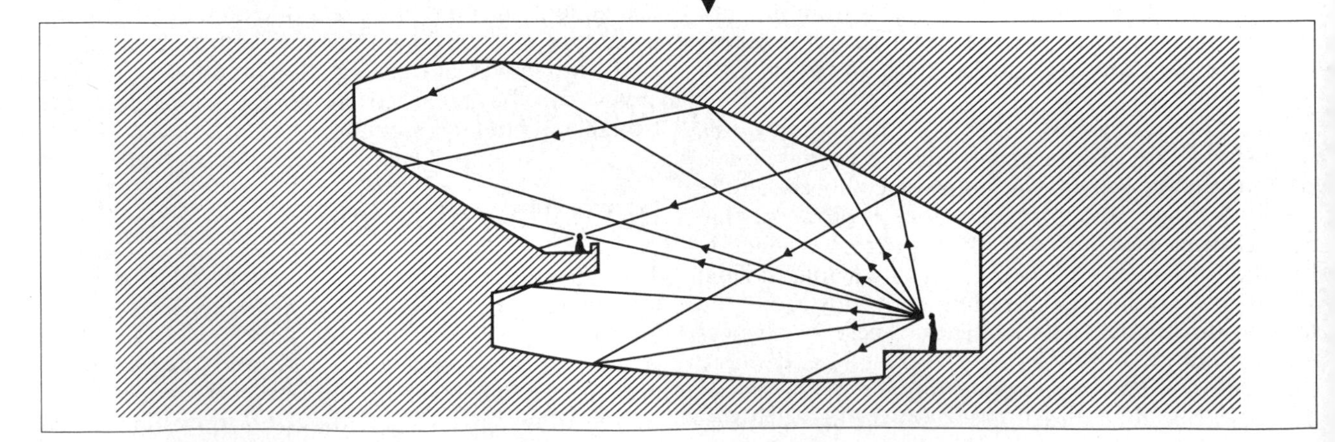

roof and the walls. The same sound arrives at the listener several times over, each one just a little later than the one before it. The result is that the speaker's voice sounds blurred. In modern halls, the walls and ceiling are made of substances which do not reflect sound but absorb it, and the echo nuisance is removed. Another way of getting rid of unwanted reflections is to stretch wires across the hall to break up the reflected waves before they reach the audience. On the other hand, in a small hall when the echoes arrive at almost the same time as the direct sound, they add to the sound received and we hear better.

Reflected sound waves are used in **echo-sounding** to tell how deep the sea bed is and detect shoals of fish and submerged wrecks. A sound transmitter is fixed to the hull and the sound echo from the sea-bed picked up by a microphone mounted close to the transmitter. Both these instruments are connected to a recorder on the bridge which measures very accurately to about $\frac{1}{1000}$ second how long the sound has taken for the double journey to the sea-bed and back, figure 26.13. Often the recorder draws a graph of the depth of the sea-bed.

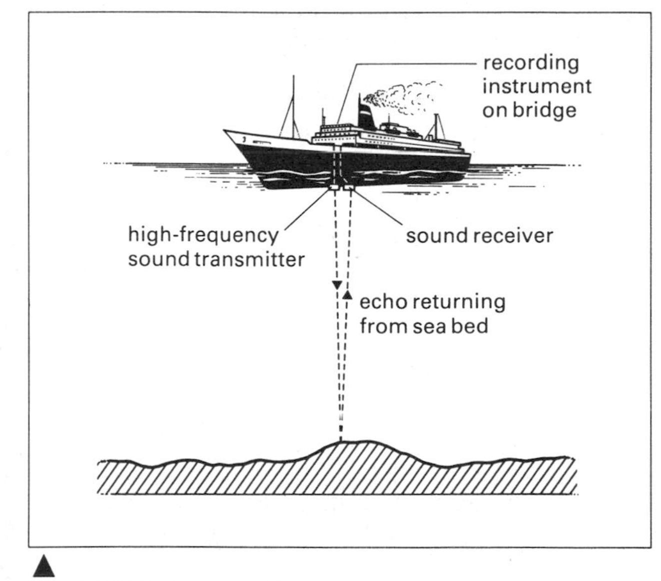

▲ fig. 26.13 Echo-sounding

fig. 26.14 Echo-sounding is also used in geological surveys ▼

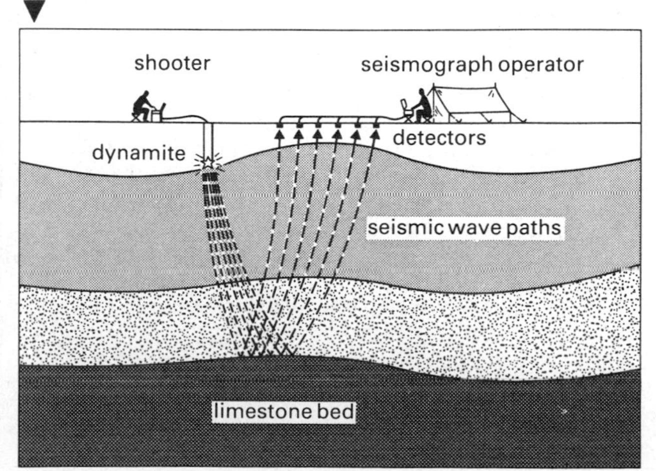

Refraction of sound

Refraction takes place when a wave passes from one medium to another in which it travels at a different speed. When the speed of sound changes there is also a change of direction. An interesting example of this occurs when different parts of the air are at different temperatures. Sound travels more quickly in warm air than in cold air, and so on passing from warm air to cold air the sound will be bent towards the normal in just the same way as light is bent on passing from air (higher speed) into glass (lower speed). The temperature changes in the atmosphere are not sudden but gradual, and this makes the path of the sound curve round gradually instead of there being a sharp change in direction.

On a warm sunny day the air next to the ground is warmer than that higher up. Sound is bent upwards, figure 26.15a, and the person would not hear the sound because it passes over his head. At night the ground cools quickly and the air near it is cooler than that above. The sound is bent downwards and can be heard at greater distances, figure 26.15b.

The bending of the sound paths can also be caused by winds moving at different speeds at different heights above the earth, and this causes sound to play rather queer tricks. Large explosions have been unheard fairly close to, but heard at much greater distances. This 'silent zone' appears because the sound has been bent upwards and then bent downwards again by some change in atmospheric conditions.

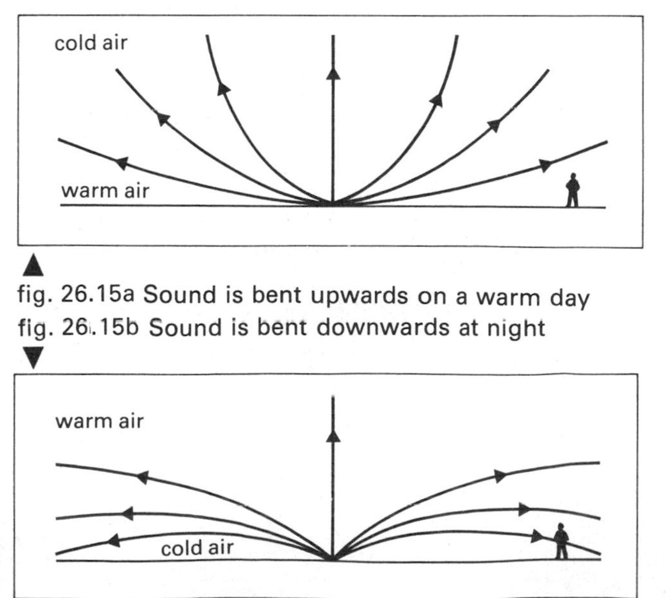

▲ fig. 26.15a Sound is bent upwards on a warm day

fig. 26.15b Sound is bent downwards at night ▼

EXERCISE 26

In questions 1–4 select the most suitable answer.

1 Which of the following does *not* apply to sound waves?
- **A** They transmit energy.
- **B** They result from vibration.
- **C** They are propagated by a series of compressions and rarefactions
- **D** They travel fastest in a vacuum.
- **E** They can be diffracted.

2 A common experiment with sound waves is to place a ringing bell in a bell jar and evacuate all the air. As the air is pumped out the sound of the bell dies away. The most likely reason for this is that
- **A** a bell striker cannot move in a vacuum
- **B** sound waves cannot travel through a vacuum
- **C** sound waves travel much faster through a vacuum and so cannot be heard
- **D** the frequency of the sound waves is increased above the level which can be heard
- **E** the wavelength becomes too small to be heard

3 A person speaking at one end of a long tube may be heard distinctly by an observer with his ear to the other end of the tube because sound waves may
- **A** be amplified
- **B** be diffracted
- **C** be reflected
- **D** be refracted
- **E** interfere

4 The echo-sounder of a ship sends waves vertically down to the sea bed. The echo is received back at the ship after 4 seconds. The speed of sound in sea water is 1500 m/s. What is the depth of the sea beneath the ship?
- **A** 375 m
- **B** 750 m
- **C** 3 000 m
- **D** 6 000 m
- **E** 12 000 m

5 a A boy in front of a wall claps his hands at one-second intervals. He hears the echoes midway between claps, that is, each echo $\frac{1}{2}$ s after the sound. Calculate the distance of the boy from the wall. Take the velocity of sound in air as 340 m/s.

b Explain how the boy could hear the same effect if he moved 170 m further from the wall.

6 a Sound is transmitted as a longitudinal wave. Explain what this means.

b An electric bell is sealed in a large flask which is connected to a vacuum pump. The bell is continuously rung and the pressure of the air is gradually reduced to zero. What would you hear during the process?

c If the pressure of the air is doubled, would it make any difference?

d A boy is at one end of a long cast-iron pipe. Another boy strikes the other end of the pipe with a hammer. The first boy hears two sounds, one following the other. (i) Why is this? (ii) Which does he hear first?

7 a (i) A ship steaming through thick fog uses its fog horn and receives an echo 5 seconds later. How far away was the object causing the echo? (Take the speed of sound in air as 330 m/s.) (ii) What would be the advantage to the ship's navigator of using radar, rather than sound from the fog horn, to detect the object?

b Figure 26.16 shows an experimental arrangement for finding the speed of sound in steel, using a cathode ray oscilloscope. It also shows the *accurate* scale (graticule) of the oscilloscope. (i) Knowing that the 'spot' travels completely across the screen in 0.001 second, calculate the speed of sound through steel. (ii) What adjustments would need to be made to the oscilloscope or other apparatus to make the method suitable for the purpose of measuring the speed of sound in air?

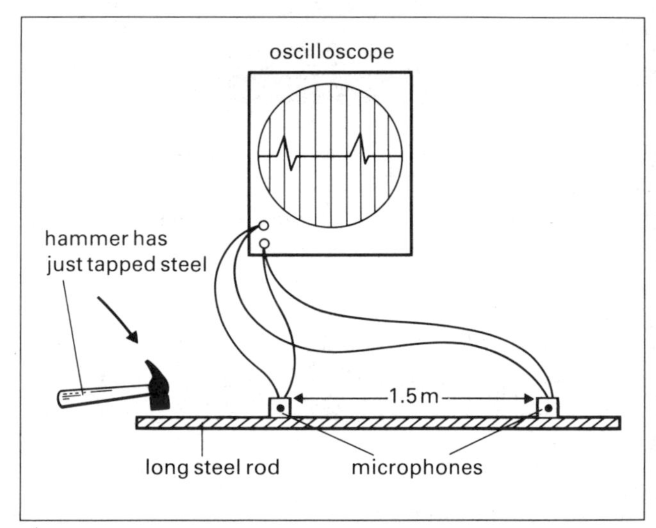

▲
fig. 26.16

8 a Describe and explain a common observation which shows that sound and light travel at different speeds. Which is the faster?

b Describe and explain (i) one feature in which the behaviour of light is similar to that of sound, (ii) one feature, other than speed, in which they differ.

c Describe briefly a method of measuring the speed of sound. State the measurements which must be taken and how they are used to obtain the result. Point out any probable sources of error and the precautions taken to reduce them.

d Explain why this method cannot be used to measure the speed of light.

27 THE STRUCTURE OF ATOMS

The idea that matter was made of atoms grew up in the nineteenth century. The atom was defined as the smallest part of an element to take part in a chemical change, and the atomic theory explained many of the laws of chemical reactions. Atoms combined to form molecules, and the behaviour of solids, liquids and gases could be explained by the kinetic molecular theory (see p. 133).

During all this time the atoms and molecules were thought of as spherical masses or particles, and no thought was given to the question whether the atoms themselves had any internal structure. During the present century scientists have shown that the atom itself has a complicated structure. It is composed of particles still smaller than itself.

In Cambridge in the 1890s Sir Joseph John Thomson showed that all kinds of matter contained small particles having a charge of negative electricity. (Some of the experiments leading to this idea are described in Chapter 38.) These particles are **electrons**. Thomson managed to measure the ratio of the charge to the mass of these particles. He showed that it was constant, and did not depend on how the electrons were produced. In 1916 Robert A. Millikan, an American physicist, measured the charge of the electron, and this has become the fundamental unit of electric charge. No fraction of this charge has ever been measured. It is, as it were, an *atom of electricity*.

The electron is a particle which is common to all matter; it has a negative electric charge of $-1{\cdot}6 \times 10^{-19}$ coulomb and a mass of $9{\cdot}1 \times 10^{-31}$ kg

If all atoms contain electrons and atoms are electrically neutral, there must be some positive charge somewhere exactly equal in magnitude to the negative charge of the electrons. Thomson's idea was that the electrons were embedded in a mass carrying an equal positive charge, figure 27.1. This was just an idea and he had no direct experimental evidence for it. The experiments of Rutherford, Geiger and Marsden in Manchester in 1909 showed that the idea was not supported by experimental evidence and that there was a positively-charged centre to the atom, the **nucleus** of the atom.

The scattering of alpha-particles by gold foil

An alpha-particle is a relatively heavy, positively-charged particle emitted from radioactive substances (see p.336). Alpha-particles are absorbed by a few cm of air, so experiments with them have to be carried out in a vacuum. Rutherford had noticed that, when alpha-particles passed through thin metal foils, some of the particles were scattered to the sides. It was

fig. 27.1 Thomson's 'currant bun' atom with electrons embedded in a positive mass

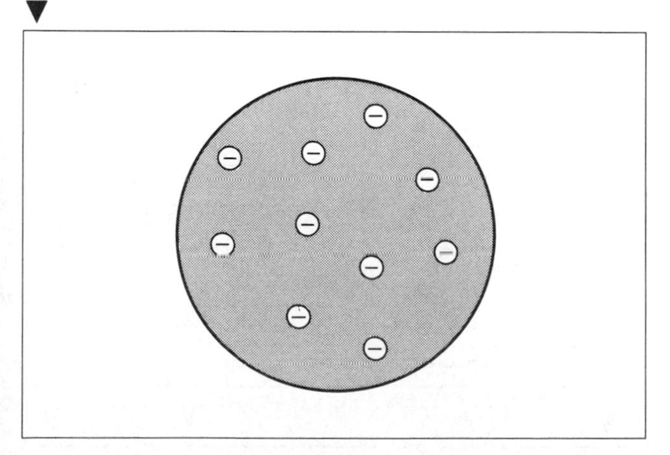

fig. 27.2 An early apparatus for measuring α-particle scattering

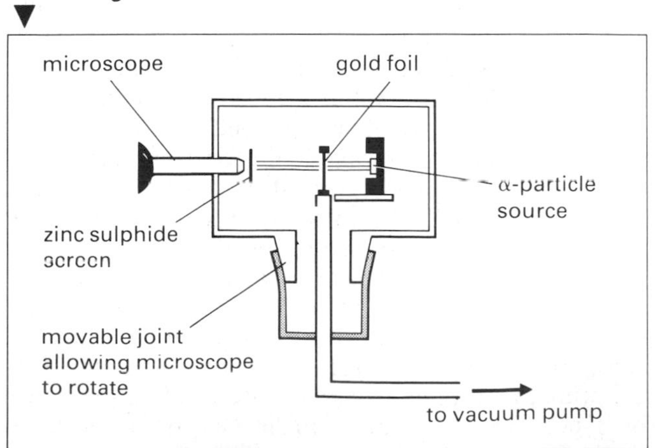

decided to investigate this scattering quantitatively.

When alpha-particles hit a zinc sulphide screen a tiny flash of light is given out, and this can be observed through a microscope, figure 27.2. There were no automatic counters in those days and all observations had to be made by eye for periods of up to twenty-four hours at a time! The numbers of alpha-particles which were scattered through different angles were counted. Most of the particles went straight through the foil, many were scattered through small angles, a few were scattered through large angles and fewer still were turned right back in the direction from which they were coming, figures 27.3a and b.

Rutherford's conclusions were

- Since most of the alpha-particles pass through the metal foil, most of an atom must be empty space.
- Of the few which were scattered back through large angles Rutherford said, 'It was about as credible as if you had fired a 15-inch shell at a piece of tissue paper and it had come back and hit you'.

This could only be explained if the alpha-particles had encountered something pretty massive which had a positive charge to repel the positive charge on the alpha-particle. There must be a positively charged nucleus to the atom.

Further, more detailed, experiments with different metal foils enabled the size and charge on the nucleus to be measured. The charge was always positive and equal to a whole number of electron charges.

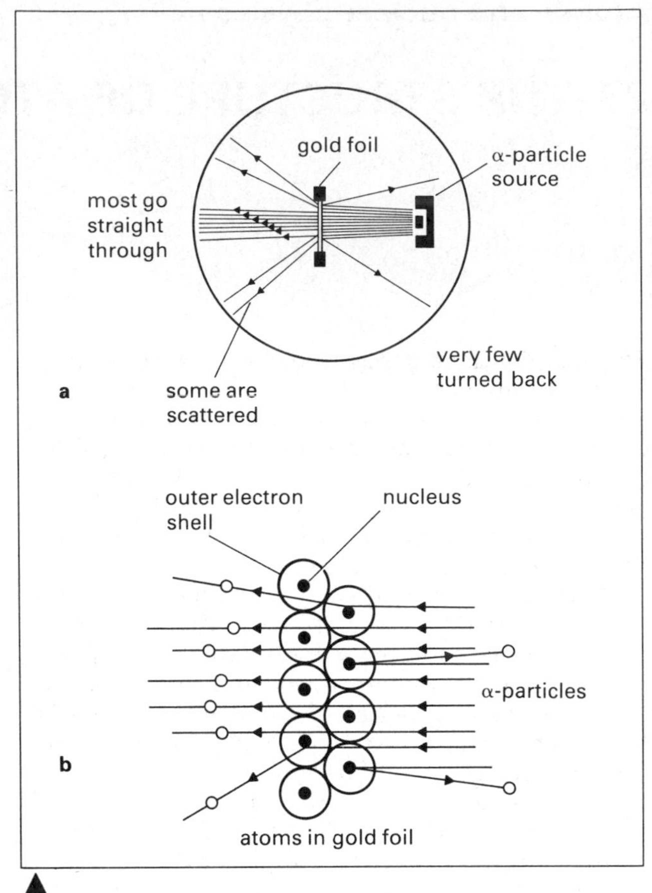

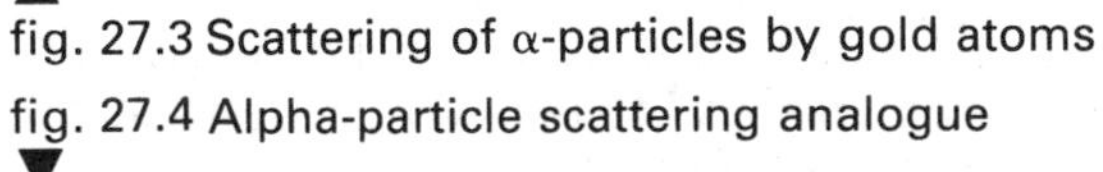

fig. 27.3 Scattering of α-particles by gold atoms

fig. 27.4 Alpha-particle scattering analogue

A 'scattering' experiment

It is impossible for us to repeat Rutherford's scattering experiment for ourselves. It is much too complicated. In a situation like this we try to imitate the conditions of the experiment in a simpler way, in an attempt to give a clearer idea of what is happening. Such a model is called an *analogue* experiment because it is analogous (similar) to the original situation.

Figure 27.4a shows a steel ball rolling down a launching slope towards a curved metal or plastic hill. The curve of the hill has been worked out so that the magnitude and direction of the force on the moving ball when it is on the hill is the same as the repulsive force between two similarly charged particles.

The steel ball is always launched from the same height on the plane, so that it always has the same velocity. The ball is sent along a series of parallel paths at different distances from the summit of the hill, and the deflected path of the ball marked on a sheet of paper, figure 27.4b. The closer the path to the summit, the greater is the deflection. A ball very close to the summit is turned back on its path. The experiment imitates the scattering of alpha-particles by a positive charge and can, in fact, be used quantitatively to produce exactly similar results.

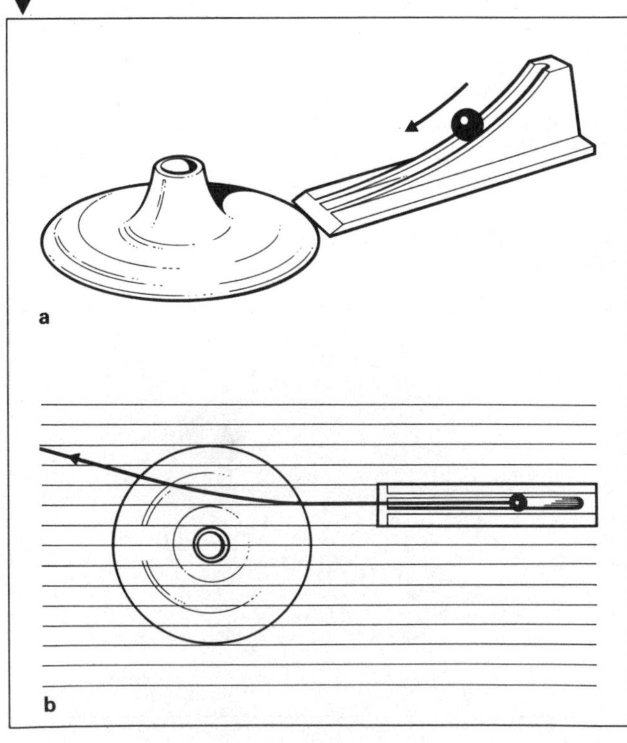

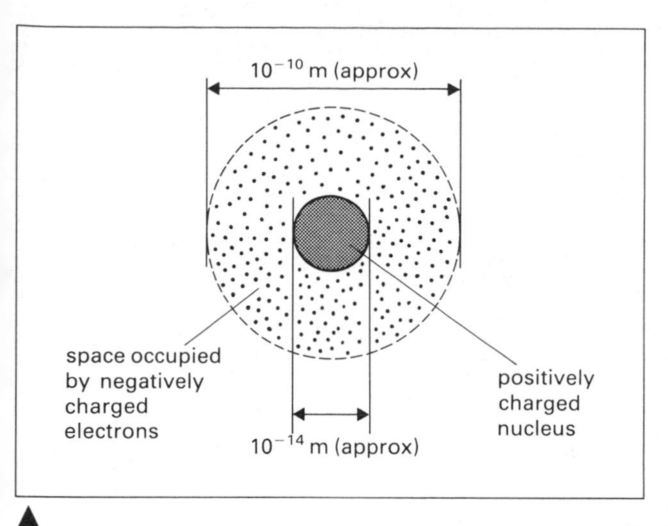

▲
fig. 27.5 Picture of an atom suggested in 1912

The nuclear atom

By about 1912 scientists had a picture of the atom, figure 27.5, as having a positively-charged nucleus with a diameter about $\frac{1}{10\,000}$ of the diameter of the atom itself. There were many more questions to be answered, but the most important ones were

- How are the electrons arranged round the nucleus?
- What gives the nucleus its positive charge?
- Are there any other constituents of atoms?

Following from the theoretical work of the Danish scientist Niels Bohr it was suggested that the electrons were arranged in specific orbits or shells round the nucleus.

The positive charge of the nucleus was explained with the identification of a positively-charged particle. This was the **proton** with a positive charge equal in magnitude to the negative charge on the electron, but nearly two thousand times heavier.

An entirely new particle, the **neutron**, was discovered by James Chadwick in 1932. The neutron has no charge at all, and a mass equal to that of the proton. The discovery of the neutron helped to clear up some difficulties about the relative masses of atoms. The table shows the relative masses of the particles in the atom.

fig. 27.6 Some examples of the structure of atoms. Note that the electron arrangements for copper and uranium have been very much simplified
▼

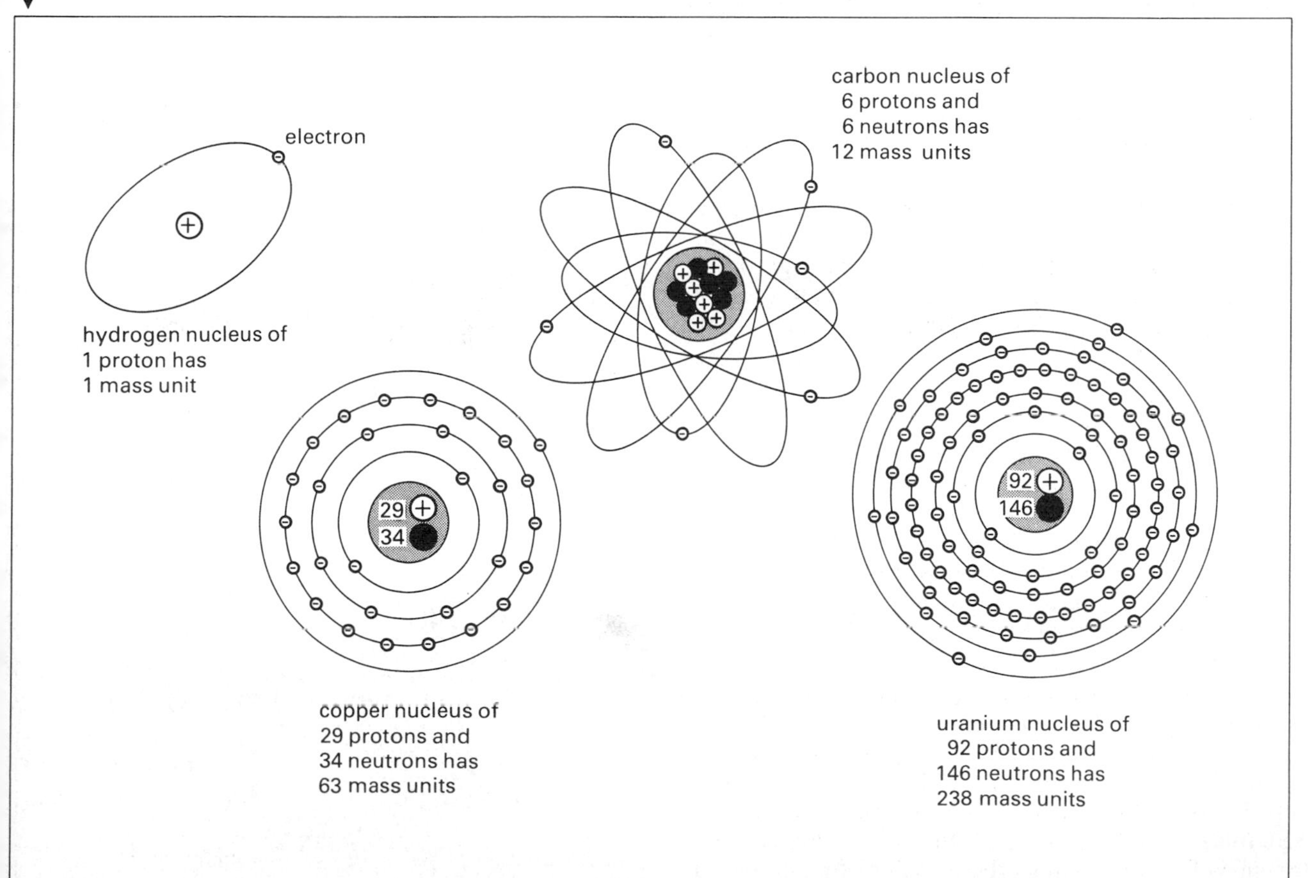

electron	**proton**	**neutron**
charge −1	charge +1	no charge
mass $\frac{1}{1840}$ unit	mass 1 unit	mass 1 unit

The electron, proton and neutron are the main building blocks of atoms. The protons and the neutrons are in the nucleus and are called **nucleons**. The number of protons in the nucleus is equal to the number of electrons surrounding the nucleus, so that the atom is electrically neutral. The balance of the mass of the atom is made up by the neutrons in the nucleus.

The **atomic number** (*Z*) (also called the **proton number**) of an element is the number of protons in the nucleus.

The **mass number** (*A*) (also called the **nucleon number**) of an element is the number of nucleons in the nucleus, i.e. the sum of the numbers of protons and neutrons.

The mass of the electrons is usually neglected compared with the mass of the nucleus.

The electrons are arranged in shells with up to 2 electrons in the first shell, up to 8 electrons in the second shell and up to 18 electrons in the third shell. After this it gets more complicated and the innner shells have to be divided into sub-shells to solve some problems about atoms. The important thing is that **the total number of electrons is equal to the total number of protons**.

The picture of the atom in figure 27.7 is the one we shall use at this stage. It must be pointed out that many more sub-atomic particles have been discovered, and that ideas about electrons have changed. It is, perhaps, more correct to think of a kind of cloud of electrons round the nucleus, with the shells defining areas where the electrons are most likely to be.

Information about the structure of an atom can be conveniently written using the chemical symbol of the atom and putting before it the mass number at the top and the atomic number at the bottom. Thus

$^{238}_{92}U$ represents an atom of uranium

there are 92 protons in the nucleus
there are 92 electrons surrounding the nucleus
there are 238 − 92 = 146 neutrons in the nucleus

Isotopes

The chemical properties of an element depend on the number of electrons in the electron cloud and on their arrangement. It is found that not all the atoms of some elements are the same. They all have the same chemical properties but different masses. Since the chemical properties are the same, the number of electrons is the same and so the number of protons is the same. The only variable which could alter the mass

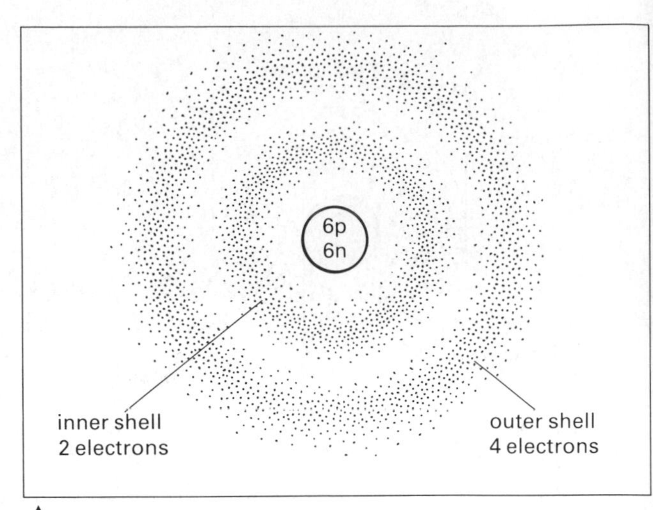

fig. 27.7 An atom of carbon

fig. 27.8 Isotopes are atoms of different masses having the same chemical properties

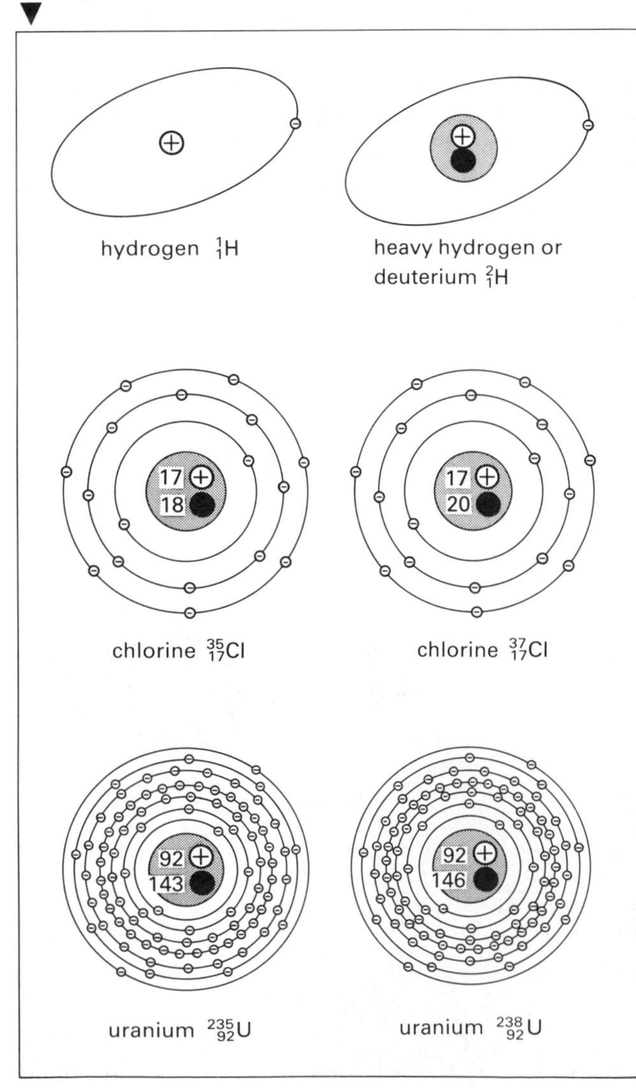

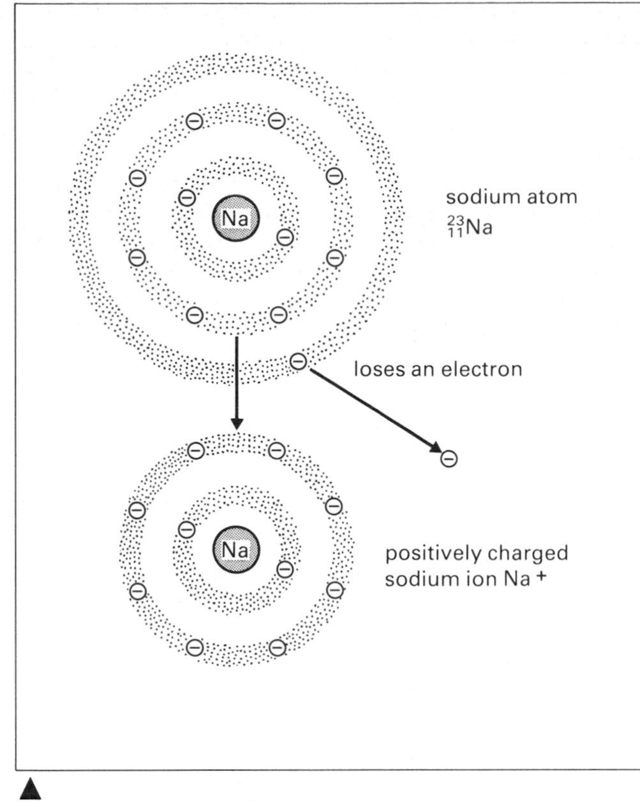

▲
fig. 27.9 Ionisation of a sodium atom

fig. 27.10 Ionisation of a chlorine atom
▼

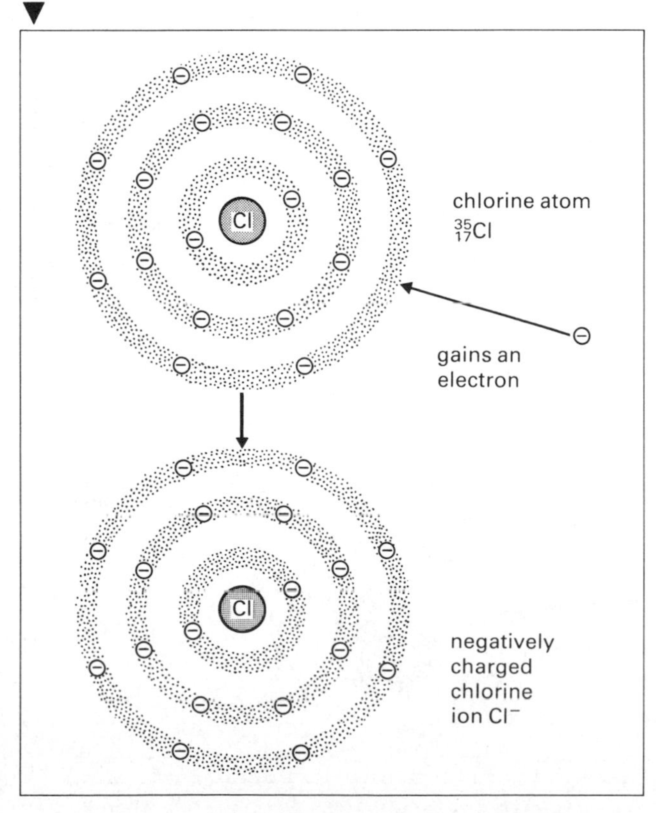

is the number of neutrons in the nucleus. Atoms with the same atomic number (therefore the same chemical properties) but different mass numbers are called **isotopes**. Isotopes have the same number of protons but a different number of neutrons, figure 27.8.

Ions and ionisation

Normal atoms are electrically neutral. The number of protons in the nucleus is exactly balanced by the number of extranuclear electrons. For some atoms, one or more of the outermost electrons can become separated from the atom if sufficient energy is supplied. The proton/electron balance is now destroyed. The atom is said to be **ionised** and the two parts are a **positive ion** and a **negative ion**. The negative ion is the electron, and the positive ion the rest of the atom which has one more proton than it has electrons, figure 27.9. The energy necessary to produce this separation is the **ionisation energy**.

In the case of some other atoms there is a gap in the electron shell which, if it were filled, would produce a more stable arrangement. A chlorine atom, which has 17 electrons arranged 2 8 7 would be more stable if the outer sub-shell had 8 electrons. When this gap is filled there is an extra electron and a negatively charged chlorine ion is formed, figure 27.10.

An ion is any charged particle, either an electron or an atom or group of atoms which has lost or gained one or more electrons

The existence of ions explains how some atoms combine to form molecules, e.g. positive sodium ions and negative chlorine ions are bound by an ionic attraction to form sodium chloride. The ionic link is broken on solution in water, and the ions are the charge carriers of the current during electrolysis (see p.267).

Gases are always ionised to some extent but the ionisation can be increased by heating them, passing X-rays through them or exposing them to strong electric fields. Flames cause ionisation of the air. Gas discharge tubes (neon tubes and fluorescent lighting tubes) conduct electricity because of the breakdown of the gas molecules into ions.

Atomic transformations

In the middle ages alchemists dreamed of changing base metals into gold. Although they did not know about atoms, what they wanted was to change one kind of atom into another kind. With our knowledge of the structure of the atom we realise that to change from one atom to another means making a change in the number of protons in the nucleus. This can be

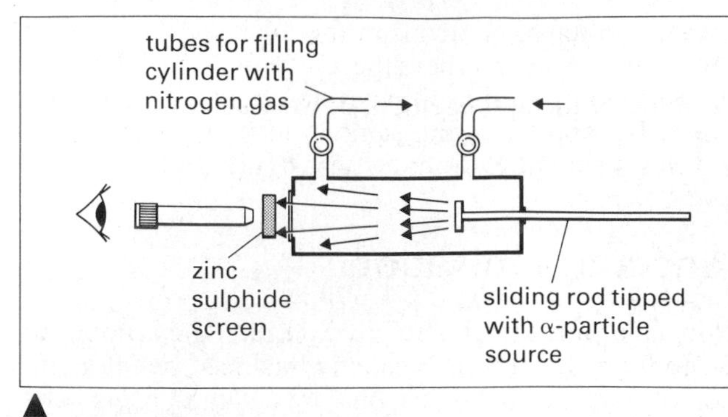

▲
fig. 27.11 Rutherford's apparatus for α-particle bombardment of nitrogen atoms

achieved by bombarding the nucleus with some kind of fast-moving particle.

In 1919 Lord Rutherford bombarded nitrogen atoms with alpha-particles given off by the radioactive element polonium, figure 27.11. Alpha-particles are the nuclei of helium atoms, and consist of two protons and two neutrons. As the result of many experiments Rutherford was able to show that a very few of the nitrogen atoms had suffered direct hits and had been converted into atoms of oxygen and hydrogen.

A reaction like that shown in figure 27.12 is called a **nuclear reaction**. Notice that the mass numbers and the atomic numbers balance on both sides of the equation.

Another important nuclear transformation was carried out by Cockcroft and Walton in 1930 when they bombarded lithium with protons accelerated by a powerful electric field. They found that alpha-particles were produced. This is the equation for the nuclear reaction:

$^1_1H + ^7_3Li \rightarrow 2\,^4_2He$

Cockcroft and Walton designed and made the first particle accelerator, figure 27.13. The particles — in this case protons — are made to accelerate by applying an electric field. The positively-charged proton will be attracted to a negatively-charged plate, and repelled by a positive one. The greater the voltage applied, and the longer the particle is in the electric field, the faster it will travel and the more energy it will have. The higher the energy of the bombarding particle, the more likely it is to enter the nuclei of the atoms being bombarded.

The voltages applied and the accelerators themselves have got bigger and bigger and more expensive. The proton synchroton at the CERN (European Organization for Nuclear Research) laboratories at Geneva cost over £20 million and works at a voltage of 30 000 million volts. The linear accelerator at Stanford in America is 3 kilometres long and works at 20 000 million volts.

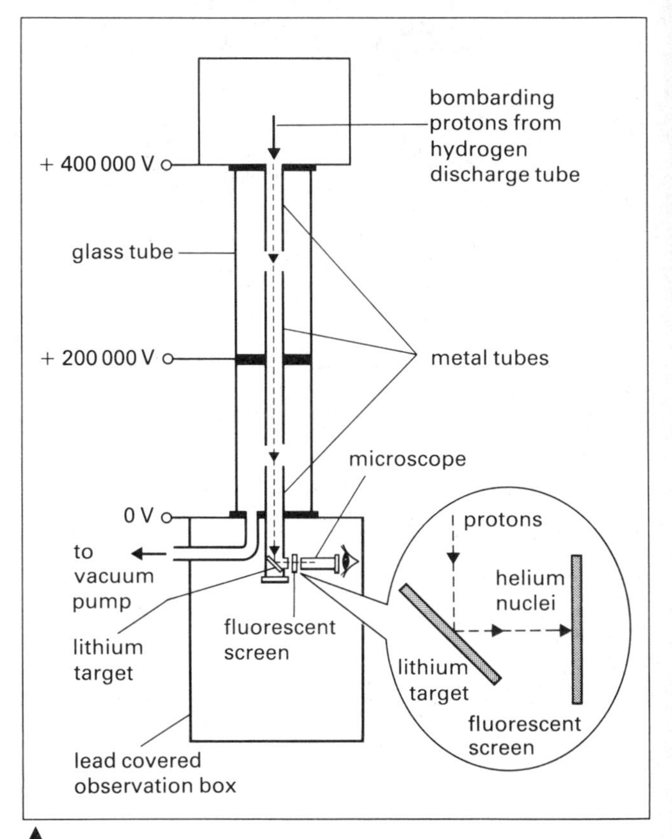

▲
fig. 27.13 Cockcroft and Walton's accelerator

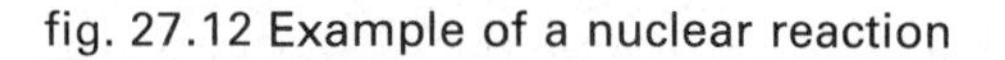
fig. 27.12 Example of a nuclear reaction
▼

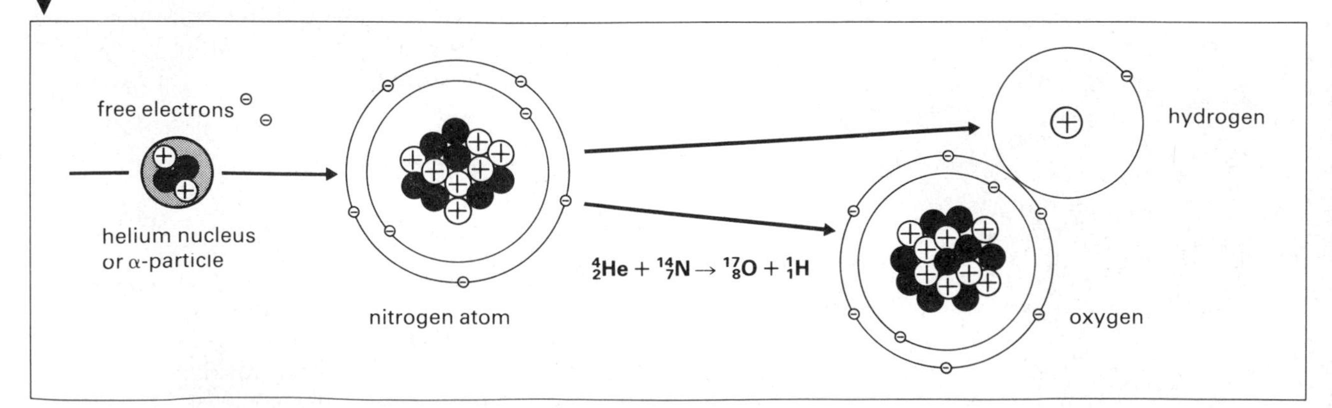

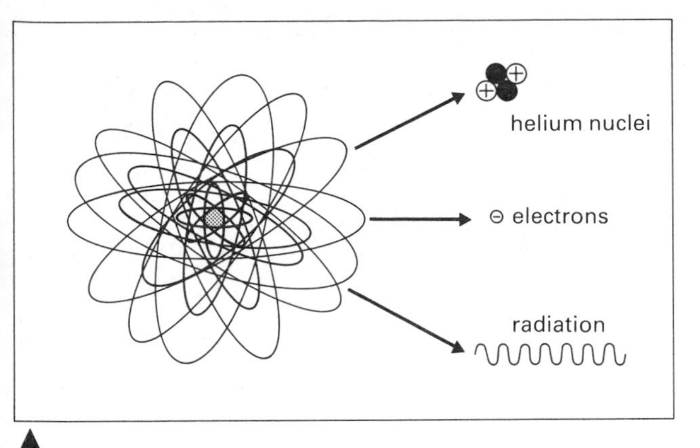

fig. 27.14 Radioactivity

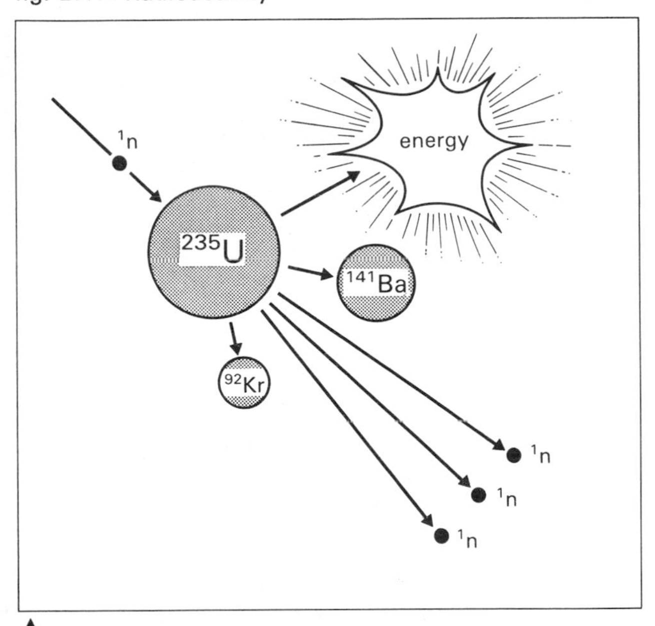

fig. 27.15 Nuclear fission

fig. 27.16 Nuclear fusion

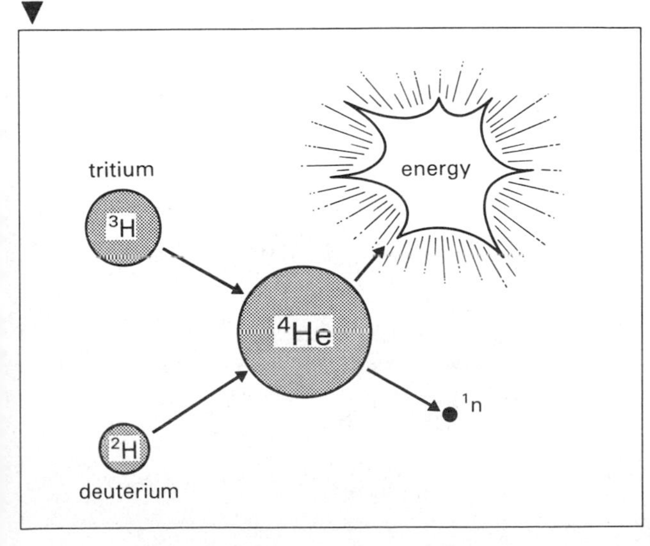

The nucleus

The study of the arrangement of the nucleons and the forces between them is the concern of nuclear physics. One obvious problem is: how do numbers of positively-charged protons manage to stay together when, normally, like charges repel each other? Possibly the nature of the forces may change when the particles are very close together, or new particles may exist which act as a kind of 'nuclear glue'. Many other sub-atomic particles are now known to exist (more than a hundred and fifty!). We shall just deal for the present with the simpler picture of the atom.

We have seen that nuclei can be changed by bombarding them. The discovery of the neutron made nuclear bombardment easier because the neutron, having no charge, would not be repelled by the nucleus. Neutron bombardment opened up an entirely new field of nuclear physics.

Radioactivity, figure 27.14, is the spontaneous disintegration of the nucleus, particularly in atoms of high atomic number, with the emission of small atomic particles (electrons and helium nuclei) and radiation (see p.336).

Nuclear fission, figure 27.15, is the breaking up of large nuclei into approximately equal parts with the emission of energy. This energy is used in 'atomic' bombs and nuclear power stations. It is often called 'atomic' energy, but this term is incorrect. Atomic energy is released during ordinary chemical reactions, such as burning, which are only concerned with rearrangements of the external electrons.

Nuclear fusion, figure 27.16, is the joining together of two light nuclei to form a more massive atom. During this process large amounts of energy are released. This is the process taking place in the sun and in the hydrogen bomb. The reaction cannot yet be controlled sufficiently for it to be used in a nuclear power station.

Einstein's mass–energy relationship $E = mc^2$

In 1905 Albert Einstein suggested that matter and energy were convertible and that the great conservation principles of the nineteenth century (conservation of energy and conservation of mass) could be combined into a single principle as the conservation of mass–energy. Einstein stated that energy and mass were related by the equation

$$E = mc^2$$

E is the energy in joules
m is the mass in kilograms
c is the velocity of electromagnetic radiation (3×10^8 metres/second)

Because c is very large, and c^2 even larger, a very small mass has a large energy equivalent. If it were possible completely to convert 1 kilogram of matter into energy it would yield 9×10^{16} joules. This is ninety thousand million million joules. In order to obtain as much energy by burning coal we should have to burn 4 000 000 tonnes!

In nuclear transformations it is found that the combined mass of the products is less than that of the original material. The figures for Cockcroft and Walton's experimental bombardment of lithium with protons are given in table 27.1. The units are atomic mass units and 1 a.m.u. is equal to $\frac{1}{12}$ of the mass of a carbon-12 atom. It will be seen that there is a loss in mass of 0.0186 a.m.u. per single atomic encounter. This mass is converted into energy according to Einstein's equation. It is the source of energy in all fission (and fusion) processes. This mass loss of 0.0186 a.m.u. may seem very small because 1 a.m.u. is equal to 1.66×10^{-27} kg but it must be remembered that this is the mass loss for only one atom of lithium. In a modern nuclear reactor millions and millions of atoms are reacting.

table 27.1 Cockcroft and Walton's experiment

Initial particles		Final particles	
$^{7}_{3}Li$	7·0160 a.m.u.	$^{4}_{2}He$	4·0026 a.m.u.
$^{1}_{1}H$	1·0078 a.m.u.	$^{4}_{2}He$	4·0025 a.m.u.
total	8·0238 a.m.u.	total	8·0052 a.m.u.

Nuclear power

A typical fission reaction occurs when uranium-235 atoms are bombarded with neutrons. The reaction was first studied by the Italian physicist Enrico Fermi in 1935. The complete explanation was later given by two German physicists, Professor Otto Hahn and Dr Fritz Strassman and two refugees from Nazi Germany, Dr Lise Meisner and Dr Otto Frisch, working in Denmark in 1939.

When a neutron hits the nucleus of the uranium atom, figure 27.17, the new nucleus formed is unstable and splits roughly in half. During this process there is release of energy (because of mass loss) and more neutrons (from one to three) are produced.

If these emitted neutrons hit other uranium-235 atoms more fissions occur. More neutrons are produced, which produce more fissions and more neutrons, and so on. Such a reaction is called a **chain reaction**. Suppose two of the neutrons produced during a fission hit other uranium atoms and produced two more fissions giving four neutrons: we should have an uncontrolled chain reaction with a multiplication factor of two, figure 27.18. If all three neutrons

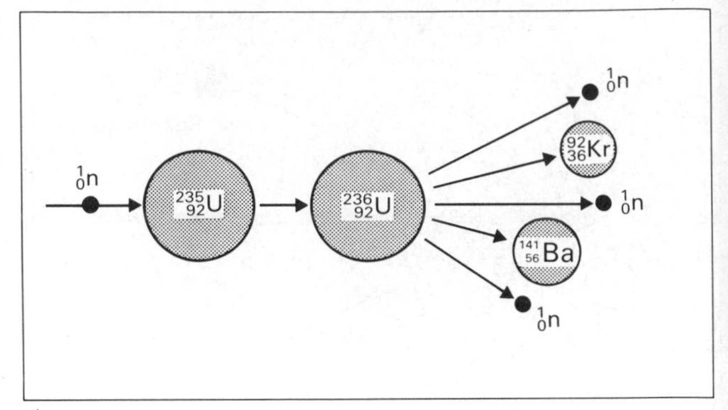

fig. 27.17 Splitting of a uranium atom

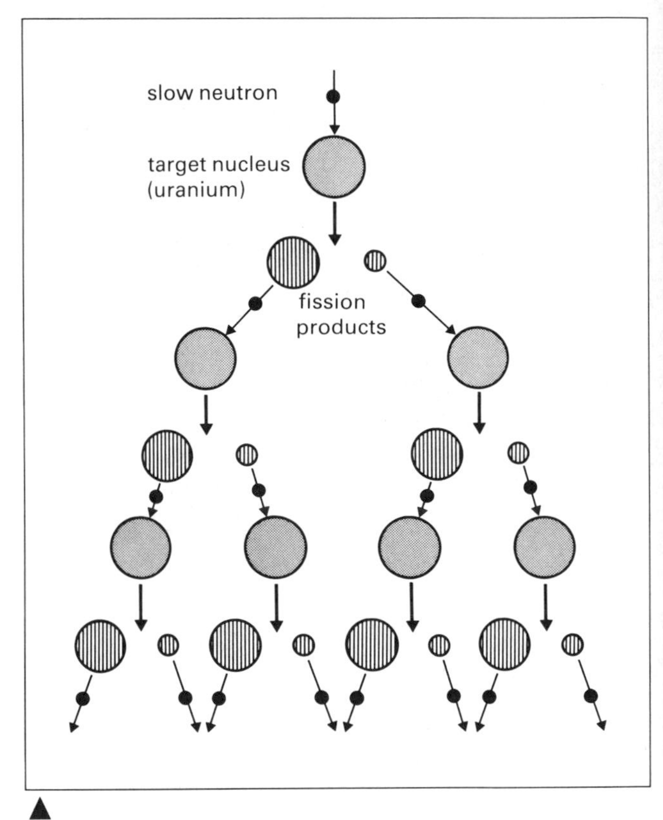

fig. 27.18 Uncontrolled chain reaction

fig. 27.19 Controlled chain reaction: only one neutron used

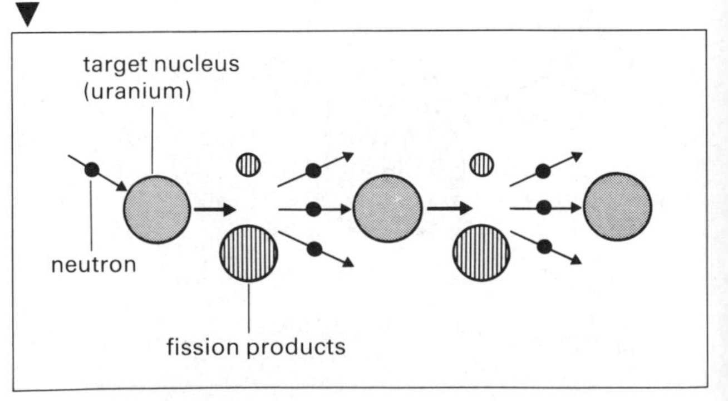

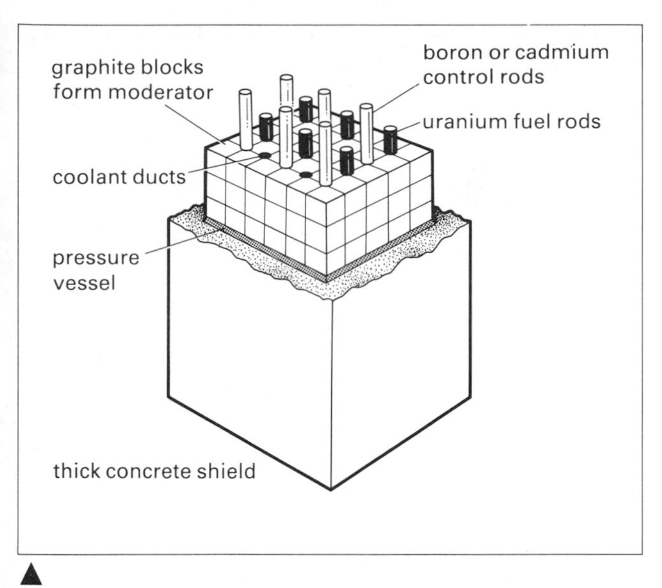

▲
fig. 27.20 General arrangement of a nuclear reactor

produced fissions the multiplication factor would be three. If we can arrange that only one neutron produces one other fission, then we have a controlled chain reaction with a multiplication factor of one, figure 27.19. An uncontrolled reaction is used in an atomic bomb. A controlled chain reaction is used in a nuclear power reactor.

There are many technical problems connected with the practical production of nuclear power.

The emitted neutrons are travelling too fast, and have to be slowed down before they will produce more fissions. This is done by a **moderator**, such as graphite. The neutrons are slowed to the best speed by repeated collisions with the carbon atoms.

The rate of the reaction is adjusted by **control rods** of cadmium or boron, which absorb neutrons. The more the control rods are pushed into the core of the reactor, the more neutrons which are absorbed and the slower the reaction. There are enough rods to stop the reaction altogether.

The products of fission are radioactive so the reactor has to be screened with thick walls of concrete and the waste material has to be safely disposed of by dumping in the sea in sealed containers or burying it deep in the ground. The dumping of radioactive waste from nuclear reactors is a subject of much concern today.

Most of the energy is released as the kinetic energy of the two large fragments which separate at high speed. This kinetic energy is shared with surrounding atoms so that the average kinetic energy of the system is increased or heat energy is produced. The heat energy is removed from the reactor by a circulating gas or liquid coolant to a heat exchanger where water is converted into steam to drive turbogenerators.

fig. 27.21 General arrangement of a nuclear reactor and its heat exchanger, in which the heat carried away from the reactor is used to generate steam
▼

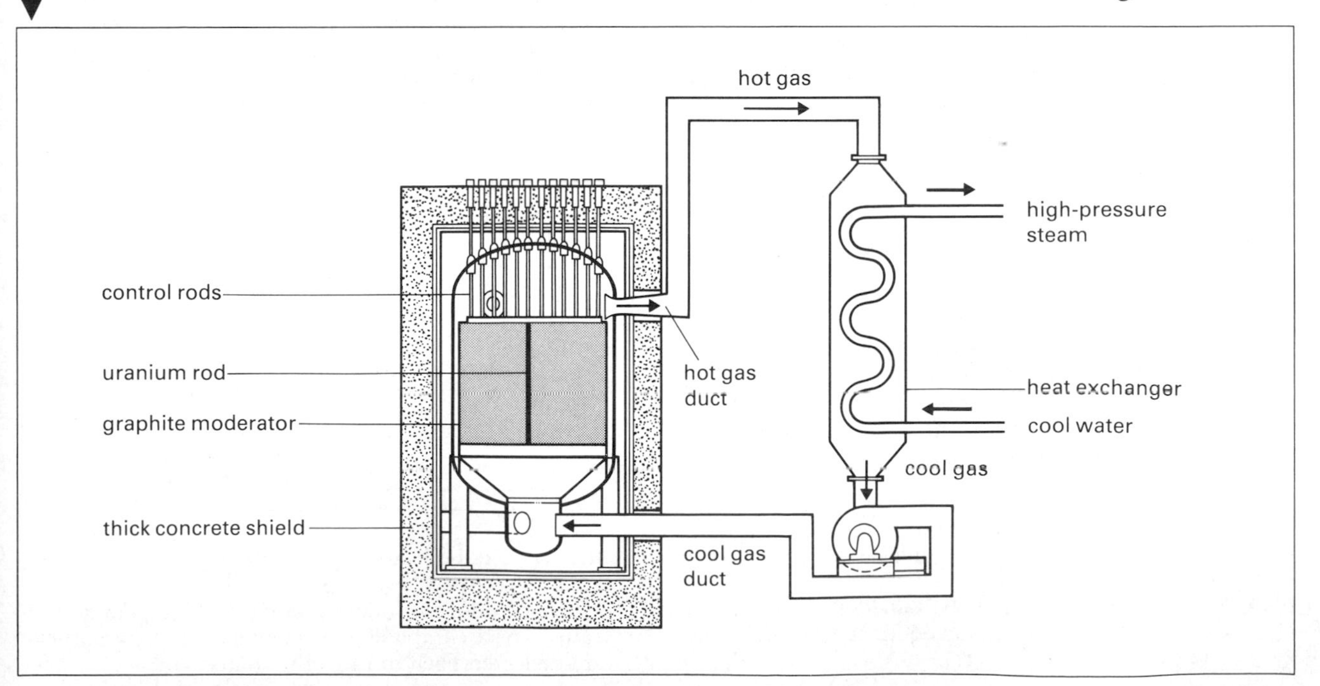

EXERCISE 27

In questions 1–6 select the most suitable answer.

1 The scattering experiments of alpha particles by gold foil suggest that
A alpha particles travel very quickly
B gold is more dense than alpha particles
C gold atoms are negatively charged
D gold atoms have very dense nuclei
E atomic nuclei contain positive and uncharged particles

2 A neutron has
A no electric charge
B a single positive charge
C a double positive charge
D a double negative charge
E a single negative charge

3 In all atoms except hydrogen you would expect to find:
A protons and electrons in orbit round a neutron
B protons and neutrons in the nucleus and orbiting electrons
C electrons and protons in the nucleus and orbiting neutrons
D electrons and neutrons in orbit and protons in the nucleus
E electrons and neutrons in the nucleus and orbiting protons

4 The nucleus of the atom of a certain element contains 5 protons and 6 neutrons. Its mass number is
A 5
B 6
C 11
D 12
E 17

5 An atom of tin has an atomic number 50 and a mass number 112. Which of the following is an isotope of tin?
A $^{112}_{51}X$
B $^{114}_{50}X$
C $^{112}_{49}X$
D $^{112}_{62}X$
E $^{51}_{112}X$

6 Isotopes of an element have
A the same number of neutrons
B different numbers of protons
C the same physical properties
D the same number of protons
E different numbers of electrons

7 a Name the particles you would expect to find (i) inside the nucleus of an atom, (ii) outside the nucleus of an atom.
b (i) Which of the particles named in **a** would you expect to find in an atom in equal numbers? (ii) Give a reason for your answer.

8 The figure represents atoms of carbon, helium and hydrogen.

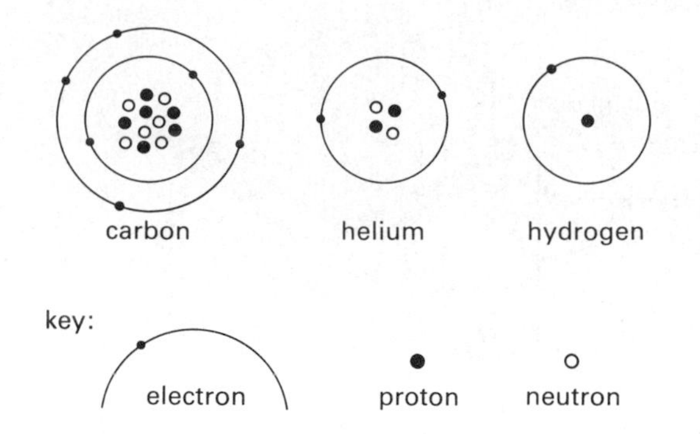

a How can you tell that the atoms shown are all neutral atoms?
b What is the atomic number of hydrogen?
c What is the atomic mass number of the isotope of carbon shown here?
d What is the symbol for the isotope of helium shown if that for the carbon atom is $^{12}_{6}C$?

9 a Copy and complete the following table.

Particle	Mass	Charge
proton		
neutron		
electron		

b How are the particles listed in the table arranged in the atom?
c Draw and label a simple picture of a helium atom that has an atomic number of 2 and a mass number of 4.
d Natural chlorine consists of two types of atom, both of atomic number 17, but one has a mass number of 37 and the other 35. (i) What name do we give to atoms which differ in this way? (ii) What accounts for the differences in the atoms? (iii) In what ways are the atoms similar?

10 Study the table and decide which pair of letters represent isotopes of the same element. Choose your answer from the following list of four possibilities: (i) P and Q, (ii) Q and R, (iii) P and S, (iv) Q and S.

Element	Number of protons	Number of neutrons
P	6	6
Q	8	8
R	7	7
S	6	8

11 Briefly describe Rutherford's experiment on the scattering of alpha-particles by gold foil.
What evidence is there that most of the gold atom is empty space? Explain why some of the alpha-particles are deflected back towards the source.

12 a The diagram represents an atom of one isotope of helium. (i) What are the names of the particles labelled X, Y and Z? The particles marked Y have no electrical charge. (ii) What is the name given to the complete group of particles in the centre of the atom?

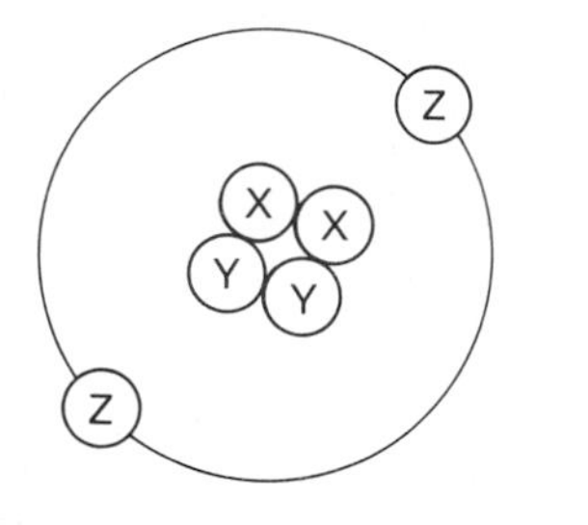

b What do you understand by the term *isotopes*? Describe an atom of another possible isotope of helium.

c An atom of uranium disintegrates during *nuclear fission*. What do you understand by the term nuclear fission? Explain why it is so important.

13 A typical nuclear equation for neutron capture and fission for uranium is given below. In reactors this process leads to a chain reaction.

$$^{235}_{92}U + ^{1}_{0}n \rightarrow ^{236}_{92}U \rightarrow ^{141}_{56}Ba + ^{92}_{36}Kr + 3\,^{1}_{0}n$$

a (i) What are isotopes? (ii) Which *two* of the materials in the above equation are isotopes of the same element?

b What is meant by the term *chain reaction?*

c Fuel rods, control rods, and moderator rods or blocks are found in a reactor. (i) Name the materials which could be used for each type of rod, and briefly describe the purpose of each. (ii) Give a labelled sketch of the likely arrangement of these rods in the reactor core.

14 a What do you understand by the terms (i) mass number, (ii) atomic number?

b Figure 27.22 shows schematically the fission of $^{235}_{92}U$. (i) What particles cause the fission? (ii) What are the products of the fission? (iii) What particles (shown as X in the diagram) are also produced in the fission? (iv) Describe what happens when these particles (shown as X) collide with other $^{235}_{92}U$ atoms? (v) What name do we give to the type of reaction shown in the diagram?

c How would you control the reaction once it has started?

d Describe briefly how the energy produced by the fission is converted into electricity.

15 Figure 27.23 shows a cross-section through a nuclear reactor.

a What is the name of the material used as a moderator?

b What is the purpose of the boron rods?

c Name a material which is used in commercial power stations to make the shield.

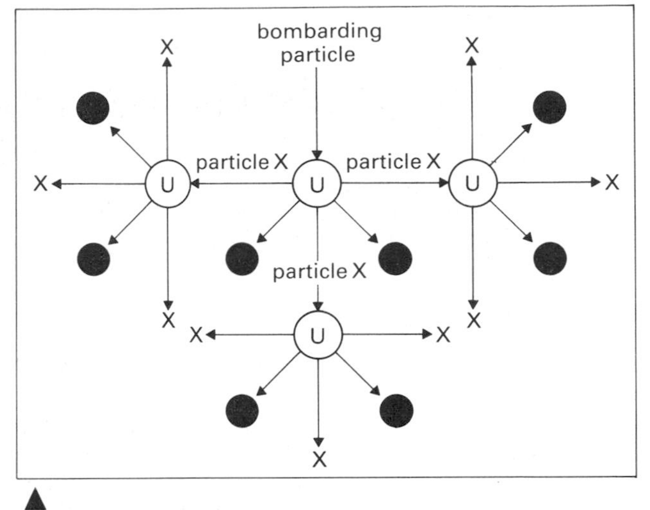

fig. 27.22

d The fuel rods are made from uranium. What are the two common isotopes of uranium and which occurs in most abundance?

e What particle is necessary to change an atom of uranium to two different atoms, and how can a chain reaction result?

f What is the name of the weapon which uses an uncontrolled fission chain reaction?

g Give two similarities in the way a coal-powered ship and a nuclear-powered ship cause the propellers to turn.

h Why would it be difficult to make a nuclear-powered aeroplane which would fly?

i Name two instruments which are sensitive to radioactivity.

j Live trees absorb carbon dioxide and some of this gas contains carbon which is radioactive. How does this fact enable scientists to make measurements which can give the age of an ancient piece of timber?

fig. 27.23

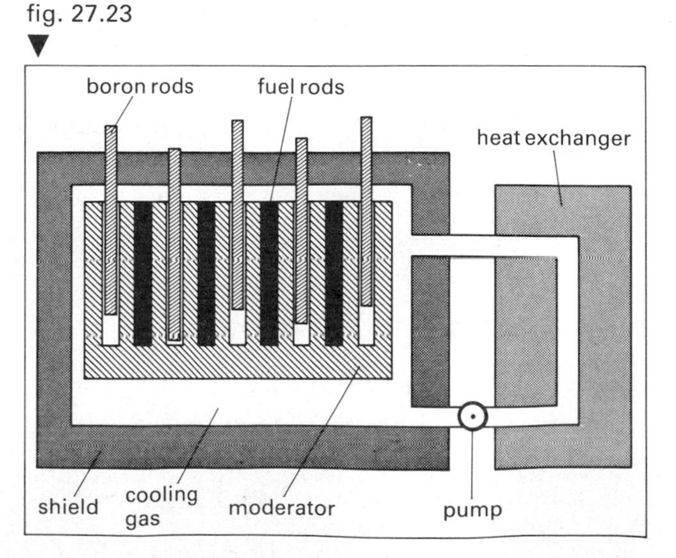

28 PROPERTIES OF MAGNETS

A **magnet** has the property of attracting iron and steel (and special alloys of iron) and to a very much less degree cobalt and nickel. These substances are known as **ferromagnetic** materials. All other substances are for practical purposes **non-magnetic**.

When a magnet is dipped in iron filings it is noticed that most filings cling to the ends of the magnet and very few to the middle. The attraction is greatest at the ends, which are called the **poles**. One of the interesting facts about magnets, which has been known from very early times, is that when they are suspended or pivoted so that they can rotate freely, they always settle down pointing in the same direction, approximately north and south. The same end always swings round to the north and this is called the **north-seeking pole**, or simply the **north pole**. The other end is the **south-seeking pole** or **south pole**, figure 28.2.

Experiments with different poles, figure 28.3, give us a result which is called the first law of magnetism:

Unlike poles attract, like poles repel

Repulsion is the only sure test for a magnet. If we want to distinguish between a magnet and an unmagnetised bar of iron, we can bring the pole of a known magnet to each end in turn of the two specimens. With the unmagnetised bar there will be attraction at both ends, but with the magnet there will be attraction at one end (unlike poles) and repulsion at the other (like poles).

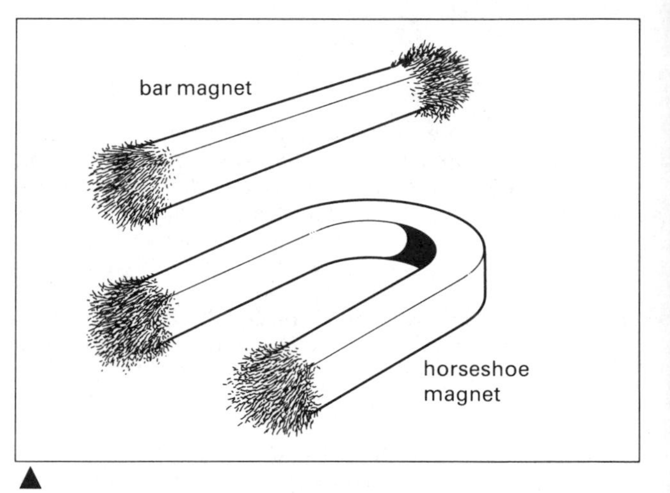

▲ fig. 28.1 Magnets with iron filings to show poles

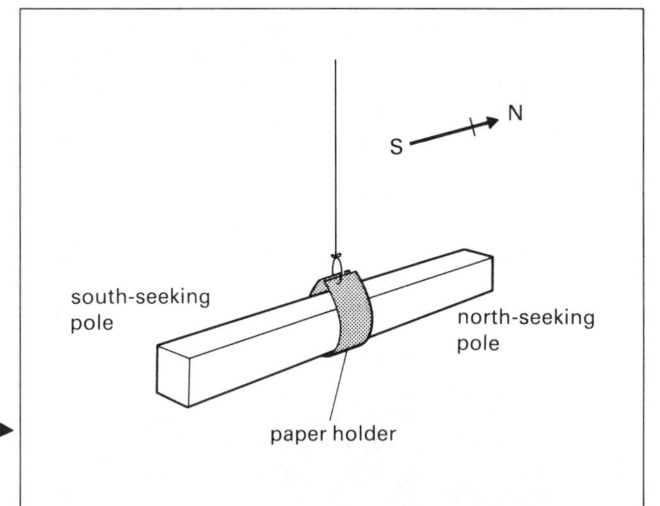

fig. 28.2 A freely suspended magnet comes to rest in an approximate N–S direction ▶

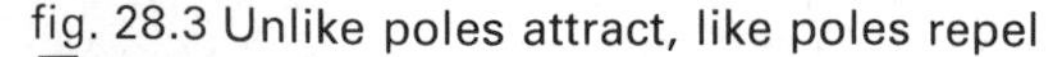

fig. 28.3 Unlike poles attract, like poles repel ▼

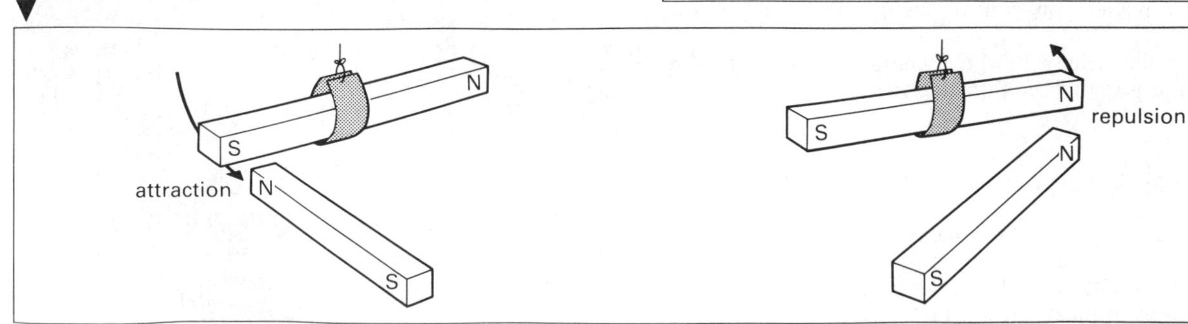

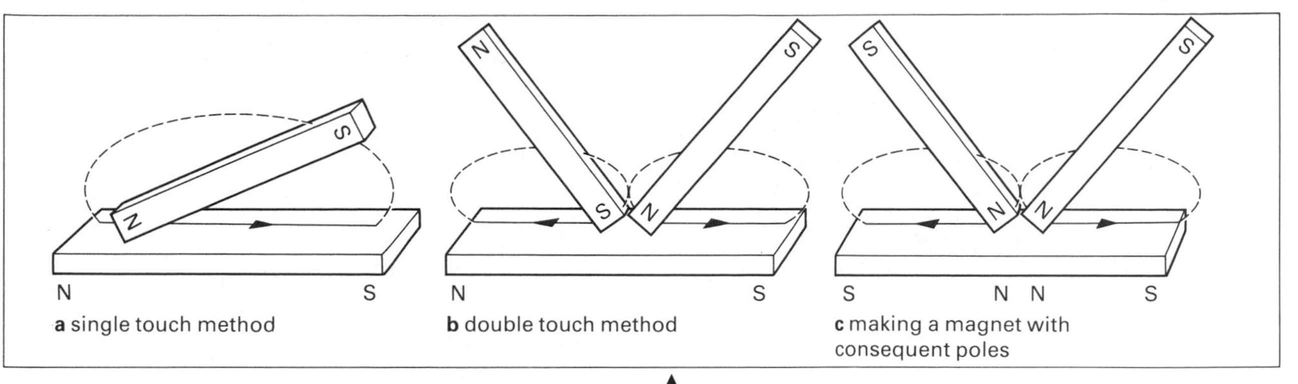

fig. 28.4 The pole at the end at which the stroke finishes is opposite to the magnetising pole

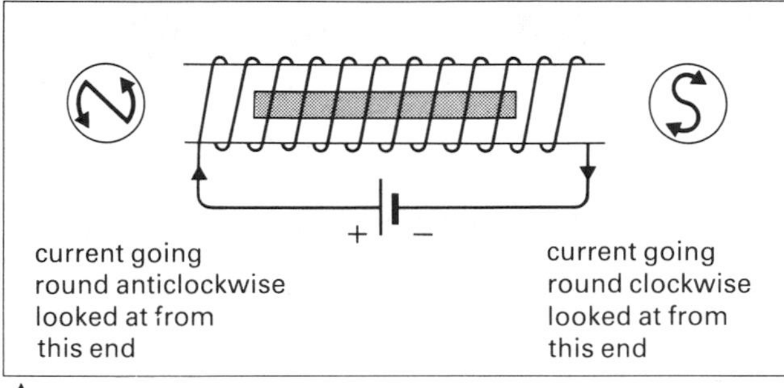

fig. 28.5 Magnetising an iron bar by placing it in a solenoid

fig. 28.6 Magnetisation by induction

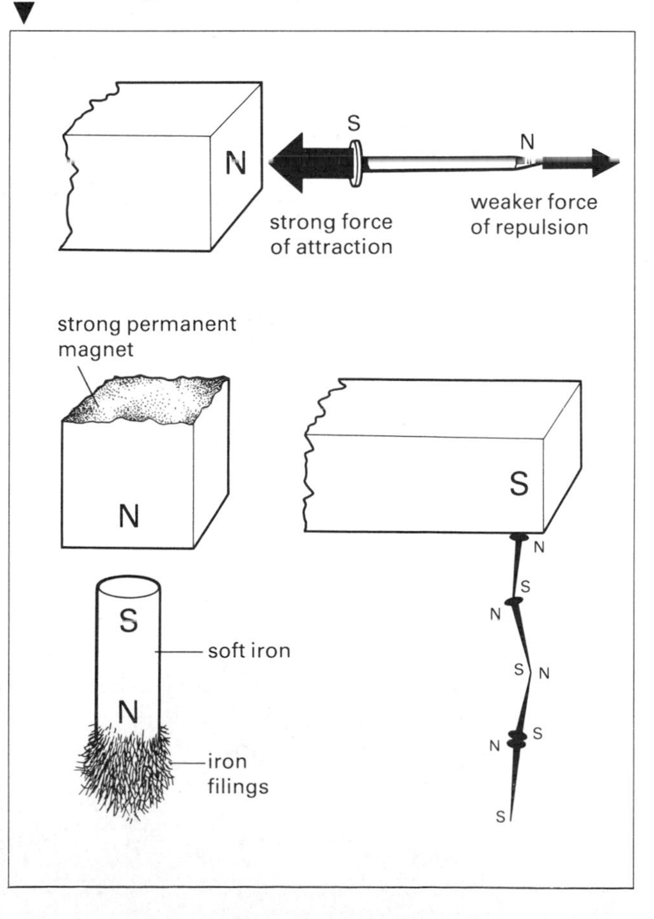

Making magnets

A piece of iron or steel rubbed on a magnet will become magnetised. If you want to magnetise a strip of steel so that there is a known pole at each end you must draw a magnet pole along it **in one direction only**, as shown in figure 28.4a and b.

There is a limit to how strong a magnet can be made. After a time further stroking would produce no further increase in strength. The magnet is said to be saturated.

If two similar magnetising poles are used in the double touch method to magnetise an iron bar we get a magnet with like poles at the ends and an unlike pole in the middle. Such a magnet is said to have **consequent poles**. The result of magnetising with two N poles is shown in figure 28.4c.

There is a close connection between magnetism and electricity (there is more about this in later sections). The best way of making a magnet is to use an electric current. The bar to be magnetised is placed in a **solenoid** (a cylindrical coil of wire) and a direct current passed through, figure 28.5. The pole produced depends on the direction in which the current is flowing round the rod. A clockwise current produces a south pole.

A piece of magnetic material becomes magnetised when placed close to a permanent magnet. We say that magnetism is **induced** in the material, or it becomes magnetised by **induction**. Whether it retains its magnetism or not depends on the material. If the permanent magnet is removed soft iron would lose its magnetism but a piece of steel in a similar situation would retain some magnetism.

Magnetisation by induction, figure 28.6, explains the attraction by a magnet of magnetic materials. Poles are induced and the attraction of the nearer unlike pole is greater than the repulsion of the more distant like pole. Chains of nails can be picked up by a magnet because they are magnetised by induction, figure 28.6.

Magnetic materials

The ferromagnetic materials are iron, cobalt and nickel and various magnetic alloys such as Alnico, an alloy of 56% iron, 19% nickel, 10% aluminium, 10% cobalt and 5% copper. There are hundreds of different magnetic alloys, each having its own special properties. Mixtures of iron and other oxides in powdered form can be magnetised under heat and pressure to form strong permanent magnets. These are ceramic magnets (e.g. ferrites) and are as brittle as pottery.

The chief magnetic properties are

- **susceptibility** or ease of magnetisation,
- **retentivity** which is the ability to retain magnetism.

Soft iron is easily magnetised, but has no retentivity, and so is suitable for the cores of electromagnets.

Steel and many alloys are not so easily magnetised but are very retentive and so are used for permanent magnets.

Figure 28.7 shows small rods of soft iron and hard steel hung from a permanent magnet and their ends dipped into iron filings. Probably more filings will cling to the soft iron than to the steel, showing it is more easily magnetised. When the permanent magnet is removed, the filings will remain clinging to the steel because of its retentivity. Practically all the filings will drop off the soft iron, showing that it has retained very little of its induced magnetism.

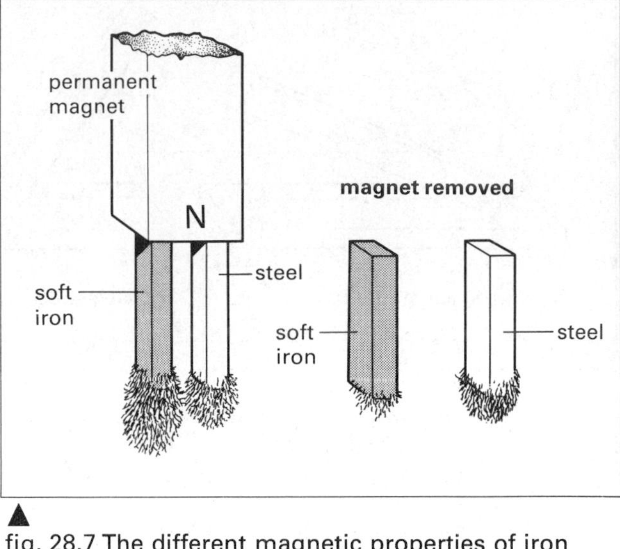

▲ fig. 28.7 The different magnetic properties of iron and steel

fig. 28.8 Two methods of demagnetisation ▼

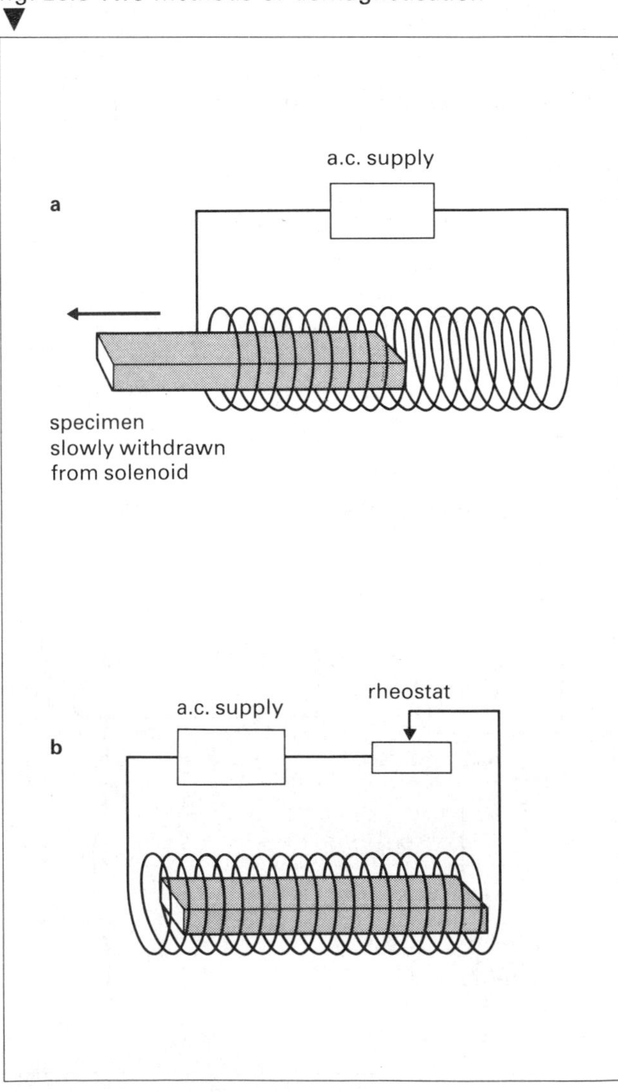

Demagnetisation

It is sometimes necessary to remove the magnetism from a specimen of magnetised material. The simplest way, probably, is to heat the material to red heat and allow it to cool down slowly away from the influence of any magnet. Since the earth itself has a magnetic influence in a north–south direction, the material should cool pointing east and west. Another method of destroying magnetism is to hammer the specimen in a field-free region. This method must obviously not be used for ceramic magnets and is not very effective for most alloys with a high retentivity.

The most satisfactory way of demagnetising is to use an electrical method. Figure 28.8a shows a specimen placed in a solenoid through which an alternating current is passed. The direction of the current is reversing one hundred times a second. The specimen is slowly withdrawn to a distance, then turned round and the process repeated. The specimen will be completely demagnetised.

Another method is to leave the specimen in the coil and slowly reduce the current to zero by means of a rheostat, figure 28.8b. The specimen is thus magnetised first one way and then the other in a gradually weakening field until it becomes demagnetised.

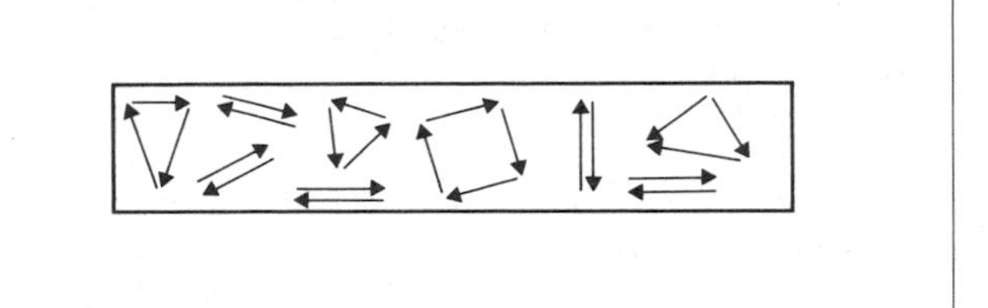

fig. 28.9a Closed chains of dipoles in an unmagnetised bar

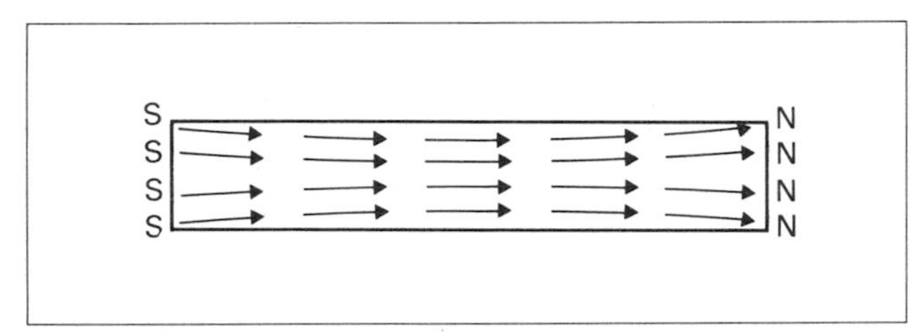

fig. 28.9b Dipoles aligned producing poles in a magnetised bar

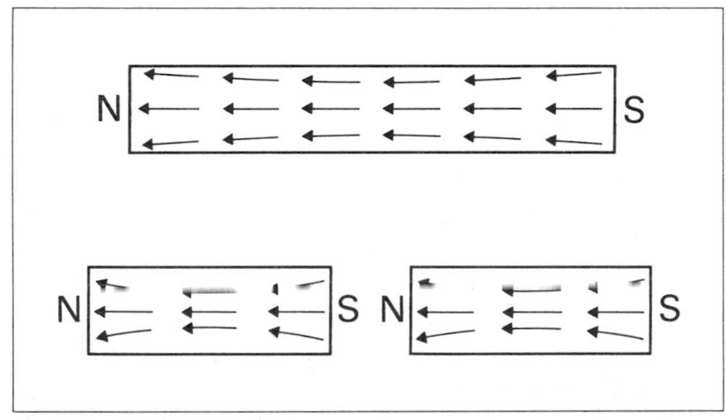

fig. 28.10 Cutting a magnet in two produces two new magnets

Theories of magnetism

In order to explain the difference between an unmagnetised and a magnetised bar of ferromagnetic material we can consider each molecule of the material as behaving like a very small magnet called a **dipole**. In an unmagnetised bar they form closed chains owing to the opposite poles attracting so there are no free poles, figure 28.9a. Or you can think of the molecular magnets as being arranged in a random manner so that there are similar numbers of north and south poles pointing in any given direction. When the bar is magnetised, the closed chains of molecular magnets are broken and the dipoles aligned along the length of the bar with free poles at the ends, figure 28.9b.

This is the simple **molecular theory** of magnetism and it explains many simple facts about magnetism such as

- The single and double touch methods of magnetisation.
- Why the poles appear at the ends of a magnet.
- The destruction of magnetism by heating and hammering.
- The phenomenon of magnetic saturation.
- When a magnet is broken in two, new poles appear on both sides of the break.

The **domain theory** is a more modern theory. A domain is a region of ferromagnetic material in which the dipoles (about 10^{15} molecules) are already aligned in one direction. The size of a domain is obviously much larger than that of a molecule and domains can be made visible under the microscope. The molecules of non-magnetic materials do not form domains. In an unmagnetised piece of iron the domains are randomly arranged, figure 28.11a. When magnetisation is complete, the dipole alignments of all the domains are in the same direction, figure 28.11b.

fig. 28.11a Randomly arranged domains in an unmagnetised bar

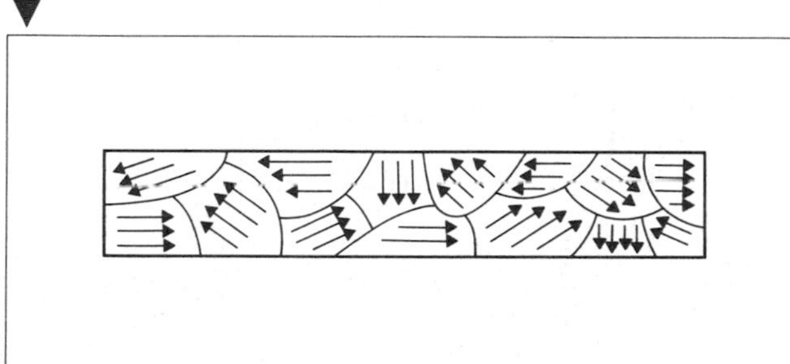

fig. 28.11b Aligned domains in a magnet

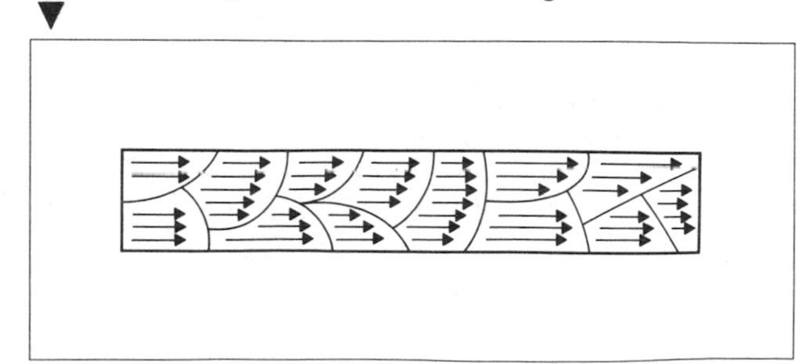

EXERCISE 28

In questions 1–4 select the most suitable answer.

1 Which of the following is a magnetic material?
A manganese
B brass
C magnesium
D iron
E copper

2 A temporary magnet should be made from
A copper
B steel
C tin
D aluminium
E iron

3 In some cases a piece of metal may be permanently magnetised by placing it in a solenoid through which a current is passed. A permanent magnet will be obtained if
A steel is used and direct current in the solenoid
B soft iron is used and alternating current in the solenoid
C copper is used and direct current in the solenoid
D soft iron is used and direct current in the solenoid
E copper is used and alternating current in the solenoid

4 A bar of metal M is suspected of being a permanent magnet. Which of the following tests would best show this?
A Suspend M freely and make it swing. It always comes to rest in the N–S direction.
B Suspend M freely and make it swing. It always comes to rest in the E–W direction.
C Suspend M freely and make it swing. Each time it comes to rest in a different direction.
D Bring one end of M near another magnet. Attraction occurs.
E Bring one end of M near a bar of copper. No attraction occurs.

5 Describe, using a diagram, a method for making a magnet.

6 Describe a method for demagnetising a magnet.

7 a Describe, with the aid of a diagram, how to make a magnet from a bar of steel by stroking with the north pole of a permanent magnet.
b Describe, using the molecular theory of magnetism, what happens inside the bar of steel as it is being magnetised.
c How does breaking a magnet into pieces support this theory?

8 Use the molecular theory of magnetism and labelled sketches to explain how heating a bar magnet destroys its magnetism.

9 Explain, with the aid of a circuit diagram in each case, how you could use a solenoid to
a magnetise a steel bar,
b demagnetise a steel bar.

10 Figure 28.12 shows two *identical* coils, carrying the same current, one with a *hard steel rod* inside and the other with a *soft iron rod* inside. State, with a reason, which of the coils has the soft iron rod inside.

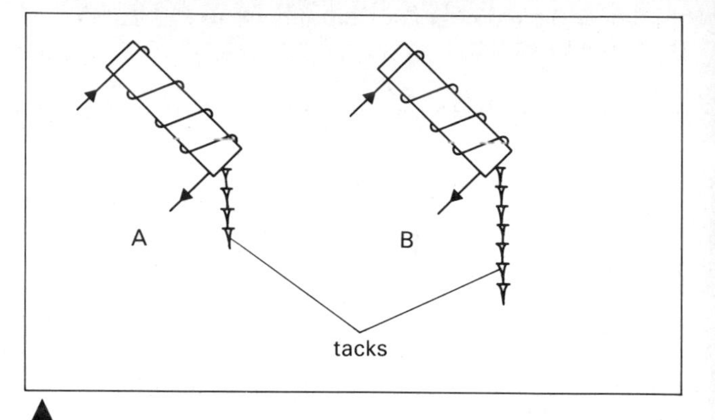

fig. 28.12

11 Pieces of soft iron and steel can be magnetised by holding them near, or touching, a permanent magnet. This is shown in figure 28.13. Assume that the permanent magnets are identical.
a What are the polarities of the ends of the piece of soft iron in A?
b Which of the two, soft iron or steel, would attract most iron filings if these are held near?
c What would happen to the iron filings on the soft iron if the permanent magnet were now taken away from it?
d What would happen to the iron filings on the piece of steel if the permanent magnet were taken away from it?

fig. 28.13

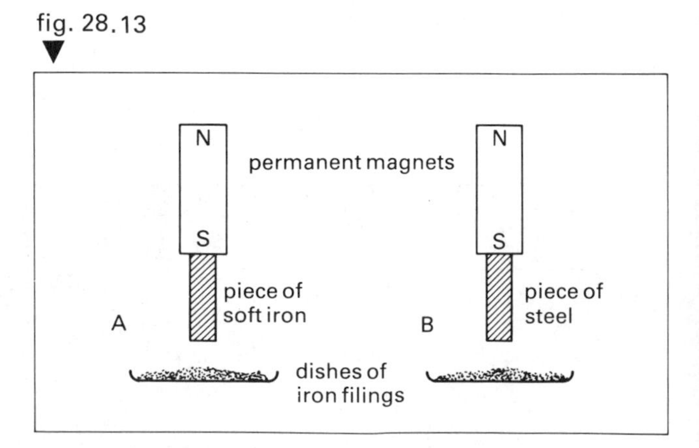

29 MAGNETIC FIELDS

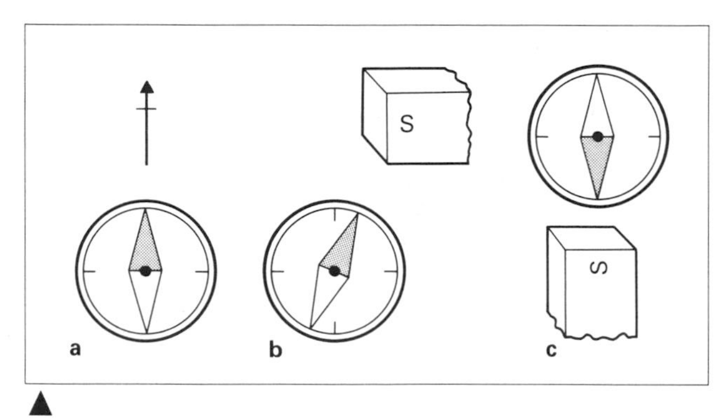

▲ fig. 29.1 Influence of a magnet on a compass needle

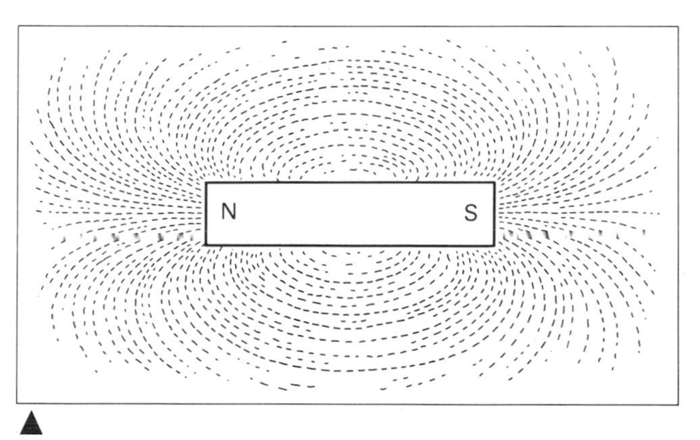

▲ fig. 29.2 A quick method of plotting a magnetic flux pattern

fig. 29.3 A more accurate method of plotting a magnetic flux pattern ▼

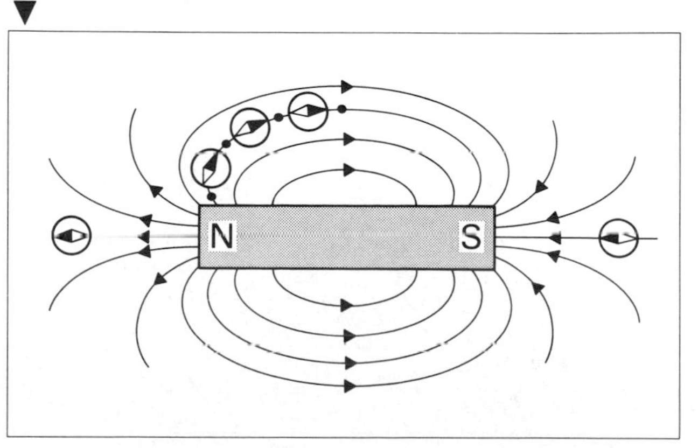

A small compass needle always settles with its north-seeking pole pointing towards the north, figure 29.1a. A magnet some little distance away will deflect it from its north–south position, figure 29.1b. If the magnet is brought closer, you can turn the compass needle round completely, figure 29.1c. The magnetic force due to the magnet is much stronger than the magnetic force due to the earth.

The space round a magnet in which it exerts a force is known as its **magnetic field**. The field is strong close to the magnet and gets weaker as we move away. At great distances it is negligible.

The existence of a magnetic field can be very simply demonstrated. Lay a magnet on a bench. Place a piece of stiff paper or card over the magnet. Sprinkle on fine iron filings from a pepper pot. Tap the card lightly so that the filings set along the lines of force. A pattern will appear as shown in figure 29.2. Note: this is not a very accurate method.

A magnetic field can be mapped out by **lines of force** which show the direction of the magnetic force, i.e. the direction in which a compass needle points. The idea of lines of force was devised by Michael Faraday and is an extremely useful one to help us in our study of magnetism and electricity. The total number of lines of force passing through any area is called the **magnetic flux** and diagrams of magnetic fields are known as **magnetic flux patterns**.

A complete line of force starts from the north pole of a magnet and ends at the south pole. The arrow on a line of force shows the direction in which an imaginary free north pole would move. Lines of force can never cross.

Magnetic fields can be plotted accurately, using a small compass, figure 29.3. Place the magnet on a sheet of drawing paper. Draw round the magnet and letter the poles. Place the compass needle near the magnet and mark its point with a dot. Move the compass so that its other end is at the dot and again mark the point. Continue until you reach the magnet again or come to the edge of the paper. Join the points with a smooth line. Draw as many lines of force as you can.

The combined fields of two magnets are shown in the diagrams in figure 29.4. The points marked X are where the forces due to the two magnets are exactly equal and opposite. They are **neutral points**.

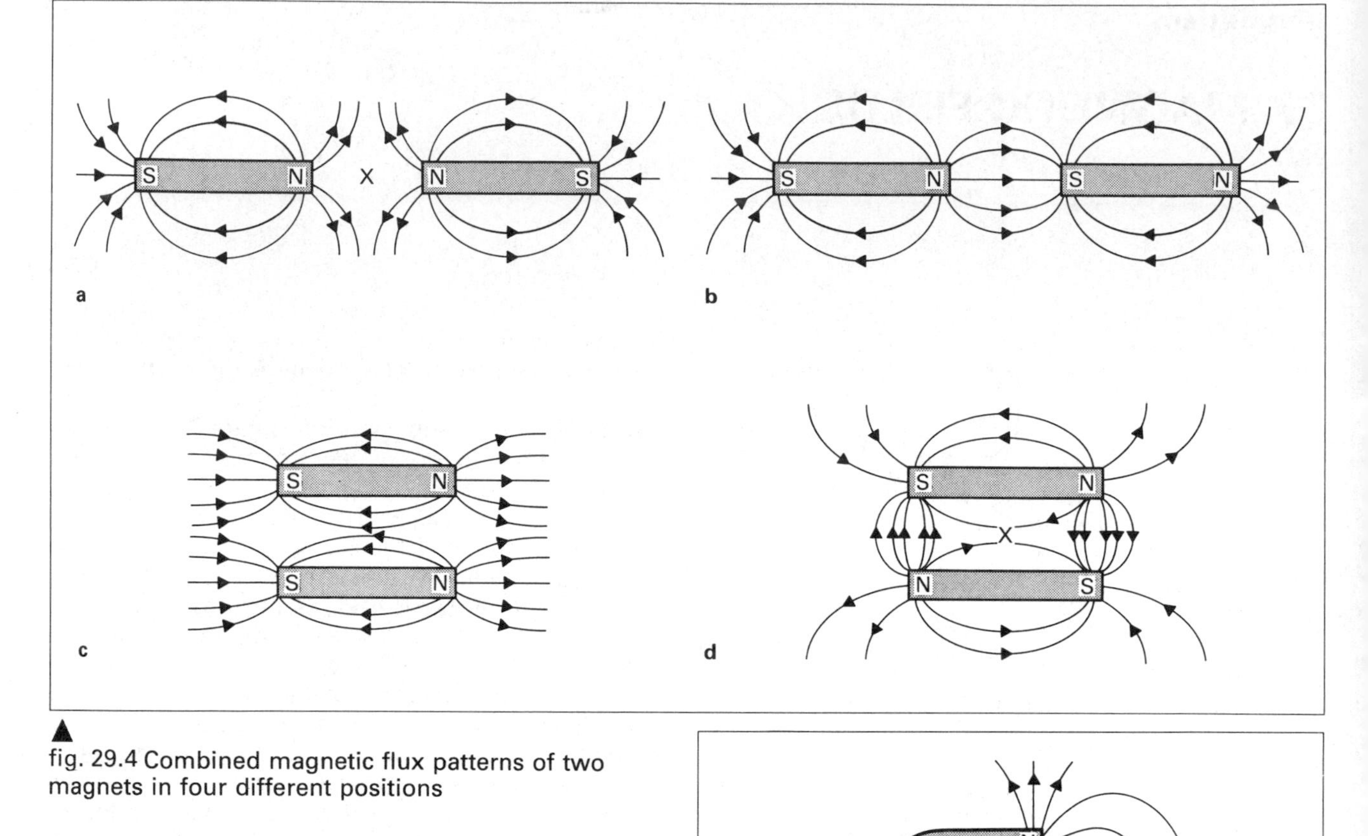

fig. 29.4 Combined magnetic flux patterns of two magnets in four different positions

A keeper is a piece of soft iron placed across the poles of a horseshoe magnet or between the unlike poles of a pair of bar magnets. The magnetic field passes through the soft iron and tends to keep the domains lined up so preventing the magnet from losing its magnetism, figure 29.5b.

fig. 29.5 Magnetic flux pattern of a horseshoe magnet without and with a soft iron keeper

fig. 29.6 Magnetic flux pattern due to magnets and pieces of iron. There is no field inside the ring, which is acting as a **magnetic shield**

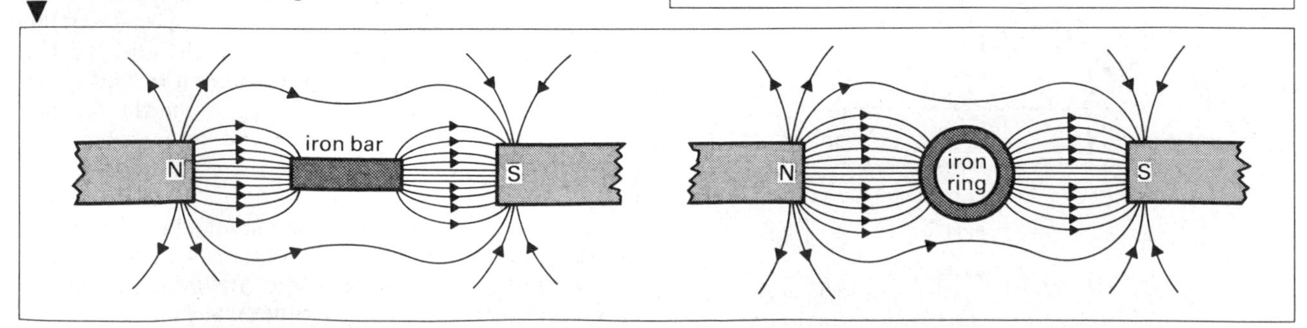

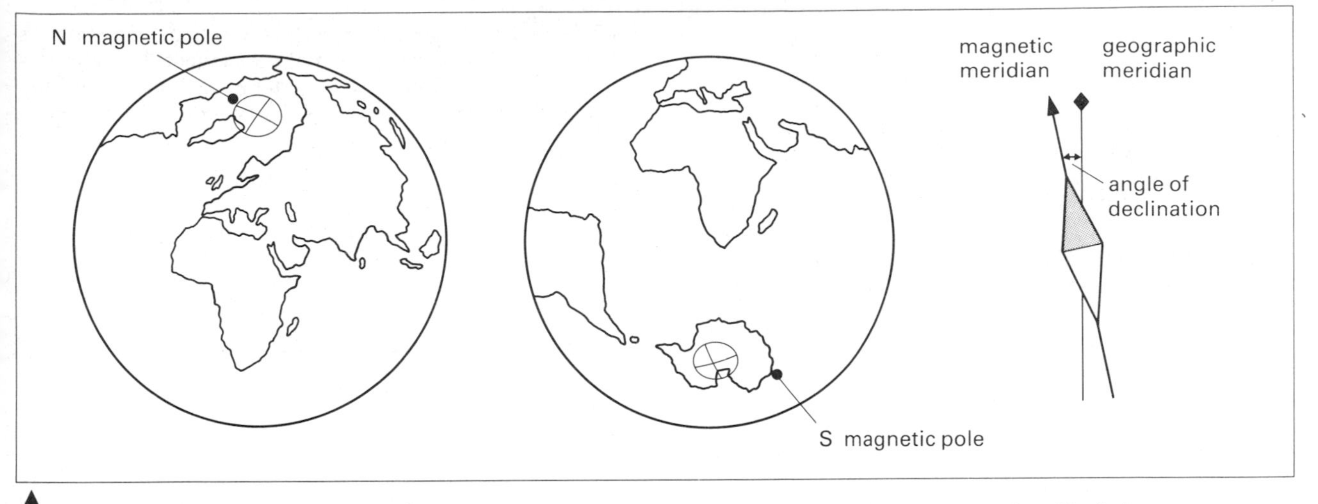

fig. 29.7 Magnetic declination

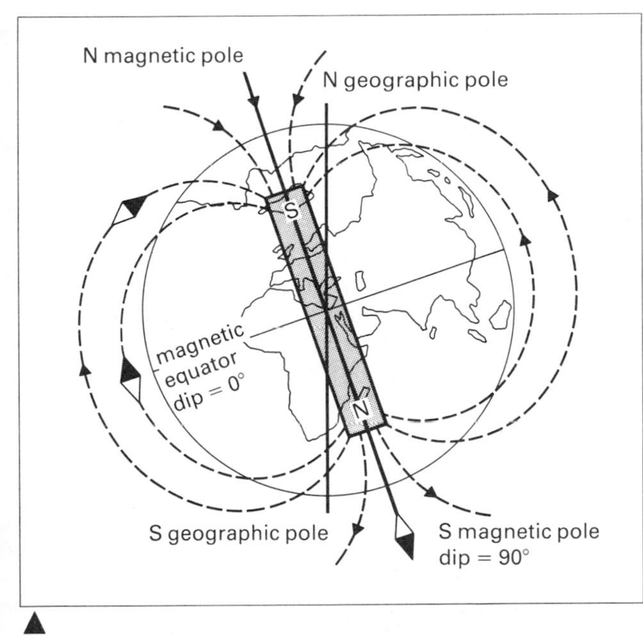

fig. 29.8 Imagine there is a magnet inside the earth

fig. 29.9 Angle of dip

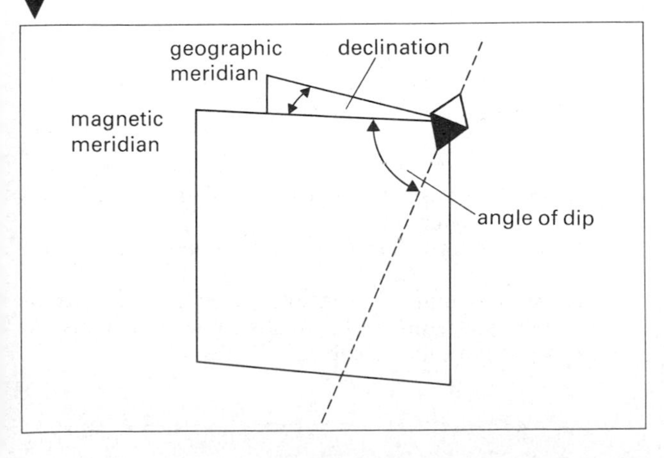

The earth's magnetic field

A pivoted magnet sets itself north and south because the earth itself has a magnetic field. If the earth's field were plotted over a small area it would be a set of parallel lines pointing to the magnetic north. The direction in which a magnet points at any place on the earth's surface is the **magnetic meridian**. The true north–south direction at a place on the earth's surface is the **geographical meridian**. At most places these two directions do not coincide. The angle between them is the **magnetic declination** (or **variation**), figure 29.7.

In England the magnetic meridian is about 4° west of the geographical meridian. At certain places in North and South America there is no variation, while on the coast of Australia the magnetic meridian is to the east of the geographical meridian. To make matters more complicated still, the value of the declination at any place changes from year to year. The navigator of a ship or an aeroplane using a magnetic compass to set his course must know the value of the declination. Magnetic maps are drawn showing the value of the declination at all points on the earth's surface. Modern methods of navigation depend on gyro-compasses which are not affected by the earth's magnetic field and on the use of radio beacons.

A simple picture of the earth's magnetic field can be obtained if we imagine that there is a large magnet situated inside the earth and making a small angle with the earth's axis, figure 29.8. This is a simple and convenient idea, and is certainly not true. The cause of the earth's magnetic field is not understood. It may be due to electric currents circulating in the earth itself.

Since a freely-suspended compass needle sets along the lines of force, you will see from the diagram that it will also tilt as well as point to the magnetic north. The angle the needle makes with the horizontal is called the **angle of dip**, figure 29.9. The value in England is about 70°.

EXERCISE 29

In questions 1–3 select the most suitable answer.

1 The magnetic field pattern shown in figure 29.10 is produced by bringing together
- **A** N pole and S pole
- **B** N pole and N pole
- **C** S pole and S pole
- **D** S pole and N pole
- **E** S pole and unmagnetised iron bar

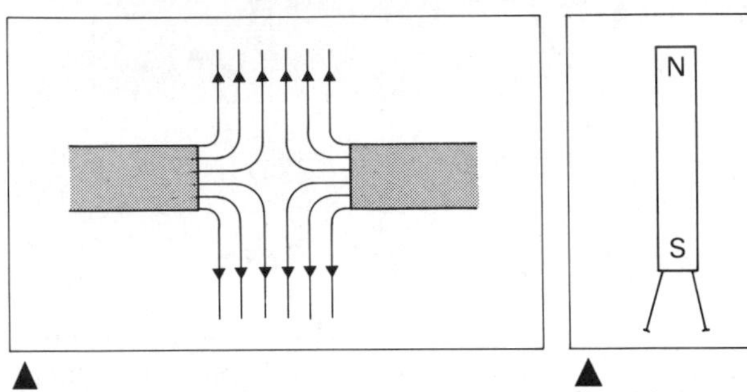

fig. 29.10

fig. 29.11

2 Figure 29.11 shows two soft iron nails hanging on the end of a magnet. Which one of the following statements about the arrangement is most likely to be correct.
- **A** The free ends of the nails are N poles.
- **B** The free ends of the nails are attracting one another.
- **C** The nails have become permanently magnetised.
- **D** The nails are induced temporary magnets.
- **E** The nails will only hang like this from the S pole of the magnet.

3 Magnetic variation (declination) is
- **A** the varying strength of the earth's magnetic field
- **B** the angle between magnetic north and true north
- **C** the angle between the earth's magnetic field and the horizontal
- **D** the angle between the earth's magnetic field and the vertical
- **E** due to the varying composition of steel

4 Figure 29.12 shows two bar magnets with the north poles marked. Draw lines of force to show the field due to the magnets. Indicate the direction of each line of force with an arrow. Dotted lines must not be used to represent lines of force. Mark any neutral points with a cross.

fig. 29.12

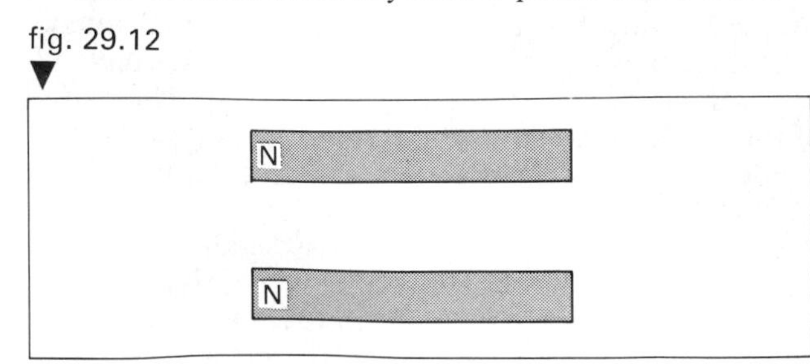

5 The circles in figure 29.13a show positions of a compass placed in the magnetic field of the magnet. In the circles mark the direction of the needle in each case. Do likewise for the coil in figure 29.13b in which the current is flowing in the direction of the arrows.

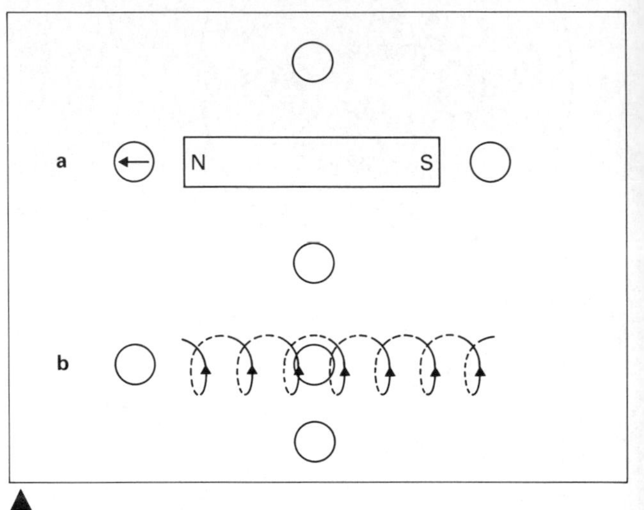

fig. 29.13

6 Figure 29.14 shows 20 turns of wire on a C-shaped core. A direct current flows through the coil.
- **a** Sketch the pattern of the magnetic field you might expect to see when iron filings were sprinkled on a piece of card placed over A and B.
- **b** What difference, if any, in the shape of the field would be noticed if the current were doubled?

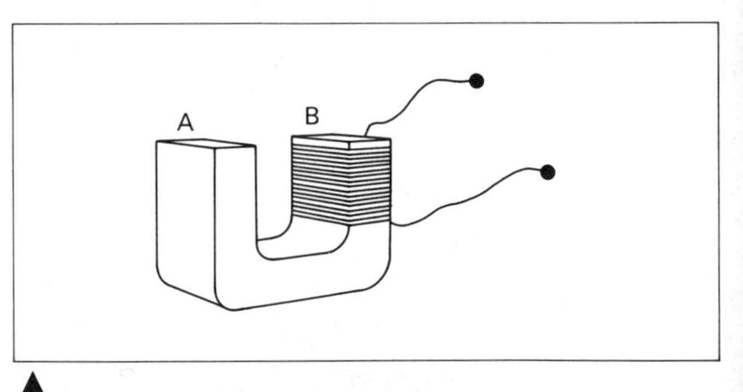

fig. 29.14

7 Explain the terms geographical meridian, magnetic meridian and declination.

How would you measure the declination in a given locality?

A ship is sailing 45° right of true geographical north at a place where the declination is 10° west. What is the reading of the ship's compass?

30 ELECTRIC CHARGES AND THEIR FIELDS

In very early times (about 600 BC) it was noticed that certain substances when rubbed with cloth or fur attracted light objects to them. One substance which showed this property strongly was amber. This is a yellow, glass-like solid formed from the resin of now extinct cone-bearing trees. Modern plastics behave in a similar way. If you rub your pen or a piece of Perspex on your sleeve it will then pick up small pieces of paper or sawdust. The Greek word for amber is *elektron*. Substances like this were said to be electrified or charged with electricity after rubbing. Since the charge stays on the rubbed rods and does not move this kind of electricity is called **static** electricity. Its study is the branch of physics known as **electrostatics.**

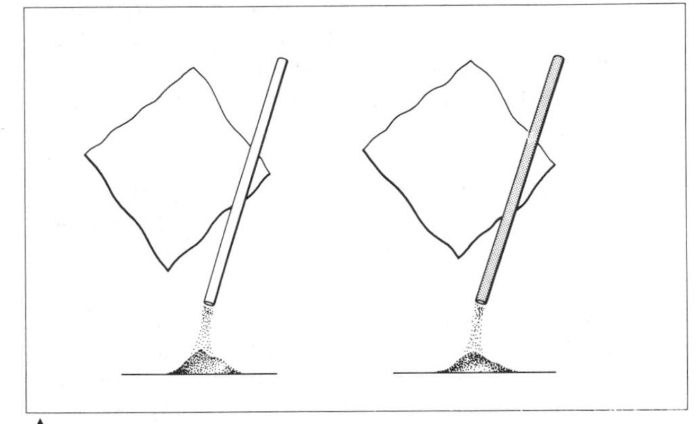

▲
fig. 30.1a A polythene rod rubbed with cloth attracts small particles of sawdust. A rubbed acetate rod also has the power of attraction

Attraction and repulsion

Rods of cellulose acetate and polythene become charged when rubbed with a dry cloth, figure 30.1a. It was found in the eighteenth century that there appeared to be two kinds of electric charge. If two charged acetate rods or two charged polythene rods are brought close together they repel, figure 30.1c, while charged acetate and charged polythene rods attract, figure 30.1b, No material has been found which when charged repels itself and also repels both acetate and polythene. We can conclude there are only two kinds of charges.

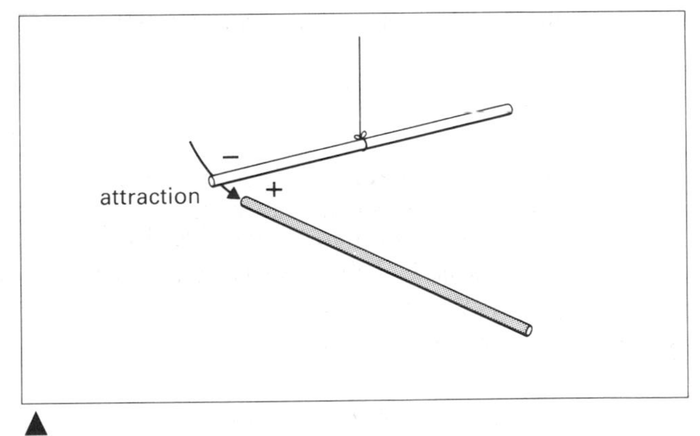

▲
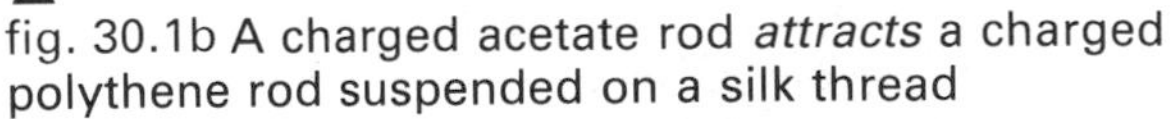
fig. 30.1b A charged acetate rod *attracts* a charged polythene rod suspended on a silk thread

Like charges repel, unlike charges attract

The rule is the same as that for magnetic poles. To distinguish between the two kinds of charge one was called positive and the other negative. The original choice was purely by chance. It turns out that acetate (or Nylon) is positively charged and polythene negatively charged.

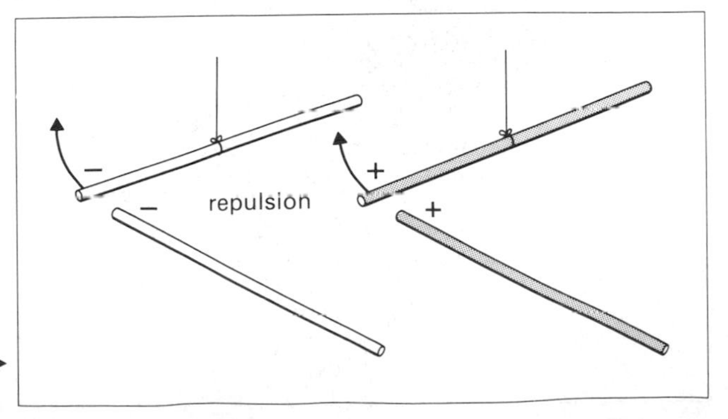

fig. 30.1c Two charged polythene rods or two charged acetate rods *repel* each other ▶

Explanation of charges in terms of electrons

Normal atoms, and the material they make up, are neutral because there are equal numbers of positively charged protons and negatively charged electrons. The outermost orbital electrons are not always very tightly bound to their atoms. When certain substances come in contact, weakly-held electrons transfer from one substance to the other where they are more strongly attracted. One substance loses electrons and the other substance gains electrons. One substance becomes positively charged (it has more protons than electrons) and the other substance becomes negatively charged (it has more electrons than protons).

In the case of the polythene rod in figure 30.2, it gains electrons from the cloth and becomes negatively charged. The cloth at the same time becomes positively charged. Since the same numbers of electrons are lost and gained the two charges are equal in magnitude.

The process used to be called (and often still is) 'charging by rubbing' or 'charging by friction', but it is really charging by contact. The process of rubbing produces more points of contact between the two substances and so increases the exchange of electrons.

Negatively-charged bodies have an excess of electrons

Positively-charged bodies have an electron deficiency

Conductors and insulators

If we try to charge a metal rod held in the hand by rubbing it with a cloth, we get no result. But if the metal is mounted on and held by Perspex, for example, it can be charged. The point is that metals are **conductors**. This means electric charges can move freely in and on them. When a metal rod is held in the hand any charge produced on it flows to the hand, itself a conductor, and through the body to earth. Conductors are materials with free electrons in them. Very little energy is required to free some electrons from their atoms and metals at ordinary temperatures have received sufficient energy to free some electrons. **Insulators** have hardly any free electrons and charges cannot move through them. A conductor which has an electron deficiency (positive charge) at one part of it can make up this deficiency by supplying electrons from elsewhere so that the charge is the same all over the conductor.

Metals, carbon, water and some other liquids are conductors. Plastics, ceramics, wax and rubber are insulators. If insulators get a film of water on the surface they will lose charge. All materials used for electrostatic experiments should be thoroughly dry.

The gold leaf electroscope

The electroscope, figure 30.3, is used for detecting charges. If a charged polythene rod touches the top metal plate (cap) both the leaf and brass stem become negatively charged and the leaf is repelled.

When a positive charge is brought near the cap of a negatively charged electroscope the deflection of the leaf is reduced. The approach of a negative charge would increase the deflection. Thus we can distinguish between positive and negative charges.

Induced charges

When a piece of soft iron is placed near a magnet, poles are induced in the soft iron. Something very similar happens with electric charges. When a nega-

fig. 30.2 The rod becomes negatively charged. The cloth becomes positively charged ▼

fig. 30.3 Gold-leaf electroscope ▼

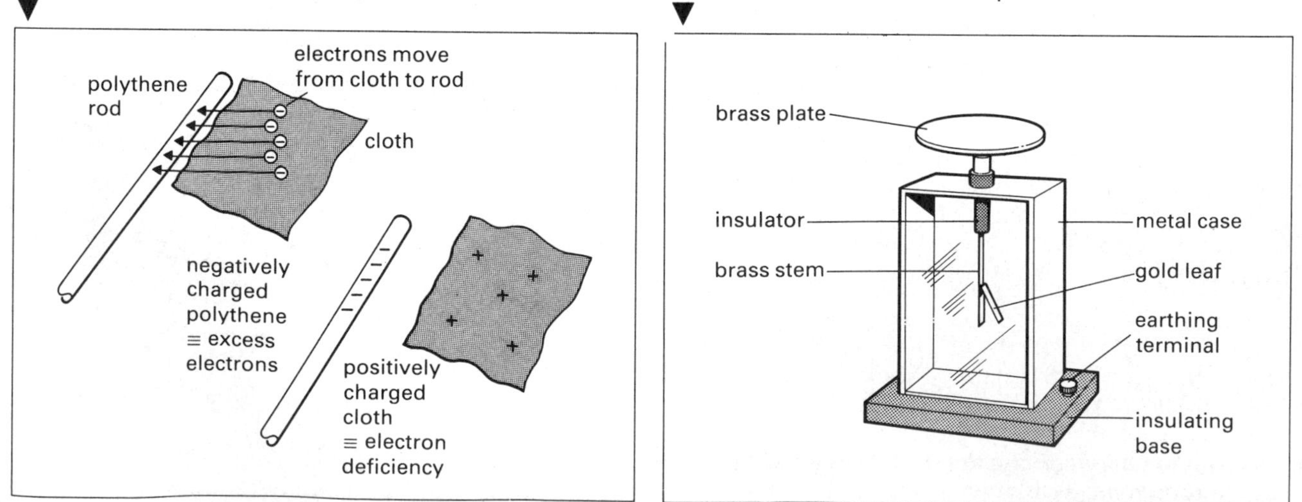

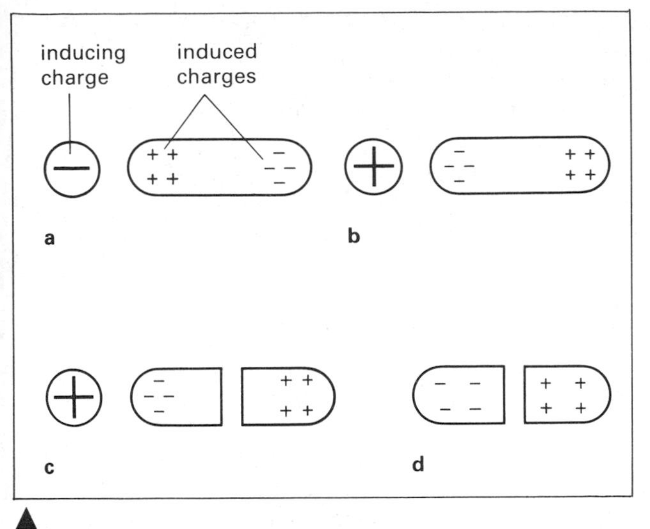

fig. 30.4 Induced charges

tive charge is placed near a conductor, free electrons in the conductor are repelled to the far end which then has an excess of electrons and so a negative charge. The deficiency of electrons at the near end is equivalent to a positive charge, figure 30.4a. The reverse occurs when a positive charge is used, figure 30.4b.

If the conductor could be cut in two, figure 30.4c, these induced charges could be separated, and the two halves would still be charged after the inducing charge was removed, figure 30.4d. The two induced charges are equal in magnitude and opposite in sign. Induced magnetic poles cannot be separated in this way because it is impossible to have an isolated north or south pole.

As with magnetism, the production of induced charge in an uncharged object takes place before the object is attracted.

Charging by induction

Experiments in electrostatics are often carried out with conductors mounted on insulating stands. The conductors could be made of wood covered with aluminium foil.

A conductor can be charged by induction in the following way. A charge is held close to the conductor. The charge could be a rubbed polythene rod. The induced charges separate to opposite ends of the conductor, figure 30.5a.

With the inducing charge still in place, the conductor is earthed. This can be done quite simply by touching the conductor with your finger. The excess electrons producing the induced negative charge are repelled to earth, figure 30.5b.

With the inducing charge still in place, the earth connection is removed, figure 30.5c.

The inducing charge is now removed. The remaining positive charge flows over the whole of the conductor, figure 30.5d.

Notice that the final induced charge is of opposite sign to the original inducing charge. If we want to charge a conductor negatively we must use a positive inducing charge, e.g. a rubbed acetate rod. Work out with a series of diagrams what happens in this case. When the conductor is earthed, electrons will flow from earth to neutralise the induced positive charge.

Electric fields

Just as there is a magnetic field round a magnet or between two magnetic poles, so there is an **electric field** round a charged body or between two charges. Electric fields are mapped out with lines of force, and the direction of the line of force is taken as the direction in which a positive charge would move, figure 30.6.

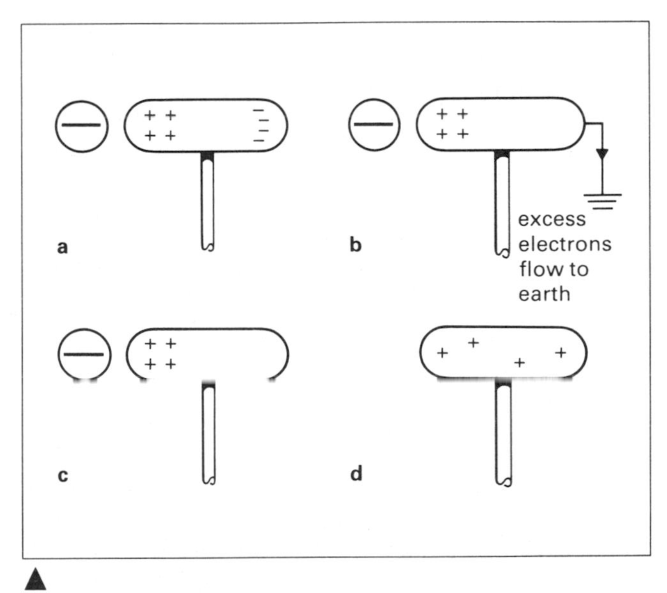

fig. 30.5 Charging by induction

fig. 30.6 The radial electric field of a negatively-charged sphere

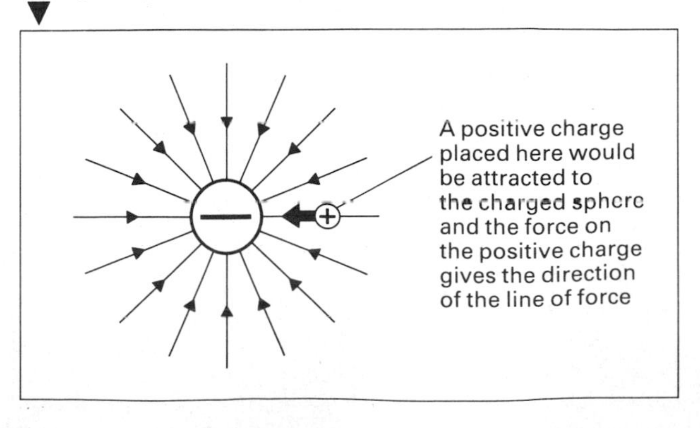

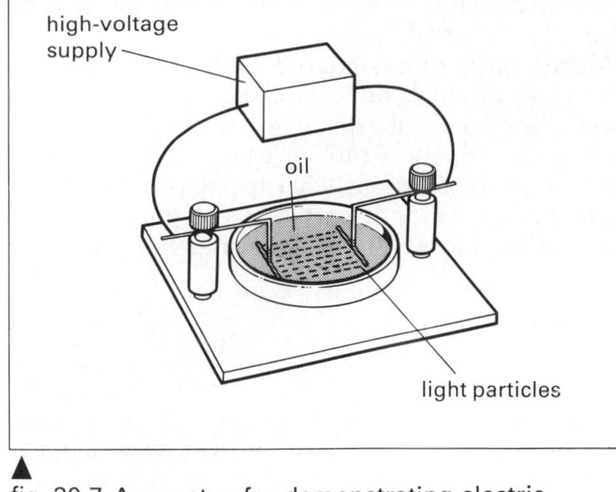

fig. 30.7 Apparatus for demonstrating electric fields. If the apparatus is placed on an overhead projector, the field can be seen on the screen

Electric fields can be demonstrated by light-weight floating particles in a liquid. Grass seeds or semolina grains floating on a thin oil can be used, figure 30.7. On connecting the electrodes to a high-voltage unit, the particles are charged inductively and set along the lines of force.

Michael Faraday, who developed the ideas of fields and lines of force in the nineteenth century, found it useful to think of these imaginary lines as having certain properties. These properties could be used to give explanations of various observations.

- Lines of force start and finish on equal and opposite charges.
- They behave as if they were trying to contract like stretched elastic.
- Lines in the same direction repel each other sideways.
- Parallel lines of force indicate a uniform field and the stronger the field the closer the lines of force.

Figures 30.8 and 30.9, showing the electric field lines for two opposite charges and two like charges, illustrate the idea of contracting lines of force as producing attraction and the sideways repulsion as producing a repelling force.

fig. 30.8 Electric field for two opposite charges

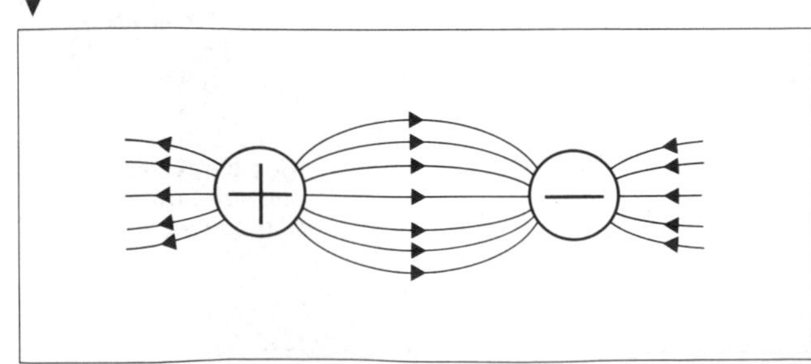

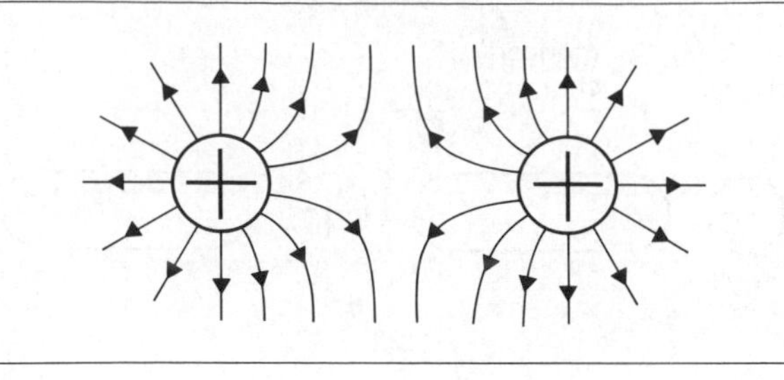

fig. 30.9 Electric field for two like charges

Potential difference

The electric field between two metal plates is uniform except at the edges. A field like this is used in the cathode ray oscilloscope (see p. 328) to deflect electrons.

If an electron were moved from plate A to plate B, in figure 30.10, work would have to be done against the force of repulsion between the negatively-charged electron and the negatively-charged plate B. This work would be stored in the electron (as long as it remains at plate B) as potential energy. Similarly, if a positive charge were moved from plate B to plate A, against the direction of the electric field, it would gain in potential energy. Because of this change in potential energy we say there is a 'difference of potential' or a **potential difference (p.d.)** between the plates.

The difference of potential between the two charged conductors is a way of measuring the strength of the electric field between them. A large potential difference produces a large electric field strength and so a large force on any charge placed in the field.

The unit in which potential difference is measured is the **volt (V)**.

There is a potential difference of 1 volt between two points when 1 joule of work has to be done to move 1 coulomb of charge between them (see p. 273)

fig. 30.10 Electric field between two metal plates

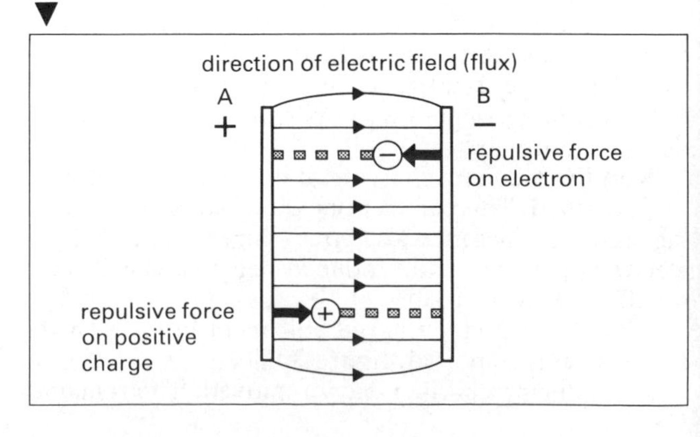

Just as heights are measured from an arbitrary zero of sea-level, and temperatures are measured from an arbitrary zero of the melting point of ice, so we have an arbitrary zero of potential. The earth is taken to be at zero potential. The idea is that the earth is such a large conductor that the comparatively small extra charges that we may give it or remove from it will make no difference to its electrical state. Positively charged bodies are above earth potential, negatively charged bodies are below earth potential. Usually we are not concerned with the absolute potential of a charge, but are more interested in the difference in potential between two charges.

The action of charged points

When a conductor is charged, the charge is distributed so that there is greater concentration at places of greater curvature, figure 30.11. The curvature at a point is very great (theoretically it is infinite), so any charge on a conductor tends to concentrate at points, figure 30.12. This high charge produces a strong electric field in the neighbourhood of the point. This strong field is often sufficient to ionise the atoms of gases in the air. Ions of the opposite charge to that of the point are attracted and neutralise the charge on the conductor, which rapidly discharges. Ions of the same charge as that of the point are repelled, and as they steam away produce a kind of 'electric wind'. This produces a bluish glow which can sometimes be seen round power line pylons on a very dark night.

If a pointed object is directed towards a charged conductor, it has the effect of discharging it, figure 30.13. Induced charges are produced with a high density of charge of the opposite sign to that of the conductor at the point. Movement of ions takes place in the strong electric field, with the result that the conductor is discharged and the pointed object itself acquires charge.

The discharging action of points is made use of in the lightning conductor, figure 30.14. During thunderstorms clouds become highly charged with electricity. The charge on the lower part of the cloud

fig. 30.11 Charge distribution on different shapes of conductor

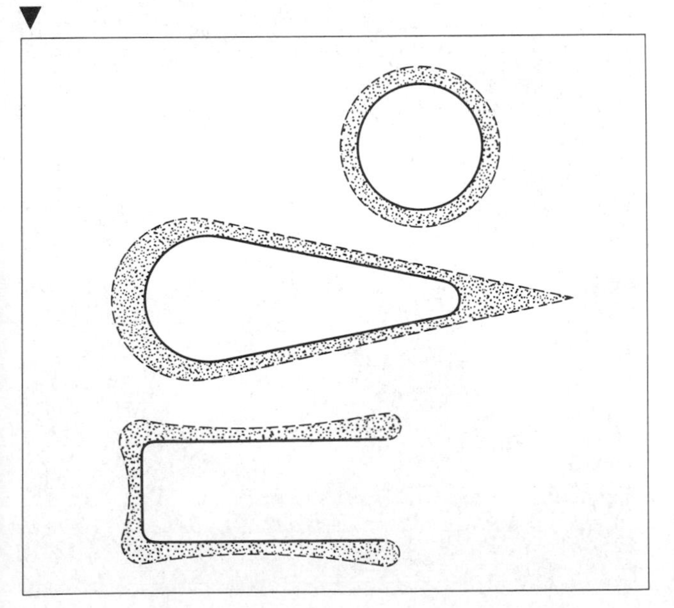

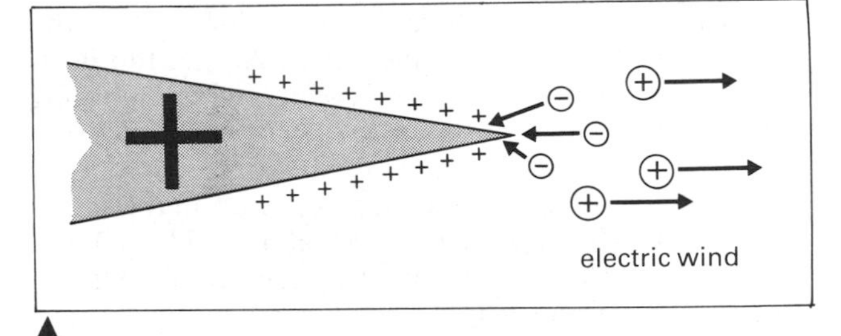

fig. 30.12 Charge concentration at a point

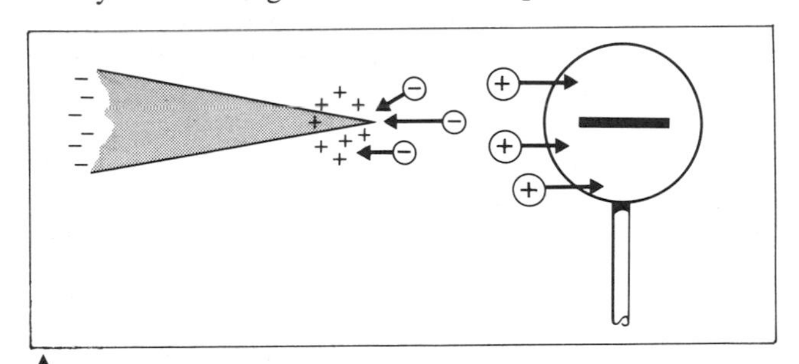

fig. 30.13 Charge collection by a point

fig. 30.14 The lightning conductor

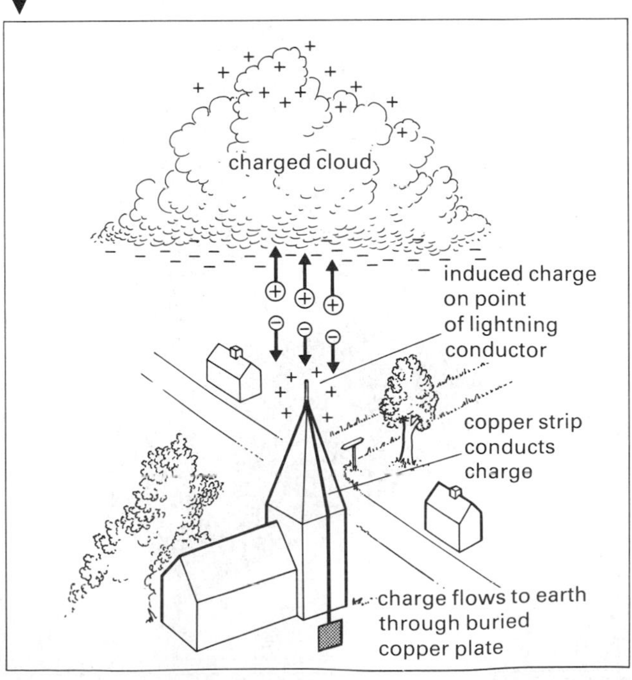

induces an opposite charge on points high above the earth's surface. After a build-up of charge, the insulation of the air can break down and a lightning 'strike' takes place. With a lightning conductor the induced charge concentrates on the points of the conductor and produces a discharging effect on the lower part of the cloud as described above. The charge collected by the conductor is conducted to a large metal plate buried in the earth.

Electrostatic generators

The idea of replacing the simple 'rubbing-a-rod' method of producing electrostatic charges by a mechanical device was thought of very early in the history of electricity. Most of the early machines consisted of glass cylinders which were rotated so that they rubbed against silk pads. One of the most successful machines used the *charging-by-induction* principle. This was the Wimshurst machine, invented about 1830, figure 30.15. The most modern of the electrostatic generators was designed by Robert Van de Graaff in America in 1931, figure 30.16. The idea was to produce very high voltages to accelerate charged particles in the 'atom smashing' experiments which were just beginning to take place.

fig. 30.15 Wimshurst machine. One knob gives a positive charge, the other a negative charge

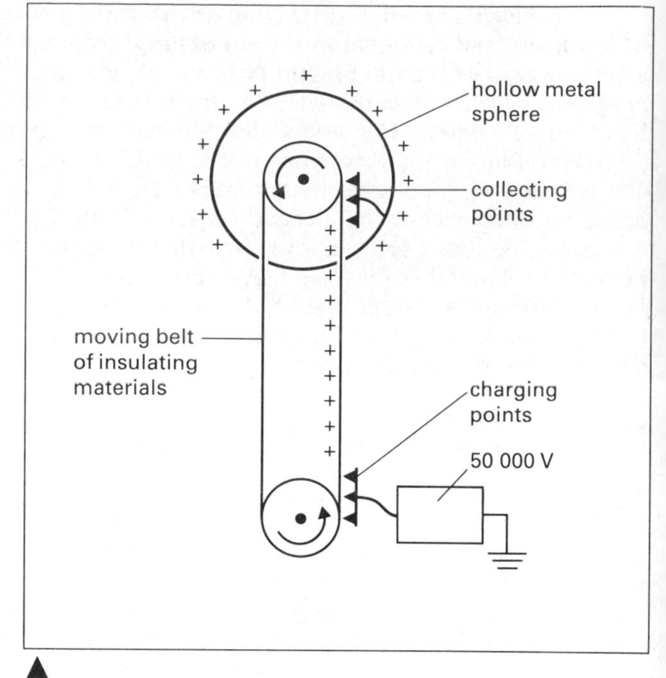

fig. 30.16 Principle of the Van de Graaff generator

fig. 30.17 Large Van de Graaff generators such as this one can generate up to 6 million volts

The generator consists essentially of an endless belt driven by an electric motor. A set of points near the bottom of the belt is connected to a high-voltage supply. Charges are sprayed on to the belt by the point action and these are carried upwards where they are collected by another set of points inside a hollow metal sphere. The charge on the sphere builds up and the final voltage may be several million. Even small school laboratory Van de Graaff machines can produce impressively large sparks, but these are not dangerous because, although the potential is high, the quantity of charge involved is relatively small.

▲
fig. 30.18 A hand-driven Van de Graaff generator for the small laboratory

Capacitance and capacitors

If the same charge is given to two differently-sized conductors they are not raised to the same potential.

Experiments with different sized conductors led to the idea of the 'capacity' of a conductor to store charge. A large conductor could hold more charge before it was raised to a given potential or had a given amount of potential energy. We now use the word **capacitance** to decribe this property.

The capacitance of a conductor is defined as the quantity of charge required to raise the potential of the conductor by one unit

$$\textbf{capacitance} = \frac{\textbf{charge}}{\textbf{potential}}$$

$$C = \frac{Q}{V}$$

The unit of capacitance is the **farad** (**F**), which is the capacitance of a conductor which requires 1 coulomb of charge to raise its potential by 1 volt. The farad is a very large unit of capacitance, and the smaller units microfarad (10^{-6} F or μF) and picofarad (10^{-12} F or pF) are more often used.

The capacitance of a conductor depends not only on its dimensions but also on the presence of neighbouring conductors. Any arrangement by which the capacitance of a conductor is artificially increased is called a **capacitor.** In its simplest form a capacitor consists of a charged plate with an earthed plate close to it. The capacitance of a capacitor depends on the area of the plates (directly), their distance apart (inversely) and on the material, or **dielectric,** between them. Capacitors form part of all electronic circuits.

In order to achieve a large plate area and have the plates close together capacitors are made from sheets of tinfoil separated by sheets of paper, mica or polyester. The sheets are then rolled into a cylinder so that they do not take up much room. In **electrolytic** capacitors the dielectric is a thin film of aluminium oxide formed from a solution of aluminium borate. The connections to an electrolytic capacitor are marked + and − and it is essential that it is connected in circuit the correct way round or the film of aluminium oxide will be destroyed.

fig. 30.19 Types of capacitor
▼

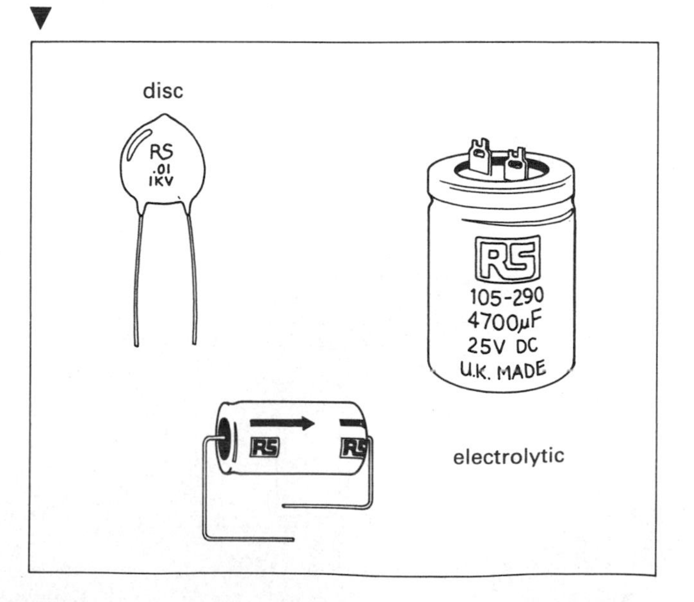

EXERCISE 30

In questions 1–4 select the most suitable answer.

1 A piece of plastic is rubbed with a cloth and is then found to be negatively charged. The most likely explanation of this is that
A cloths are always negatively charged
B plastic is always negatively charged
C the cloth rubs electrons off the plastic
D positive ions are removed from the plastic
E electrons are transferred from the cloth to the plastic

2 Two polythene rods, PQ and RS, are rubbed vigorously from end to end with a duster. PQ is suspended in a horizontal position by an insulating thread. When RS is held parallel to PQ and in the same horizontal plane, as shown in figure 30.20.
A P moves towards R and Q moves away from S
B P moves away from R and Q moves towards S
C PQ remains in its original position
D PQ moves towards RS
E PQ moves away from RS

fig. 30.20

3 Two identical, uncharged metal spheres on insulated stands are in contact, when a negatively charged polythene rod is brought close to X, figure 30.21. Whilst the charged polythene rod is held stationary, spheres X and Y are separated. What are the charges on X and Y now?

	Charge on X	*Charge on Y*
A	positive	negative
B	negative	negative
C	neutral	neutral
D	positive	positive
E	negative	positive

fig. 30.21

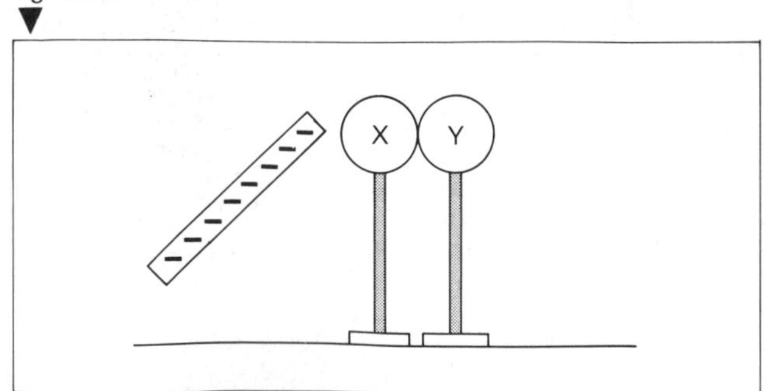

4 Metal spheres A (positively charged) and B (uncharged) are close to each other and B is touched, figure 30.22, with a finger. The finger is removed from B. After this, sphere A is taken away. Sphere B is now
A positively charged
B negatively charged
C uncharged
D negatively charged on the left side and positively charged on the right side
E positively charged on the left side and negatively charged on the right side

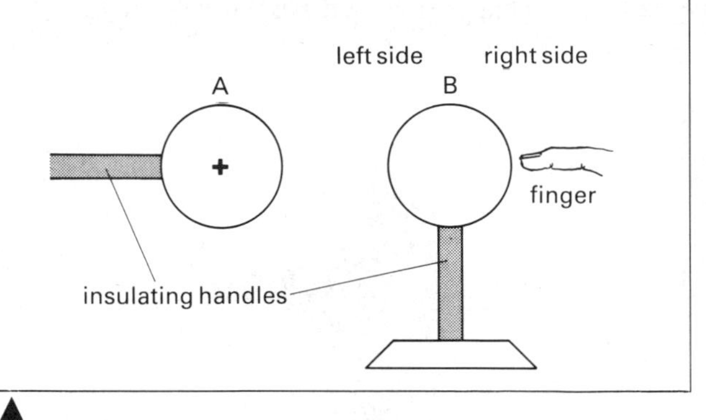

fig. 30.22

5 You are provided with two small insulated charged conductors of different shape, an electroscope and any other apparatus you may require. Describe *one* experiment, in each case, to show, preferably without discharging the conductors, that they possess
a opposite charges,
b unequal charges.

6 The diagram shows two metal plates in a vacuum viewed from above connected to a high-voltage d.c. supply.

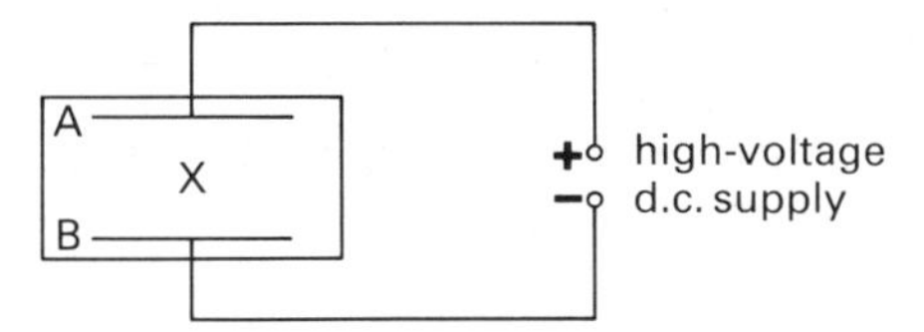

a Various particles are introduced at X. State, in each case, whether they *remain at* X, *move towards* A, *move towards* B, or *oscillate* if the particle at X is (i) a neutral atom, (ii) an electron, (iii) a proton.
b If the atom was carrying a negative charge how would its behaviour differ from an electron in the above experiment?
c What would be the effect on an electron of (i) decreasing the voltage (potential difference) between the plates? (ii) using an alternating high voltage in place of the high d.c. voltage?

7 Two uncharged metal spheres, X and Y, each having an insulated handle, are held in contact, figure 30.23a. A negatively charged rod is then brought near to metal sphere X, figure 30.23b. While the charged rod is held in this position, the two spheres are separated, figure 30.23c.

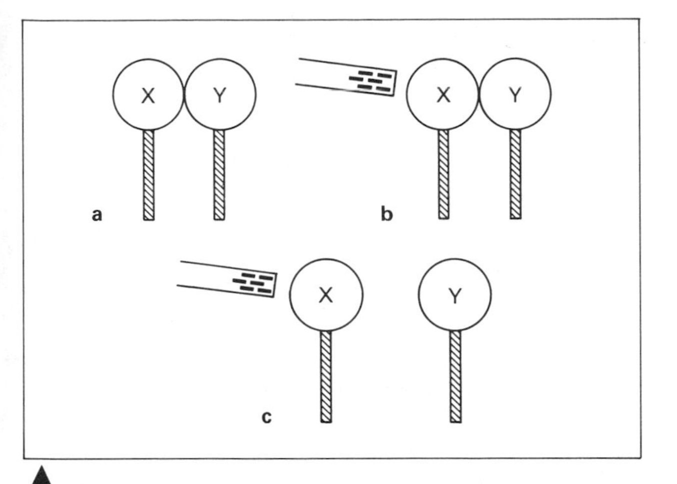

fig. 30.23

a In figure 30.23c, will sphere X be positively charged, negatively charged or uncharged?
b In figure 30.23c, will sphere Y be positively charged, negatively charged or uncharged?
c Describe how you could use an uncharged electroscope and the negatively-charged rod to find out if your statement about the charge on sphere Y is correct.

8 Figure 30.24 shows two metal plates X and Y in a vacuum connected to an e.h.t. supply. A particle P is positioned between the plates.

fig. 30.24

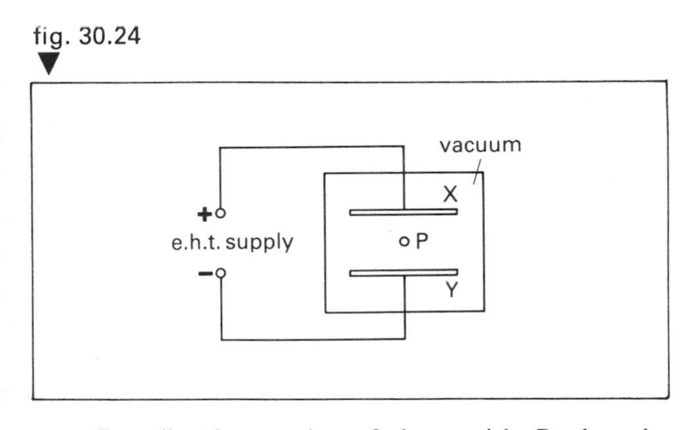

a Describe the reaction of the particle P when the switch is closed if P is (i) an electron, (ii) an atom, (iii) a negative ion, (iv) a positive ion.
b Explain how and why the reaction of the electron differs from the negative ion.
c What would be the effect on any motion of the particle P if air were introduced between X and Y?
d What would be the effect on any motion of the particle P if the potential difference between the plates were (i) increased, (ii) decreased?

9 a A student performed the following operations in the sequence described. (i) Brought a negatively-charged polythene rod *near* to the cap of an uncharged leaf electroscope. (ii) Touched the cap of the electroscope momentarily with a finger. (iii) Removed the rod.

Draw diagrams showing the charge distribution on the cap and the leaf of the electroscope and the position of the leaf after each of the above operations. What would be the effect of removing the rod before removing the finger?

Another student performing the same experiment allowed the polythene rod to rest on the cap of the electroscope in operation (i). State, giving reasons, whether or not you would expect any marked difference in the results obtained by the two students.

b A manufacturer of Nylon thread put heavy rubber mats under his spinning machines, which were made of metal, to deaden the noise. The following effects were subsequently noted. (i) The workers sometimes received an electrical shock when touching the machines. (There was no leak from the mains cable.) (ii) Small bits of Nylon fluff stuck to the thread, but this could be overcome by keeping the air in the workshop moist. State and explain the physical reasons for (i) and (ii), and say how you could overcome (i).

10 The diagram shows a charged pear-shaped conductor on an insulating stand.

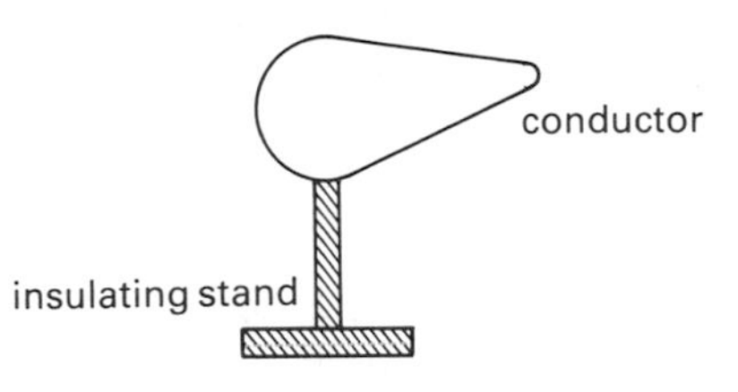

a (i) Of what material is the conductor most likely to be made? (ii) Of what material is the insulating stand most likely to be made?
b Copy the diagram and mark on it with the letter A a place where the charge density is likely to be greatest.
c What shape would a charged conductor have if it were to have a uniform charge density over its surface?

31 THE ELECTRIC CIRCUIT

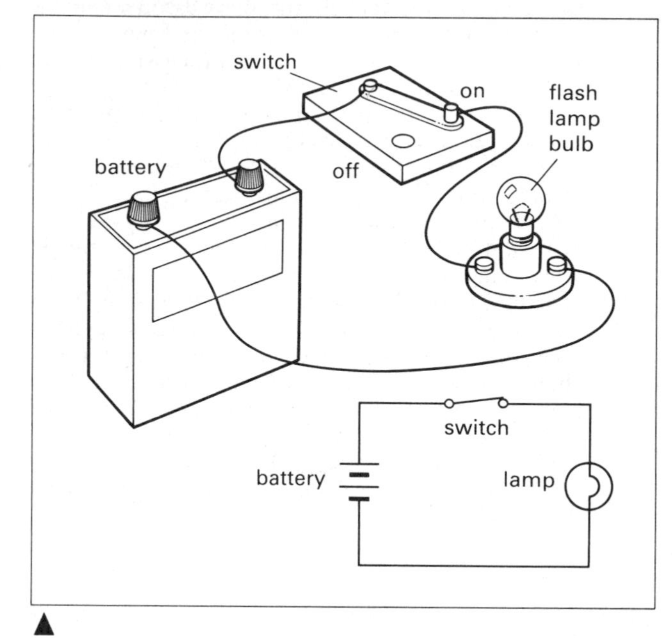

▲
fig. 31.1 A simple circuit and how it is drawn using circuit symbols. (See p. 351 for table symbols)

In order to use electricity to light a lamp, we must have a source of electricity. This could be a battery or it could be a dynamo. The lamp must be connected to the battery in some way and it is convenient to have a switch so that the lamp can be turned on and off. These four objects — battery, switch, lamp and connecting links — are arranged to form a simple **circuit**. For an electric current to flow there must be a complete path for it, with no gaps.

The circuit shown in figure 31.1 is a simple **series** circuit. The electric current flows through the series of three things connected one after the other. There are no 'side-paths' for the current to follow. The order in which the pieces of equipment are connected does not matter. The switch will work just as well on the other side of the battery.

Circuit boards on which various components can be clipped between metal pillars are a convenient way of building up different circuits. The torch cells which fit between the clips are the source of electricity.

fig. 31.2 Circuit board with a simple rheostat circuit connected up
▼

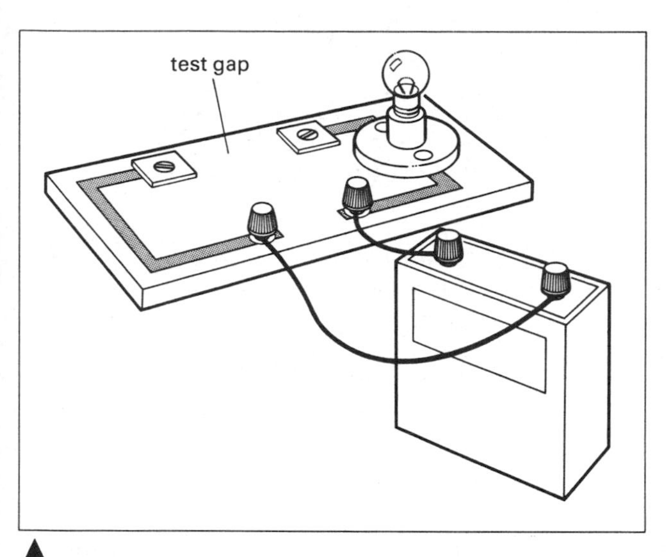

fig. 31.3 Testing for conductors and insulators

Conductors and insulators

Not all substances allow electricity to flow through them. Substances which do are called **conductors**. Materials which do not allow a current to flow through them are **insulators**. Different materials can be tested by placing them across the test-gap of the apparatus in figure 31.3 and seeing whether the lamp lights.

Most wires used for making electrical connections are covered with a layer of insulating material so that if two wires touch they will not cause a **short circuit**. Metals are good conductors, while non-metallic substances are usually insulators. Rubber and plastics are good insulators.

The electric current

The word current means a flow or movement. The flow of an electric current round a circuit is often compared with the flow of water through a pipe system. The pump produces a pressure which drives the water through the pipes, figure 31.4a, and the battery produces an 'electrical pressure' which causes the electric current, figure 31.4b. Both the moving water and the moving electricity can do work such as driving a turbine or an electric motor.

An electric current is a flow of electrically-charged particles. In the nineteenth century, before the discovery of the electron, it was decided that the particles were positively charged and moved from the positive terminal of the battery to the negative. This is the **conventional electric current**. We now believe that the current consists of negatively-charged electrons moving in the opposite direction. This is the **electron current**. In spite of this new idea, we still tend to label diagrams with the conventional current direction which has been in use for over a hundred years, figure 31.5.

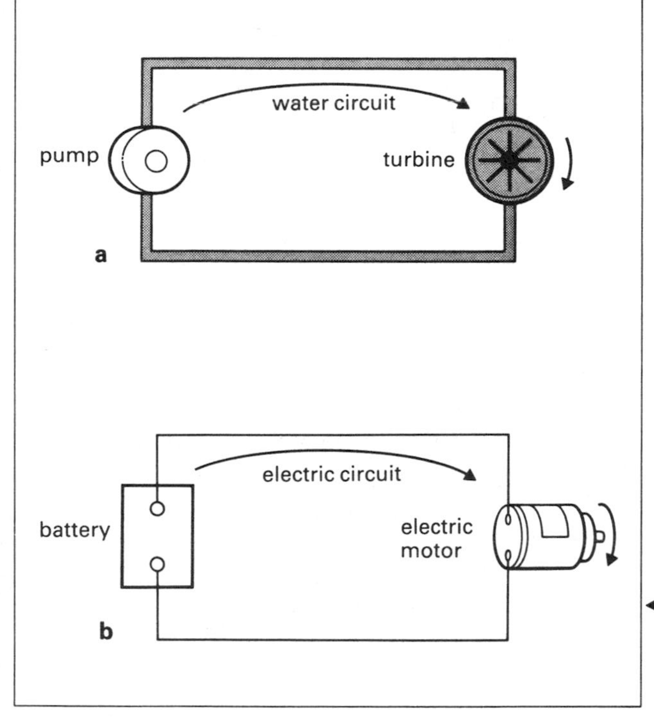

fig. 31.4 The water flow analogy

fig. 31.5 The conventional current and the electron current

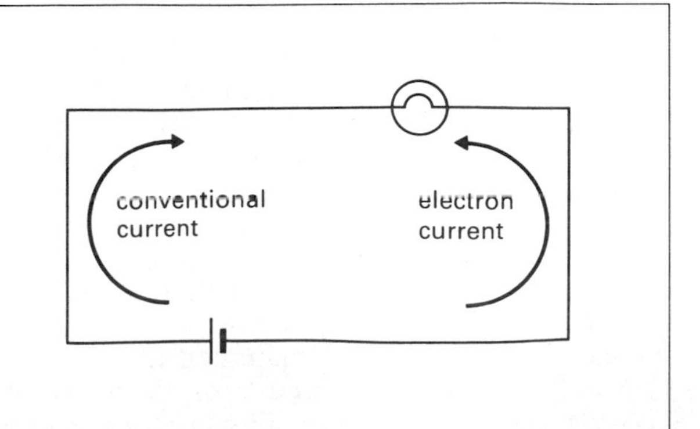

Electrical conduction in metals

The atoms of metals are arranged in a regular pattern and are held in position by electrical forces. The loosely-held outer electrons are shared between the atoms, and these free electrons move at random within the metal in the same way as the molecules of a gas move within the confines of the container. When the ends of a metallic conductor are connected to a battery, the negatively-charged electrons move towards the positive end, figure 31.6. This drift of electrons is what we mean by an electric current. Each electron carries a small negative charge, so an electric current is a movement of charge. The rate of movement of charge is a measure of the magnitude of the current.

Metals are good conductors of electricity because they contain many free electrons. In a cubic centimetre of copper there are about 10^{23} free electrons. Insulators contain no, or very few free electrons. In between the conductors and the insulators is a group of materials containing very many fewer free electrons than conductors. These are called **semiconductors**.

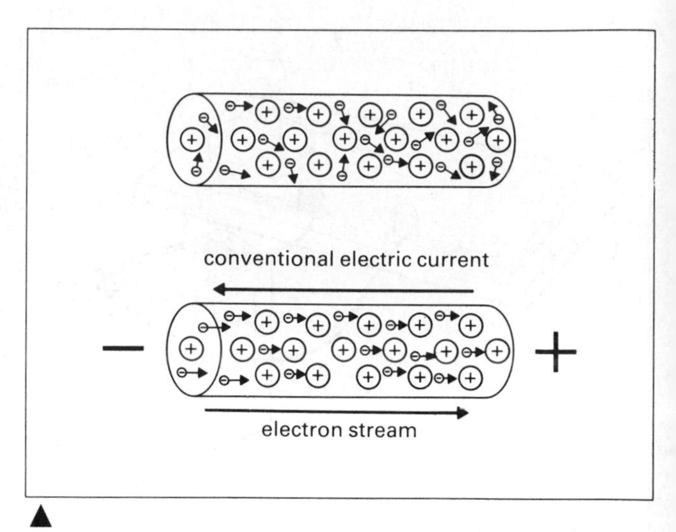

▲ fig. 31.6 How electricity is conducted along a wire

Measuring the current

The instrument used to measure the magnitude of an electric current is the **ammeter**. The unit is the **ampere** (**A**), named after the French scientist, André Marie Ampère, 1775–1836. (The ampere is defined precisely on p.272.)

The ammeter is connected in the circuit so that the current to be measured flows through it. In a simple series circuit, the current is the same at all points in the circuit, figure 31.7. Both experimental observations and comparison with the 'water circuit' show that this must be so. The rate of flow of water in the pipe system must be the same at all points, otherwise the water would pile up at some point in the system and elsewhere there would be little or no water.

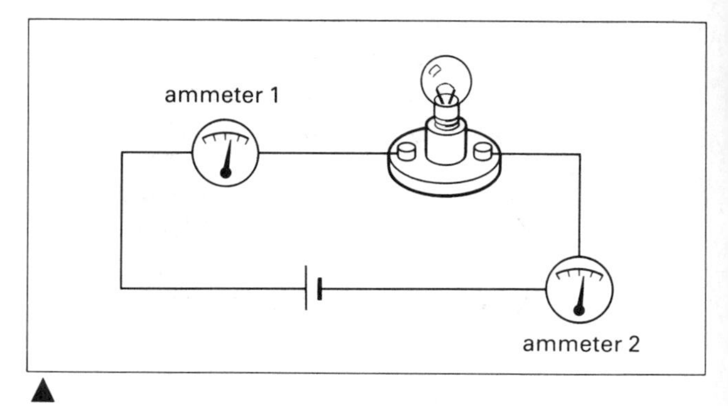

▲ fig. 31.7 Both ammeters show exactly the same current reading

Series and parallel circuits

Cells and lamps can be connected in a variety of ways. Two examples are shown in figure 31.8. In figure 31.8a the lamps are connected in **series** and the same current flows through each lamp. In figure 31.8b the lamps are in **parallel**.

Such circuits can be connected up on the circuit board. If one lamp is removed from its holder in the series circuit, the other lamps go out because the circuit is broken. If a lamp is unscrewed in the parallel circuit, the other lamps remain alight.

A cell (or any other source of electricity) which converts one form of energy into electrical energy has

fig. 31.8 Bulbs connected in series and in parallel ▼

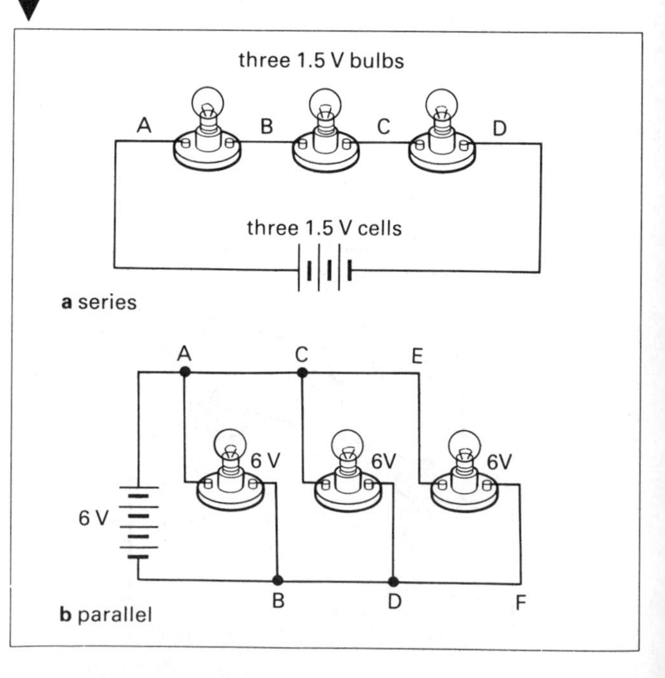

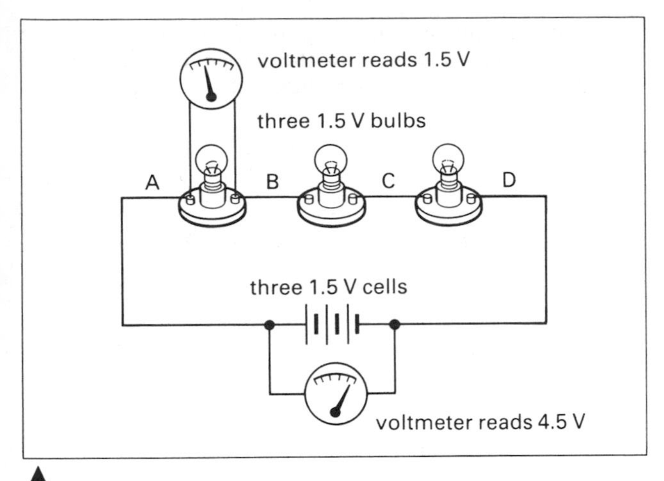

fig. 31.9 Lamps and cells in series. Each lamp has its correct voltage and lights normally

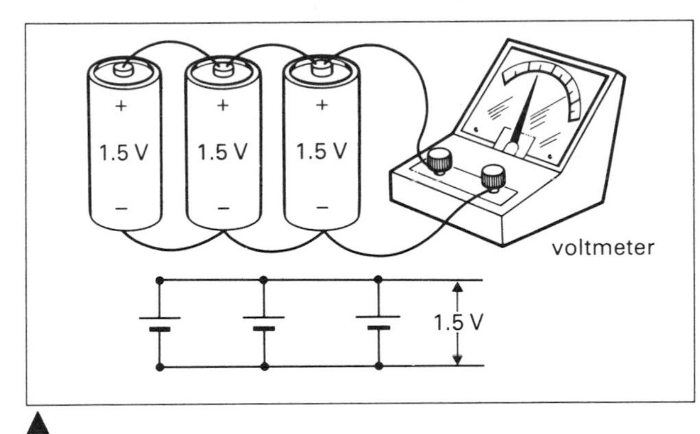

fig. 31.10 Cells in parallel

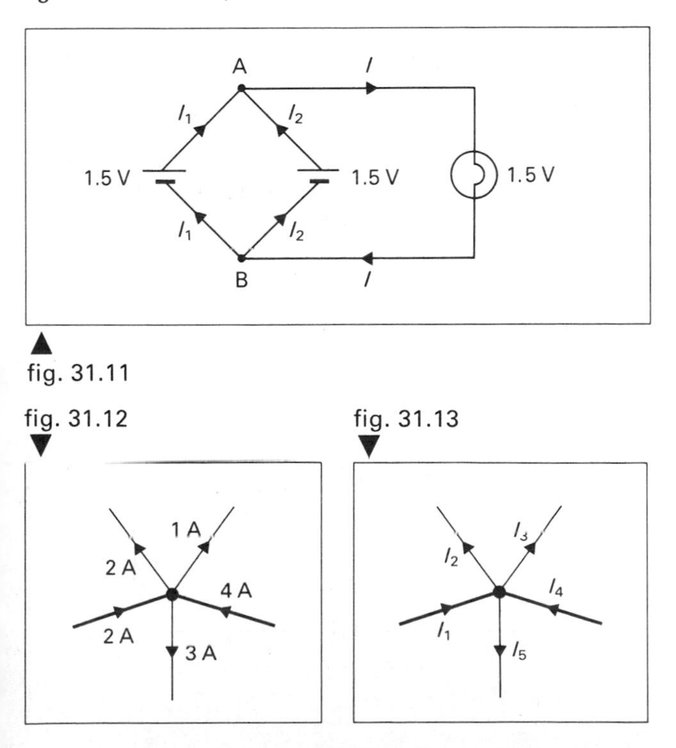

fig. 31.11

fig. 31.12

fig. 31.13

an **electromotive force (e.m.f.)**, which is a measure of its ability to drive a current through a circuit. The electromotive force is measured with a **voltmeter**. The unit of electromotive force is the **volt (V)**, named after the Italian Alessandro Volta (1745–1827), who developed the first electric cell. (The volt is defined precisely on p.244.)

We often speak loosely of the e.m.f. of a cell as its 'voltage', just as we use the word 'mileage' sometimes when we mean distance. The e.m.f. of a cell is also called the **potential difference (p.d.).** When a current flows between two points in a circuit, we say there is a difference in potential between them.

A lamp is designed to work at a particular voltage and this should match the electromotive force supplying it. In a series circuit of lamps and cells, the voltages add together, figure 31.9.

When cells are joined in parallel, figure 31.10, the voltage of the battery is the same as that of each of the single cells. The battery has three times the capacity of a single cell, but its e.m.f. is no greater.

Figure 31.11 shows two 1.5 V cells connected in parallel and supplying current to a 1.5 V lamp. Both cells provide currents I_1 and I_2 and these join together at the point A to form the current I in the lamp circuit. At B, the main current I divides into I_1 and I_2. At both these points

$$I_1 + I_2 = I$$

If this were not so, then either more electrons would be arriving at a point than were leaving it, so charge would be accumulating at the point, or more electrons would be leaving than were arriving.

This is really a statement of Kirchhoff's first law of electric circuits which states

The sum total of the current flowing towards a junction in an electric circuit is equal to the sum total of the current flowing away from the junction

or

The algebraic sum of the currents meeting at a point is zero

In figure 31.12

sum of currents towards junction = sum of currents away from junction

$$2 + 4 = 2 + 1 + 3$$

In figure 31.13, calling the currents towards the junction positive and those away from the junction negative, then

$$I_1 + I_4 - I_2 - I_3 - I_5 = 0$$

EXERCISE 31

In questions 1–5 select the most suitable answer.

1 Metals allow electric currents to flow through them easily because they
- **A** have more atoms
- **B** have heavier atoms
- **C** have free electrons
- **D** are good conductors of heat
- **E** have high densities

2 Which of the following instruments is used to measure an electric current?
- **A** voltmeter
- **B** electrometer
- **C** ohm-meter
- **D** ammeter
- **E** micrometer

3 S_1, S_2 and S_3 are switches in the circuit shown in figure 31.14. To make lamp L as bright as possible close
- **A** S_1 only
- **B** S_2 only
- **C** S_3 only
- **D** S_1 and S_2 only
- **E** S_1 and S_3 only

fig. 31.14

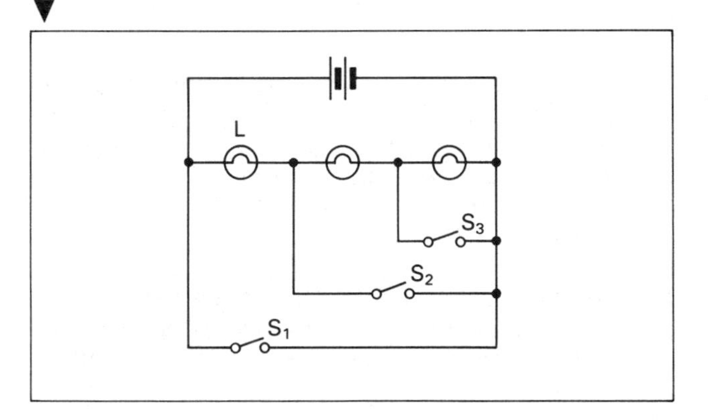

4 The current in the conductor YZ in figure 31.15 will be
- **A** 2 amperes
- **B** 3 amperes
- **C** 7 amperes
- **D** 8 amperes
- **E** 10 amperes

fig. 31.15

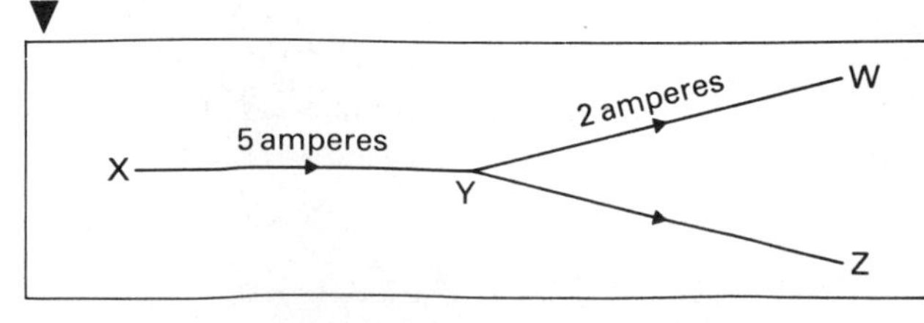

5 In the circuits shown in figures 31.16 **A** to **E** all the cells are identical and all the lamps are identical. In which circuit will the lamp marked L be brightest?

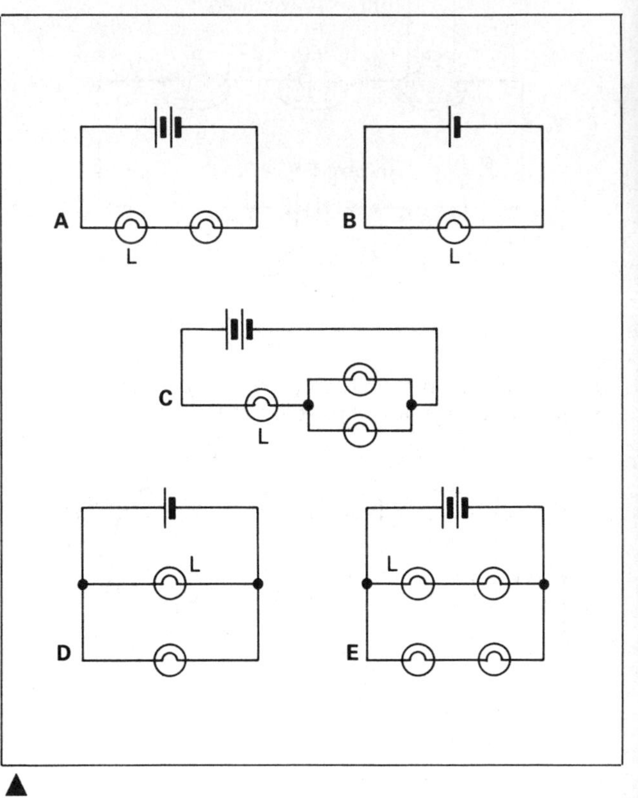

fig. 31.16

6 Figure 31.17 shows three identical bulbs L_1, L_2 and L_3 which shine with normal brightness when 2 V is applied to each of them.
- **a** What is the potential difference between AB when the switch is open?
- **b** When the switch S is closed, what will happen to the brightness of: (i) bulb L_1, (ii) bulb L_2?
- **c** What is the potential difference across AB when the switch is closed?

fig. 31.17

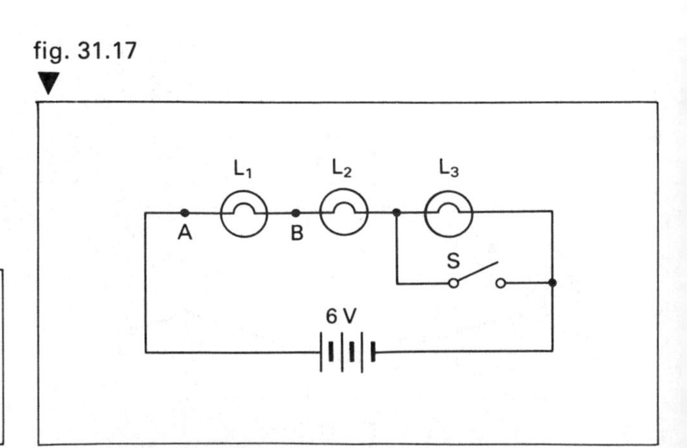

7 a From the list of substances below pick *three* which are good conductors of electricity:
bakelite, tin, paraffin wax, acidified water, mercury, sulphur, wood.

b From the list of substances below pick *three* which are good electrical insulators:
aluminium, mercury, distilled water, glass, platinum, polythene, gold.

c In general what broad class of substances includes the best conductors of electricity?

8 a Draw a diagram to show a circuit containing two lamps in parallel, and another circuit containing the same two lamps in series.

b In which of the above circuits is the current through both the lamps the same as the current from the power supply?

c If the lamps in part **a** are identical 12 volt lamps, what voltage supply will be required to light them to normal brightness when they are (i) in parallel? (ii) in series?

d Look at figure 31.18 and say which lamp or lamps are controlled by each of the switches. Answer this question by rewriting and completing the following sentences:
Switch X controls lamp(s) ....
Switch Y controls lamp(s) ....
Switch Z controls lamp(s) ....

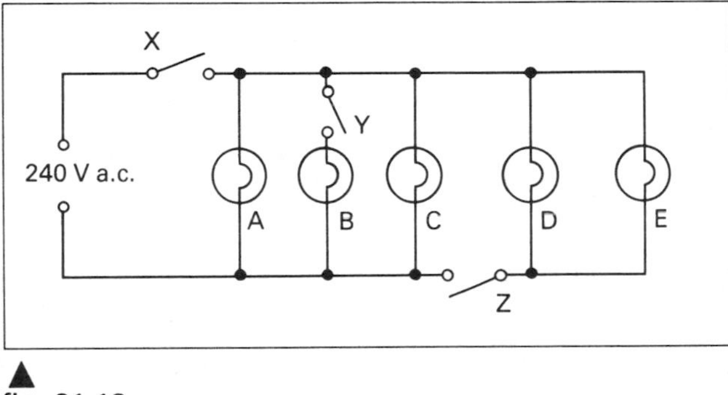

fig. 31.18

9 Figure 31.19 shows two circuits using the same cells and bulb. State, with reasons, in which circuit the bulb will be brighter.

fig. 31.19

10 In which one of the circuits shown in figure 31.20 do all the bulbs go out if any one of the bulbs fuses?

fig. 31.20

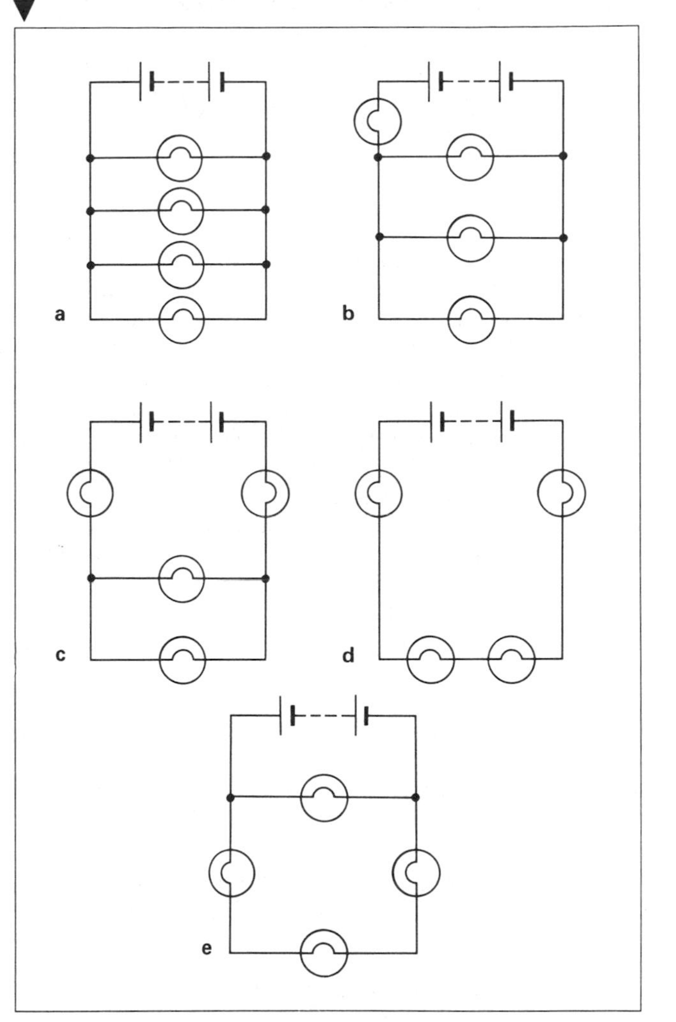

11 Five cells, each of e.m.f. 1.5 V, are connected as shown in figure 31.21. What would a voltmeter read when connected across

a A and B?
b B and C?
c A and C?

fig. 31.21

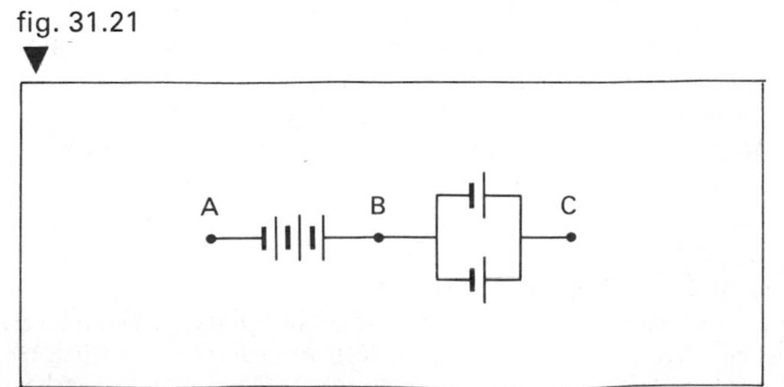

32 ELECTRIC CELLS

An electric **cell** is a device for converting some form of energy into electrical energy. Usually, chemical energy is converted into electrical energy, but in other types of cell light energy or heat energy can be made to produce electricity. A cell which produces an electric current from chemical changes or directly from light or heat is a **primary** cell. When the chemicals are used up the cell will provide no more current. The process is irreversible. In a **secondary** cell the energy conversion can be reversed. Accumulators or storage batteries are secondary cells. The accumulator can be charged by supplying electricity from another source. This electricity produces chemical changes in the cell. When the cell supplies a current the chemical changes are reversed. The accumulator can be recharged.

During the years 1790–1800 two Italian scientists, Luigi Galvani and Alessandro Volta, discovered that when two different metals are placed in certain solutions electricity flows along a wire connecting the two metals. Volta's pile, figure 32.1, was the first battery and consisted of about a hundred cells joined in series.

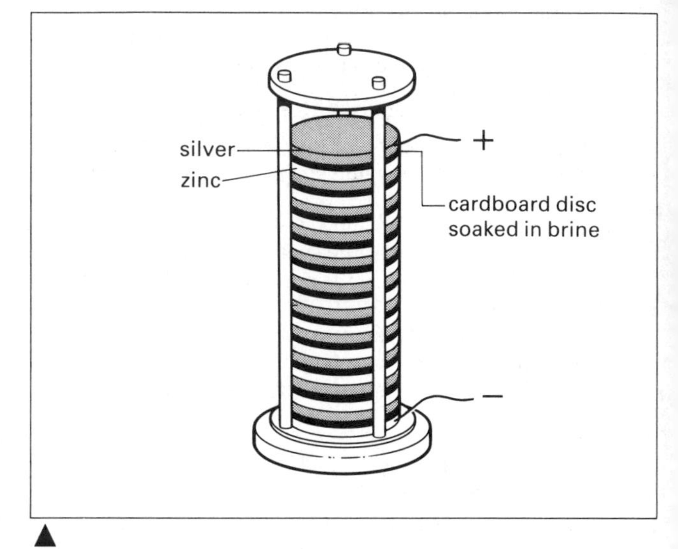

▲ fig. 32.1 Volta's pile consisted of discs of zinc and silver separated by cloth or cardboard discs dipped in salt solution

fig. 32.2 Simple cell ▼

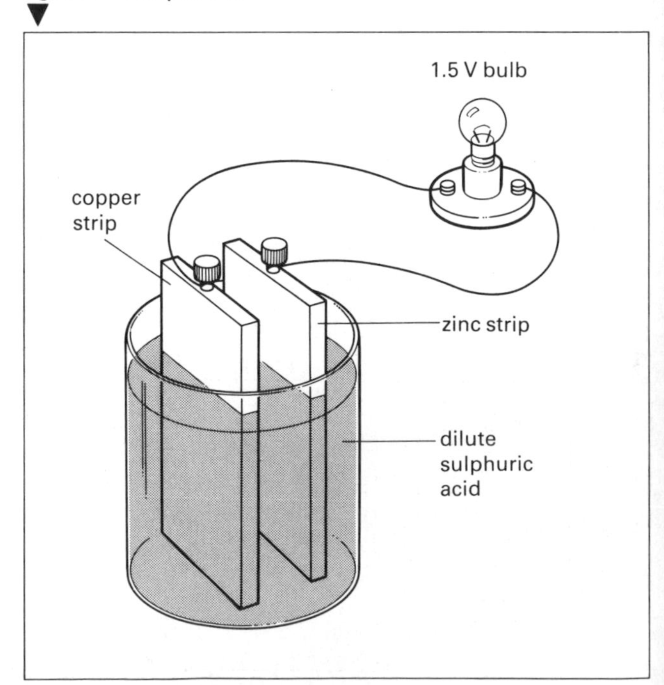

The simple cell

When two different metals, for example copper and zinc, are placed in a liquid which conducts electricity (an **electrolyte**) a current flows through a wire joining the metals. The arrangement is a simple cell, figure 32.2. A bulb connected to the cell goes out after a very short time, and a voltmeter connected across the cell shows that the voltage drops quickly with time. While this is happening it will be observed that bubbles of gas are collecting on the copper plate. The gas on collection and testing is shown to be hydrogen. This collection of hydrogen on the copper plate is known as **polarisation** and is the cause of the fall in voltage shown on the graph in figure 32.3. The hydrogen layer replaces the copper, as it were, and a zinc–hydrogen cell is formed which has a lower e.m.f. than the zinc–copper cell.

Another problem is that the zinc plate dissolves in the acid when the cell is not in use. This solution takes place mainly where there are impurities in the

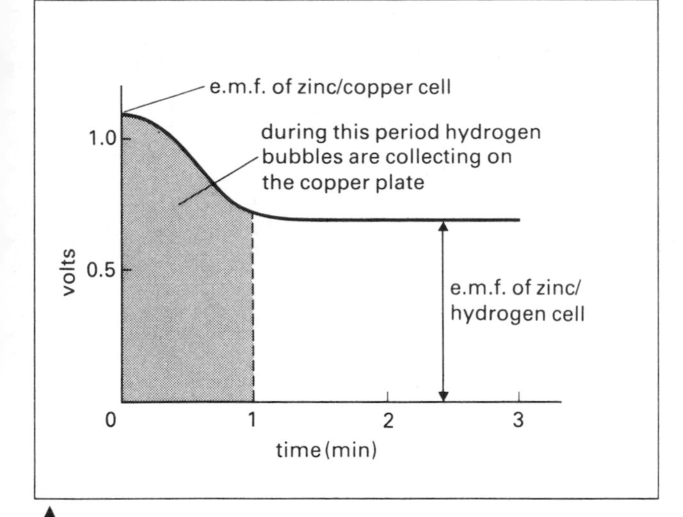

fig. 32.3 Polarisation and the simple cell

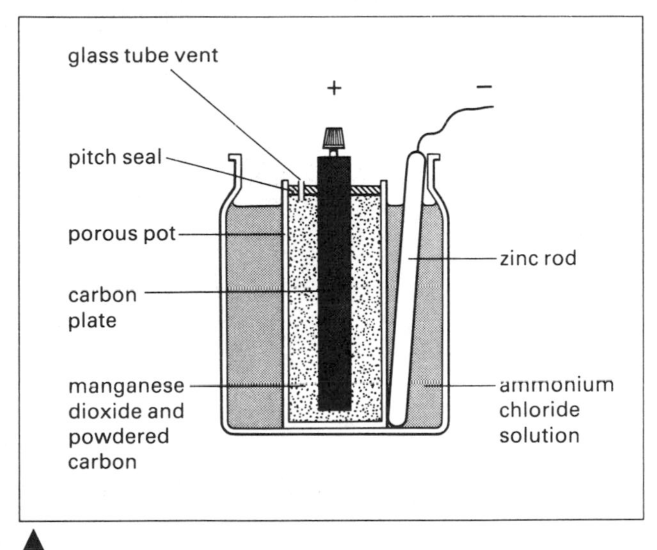

fig. 32.4 Leclanché cell

fig. 32.5 Single dry cell (Leclanché type)

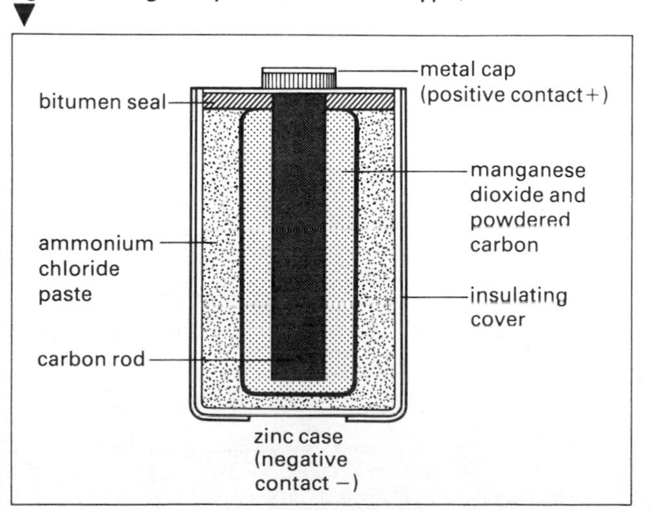

metal and the defect is called **local action**. If some practical form of cell is to be designed these two defects must be overcome.

Local action can be cured by rubbing the zinc with mercury which dissolves the pure zinc to form an amalgam, or by changing the electrolyte.

Polarisation can be overcome by changing the materials so that hydrogen is not formed, or by adding an oxidising agent so that the hydrogen is converted into water.

Leclanché cell and the dry cell

Many different combinations of materials were tried out in the construction of cells. Probably the most successful was the **Leclanché cell** designed in 1868, figure 32.4. We still use the same principle in the **dry cell**, figure 32.5. The electrodes are zinc and carbon, the electrolyte is ammonium chloride solution and manganese dioxide is used as a depolariser.

The dry cell is not really 'dry' because chemical changes do not take place in the absence of water, but the ammonium chloride solution is replaced by a paste. The zinc case acts as the negative terminal. The e.m.f. of the Leclanché cell is 1.5 volts whatever its size. Larger cells contain more chemicals and last longer. The manganese dioxide is a rather slow oxidising agent so the cell is best used for short intervals with periods to recover in between, but for the small currents required for transistor radios it is ideal.

Mallory cell

The **Mallory cell** has electrodes of zinc and mercury oxide with potassium hydroxide for electrolyte. The e.m.f. is 1.35 volts. The Mallory cell can deliver larger currents than the Leclanché cell without polarising and does not deteriorate when stored. It has what the trade call a long shelf-life. The cell can be made very small and is used in watches, cameras and hearing aids. Several cells are joined together to form a battery if a larger e.m.f. is required.

fig. 32.6 Mallory cell

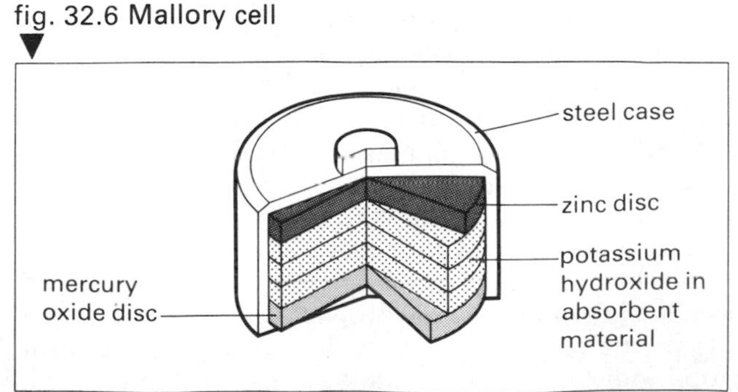

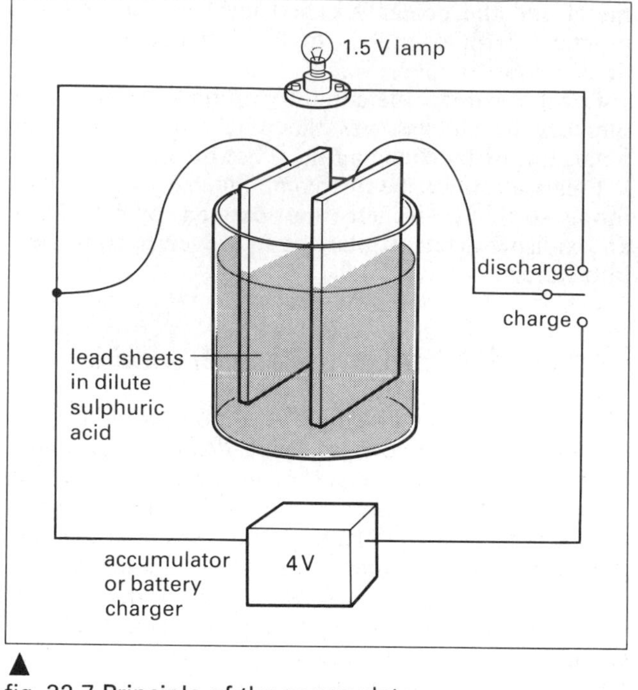

▲ fig. 32.7 Principle of the accumulator

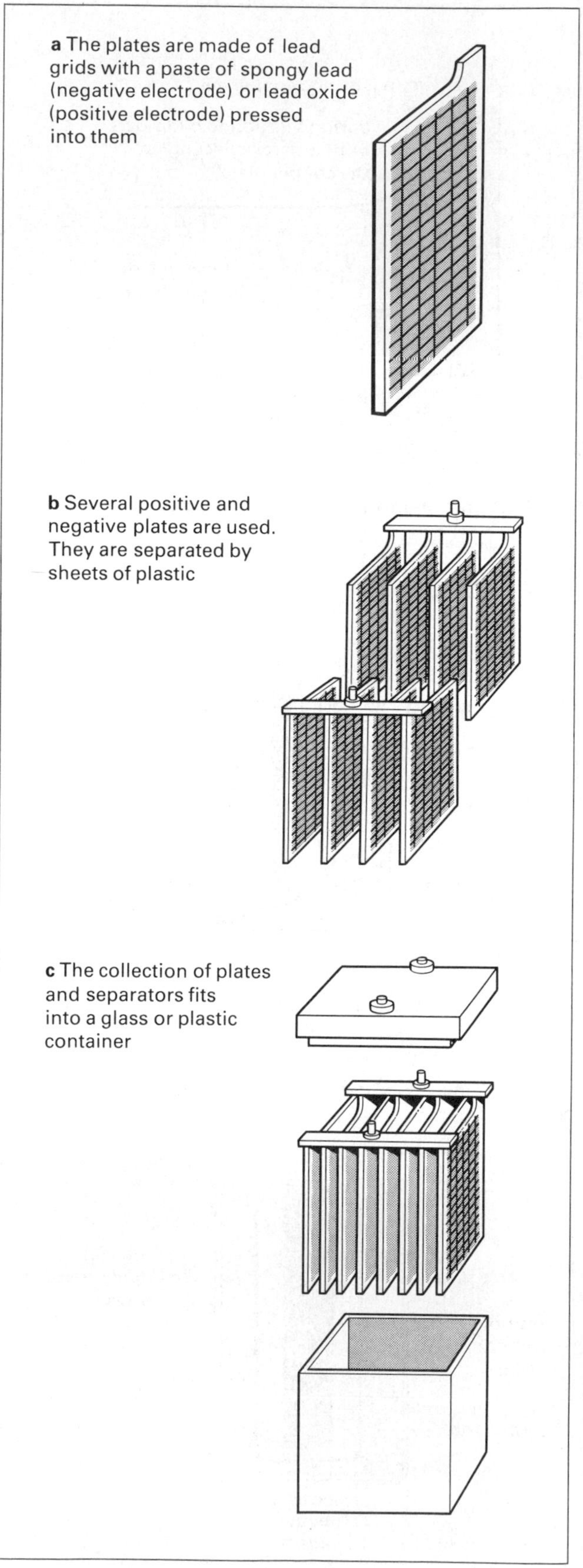

fig. 32.8 General construction of the lead–acid accumulator ▶

a The plates are made of lead grids with a paste of spongy lead (negative electrode) or lead oxide (positive electrode) pressed into them

b Several positive and negative plates are used. They are separated by sheets of plastic

c The collection of plates and separators fits into a glass or plastic container

Secondary cells — accumulators

When the two electrodes and the electrolyte of a primary cell are put together an electromotive force is produced. The cell is a source of electrical energy. A secondary cell has to be charged before it can be used. During the charging process electrical energy is stored as chemical energy, and the reverse change (chemical energy → electrical energy) occurs during discharge.

The principle of the accumulator can be demonstrated with the apparatus in figure 32.7. The switch is put in the 'charge' position for a few minutes. When the switch is put to 'discharge' the lamp lights for a short time, during which the electrical energy stored during the charging process is converted to heat and light. When the bulb goes out the switch can be put to 'charge' and the reversible process repeated.

In the **lead–acid accumulator** the plates are made of lead grids, with a paste of brown lead dioxide pressed in for the positive plate, and grey spongy lead for the negative plate, figure 32.8. The electrolyte is dilute sulphuric acid. During discharge both plates are converted to lead sulphate and water is produced. The reverse changes take place during charging. Because of the water produced during discharge the specific gravity of the electrolyte is reduced. The testing of the specific gravity with a hydrometer gives a good indication of the state of charge of the cell.

When fully charged the specific gravity is 1.25 and when discharged it drops to about 1.11. The e.m.f. of a single cell is about 2 volts. A car battery usually has six cells in series giving a total e.m.f. of 12 volts.

Other types of storage cell use an alkali (potassium hydroxide) for the electrolyte, and plates of nickel and iron or cadmium and iron or nickel and cadmium. The alkali cells have an e.m.f. of about 1.3 volts. They are much lighter than the lead–acid cells and have a longer working life. They are used in electric vehicles such as milk floats.

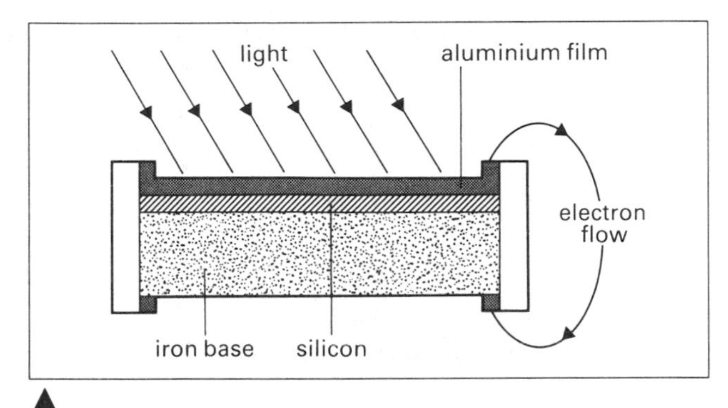

▲ fig. 32.9 The photoelectric effect

The photoelectric effect — solar cells

Many metals and semiconductors (such as selenium and silicon) emit electrons when light of certain wavelengths falls on them. The released electrons flow through an external circuit and produce a current, figure 32.9. This is the **photoelectric effect** and the photoelectric cell converts light energy directly into electrical energy.

Photocells or solar cells consist essentially of a slice of semiconductor material sandwiched between an iron base and a very thin metallic film of aluminium which will allow light to pass through it. When light falls on the cell, electrons are released from the boundary between the aluminium and the silicon and flow through an external circuit to the iron base. The arrangement acts as a cell of which iron is the positive terminal and aluminium is the negative terminal. The e.m.f. is about 0.5 volts, and the current produced depends on the surface area and the intensity of the light. Solar batteries used in satellites consist of thousands of solar cells connected in series.

fig. 32.10 Technicians checking the array (wing) of solar cells of the Landsat 4 earth-resources spacecraft prior to launch ▼

EXERCISE 32

In questions 1–2 select the most suitable answer.

1 A simple voltaic cell was made by dipping a copper plate and a zinc plate into dilute sulphuric acid. The zinc plate was removed and replaced by another plate X but no e.m.f. was obtained, figure 32.11. Which of the following substances was the plate X made of, assuming that all the connections were still good?
A carbon
B copper
C iron
D lead
E steel

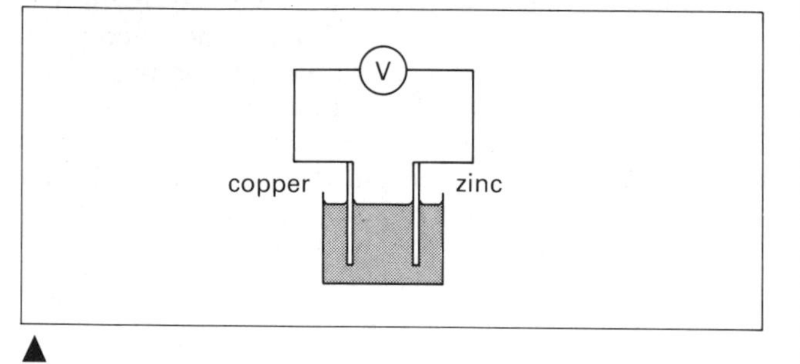

fig. 32.11

2 One advantage lead–acid batteries have over dry-cell batteries for use in a car is that
A topping up with distilled water restores the voltage
B they have the right voltage
C they have a larger e.m.f.
D it does not matter if they are treated roughly
E they can give a larger current

3 Make a labelled diagram of the simple cell.
a What are the *two* main disadvantages of this cell?
b How may one of these disadvantages be overcome?

4 Draw a labelled diagram through a *dry cell*. Indicate which is the positive pole.

5 State two *advantages* and two *disadvantages* the lead-acid accumulator has when compared with a dry cell.

6 A section of a secondary cell suitable for a car battery is shown in figure 32.12. Alternate plates are connected together in the cell which is filled with a suitable electrolyte to provide a powerful source of electrical energy.
a Identify the chemicals on each set of plates of a new fully charged accumulator and state which set of plates is positive. What electrolyte is normally used in this cell?
b How could you test whether the accumulator is fully charged?
c Figure 32.13 shows a number of items that can be connected together to enable two 12 V car batteries to be charged slowly at the same time. Using all these items, draw a circuit diagram which will enable a suitable charging current to pass through both batteries.

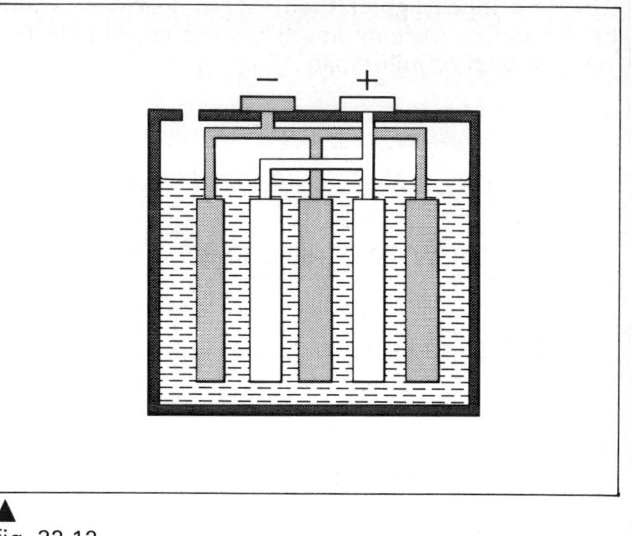

fig. 32.12

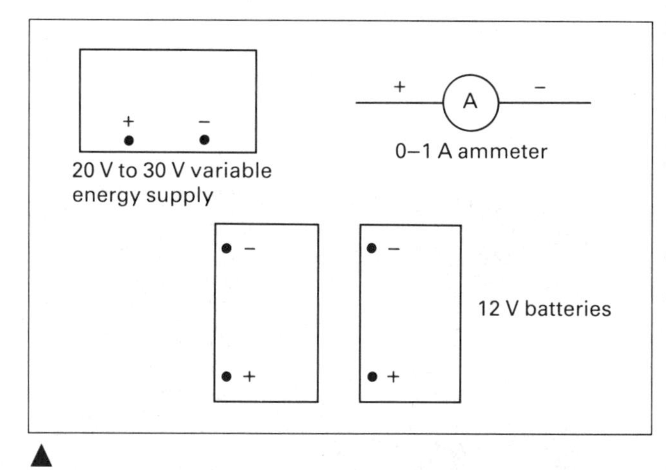

fig. 32.13

7 Describe the chemical changes which occur during the charging and discharging of an accumulator.
Which would you use for testing the state of charge of an accumulator — a voltmeter or a hydrometer? Give reasons for your choice.

33 THE EFFECTS OF AN ELECTRIC CURRENT

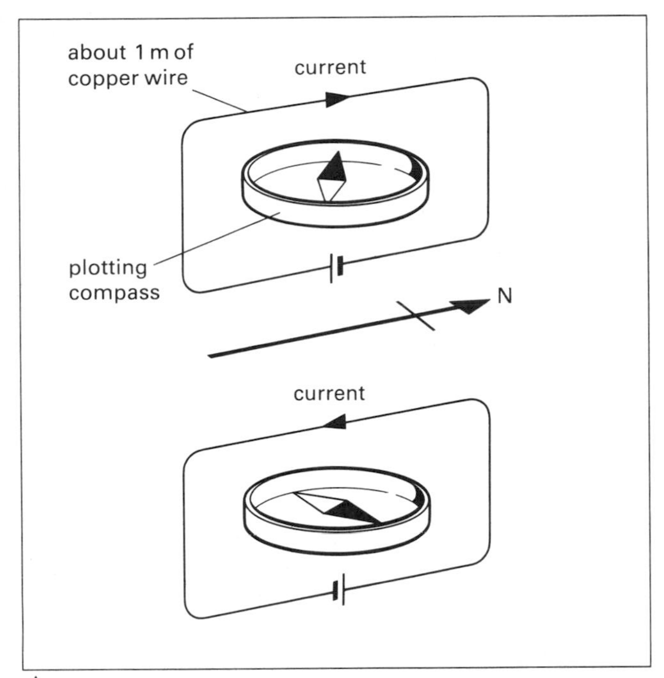

▲
fig. 33.1 Oersted's experiment showing the presence of a magnetic field near a current-carrying wire

fig. 33.2 Plotting the flux pattern round a straight current-carrying wire
▼

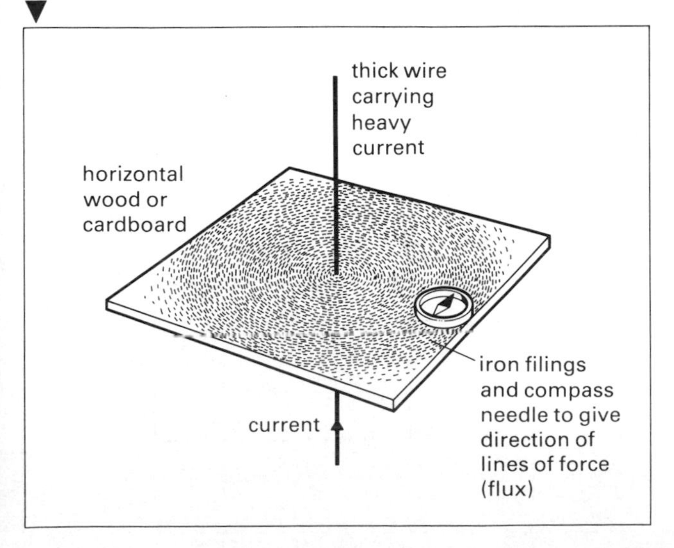

When an electric current flows in a conductor it produces certain changes in the space around the conductor and in the conductor itself. An electric current has

- a magnetic effect,
- a heating effect,
- a chemical effect if the conductor through which the current is passing is a solution.

An electric current produces a magnetic field

In 1819 a Danish scientist, Hans Christian Oersted, discovered that a wire carrying an electric current could deflect a compass needle, figure 33.1. When the direction of the current is reversed the deflection of the compass needle is in the opposite direction. The shape of the magnetic field can be mapped with iron filings, figure 33.2. The lines of force are a series of concentric circles with their common centre at the wire. The direction of the lines of force can be remembered by using one of the following rules:

Clockwise rule Looking in the direction of the current, the lines of force are going round clockwise, figure 33.3.

Corkscrew rule If a right-handed corkscrew is screwed in the direction of the current, the hand turns in the direction of the lines of force, figure 33.4.

The field due to a single loop of wire is shown in figure 33.5. It is often useful to draw a cross-section of a wire carrying a current. The convention used to show the direction of the current is

current coming towards you ⊙

current going away from you 

The dot and the cross represent the point and the tail of an arrow.

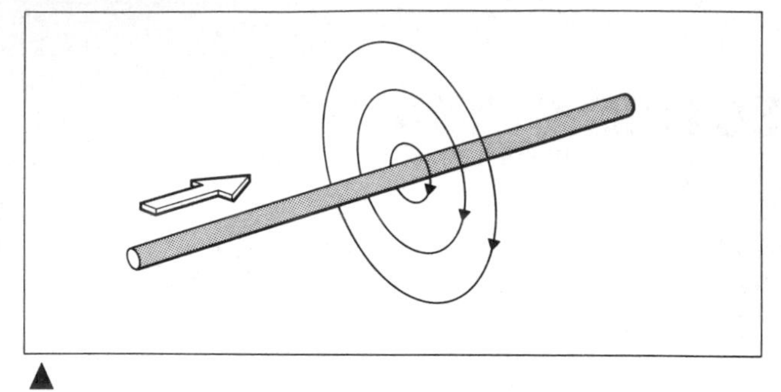

fig. 33.3 The clockwise rule

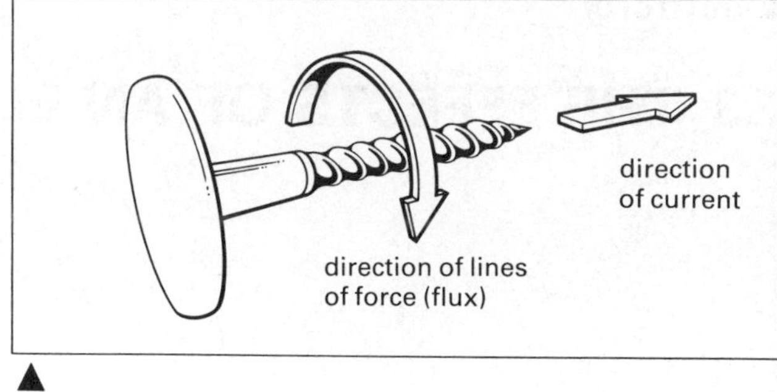

fig. 33.4 The corkscrew rule

When a current is passed through a solenoid, the magnetic fields of the individual turns combine to form a field which is exactly like that of a bar magnet, figure 33.6. The polarity rule is exactly the same as the one we used when making a magnet using an electric current (see p. 233). When a rod of soft iron is inside the solenoid, the magnetic field is more concentrated and we have an **electromagnet**.

Forces between electric currents

Just as two magnets exert forces on each other because of the interaction of their magnetic fields, so do two current-carrying conductors. Experiments with strips of aluminium tape demonstrate the direction of the forces, figure 33.7. The strips are pinned to blocks of wood and the current is supplied by a low-voltage unit. The blocks are arranged so that the strips are not stretched taut and are capable of movement sideways. When the directions of the currents are as shown in figure 33.7. the strips either attract or repel, showing that forces are acting on them.

Currents in opposite directions repel each other

Currents in the same direction attract each other

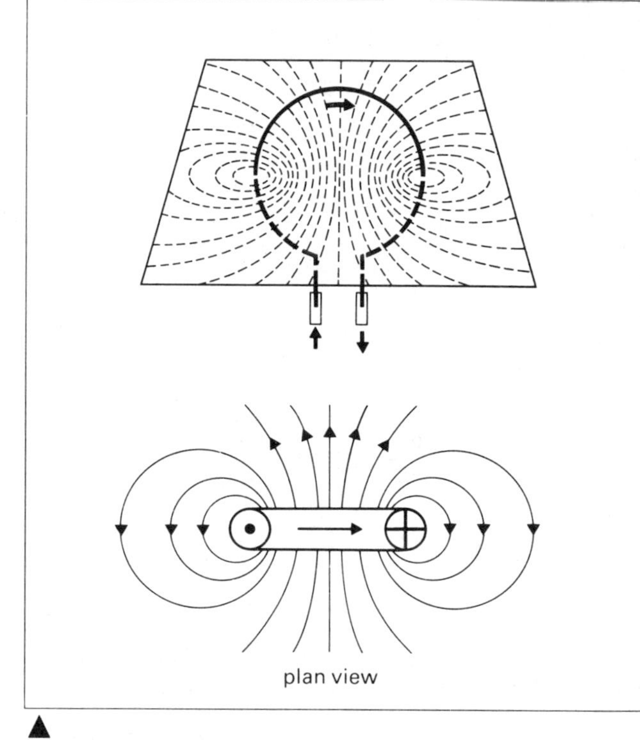

fig. 33.5 The magnetic flux pattern through a circular loop carrying a current

fig. 33.6 Coil of wire (solenoid) carrying a current has a magnetic field like a bar magnet

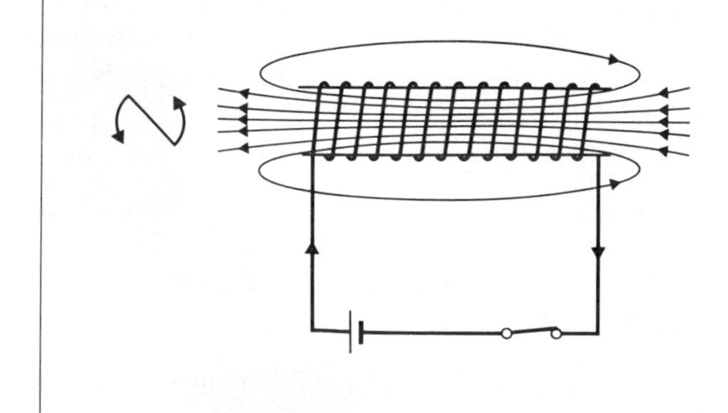

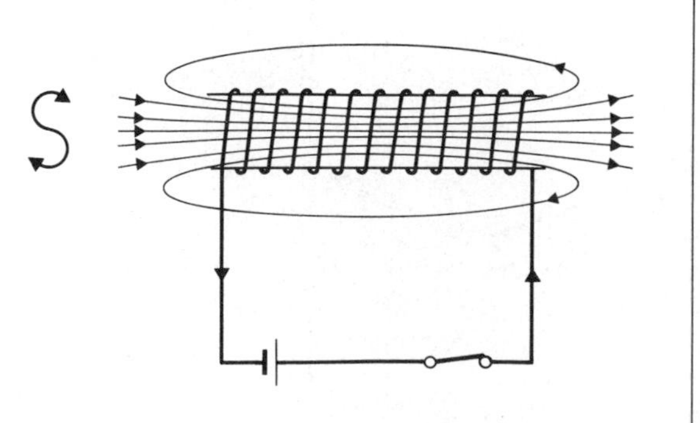

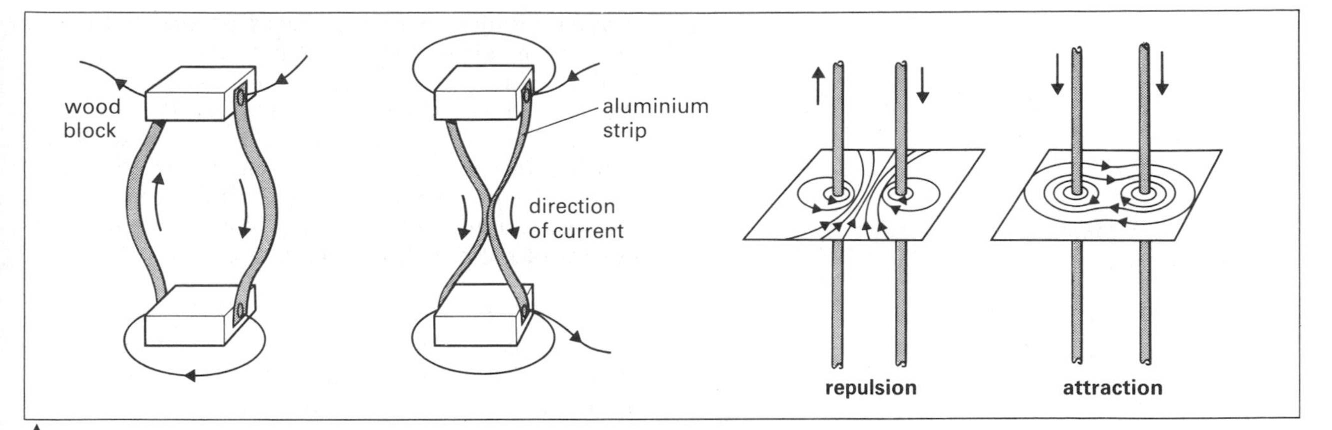

fig. 33.7 Interaction of current-carrying conductors

fig. 33.8 Some examples of electromagnets

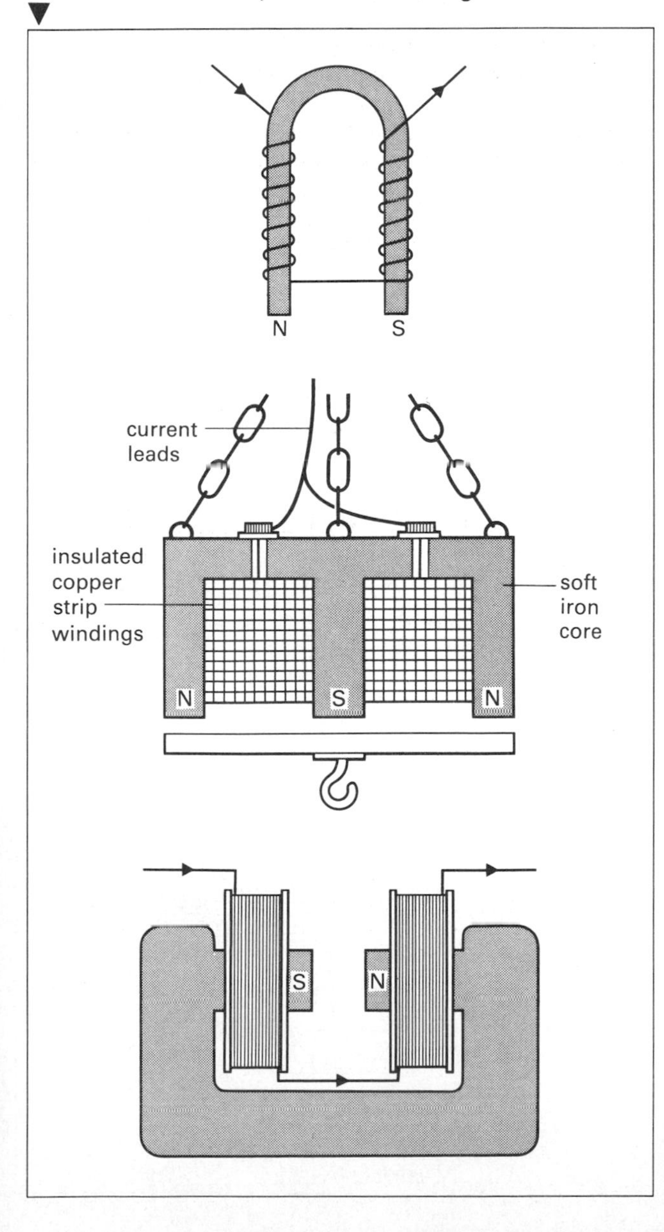

The combined fields of the two conductors explain the directions of the forces. This is only one example of the forces produced between currents and magnetic fields. The problem will be discussed in greater detail in Chapter 36.

Electromagnets

An electromagnet consists of a coil of wire round a soft iron core. When a current flows through the coil, the soft iron core becomes a magnet. When the current is switched off, the soft iron core loses its magnetism because the soft iron has little or no retentivity for magnetism. The ability to switch the magnetism on and off is extremely useful and leads to many applications of electromagnets, e.g. the lifting magnet in figure 33.9. An important advantage is that they can be operated from a distance.

fig. 33.9 Electromagnet lifting a 12 tonne slab

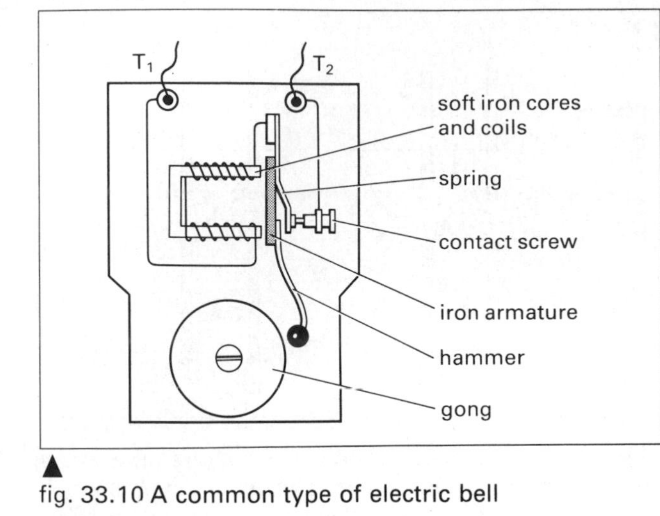

fig. 33.10 A common type of electric bell

The **electric bell**, figure 33.10, makes use of an electromagnet. Trace out the path of the current from the terminal T_1 through the coils to T_2. When the current is switched on by pressing the bell-push, the soft iron cores become magnetised. The armature is attracted to the cores, so pulling the spring away from the contact screw. The circuit is broken and the hammer hits the gong. When the current no longer flows, the cores lose their magnetism. The armature is pulled away by the spring and the circuit is re-made. The process is now repeated for as long as the bell push is pressed. The spring and the contact screw act as a 'make-and-break' for the current. Without the hammer and gong, the armature vibrates as the circuit is made and broken and we have a buzzer.

An electromagnet is used in a **relay**, which is really an electromagnetic switch. A small current can be used to switch on or off a second circuit in which a larger current is flowing. In figure 33.11 when the switch S is closed, the electromagnet is energised and attracts the armature to it, so closing the contacts at R and switching on the equipment E. In an electric organ, for instance, the switch S is the key on the keyboard which in turn switches on the mechanism for allowing air to flow through a certain pipe. There is a relay in the starter circuit of a motor car. When the ignition key is turned, a small current of a few amperes flows through the relay solenoid, which switches on the starter motor circuit, which has to carry a heavy current of over a hundred amperes. Relays are used to operate the switching in the earlier designs of the telephone exchange.

fig. 33.11 Principle of the relay

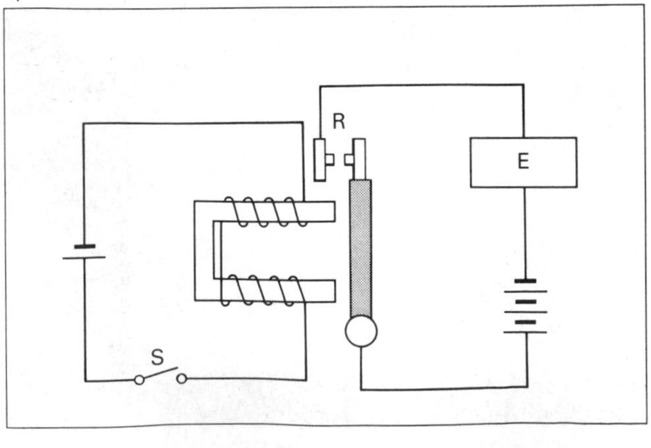

Strength of an electromagnet

The strength of an electromagnet is most simply measured by finding the mass of iron it will attract. There are three variables involved:

- the current,
- the number of turns of wire,
- the size and material of the core.

Keeping two of these factors constant, we can investigate how the strength of the magnet depends on the third factor.

Using the circuit shown in figure 33.12, the number of turns is kept constant and the number of nails which the magnet will pick up is found for different current values. A typical graph of the results is also shown in figure 33.12. As the current is increased,

fig. 33.12 Finding the strength of an electromagnet

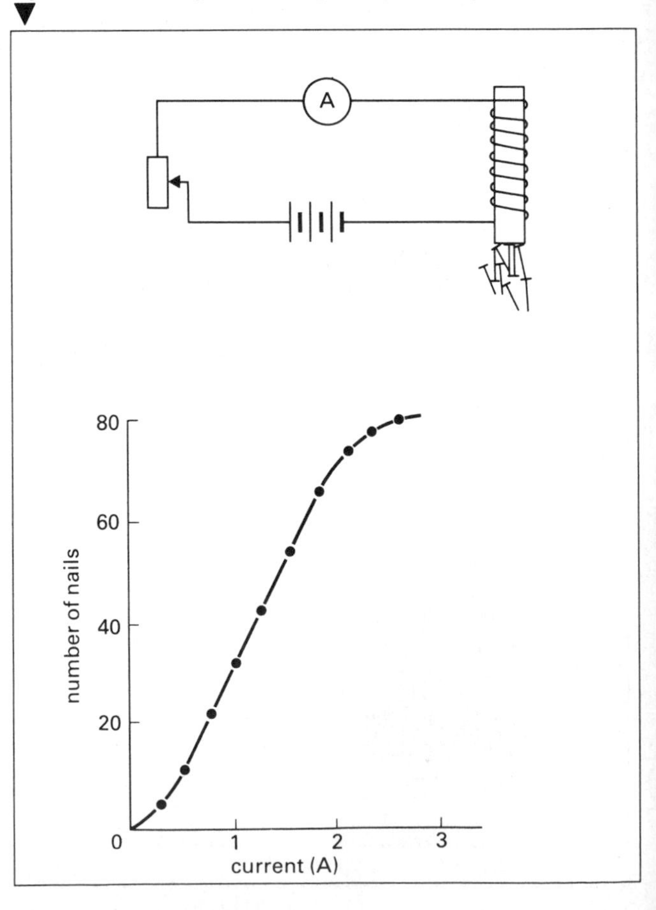

more and more domains are lined up until **magnetic saturation** is reached. Over the greater part of the graph the strength of the magnet is proportional to the current strength. A similar graph would be obtained if the current were kept constant and the number of turns varied.

Putting the two results together the strength of an electromagnet depends on the number of **ampere-turns**. A magnet coil with 200 turns and a current of 2 amperes has 400 ampere-turns, and is just as strong as a magnet coil of 400 turns and a current of 1 ampere. There is a limit, however, to the magnetic strength which can be produced with a given piece of iron. The iron reaches magnetic saturation and increasing the ampere-turns will not make it any stronger.

The heating effect of an electric current

Since electricity is a form of energy we should expect to be able to change it into other forms of energy, such as heat energy.

When the current supplying an electric radiator is switched on, the coil of wire in the radiator (called the element) grows red hot. Electrical energy is being converted into heat energy in the wire. At the same time the cable connecting the radiator to the power-plug remains cold. Some wires get hot while others do not.

An experimental investigation of the heating effect can be carried out as follows. Using the circuit shown in figure 33.13, set the variable resistance to give minimum current. Connect wires of different materials and diameters across the two terminals screwed into the piece of wood. Increase the current in stages and note what happens for different ammeter readings.

We see that the heating effect depends on how big the current is, on the thickness of the wire and also on the material of which the wire is made. Wires which produce a large amount of heat when a given current passes through them are said to have a large **resistance**. Electrons do not move through them easily. Thick copper wire offers little resistance. Nichrome wires have a high resistance and are used in heating elements. Thin wires of tin and special alloys of tin and lead have a high resistance and also a low melting point. They melt and break the circuit when too large a current flows through them.

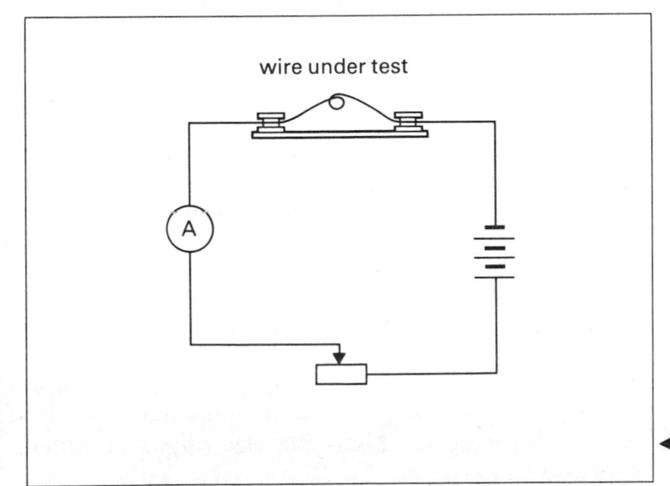

◀ fig. 33.13 Simple arrangement for investigating the heating effects of an electric current

Fuses

A **fuse** is a safety device for ensuring that the current in any particular circuit does not rise above a certain value and either damage equipment or cause wiring to overheat and start a fire. A fuse is a thin piece of tinned copper wire mounted in some suitable holder. Fuses are carried in the three-pin plugs which connect appliances to the mains, figure 33.14. It is obviously important that the correct value of fuse should be used, so that when a safe value of the current is exceeded the fuse wire melts and breaks the circuit.

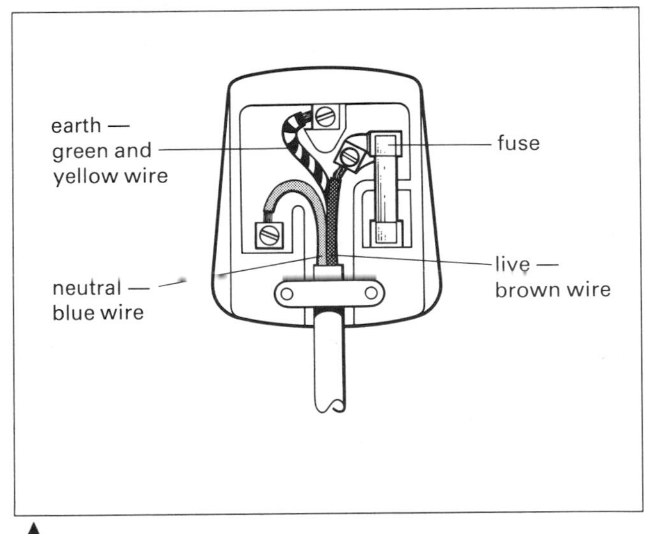

▲ fig. 33.14 Fused three-pin plug for use in a domestic ring main circuit. Note that the live wire (brown) is connected to the fuse terminal

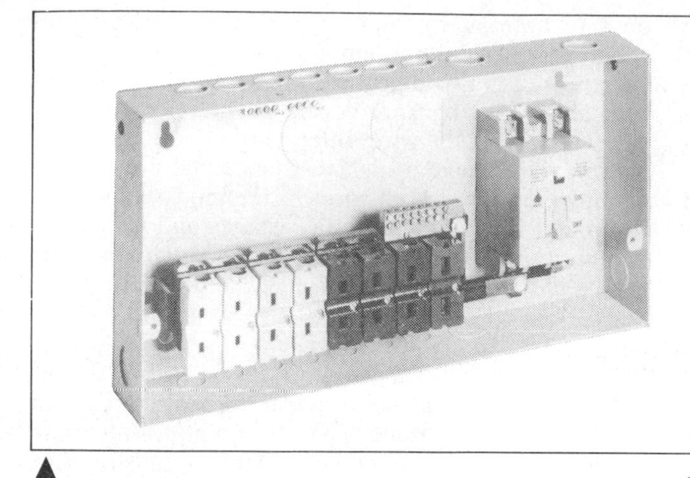

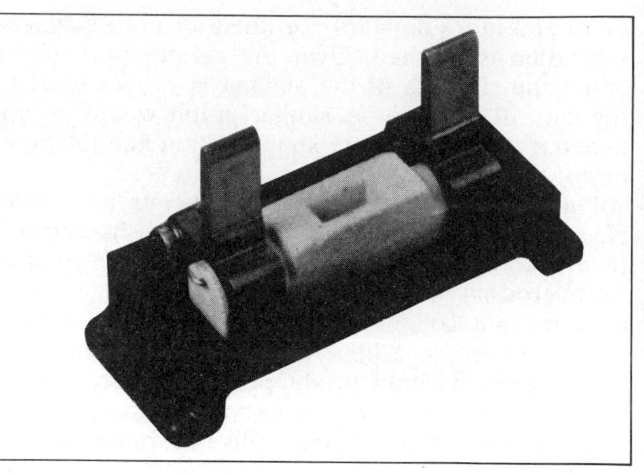

fig. 33.15 Consumer unit with its fuses before installation and, right, a rewirable fuse. When replacing a burnt-out fuse, use the correct length of the correct strength of fuse wire. Always **switch off** first

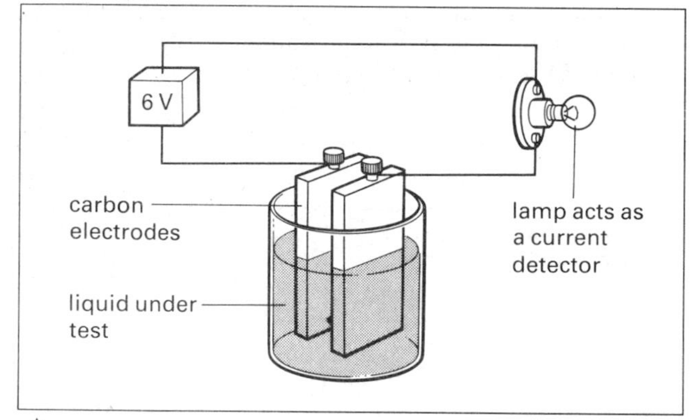

fig. 33.16 Conductivity tester for liquids

fig. 33.17 Electrolysis of water

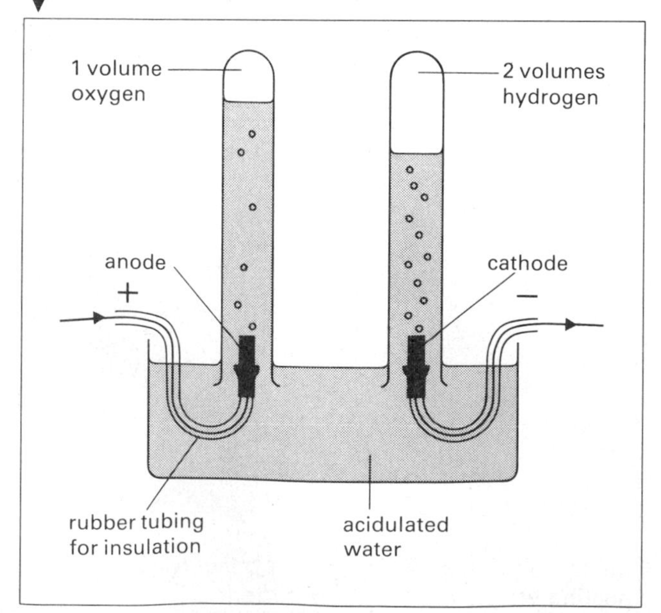

The chemical effect of an electric current

In an electric cell chemical changes take place and electricity is produced. Chemical energy is being converted into electrical energy. Since many energy changes of this kind are reversible, we might expect that an electric current could bring about chemical changes — and it does.

The conductivities of various liquids can be tested as shown in figure 33.16. Using rods of carbon (**electrodes**) dipping into the liquid under test, see whether the lamp lights brightly or dimly, and whether any chemical action as indicated by bubbles of gas or changes at the carbon rods is taking place.

We conclude that solutions of acids, alkalis (such as caustic soda) and salts (such as common salt and copper sulphate) conduct electricity. Water is a very poor conductor of electricity. Liquids which conduct electricity are called **electrolytes**. Oils and solutions of substances such as sugar do not conduct electricity. Oil, in fact, is a good insulator and some electrical equipment is oil-filled for this reason. One pure liquid is a very good conductor of electricity, but this is because it is a metal — mercury.

When an electric current is passed through an electrolyte, a chemical change takes place and the process is known as **electrolysis**.

The electrode at which the current enters the liquid is the **anode**.

The electrode at which the current leaves the liquid is the **cathode**.

The apparatus in which electrolysis is carried out is called a **voltameter**.

The electrolysis of water

The chemical effect of a current can be used to find out the composition of different compounds. An important example of this is the **electrolysis of water**, figure 33.17.

Pure, distilled water is only a feeble conductor of electricity, so an acid (a few cubic centimetres of concentrated sulphuric acid is sufficient) is added, to increase the conducting power and allow more electricity to pass through and produce the effect more quickly. It could be argued that the acid is being split up rather than, or perhaps as well as, the water. But it can be shown experimentally that the amount of acid present is the same at the end as at the beginning, so we know that none can have been decomposed.

As soon as the current is switched on, bubbles of gas appear at each electrode. When a sufficient quantity of each gas has been collected, it can be tested with a lighted splint. In one case the gas burns with a characteristic 'pop', and in the other the splint burns more brightly. The gas produced at the cathode is hydrogen and the gas at the anode is oxygen. This shows that water contains oxygen and hydrogen. The chemical name for water is hydrogen oxide.

Fuel cells

In the electrolysis of water, electrical energy produces a chemical change and two substances, hydrogen and oxygen, are formed. If the process could be reversed it might be possible, under the right conditions, to feed hydrogen and oxygen to a cell and produce electricity. This is the principle of the **fuel cell**, in which fuels such as hydrogen, natural gas and carbon monoxide are used with oxygen to produce electricity directly.

In figure 33.18 the cell is continuously supplied with hydrogen and oxygen under pressure. The gases diffuse through the special nickel electrodes and, on reaching the hot concentrated potassium hydroxide solution, react to form water and produce electrons which can flow through an external circuit as an electric current. Fuel cells have been used in spacecraft, but are not so far in common use.

fig. 33.18 Simple fuel cell

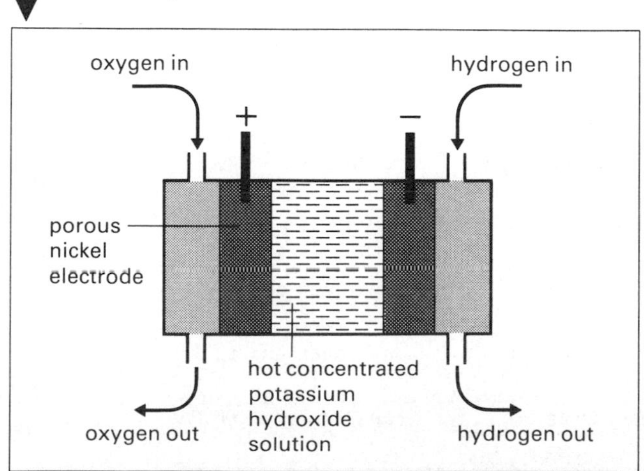

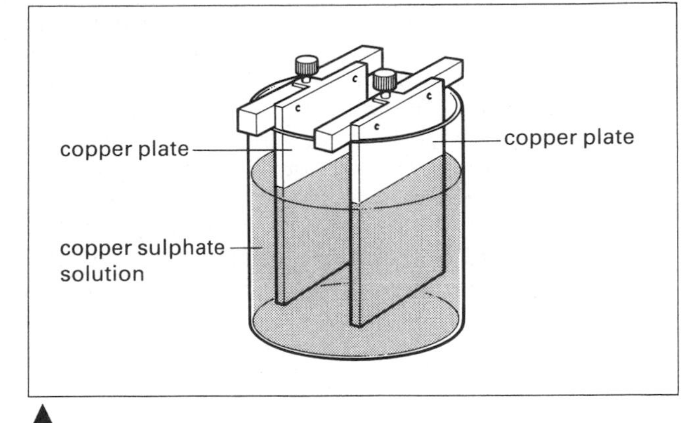

fig. 33.19 Copper voltameter

The electrolysis of copper sulphate

A copper voltameter consists of two copper plates with terminals attached, supported in a solution of copper sulphate, figure 33.19. To carry out electrolysis both plates are weighed and a current is passed for some time. When the plates are examined, a new coating of copper is observed on the cathode. Both plates are re-weighed. The gain in mass of the cathode is found to equal the loss in mass of the anode. No copper has been taken out of the solution. The net result of the electrolysis is a transfer of copper from anode to cathode.

The ionic theory

When copper sulphate (or any salt, acid or alkali) dissolves in water its molecules split up (**dissociate**) into two parts. These parts have either more or fewer electrons than the normal atoms, and so are negatively or positively charged. Charged atoms or groups of atoms are called **ions**.

fig. 33.20 Dissociation of copper sulphate

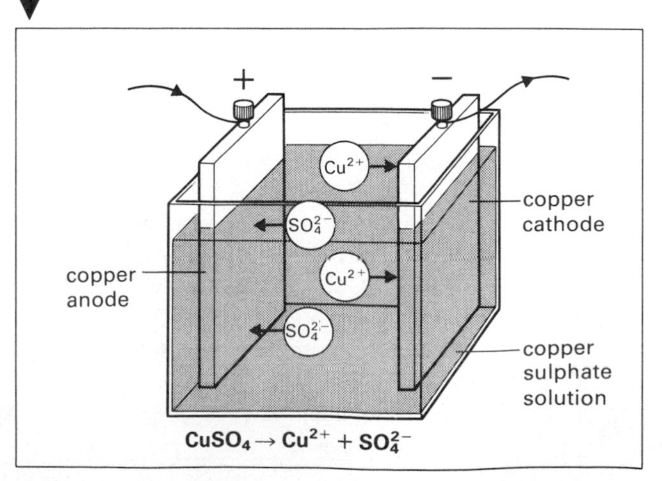

For example, copper sulphate dissociates into copper ions (Cu^{2+}) and sulphate ions (SO_4^{2-}), figure 33.20. When the current is passed through the solution, the charged ions are attracted to the electrodes. The positively-charged copper ions are attracted to the negatively-charged cathode, and the negatively-charged sulphate ions are attracted to the positively-charged anode. This movement of charged atoms within the liquid is equivalent to an electric current.

Two points must be stressed. The dissociation into ions takes place on solution and is quite independent of the presence of an electric current. Conduction in metals is brought about by a *one-way* flow of electrons. Conduction in electrolytes is brought about by a *two-way* flow of positive and negative ions.

What actually happens when the ions reach the electrodes depends on the nature of the ions and the material of the electrodes. During the electrolysis of copper sulphate solution with copper electrodes:

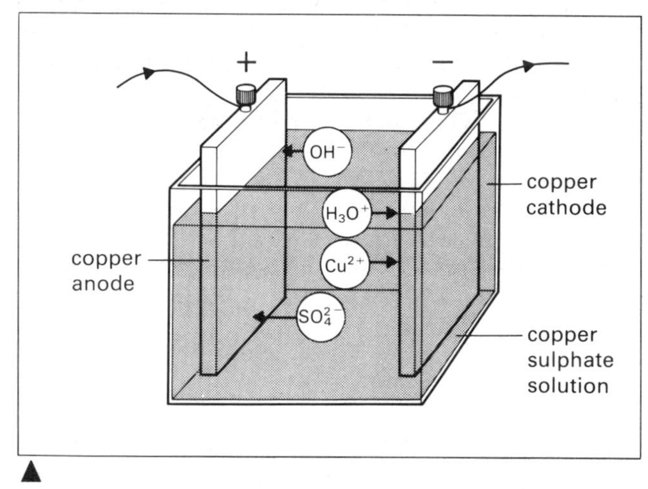

▲ fig. 33.21 In copper sulphate solution, four different kinds of ion are formed: Cu^{2+}, H_3O^+, OH^- and SO_4^{2-}

at the anode: SO_4^{2-} ions give up their two extra electrons and new copper ions are formed from the copper anode.

at the cathode: Cu^{2+} ions gain two electrons and the neutral copper atoms so formed are deposited on the cathode.

This simple explanation does not really cover all the facts. Water itself dissociates into hydroxonium ions (H_3O^+) and hydroxyl ions (OH^-):

$$2H_2O = H_3O^+ + OH^-$$

In copper sulphate solution we have four different kinds of ions, figure 33.21. Positive ions travel to the cathode and negative ions travel to the anode. At the electrodes the possibilities are that one or other of the ions will discharge, or that the atoms of the electrode will dissolve to form ions. Which of these actions takes place depends on which requires the least energy. At the anode, less energy is required to form copper ions than to discharge either sulphate or hydroxyl ions, so the copper anode dissolves. At the cathode, less energy is required to discharge copper ions than hydroxonium ions, so copper is deposited.

Ions can be arranged in descending order of the energy required to discharge them. This order or list is known as the **electrochemical series**.

The results obtained in some common cases of electrolysis are given in table 33.1.

Electroplating

The chemical effect is used in electroplating, in which a thin layer of one metal is deposited on another metal (or sometimes plastic) in order to give a more attractive appearance, to prevent corrosion, or to increase mechanical strength. The object to be plated is made the cathode of an electrolytic cell. In some cases, the metal to be deposited comes from a pure

table 33.1 Some examples of electrolysis

Electrolysis	At anode	At cathode
water with carbon or platinum electrodes	oxygen given off	hydrogen given off
copper sulphate solution with copper electrodes	copper dissolves	copper deposited
copper sulphate solution with carbon or platinum electrodes	oxygen given off	copper deposited
sodium chloride solution with carbon or platinum electrodes	chlorine given off	hydrogen given off
sodium sulphate solution with carbon or platinum electrodes	oxygen given off	hydrogen given off
fused aluminium oxide with carbon electrodes	oxygen given off	aluminium deposited

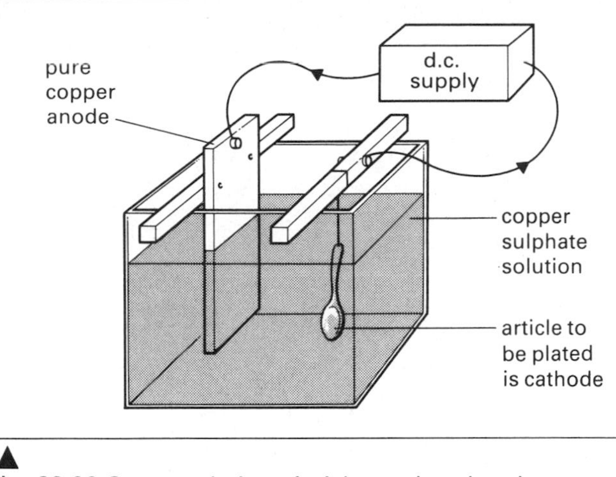

fig. 33.22 Copper plating. Articles to be plated must be clean and free from grease. A useful solution for copper plating is 150 g of copper sulphate crystals with 25 cm^3 of concentrated sulphuric acid in 1000 cm^3 of water

metal anode (copper, nickel and silver); in other cases, the metal ions are taken out of solution (chromium). Records are produced from master discs made by electroplating. Metal parts of cars are electroplated with nickel to prevent rusting and with chromium to produce a non-tarnishing finish.

Extraction and purification of metals

The chemical industry uses electrolysis to purify metals and extract them from their ores. Copper obtained by smelting contains about 5% impurities of iron, gold and silver which were present in the original ore. The impure copper is made the anode of a copper voltameter, and 99.9% pure copper is deposited on a pure copper cathode. Some impurities remain in solution and some fall to the bottom of the cell as a kind of sludge.

Aluminium is produced from its ore, bauxite (aluminium oxide), in an electrolytic cell.

Faraday's laws of electrolysis

Michael Faraday (1791–1867) investigated the process of electrolysis quantitatively and in 1834 discovered two laws about the masses of materials liberated during electrolysis.

Law 1 The mass of a particular element liberated during electrolysis is proportional to the magnitude of the current and to the length of time for which it flows through the electrolyte:

mass $\propto$ current $\times$ time
or mass = constant $\times$ current $\times$ time

$$m = zIt$$

The constant z is the **electrochemical equivalent** of the element.

The product of the current (amperes) and the time (seconds) is the quantity of electricity which has passed through the circuit. The unit of quantity is the **coulomb (C)**, which is the charge passing when a current of 1 ampere flows for 1 second.

$$Q \text{ (coulombs)} = I \text{ (amperes)} \times t \text{ (seconds)}$$

Faraday's first law could be rewritten as: the mass of a particular element liberated during electrolysis is proportional to the charge which passes through the electrolyte. That is

$$m = zQ$$

Law 2 The mass of a substance liberated by a given quantity of charge is proportional to the chemical equivalent of the substance.

If the same current, I, is passed for the same time (i.e. the same charge passes) through the series of voltameters shown in figure 33.23, the masses of the elements set free are in the ratio of their chemical equivalents.

fig. 33.23 Illustrating Faraday's second law of electrolysis

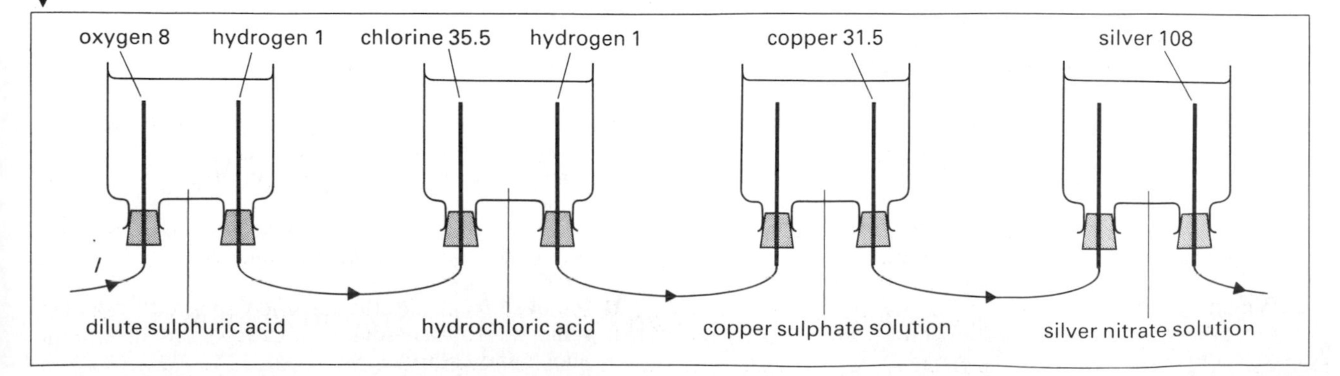

From the equation $m = zIt$, it follows that the electrochemical equivalent of a substance is the mass deposited by a current of 1 ampere flowing for 1 second, or by a charge of 1 coulomb.

The electrochemical equivalent of copper = 0.000 33 gram per coulomb

The electrochemical equivalent of silver = 0.001 12 gram per coulomb

EXAMPLES

1 When a current is passed through a silver plating bath for 25 minutes it is found that 0.4 gram of silver is deposited. Calculate the average value of the current.

$$m = zIt \quad \text{or} \quad I = \frac{m}{zt}$$
$$= \frac{0.4}{0.001\ 12 \times 25 \times 60}$$
$$= 0.24 \text{ ampere}$$

2 A sheet of area 100 square centimetres is to be copper-plated on both sides with a layer of copper 0.05 centimetre thick, using a current of 5 amperes. The density of copper is 8.8 grams/cubic centimetre. How long should the sheet be left in the plating bath?

volume of copper deposited $= 2 \times 100 \times 0.05 \text{ cm}^3$
$= 10 \text{ cm}^3$
mass of copper deposited $=$ volume $\times$ density
$= 10 \times 8.8$
$= 88$ g

$$m = zIt \quad \text{or} \quad t = \frac{m}{zI} = \frac{88}{0.000\ 33 \times 5}$$
$$= 53\ 333 \text{ seconds}$$
$$= 14.8 \text{ hours}$$

EXERCISE 33

In questions 1–4 select the most suitable answer.

1 The strength of the magnetic field between the poles of an electromagnet would be unchanged if the
A current in the electromagnet windings were doubled
B direction of the current in the electromagnet windings were reversed
C distance between the poles of the electromagnet were doubled
D material of the core of the electromagnet were changed
E number of turns in the electromagnet windings were doubled

2 Which of the diagrams in figure 33.24 correctly represents the magnetic field round a wire X carrying current into the plane of the paper?

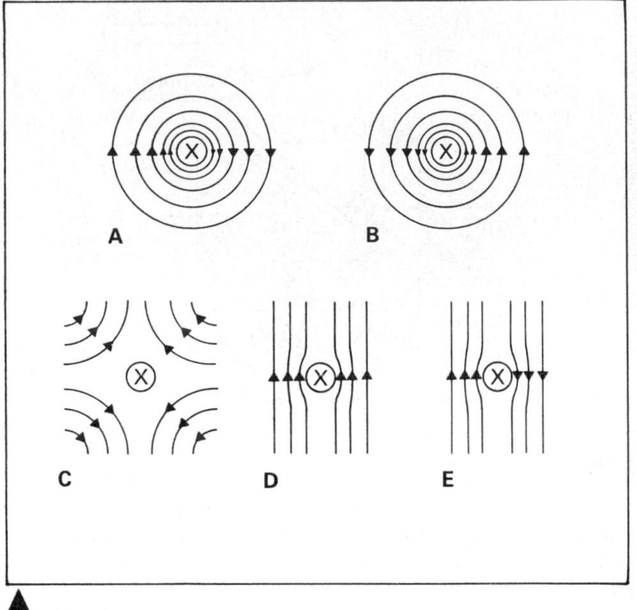

fig. 33.24

3 Figure 33.25 shows a circuit which is an attempt to make an electric bell work. The bell will *not* ring because
A the core is made of soft iron instead of steel
B the armature is made of soft iron instead of steel
C there should be a coil on only one arm of the U core
D the coil should be continued around the end of the U core
E the circuit is wired incorrectly

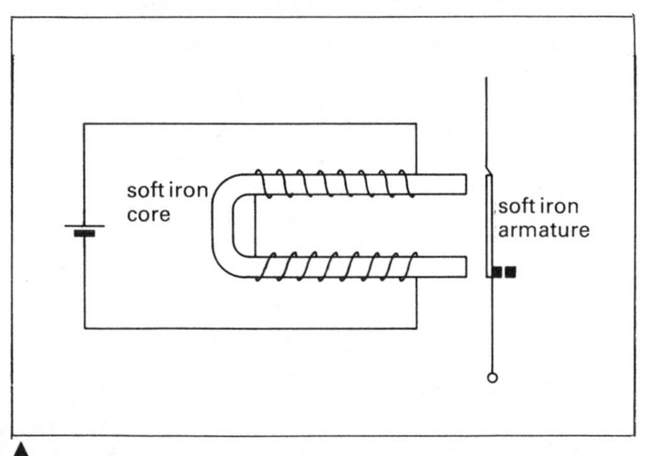

fig. 33.25

4 When a current is passed through copper (II) sulphate solution, using copper electrodes, copper is
A removed from the anode and deposited on the cathode
B removed from the cathode and deposited as sludge
C removed from the cathode and deposited on the anode
D removed from the solution and deposited as sludge
E removed from the solution and deposited on both the anode and cathode

5 Describe how an electric current may be used
a to magnetise a steel bar,
b to demagnetise it.

Draw a diagram of the magnetic field produced by a current flowing in a long straight wire in a plane at right angles to the wire. State the rule which gives the relation between the direction of the current and that of the field.

A long vertical wire carrying a current passes through a horizontal bench. Draw a diagram of the resultant magnetic field on the surface of the bench around the wire due to the current and the earth. Mark the positions of any neutral points formed in this field.

6 The diagram shows a circuit in which an electric current flows through two solenoids connected in series.

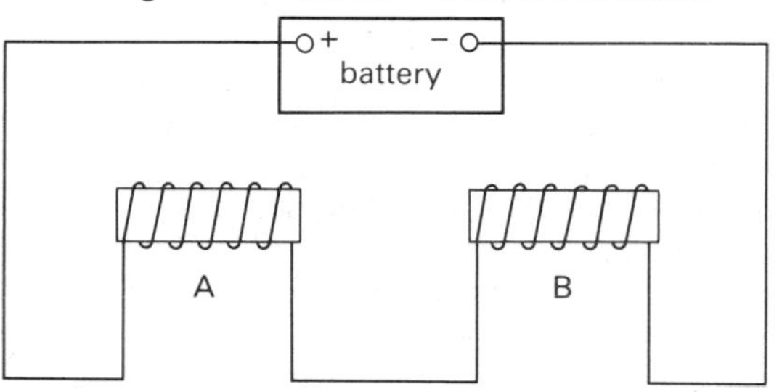

a Sketch the magnetic field between the two solenoids and indicate the polarity.
b What, if any, would be the effect on the force between the two solenoids of (i) increasing the current, (ii) increasing the number of turns in the coils, (iii) introducing soft iron cores into the coils, (iv) reversing the direction of the current in coil A, (v) reversing the direction of the current in both coils, (vi) increasing the distance between the coils.

7 Describe how the electric bell works, and state the effect on its working of
a altering the position of the contact screw, and
b increasing the size of the current passing through the electromagnet.

8 Describe briefly one experiment in each case to illustrate the heating, chemical and magnetic effects of an electric current.

9 Figure 33.26 shows an electromagnetic relay.
a Name the parts labelled, A, B, C and D.
b What material is B made from?
c Why is this material used?
d Describe briefly what happens when the current is switched on.

fig. 33.26

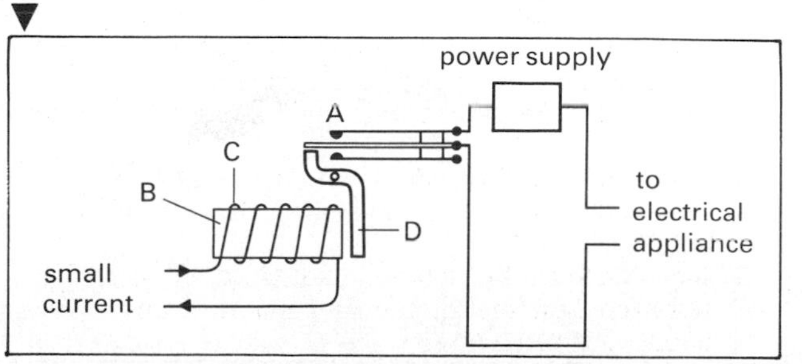

10 Figure 33.27 shows an electromagnet.
a Complete the circuit to which it must be connected in order to make the electromagnet work. The circuit should contain a cell, a switch and a variable resistance (in order to vary the current).
b When no current passes through the electromagnet, would the piece of soft iron, AB, be attracted? Give a reason.
c With the circuit connected and a small current flowing, the maximum weight which can be supported on the hook is 10 newtons. Give *two* alterations each of which could be made in order to make the electromagnet stronger.
d What difference would it make to the electromagnet if the coil were wound on a *wooden* core instead of a soft iron one?
e If the electromagnet were wound on a steel core and the current were switched on, would AB still be attracted?
f What would happen if the current were then switched off again?
g Give *two* reasons why the core of an electromagnet is usually made of soft iron.

fig. 33.27

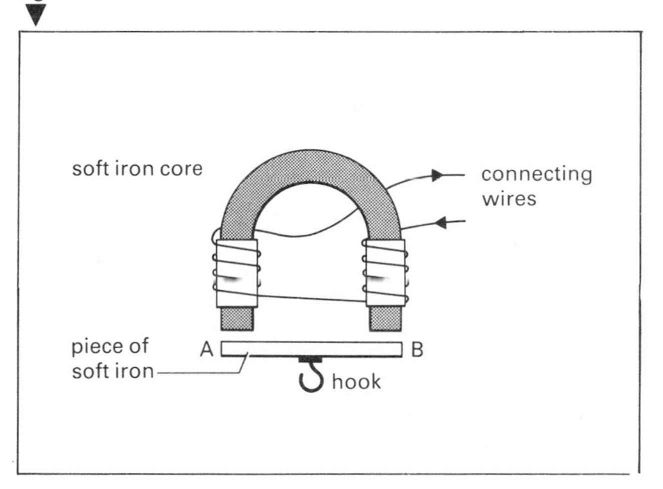

11 a Draw a clearly-labelled diagram to show the structure of an electric fire.
b Draw a three-pin plug which could be used to plug the fire into the mains, and indicate on it (i) the names of the three pins, (ii) the colours of the three wires, (iii) the position of the fuse.
c What is the purpose of the fuse?
d How does the fuse fulfil this purpose?

12 a In the process of electroplating what name is given to (i) the positive electrode? (ii) the negative electrode? (iii) the solution between the electrodes?
b What type of current is used in electroplating?
c What name is given to the electric 'charge carriers' in the solution?
d Name a material which would be suitable for electroplating.
e To which terminal of the electrical supply is an object connected when it is to be electroplated?

34 ELECTRICAL UNITS AND MEASUREMENTS

We have already used certain electrical quantities and the units in which they are measured. In this section we shall define these units and quantities precisely and also investigate the relationships between them.

The fundamental electrical quantities are

- **current** which is the flow of electrons carrying a charge,
- **charge** which measures the quantity of electricity,
- **potential difference** which is the property which determines whether or not a current will flow, and in which direction,
- **resistance** which is the opposition offered by a conductor to the flow of current.

The unit of current — the ampere

The strength of a current is measured by using the magnetic effect. In Chapter 33 it was shown that there are forces exerted between two current-carrying conductors. Currents in the same direction attract, and currents in opposite directions repel. The larger the currents, the greater the forces.

The ampere (A) is defined as the constant current which, flowing through two infinitely long, straight, parallel conductors placed 1 metre apart in a vacuum, produces a force between them of 2×10^{-7} newton per metre length of the conductor

fig. 34.1 Principle of the current balance

The definition may seem complicated, but the ampere is the fundamental unit of electrical measurement, so it is important to define it very precisely. The conditions (e.g. infinitely long wires) stated in the definition cannot be realised practically. The definition is used to work out the forces exerted between coils of current-carrying wire and these forces can be measured using a **current balance**.

The current balance, figure 34.1, consists of two solenoids X and Y which are suspended in equilibrium from the arms of the balance and can move inside two pairs of fixed solenoids A, B and C, D. The current to be measured is passed through all six solenoids in such a way as to displace X and Y. For example, in one arrangement the current flows in the same direction in X, Y, A and D but in the opposite direction B and C. This will force X upwards and Y downwards. The force acting on X and Y is found by adding weights to counterbalance it. The current can then be calculated from the value of the counterbalancing weights and the geometry of the solenoids. Thus the current is determined in terms of the fundamental units of length, mass and time.

Current balances are used in standardising laboratories to calibrate precision ammeters.

fig. 34.2 The ampere

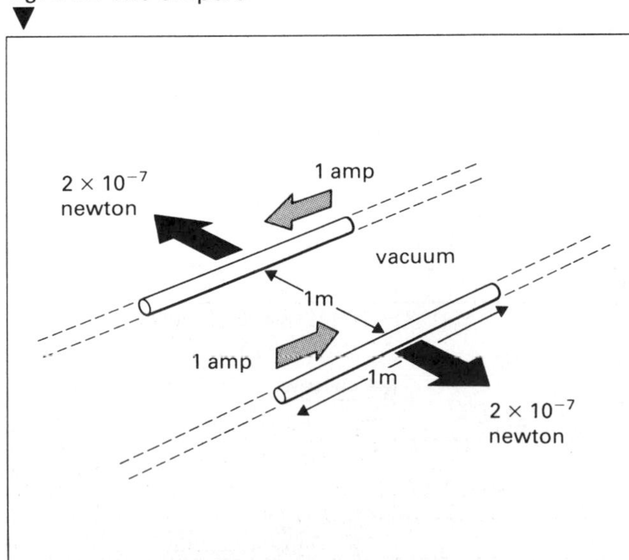

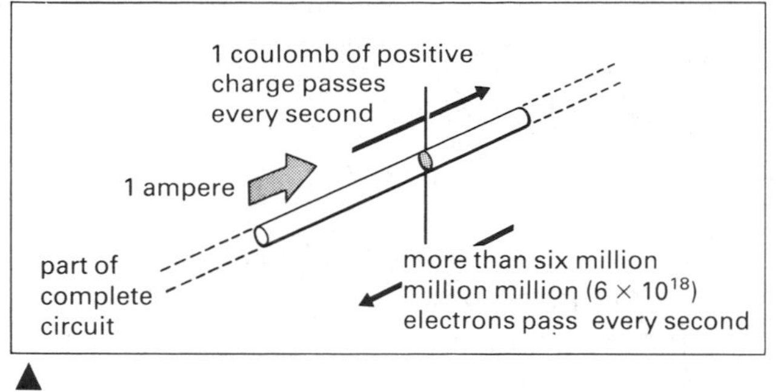

fig. 34.3 The coulomb

The unit of charge — the coulomb

A current is a flow of charge and the larger the current the greater the amount of charge which is passing any point in a circuit per second. Current is rate of flow of charge.

The coulomb (C) is the quantity of charge involved when a current of 1 ampere flows for 1 second

1 ampere for 1 second, charge = 1 coulomb

I amperes for t seconds, charge = It coulombs

$$Q = It$$

coulombs = amperes × seconds

or

$$I = \frac{Q}{t}$$

$$\textbf{amperes} = \frac{\textbf{coulombs}}{\textbf{seconds}}$$

One electron has a negative charge of 1.6×10^{-19} coulomb, so that to make up 1 coulomb of charge we should require

$$\frac{1}{1.6 \times 10^{-19}} = 6.25 \times 10^{18} \text{ electrons}$$

In other words, when there is a current of 1 ampere in a conductor, 6.25×10^{18} electrons pass any point in the conductor every second.

The unit of potential difference — the volt

In order to move a charge of electricity along a conductor, work has to be done. The situation is very similar to lifting a mass from one level to another. The work done depends on the mass lifted and the difference in height between the starting and finishing points. In electricity, the quantity which corresponds to difference in height is potential difference (measured in volts).

The volt (V) is the potential difference between two points when 1 joule of energy is required to transfer 1 coulomb of charge from one point to the other

1 volt = 1 joule/coulomb

$$\textbf{volts} = \frac{\textbf{joules}}{\textbf{coulombs}}$$

An electric current in a conductor has already been compared with the flow of water through a pipe. The analogy is a useful one. A difference in pressure produces a water flow. A difference in potential produces an electric current. Georg Simon Ohm (1787–1854), the German scientist, modelled his ideas about electric currents on the flow of heat along a metal bar heated at one end. A temperature difference produces a heat flow. This is illustrated graphically in figures 34.4a–c.

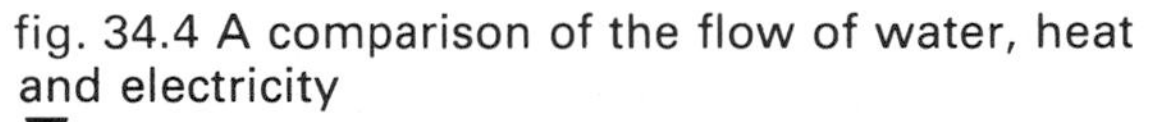

fig. 34.4 A comparison of the flow of water, heat and electricity

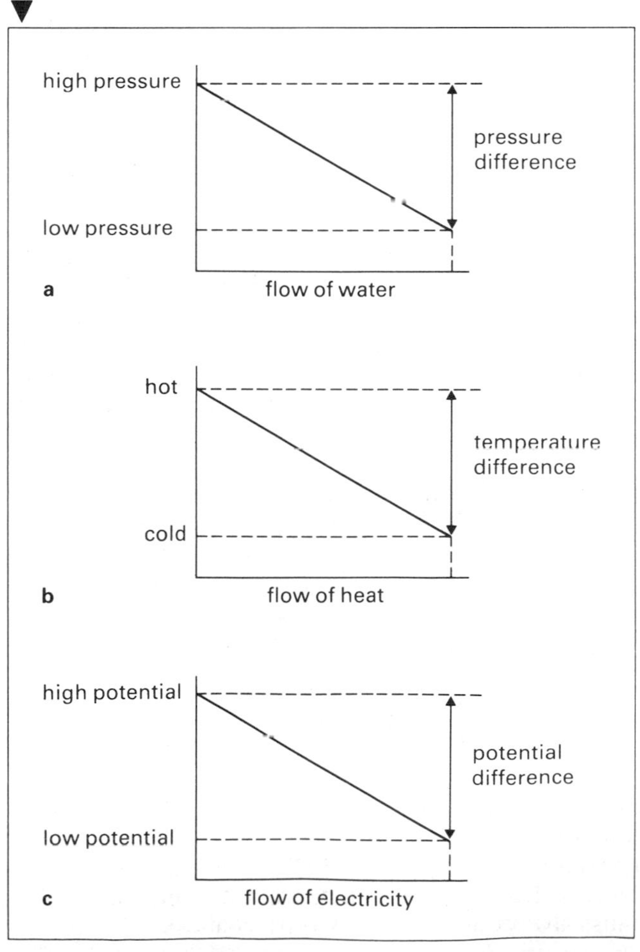

Ohm's law

About 1826 Ohm made a large number of investigations to see how the current flowing through a conductor was related to the magnitude of the potential difference between the ends of the conductor. At that time it was difficult to obtain reliable cells, and ammeters as we have them today did not exist. Nevertheless, Ohm found from his observations the relationship which is known as Ohm's law,which states that

The current flowing through a conductor is directly proportional to the potential difference across its ends provided the temperature is constant

$$I \propto V \quad \text{or} \quad \frac{V}{I} = \textbf{constant}$$

We can easily investigate and verify Ohm's law for a conductor such as a length of wire (about 1 metre of 24 s.w.g. Eureka wire would be suitable). The potential difference is measured by a voltmeter across the ends of the wire. An ammeter measures the current. The potential difference is varied by means of a rheostat, figure 34.5. A series of readings of p.d. and corresponding current are taken. These are given in table 34.1.

It can be seen from the table that the ratio V/I is constant and that a graph of I against V is a straight line through the origin, figure 34.6. Both of these show the direct-proportion relationship between current and potential difference.

The constant quantity V/I is the **resistance** of the conductor. A high resistance will allow only a small current to flow through it, i.e. the ratio V/I is large. If the ratio V/I is small, it means a large current is flowing, and so the resistance of the particular conductor is small. The unit of resistance is the **ohm** which is defined as follows.

A conductor has a resistance of 1 ohm when a p.d. of 1 volt across its ends produces a current of 1 ampere

The symbol for the ohm is Ω. This is the Greek letter omega.

The equations

$$R = \frac{V}{I} \qquad I = \frac{V}{R} \qquad V = IR$$

are all mathematical statements *derived* from Ohm's law for a conductor provided the resistance remains constant. They are very useful in solving problems on electrical circuits. But they are not Ohm's law. This must always be stated in words as above.

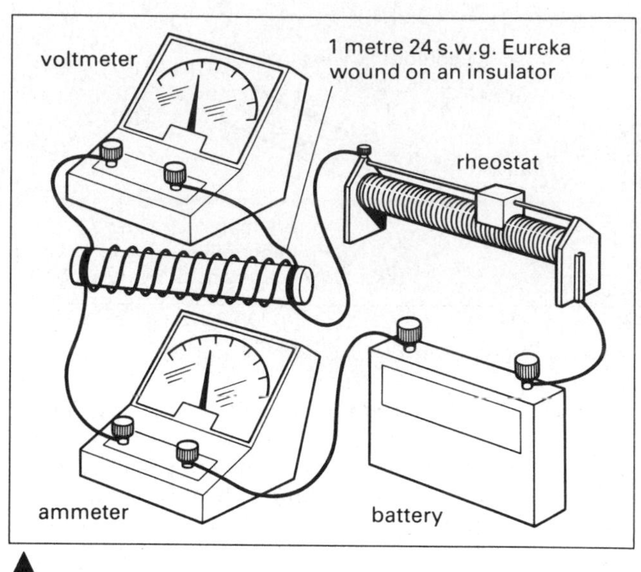

▲
fig. 34.5 Verifying Ohm's law

table 34.1

p.d. V (volts)	Current I (amps)	$\frac{\text{p.d.}}{\text{Current}}$
0.20	0.11	1.80
0.30	0.17	1.76
0.50	0.28	1.78
0.75	0.41	1.82
1.00	0.56	1.79
1.40	0.79	1.77
1.90	1.06	1.79

fig. 34.6 Relationship between current and p.d. for a wire, from the values in table 38.1
▼

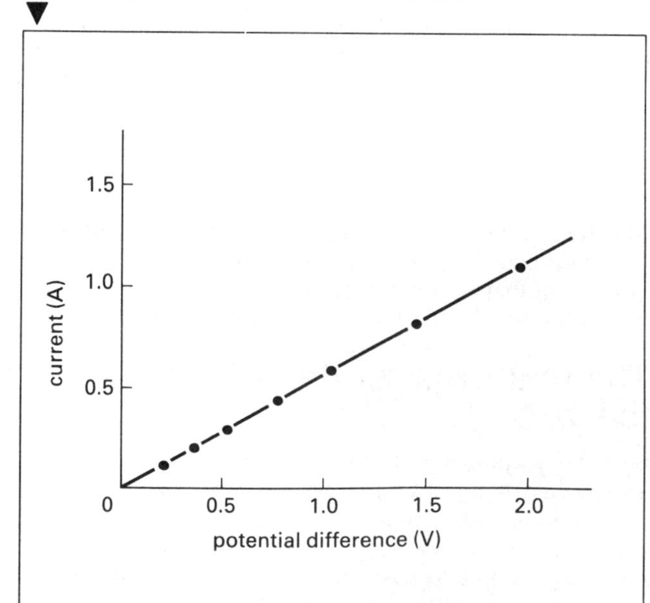

EXAMPLES

1 A 12 volt car headlamp takes a current of 3 amps. What is its resistance?

$$R = \frac{V}{I} = \frac{12}{3} = 4\ \Omega$$

2 An electric iron for use on a 240 V circuit has a resistance of 80 ohms. What current does it take?

$$I = \frac{V}{R} = \frac{240}{80} = 3\ \text{A}$$

3 The working current for an electric soldering iron of resistance 500 ohms is $\frac{1}{2}$ amp. What is the correct voltage to use?

$$V = IR = \frac{1}{2} \times 500 = 250\ \text{V}$$

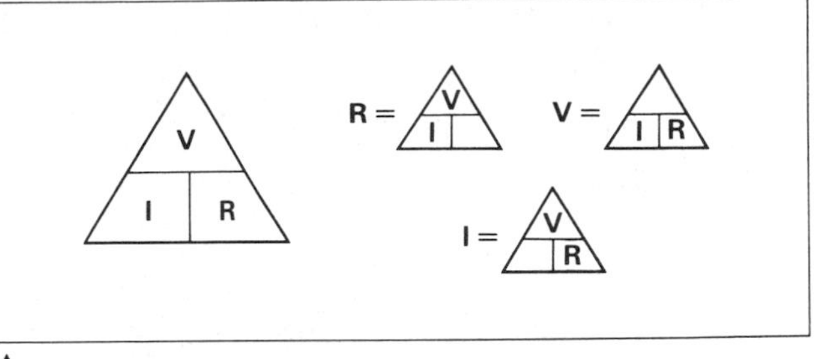

▲ fig. 34.7 How to remember the Ohm's law formula. Cover up the quantity you wish to find and the remaining symbols are arranged to give the formula required

fig. 34.8 Relationship between current and p.d. for four different components ▼

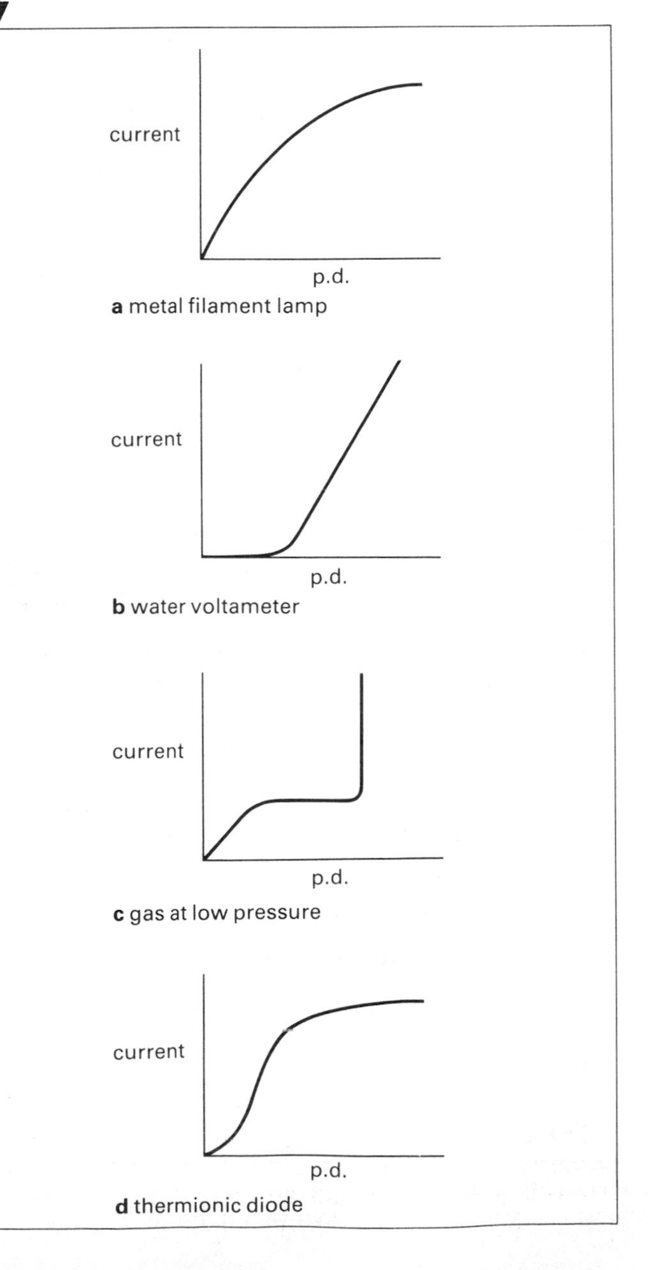

The limitations of Ohm's law

Ohm's law has been described as the most important single generalisation in the whole of electrical circuit theory. It is true that we can use the formulae derived from the law to solve many problems. However, we must be sure that the law really applies to a particular case before we attempt to use these formulae. If the current/potential difference relationship is investigated for different materials, and in different sets of circumstances, we shall not always get the simple straight-line graph that we obtained for a metallic conductor at constant temperature.

The results of various investigations are shown in the graphs in figure 34.8.

Clearly, none of the graphs in figure 34.8 shows a simple straight line, direct-proportion relationship. In the case of the metal filament lamp, the ratio V/I is increasing, which means that the resistance of the filament must be increasing. As the current is increased the filament gets hotter and hotter, so that the 'constant temperature' condition of the law does not hold. We shall learn later on that resistance does, in fact, increase with increasing temperature. In the other examples, too, Ohm's law is not obeyed over the whole range of the graph.

The use of Ohm's law must be limited to cases where the resistance of the conductor remains constant. This means metallic conductors at constant temperature. Such conductors are called **resistors**.

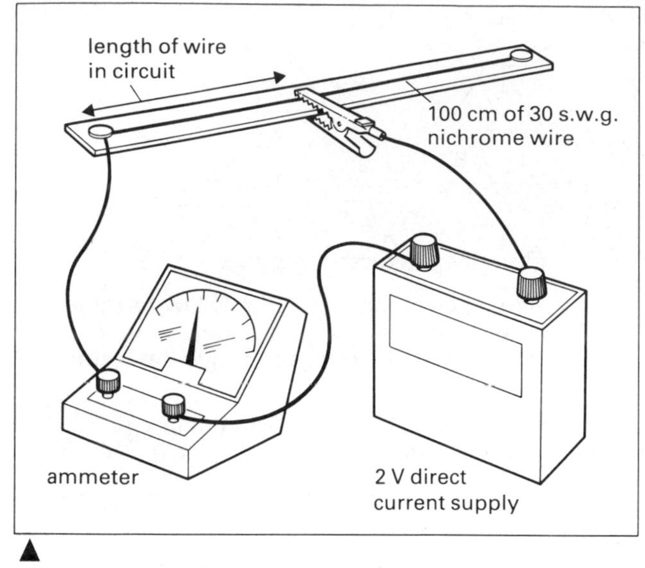

fig. 34.9 Demonstrating the dependence of resistance on length

Resistance and resistors

Figure 34.9 shows 100 cm of 30 s.w.g. Nichrome wire fastened to a metre ruler. A crocodile-clip connection enables different lengths of wire to be included in the circuit. Adjust the length of wire in the circuit so that the ammeter reads 2 amperes. Now double the length of wire in the circuit and take the ammeter reading. Now double the length again and see what the current is.

When the length is doubled the current is halved, and when the length is doubled again the current falls to a quarter of its original value. This means that when you double the length of a wire you also double its resistance. Which is what we should expect. The resistance of a wire varies directly as its length.

A similar experimental approach, with constant lengths of wires of different diameters, shows that thin wires have a greater resistance than thick wires, and that the resistance varies inversely as the cross-sectional area of the wire. If the cross-sectional area is doubled the resistance is halved, and so on.

Combining these two ideas we have

$$\textbf{resistance} \propto \frac{\textbf{length}}{\textbf{cross-sectional area}}$$

or

$$R = \rho\frac{l}{a}$$

The constant ρ is called the **resistivity**, and is a property of the material of which the wire is made. If the length is 1 m and the cross-section 1 square metre, then $R = \rho$ and the definition of resistivity is

The resistivity of a material is numerically equal to the resistance of a specimen 1 metre long and with 1 square metre cross-section

The unit of resistivity is the **ohm-metre**.

The resistivity of the good conductors, such as copper and aluminium, is low. Insulators have a high resistivity, about a million million times greater than that of the metallic conductors. In between these two extremes are the **semiconductors**, such as silicon and germanium, which are used in the manufacture of transistors.

Resistance and temperature

The resistance of a conductor depends on its length, its area of cross-section and the material of which it is made. Resistance also depends on temperature. Ohm's law speaks of 'constant temperature', and in the section on the conditions under which the law holds we saw that the resistance of a lamp filament increases as it heats up.

In metallic conductors resistance increases with increasing temperature

The resistance of an electric lamp filament at its working temperature is several times its resistance when cold. This means that the current through the cold filament is much greater than when the lamp is working normally. Fuses in circuits have to be large enough to stand this initial current surge. Lamps most often burn out as soon as they are switched on.

In semiconductors the resistance decreases with increasing temperature

The change in resistance of semiconductors with temperature is much greater than that of conductors. The resistivity of some semiconducting materials drops to half its value for a rise of 20 °C. Such materials are used in **thermistors**, which are used as thermometers and in t.v. circuits as current-controlling devices. Other uses are illustrated on pp. 319–320.

According to the kinetic theory, a temperature rise increases the kinetic energy of the atoms and makes them vibrate more vigorously. The electrons moving through the conductor will experience more violent and frequent collisions, and so the electron flow is slowed down as the resistance to the flow increases. The change in resistivity for semiconductors is due to a change in the number of free electrons, which increases with increase in temperature.

When metals are cooled to nearly absolute zero their resistance to a flow of current vanishes and they become **superconducting**. A current started in a superconducting circuit would theoretically go on for ever without losing energy.

Light dependent resistors (LDRs)

Another factor which affects the resistance of certain semiconductors (such as cadmium sulphide) is the intensity of the light falling on them.

In light dependent resistors resistance decreases with increasing light intensity

The resistance can drop from a few megohms to about a hundred ohms when the LDR is taken from complete darkness to bright sunlight.

LDRs are used in light meters and in automatic cameras for controlling the aperture. More examples of their use are given on pp. 319–320.

The behaviour of metals, semiconductor thermistors and light dependent resistors with changes in temperature and light is summarised in table 34.2.

table 34.2.

	Cold	Hot
metallic conductors	low resistance	high resistance
semiconductors, thermistors	high resistance	low resistance
	Dark	**Light**
light dependent resistors	high resistance	low resistance

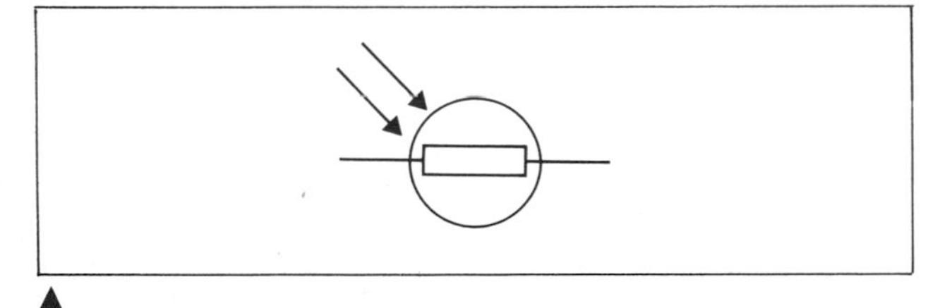

▲ fig. 34.10 The symbol for a light dependent resistor

Resistors in series and parallel

Like cells and lamps, resistors can be connected in series and parallel or in a combination of both systems.

When resistors are joined in **series** the same current I flows through them all, figure 34.11.

The potential differences across individual resistors are

$$V_1 = IR_1 \qquad V_2 = IR_2 \qquad V_3 = IR_3$$

The total drop in potential $V = IR$, where R is the total equivalent resistance.

$$\text{Since } V = V_1 + V_2 + V_3$$
$$IR = IR_1 + IR_2 + IR_3$$

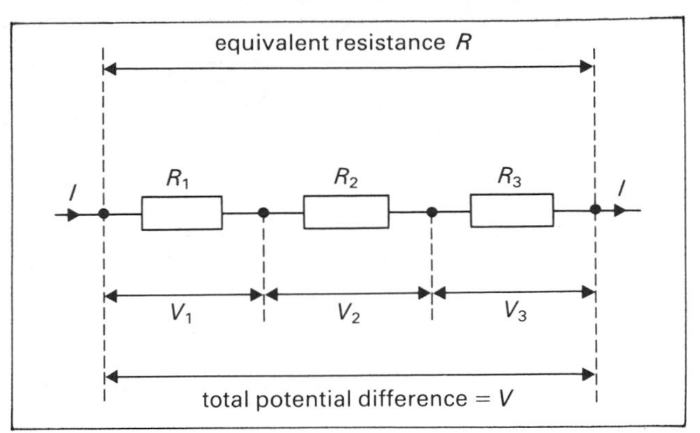

fig. 34.11 Three resistors in series

$$\text{Therefore } \boldsymbol{R = R_1 + R_2 + R_3}$$

When resistors are in series, the equivalent resistance is equal to the sum of the individual resistances.

When resistors are in **parallel**, the current in the main circuit divides, part passing through each resistor. The potential difference across all resistors is the same, figure 34.12. Therefore,

$$I_1 = \frac{V}{R_1} \qquad I_2 = \frac{V}{R_2} \qquad I_3 = \frac{V}{R_3}$$

Since $I = I_1 + I_2 + I_3$

$$\frac{V}{R} = \frac{V}{R_1} + \frac{V}{R_2} + \frac{V}{R_3}$$

$$\text{Therefore } \boldsymbol{\frac{1}{R} = \frac{1}{R_1} + \frac{1}{R_2} + \frac{1}{R_3}}$$

fig. 34.12 Three resistors in parallel ▼

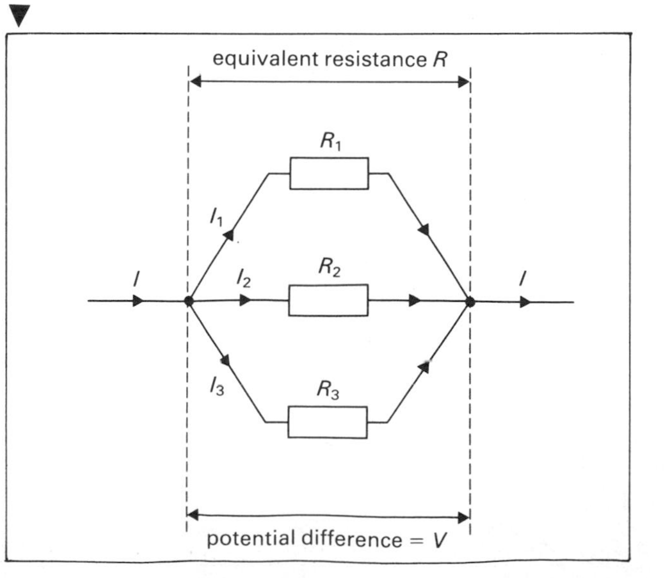

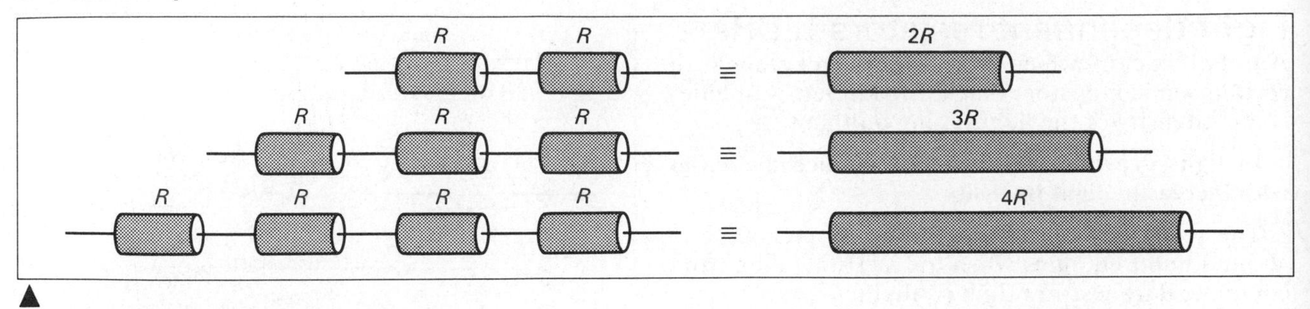

▲
fig. 34.13 When resistors are joined in series their combined resistance is *greater* than the resistance of any of the resistors

fig. 34.14 When resistors are joined in parallel their combined resistance is *less* than the resistance of any of the resistors
▼

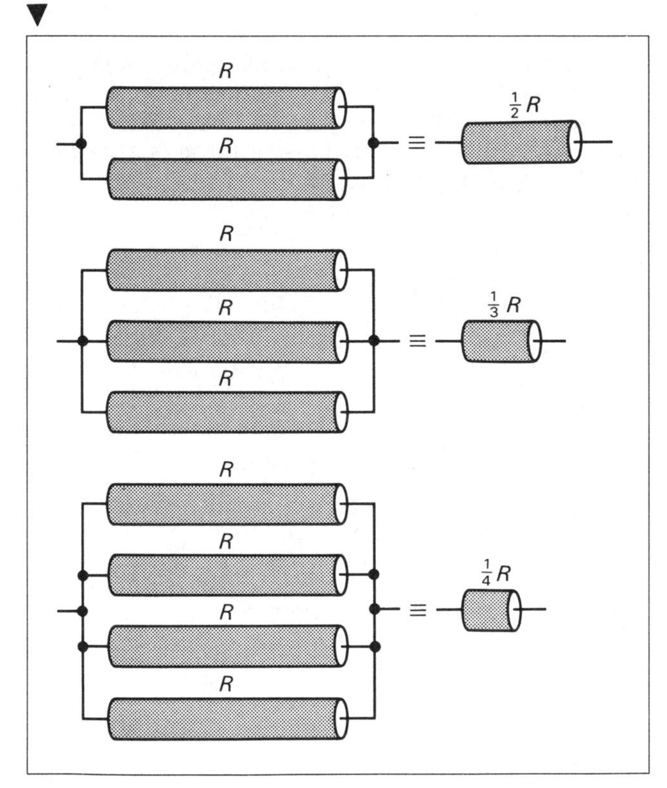

fig. 34.15
▼

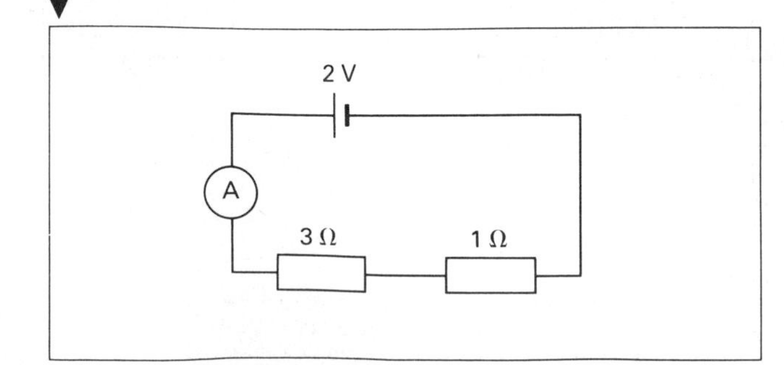

When resistors are in parallel, the reciprocal of the equivalent resistance is equal to the sum of the reciprocals of the individual resistances.

Ohm's law for a complete circuit

When applying Ohm's law to a single conductor, we use the relationship $I = V/R$. If we wish to apply the law to a complete circuit consisting of batteries and resistors, we use the relationship:

$$\textbf{current} = \frac{\textbf{total e.m.f.}}{\textbf{total resistance}}$$

$$\boldsymbol{I = \frac{E}{R}}$$

where R is the total resistance.

It is best to calculate the total e.m.f. of the batteries and the total resistance of the circuit separately, and then insert the values in the expression for the current.

EXAMPLES

1 In figure 34.15 calculate the current reading of the ammeter (A).

Since no value is given for the internal resistance of the cell, we assume it to be negligible.

total e.m.f. = 2 V
total resistance $R = R_1 + R_2$
$= 3 + 1 = 4\ \Omega$

$$\text{current } I = \frac{\text{total e.m.f.}}{\text{total resistance}}$$
$$= \frac{2}{4}$$
$$= 0.5\text{ A}$$

2 In figure 34.16 calculate the current reading on the ammeter (A).

total e.m.f. = 2 + 2 = 4 V (cells in series)

$$\text{total resistance } \frac{1}{R} = \frac{1}{R_1} + \frac{1}{R_2}$$

$$= \frac{1}{4} + \frac{1}{4} = \frac{2}{4}$$

$$R = \frac{4}{2} = 2\ \Omega$$

$$\text{current } I = \frac{\text{total e.m.f.}}{\text{total resistance}}$$

$$= \frac{4}{2}$$

$$= 2\text{ A}$$

fig. 34.16

3 In figure 34.17 calculate the current in the circuit and the potential difference across the 4 Ω resistor.

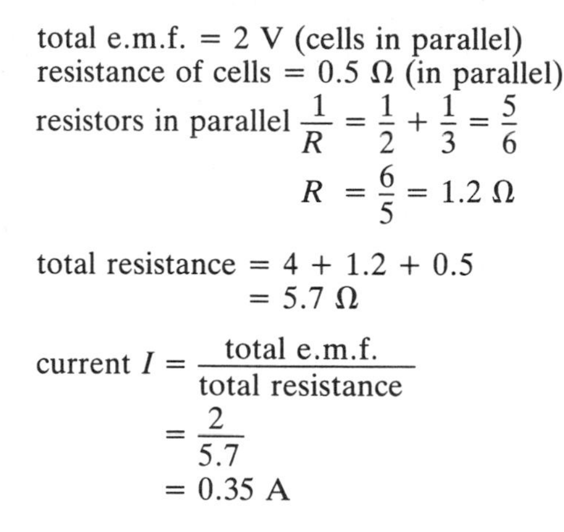

total e.m.f. = 2 V (cells in parallel)
resistance of cells = 0.5 Ω (in parallel)

$$\text{resistors in parallel } \frac{1}{R} = \frac{1}{2} + \frac{1}{3} = \frac{5}{6}$$

$$R = \frac{6}{5} = 1.2\ \Omega$$

total resistance = 4 + 1.2 + 0.5
= 5.7 Ω

$$\text{current } I = \frac{\text{total e.m.f.}}{\text{total resistance}}$$

$$= \frac{2}{5.7}$$

$$= 0.35\text{ A}$$

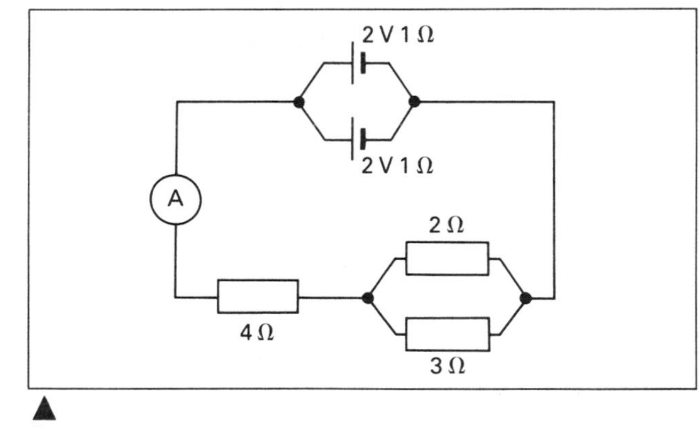

fig. 34.17

Using the simple Ohm's law formula for the 4 Ω resistor:

$$V = IR$$

$$= 0.35 \times 4$$

$$= 1.4\text{ V}$$

fig. 34.18 Arrangement of coils in a resistance box. The middle coil is shown in circuit

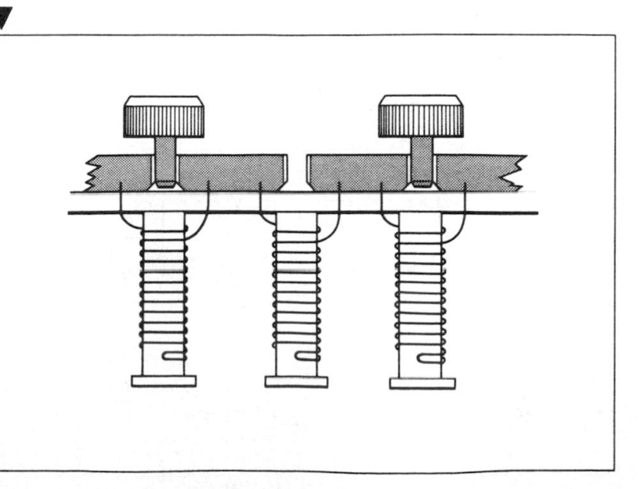

Measurement of resistance

One way to measure an unknown resistance is to compare it with a known standard resistance using the **substitution method**. The idea is that, if the known and unknown resistances are the same, they will allow equal currents to flow when connected in turn in the same circuit. A **resistance box** consists of a series of coils of known resistances connected across gaps in thick brass bars. The gaps can be closed with well-fitting tapered brass plugs. When a plug is in position, it offers little or no resistance to a current. When a plug is removed, a particular coil is put into circuit. Adding up the values of the 'gaps' gives the total

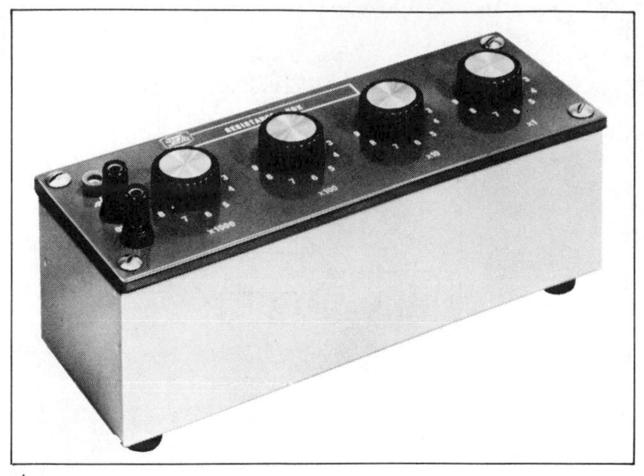

fig. 34.19 A 'dial' resistance box

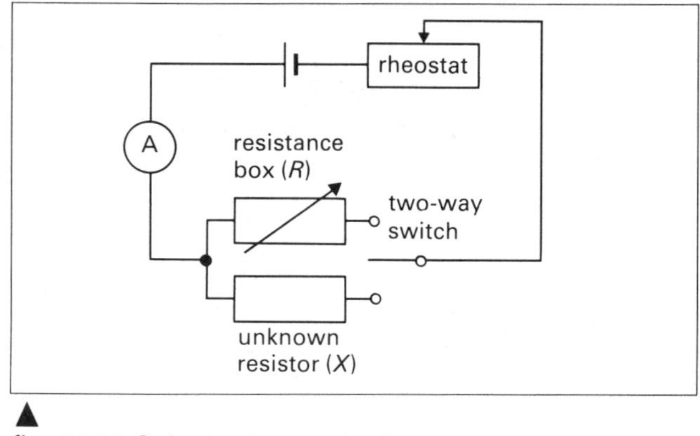

▲
fig. 34.20 Substitution method

fig. 34.21 Ammeter–voltmeter method
▼

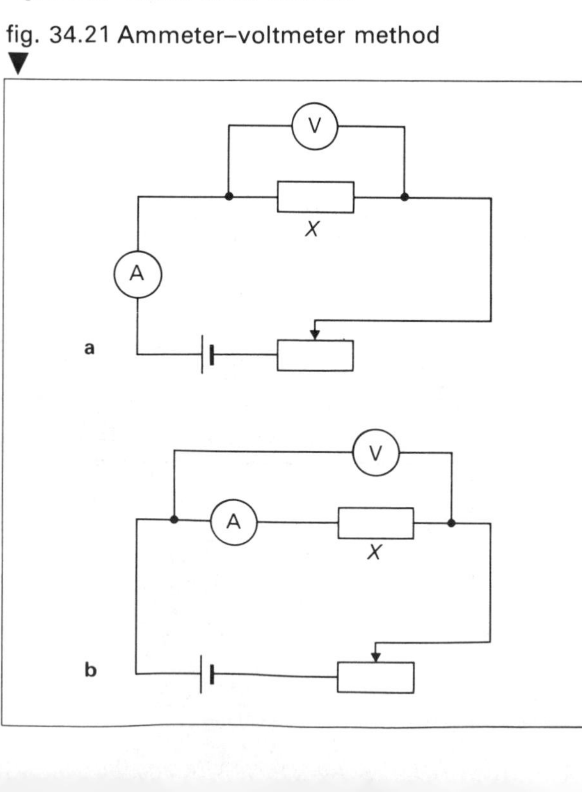

resistance in circuit. With coils of values 1, 2, 2, 5, 10, 20, 20, 50 ohms, for example, any resistance value between 1 ohm and 110 ohms in steps of 1 ohm can be obtained. Resistance boxes are often of the 'dial' type, in which the sum of the readings on the various dials indicates the total resistance in circuit. Resistance boxes must only be used with small currents (stated on the box); otherwise the coils heat up and their resistance changes.

The known and unknown resistors are connected in circuit with a two-way switch, a cell and an ammeter, figure 34.20. The unknown resistor (X) is switched in circuit and the ammeter reading observed. The standard resistor is now switched in circuit and its value (R) adjusted until the same reading is obtained; then $X = R$. A large rheostat is included in the circuit in order to prevent too large a current flowing.

Another method of measuring resistance is the **ammeter–voltmeter method**. If the current through a resistor and the potential difference across it are known, then the resistance can be worked out using the Ohm's law formula. If a rheostat is included in the circuit, a series of readings can be taken and the average value of the ratio of V to I found. An example of such a series of readings is given in table 34.3. Or a graph of V against I can be drawn and the slope of the best straight line will give the resistance.

table 34.3

p.d. (volts)	0.95	1.15	1.50	1.75	2.15
Current (amps)	0.30	0.40	0.50	0.60	0.70
$R = \frac{\text{p.d.}}{\text{current}}$ (Ω)	3.16	2.88	3.00	2.92	3.07

Average value for resistance = 3.00 Ω

There are two ways in which this circuit can be connected, as shown in figures 34.21a and b. Ideally, a voltmeter should have an infinite resistance so that no current flows through it, and an ammeter should have zero resistance so that there is no potential difference across it. But obviously these conditions cannot be realised in instruments which depend for their working on a current passing through a coil situated in a magnetic field. So some compromise has to be made. In the first circuit (a), the ammeter measures the current passing through the resistor and the voltmeter in parallel. In the second circuit (b), the voltmeter measures the p.d. across the ammeter and resistor in series. Which method is chosen depends very much on the instruments available, but in general the first method is suitable when the unknown resistor has a low value and the second method will be more accurate when the unknown resistor has a high value.

The potentiometer or potential divider

A length of uniform resistance wire AB has a potential difference V between its ends, figure 34.22a. Since the wire is uniform there will be a steady drop in potential along it, represented graphically in figure 34.22b. If we make contact with the wire at some point C then the p.d. between A and C can be found as a fraction of the total p.d.

$$\frac{v}{V} = \frac{l}{AB} \qquad v = \frac{l}{AB}V$$

As the point C is moved along the wire the value of v changes and we have, in effect, a source of potential difference which is continuously variable between the values of zero and V.

Two practical forms of potentiometer are shown in figures 34.23. The sliding type with the wire wound in an open coil can deal with large currents while the enclosed type is for small currents such as those in radios.

We can look at the potential divider in another way. Since we are dealing with a uniform wire the resistance is proportional to the length and the above expression for v can be written

$$v = \frac{R_{AC}}{R_{AB}} V$$

The wire could be replaced by two resistors, one of which is variable, figure 34.24.

$$R_{AC} = R_2 \text{ and } R_{AB} = R_1 + R_2$$

$$\text{so } v = \frac{R_2}{R_1 + R_2} V$$

As R_2 is changed v changes. When R_2 is small v is small. When R_2 increases v increases.

The idea of the potential divider is used in the transistor switching circuits described in chapter 38.

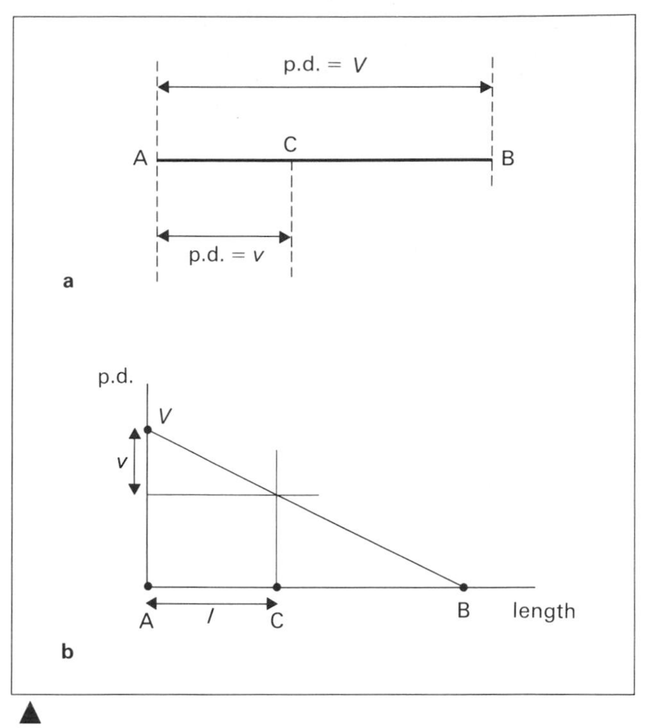

fig. 34.22 Principle of the potentiometer

fig. 34.23 Sliding and rotary type potentiometers

fig. 34.24 The potential divider

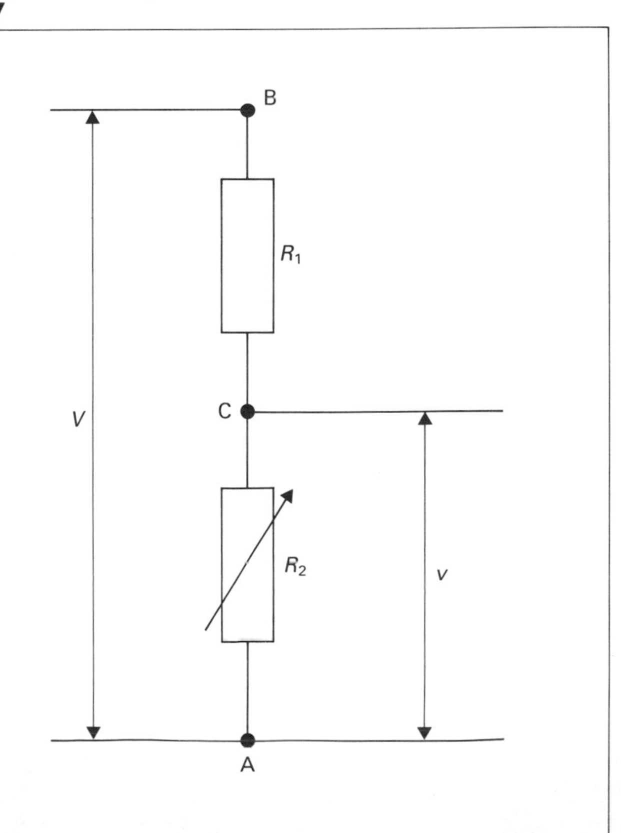

EXERCISE 34

In questions 1–5 select the most suitable answer.

1 Electric charge is measured in
A amperes
B volts
C joules
D watts
E coulombs

2 Ohm's law states that provided the temperature of a conductor is constant
A its resistance is proportional to the current through it
B the current through it is constant
C the current through it is proportional to the potential difference across it
D the current through it is inversely proportional to the potential difference across it
E the potential difference across it is proportional to its resistance

3 A volt may be defined as
A a coulomb per second
B a joule per second
C an ampere per coulomb
D a coulomb per joule
E a joule per coulomb

4 The circuit in figure 34.25 shows a 9 V supply connected across three resistors in series. What is the p.d. across the 3 ohm resistor?
A 2 V
B 3 V
C 4 V
D 6 V
E 9 V

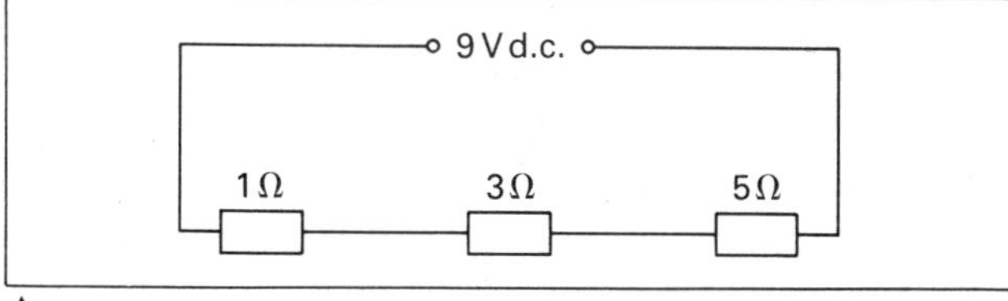

fig. 34.25

5 Figure 34.26 shows 3 resistors connected between the points X and Y. What is the total resistance between X and Y?
A 1.6 ohms
B 9.5 ohms
C 11.0 ohms
D 13.5 ohms
E 18.0 ohms

fig. 34.26

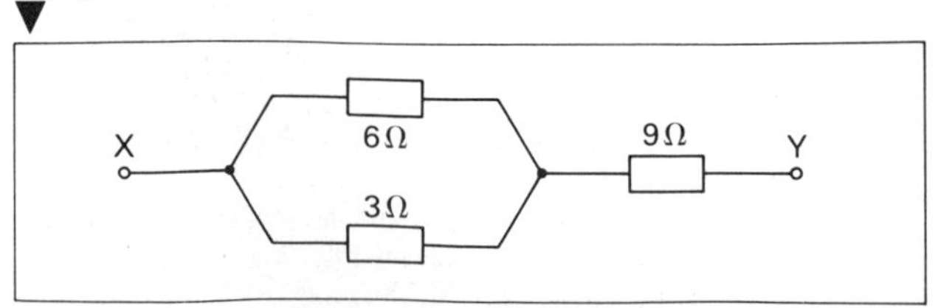

6 Ohm's law states: 'The current flowing through a conductor is proportional to the potential difference applied, provided physical conditions remain constant.' Explain in your own words what is meant by the following words or phrases, which come from the above statement:
a current,
b conductor,
c potential difference,
d physical conditions,
e is proportional to.

7 This question is concerned with simple resistance networks. Read each part and then decide which of the arrangements in figure 34.27 provides the correct answer.
a In which arrangement is there a current of 3 A in a 2 Ω resistor?
b In which arrangement is there a current of 1 A in a 4 Ω resistor?
c In which arrangement is there a potential difference of 4 V across a 4 Ω resistor?
d In which arrangement is there a potential difference of 24 V across a 4 Ω resistor?

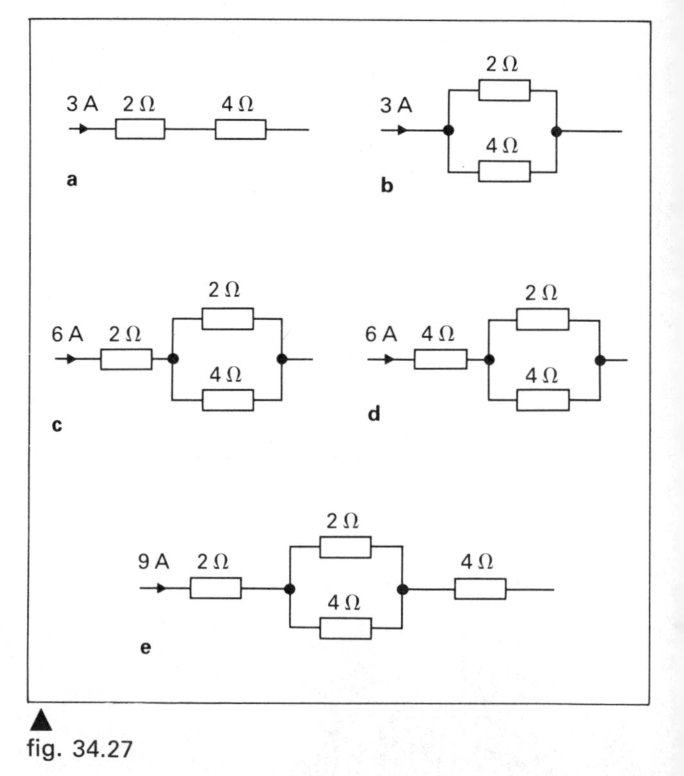

fig. 34.27

8 a You have been provided with three resistors of value 2 Ω, 3 Ω and 6 Ω. (i) Draw a diagram of these three resistors arranged in series and then calculate the value of the total resistance. (ii) Draw a diagram of these three resistors arranged in parallel and then calculate the total resistance of this arrangement.
b What resistor must be placed in parallel with a 1 Ω resistor to create a total resistance of 0.5 Ω?

9 A portable radio can be powered by a 9 V dry battery, but the electrical system of a car uses a 12 V lead–acid accumulator.

- **a** How many 1.5 V cells are there in the 9 V battery? Draw a sketch to show how they are connected.
- **b** What is the e.m.f. of a single lead–acid cell? How many would be required to make a 12 V battery?
- **c** Give *two* reasons why a dry-cell battery would be unsuitable as a car battery.
- **d** Give *two* reasons why accumulators are not usually used for radios.
- **e** A 12 V accumulator is connected in the circuit shown in figure 34.28 and supplies a current of 2 A. What voltage is required between A and B to get this current flowing through the 4 Ω resistor?
- **f** What voltage remains across B and C?
- **g** What current flows through the 6 Ω resistor?
- **h** What current flows through the resistor R?
- **i** What is the value of the resistor R?

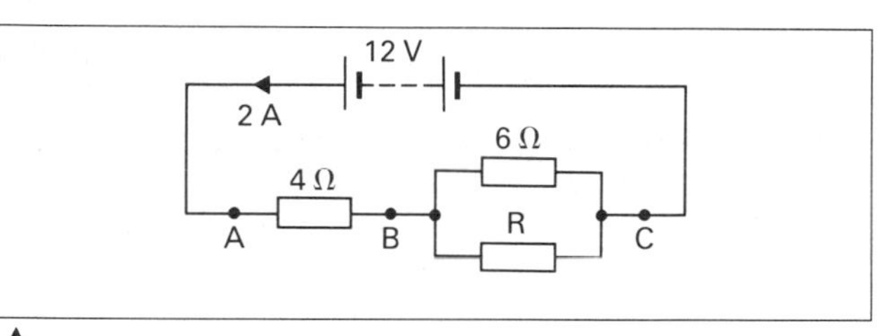

fig. 34.28

10 Figure 34.29 shows a simple circuit. Calculate

- **a** The resistance of the 5 Ω and 1 Ω resistors combined as shown.
- **b** The resistance between A and B.
- **c** The current flowing through the battery, assuming the battery has no resistance.
- **d** The current flowing through the 5 Ω resistor.
- **e** The current flowing through the 1 Ω resistor.

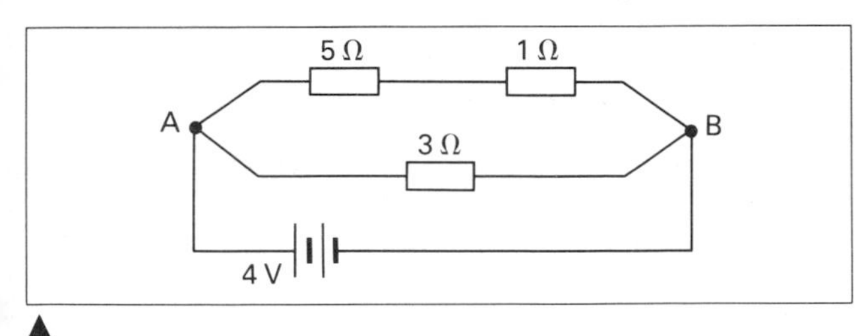

fig. 34.29

11 a (i) In figure 34.30a what is the value of the current flowing when the switch is open? (Show all working.) (ii) What current flows when the switch is closed?

b (i) In figure 34.30b what current flows through the ammeter when the switch is closed and the variable resistor set at zero? (ii) What current flows when the switch is closed and the variable resistor set at 20 Ω?

c (i) Using figure 34.30c explain why no current will flow when the switch is closed. (ii) What effect will there be on the circuit if the switch remains open but A is joined to B? (iii) What effect will there be on the circuit with the switch closed with A joined to B?

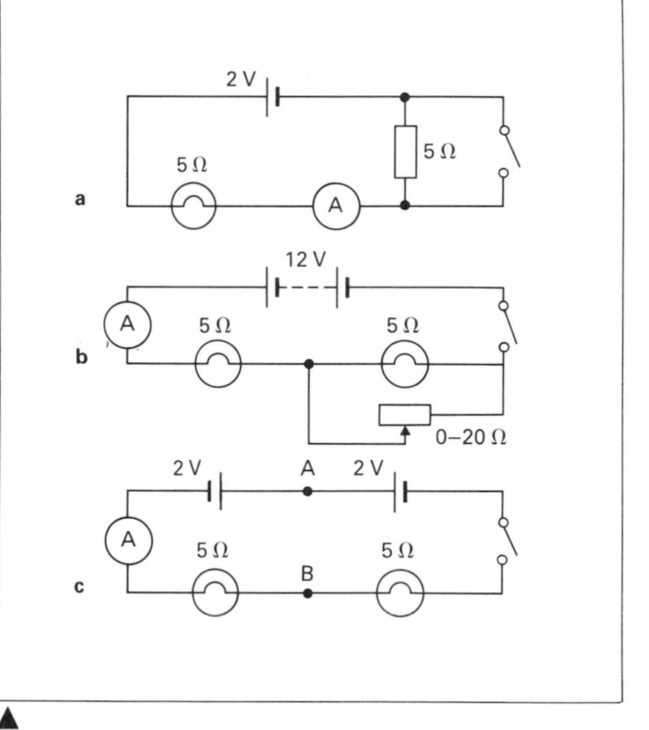

fig. 34.30

12 State in each case whether the electrical resistance of a length of copper wire will increase, decrease or remain unchanged if

- **a** its length is doubled,
- **b** its cross-sectional area is doubled,
- **c** the copper wire is replaced by an iron one,
- **d** the temperature of the wire is raised.

13 As the current through a tungsten filament lamp was increased, the potential difference across it varied as given in the table below

- **a** Sketch a suitable circuit for obtaining these data.
- **b** Plot a graph of potential difference against current.

Current I (amperes)	Potential difference V (volts)
0	0
0.2	1.3
0.4	2.6
0.6	4.0
0.8	5.4
1.0	7.0
1.2	9.0
1.4	12.0

- **c** Using the graph, what is the resistance of the filament when current I = 0.5 ampere?
- **d** What happens to the temperature of the filament as the current increases?
- **e** What happens to the resistance of the filament as the temperature rises? Explain how your answer may be deduced from the graph.

35 ELECTRICAL POWER AND ITS DISTRIBUTION

An electric current is a very convenient way of transmitting energy from place to place; for example, from a battery or generator to where it is needed to light a lamp, heat a radiator or drive a motor. The energy is carried by the moving charges (electrons) in the conductor and these charges require a potential difference between the ends of the conductor to keep them moving. We can calculate the quantity of energy transferred from the fundamental definitions of charge and potential difference (p. 273).

Electrical energy

One joule of energy is required to transfer one coulomb of charge between two points having a potential difference of one volt. For Q coulombs and a potential difference of V volts, the energy required would be VQ joules.

$$\boldsymbol{E = VQ}$$

joules = volts × coulombs

The coulomb is defined as the quantity of charge which passes when a current of one ampere flows for one second. A current of I amperes flowing for t seconds involves a movement of It coulombs.

$$\boldsymbol{Q = It}$$

coulombs = amperes × seconds

Substituting the value for Q in the first equation, we get

$$\boldsymbol{E = VIt}$$

joules = volts × amperes × seconds

This relationship enables us to calculate the energy transfer in any given situation.

EXAMPLE

A current of 4 A flows through the element of a radiator working on a 250 V supply for half an hour. How much heat energy is produced?

$$\begin{aligned} \text{energy transfer} &= VIt \\ &= 250 \times 4 \times 30 \times 60 \\ &= 1\,800\,000 \text{ joules} \\ &= 1800 \text{ kJ} \quad \text{or} \quad 1.8 \text{ MJ} \end{aligned}$$

The energy equation

$$E = VIt \qquad (1)$$

can be rewritten to include resistance R by using the Ohm's law relationship, for example:

$$\text{Using } V = IR \qquad \boldsymbol{E = I^2Rt} \qquad (2)$$

$$\text{Using } I = \frac{V}{R} \qquad \boldsymbol{E = \frac{V^2t}{R}} \qquad (3)$$

These are alternative forms of the energy equation and in each we assume that the resistance R is constant, which it is not if the temperature changes. But they are useful if applied in the correct conditions. Equation (2) shows that, for constant resistance, the energy transfer is proportional to the square of the current. If the current is doubled, the energy transfer is four times as great, and so on.

Electrical power

Power measures the rate at which energy is transferred or converted. A rate of one joule per second gives a power of one watt (see p. 47).

$$\frac{E}{t} = VI$$

$$\frac{\text{joules}}{\text{seconds}} = \text{volts} \times \text{amperes}$$

or

$$\text{power } \boldsymbol{P = VI}$$

watts = volts × amperes

Electrical equipment is usually labelled to show the correct voltage and power rating, figure 35.1.

fig. 35.1 All reputable electrical equipment is marked with the voltage and power rating. This notice, from the back of a portable air filter, also carries a warning

EXAMPLES

1 An electric kettle uses power at the rate of 600 watts. What current does it take on a 240 volt circuit?

$$\begin{aligned} \text{watts} &= \text{volts} \times \text{amperes} \\ 600 &= 240 \times I \\ I &= \frac{600}{240} \\ &= 2\tfrac{1}{2}\ \text{A} \end{aligned}$$

2 Each element of a radiator is rated at 1 kW. What current does it take when both elements are in use on a 240 volt circuit?

$$\begin{aligned} \text{watts} &= \text{volts} \times \text{amperes} \\ 2000 &= 240 \times I \\ I &= \frac{2000}{240} \\ &= 8\tfrac{1}{3}\ \text{A} \end{aligned}$$

3 An electric cooker with the oven and two hot plates in use takes a current of 25 amps, on a 250 volt circuit. What is its power rating?

$$\begin{aligned} \text{watts} &= \text{volts} \times \text{amperes} \\ &= 250 \times 25 \\ &= 6250 \\ \text{power rating} &= 6\tfrac{1}{4}\ \text{kilowatts} \end{aligned}$$

Two further expressions for electrical power can be deduced from the energy conversion expressions:

$$P = I^2R \quad \text{and} \quad P = \frac{V^2}{R}$$

but it is probably easier to remember $P = VI$ and use the Ohm's law relation $I = V/R$ to solve problems.

Paying for electricity

When we use electrical energy to produce light or heat or drive a motor, the amount we have to pay must obviously depend on the amount of energy used. The electricity generating boards could charge so much per joule or, since this is a very small unit of energy, per kilojoule or megajoule. But, since energy = power × time, the unit chosen is the **kilowatt-hour (kWh)**. A kilowatt-hour is the amount of electrical energy used when a piece of electrical equipment rated at 1000 watts is used continuously for one hour.

1 kilowatt = 1000 watts = 1000 joules/second
1 hour = 60 × 60 seconds
1 kilowatt-hour = 1000 × 60 × 60 joules
= 3 600 000 joules
= 3.6 MJ

If a 2 kW radiator is switched on for 3 hours it uses 2 kW × 3 hours = 6 kWh, or 6 'units' of electricity. At 5.85 p (1983 price) per unit the electricity consumed would cost 35.1 p.

An electric meter records the kilowatt-hours used. It is driven by a small electric motor, the speed of which depends on both the voltage of the mains and the total current being used, so that its speed measures the watts. The longer it is in use, the more revolutions it makes, so that it actually records watts × time. The motor is geared to dials (or a digital display) which show the kilowatt-hours (units) used, figures 35.2a–c.

The home wiring system

It is desirable to have electricity available for lighting and heating and running other electrical appliances at many different points in a home. Since all the appliances are designed to run on the same voltage they must all be connected in parallel. It should be pointed out here that the domestic supply is **alternating current** from a generator, not direct current (as from a battery) that we have up to now been considering. The reason for using a.c. and not d.c., and the special properties of a.c., are discussed in detail in Chapter 37. Instead of speaking of positive and negative wires, with a.c. we speak of **live** and **neutral** wires.

fig. 35.2a Old-style electric meter

fig. 35.2c Digital-display electric meter

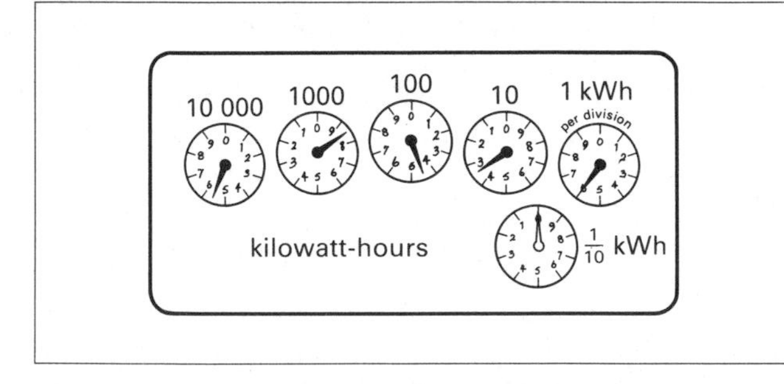

fig. 35.2b Reading = 58 436 units. When the pointer is between the figures, read the smaller figure. When between 9 and 0, read 9

Power up to 3kW (13A) is supplied on the **ring main circuit**, figure 35.3. This is a three-core cable (live, neutral and earth wires) which starts from the consumer unit, runs right round the home and returns to the unit, and into which are connected standardised 13A wall socket outlets (power points). Remote sockets are fed by spur cables which branch off the ring. The ring is protected by a 30 A fuse at the consumer unit. Only one type of plug is used, fused according to its appliance (e.g. 3 A for 700 W, 5 A for 1000 W and 13 A for 3 kW). For a large house, two rings may be needed.

The advantages of the ring main are

- There are two paths to any appliance, each of which carries half the current needed. Therefore, thinner cable can be used.
- One cable supplies many outlets.
- Standardisation of sockets, plugs and fusing.

Lighting is on a separate set of circuits (*not* a ring). They are fed from 5 A fuseways and earthed.

Safety and earthing

It is important to realise the difference between the live and neutral wires carrying an alternating current. The neutral wire is earth-connected at the electricity substation. Figure 35.4b shows that a lamp would light if one contact were connected to the live wire and the other contact to earth, since there would be a complete circuit back to the substation transformer. So, if you touch just *one* bare wire (when the switch is on), and it happens to be the live wire, you will get a dangerous shock because the current passes through your body to earth. If you were standing with wet feet on the bathroom floor the shock would be fatal. It is essential that the switch is put in the live wire. If the electrician made a mistake and put the switch in the neutral wire you would get a shock from the live wire even if the switch were off.

Never touch any wires unless the supply is switched off at the switchboard (consumer unit)

All electrical apparatus made of metal must be **earthed** by using a three-pin plug and three-core wire. If an insulation fault develops in an electric kettle, for example, and the element comes in contact with the metal, a heavy current flows straight to earth and blows the fuses, figure 35.5. If an earth wire is not included and the insulation is faulty, anyone touching the kettle would receive a shock.

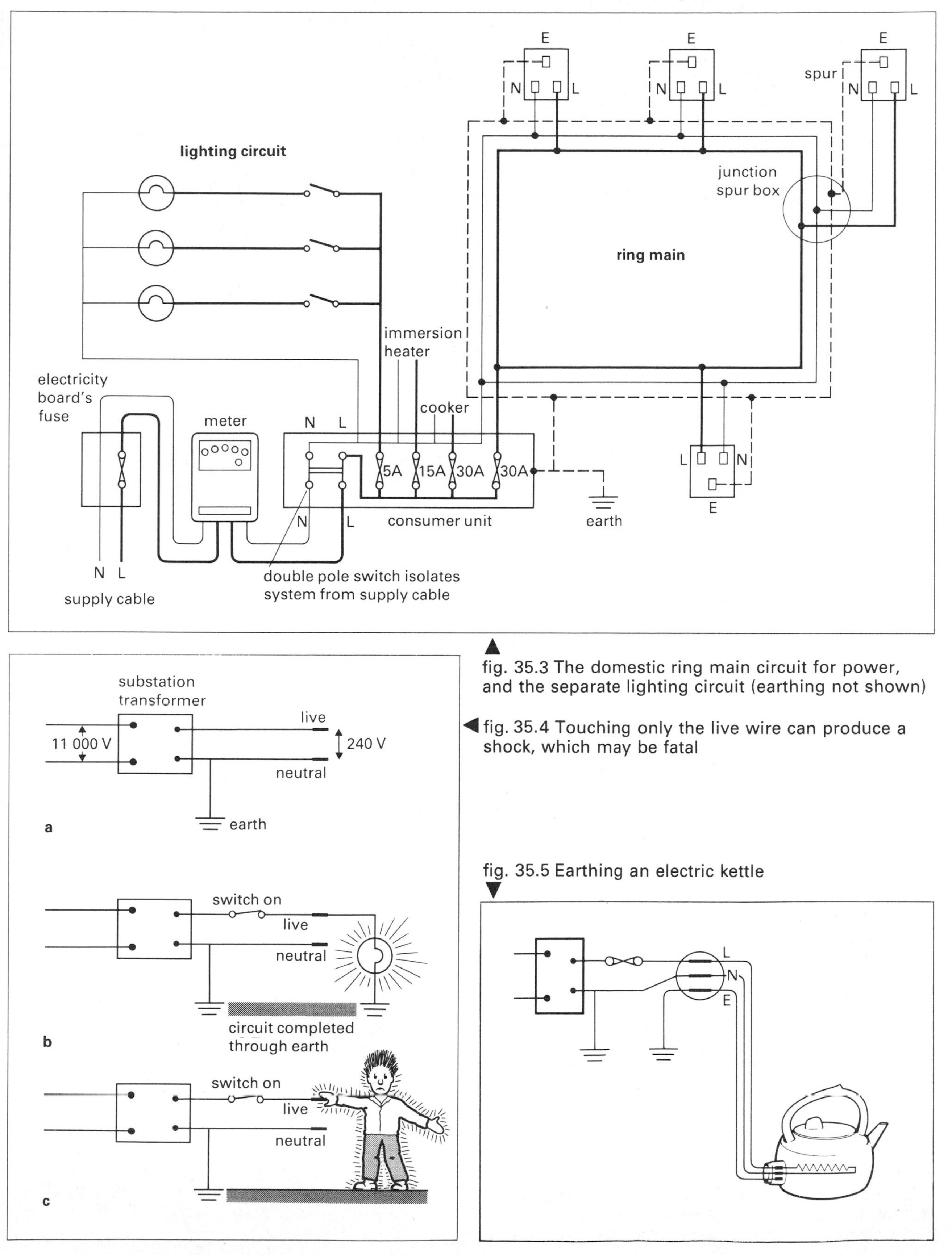

▲ fig. 35.3 The domestic ring main circuit for power, and the separate lighting circuit (earthing not shown)

◀ fig. 35.4 Touching only the live wire can produce a shock, which may be fatal

fig. 35.5 Earthing an electric kettle ▼

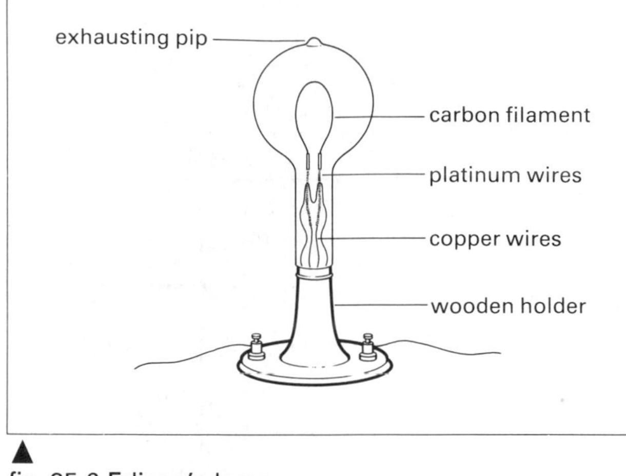

fig. 35.6 Edison's lamp

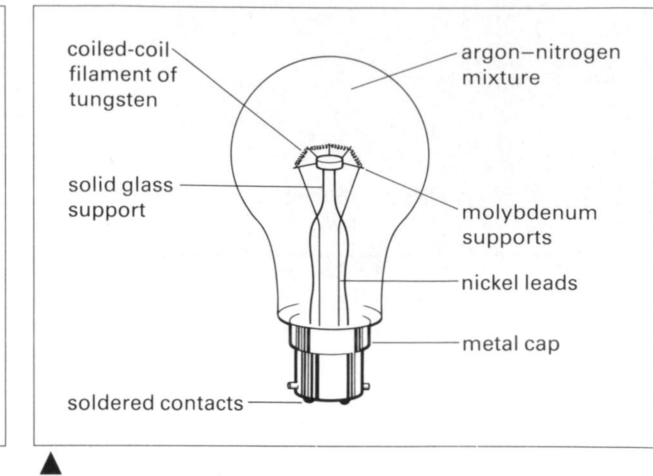

fig. 35.7 Gas filled coiled-coil lamp

Electric lamps

Electric lamps were developed by Joseph Swan in England and Thomas Edison in America in 1878-9. When a current passes through a thin wire, the wire gets hot, and if the temperature is high enough light is produced. At red-heat (about 800 °C) infrared radiation and visible reddy-orange light are produced. As the temperature rises, the remaining colours of the spectrum are added until we get white light emitted at about 1300 °C. Most of the energy supplied to an electric lamp is converted to heat and only about 2% of the energy is emitted as light.

The early lamp filaments worked in a vacuum, figure 35.6, but modern lamps are gas-filled, figure 35.7. This prevents the metal of the filament evaporating and blackening the inside of the glass, thus cutting down the light emitted. The area occupied by the filament is made as small as possible by coiling the filament and then coiling again to give a 'coiled-coil' filament, figure 35.8. The smaller area cuts down heat losses by radiation and makes the lamp more efficient. A coiled-coil lamp gives more light for the same power input than a single-coil lamp.

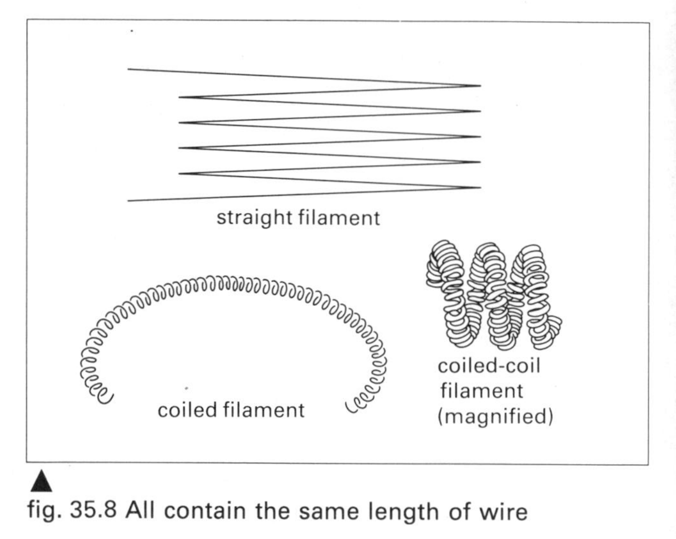

fig. 35.8 All contain the same length of wire

Power loss during transmission of electricity

When a current flows through a resistor heat is produced and given off to the surroundings. This happens in the transmission cables from the power station and produces a loss of power, figure 35.9. The power output at the end of the transmission line is equal to the power input minus the power lost as heat. In order to make the system as efficient as possible, the power loss must be kept to a minimum.

To reduce the loss during transmission, I^2R must be kept small. There is a limit to how small the resistance R can be made; so the best thing to do is to reduce

fig. 35.9 Power loss on transmission lines

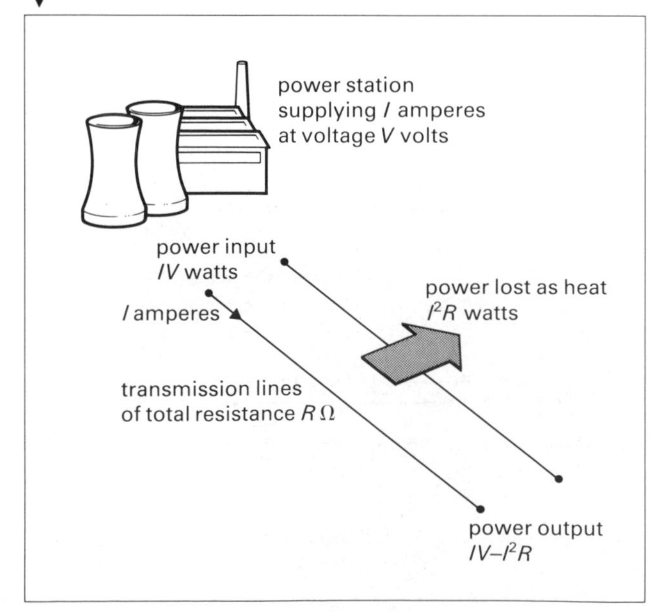

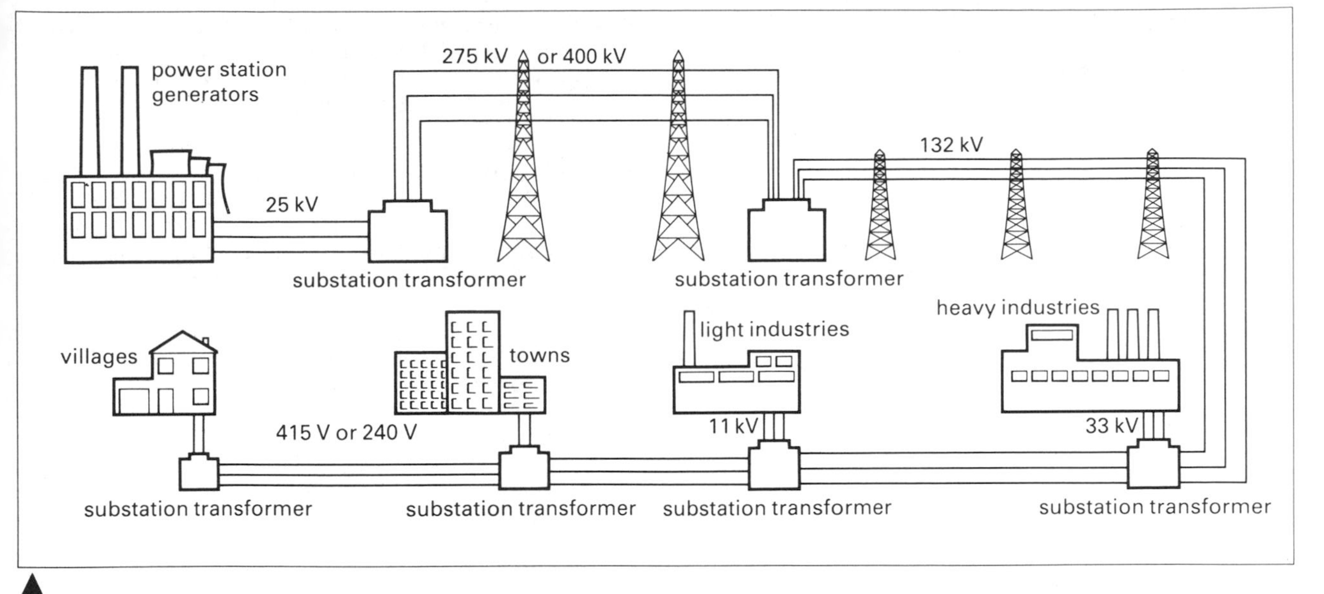

fig. 35.10 Distribution of electric power: the grid system

fig. 35.11 'Pole' transformer for isolated dwellings such as farms

the current I. This can be done by increasing the voltage V because, if a certain quantity of power (IV) has to be transmitted, we can choose whatever values of I and V we like so long as their product remains the same. 1000 kW of electrical power can be transmitted at 100 kV and 10 A or at 10 kV and 100 A. The power loss at the higher voltage and smaller current will be less. An efficient transmission system uses as high a voltage as is consistent with safety. The main problem is that the higher the voltage, the better the insulation has to be.

The advantage of using alternating current is that the voltage can be changed using a transformer (see p. 304). A transformer does not work with direct current.

The grid system

The grid system for the distribution of electrical power uses a few very large power stations and joins them up with a cable system to supply the whole country. The word 'grid' is used meaning a network of cables, figure 35.10. This use of longer cables means that the power losses in the cables has to be as small as possible, hence the use of high-voltage transmission as described above. Large power stations are more economical to run than small ones. Since all the grid stations are connected, current from one region can be supplied to another region in the event of a large demand, or a breakdown. The substations and grid supply points house the transformers which step up or step down the voltage to the appropriate value.

The grid cables are made of aluminium wire wound on to a steel core to give strength. They are about 2 cm in diameter. Long porcelain insulators give a large surface area and the 'umbrella' shape used keeps the lower parts dry to some extent.

EXERCISE 35

In questions 1–4 select the most suitable answer.

1 The power of an electrical apparatus is measured in
A amperes
B coulombs
C volts
D watts
E watt-hours

2 An electric oven is rated at 12 000 W for use on a 240 V supply. Select from the ratings given below the minimum current rating of supply cable for the oven to operate safely.
A 15 A
B 30 A
C 45 A
D 60 A
E 120 A

3 In the circuit shown in figure 35.12 the supply voltage is 24 V d.c. What is the reading on the ammeter?
A 0.25 A
B 0.50 A
C 2.0 A
D 4.0 A
E 8.0 A

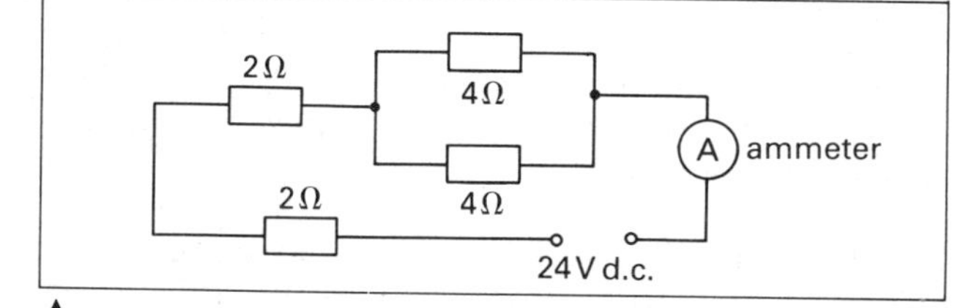

▲
fig. 35.12

4 The equivalent resistance of the arrangement of the four 2-ohm resistors in figure 35.13 is
A $\frac{1}{2}$ ohm
B $1\frac{1}{2}$ ohms
C $4\frac{1}{2}$ ohms
D 5 ohms
E 8 ohms

fig. 35.13
▼

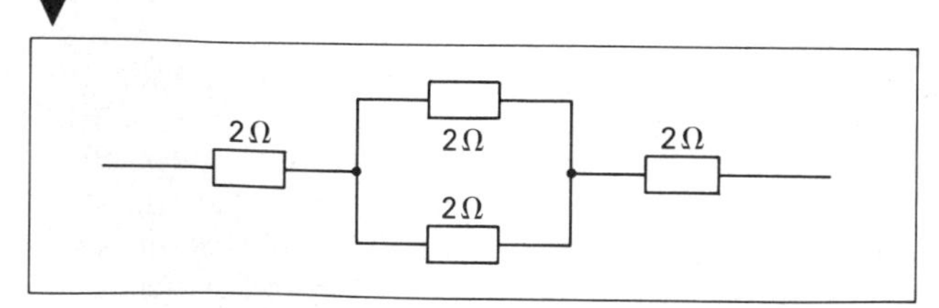

5 a (i) Define the term *power* as applied to an electrical component. (ii) What unit is used to measure electrical power?
b A 750 W heating element of an electric fire is used on a 250 V mains supply. Calculate the current flowing in the element.

6 A light bulb is marked '100 W'. What does 100 W mean? This bulb is used an average of 10 hours a week for 50 weeks of the year. How many kilowatt-hours of electrical energy has the bulb used?

7 a The starter motor of a car has a power rating of 960 W. If it is switched on for 5 seconds, how much energy does it use?
b The same starter motor is powered by connecting it to a 12 V car battery. How much current does it use?

8 a Calculate the current which passes through the following electrical devices operating on 240 V a.c.: (i) radiant heater rated at 1.5 kW; (ii) electric clock rated at 2 W.
b How many 240 V, 100 W bulbs can be connected in parallel before a current of 5 A would be drawn from the electrical supply?
c How many joules of energy are used when five 100 W bulbs are left switched on for one hour?

9 a What physical quantity is measured in megajoules?
b A 1 kW electric fire and a 3 kW immersion heater are allowed to remain switched on for five hours. Calculate the cost of electricity used if one megajoule costs 1p.

10 a A series circuit of 12 fairy lights on a Christmas tree operates on 240 V mains supply. The lamps are rated at 3 W each. (i) What is the voltage drop across each lamp? (ii) What current will be taken by each lamp? (iii) What current will be taken by the whole circuit? (iv) If one lamp fails, the string of lights will usually all go out. Why is this?
b If the string of fairy lights is allowed to operate continuously for 500 hours over the Christmas period, how much will it cost to run them at 2p per megajoule?

11 Figure 35.14 shows a light bulb.
a Of what material is the filament made? Why is this material good for this purpose?
b What gas is contained in the bulb? Why is this gas used rather than air?
c Give a reason why glass is a good material for the stem to which the filament supports are connected.

fig. 35.14
▼

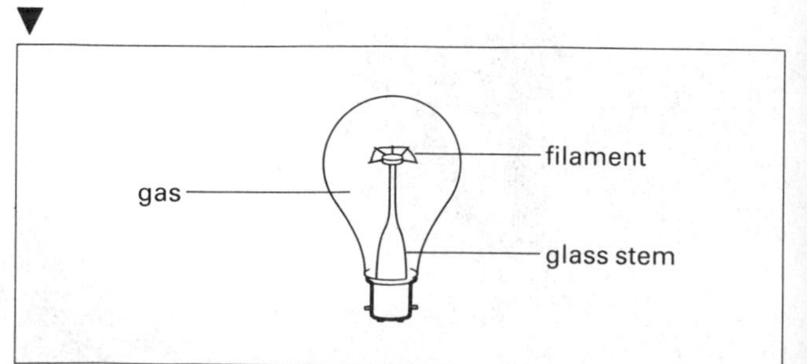

12 A large battery is connected as shown in figure 35.15 to a resistor of resistance 1000 Ω. The potential difference across the resistor is 50 V.

a What is the reading on the ammeter in the circuit?
b How many coulombs of charge are supplied by the battery in 1 minute?
c Calculate the energy dissipated by the resistor in this time.
d What is the power dissipated by the resistor?

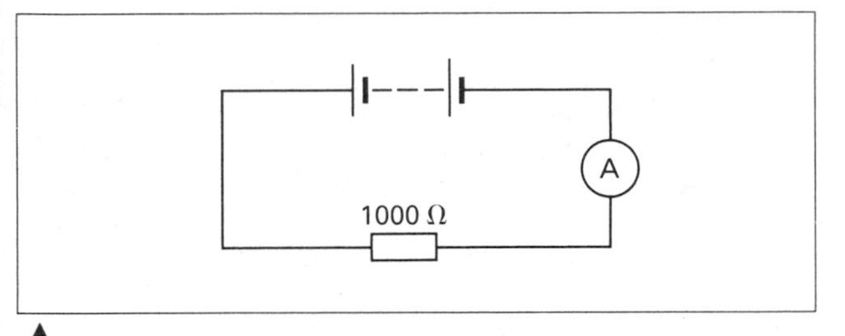

fig. 35.15

13 a Draw a labelled diagram of a ring main circuit showing four sockets, three on the ring main and one on a spur.
b Explain why two electric fires, each taking a current of 12 A, would both work satisfactorily when plugged into any two sockets on the ring main. In your answer refer to the fuse ratings which would normally be used. Briefly state why it would be unwise to plug both fires into the same socket using an unfused multi-socket adapter.

14 Figure 35.16 shows a common switching arrangement. Explain how switches G and H control the lamp. Give an example of where such a system is used.

a If the lamp is a 60 W lamp what is the value of the current flowing through it when it is switched on?
b What is the resistance of the lamp when it is working?
c The resistance of the bulb when cold is found to be very different. Is the resistance when hot higher or lower than the resistance when cold? Explain briefly why this is so.

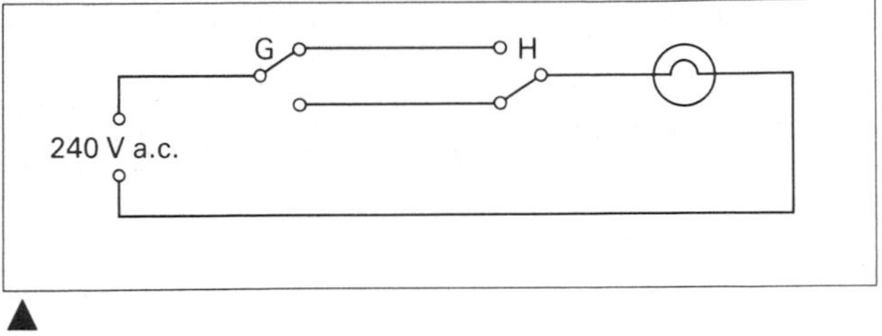

fig. 35.16

15 a State the advantages of using (i) a.c. rather than d.c. for power transmission, (ii) high-voltage a.c. rather than low-voltage d.c.
b A person connecting a 13 A plug to a standard 3-core lead attached to a metal-framed washing machine did not bother to read the instructions. The final connections made were the brown-covered wire to the neutral terminal, the green and yellow-striped covered wire to the live terminal, and the blue-covered wire to the earth terminal. (i) State what the correct connections should have been. (ii) Explain carefully why this connection could be very dangerous, even when the switch on the washing machine was in the off position.

16 Electrical energy is transmitted across the country in the United Kingdom grid system at a voltage of 400 kV using alternating current in transmission lines carried overland on pylons.

a Why is such a dangerously high voltage used in the grid system?
b Why is alternating current used in preference to direct current?
c Although this is a fairly efficient method of transmitting energy, how is some electrical energy nevertheless lost in the grid system?

36 ELECTRICITY PRODUCES MOVEMENT

Several examples of an electric current producing movement have already been described. A current-carrying conductor can move a compass needle placed near it (p. 261). Two current-carrying conductors attract or repel each other (p. 262). The hammer of an electric bell moves when the current is switched on (p. 264). In all these cases, the electric current is producing a force, and the force is producing movement. Electrical energy is being converted into kinetic energy. In this section the forces produced and their applications are discussed in detail.

Force, current and magnetic field

The basic fact is that a current experiences a force when placed in a magnetic field. There are several ways of demonstrating this. The essentials of all the demonstrations are a conductor capable of movement, ways of reversing and varying the current and a fairly strong magnetic field. When the current is switched on the conductor moves, showing that a force is acting. The greater the current, the greater the force. Reversing either the current or the magnetic field direction reverses the direction of the force. In the experiments shown in figures 36.1–36.3 the current and the magnetic field are at right angles. It will be found that the force produced will be mutually at right angles to both the current and the field. The magnitude of the force is proportional to both the current and the strength of the magnetic field.

fig. 36.1 Bent copper wire swings from mercury cups between the poles of a horseshoe magnet

▼

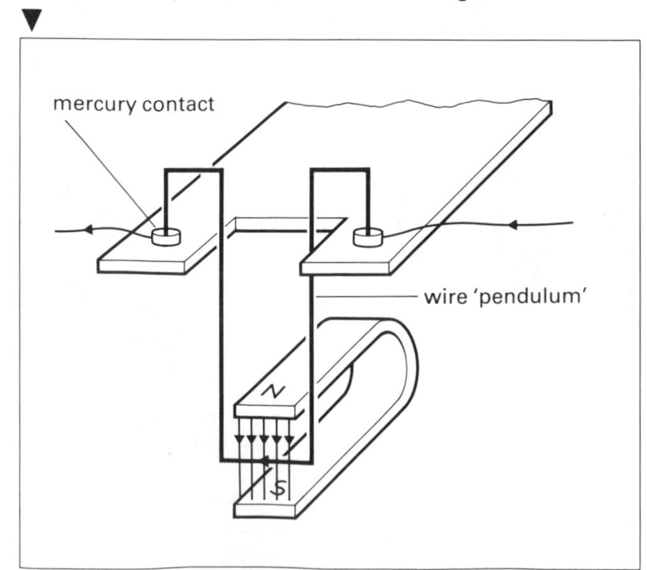

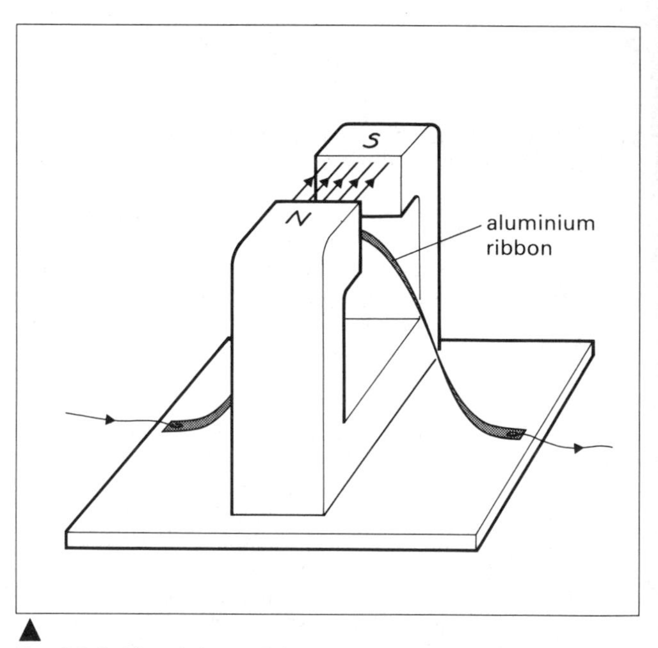

▲

fig. 36.2 Aluminium ribbon, pinned at the ends, lies slackly between the magnet poles until the current flows through the ribbon

fig. 36.3 Brass rod rolling on brass rails carrying a current. The rod is between the poles of a powerful magnet

▼

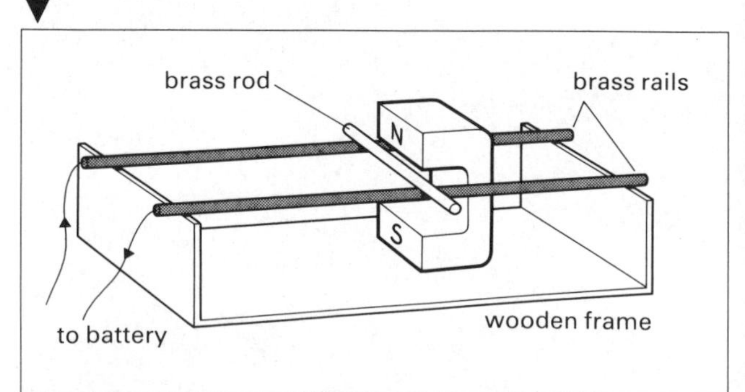

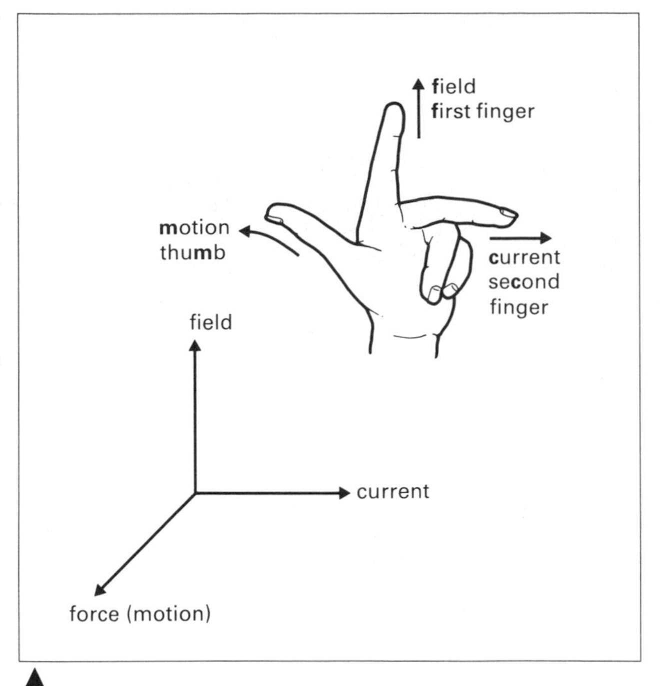

fig. 36.4 Fleming's left-hand rule

A simple way of remembering how these three quantities are arranged is given by **Fleming's left-hand rule**, figure 36.4. You must use your left hand or you will get the directions wrong. Extend the first and second fingers and the thumb of the left hand so that they are as nearly at right angles to each other as you can get them. Set the first and the second fingers in the directions of the field and the current and the thumb points in the direction of the motion.

The deflection of an electron beam

The force on a current-carrying conductor in a magnetic field is exerted on the electric charges themselves — the moving electrons which constitute the current. So a stream of electrons independent of any metallic conductor will be deflected. This can be demonstrated using a cathode ray tube. In the cathode ray tube a stream of electrons from an electron gun is directed on to a fluorescent screen at the end of the tube to produce a bright spot of light. When a magnet approaches the tube, the beam of electrons is deflected and the spot of light moves, figure 36.5a.

Remembering that the electron stream moves in a direction opposite to the conventional current, the left-hand rule can be used to predict the direction of the deflection, figure 36.5b.

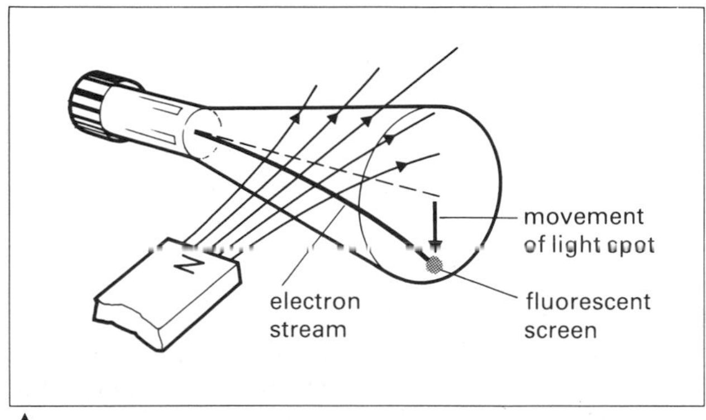

fig. 36.5a Deflection of an electron beam by a magnetic field

fig. 36.5b Path of an electron is circular while it is in the field because the force is at right angles to the motion

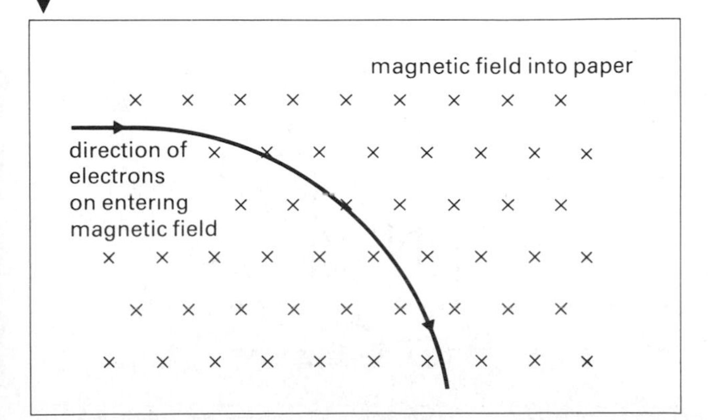

fig. 36.6 Interaction of currents

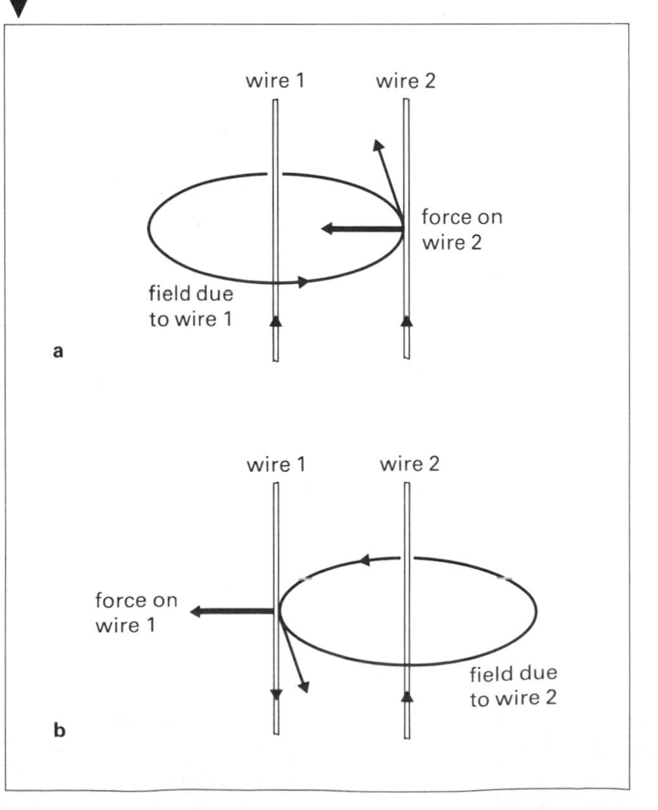

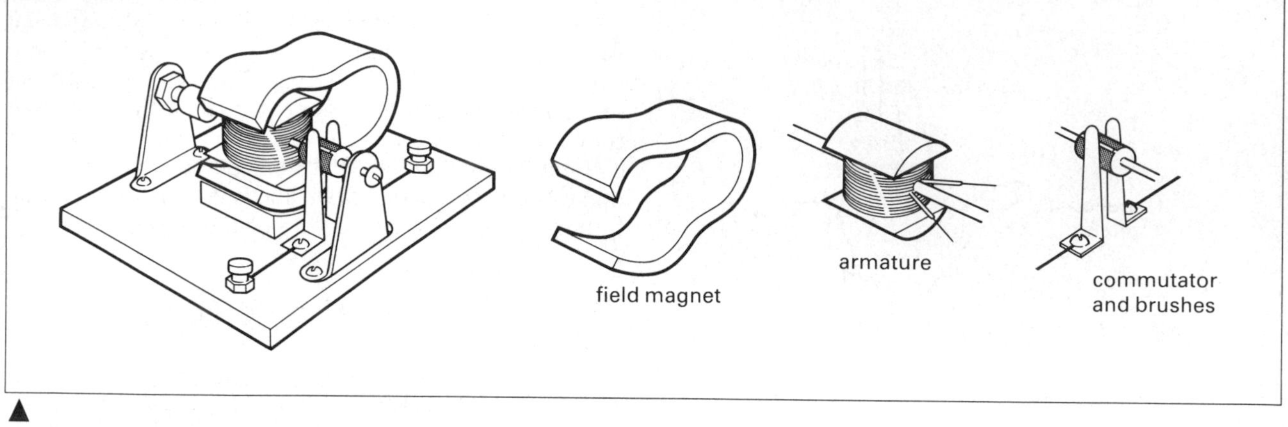

fig. 36.7 Model electric motor

Forces between currents

A current produces a magnetic field around it. A second current placed in this magnetic field will experience a force. Similarly, the first current will be in the magnetic field of the second current and experience a force.

These mutual forces between currents have already been described on p.263 and form the basis for the definition of the ampere on p.272.

The attraction of currents in the same direction, and repulsion of currents in opposite directions, fit in with Fleming's left-hand rule, as shown in figures 36.6a and b. Only the field due to one wire and the force on the other wire are shown for simplicity, but remember that there are forces on both wires.

The direct current electric motor

An obvious application of the force on a current-carrying conductor placed in a magnetic field is the **electric motor**, in which electrical energy is converted into mechanical energy.

A simple direct-current motor, figure 36.7, consists essentially of three main parts:

- A magnet. This may be a permanent magnet or electromagnet and is called the **field magnet**, because it produces the magnetic field of the motor.
- The rotating part or **armature**, which consists of a coil of wire wound on a soft iron core.
- The contacts by which the current passes through the armature coils. These are made up of the **commutator** and **brushes**.

Using the left-hand rule and with the help of figure 36.8, we can see how it is that the coil of an electric motor turns when the current is switched on, and also that it continues to turn in the same direction.

At first the force on AB is upward. When the coil is vertical, the commutator segments are changing from one brush to the other. This reverses the current in AB and the force on it is now downwards. The coil continues to turn in a clockwise direction. Trace out the direction of the current in the coil in the various positions, and work out the directions of the forces on the two sides.

An electric motor with a single-coil armature would

fig. 36.8 Action of a d.c. motor

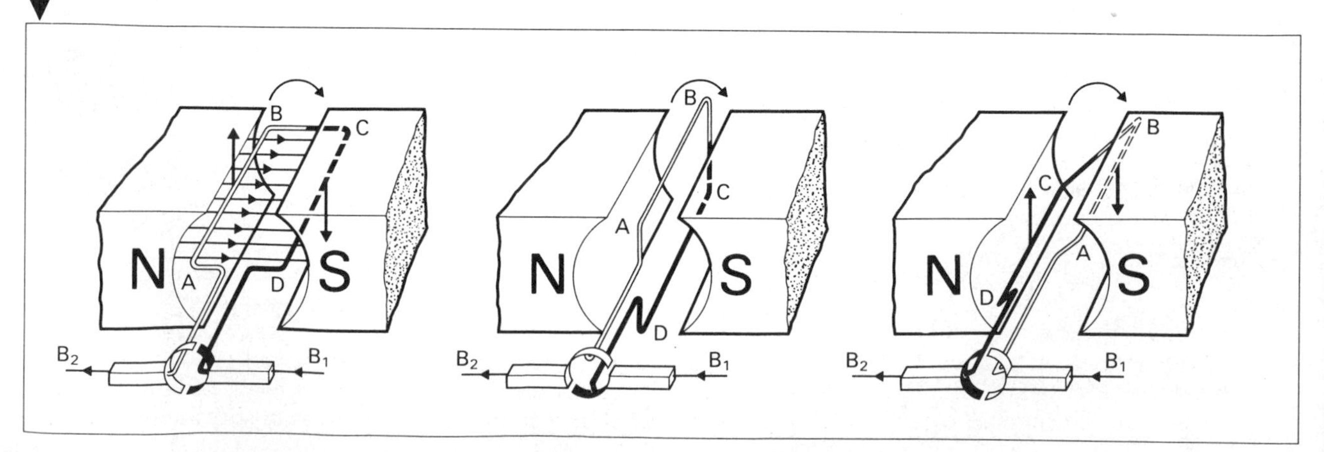

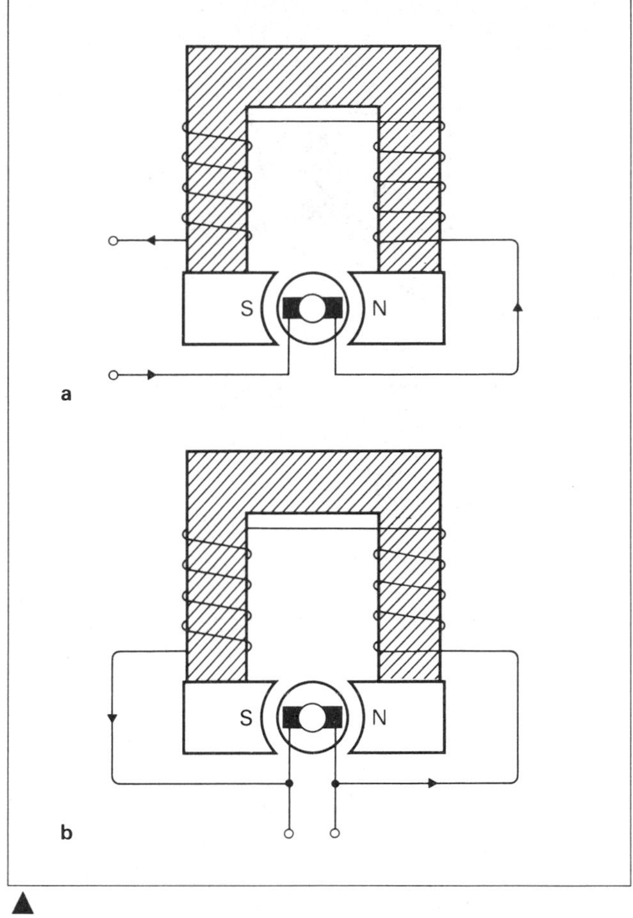

fig. 36.9 Series and shunt-wound electric motors

not work well in practice because the turning force varies with the position of the coil. The turning force (moment or torque) is greatest when the coil is horizontal (see figure 36.8) and zero when the coil is vertical. Practical motors overcome this disadvantage by using many coils, set at a small angle to each other, in the armature. The field magnets are constructed with several pairs of poles. This makes for smoother running and a constant turning force is exerted by the motor in all positions.

The field coils of an electric motor may be connected in series or parallel (shunt) with the armature coils. In the series-winding the same current flows through both sets of coils, figure 36.9a. In the shunt-winding some of the current goes through the field coils, but most of it goes through the armature coils which have a much lower resistance, figure 36.9b. A series-wound motor will run on a.c. or d.c. Series-wound motors speed up quickly, but tend to 'race' when the load is small. Shunt-wound motors run at a more or less steady speed. Many motors have a mixture of both windings, and combine the advantages of both types of machine.

An electric motor is a machine for converting electrical energy into mechanical energy. Like any other machine its efficiency is measured by comparing its energy output with its energy input.

$$\textbf{efficiency} = \frac{\textbf{energy output}}{\textbf{energy input}} \times \mathbf{100\%}$$

Suppose that a small motor has an output of $\frac{1}{2}$ horse-power and that when running on 240 V mains it takes a current of 2 A. We can find its efficiency as follows (1 h.p. = 746 watts):

$$\text{energy output} = \tfrac{1}{2}\text{ h.p.} = \tfrac{1}{2} \times 746\text{ W} = 373\text{ W}$$
$$\text{energy input} = \text{V} \times \text{A} = 240 \times 2 = 480\text{ W}$$

$$\text{efficiency} = \frac{\text{energy output}}{\text{energy input}} \times 100 = \frac{373 \times 100}{480} = 77.7\%$$

A large, well-designed electric motor may have an efficiency as high as 95%.

The moving-coil galvanometer

The moving-coil galvanometer, figure 36.10, works on the same principle as the electric motor. A coil rotates between the poles of a horse-shoe magnet when a current is passed through it. The rotation is balanced by spiral control springs, figure 36.10b. The greater the current, the greater the turning force acting on the coil and the more the control spring is wound up, until an equilibrium position is reached where the deflecting moment due to the current is equal to the restoring moment of the control spring. The scale can be calibrated by using known currents measured on a current balance.

The current is led into and out of the coil by control springs at the top and bottom. The special shape of the pole pieces of the magnet and the cylinder of soft iron round which the coil rotates produce a very uniform radial field, with the result that the scale is uniform.

The **sensitivity** of a galvanometer is measured by the ratio of the *deflection* produced to the *current* passing through the coil. Thus, a very sensitive instrument produces a large deflection for a small current. A very sensitive instrument should have

- a strong magnet,
- a large coil with many turns,
- a weak controlling spring,
- a long, light pointer.

Some of these requirements are mutually impossible to achieve. For example, a large coil with many turns would be heavier and require a larger force to move it, so in practice a compromise has to be made. In very sensitive instruments a beam of light is often used as a pointer. There are many advantages to the

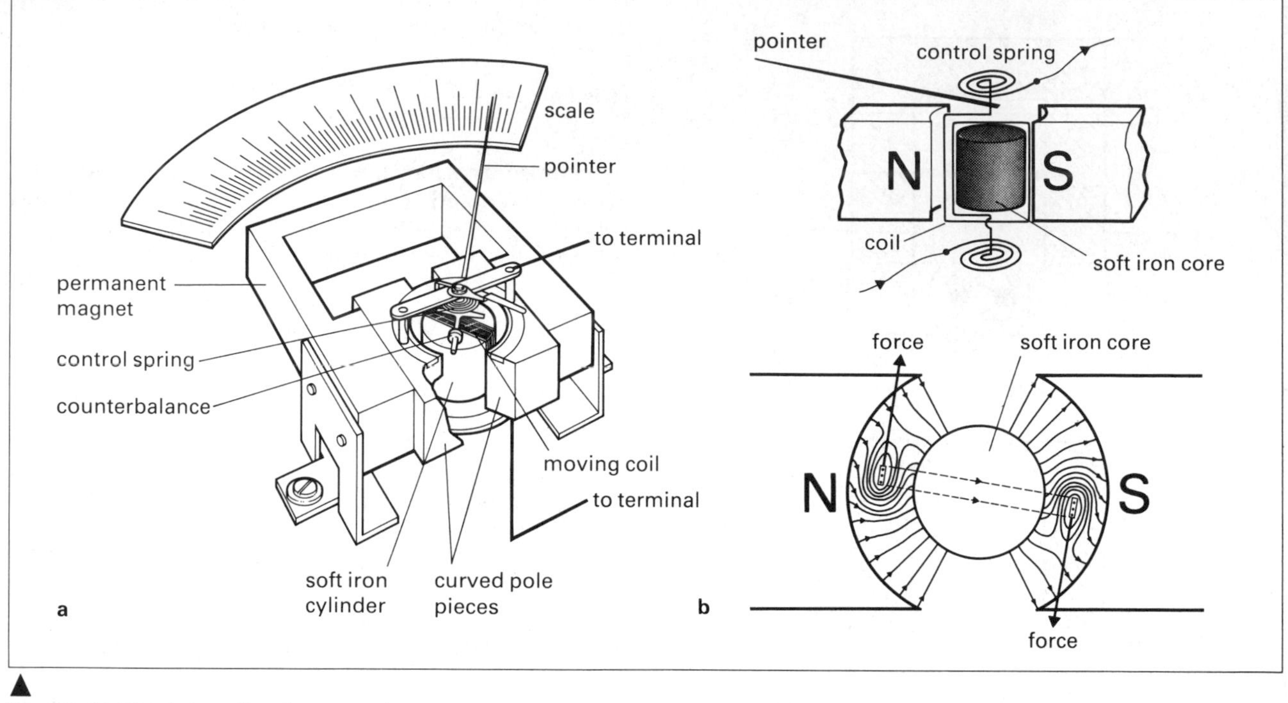

fig. 36.10 Moving-coil galvanometer

moving coil galvanometer, but one disadvantage is that it *cannot* be used to measure an alternating current.

The moving-coil galvanometer can be converted for use as an ammeter or a voltmeter.

EXERCISE 36

In questions 1–4 select the most suitable answer.

1 A length of insulated wire is held between the poles of a strong horse-shoe magnet and a current is passed through the wire in the direction shown in figure 36.11. The wire will experience
A a force in the direction W (upwards)
B a force in the direction X (towards the south pole)
C a force in the direction Y (downwards)
D a force in the direction Z (towards the north pole)
E no force because of the effect of the insulation

fig. 36.11

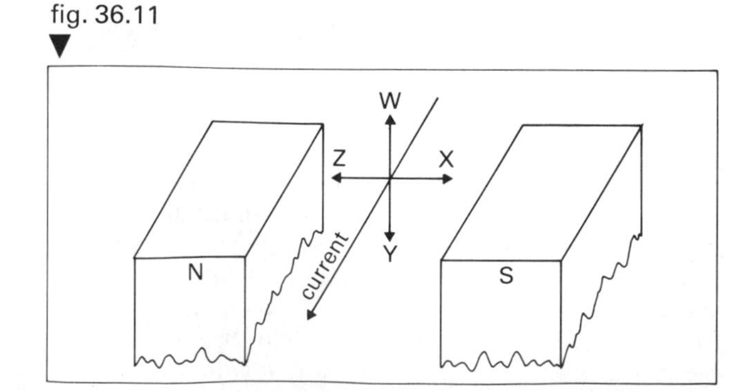

2 A stream of positively charged particles is projected between two magnetic poles as shown in figure 36.12. Under the influence of the magnetic field the particles will
A accelerate without change of direction
B be deflected towards the north pole N
C be deflected towards the south pole S
D be deflected into the paper
E be deflected out of the paper

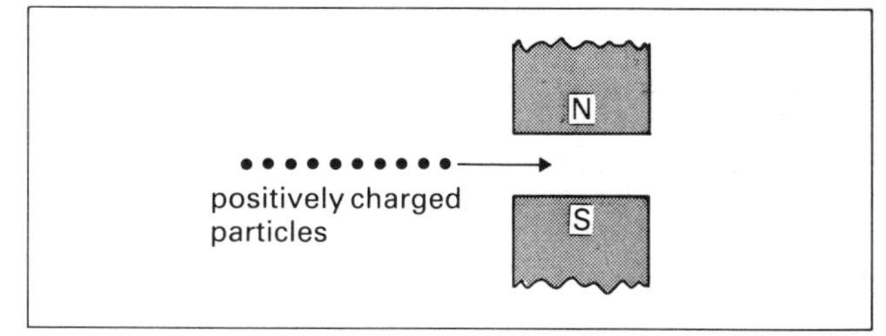

fig. 36.12

3 The purpose of the commutator in a simple d.c. motor is to
A provide an a.c. output from the motor
B reverse the direction of the current in the coil twice in each revolution
C reverse the direction of rotation of the armature twice in each revolution
D reverse the polarity of the magnet twice in each revolution
E stop the armature getting overheated

4 The purpose of a moving-coil meter is to
A measure the value of a direct current
B measure the peak value of an alternating current
C convert alternating current into direct current
D convert direct current into alternating current
E increase the value of an alternating current

5 Two powerful magnets are placed alongside each other as shown in the figure. Note that they are magnetised so that their sides are the poles.

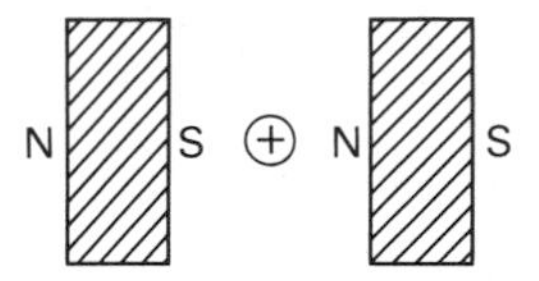

a Copy the diagram, and draw in at least six field lines to represent the field between the two magnets.
b ⊕ represents the end view of a wire connected to a cell and arranged at right angles to the page. When the current is switched on its direction is into the page. Which way would the wire move?

6 The figure shows a coil free to rotate about the axis shown between the poles of a strong magnet. What will happen to the coil

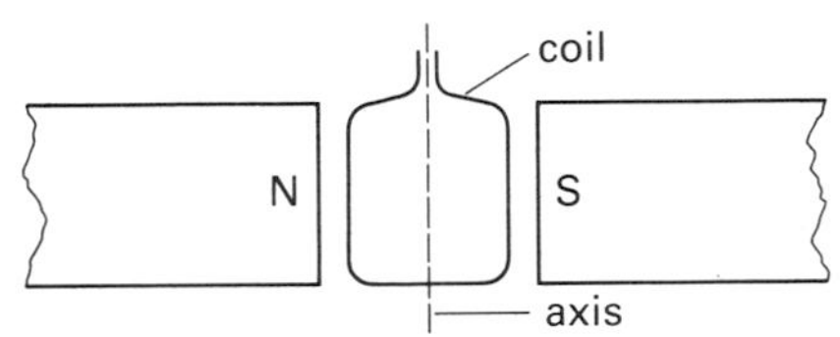

a When a current is passed through the coil?
b When the same current is passed through the coil but the poles are interchanged?

7 Equal currents flow in two long, straight parallel wires. Draw the magnetic field patterns when
a the currents are in the same direction,
b the currents are in opposite directions.
Show the directions of the forces acting on the conductors in each case.

8 Draw a diagram showing the lines of magnetic force due to a current in a straight conductor, in a plane perpendicular to the conductor.

Describe an experiment to show that a current-carrying conductor which is perpendicular to a magnet field experience a mechanical force. Show the directions of the current, the magnetic field and the force.

Describe a simple electric motor and explain why a commutator is necessary to obtain continuous action.

9 a Name *two* domestic appliances which contain an electric motor.
b The graph in figure 36.13 shows how the current through a 240 V commercial motor varies with its speed. The current flowing when the motor is running on full load is found to be 2 A. At what speed is the motor running when on full load?

fig. 36.13

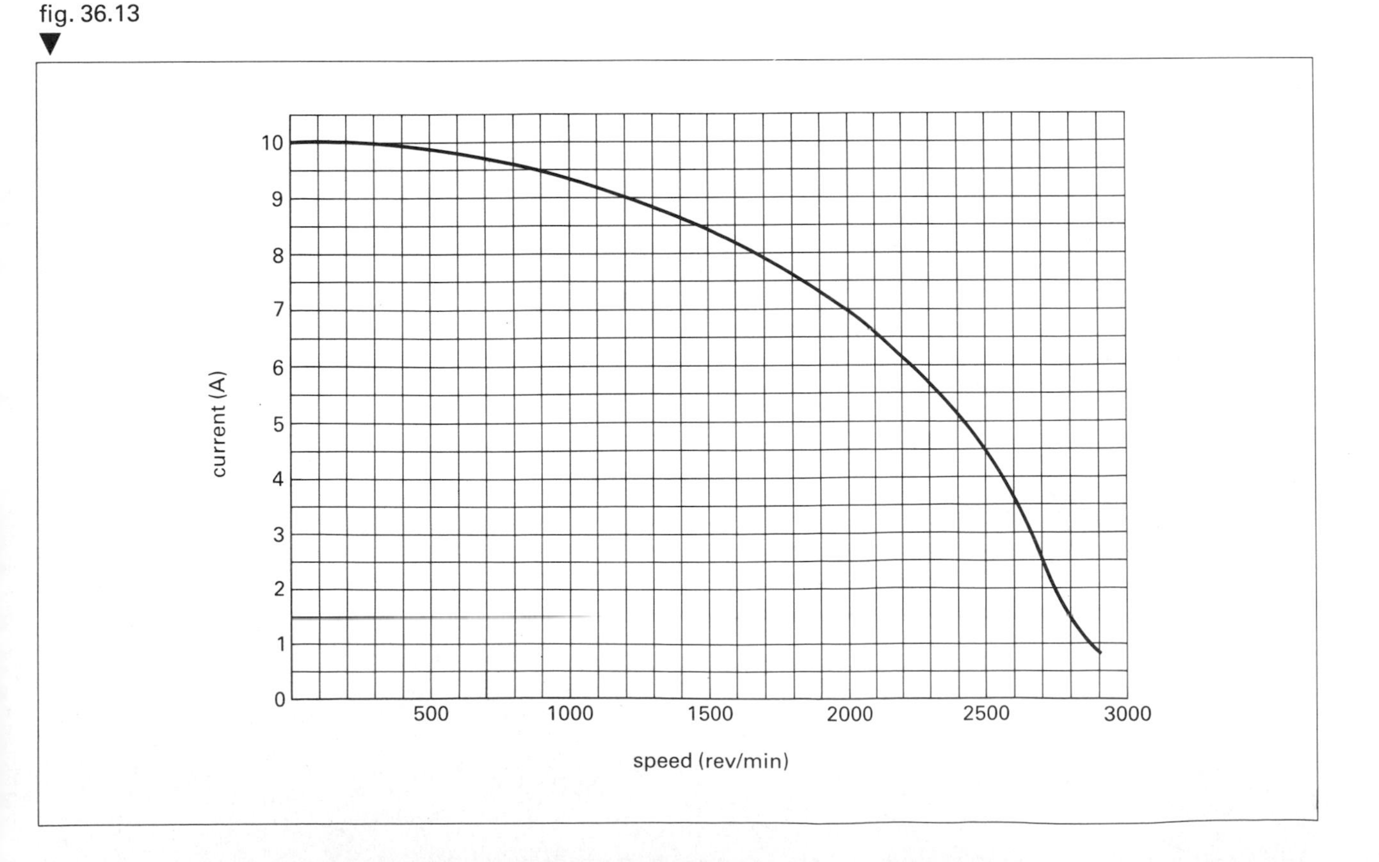

c What is the maximum speed at which the motor can run?

d What is the maximum current which can pass through the motor?

e At what speed is the motor rotating when this maximum current flows?

f What would be a suitable fuse to use in the electrical lead to the motor? Choose one of the following: 13 A, 10 A, 5 A or 250 mA.

g In an experiment to plot the graph in Figure 36.13 a mechanical brake was arranged to control the speed of the motor. (i) How could the current be measured? (ii) How could the speed be measured in revolutions per minute?

h The brake will slow the motor down. What happens to the mechanical energy which the motor transfers to the brake?

i What power in watts is being supplied to the motor when it is running on full load?

j Why is it likely to be necessary to make a costly repair if at any time the motor is accidentally jammed so that it will not turn?

10 Figure 36.14 shows a simplified diagram of the essential parts of a moving-coil meter.

a What is the purpose of the spring?

b Explain why the pointer moves when a current flows.

c The moving-coil meter is suitable only for measuring direct current, and cannot measure alternating current. Explain why.

d The meter should have a high resistance if it is to be used as a voltmeter. Explain why.

e The meter should have a low resistance if it is to be used as an ammeter. Explain why.

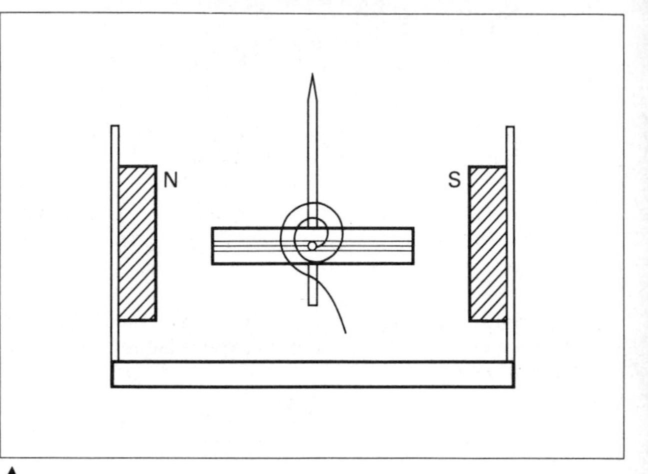

fig. 36.14

37 ELECTROMAGNETIC INDUCTION

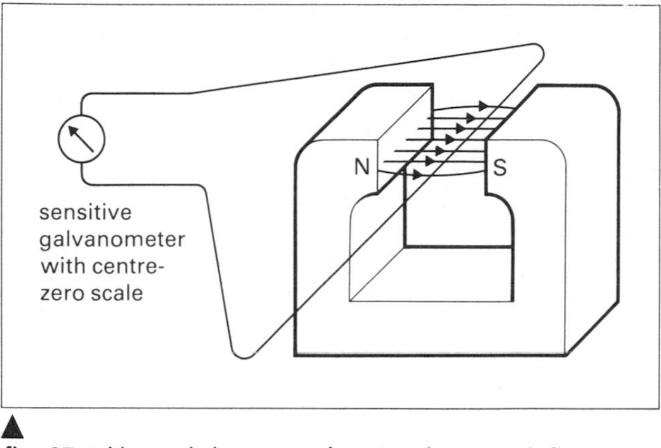

fig. 37.1 Up and down motion produces a deflection on the galvanometer. Sideways motion does not

fig. 37.2 Electromagnetic induction in a coil

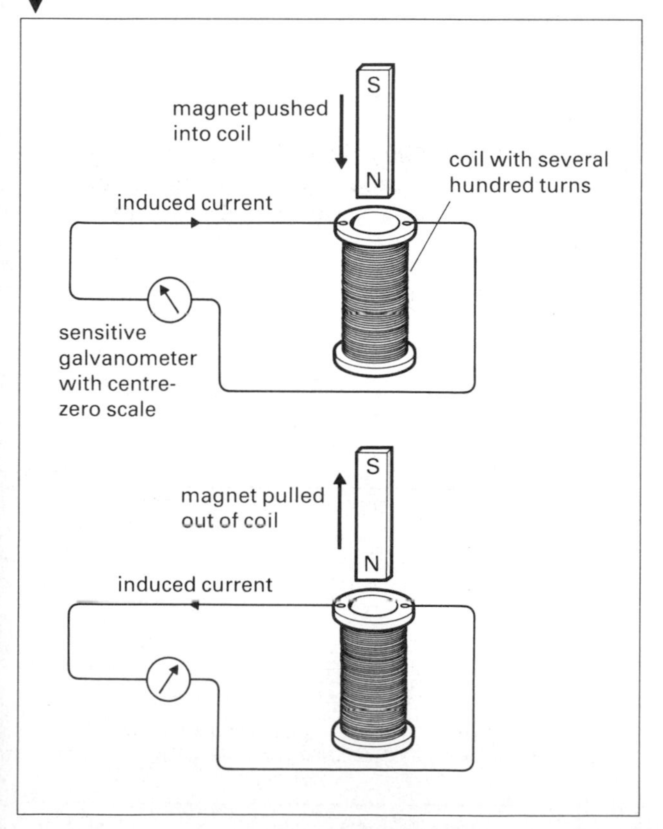

Many physical effects are reversible. For example, if heat is supplied to a fixed mass of gas, the pressure increases. In reverse, if the pressure on a gas is increased, the gas gets hot. Electrical energy produces a chemical change when water is split into hydrogen and oxygen during electrolysis. In reverse, electrical energy is produced by the combination of hydrogen and oxygen in the fuel cell. When a current-carrying conductor is in a magnetic field it experiences a force and if free to do so it moves. So we ask if we move a conductor in a magnetic field, will it cause a current to flow through the conductor? A series of experiments by Michael Faraday in the first half of the nineteenth century answered the question. These are outlined below.

Experiments on electromagnetic induction

A loop of wire is connected to a sensitive galvanometer with a centre zero, figure 37.1. A section of the loop is moved downwards between the poles of a strong magnet so that it cuts through the lines of force. The galvanometer shows a deflection while the wire is moving in the magnetic field. When the wire is moved upwards the deflection is in the opposite direction. So the reverse of the motor effect *is* true. When a conductor is moved in a magnetic field a current is produced. Such a current is called an **induced current** and the effect is known as **electromagnetic induction**.

Figure 37.2 shows a coil connected to a galvanometer. A magnetic field is introduced by pushing a bar magnet into the coil. When the magnet is pushed into the coil there is a deflection on the galvanometer, but this deflection ceases when the magnet is stationary inside the coil. When the magnet is withdrawn there is again a deflection in the opposite direction. If the other pole of the magnet is used, the deflections are in the opposite directions. You can investigate what happens if a stronger magnet is used and what happens if the magnet is moved more quickly.

The results of these investigations can be summed up as follows:

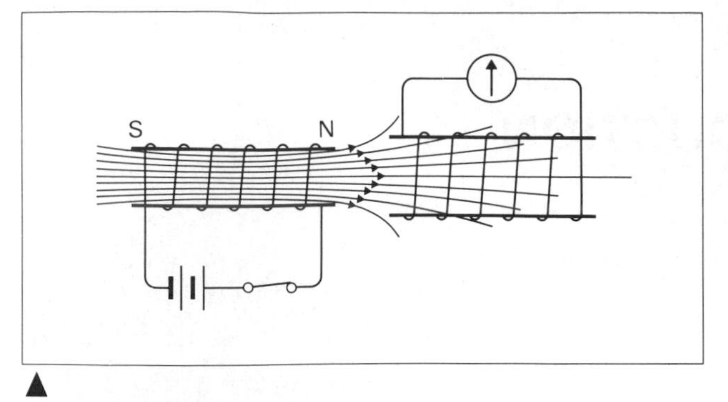

▲
fig. 37.3 Induction using a current-carrying coil

- An induced current is only produced when there is movement or a change of some kind. Either the magnet or the coil can move.
- The direction of the induced current depends on which pole of the magnet is used and in which direction it is going.
- A stronger magnet, a quicker movement and a coil with more turns, all produce a larger induced current, as shown by the 'kick' of the galvanometer.

The changing magnetic field can be supplied by a coil of wire with a current passing through it, instead of by a moving permanent magnet, figure 37.3. An induced current is only produced in the second coil when the current in the first coil is switched on or off or is increased or decreased or if the coils are moved relatively to one another when a steady current is flowing in the first coil. In all these cases it will be realised that there is a change in the magnetic flux linking with the second coil.

Faraday found that this effect was greatly increased if the coils were linked together with iron. He used an iron ring with two coils wound on it, as shown in figure 37.4.

fig. 37.4 Faraday's iron ring experiment
▼

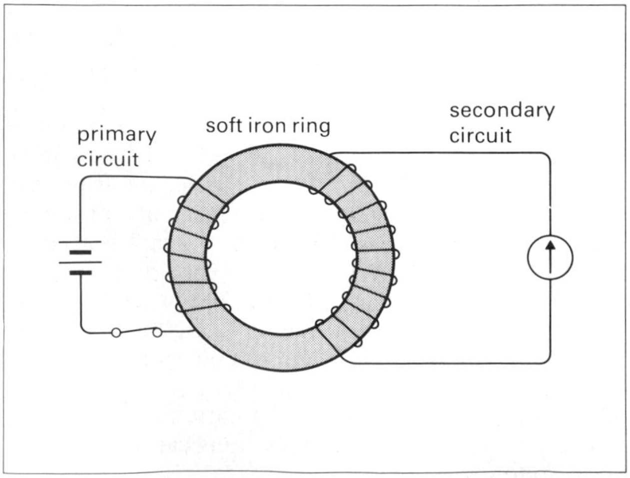

Faraday gave his explanation of electromagnetic induction in terms of the changing magnetic field. Whenever a conductor is moved in a magnetic field so that it cuts across the lines of force, or when the field round a conductor is changed in any way, an induced current is produced. The size of the induced current increases as the speed of the change is increased.

The induced electromotive force

Although we have talked so far about induced currents it is important to realise that what is really produced is an e.m.f. Consider what happens when the conductor AB in figure 37.5 is moved down the page in a magnetic field which is directed into the paper. Remember that a force is exerted on electrons moving in a magnetic field. The left-hand rule shows that this force is from A to B, so that the free electrons in the conductor move towards end B where there will be excess electrons, leaving an electron deficiency at A. The two ends of the conductor will be at different potentials, and this potential difference constitutes an e.m.f. between the ends of the wire. If the wire is part of a complete circuit, then a current will flow and this is the induced current shown by the galvanometers in the introductory experiments.

Faraday's law of electromagnetic induction

Faraday's findings and explanation can be summed up as follows:

When the magnetic flux through a circuit changes, an induced e.m.f. is produced and the magnitude of the induced e.m.f. is proportional to the rate of change of the flux

fig. 37.5 How an e.m.f. is induced
▼

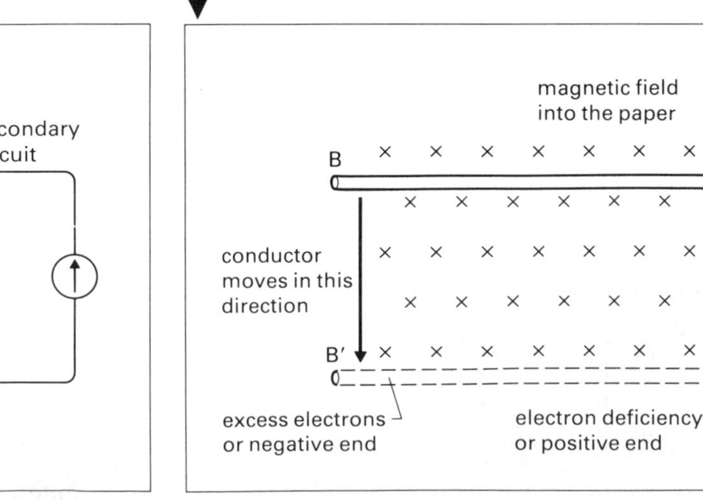

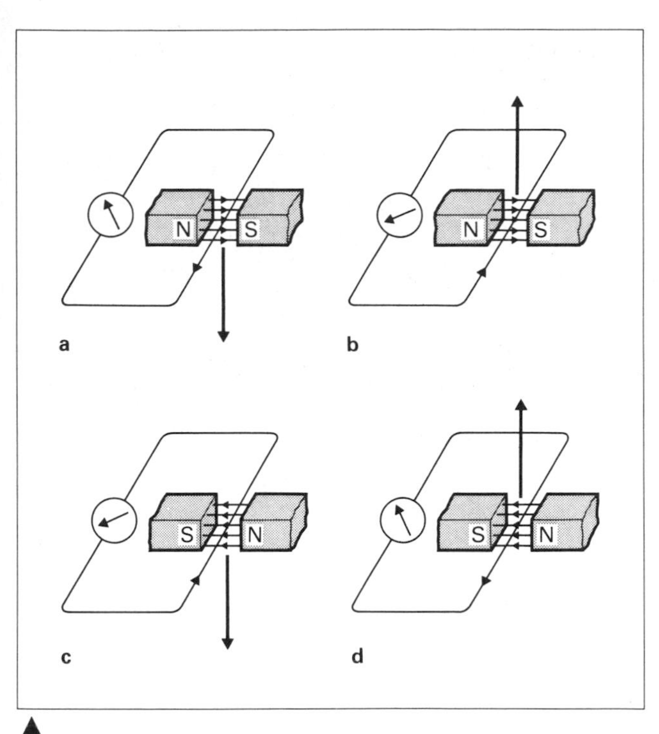

fig. 37.6 The direction of the induced current

fig. 37.7 Fleming's right-hand rule for induced currents

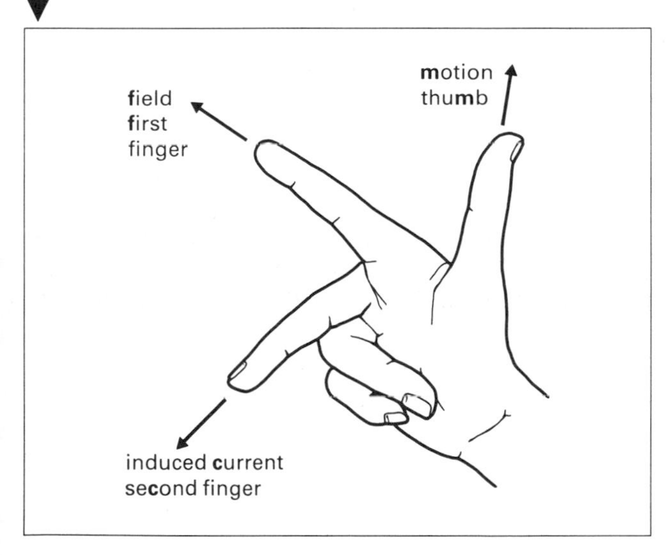

The direction of the induced e.m.f. and induced current

The first of the electromagnetic induction experiments is repeated with a galvanometer which will indicate the direction of the current. A preliminary experiment, connecting the galvanometer in series with a cell and a high resistor, may be necessary to find out in which direction a current must flow to produce a deflection to the right, say. The direction of the induced current corresponding to a particular movement and a particular field can then be found. Four cases are shown in figures 37.6a–d.

We shall consider only those cases where the field, motion and induced current are mutually at right angles. The relative directions can be remembered using **Fleming's right-hand rule**, figure 37.7.

Extend the first and second fingers and the thumb of the right hand so that they are at right angles to each other. Set the first finger in the direction of the field, the thumb in the direction of the motion and the second finger gives the direction of the induced current.

A much more fundamental way of considering the direction of the induced e.m.f. and current is given by **Lenz's law** which states

The direction of the induced e.m.f. is such as to produce a current which opposes the change to which it is due

For example, in figure 37.8, when a N pole is being pushed towards a coil, the current flows in the direction to produce a N pole at the near end, and repel the approach of the magnet. When the N pole is being withdrawn, a S pole is induced at the near end to attract it.

Lenz's law is really a special case of the conservation of energy. In order to push the N pole into the coil, work has to be done to overcome the force of repulsion and this work appears as electrical energy.

fig. 37.8 The induced current direction is such as to produce at the end nearest the magnet a pole which opposes the movement of the magnet

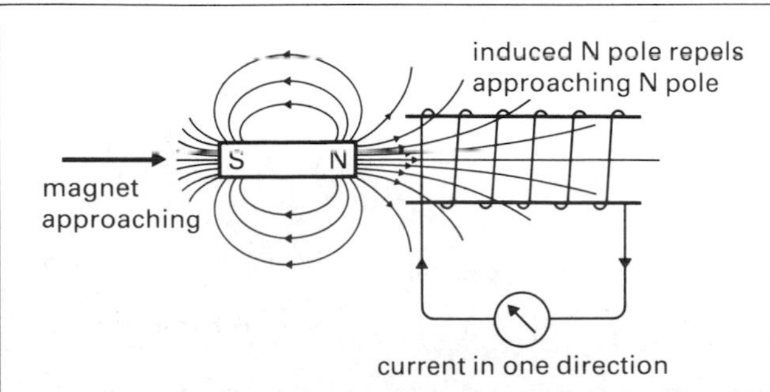

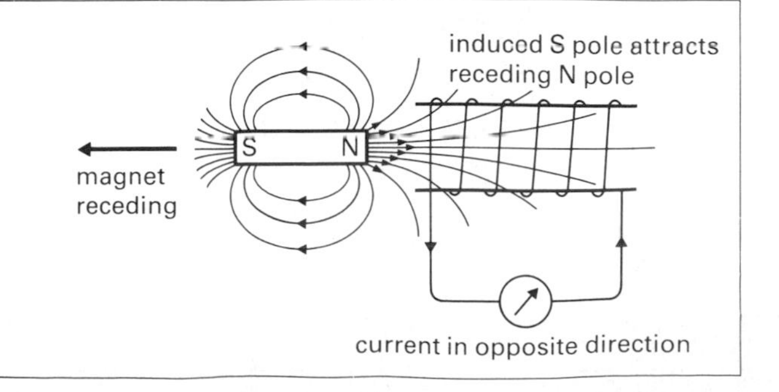

Electric generators

A generator or dynamo converts mechanical energy into electrical energy. It is therefore the opposite of the electric motor.

In order to produce an induced current, movement between the coil of wire and the magnetic field is essential. If a magnet suspended by a spring is allowed to move up and down as it dips into a coil connected to a centre-zero galvanometer, a current is recorded as long as the magnet is in motion, figure 37.9. The galvanometer needle moves over to the right (say), then back to zero and over to the left, back to zero, and so on. The movement of the needle to left and right shows that the current is changing its direction. A graph of current against time is shown in figure 37.10.

A current which behaves in this way is called an **alternating current**. The number of complete cycles made in one second is known as the **frequency** of the particular generator producing the current. The normal frequency of our a.c. mains is 50 hertz (cycles per second). The sign '50~' seen on electrical apparatus means that it is for use on a 50 Hz a.c. supply.

It is always easier to produce movement by spinning than by a backward and forward motion. In an electric generator or dynamo, a coil of wire is spun between the poles of an electromagnet which supplies the magnetic field, figure 37.11.

The direction of the current changes when the coil is at right angles to the magnetic field. The current is a maximum when the coil is in line with the field and is cutting directly through the lines of force. Direct current can be obtained from a simple generator by using a split-ring commutator, as in the electric motor, figure 37.12b. The brushes must change from one segment of the commutator to the other just when the direction of the current is changing, that is when the coil is at right angles to the field. A single coil

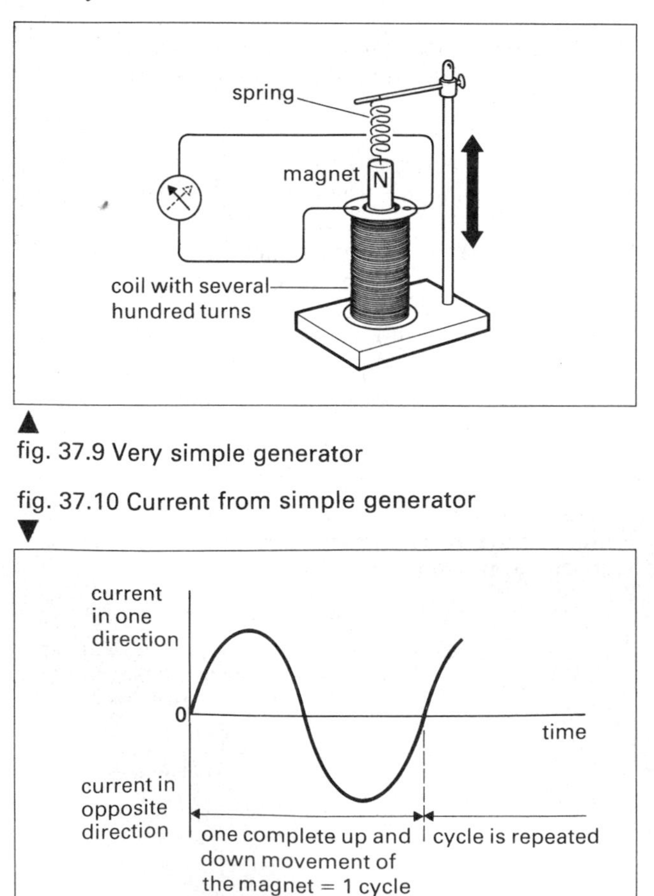

▲ fig. 37.9 Very simple generator

fig. 37.10 Current from simple generator ▼

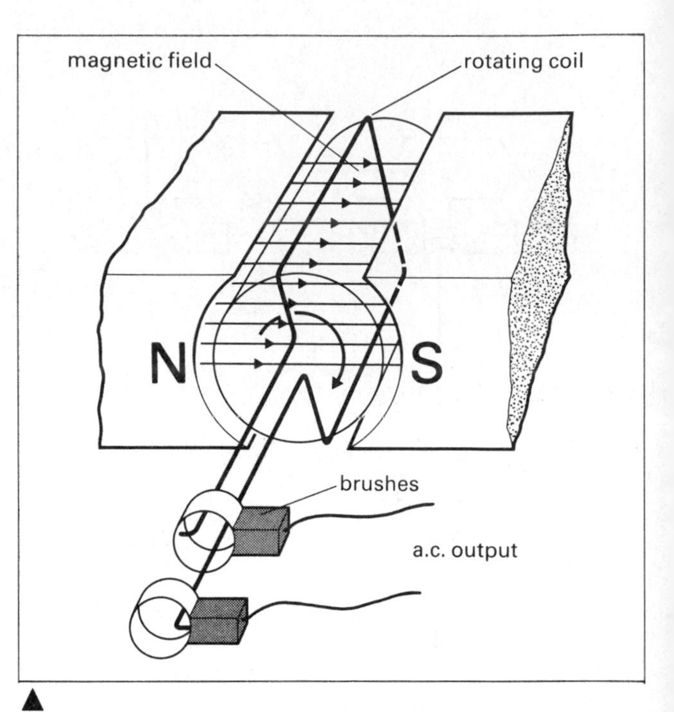

▲ fig. 37.11 Simple generator giving a.c.

fig. 37.12 Connections to single-coil a.c. and d.c. generators ▼

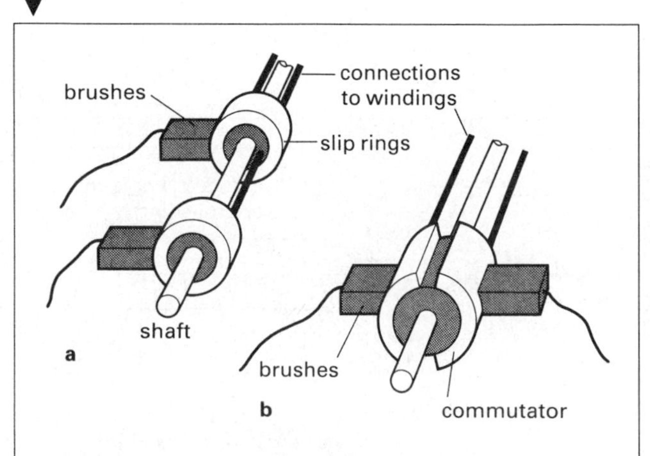

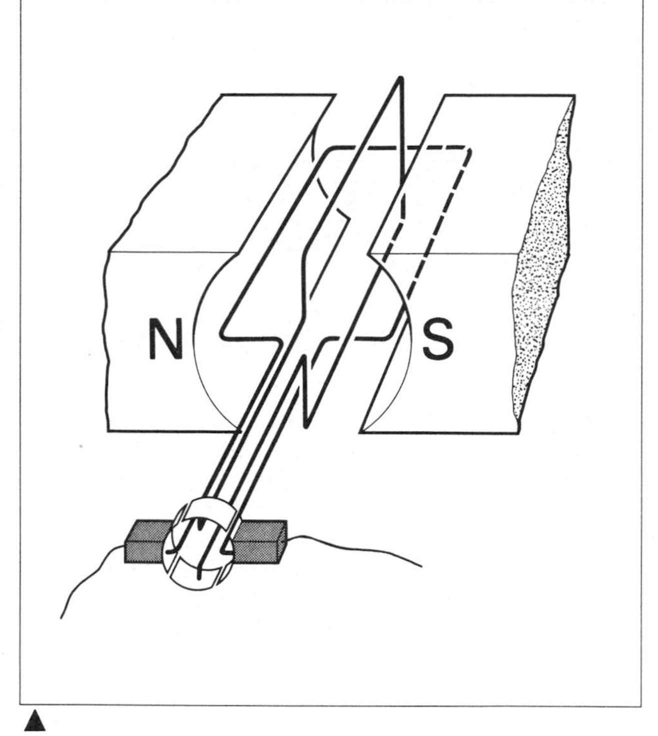

fig. 37.13 A two-coil dynamo producing direct current

fig. 37.14 Comparison of the current outputs of three types of generator

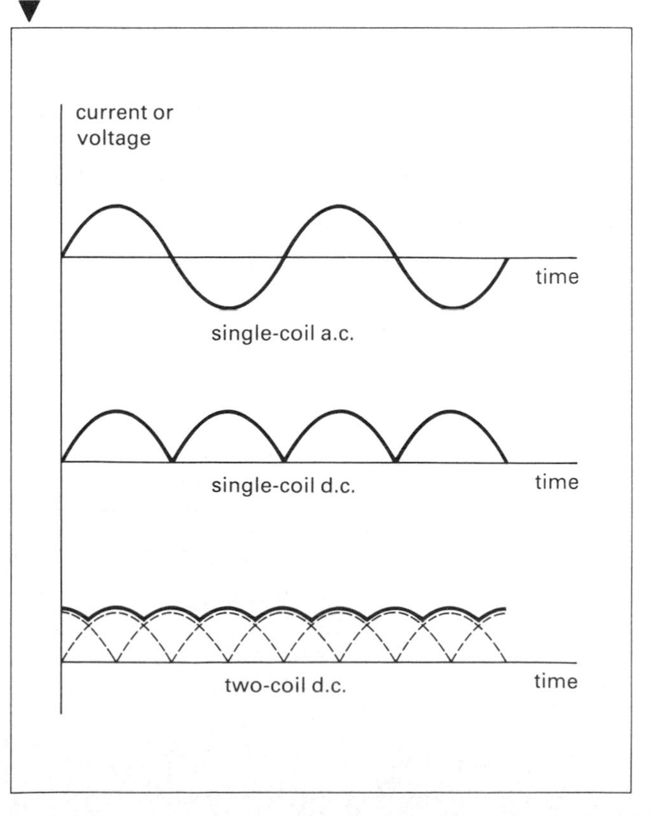

would produce a direct current which is varying in size. To produce a smoother current, two coils could be used with a four-segment commutator, figures 37.13 and 37.14. Each coil is used for only a quarter of a revolution at a time. When twenty or more coils are used, an almost steady current can be produced. In an actual generator, all the coils are joined in series, so all are contributing to the generation of electricity all the time. The arrangement is too complicated to be described here.

According to Faraday's law, the magnitude of the induced e.m.f. depends on the rate of change of flux. To produce a high voltage a strong magnetic field, a large number of turns on the rotating coil and a high speed of rotation are necessary. It should be noted, however, that the higher the speed of rotation, the higher the frequency if a.c. is being produced.

The efficiency of a generator is the ratio of the electrical energy supplied by the machine to the mechanical energy put into it. Losses of energy are due to friction and to the heat generated when the current flows through the coils. A well-designed generator may have an efficiency of over 90%.

Alternating current

An alternating current is a current which changes its direction at regular intervals. The magnitude of the current is also changing with time. The current grows to a peak value, diminishes to zero, grows to a maximum in the opposite direction and then decreases to zero again. We picture a direct current as a continuous drift of electrons along the conductor. With an alternating current, we must picture the electrons

fig. 37.15 In a cycle dynamo it is simpler to make the magnet rotate and keep the coil stationary so that no brushes are needed

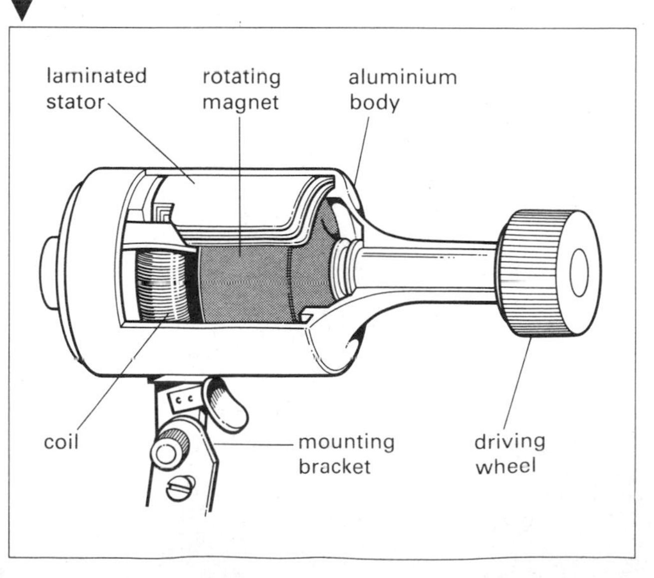

vibrating backwards and forwards with the frequency of the alternating current.

The following experiment demonstrates that a change of direction is really occurring with a.c. A solution of sodium sulphate has phenolphthalein added to it. Phenolphthalein is an indicator which is colourless with acids and red with alkalis. When this solution is electrolysed the resulting alkalinity at the cathode turns the phenolphthalein red. Using d.c., and making the screwdriver in figure 37.16a the cathode, you can draw a continuous red line on the blotting paper. With a.c. a dotted line results, figure 37.16b, showing that the screwdriver is alternately negative and positive. If you drew the screwdriver across the paper for one second you could count the number of dots and so find the number of changes per second. With 50-cycle a.c. there are 100 changes of direction every second.

The measurement of alternating currents and voltages presents a problem. When a current or voltage is changing all the time, which value should we measure? When we talk of 240 V a.c. mains, what do we mean? To solve this it has been decided to measure a kind of average value called the **effective value**. An alternating current with an effective value of 1 ampere produces the same amount of heat in the same time as a direct current of 1 ampere, figure 37.17. This kind of average value is found by taking the square root of the average value of the current squared and is called the **root mean square current** or r.m.s. current. It turns out that the effective value is equal to the peak value divided by $\sqrt{2}$ or 1.41.

peak value = 1.41 × effective value

A 240 volt circuit has a peak value of $1.41 \times 240 = 338.4$ volts. This is the reason why a shock from a 240

fig. 37.16 Demonstrating the nature of a.c.

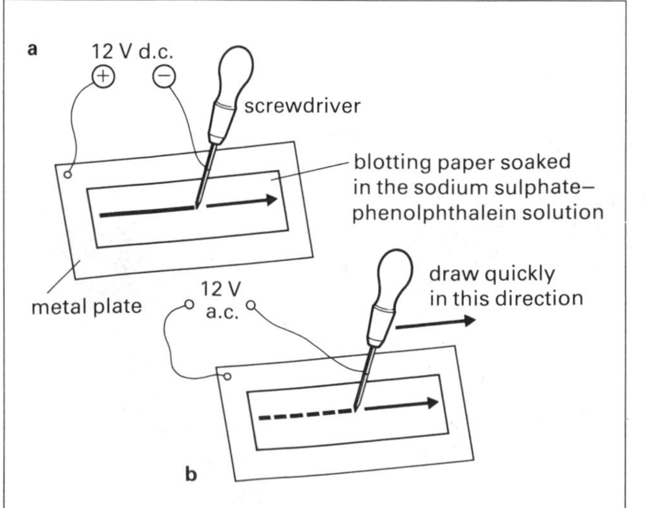

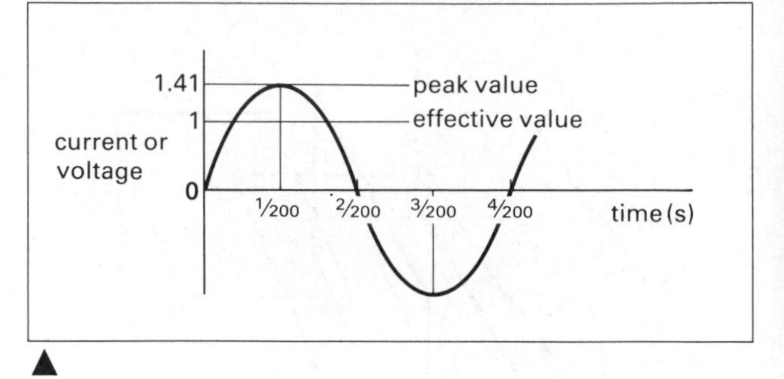

fig. 37.17 Difference between peak value and effective value

volt a.c. circuit is more dangerous than one from a 240 volt d.c. circuit. The scales on alternating current instruments are arranged to read effective or r.m.s. values.

Alternating current produces a heating effect and so can be used for heating and lighting. Alternating current can also be used to make an electromagnet, although the polarity is constantly changing. But an alternating current cannot be used for electroplating, which needs a direct current to bring about a one-way transfer of material. The great advantage of alternating current is that it can be 'transformed', that is, the voltage can be altered using a transformer.

Transformers

In one of his experiments on induced currents, Faraday used an iron ring with two coils of wire wound on it (see p. 300). The coil connected to the battery and switch is called the **primary coil** and the coil connected to the galvanometer is the **secondary coil**, figure 37.18. (The term **winding** is often used instead

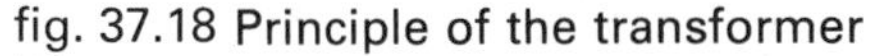

fig. 37.18 Principle of the transformer

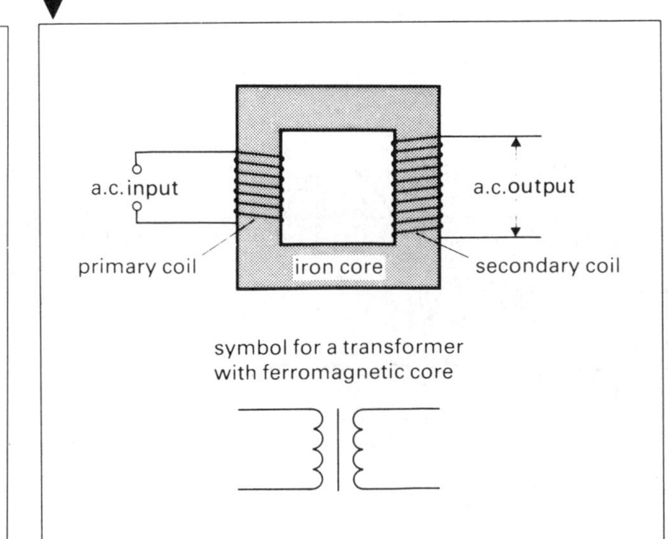

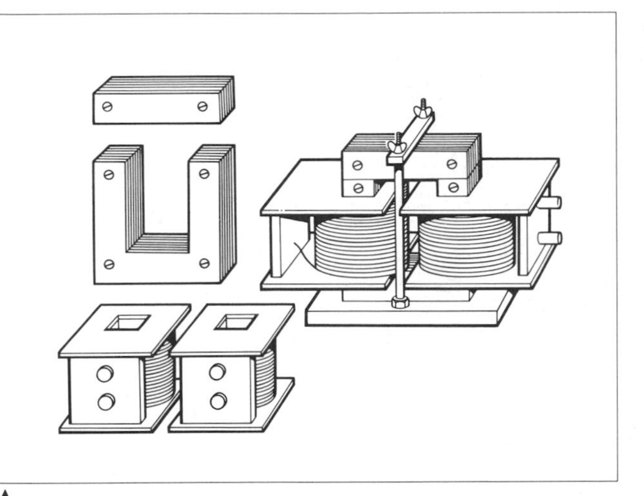

fig. 37.19 Experimental transformer with removable windings

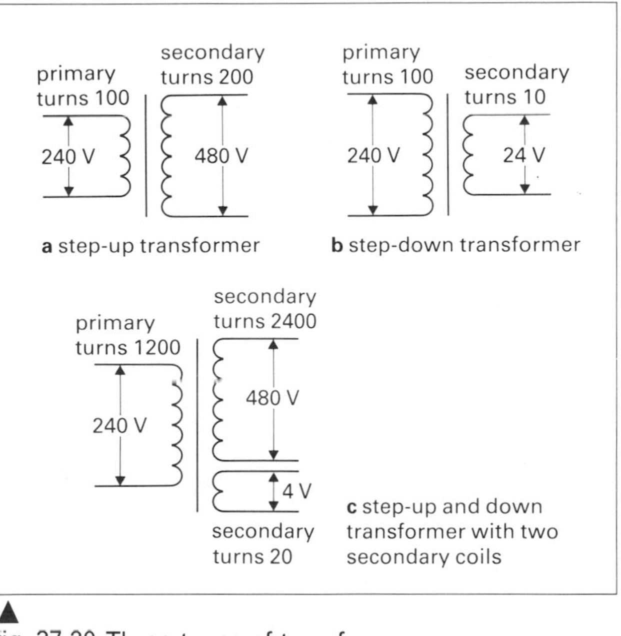

fig. 37.20 Three types of transformer

of coil.) When the primary current is switched on and off, a current is induced in the secondary coil, as shown by the kick of the galvanometer needle. But the current only flows in the secondary when the primary current is changing. Now, an alternating current is changing all the time. If we connect the primary coil to a source of alternating current there will always be an induced current in the secondary coil, and this induced current will also be changing in the same way and at the same rate as the primary alternating current. This arrangement is the basis of the **transformer**, which consists of a primary coil, an iron core and a secondary coil.

The purpose of a transformer is to change the voltage. The alternating voltage across the terminals of the secondary coil depends on the voltage applied to the primary, and also on the number of turns on each coil. The ratio of the voltages is the same as the ratio of the numbers of turns on the two coils. This can be verified by using an experimental transformer with secondary coils having different numbers of turns and measuring the secondary voltage, figure 37.19.

$$\frac{\textbf{secondary voltage}}{\textbf{primary voltage}} = \frac{\textbf{number of turns on secondary coil}}{\textbf{number of turns on primary coil}}$$

It looks at first sight as if the step-up transformer is producing volts from nowhere! But if the secondary voltage goes up, the current supplied by the secondary coil must come down. The power (in watts) taken from a transformer can never be greater than the power supplied to it, and in practice it is always less because the transformer is not one hundred per cent efficient. So at the very best

primary volts × primary amperes

= secondary volts × secondary amperes

Transformers are used to supply low voltages for laboratories and train-sets and high voltages for radios and television sets and on the grid system.

fig. 37.21 Modern transformer design

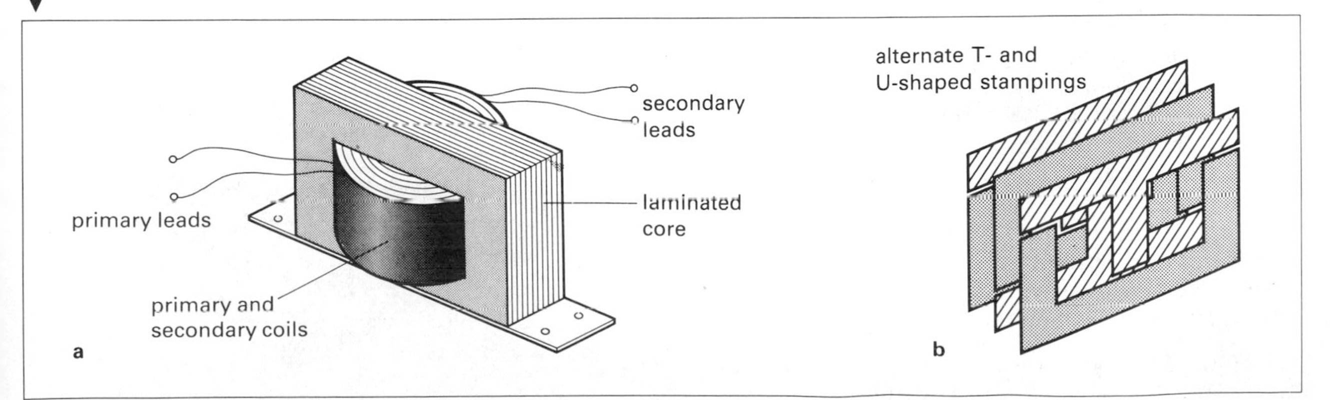

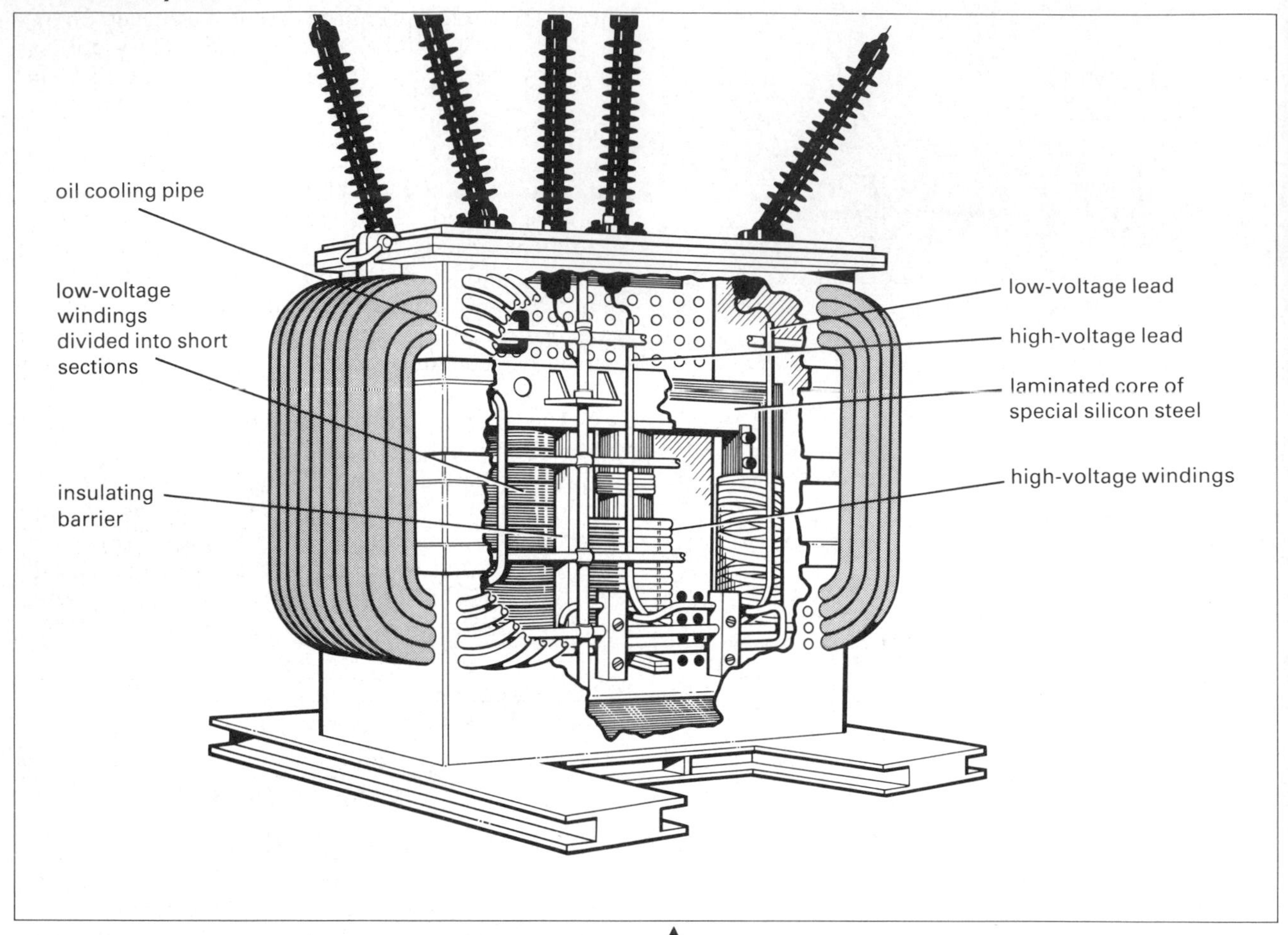

fig. 37.22 General arrangement in a large grid transformer

In practical transformers some of the input energy is lost and the efficiency of the transformer is less than 100%. Energy loss in a transformer occurs as follows:

- Heat is produced as current flows through both windings. The resistance of the windings should be kept as low as possible.
- Currents are induced in the iron core itself. These are called **eddy currents** and, since the core has a low resistance, the eddy currents could be large and produce a lot of heat. The eddy currents are reduced by **laminating** the core, figure 37.21. That is, making it of thin stampings of iron interspersed with insulating layers of varnish.
- The core is continually being magnetised and demagnetised by the alternating current, and the reversal of the molecular magnets requires energy. The core should be of soft iron with a low susceptibility.
- All the magnetic field produced by the primary does not link with the secondary; there is **flux leakage**. The coils are wound in alternate layers and the core forms a complete magnetic circuit.

In spite of all these possible losses, a well-designed transformer can have an efficiency of up to 99%.

fig. 37.23 High-temperature transformer for aircraft systems

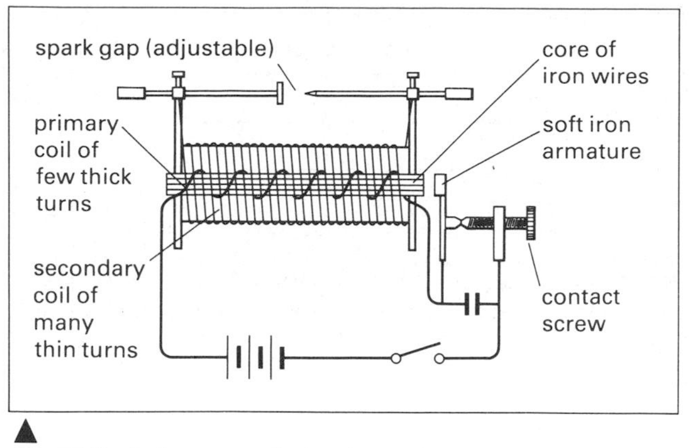

▲ fig. 37.24 Laboratory induction coil

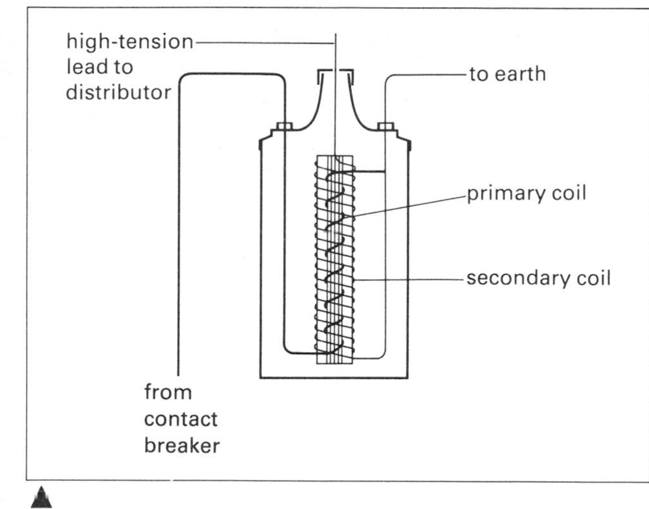

▲ fig. 37.25 Car ignition coil

The induction coil

A transformer cannot be used with direct current because the electromagnetic induction effect depends on there being a change of magnetic flux, as for instance when the current through a coil is switched on and off. The **induction coil** is really a transformer in which the primary current is automatically switched on and off by a make-and-break, as used in the electric bell, figure 37.24. When the primary current is switched on, the soft iron core becomes magnetised and attracts the soft iron armature, thus breaking the circuit again. The result is that the primary current is being continually switched on and off, and on each of these occasions an e.m.f. is induced in the secondary coil. Since the primary coil consists of a few turns of thick wire and the secondary coil has many turns of thin wire, the induction coil acts as a step-up transformer and a high voltage, up to many thousands of volts, is produced. These high voltages produce large sparks. The magnitude of the induced e.m.f. depends on the rate of change of flux and, since the sudden break of the circuit takes place more quickly than the comparativcly slow build-up of current on switching on, the e.m.f. on the break is very much larger than that on the make. The induced e.m.f. is intermittent but mainly undirectional.

The spark to fire the petrol-and-air mixture in a car engine is produced by a small induction coil, figure 37.25. This make-and-break is driven mechanically and the high voltage sent to the sparking plugs in turn by the distributor, figure 37.26.

fig. 37.26 Ignition system of a motor car ▼

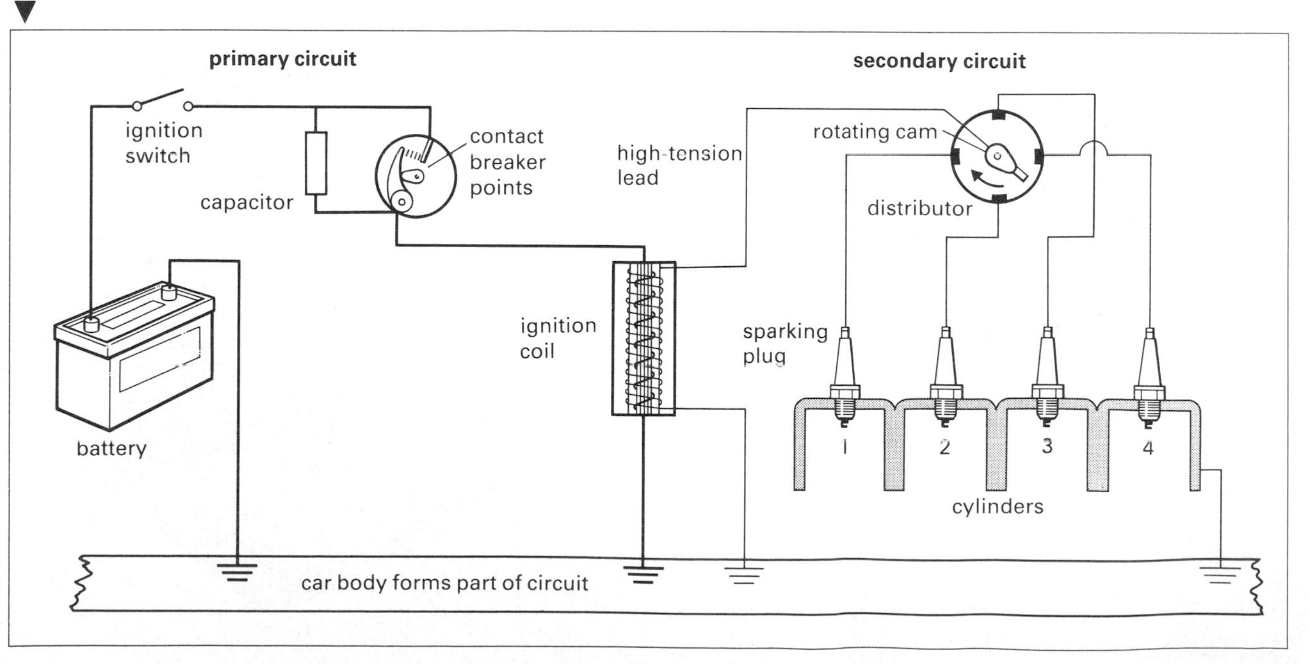

Rectification of a.c.

Although alternating current is easy to generate and transmit and can be used for most domestic and industrial purposes, some electrical devices, e.g. radio and television sets, require a direct current. An alternating voltage can be converted to a direct voltage using a **rectifier**, figure 37.27.

Rectifiers have a low resistance for current flow in one direction (forward direction) and a high resistance in the opposite direction (reverse direction). Rectifiers will only allow a one-way current. The effect of including a single rectifier in an a.c. circuit is to cut out the negative half-cycles and produce what is called **half-wave rectification**. Half of the original wave is wasted, figure 37.28.

Using two rectifiers and a connection in the middle of the transformer secondary produces **full-wave rectification**, figure 37.29. More often four rectifers arranged in a **bridge** are used to produce full-wave rectification, figure 37.30.

Although the output from the rectifiers is in one direction, it is not a steady voltage. Used in a radio set, for example, this variation in voltage would produce an unpleasant humming sound. The output has to pass through a **smoothing** circuit before it can be used.

There are three types of rectifier:

- metal rectifiers,
- thermionic diodes (valves),
- semiconductor diodes.

More details about valves and semiconductor diodes are given in Chapter 38.

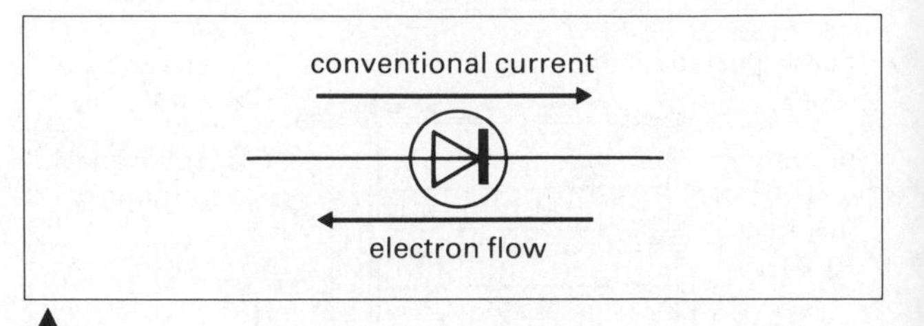

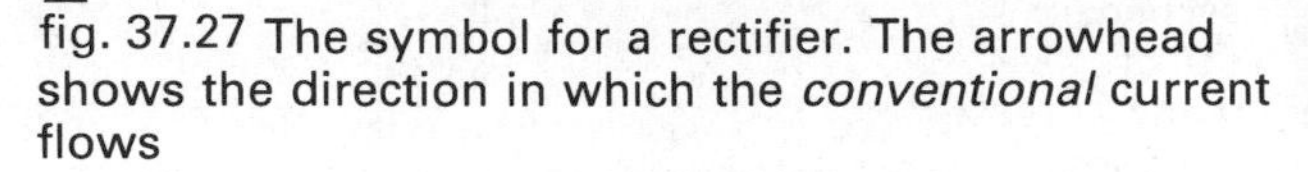

fig. 37.27 The symbol for a rectifier. The arrowhead shows the direction in which the *conventional* current flows

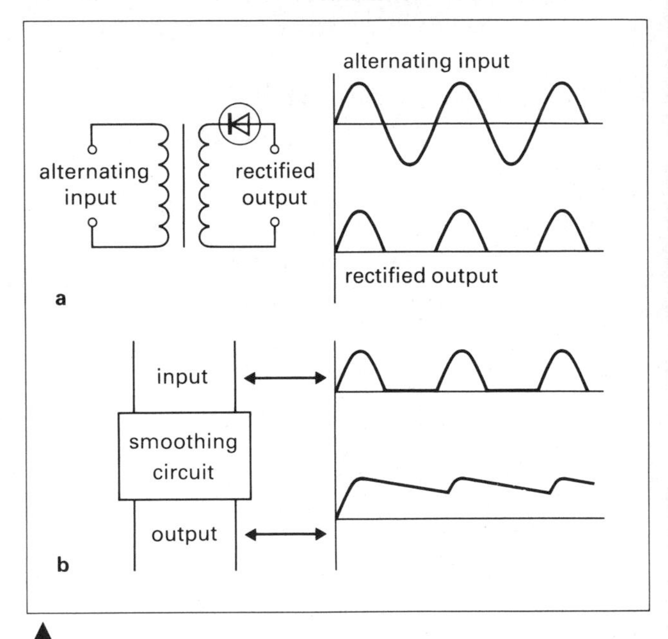

fig. 37.28 Half-wave rectification

fig. 37.29 Full-wave rectification

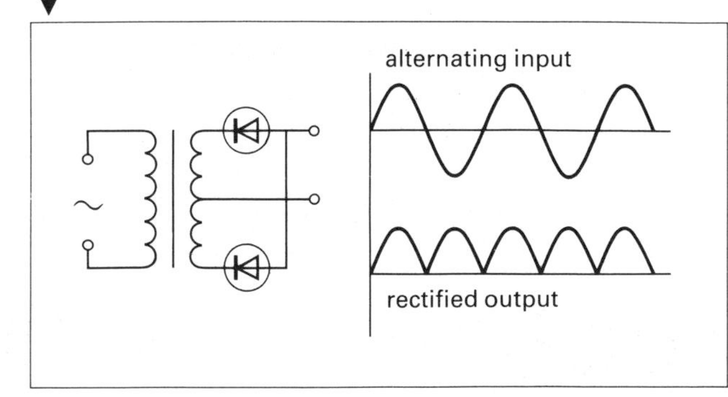

fig. 37.30 Bridge rectifier

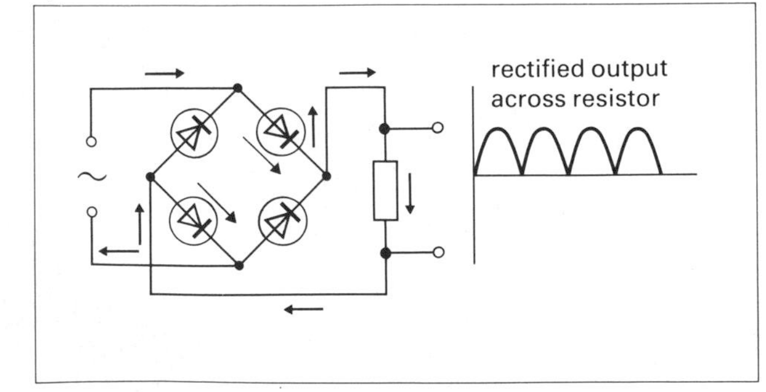

EXERCISE 37

In questions 1–6 select the most suitable answer.

1 Figure 37.31 shows a coil connected to a sensitive centre reading galvanometer, with a magnet near by. Which of the following statements is true?
- **A** The meter pointer moves when the magnet is moving towards the coil, but not when it is moving away from the coil.
- **B** The meter pointer moves if the magnet is moved quickly towards the coil.
- **C** The meter pointer does not move if the coil is moved towards the stationary magnet.
- **D** The biggest movement of the meter pointer is obtained when the magnet is stationary in the coil.
- **E** The movement of the meter pointer is in the same direction regardless of the direction in which the magnet is moving.

fig. 37.31

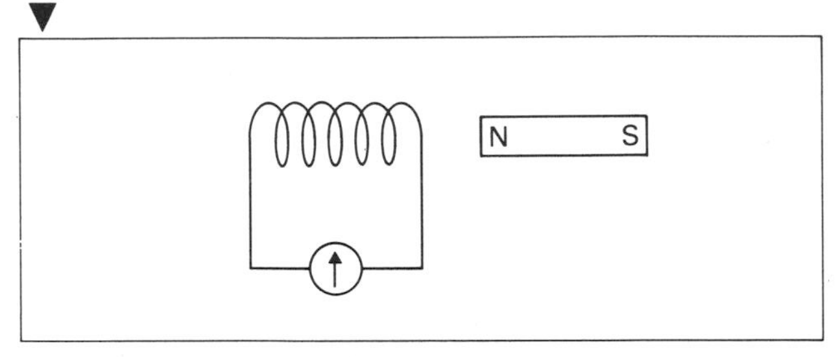

2 In which direction must the wire be moved in order to induce current in the direction shown in figure 37.32?
- **A** upwards
- **B** downwards
- **C** towards the north pole
- **D** towards the south pole
- **E** in some other direction

fig. 37.32

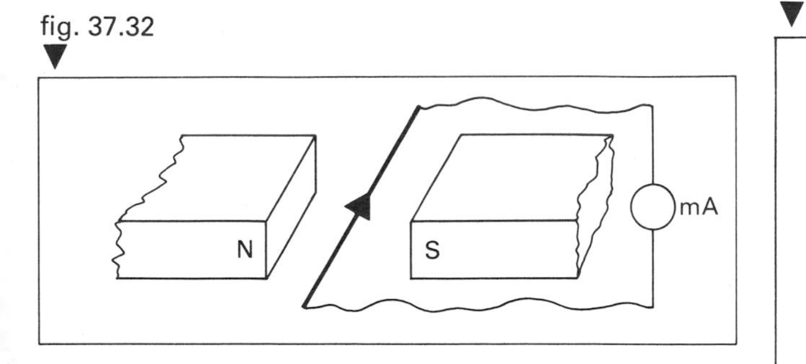

3 The step-up transformer
- **A** operates on direct current
- **B** has a higher current flowing in its secondary circuit than in its primary circuit
- **C** has a higher potential difference across its secondary coil than across its primary coil
- **D** has its coils wound on an insulating core
- **E** does not depend for its action on the principle of electromagnetic induction

4 A mains transformer working on 240 V has 4000 turns on the primary and 100 turns on the secondary. Assuming no power loss, the secondary output will be
- **A** 6 V
- **B** 12 V
- **C** 24 V
- **D** 960 V
- **E** 1440 V

5 Which of the following would work only on a d.c. supply and not on an a.c. supply of suitable voltage?
- **A** a filament lamp
- **B** an electric iron
- **C** an electric water heater
- **D** an electroplating machine
- **E** a fluorescent lamp

6 Electrical energy is transmitted at high voltages in the grid system because
- **A** the power station generator rotates at high speed
- **B** some factories need high voltages
- **C** it is easier to generate high voltages
- **D** energy losses in the cables are less at high voltages
- **E** the resistance of the cables is less at high voltages

7 A bar magnet is attached to one end of a long wooden rod AB which is hung from a pivot so that the magnet can swing in and out of the coil PQ, figure 37.33.

The magnet is pulled to the right and released. It makes ten swings before coming to rest. Describe the changes which take place in the voltmeter reading during this time.

What difference will there be in the voltmeter readings if the experiment is repeated but with
- **a** the coil PQ replaced by one of more turns of the same wire,
- **b** the magnet reversed?

fig. 37.33

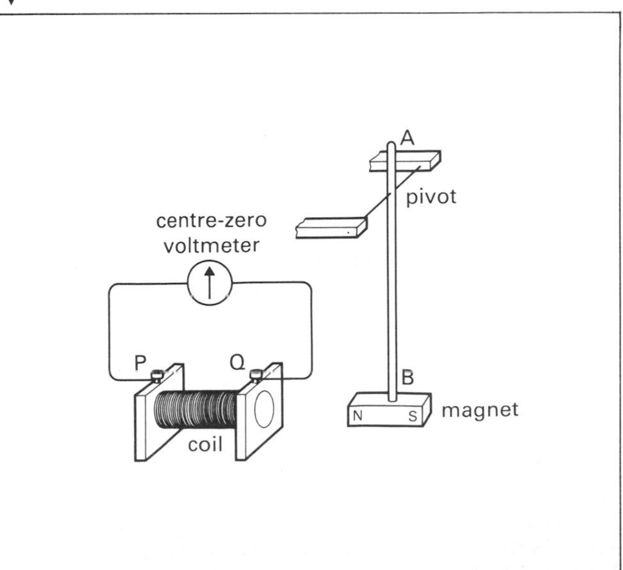

8 Figure 37.34 shows a transformer in which the primary coil is connected to a cell and a switch and the secondary coil is connected to a centre-zero galvanometer.

a What is observed when (i) the switch is closed? (ii) the switch remains closed so that a steady current flows through the primary coil? (iii) the switch is opened?

b The cell in the diagram is replaced by an a.c. supply, and the galvanometer is replaced by a filament lamp. What is observed when (i) the switch is closed? (ii) the switch remains closed? (iii) the switch is now opened? (iv) Give *three* ways in which the brightness of the lamp in this transformer circuit could be increased.

c You are given a transformer designed to operate a 12 V train-set from the 240 V mains. You find that the secondary coil of the transformer is damaged. How many turns would you need to wind on a new secondary coil if you find the primary coil has 1500 turns?

d Explain why the national grid uses step-up transformers at the power station and step-down transformers at the consumer end.

fig. 37.34

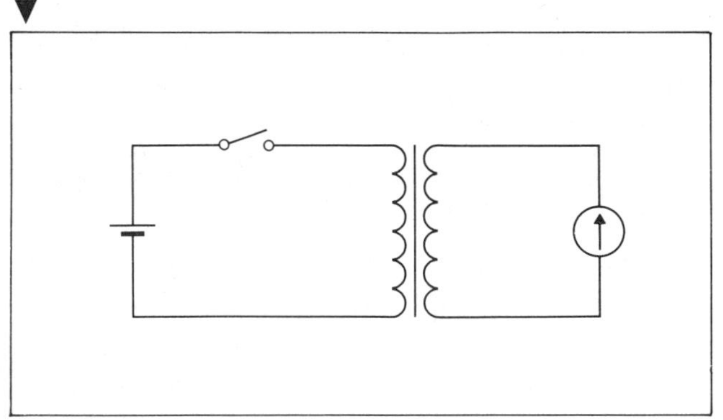

9 This question is about electromagnetic induction and transformers.

a In figure 37.35 P and Q are two coils of insulated wire, coil P being connected to a 36 W, 12 V lamp. It is observed that the lamp will light when an alternating current supply is connected to coil Q but not when a direct current supply is connected to Q. (i) Sketch the magnetic flux pattern produced by coil Q when the direct current supply is used. (ii) Account for the observations described above. (iii) With the alternating supply connected to Q, why will the brightness of the lamp decrease if P is moved away from Q? (iv) With the alternating supply connected to Q a rod of aluminium, held along the axis of P and Q, becomes warm but a rod of plastic does not. Suggest a reason for this. (v) With the lamp glowing at its normal brightness, calculate the current in the lamp, and then in Q, assuming the arrangement is 60% efficient.

b When electricity is transferred by cable from one part of the country to another, the voltage is stepped up to 275 000 V; it is then transformed down to 240 V before domestic use. What is the advantage of doing this?

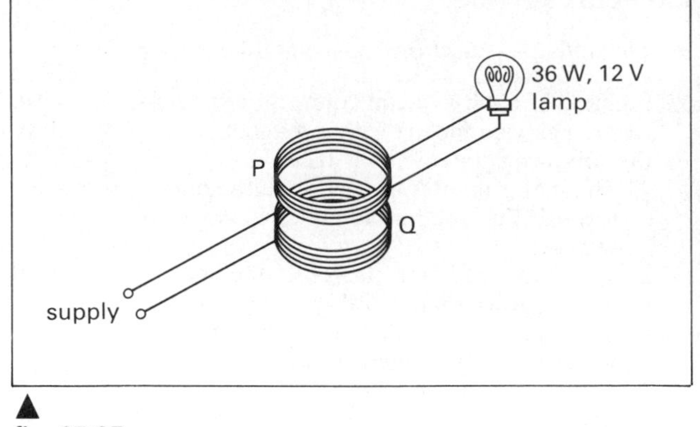

fig. 37.35

10 a It is required to run a 6 V, 24 W lamp from a 240 V a.c. mains using a transformer. (i) Calculate the current that would be taken by the lamp when operating normally. (ii) Calculate the turns ratio of the transformer you would use. (iii) Calculate the current taken by the primary coil of the transformer, assuming it to be 100% efficient. (iv) Why, in practice, is the efficiency of the transformer less than 100%?

b Alternatively the 6 V, 24 W lamp can be operated normally from a 240 V d.c. supply using a suitable fixed series resistor. (i) What is the resistance of the lamp? (ii) What is the p.d. across the resistor? (iii) What is the resistance of the resistor? (iv) How much energy is lost in it in 1 s?

c Why may the method used to light the lamp described in **a** be preferable to that described in **b**?

11 Figure 37.36 shows the circuit of the electrical system of a four-cylinder four-stroke spark ignition engine.

a Name, describe and explain the action and purpose of the parts indicated.

b What routine maintenance must be given to parts A and B to keep the engine in good running order?

fig. 37.36

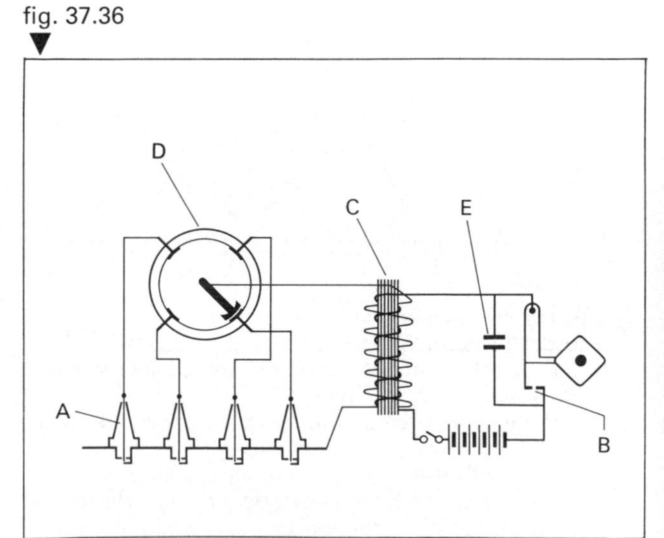

38 ELECTRONIC DEVICES

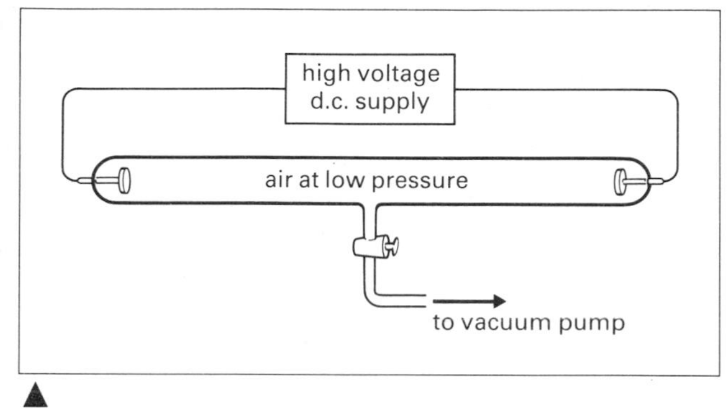

fig. 38.1 Electric discharge tube

The electron was the first of the sub-atomic particles to be identified. It was discovered by Sir Joseph John Thomson in 1897. His experiments were concerned with the passage of electricity through gases at low pressures. The development of vacuum pumps and the techniques of sealing metal electrodes into glass tubes had grown out of the manufacture of electric lamps, which began in the 1880s. These developments made it easier to carry out experiments with gas discharge tubes, and are an example of how scientific research often depends on technological progress.

Cathode rays

When a high voltage (about 5000 V) is connected to two metal electrodes at the ends of a glass tube full of air and the pressure is gradually reduced, figure 38.1, the air becomes conducting at about 10 mm of mercury pressure and a pink glow appears. As the pressure is further reduced, various changes take place in the glow until (at about 0.02 mm mercury) the glow disappears and the glass of the tube *fluoresces*. It was investigations of the tube at this stage which led to the idea that something was travelling from the cathode towards the anode; the name **cathode rays** was given to the phenomenon.

The cathode rays travel in straight lines because if an object is placed in their path a geometrically similar shadow is produced on a screen at the end of the tube. The experiment is usually carried out with a tube containing a metal Maltese cross, figure 38.2.

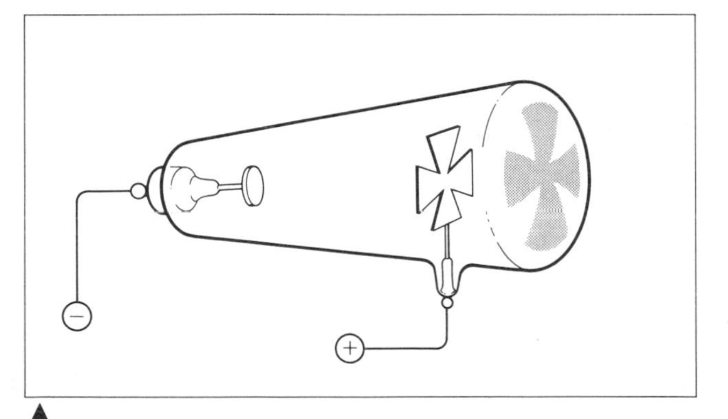

fig. 38.2 Maltese-cross tube shows cathode rays travel in straight lines

fig. 38.3 Deflection of cathode rays by an electric field

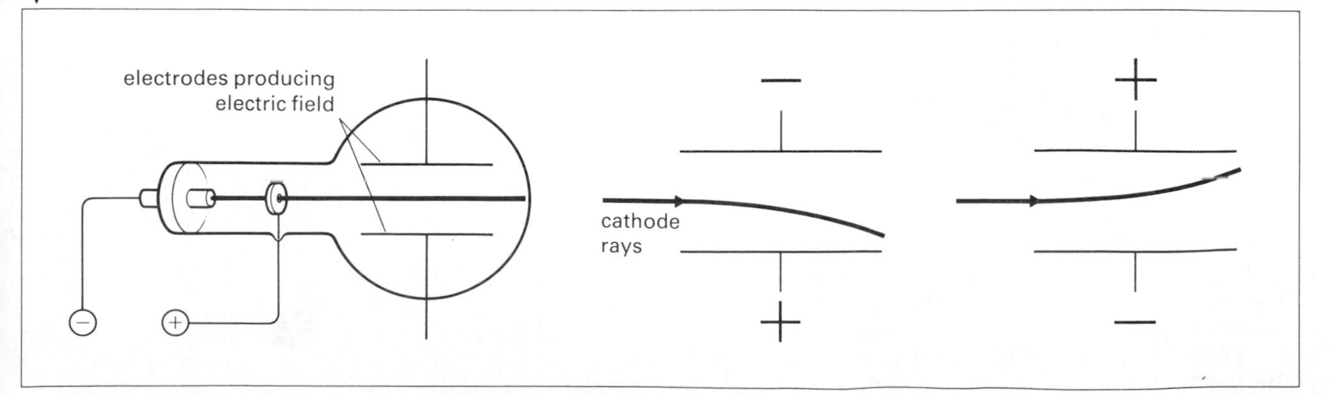

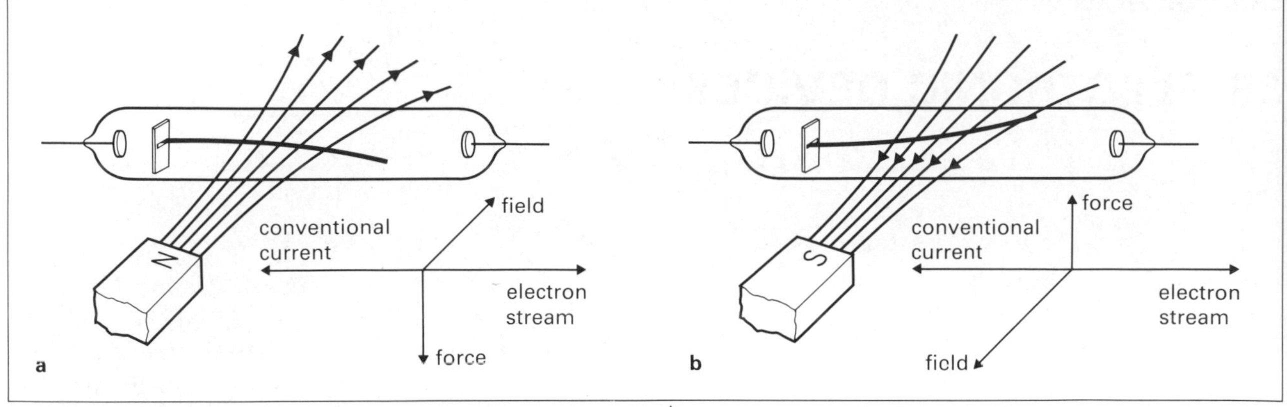

fig. 38.4 Deflection of cathode rays by a magnetic field. The direction of the deflection can be worked out using Fleming's left-hand rule

The cathode rays can be deflected by an electric field. When the rays are passed between two metal plates to which a potential difference is applied, they are deflected upwards or downwards depending on the direction of the field, figure 38.3. Observation shows the rays are attracted by the positively-charged plate and so must be negatively charged.

The cathode rays can be deflected by a magnetic field in a similar way to a conductor carrying an electric current, figure 38.4, and this suggests the cathode rays are moving charges. The direction of the deflection again shows the charges to be negative.

The negative charge carried by the cathode rays was confirmed by collecting the rays in a metal cylinder connected to an electroscope. The deflection of the electroscope could be shown to correspond to an increase in its negative charge.

All these observations suggested that the cathode rays consisted of a stream of negatively-charged particles travelling from the cathode to the anode.

The ratio of the charge e to the mass m was measured by Sir Joseph Thomson in 1897 and the separate measurement of the charge was made by R. A. Millikan in America from 1909 to 1916 (see p. 221). These negatively-charged particles are **electrons**. They are essential constituents of all atoms (see p. 221). The charge of an electron is 1.6×10^{-19} C, and its mass is 9.1×10^{-31} kg.

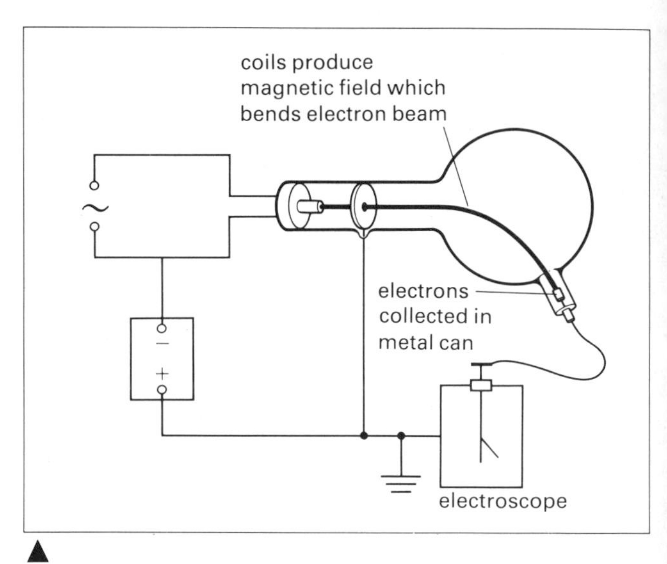

fig. 38.5 Using a Perrin's tube to show that cathode rays carry negative charge

The thermionic effect

About the beginning of this century it was discovered that heated metals give off electrons. The free electrons in a metal are attracted by the positive ions of the metal lattice (see p. 252), but if they are given sufficient energy they can overcome the attractive forces and escape, figure 38.6. The process is similar to the evaporation of molecules from the surface of a liquid when it boils. To make the liquid boil requires latent heat of vaporisation. The energy required to free an electron from the metal is called the **work function**, and this depends on the nature of the metal.

fig. 38.6 Thermionic emission

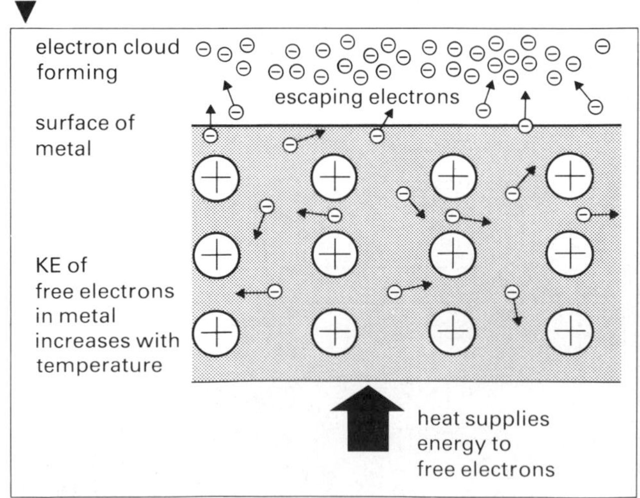

Thermionic emission is the emission of electrons from a heated substance (usually a metal or a metal oxide). The number of electrons emitted depends on the temperature (the number increases very rapidly as the temperature rises) and the nature of the emitting substance. Different substances have different work functions. Some materials—for instance a mixture of barium and strontium oxides—emit electrons very easily, but they have to be heated indirectly by a separate heating filament because they do not conduct electricity.

Semiconductors

Semiconductors are solids and electrons move between the atoms of the crystal lattice. Their electrical conductivities lie between those of metallic conductors and insulators. Silicon and germanium are typical semiconductors.

Silicon $^{28}_{14}$Si has 14 protons in the nucleus and 14 electrons around the nucleus. These are arranged in shells with 2, 8 and 4 electrons in them. The outer four electrons are the valency electrons concerned in chemical compounds. Germanium $^{74}_{32}$Ge has 32 electrons in shells with 2, 8, 18, 4 electrons in them. Both silicon and germanium are quadrivalent.

In pure silicon crystals the valency electrons bond with neighbouring atoms to form a stable lattice, figure 38.7. A few bonds break down to produce a few free electrons. As the temperature is raised, more and more electrons receive sufficient thermal energy to break away from the lattice, so the conductivity of semiconductors increases with increasing temperature. Remember that the reverse is true of metallic conductors. Above about 100 °C the structure of the lattice changes and so semiconductor devices (such as transistors) can only work at comparatively low temperatures.

The conductivity of semiconductors can be increased without raising the temperature by the introduction of certain impurity atoms into the lattice. Suppose a pentavalent atom, such as arsenic or antimony, is used. A pentavalent atom has five valency electrons. Four of these electrons bond with four neighbouring silicon atoms and one electron is left over to act as a conductivity electron. This produces ***n*-type silicon,** (*n* for negative). The impurity atom is called a **donar,** since it donates one electron, figure 38.8.

If a trivalent atom, such as aluminium or indium, is introduced there are only three valency electrons to bond with the surrounding silicon atoms. The position of the missing electron is called a **hole.** Any free electron in the silicon can fill this hole, but this will leave another hole in another atom. This new hole can be filled and will leave another hole, and so on. As electrons move in one direction holes move in the opposite direction, so the holes are effectively positive charge carriers. This type of silicon is ***p*-type** (*p* for positive). The impurity atoms are called **acceptors** since they can accept electrons from other atoms, figure 38.9.

Although there are negative charge carriers in *n*-type semiconductors and positive charge carriers in *p*-type semiconductors, any piece of material is uncharged because the donor and acceptor atoms have opposite balancing charges to those of the carriers they supply, figure 38.10.

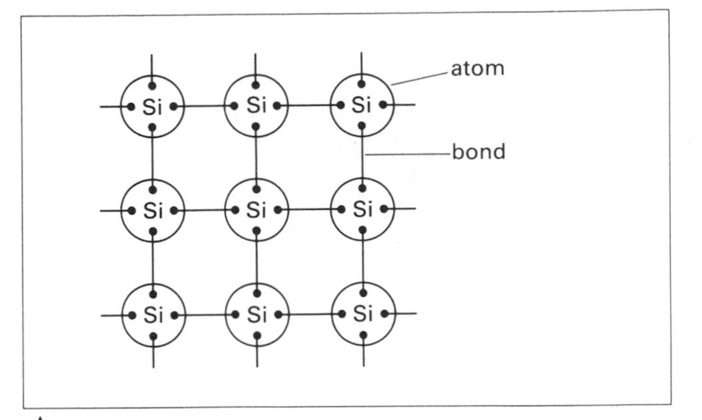

fig. 38.7 Normal silicon crystal

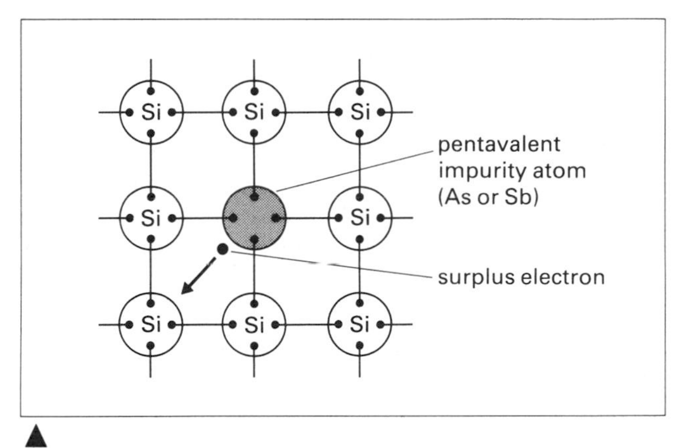

fig. 38.8 Surplus electrons in *n*-type silicon

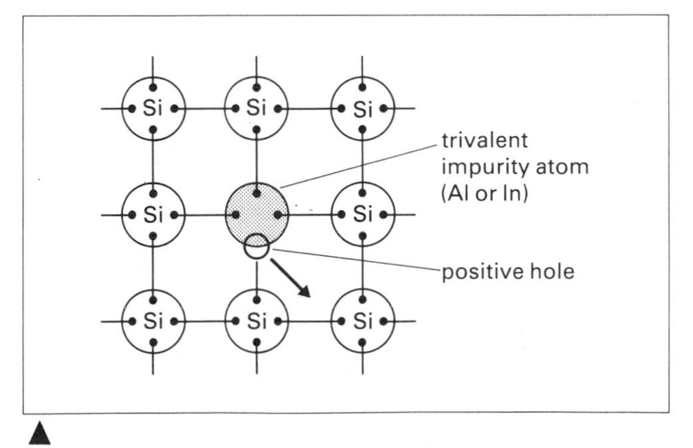

fig. 38.9 Positive holes in *p*-type silicon

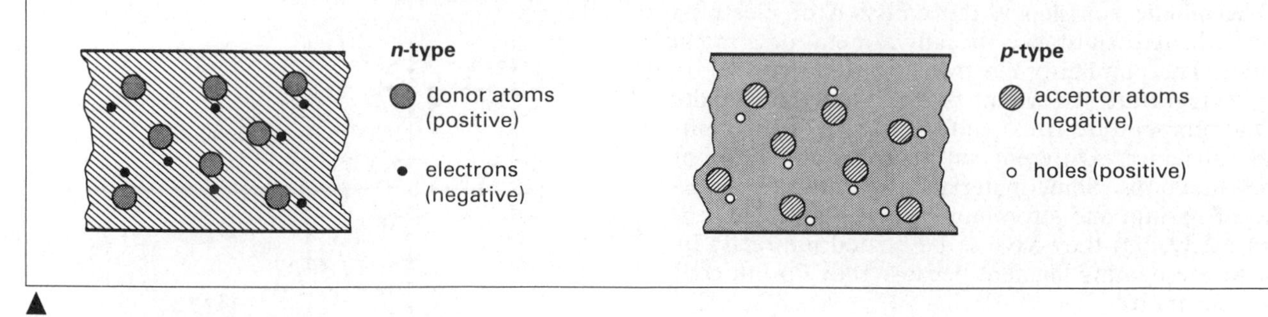

fig. 38.10 The charges balance in a semiconductor

The p-n junction diode

At the junction of a *p*-type with an *n*-type semiconductor, the diffusion of electrons and holes takes place, figure 38.11. But this process soon stops because, as the *n*-type loses electrons, it becomes positively charged (there are more positive donor atoms than electrons). This positive charge tends

- to hold on to electrons and stop them diffusing,
- to repel the positive holes diffusing across from the *p*-type.

In the same way, the *p*-type becomes negatively charged (excess of negative acceptor atoms) and this tends to repel diffusing electrons and attract holes. A barrier layer about 10^{-4} cm wide is set up, and across this barrier layer there is a potential difference (the *n*-type is positive and the *p*-type is negative) of a few tenths of a volt. The movement of electrons and holes ceases.

If a battery whose e.m.f. exceeds the barrier potential is connected across the junction with the positive terminal to the *p*-type and the negative terminal to the *n*-type, figure 38.12a, the potential barrier is overcome and charge carriers can move across the boundary. The junction is said to be **forward biased** by the battery. If the battery is connected the other way round, figure 38.12b, with the positive terminal to the *n*-type and the negative terminal to the *p*-type, the barrier potential is increased. The battery and the barrier potential are both acting in the same direction. The junction is now **reverse biased** and practically no current flows. Perhaps a few microamperes flow due to a very small number of electrons produced by ionisation.

It will be seen that the junction diode acts as a rectifier. It will allow a current to flow from *p* to *n* when it is forward biased, but not from *n* to *p* when it is reverse biased. The *p-n* junction can be used in the rectifier circuits described on p. 308.

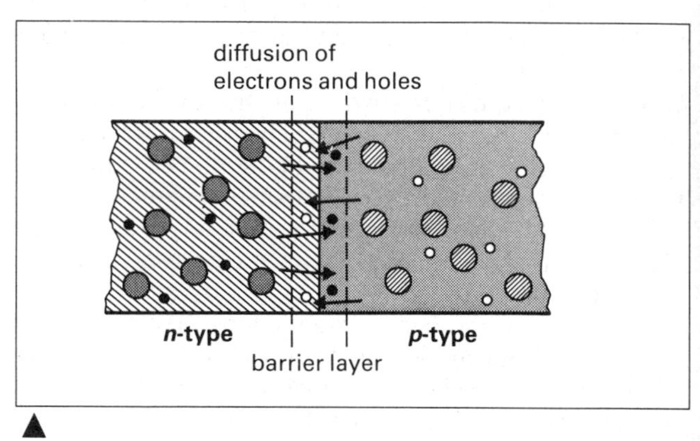

fig. 38.11 The *p-n* junction

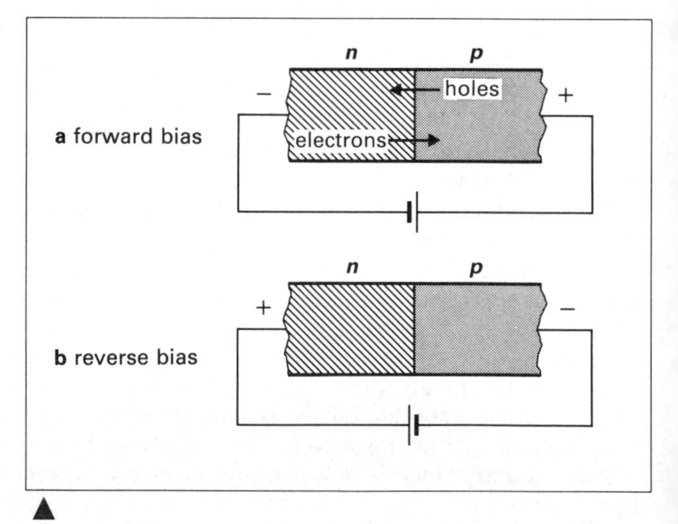

fig. 38.12 Forward and reverse bias

It must be pointed out that a diode cannot be made just by putting pieces of *p*- and *n*-germanium together. The crystal must be grown with a continuous lattice. The *p-n* junction is formed by melting a small pellet of indium into a slice of *n*-type germanium or silicon. The pellet produces a layer of *p*-type in contact with the *n*-type and this forms the junction.

fig. 38.13 Compact bridge assemblies using semiconductor rectifiers

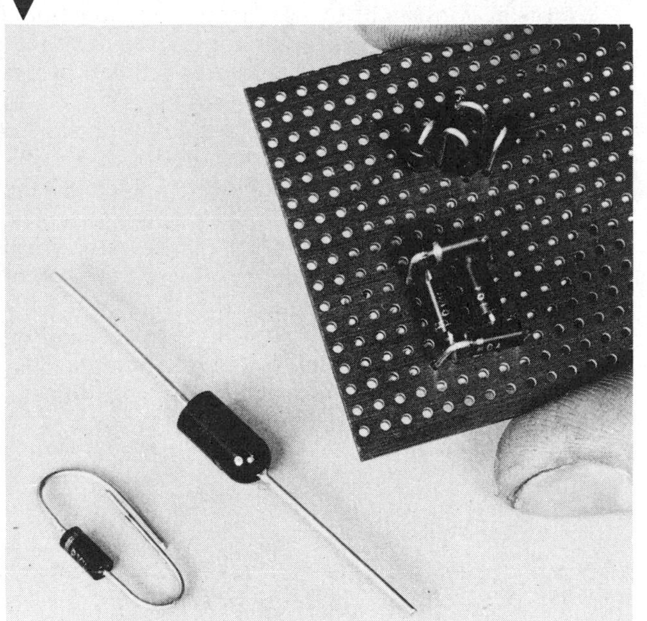

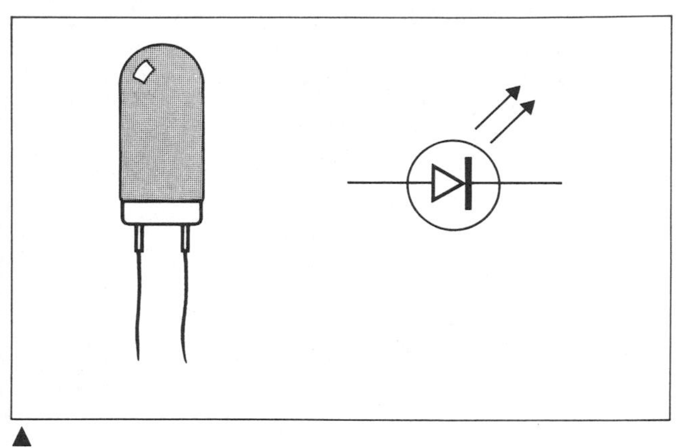

fig. 38.14 An LED and its symbol

The light emitting diode (LED)

The light emitting diode is a semiconductor diode which emits light when a current is passed through it. In diodes using silicon and germanium semiconductors, the electrical power is dissipated as heat. In diodes made from compounds of gallium about 1% of the power is emitted as red light. Other compounds produce green and yellow light.

A light emitting diode will only work satisfactorily if

- the current is in the forward biased direction,
- the applied voltage is correct (usually between 2 V and 3 V),
- the current does not exceed a certain value, otherwise the diode is damaged. A series resistor is used to limit the current.

The value of the series resistor in any particular case can be easily calculated. An LED rated at 2.3 V and 10 mA is to work on a 6 V supply. What is the value of the series resistor?

The voltage drop across the diode = 2.3 V

The voltage drop across the resistor = 6 −2.3 = 3.7 V

The current through the resistor = 0.01 A

$R = \frac{V}{I} = \frac{3.7}{0.01} = 370\ \Omega$

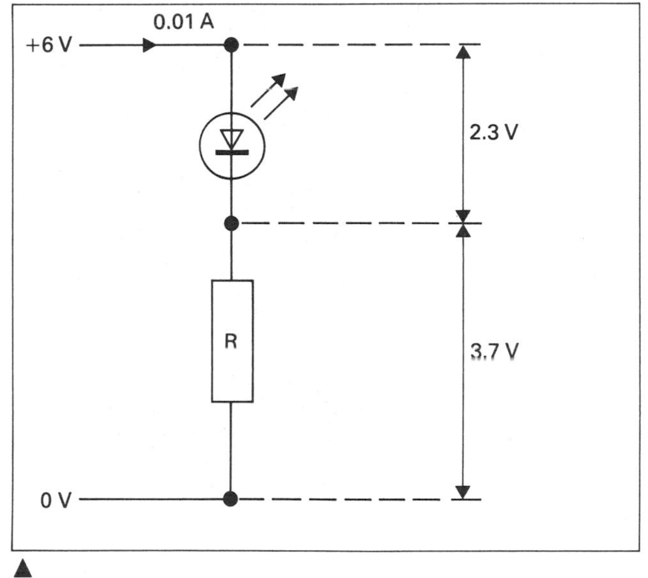

fig. 38.15

Light emitting diodes are useful as indicator lamps as they only need small currents and last much longer than filament lamps. They are used in the digital displays of clocks, calculators and cash registers. A seven segment digital display is shown in figure 38.16. By illuminating different combinations of segments all the numbers and some letters can be produced.

The LED has been replaced in quartz controlled watches by the liquid crystal display (LCD) which uses considerably less power, so the battery lasts much longer.

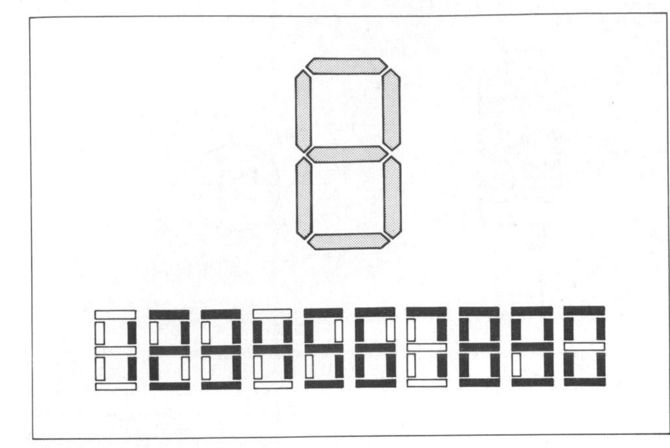

fig. 38.16

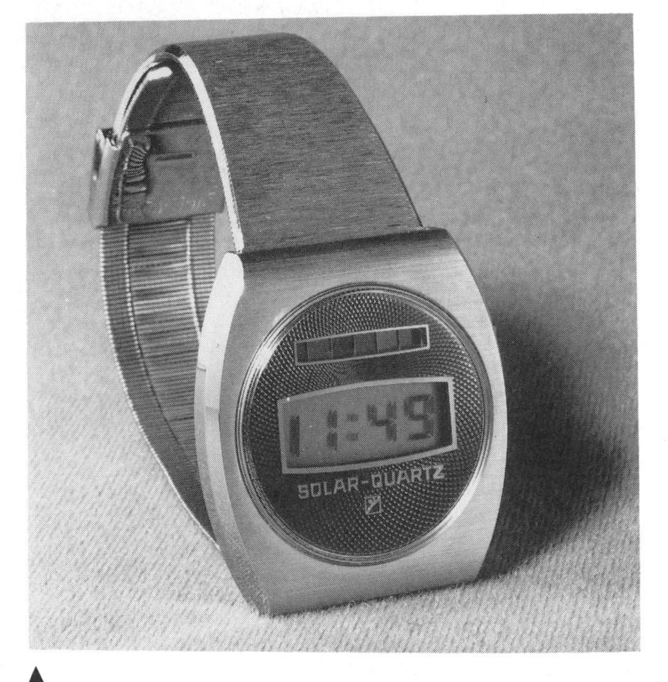

fig. 38.17 An early digital watch using five LED displays

The transistor

A **transistor** consists of two semiconductor junctions, e.g. a slice of *n*-type semiconductor between two pieces of *p*-type or a slice of *p*-type between two pieces of *n*-type. Thus, we can have a *p-n-p* transistor or an *n-p-n* transistor, figure 38.18. The middle portion is a very thin slice. It is about a hundredth of a millimetre thick. The three portions are known as the **emitter**, the **base** and the **collector** respectively. The transistor is sealed in a metal or plastic case which protects the junctions from heat and light, to both of which it is very sensitive, figure 38.19.

The transistor was developed in 1948 in the Bell Telephone Laboratories in New York by Schockly, Barden and Brittan. These scientists received the Nobel Prize in 1956 in recognition of their work.

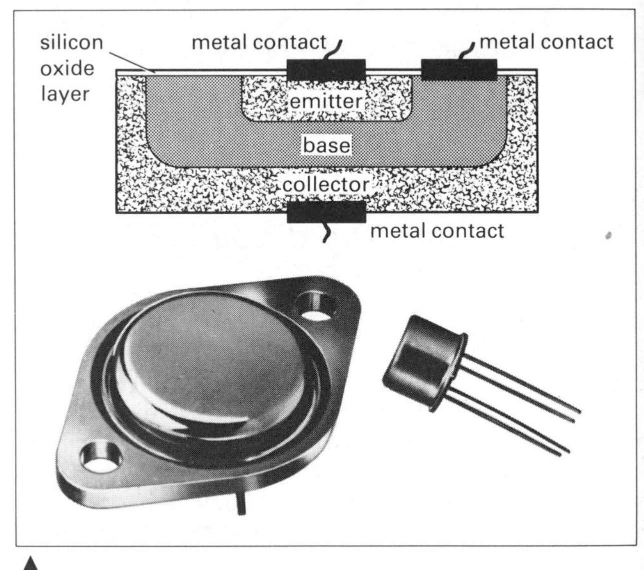

fig. 38.19 General construction of a transistor and two common designs

fig. 38.18 *p-n-p* and *n-p-n* transistors

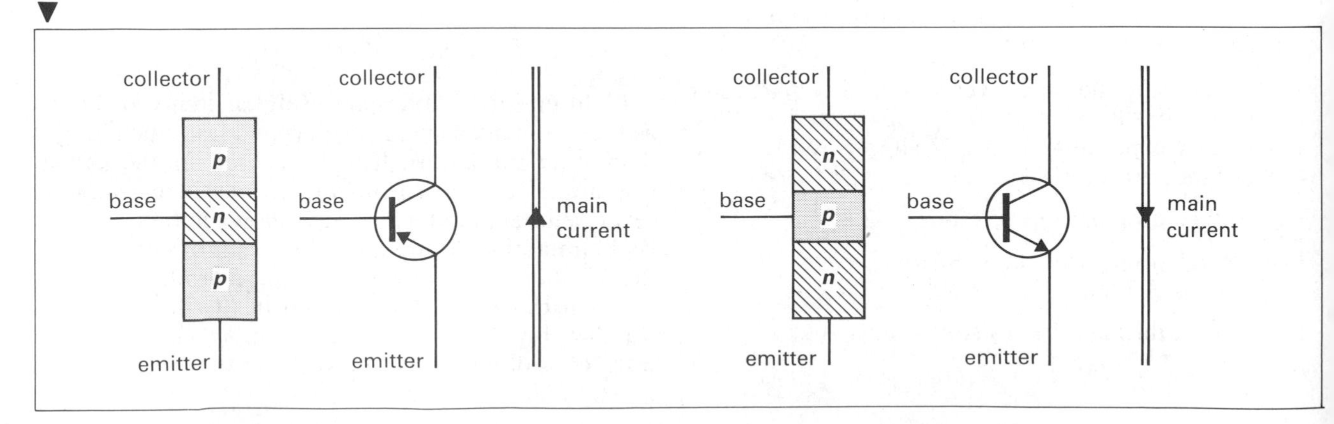

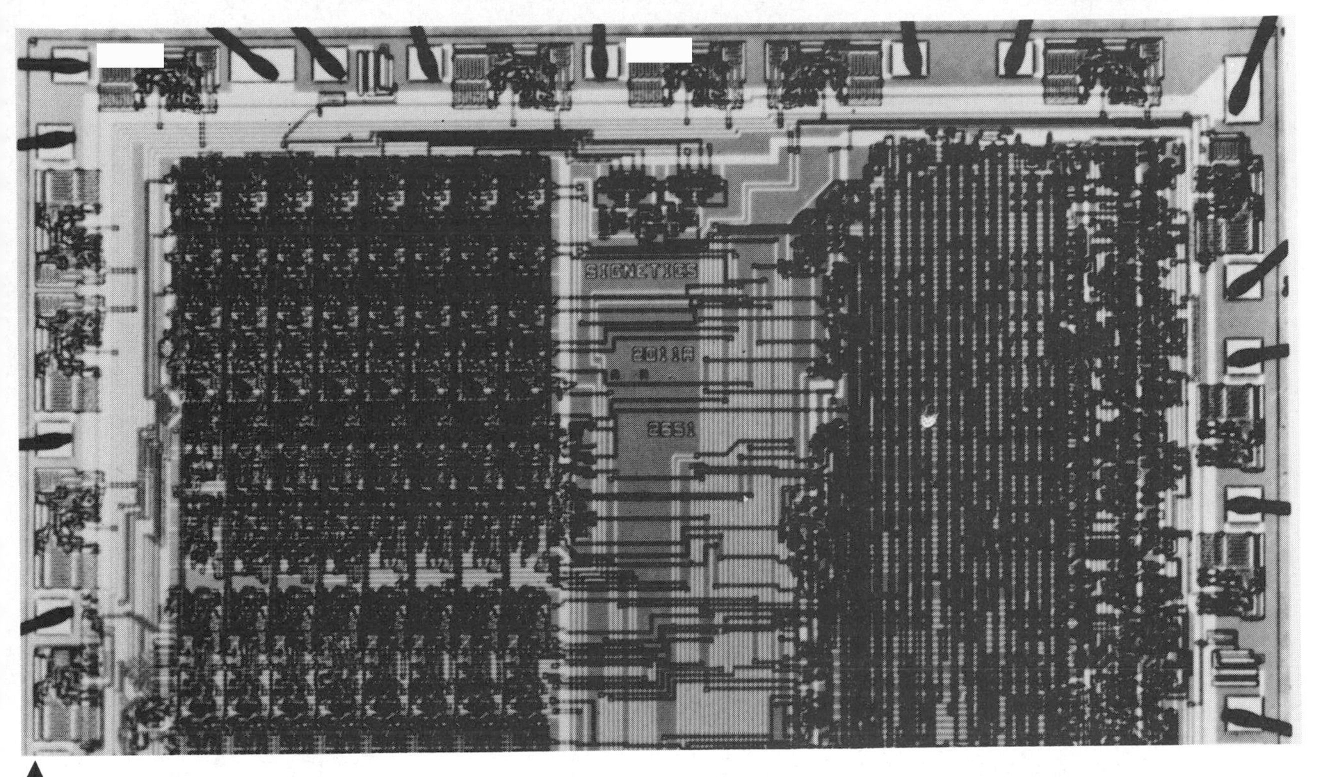

fig. 38.20 Part of a silicon chip integrated circuit for a microprocessor. The whole chip is some 6 mm x 4 mm

Transistors use much less power than valves, are cheaper and are more reliable. They are damaged if connected the wrong way round, and are very sensitive to temperature changes. They have to be soldered very carefully so as not to damage them. Their great advantage is that they can be made very small indeed, so that tens of thousands of components and their circuits can be produced on a single 'silicon chip' no bigger than a few millimetres square. This is why computers are getting smaller and smaller and cheaper and cheaper, figure 38.20.

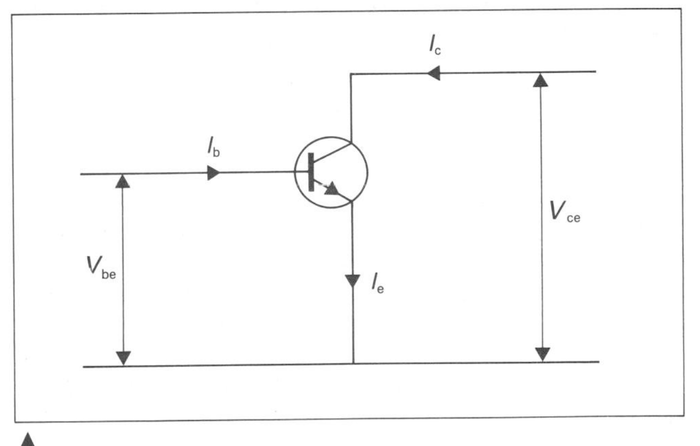

fig. 38.21

fig. 38.22

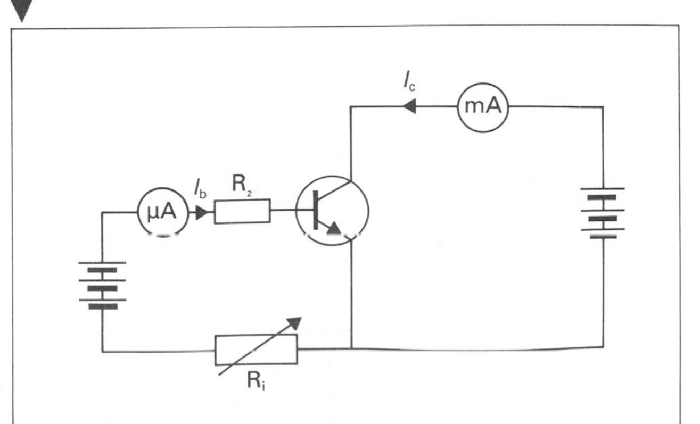

The characteristic currents and voltages of a transistor

The characteristic electrical quantities associated with a transistor are

- the base current I_b,
- the potential difference producing the base current V_{be},
- the collector current I_c,
- the potential difference producing the collector current V_{ce},
- the emitter current I_e.

These quantities are shown in figure 38.21.

To find what relationships exist between these quantities we concentrate on pairs of quantities. For example using the circuit in figure 38.22 we could find out how the collector current, measured by the milliammeter, varied with base current measured by the microammeter. A typical result is given in figure 38.23.

The base current is varied by using the variable resistor R_1. The resistor R_2 is to protect the transistor and has a value (usually about 10 kΩ) such that when R_1 is zero I_b does not exceed the maximum permissable value for the particular transistor being used.

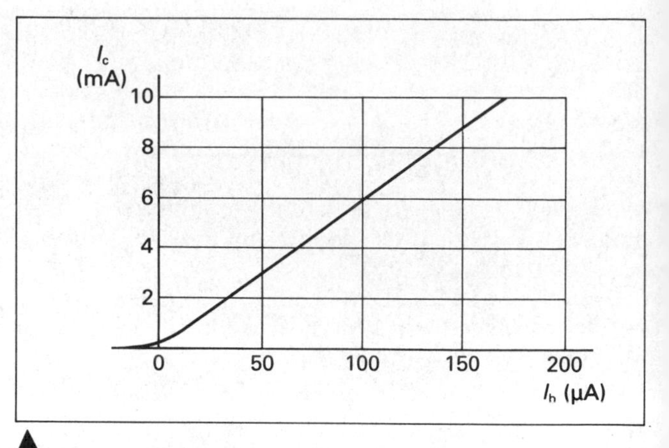

fig. 38.23 Variation of collector current with base current for a transistor

The graph clearly shows that

- when the base current is altered then the collector current changes,
- a small change in I_b produces a much larger change in I_c.

For example a change in the input current of 100 μA produces a change in the output current of 6 mA. The transistor acts as a **current amplifier.** The current amplification or **gain** (I_c/I_b) in this case is 60 times. The extra current is supplied by the battery in the collector/emitter circuit. In order to produce voltage amplification the output current is passed through a resistor in the collector circuit and the voltage is tapped from the ends of the resistor.

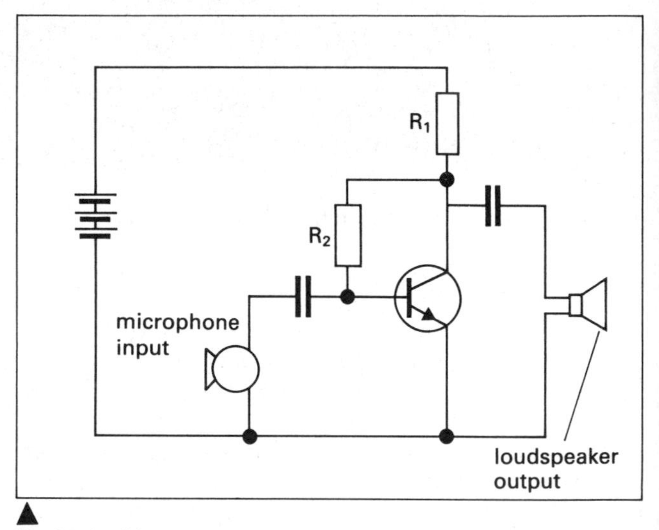

fig. 38.24 Single stage amplifier circuit. Appropriate values of R_1 and R_2 allow the same battery to be used for the input and output circuits.

The transistor as an amplifier

Radios, cassette players and all other types of electronic communication systems use transistors as amplifiers. The varying current produced by a microphone acts as the input (I_b) and the output current (I_c) is supplied to the loudspeaker. A single stage amplifier circuit is shown in figure 38.24. Additional stages of amplification can be added by connecting the output of the first stage to the input of the second stage and so on.

fig. 38.25 **a** Connections of a typical op-amp

b Symbol for an op-amp

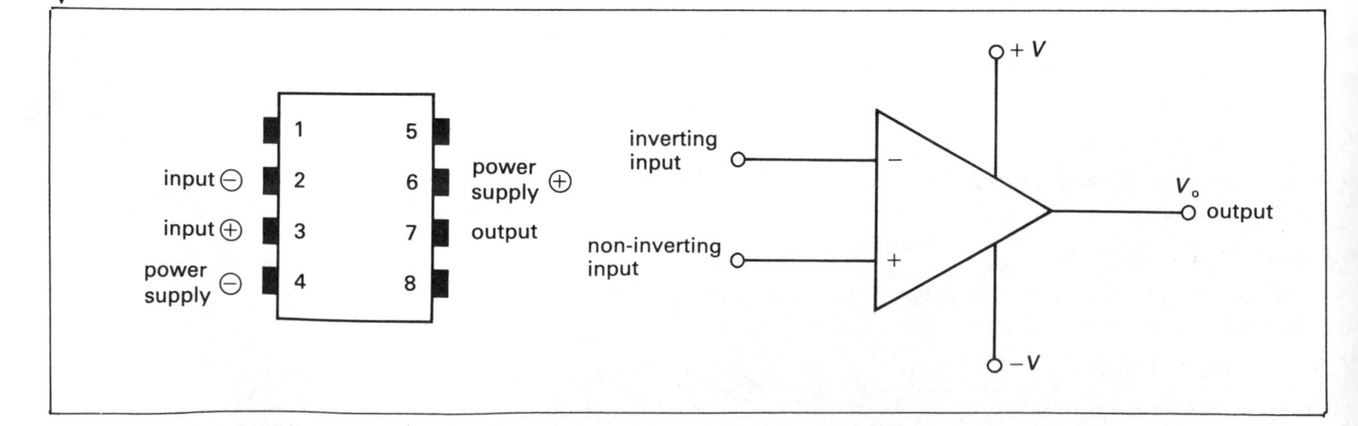

An **operational amplifier** is an integrated circuit containing a series of transistor amplifiers, each one amplifying the output of the previous one. A simple 'op-amp' could contain twenty transistors with resistors and capacitors in a single chip about 3 mm square. The op-amp is used in calculators and computers as a high gain voltage amplifier. The connections of a typical op-amp and the symbol for it are shown in figure 38.25.

There are two input connections. The input terminal marked pin 2 is an 'inverting' input – a negative input here will produce a positive output and vice versa. The input terminal marked pin 3 does not produce this inverting effect. For use as an inverting amplifier the op-amp is connected as shown in figure 38.26 using a **feedback** resistor R_f which 'feeds back' some of the output to the input side of the device.

The output voltage is larger than the input voltage and is the opposite way round (inverted). The voltage gain is equal to V_o/V_i and for all practical purposes this ratio is equal to R_f/R_i. If R_i = 10 kΩ and R_f = 1 MΩ, the voltage gain would be 100 times. Op-amps can give gains of up to 1000 times.

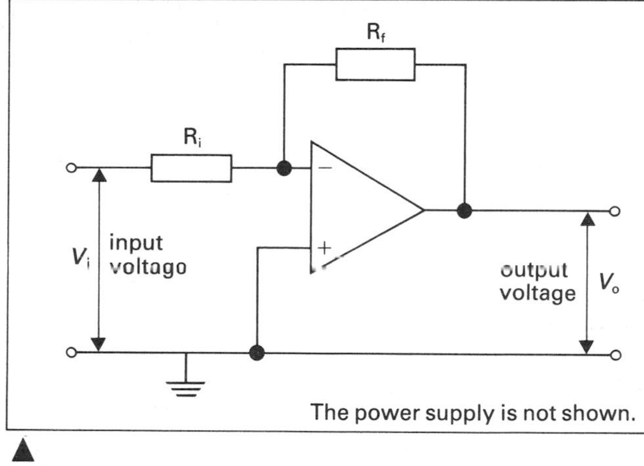

fig. 38.26

The transistor as a switch

When the base current is very small there is virtually no current in the collector/emitter circuit: the transistor is **off.** When the base current is sufficiently large to give a collector/emitter current, the transistor is **on.** The minimum base/emitter voltage to switch a silicon transistor on is about 0.6 V.

Another way of looking at this behaviour of a transistor is to realise that when no base current flows the collector/emitter circuit has a very high resistance and when the base current is increased the collector/emitter circuit resistance is low. The use of a transistor as a switch is illustrated by using the circuit shown in figure 38.27.

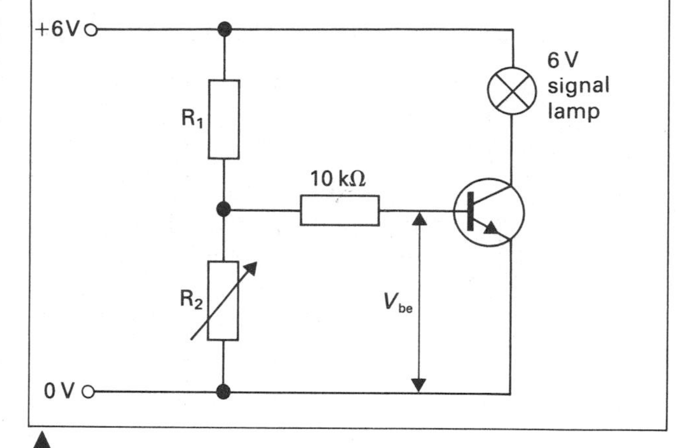

fig. 38.27

The 6 V supply is connected to a potential divider (R_1 R_2). When R_2 is very small the base/emitter voltage V_{be} will be very small and the transistor switch is **off.** As R_2 is increased and the value of V_{be} becomes greater than about 0.6 V the transistor switch is **on** and the signal lamp lights.

Suppose we replace R_1 of the potential divider by a thermistor, figure 38.28, and adjust R_2 so that the lamp does not light. This means V_{be} is less than 0.6 V. When the temperature of the thermistor increases, its resistence decreases. The voltage drop across the thermistor decreases and the voltage across R_2 increases. When it reaches 0.6 V the lamp will light. We now have a heat operated switch or a fire alarm.

In figure 38.29 the thermistor is replaced by a light dependent resistor. When light shines on the LDR its resistance decreases and at some stage V_{be} will be

fig. 38.28

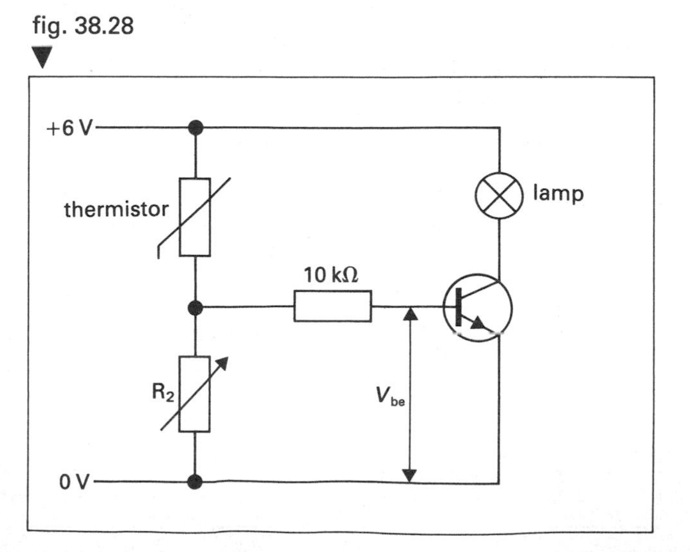

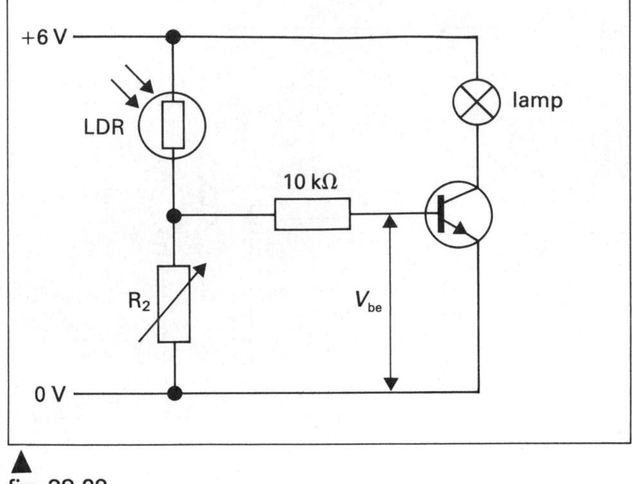

fig. 38.29

large enough to 'trigger' the transistor switch so that the light is switched on. This is a light operated switch. The switch is open in the dark and closed in the light.

Rather than draw out the whole circuit every time, the idea behind the system can be represented by a block diagram. The two circuits above could be represented quite simply as

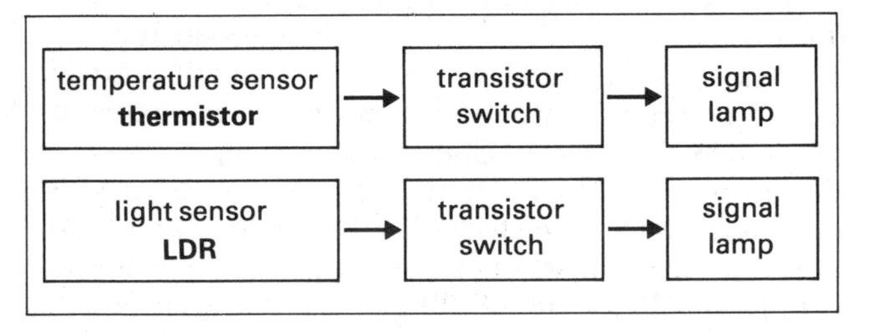

Consider the two circuits shown in figure 38.30 where the positions of R_2 and the thermistor and LDR have been interchanged. We still have heat operated and light operated switches, but they work the other way round. When the temperature falls or the light intensity decreases the signal lamp will light. For what purposes might these switches be used?

Suggest uses for the systems represented by the following block diagrams.

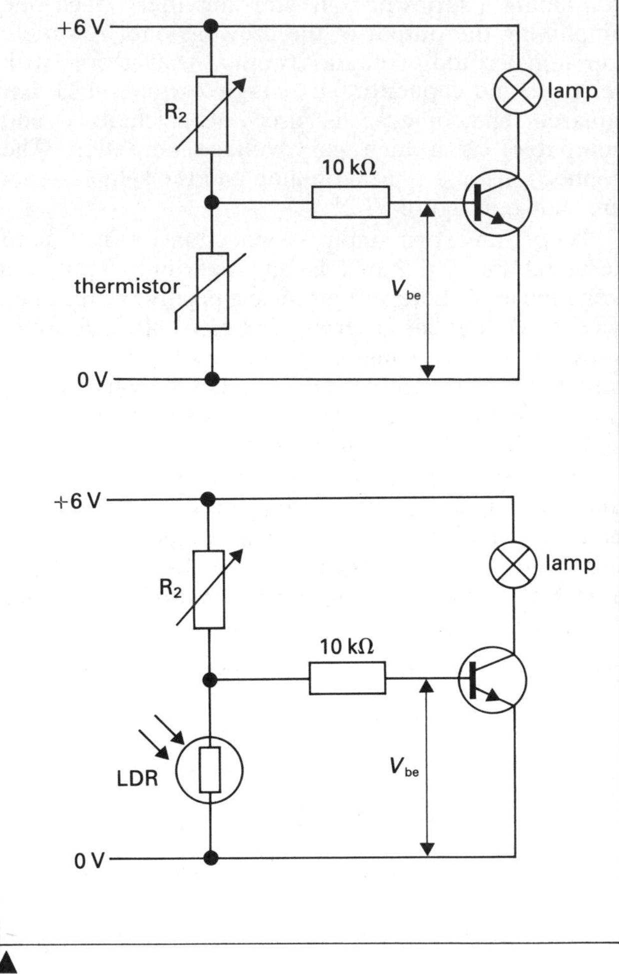

fig. 38.30

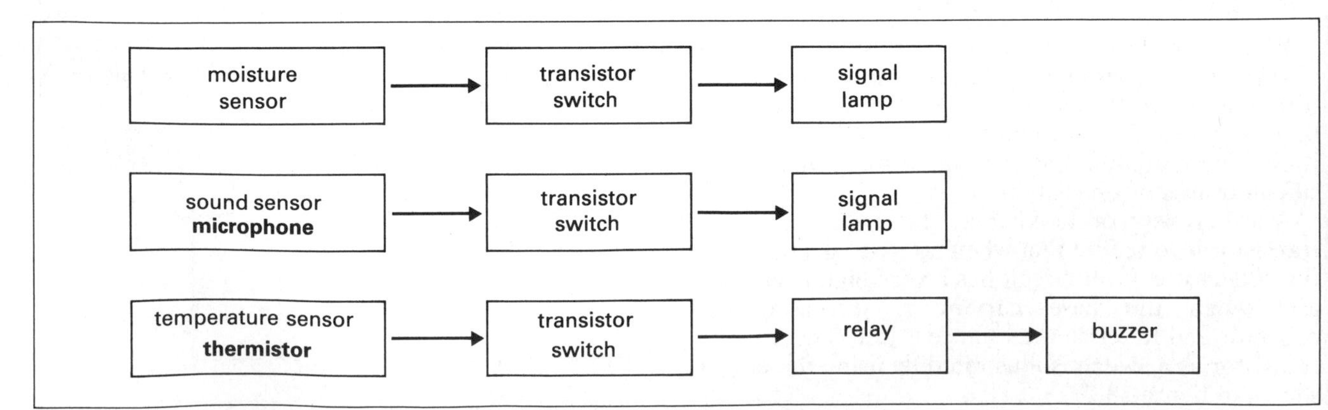

Logic gates

Electronic components in quartz watches, electronic robots, calculators and computers are connected together by systems of switches which will only pass on a signal (an electric current) if they receive the correct input signals. These systems of switches are called **logic gates.** These devices are of no use by themselves, they must be supplied with input voltages so that the output can be used to trigger other devices.

Let us first consider a simple circuit using mechanical switches (figure 38.31). A series circuit consists of two switches, a battery and a lamp. It is clear that the lamp will only light if both switch A *and* switch B are closed. This system is basically an **AND** gate. The behaviour of the system is summarised in table 38.1.

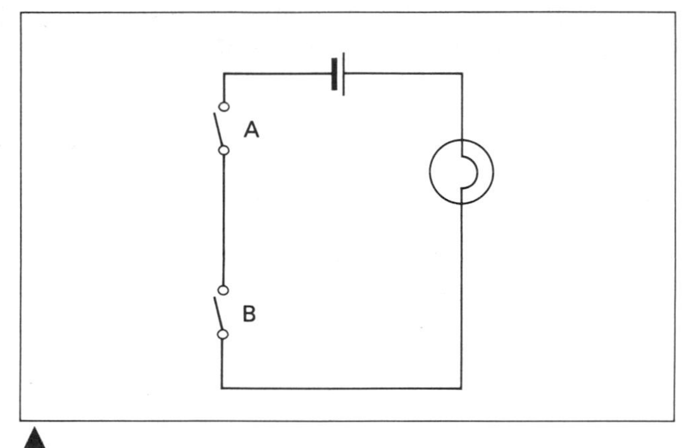

fig. 38.31

A switch can be in one of two states, it is either *off* or *on*. These two states are represented by the numbers 0 and 1 respectively (referred to as logical 0 and logical 1). In the same way, a lamp can be either off or on and these two states are represented by 0 and 1 respectively. When we rewrite the table using numbers instead of words we have what is called the **truth table** for the gate, table 38.2. It is called a 'truth' table because in logical arguments we are often concerned with whether a particular statement is true or false. For an AND gate the output is only true if both inputs are true. Such a switching system could be used in an electric washing machine or a spin drier. The power is only supplied to the motor when both the main switch AND the switch operated by the door

table 38.1 Summary table for an AND gate

Switch A	Switch B	Lamp
off	off	not lit
on	off	not lit
off	on	not lit
on	on	lit

table 38.2 Truth table for an AND gate

Input A	Input B	Output
0	0	0
1	0	0
0	1	0
1	1	1

are closed. The mechanical switches can be replaced with either relays or transistors.

The lamp is figure 38.32 will only light when the coils of both relays are energised. If the relay will only operate when the applied voltage $+V$ is greater than 6 V, then we would represent a voltage less than 6 V by 0 and a voltage greater than 6 V by 1 in the truth table.

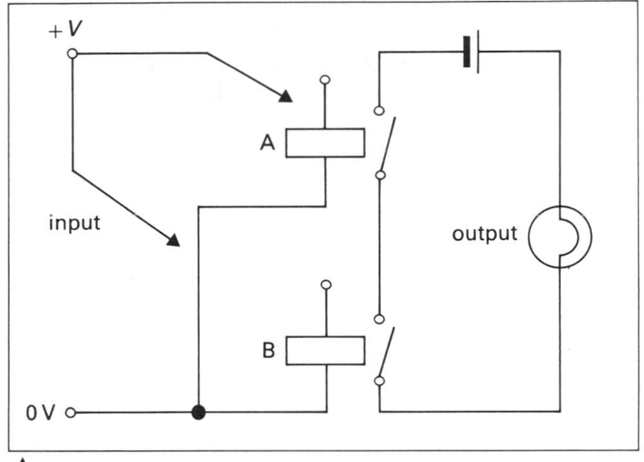

fig. 38.32

The transistor switches are included in an integrated circuit which could contain several AND gates (or other types of gates to be described) in a single chip. An AND gate is used when two conditions have to be satisfied at the same time before a specified operation is carried out. For example we could arrange to open the windows of a greenhouse in daylight when the temperature reached a certain value. The block diagram for the system would be as shown in figure 38.34.

If the LDR is in the dark its resistance is high and the voltage to the first input of the AND gate is logical 0. If the thermistor is below a preset temperature its resistance is high and the voltage to the second input of the AND gate is also logical 0. The output of the gate is zero.

fig. 38.33 The symbol for an AND gate

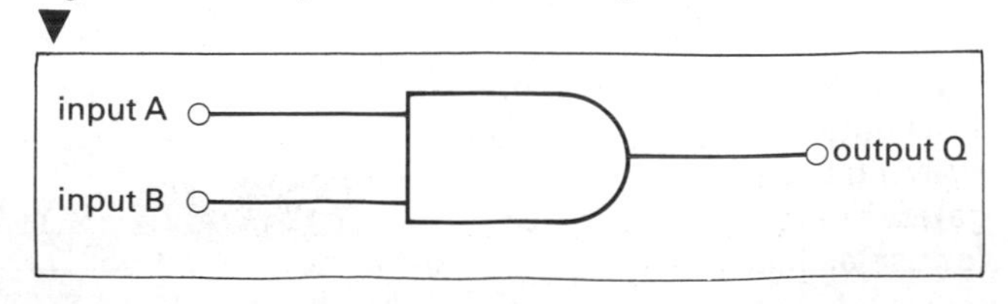

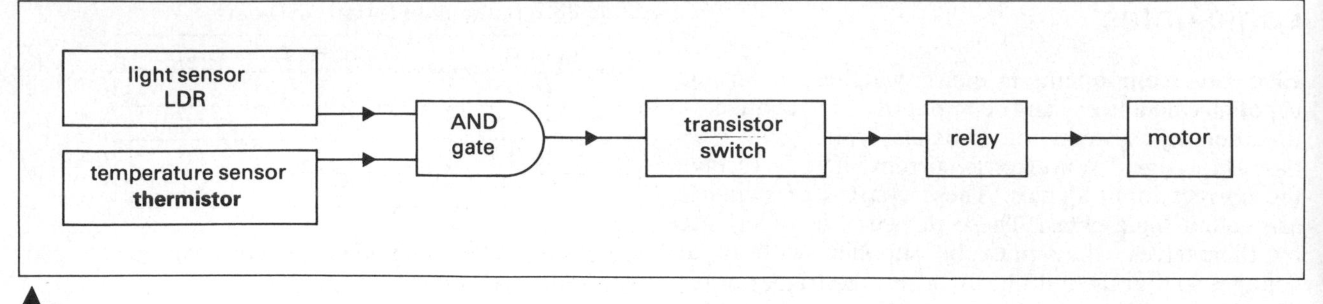

fig. 38.34

When the light falling on the LDR reaches a certain intensity its resistance decreases and a voltage is supplied to the first input of the gate (logical 1). As the temperature rises the resistance of the thermistor decreases. A voltage is supplied to the second input of the gate (logical 1) and the gate 'opens' allowing the transistor switch to pass a current through the relay and start the motor to open the window. A relay has to be used because the current needed to run the motor would damage the transistor.

The circuit diagram for the 'greenhouse' example (figure 38.35) shows that it essentially consists of two **potential dividers.** The potential difference (p.d.) across R_1 is controlled by the LDR and the p.d. across R_2 by the thermistor. These p.d.s are applied to the inputs of the AND gate.

The experimental investigation of the properties of logic gates is most simply carried out using specially designed circuit boards such as that shown in figs. 38.36 – 38.38.

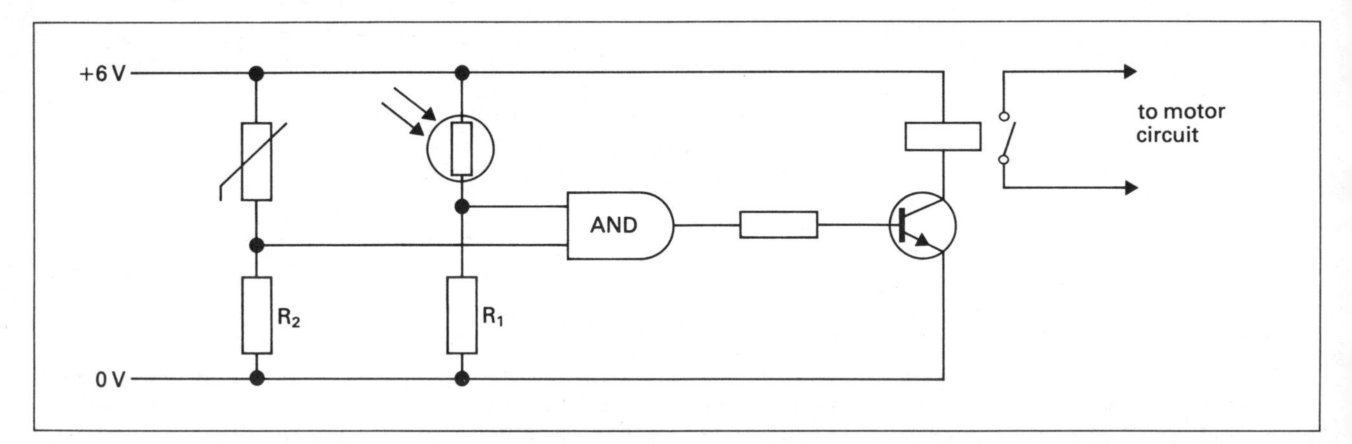

fig. 38.35

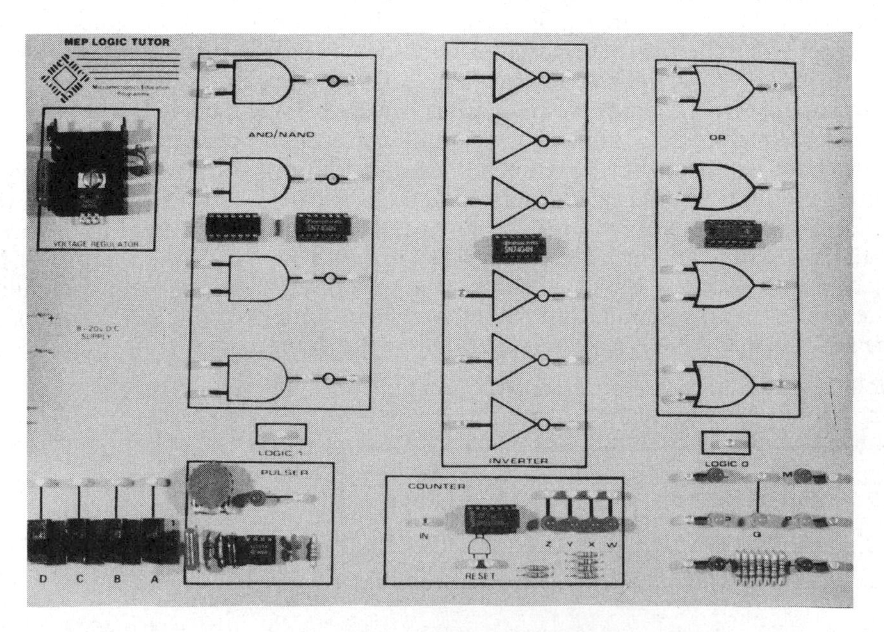

Fig. 38.36 This board has four AND gates, four OR gates, four NAND gates six NOT gates. There are seven LCDs for use as indicators

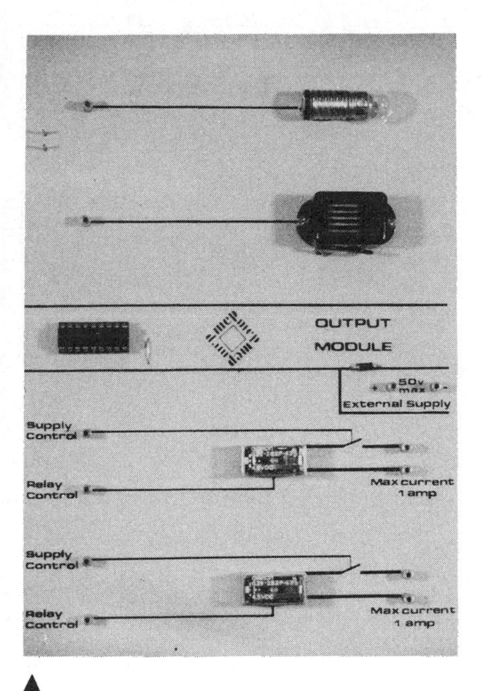

▲
fig. 38.37 Various input transducers: a switch, moisture, light and temperature sensors

fig. 38.38 Outputs include lamp, buzzer and two channels for driving motors
▼

More logic gates

The OR gate

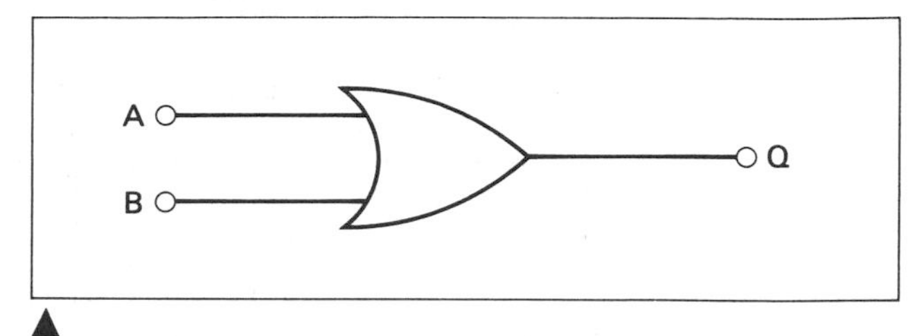

▲
fig. 38.39 Symbol for an OR gate

The basic circuit of the OR gate is illustrated using mechanical switches in figure 38.40. The lamp will light if *either* switch A *or* switch B is on, or *both* are on. The OR gate is used when either (or both) of the two conditions is satisfied before a specified operation is carried out.

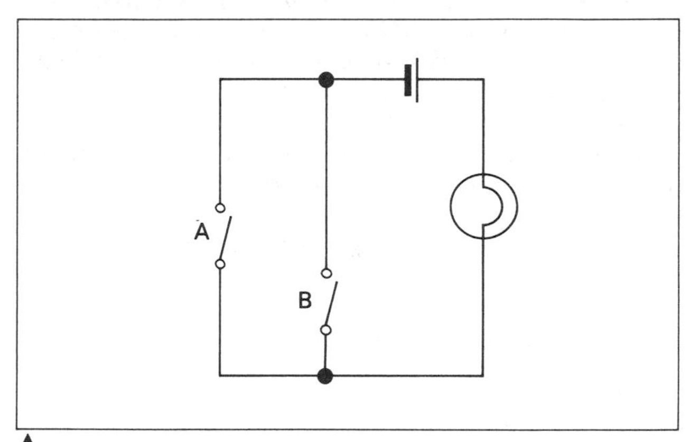

▲
fig. 38.40

table 38.3 Truth table for an OR gate

Switch A	Switch B	Lamp
0	0	0
1	0	1
0	1	1
1	1	1

An OR gate could be used to add a test button to the light/temperature sensor system for the greenhouse illustrated in figure 38.41.

The motor will start if either the output from the AND gate is logical 1 *or* the test button is pressed.

The NOT gate

The circuit of a NOT gate is arranged so that when the input is low the output is high, and when the input is high the output is low. In other words it is an **inverter:** the output is always opposite to the input.

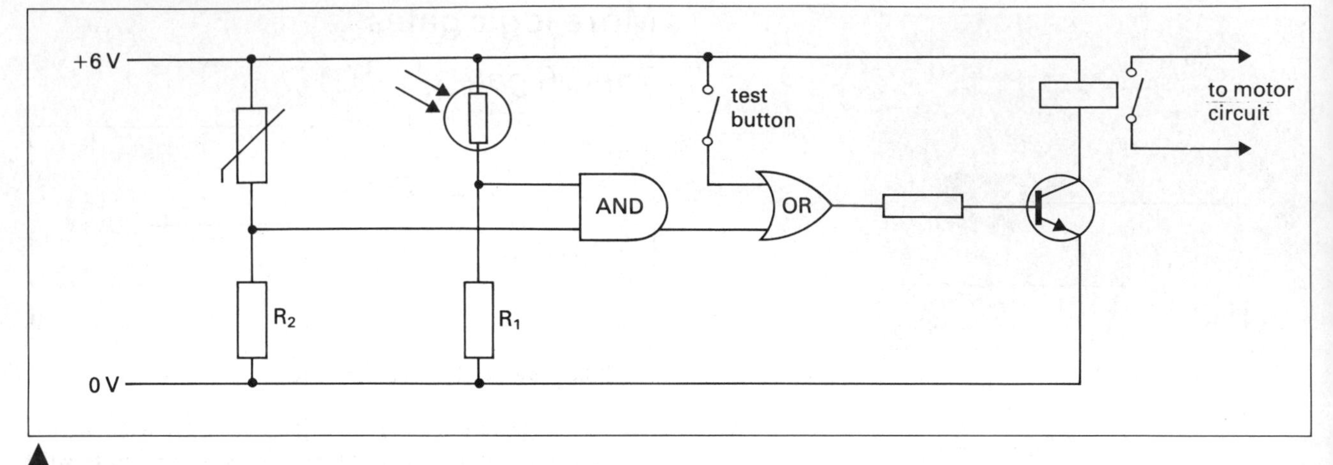

fig. 38.41

A NOT gate can be used to reverse a particular condition. The first logic system described, figure 38.34, was designed to work when it was light and warm. If we insert a NOT gate after the temperature sensor the system will work when it is light and cold. The relay could then be used to switch on a heating system, figure 38.43.

A similar result could be obtained by redesigning the original circuit and reversing the positions of the thermistor and R_2. The circuit in figure 38.44 is arranged so that the filament bulb lights when the LDR is in darkness.

table 38.4 Truth table for a NOT gate

Input	Output
0	1
1	0

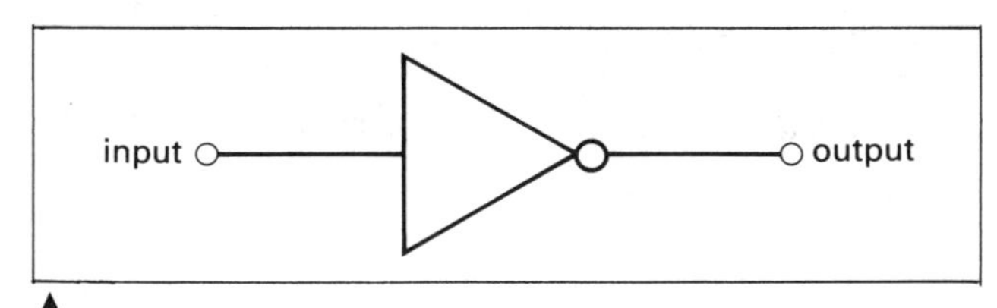

fig. 38.42 Symbol for a NOT gate

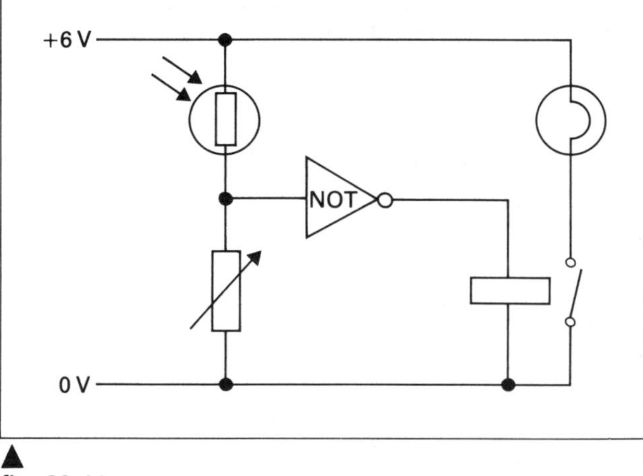

fig. 38.44

fig. 38.43

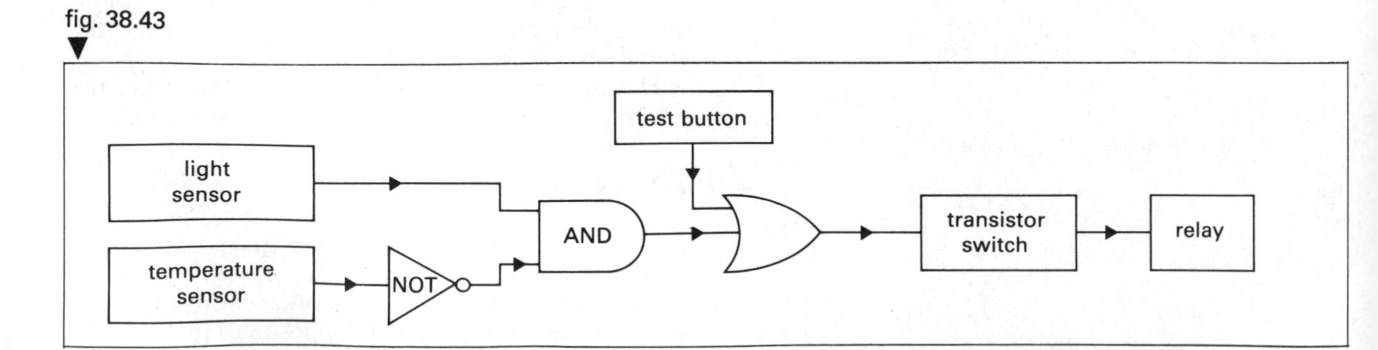

Combinations of gates

The electronic circuits of logic gates can be combined to produce new switching patterns. An AND gate and a NOT gate combine to give a NAND gate. The truth table of this gate is given in table 38.5. If you compare this table with that of an AND gate you will see that the resulting outputs have been reversed.

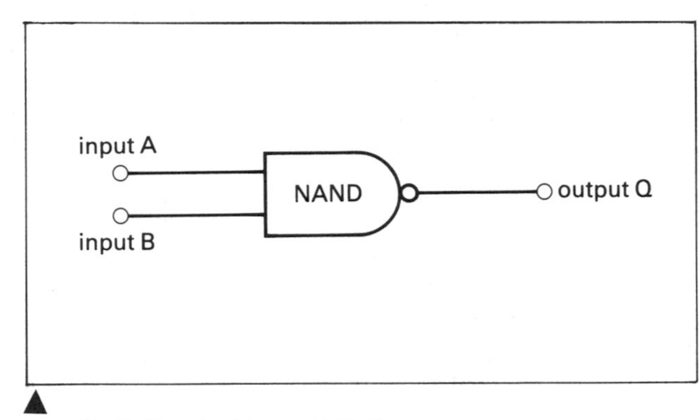

fig. 38.45 Symbol for a NAND gate

table 38.5 The truth table for a NAND gate

The NAND gate
AND and NOT = NAND

Input A	Input B	Output
0	0	1
1	0	1
0	1	1
1	1	0

In a similar way OR and NOT gates combine to give a NOR gate. Comparing the truth table with that of an OR gate you will once again see that the outputs have been reversed.

fig. 38.46 Symbol for a NOR gate

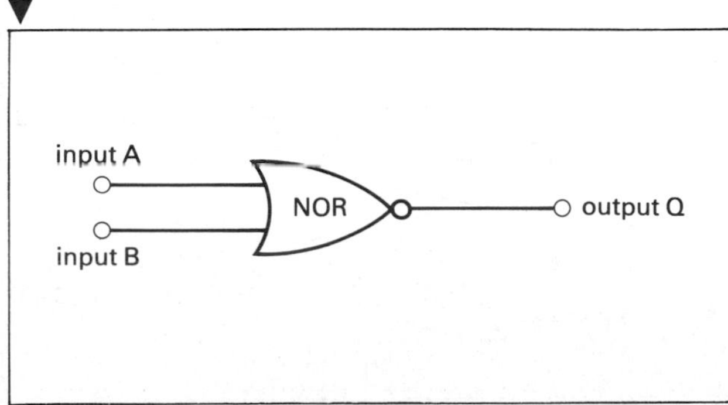

table 38.6 The truth table for a NOR gate

The NOR gate
OR and NOT = NOR

Input A	Input B	Output
0	0	1
1	0	0
0	1	0
1	1	0

The truth table for any particular combination of gates can be worked out systematically. First of all write down all the possible combinations of the inputs. For two inputs there are four possibilities (four rows in the truth table); for three inputs there are eight possibilities. Then write down the outputs of the gates; these become the inputs for the next gates, and so on.

For the system shown in figure 38.47 the result at C is the output for an AND gate. The result at D will be opposite to that at B since a NOT gate acts as an inverter. The results at C and D are used as inputs for the OR gate and give rise to the final output Q. Check the res'lts in table 38.7.

fig. 38.47

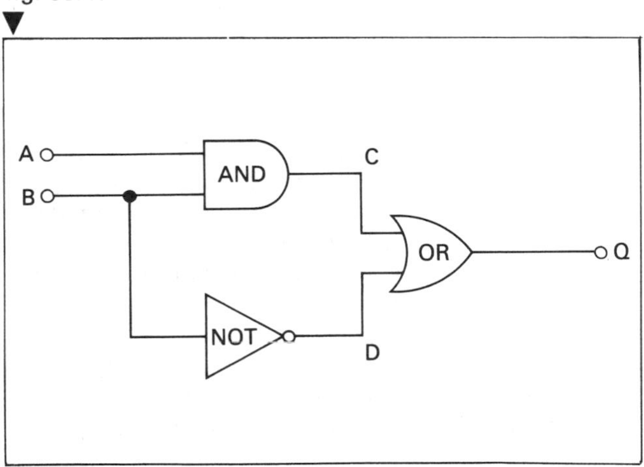

table 38.7

A	B	C	D	Q
0	0	0	1	1
1	0	0	1	1
0	1	0	0	0
1	1	1	0	1

The bistable combination of gates

A special and very useful combination of gates is the **bistable:** also called a multivibrator, a flip-flop or a latch. It consists of two gates cross-connected so that there is positive feedback from the output of each gate to one input of the other gate. The system has two stable states (depending on the input signal) either of which can be indefinitely maintained. In other words the system 'remembers' the state it is in until another input signal is received. The inputs are called SET (S) and RESET (R).

A bistable consisting of two NAND gates is shown in figure 38.48. When input S is logical 0 (low or zero potential) and input R is logical 1 (high potential) the output Q is logical 1 and output $\bar{Q}$ is logical 0. Check this with the truth table for a NAND gate. This state will continue when S is returned to logical 1. The bistable is then in its 'store' or 'rest' state (when both its inputs are at logical 1) but it remembers which input was last placed at logical 0. If R is now briefly set at logical 0 the outputs Q and $\bar{Q}$ change over (Q goes to logical 0 and $\bar{Q}$ goes to logical 1) and this new state is remembered when the bistable returns once more to its store state.

A bistable can be made from two NOR gates and also from two NOT gates; the first of these systems is illustrated in figure 38.49. Check its working with the truth table for a NOR gate.

The bistable responds permanently to a brief input signal and remains in this state when the signal has gone; the bistable returns to its store state only when it is reset by a new input.

A simple use of a bistable circuit is in railway signalling, figure 38.50. A magnet mounted underneath the train passes over reed switch A on the track. The switch is momentarily closed and input S is logical 1. In this state $\bar{Q}$ is logical 1 and the red signal light is switched on and remains on after the train has passed. When the train passes over the second reed switch B placed at a distance down the track, input R is logical 1 and output Q is logical 1. The green signal light is switched on showing the track ahead is clear and remains on until another train comes along.

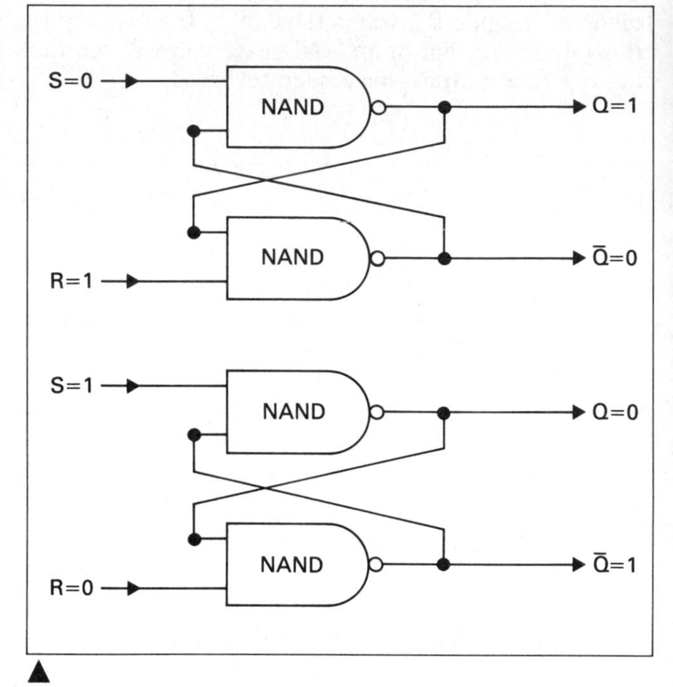

fig. 38.48 A bistable made from two NAND gates

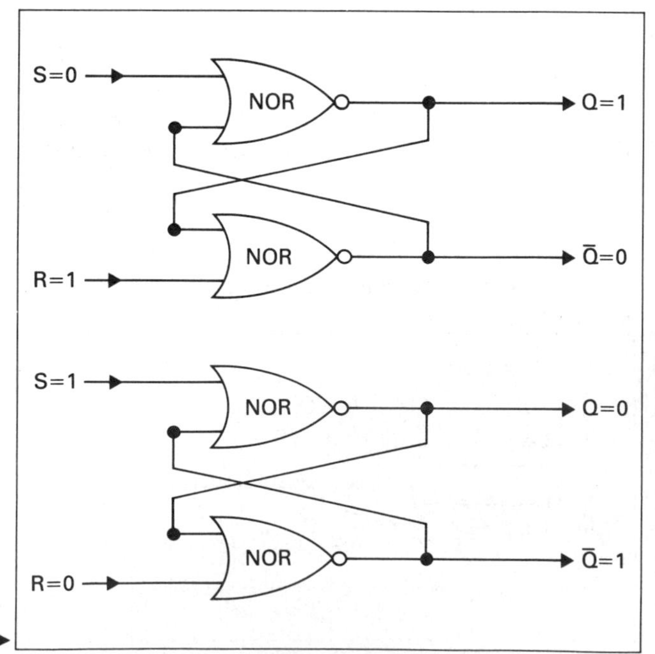

fig. 38.49 A bistable made from two NOR gates

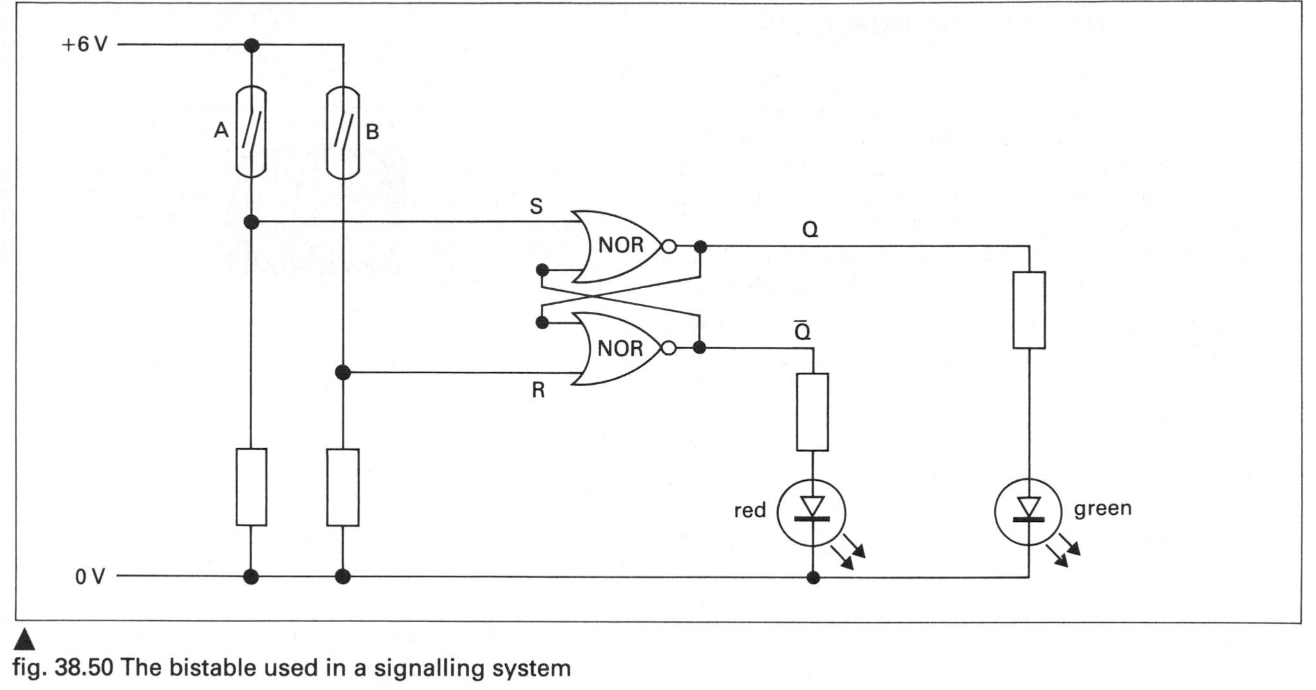

fig. 38.50 The bistable used in a signalling system

fig. 38.51 The bistable in this circuit is used to reverse the direction of a motor. When reed switch A is closed the motor rotates in one direction; when reed switch B is closed the direction of rotation reverses

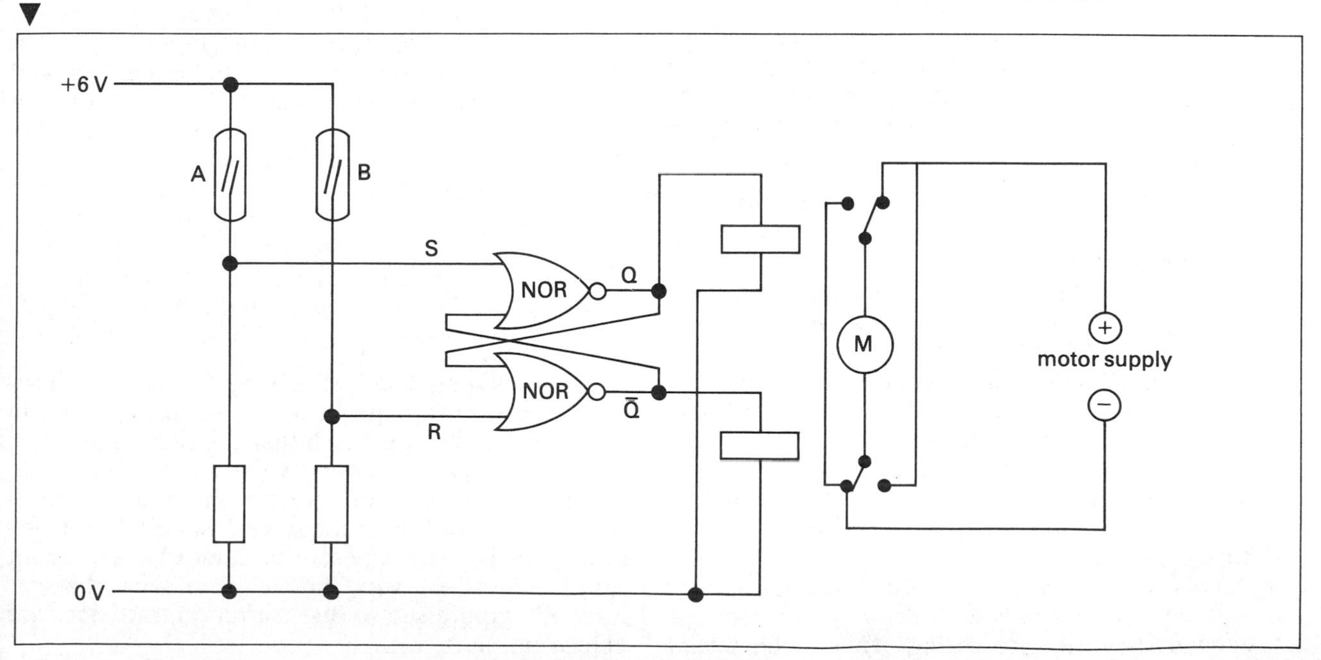

The cathode ray oscilloscope

The cathode ray tube combines two of the discoveries we have already discussed — the thermionic production of electrons and the deflection of electrons in an electric field. The basic principles of the tube are essentially those of a television tube. The cathode ray oscilloscope (CRO) is a very important scientific measuring instrument and research tool, figure 38.52.

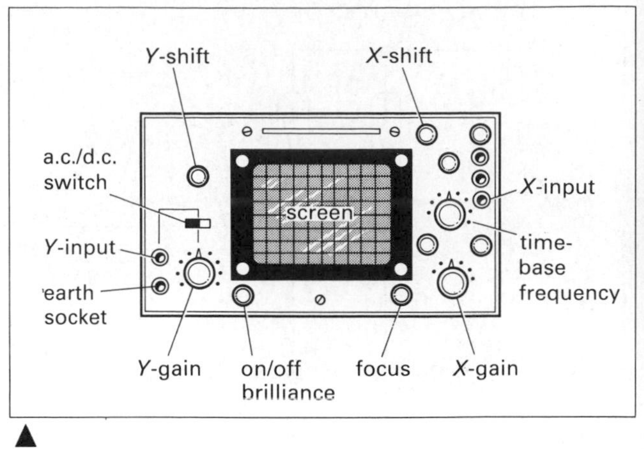

fig. 38.52 Controls on a basic CRO

fig. 38.53 Cathode ray tube

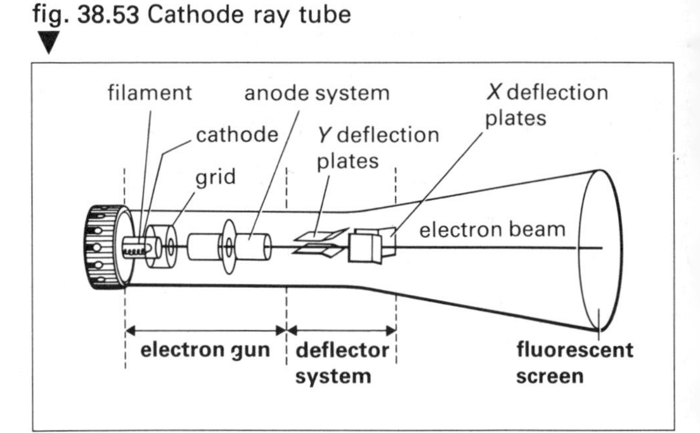

The cathode ray tube, figure 38.53, consists of

- An **electron gun** which produces a narrow beam of electrons from a heated cathode and has controls for altering the intensity (brightness) and for focusing.
- A **deflecting system** so that the beam can be deflected upwards and downwards and from side to side.
- A **fluorescent screen** on which the electrons produce a spot or line of light.

All the components are contained in a highly-evacuated glass tube.

The electrons are produced by an indirectly-heated cathode. The potential of the grid cylinder is negative with respect to the cathode and so controls the number of electrons leaving the cathode (and so the brightness of the spot on the screen). It also concentrates the electrons into a beam. The anodes are positive with respect to the cathode. They accelerate the electrons so that they have sufficient energy to produce light when they hit the fluorescent screen. The anodes control the fine focusing of the beam.

When a potential difference is applied to the *Y*-plates the beam is moved upward or downward. When a potential difference is applied to the *X*-plates the beam is moved either left or right.

The controls of the CRO can be very complex indeed. It is essential to understand the particular instrument you have before attempting to use it. The following controls will probably be present in a typical CRO figure 38.52.

- An **on/off** switch incorporated in the **brilliance** knob.
- A **focus** control.
- The ***X*-shift** and ***Y*-shift** controls enable the operator to centre the spot on the screen before starting an experiment.
- The ***X*-gain** and ***Y*-gain** are connected to amplifiers between the input terminals and the deflector plates. These controls are used so that applied voltages of various sizes can produce deflections, which are still on the screen in the case of large applied voltages, and can be magnified in the case of very small applied voltages.
- An **a.c./d.c. switch.** On the d.c. setting applied voltages are displayed directly on the screen, even if they are alternating. This is the setting used for all the basic applications of the CRO described in the following sections. The a.c. setting is used for more complex inputs in which a steady voltage component has to be removed and only the varying component displayed on the screen.
- A **time-base** control. The time-base applies a steadily changing voltage to the *X*-plates so that the beam is swept across the screen from left to right and then very quickly returned to the left to start again. This produces a line on the screen. The frequency of the time-base circuit can be changed, and there are usually coarse and fine adjustments.

The oscillating circuit supplying the time-base produces a sawtooth voltage, as shown in the graph in figure 38.54. When the right-hand *X*-plate is negative the beam is pushed over to the left. As the plate becomes more and more positive the beam is moved over to the right side of the screen. Since the voltage–time graph is linear, the spot must move at a constant speed. During the very short fly-back time, the spot is usually suppressed so that the return trace does not appear on the screen.

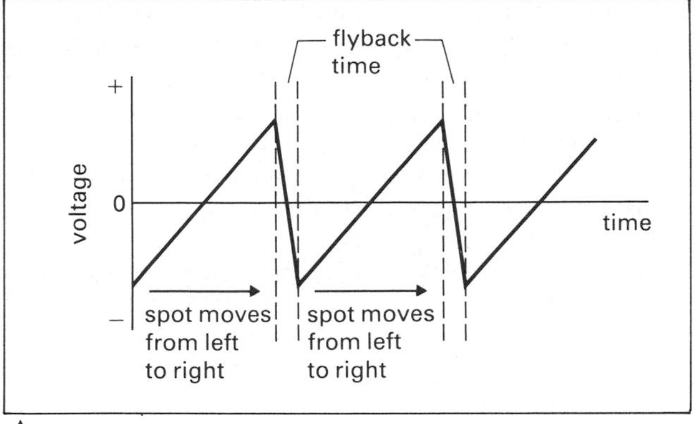

fig. 38.54 Time-base voltage (sawtooth waveform)

Using the CRO

The CRO is essentially a voltage-measuring device. A deflection occurs whenever there is a potential difference between the deflecting plates. Its great advantage is that the only moving part is the electron beam. This responds instantly to any deflecting force and returns instantly to its zero position when the force is removed. The resistance of the instrument is virtually infinite, so it acts as a perfect voltmeter, allowing no current to flow through it.

When the brightness and focus controls are properly adjusted and the spot is centred using the X- and Y-shifts, the instrument is ready for use. In practice the X_2 and Y_2 plates are connected to earth and the potentials under investigation connected to X_1 or Y_1, figure 38.55.

With the *time-base switched off*, the deflections shown in figures 38.56a–h could be obtained.

With the *time-base switched on*, the deflections shown in figures 38.57a and b could be obtained.

fig. 38.55 Deflector plate connections

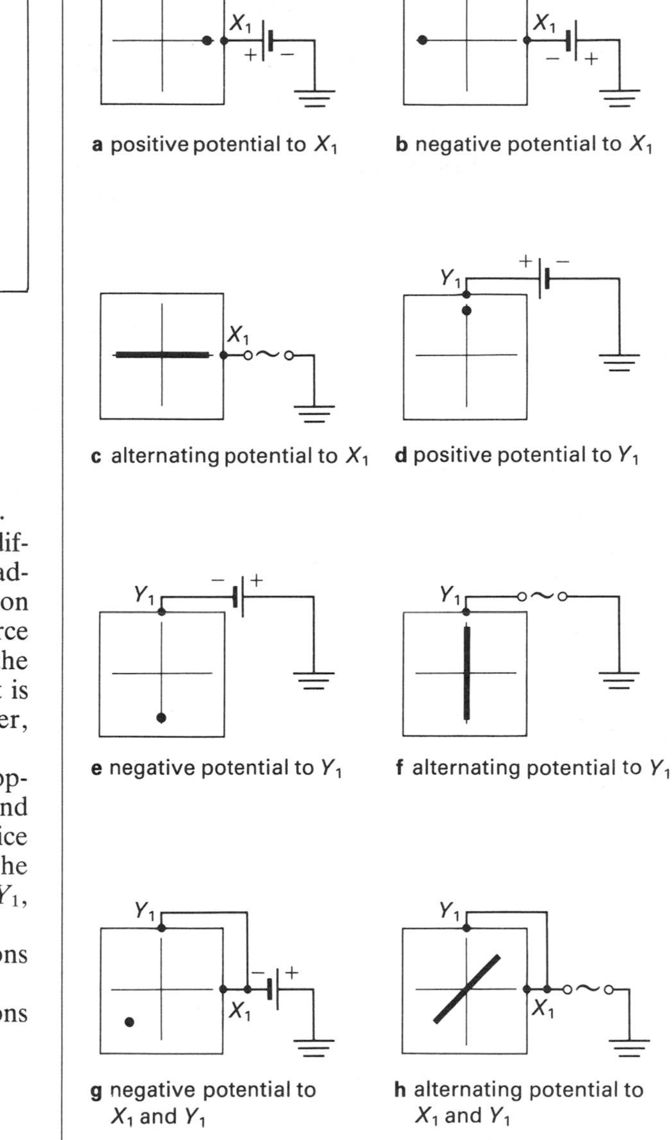

fig. 38.56 Deflections produced by various d.c. and a.c. inputs to an oscilloscope with the time-base off

fig. 38.57 Two traces produced when a linear time-base is on the X-plates

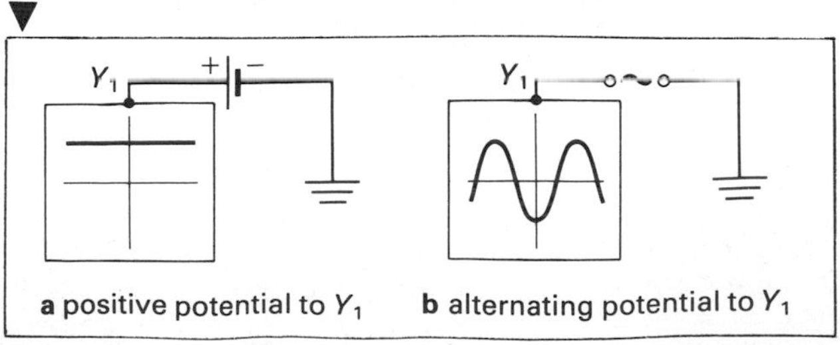

If the CRO is to be used as a voltmeter it has to be calibrated. This can be done using standard cells of known voltage, or the instrument can be checked against an accurate voltmeter. If there is no scale on the face of the tube, a strip of millimetre graph paper can be stuck on, figure 38.58. A graph of deflection against applied voltage is a straight line, so that the deflection per volt can be worked out, figure 38.59. This will have a different value for each setting of the *Y*-gain control. On some instruments the gain control is marked in volts per centimetre.

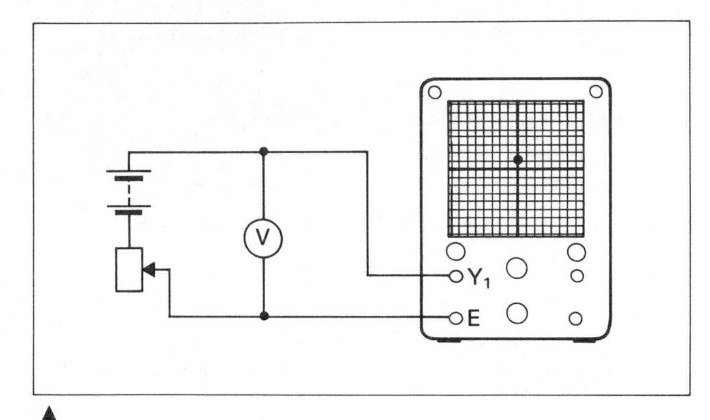

fig. 38.58 Calibrating a CRO

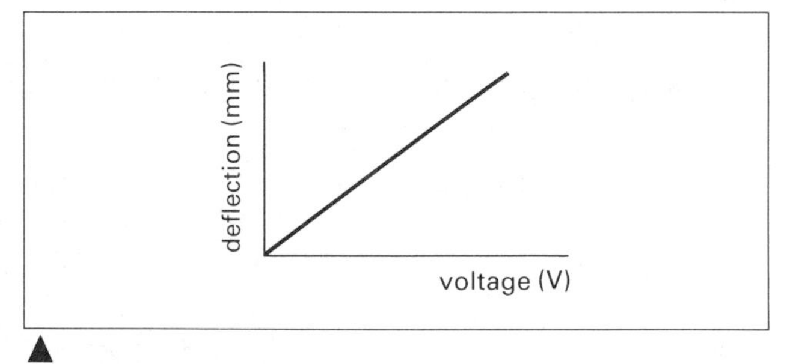

fig. 38.59 Calibration graph

fig. 38.60 Effect of different input frequencies

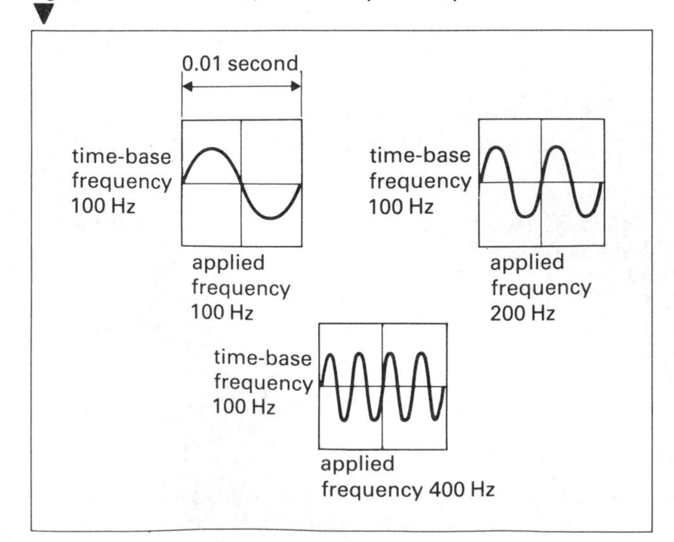

If an alternating potential difference is applied, the length of the line gives the peak-to-peak-value of the voltage. The root-mean-square value then has to be worked out by dividing the peak value by $\sqrt{2}$.

When the time-base is used and an alternating p.d. applied to the *Y*-plates, the waveform seen on the screen depends on the relative frequencies of the time-base and the applied p.d.

For example, suppose the time-base frequency is 100 Hz. This means that the spot travels across the tube 100 times a second, or the length of the trace is equivalent to $\frac{1}{100}$ second. If the applied p.d. has a frequency of 100 Hz, figure 38.60, then one complete wave will be seen since one complete wave is produced every $\frac{1}{100}$ second. If the applied p.d. has a frequency of 200 Hz, figure 38.60, then two complete waves will be seen. If four complete waves were seen on the screen, the frequency of the applied p.d. would be 4 waves in $\frac{1}{100}$ second or 400 Hz.

In order to get a stationary pattern on the screen, the applied frequency and the time-base frequency must begin at the same time. Varying the control marked 'sync' (synchronisation) will lock the two signals together and give a stationary pattern.

The electron microscope

So far we have regarded the electron as behaving like a particle of mass m and having a negative charge e. It also now turns out that a moving electron can behave like a wave. In 1927 Davisson and Germer in America used the atoms of a nickel crystal to produce diffraction effects with a beam of electrons and the following year G. P. Thomson (son of J. J. Thomson) produced the same diffraction effects by passing a beam of electrons through a very thin metal foil, figure 38.61. We have the same kind of wave/particle behaviour for moving electrons as for light. The faster the electrons move, the shorter the wavelength of the electron beam.

The usefulness of the light microscope is limited by the wavelength of light. It is impossible to distinguish between two objects which are less than about half a wavelength apart. The diffraction patterns overlap and produce a blurred image. If shorter wavelengths can be used, more detail should be revealed. In other words, a greater magnification could be produced. Fast-moving electrons have a wavelength about a hundred thousand times shorter than that of visible light, so an electron microscope can produce magnifications up to a million times.

Both magnetic and electric fields can deflect electrons, so that either could be used as electron 'lenses' to focus the electron beam. Most electron microscopes use magnetic lenses and the final image is projected on to a fluorescent screen or a photographic plate, figures 38.62–38.64.

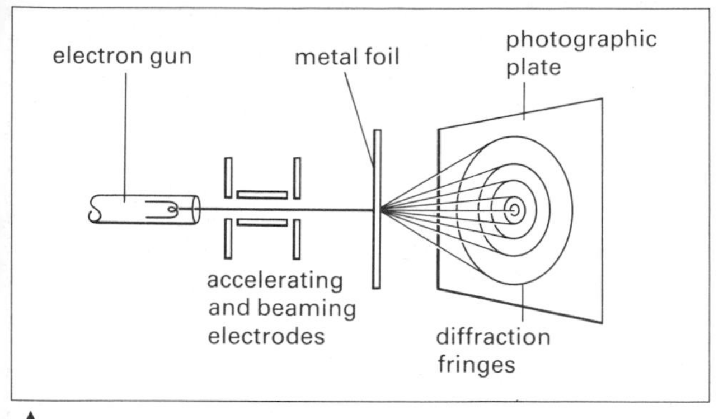

fig. 38.61 Thomson's experiment on electron diffraction

fig 38.62 Comparison of the electron microscope with the optical microscope

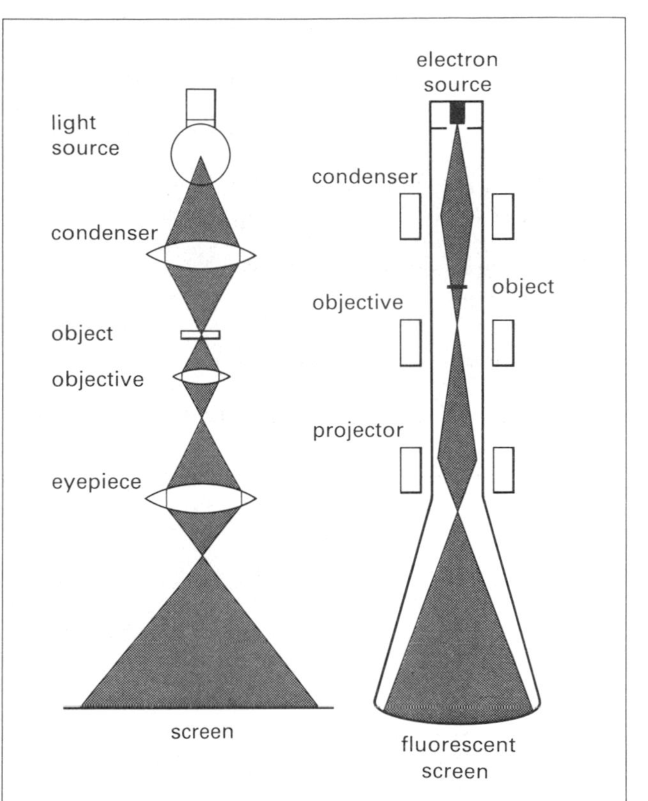

fig. 38.63 High-voltage analytical transmission electron microscope

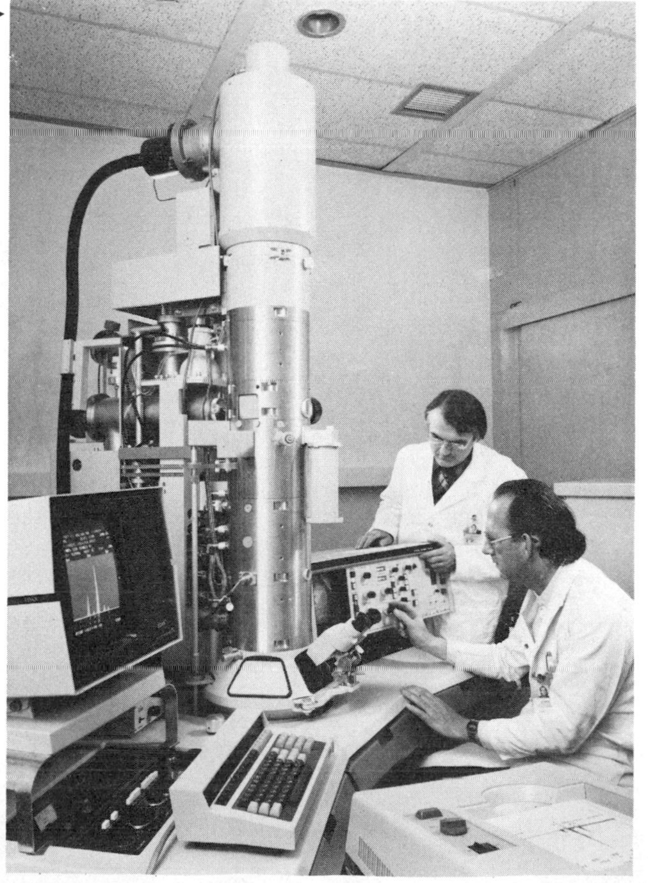

fig. 38.64 Electron micrograph of a fractured surface of an irradiated uranium dioxide fuel pellet, showing fission gas-filled cavities (×12 000 magnification)

EXERCISE 38

In questions 1–4 select the most suitable answer.

1 Which of the following is *not* a correct statement about cathode rays?
A They are deflected by an electric field.
B They are deflected by a magnetic field.
C They travel in straight lines.
D They are streams of electrons.
E They have a positive charge.

2 In a cathode ray tube a beam of cathode rays is most likely to cross the tube if
A the filament is heated and the anode voltage is switched off
B the filament is heated and the anode is negative with respect to the cathode
C the filament is switched off and the anode is negative with respect to the cathode
D the filament is heated and the anode is positive with respect to the cathode
E the filament is switched off and the anode is positive with respect to the cathode

3 When cathode rays pass through, and at right angles to, a uniform magnetic field they will experience
A a force towards the north pole
B a force towards the south pole
C a force along the direction of motion
D a force at right angles to the direction of motion
E no force at all

4 Symbols for five electrical devices are shown below. Which one converts electrical energy into light energy? Give its name.

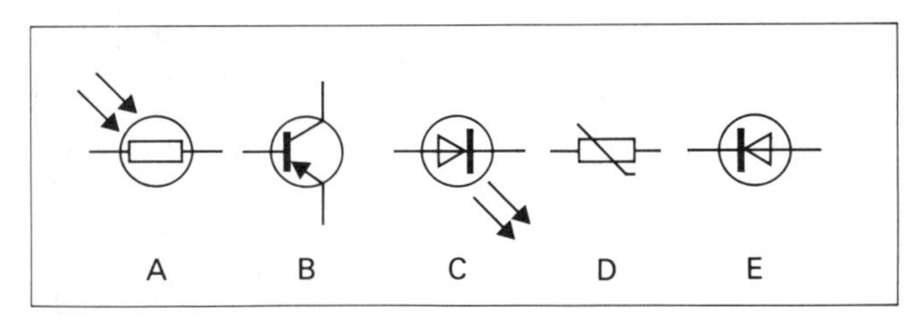

5 Figure 38.65 shows a simple rectifier circuit using a diode. When an oscilloscope is placed across the resistor and suitably adjusted, which of the traces **A** to **E** in figure 38.66 would you expect to see on the screen?

fig. 38.65

a.c. supply
diode
resistor

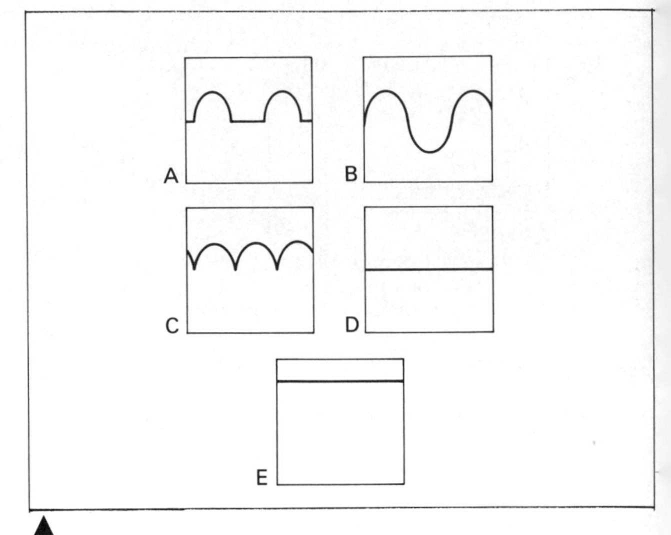

fig. 38.66

6 Figure 38.67 shows a simplified version of a tube designed to illustrate the effect of an electrostatic field on a stream of electrons. Explain carefully
a how A is made to emit electrons, naming a suitable substance for A and giving a reason for your choice,
b the purpose of B and how this purpose is achieved,
c how you would establish the nature of the charge on the electrons,
d why the air must be removed from the tube,
e how the path of the electron beam is made visible.

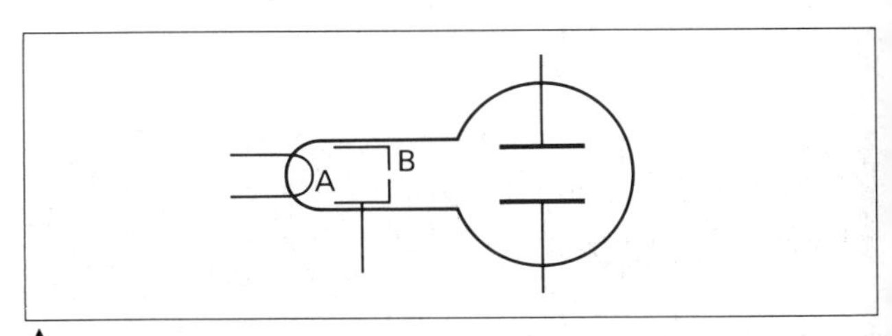

fig. 38.67

7 a Describe the differences between *n*-type and *p*-type semiconductors.
b Describe, with the aid of a diagram, the structure of a semiconductor diode.
c Draw circuit diagrams showing how semiconductor diodes may be used to give: (i) half-wave rectification; (ii) full-wave rectification.

8 a A *p-n-p* transistor consists of two layers of *p*-type semiconductor with a thin layer of *n*-type semiconductor between. Give the circuit symbol for a *p-n-p* transistor and label the base, collector and emitter.
b When a transistor such as described in **a** is connected into a circuit to determine the characteristic curve,

the results given in the table are obtained. (i) Plot the graph of I_c against I_b (I_c on the y-axis). (ii) Draw the best straight line through the points. (iii) Calculate the gradient I_c/I_b. (iv) What property of a transistor does this illustrate?

I_b (μA)	20	40	60	80	100
I_c (mA)	1.1	2.3	3.2	4.4	5.5

c Draw a circuit which can be used to produce the results in **b**.

9 Draw block diagrams of systems which could be used
a to switch on a heater if it gets too cold,
b to light a warning lamp when an oven exceeds a certain temperature,
c to open a window when it is warm in daylight
d to close a window when it is cold and wet at night.
Draw the transistor circuits for systems **a** and **c**.

10 Copy and complete the truth table for the system of logic gates shown.

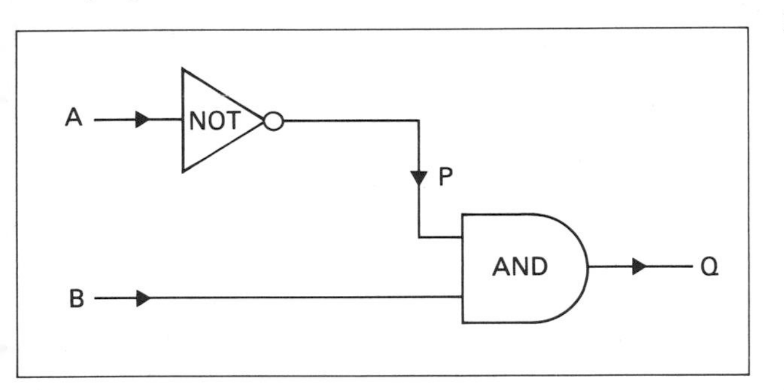

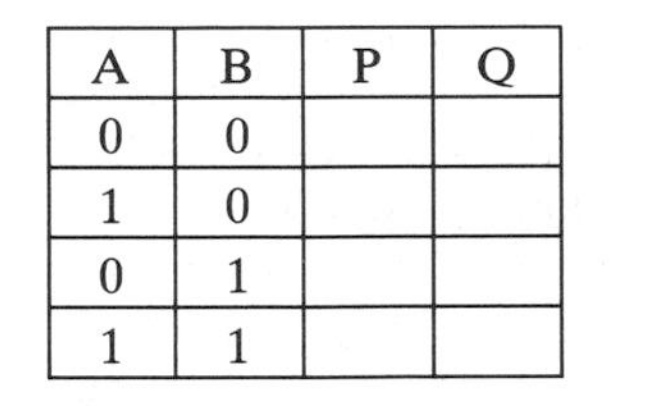

A	B	P	Q
0	0		
1	0		
0	1		
1	1		

11 a Figure 38.68 shows the main features of an oscilloscope. (i) What particles are emitted from the cathode? (ii) How are the electrons accelerated? (iii) To which plates is the time-base connected? (iv) Why must the cathode be heated? (v) What device is necessary (not shown in the diagram) to control the brightness of the electron beam, and where should it be located in the oscilloscope?

b Sketch the trace that would be obtained on the oscilloscope screen (i) with the time-base alone connected; (ii) with an alternating potential difference acting across the Y-plates but without the time-base operating; (iii) with an alternating potential difference acting across the Y-plates with the time-base operating.

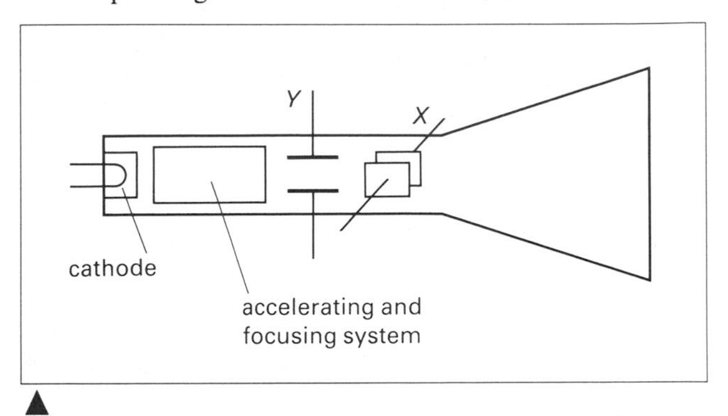

fig. 36.68

12 a With the aid of sketches, describe the action of the variable time-base of a cathode ray oscilloscope on the beam of electrons. If the time-base is switched off and a 50 Hz a.c. voltage connected to the X-plates, how will the movement of the beam differ?

b The traces shown in figures 38.69a–d were obtained on a cathode ray oscilloscope with the time-base on by connecting four voltage supplies in turn without altering any other settings on the oscilloscope. Describe each voltage supply, and compare those shown in (b), (c) and (d) with that shown in (a).

fig. 38.69

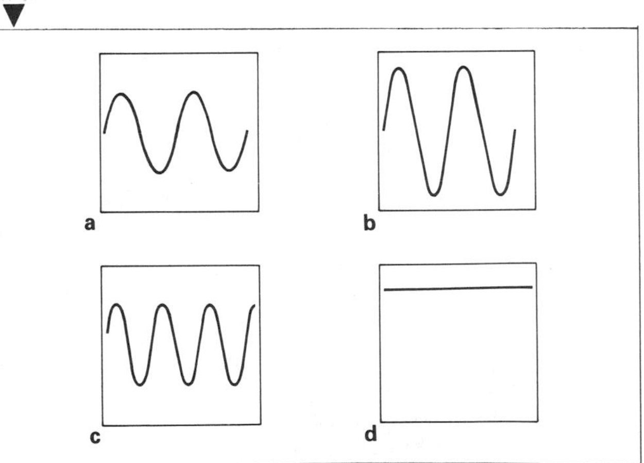

39 X-RAYS AND RADIOACTIVITY

One very important scientific advance which arose from the study of electric discharge in gases was the discovery of the electron, as described in the last chapter. Equally important discoveries made at about the same time and linked with the original investigations were the discovery of X-rays by Wilhelm Conrad Röntgen in Germany in 1895 and the discovery of radioactivity by Antoine Henri Becquerel in France in 1896.

Röntgen noticed that some crystals of barium platino-cyanide fluoresced when near an electric discharge tube, even when the tube was covered with black paper. Something was being given out by the tube. Many scientists must have noticed similar effects when working with cathode rays, but did not investigate them further. Sir William Crookes had observed that photographic plates stored near discharge tubes became fogged and had once returned the plates to the makers saying they were unsatisfactory! Röntgen not only noticed something interesting, but was also sufficiently curious to carry out further experiments on the new kind of rays, which he called X-rays because the nature of the rays was uncertain.

Properties of X-rays

It was soon discovered that the point of origin of the X-rays was the fluorescent spot where the cathode rays hit the glass at the end of the tube. Some of the first properties to be discovered are shown in figure 39.1.

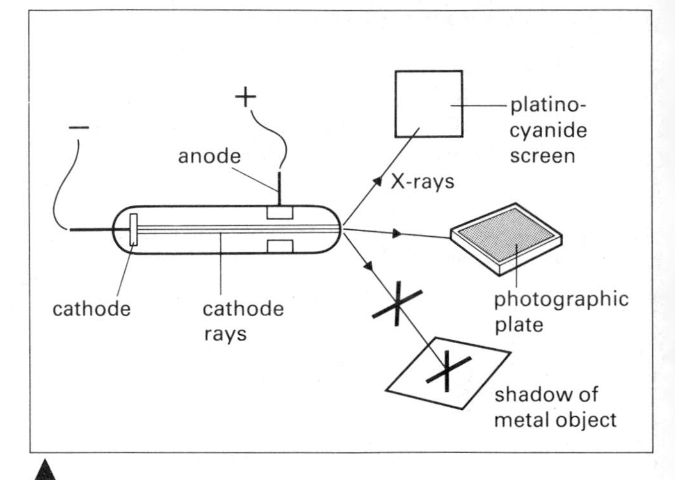

▲
fig. 39.1 Some properties of X-rays

X-rays cause certain substances, e.g. barium platino-cyanide, to fluoresce. They affect the emulsions of photographic plates just as if the plates had been exposed to light. The rays can penetrate matter, but are absorbed by dense materials. The shadow of a metal object placed in the path of the rays showed that the rays travelled in straight lines.

The question to be settled was whether X-rays were particles or waves. It was shown that the rays could not be deflected by electric or magnetic fields, so they did not behave like charged particles. On the other hand, the rays could discharge electroscopes, so that it seemed that they could ionise the air and cause the charge on the electroscope to leak away. In 1912 Max von Laue suggested that the regular spacing of atoms in crystals might produce a diffraction effect. This proved to be true, and not only established the wave nature of the rays but also enabled the wavelength to be measured. X-rays are electromagnetic waves with wavelength shorter than ultraviolet light (see the electromagnetic spectrum p. 197).

X-rays have a destructive effect on living tissues. Their high energy upsets the chemical balance of living cells. Many of the early experimenters suffered severe damage because they were exposed to the rays for long periods. Now, protective clothing is worn by people working with X-rays and regular checks are made to determine the dose of radiation and ensure that it remains at a safe level. A small piece of film is worn in a badge. When the film is developed the density of the blackening is a measure of the radiation dosage. The rays have a more damaging effect on cancer cells than on healthy tissue and this is the basis of the treatment of the disease by radiotherapy.

The production of X-rays

The first X-rays were produced when the cathode rays hit the end of the glass tube. In modern terms, X-rays are produced whenever fast-moving electrons hit a solid, usually a heavy metal. The electrons are pro-

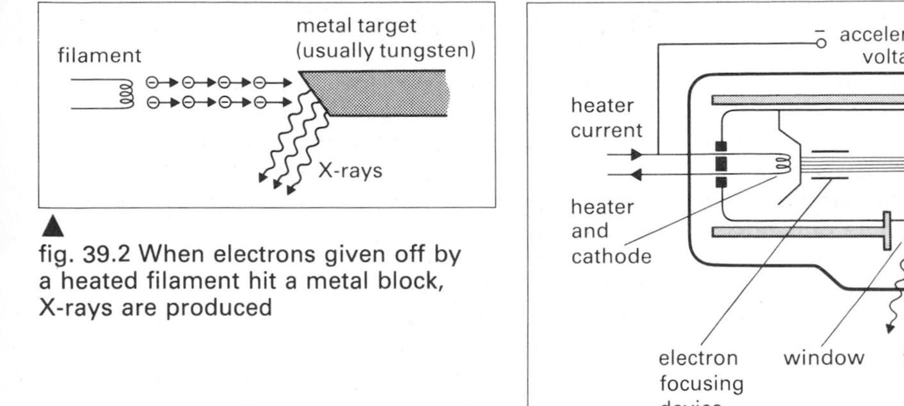

fig. 39.2 When electrons given off by a heated filament hit a metal block, X-rays are produced

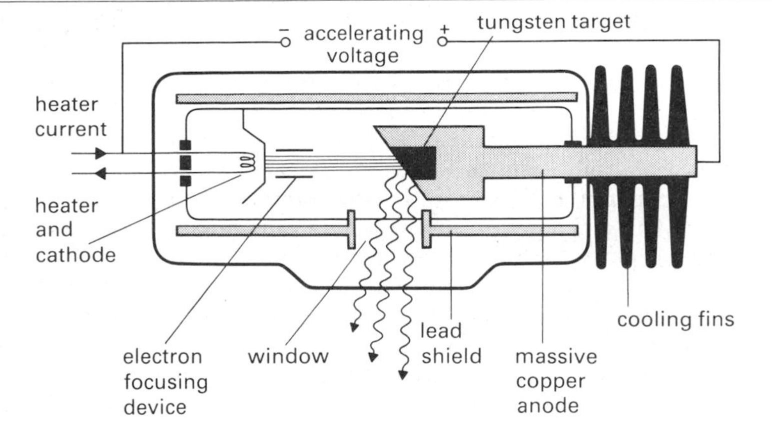

fig. 39.3 The parts of an X-ray tube

duced by thermionic emission from an indirectly-heated cathode and are accelerated towards the anode, which is also the target, by a potential difference of several hundred kilovolts. The essentials of an X-ray tube are shown in figure 39.3.

The **intensity** of the X-rays produced depends on the number of electrons hitting the target per second and this, in turn, depends on the value of the heater current.

The **quality** or **penetrating power** of the rays depends on the accelerating voltage and to some extent on the material of the target. The higher the voltage the greater the energy of the bombarding electrons and the shorter the wavelength of the X-rays emitted. Short-wave X-rays are more penetrating than the longer waves, and are called 'hard' X-rays. 'Soft' X-rays can penetrate flesh and are used in medicine for photographs of broken bones, while 'hard' X-rays can penetrate metals, figures 39.4 and 39.5.

A large amount of energy is supplied to an X-ray tube. Suppose the tube is working at 100 000 volts and the current is 10 milliamperes. The energy supplied is 1000 watts (watts = volts × amperes). Less than 1% of this energy is converted into X-rays. The rest is produced as heat. So the tube is producing as much heat as a one-bar radiator! Very efficient cooling is necessary. Large tubes are oil-cooled. In smaller tubes, the target is rotated rapidly, so that the intense beam of electrons does not hit the same spot for very long.

fig. 39.4 X-ray photographs are used to reveal hidden defects in welds and castings. This is a radiograph of welded stainless steel piping

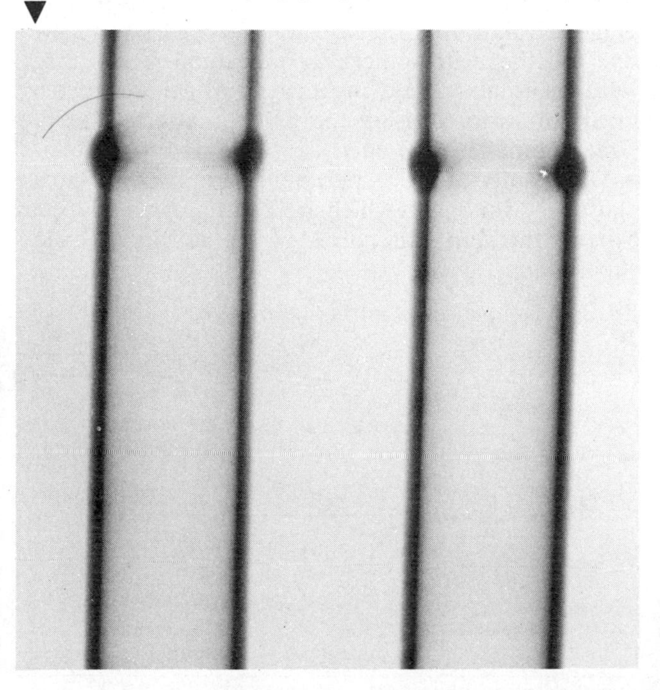

fig. 39.5 Radiograph of a forearm in which both the ulna and radius have been broken. Note how the details of the wrist bones have been revealed

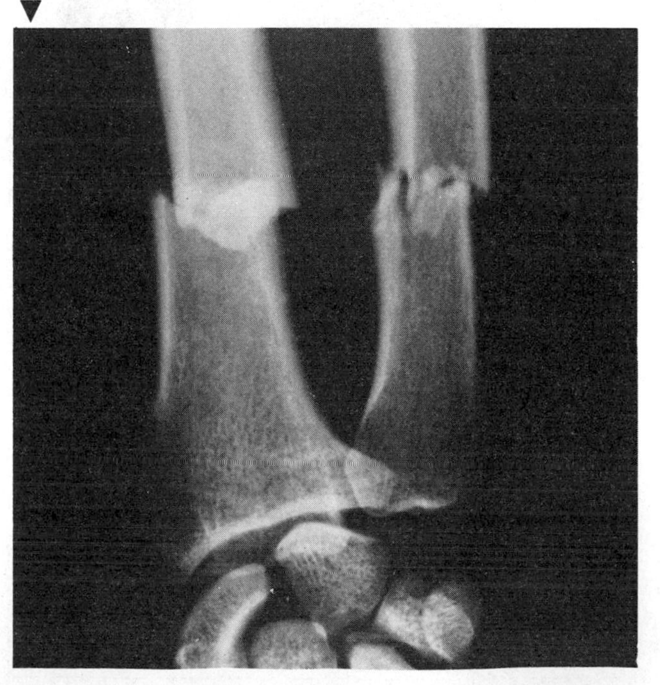

Radioactivity

While Sir Joseph John Thomson was investigating the properties of cathode rays in Cambridge, and Wilhelm Conrad Röntgen discovered X-rays in Germany, in France Antoine Henri Becquerel became interested in phenomena associated with compounds of uranium. These compounds fluoresced in sunlight and had the property of blackening photographic plates even when they were covered with black paper. It looked as if the uranium salts were emitting some kind of radiation. This was in 1896.

A great deal of research followed and the namc **radioactivity** was first used in 1898 by Madame Curie who, with her husband Pierre, discovered that radium and polonium compounds behaved in the same way as those of uranium.

Radioactivity is the spontaneous disintegration of unstable nuclei of atoms of heavy elements (particularly those with atomic numbers greater than 84). During the disintegration of the nucleus, certain particles and radiation are given off and a new nucleus (which means a new element) is formed.

Some of the research experiments which unravelled the problems connected with radioactivity will be described and the main facts about radioactivity summarised.

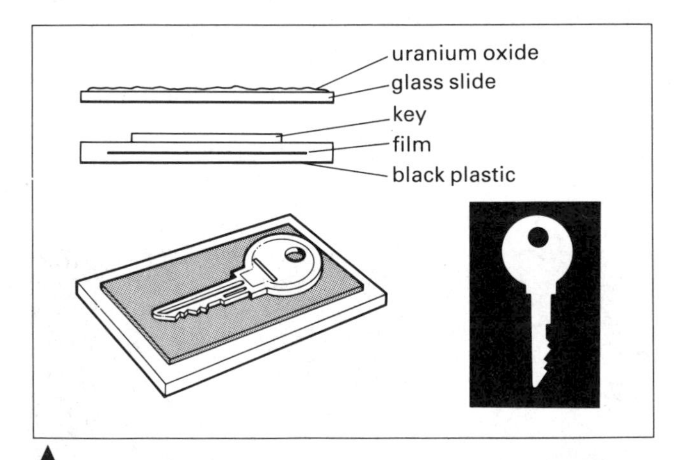

▲ fig. 39.6 Making a radiograph

fig. 43.7 Deflection in an electric field ▼

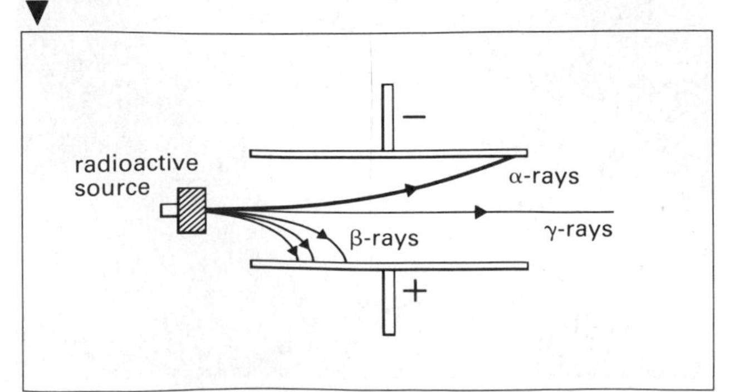

Making a radiograph

The radiation from a radioactive source is dangerous to living material, and all experiments have to be carefully controlled and carried out under supervision. Weak radioactive sources can be used in school experiments if they are handled properly by qualified persons.

One of the early experiments can be repeated, figure 39.6. A strip of fast industrial X-ray film is sealed in a black paper or plastic envelope and a metal object such as a key placed on top. A layer of uranium oxide on a glass slide is placed on the top and left for several days. On developing the film a clear 'shadow' of the key appears, figure 39.6.

The experiment shows that the uranium oxide is emitting some kind of radiation which affects the photographic emulsion and that the radiation does not pass through metal.

Three kinds of radioactive emission

The nature of cathode rays was established by observing their behaviour in electric and magnetic fields. Similar experiments were done with the radioactive rays. In both cases the results showed that there were three kinds of emission. In an electric field, something was attracted to the positive plate, something was attracted to the negative plate and something was undeflected, figure 39.7.

In a magnetic field, something was deflected in one direction, something in the opposite direction and something was undeflected, figure 39.8. This strongly suggested that oppositely charged particles were present together with something uncharged. These three were called **alpha rays**, **beta rays** and **gamma rays** by Lord Rutherford, using the first three letters of the Greek alphabet α, β and γ.

The results of the experiments are summed up in figures 39.7 and 39.8 which are not diagrams of actual experiments but summaries of the results of many experiments.

fig. 39.8 Deflection in a magnetic field ▼

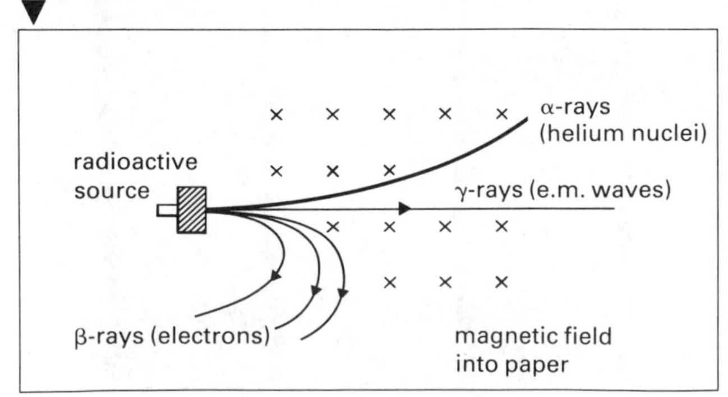

The deflections in the electric field show that the α-rays carry positive charges, the β-rays carry negative charges and the γ-rays are uncharged.

When Fleming's left-hand rule is applied to the deflecting forces acting in a magnetic field, figure 39.8, it shows again that the α-rays act like moving positive charges and the β-rays like moving negative charges, while the γ-rays are again uncharged.

The nature of the radiations

The determination of the ratio of charge to mass and of the actual magnitude of the charge of the alpha particle by Rutherford and Geiger (from 1903 to 1908), showed that the charge was twice that of an electron and the mass was four times that of a hydrogen nucleus. These numbers fitted in with the alpha particle being the nucleus of a helium atom. Direct evidence of the connection between alpha particles and helium came from an experiment of Rutherford and Royds in 1909. A radioactive gas, radon, was compressed into a thin-walled tube which was surrounded by an evacuated discharge tube, figure 39.9. After a week the mercury level was raised, compressing any gas into the narrow part of the discharge tube. When an electric charge was passed through it and the light emitted examined with a spectrometer, it was seen to give the characteristic spectrum of helium.

Alpha particles are the nuclei of helium atoms. They consist of two protons and two neutrons joined together. They have a charge of plus two and a mass number of four, i.e. $^{4}_{2}He$.

When the ratio of charge to mass for the beta rays was measured it was found to be of the same order as that of the cathode rays and β-particles are, in fact, fast-moving electrons.

Gamma rays are not particles and carry no charge. It was thought that they could be of the same nature as X-rays and when they were passed through a crystal they were diffracted showing their wave nature. The wavelength turned out to be that of very short-wave X-rays (about 10^{-11} m), so γ-radiation is very short-wave electromagnetic radiation.

A summary of the nature and properties of the three types of radioactive radiation is given in table 39.1

fig. 39.9 Rutherford and Royds' apparatus used to show that α-particles are helium nuclei

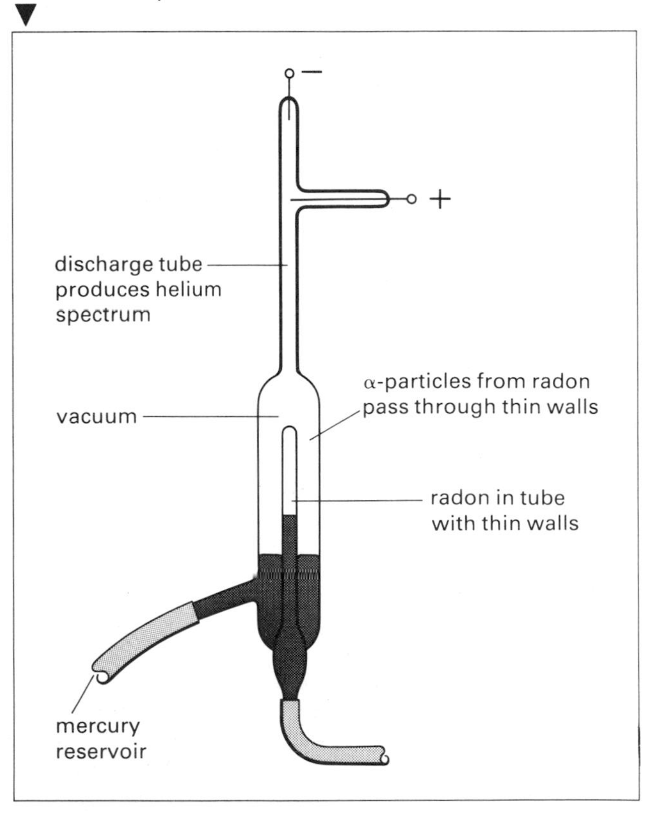

table 39.1 Properties of α-, β- and γ-rays

Radiation	Charge	Mass	Nature	Penetrating power	Range in air	Ionising power
α-rays	positive (+2) deflected by electric and magnetic fields	4 a.m.u.	helium nuclei	absorbed by skin or a sheet of paper	a few cm	very great 10^5 ion pairs per cm
β-rays	negative (−1) deflected by electric and magnetic fields	1/1840 a.m.u.	electrons	absorbed by a few mm of aluminium	a metre or so	much less than α-particle 10^2 ion pairs per cm
γ-rays	none undeflected by electric and magnetic fields	none	electro-magnetic radiation of very short wavelength	very penetrating absorbed by thick lead or concrete	very high	very low

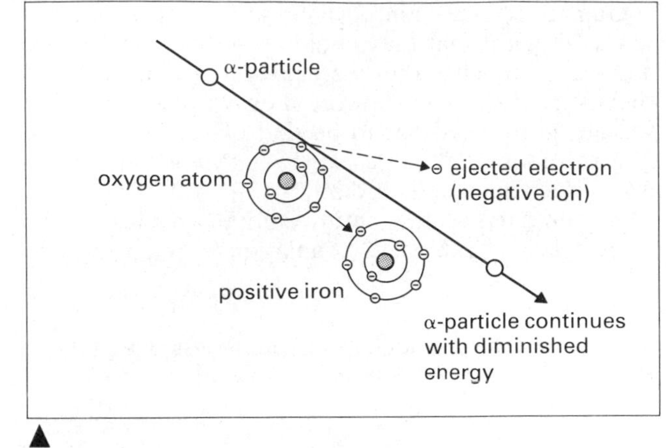

fig. 39.10 Formation of an ion pair

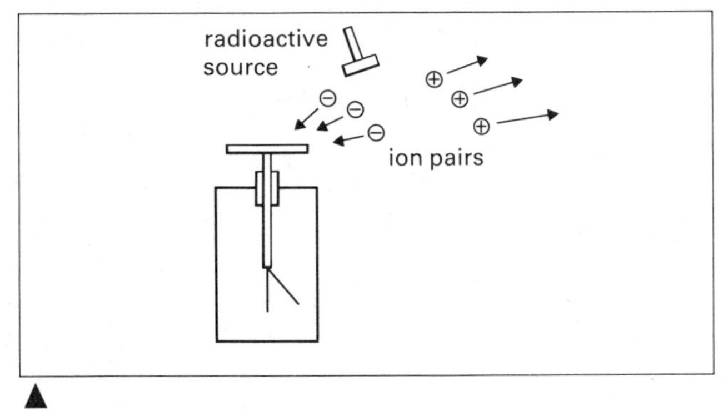

fig. 39.11 Detecting α-particles by electroscope

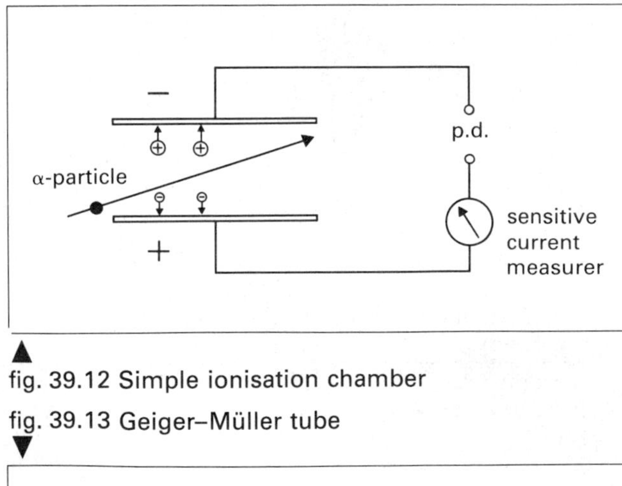

fig. 39.12 Simple ionisation chamber

fig. 39.13 Geiger–Müller tube

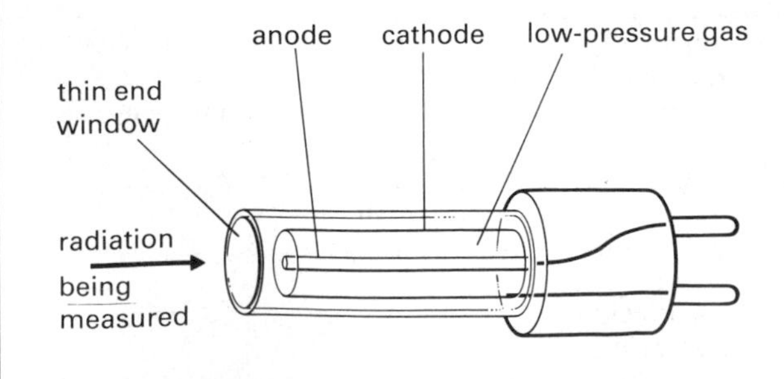

Detecting radioactive radiations

The production of ionisation is an important property of radioactive radiations and is used in devices to detect the radiations. The α- and β-particles both have mass and velocity and therefore possess kinetic energy. When they collide with atoms or molecules of the gas through which they are travelling, they may have sufficient energy to remove an outer electron and so produce an electron and a positive ion — an **ion pair**, figure 39.10. α-particles produce a lot of ion pairs over their comparatively short range. β-particles are travelling much more quickly and so arc ncar to the gas molecules for a much shorter time. There is less chance of an ion pair being formed, so the number of ion pairs per centimetre is less. Since the range of the β-particles is much greater than that of α-particles, the total number of ion pairs formed is about the same if the particles have the same energy. The α- and β-particles gradually lose their kinetic energy in a series of ionising encounters. γ-rays produce a smaller amount of ionisation because they eject few electrons from atoms.

The production of ions is responsible for the discharge of an electroscope when a radioactive source is placed near it. In figure 39.11 the negative ions are attracted to the positively-charged electroscope and neutralise the charge. This was an early method of detecting α-particles. The rate at which the leaves fall is a measure of the strength of the radioactive source.

A basic piece of apparatus for the detection of radioactive radiations is the **ionisation chamber**, figure 39.12. This consists essentially of two electrodes with a potential difference between them. When, for example, an α-particle passes between the plates, ion pairs are formed and these are attracted to the electrodes — the electrons move to the positive plate, and the positive ions to the negative plate. This movement of ions forms an ionisation current which can be detected by a sensitive galvanometer included in the circuit. The size of the ionisation current (for a given ionisation chamber) depends on the nature of the radiation and on the rate at which the particles enter the chamber, i.e. on the intensity of the radiation. Very often the chamber is in the form of a cylinder of metal with a central electrode. The very small ionisation current is passed through an amplifier so that it can be measured on a milliammeter.

A special type of ionisation chamber is the **Geiger–Müller tube**, figure 39.13. This can be designed to detect α-, β- and γ-radiation and is the most popular form of radiation detector. It consists of a cylindrical cathode with a central anode mounted in a glass tube. The tube is filled with a mixture of gases at low pressure which is designed

- to produce ion pairs in large numbers,
- to 'quench' the discharge quickly, so that the next particle entering the tube can be detected.

A potential difference of 300–400 V is maintained between anode and cathode. At the end of the tube is a window through which the radiation can enter. In the case of α-particles it has to be very thin.

When an ionising radiation passes through the gas in the tube, the ion pairs are produced and are attracted to the appropriate electrodes. As the electrons are accelerated to the anode they acquire sufficient energy to ionise more atoms and the new electrons can produce more ionisation and so on. An 'avalanche' of electrons reaches the anode and so a large pulse of current is produced, making the Geiger–Müller tube a very sensitive detector of radiation.

The current pulse is passed through a resistor, so producing a voltage pulse, i.e. a potential difference between the ends of the resistor. This voltage pulse can be fed

- to a **loudspeaker** to produce a loud click,
- to a **scaler** which gives a record of the total number of pulses it receives, or
- a **ratemeter** which indicates the number of counts per second received.

The scaler measures the total amount of radiation entering the tube in a given time. The ratemeter measures the rate at which radiation is entering the tube.

fig. 39.14 A GM tube may be used with a loudspeaker, scaler or ratemeter

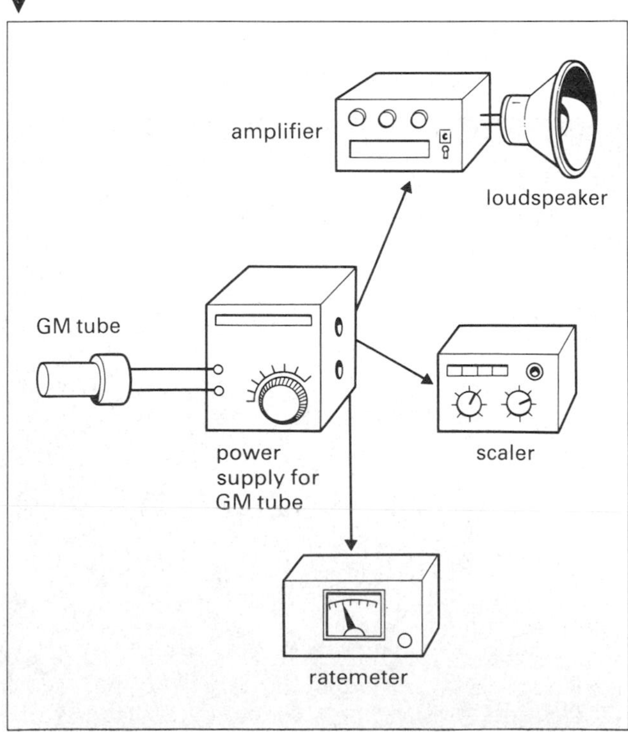

The cloud chamber

An important method of detecting radioactive radiations was devised by C. T. R. Wilson at Cambridge in 1911. It is called a **cloud chamber** and with it the tracks of the radiations can be observed and photographed, figures 39.15 and 39.16. The method depends on producing a space filled with a saturated or supersaturated vapour. Wilson originally used water vapour but the type of chamber usually used in schools has saturated alcohol vapour in it. When a saturated vapour is cooled it will condense on any particles present to produce droplets. The particles act as nuclei on which the droplets can form.

When a radioactive particle passes through the air

fig. 39.15 An α-particle track (left) and two β-particle tracks in a cloud chamber. The superior ionising power of the α-particle is shown by the strength of its track

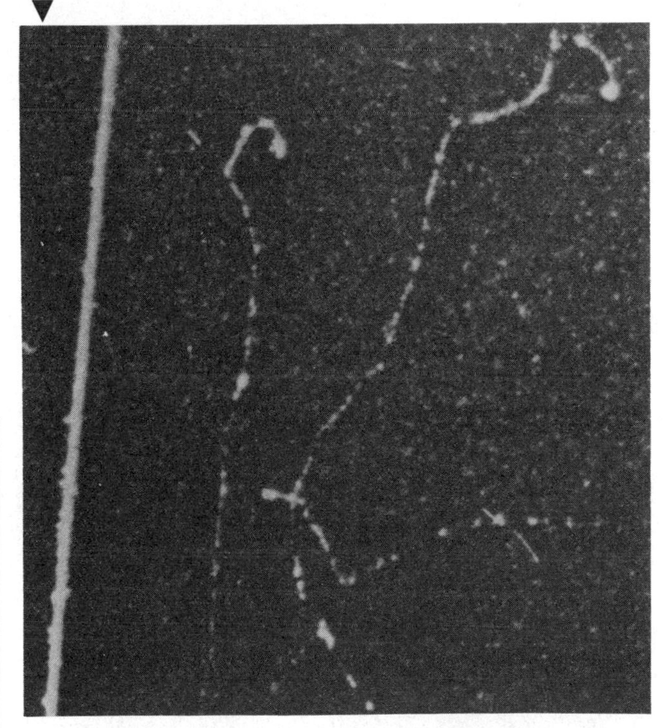

fig. 39.16 Wilson cloud chamber

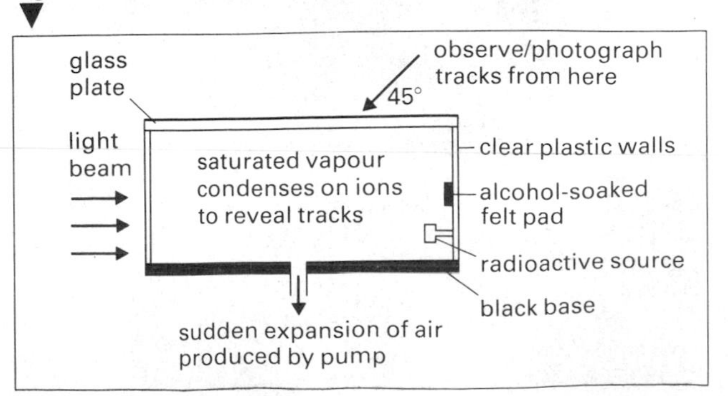

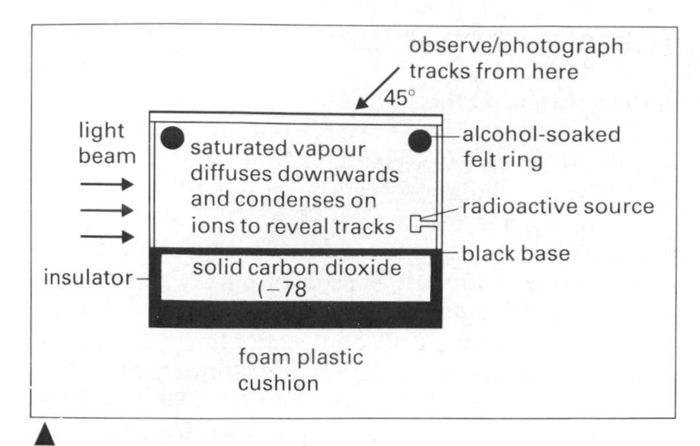

fig. 39.17 Diffusion cloud chamber

fig. 43.18 This 1.5 m bubble chamber was used with the UK Nimrod accelerator and the CERN proton synchrotron. Tracks produced by particles, which entered from the left, are illuminated through the rear window of the chamber and photographed by a camera mounted on the front window

pairs of ions are produced when electrons are knocked off the gas molecules. These ions act as nuclei on which the saturated vapour can condense. The path of the radiation is marked by droplets condensed on the ions. The condensation is produced by cooling the vapour. In the original apparatus the cooling was produced by a sudden expansion. In the more modern **diffusion cloud chamber**, figure 39.17, the cooling is produced continuously by solid carbon dioxide at the base of the chamber. The chamber is illuminated from the side so that the tracks show up clearly.

Alpha-particles produce heavy ionisation and travel in straight lines of fixed length. Thc lcss massive beta-particles produce much less ionisation and change course frequently when they collide with gas atoms. The beta-particle tracks are thin and irregular. The tracks in the cloud chamber are like the vapour trails produced by high flying aircraft on a clear day. The most up-to-date method of studying particle tracks is to use a **bubble chamber** in which bubbles form on the ions as the particle passes through liquid hydrogen, figure 39.18.

Cloud chamber photographs play a very important part in studying collisions between particles. The products formed in the collisions also produce tracks, the nature and length of which can give an indication of the nature of the products. For example, particles with more energy produce longer tracks. When a magnetic field is applied the direction of curvature of the track will show whether the particle is positively or negatively charged. Many new kinds of particles have been discovered from the study of cloud chamber photographs, figures 39.19–39.22.

fig. 39.19 Collision of an α-particle with an oxygen nucleus. The short track is the recoiling nucleus

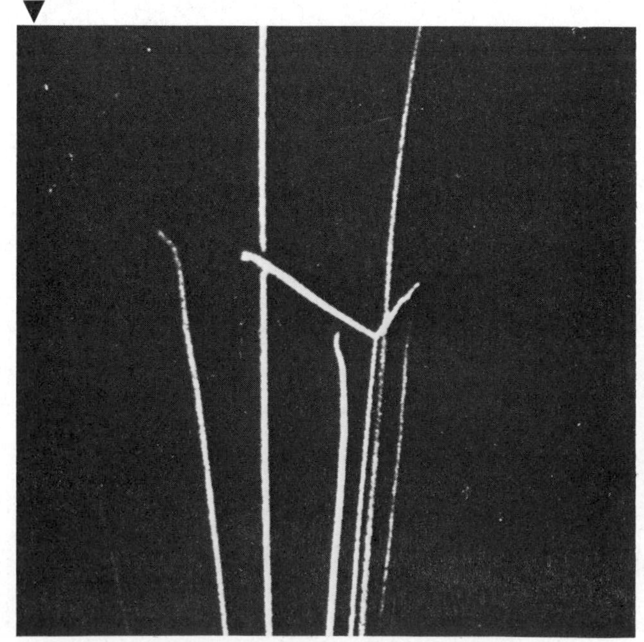

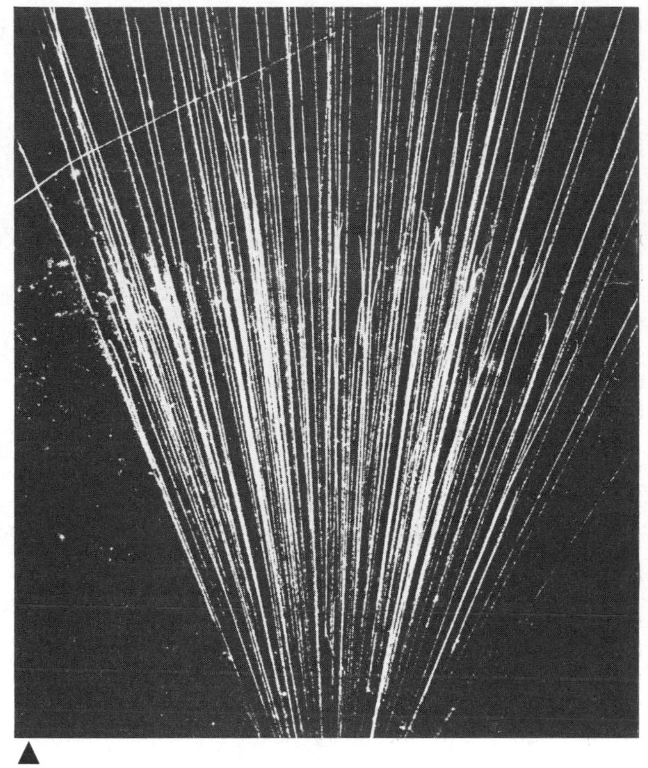

fig. 39.20 α-particle collision in nitrogen. The straight tracks are of two different lengths, showing α-particles of two different energies

fig. 39.21 β-particle tracks curved by a magnetic field. The particles are travelling from the left and the field is directed out of the page

fig. 39.22 This picture was taken in the CERN 2 m bubble chamber. Each particle, entering from the left, leaves a trail of bubbles along its path. The pairs of tracks looking like V's on their sides are due to neutral particles, which leave no track until they decay into pairs of charged particles. The spirals are due to electrons

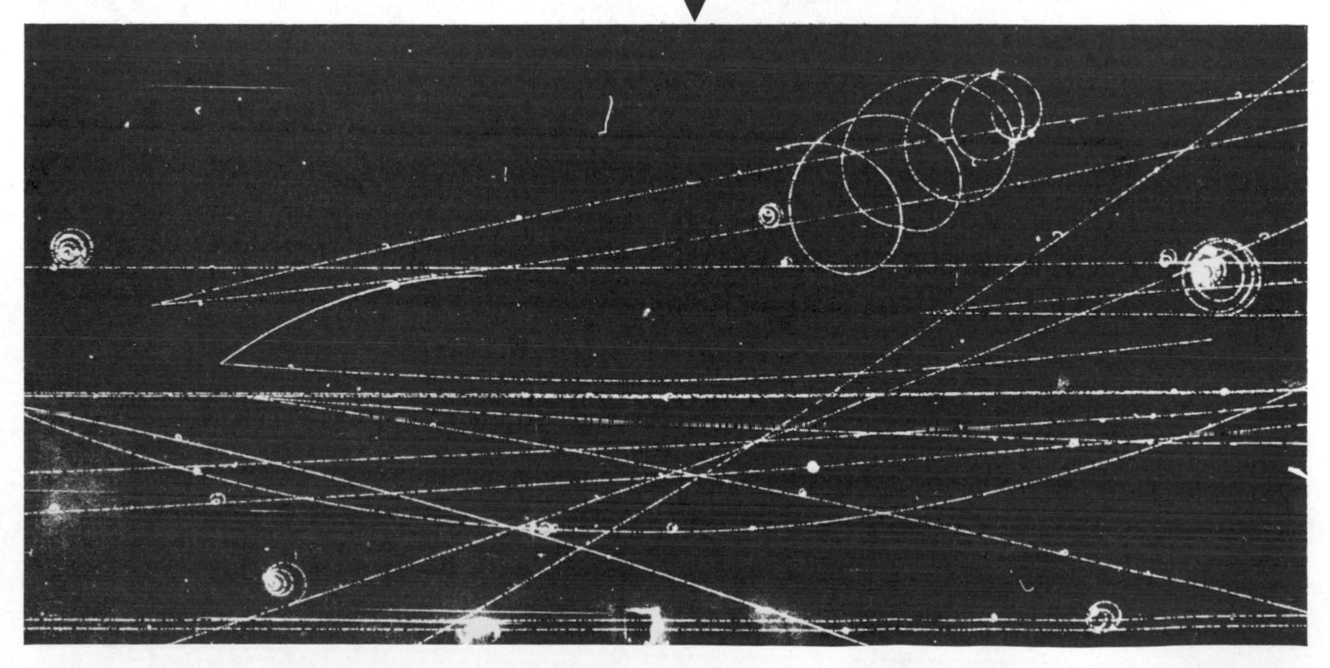

▲
fig. 39.23 The international symbol for radioactive material

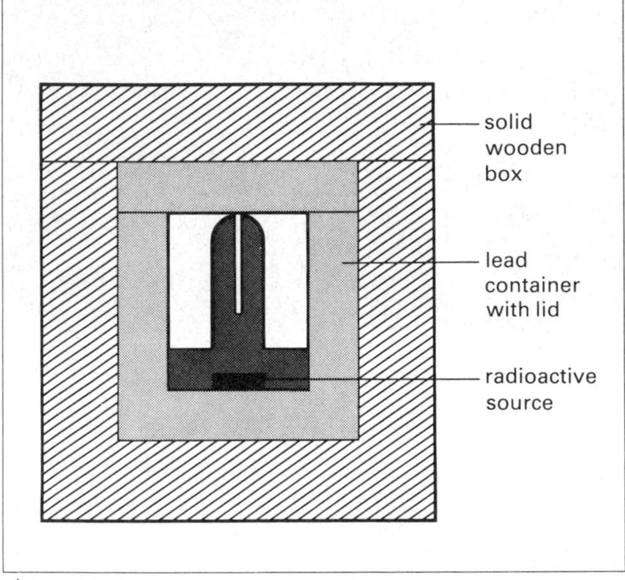

▲
fig. 39.24 Radioactive source for school use. The active material is enclosed in a lead container

fig. 39.25 Long tongs for handling radioactive sources
▼

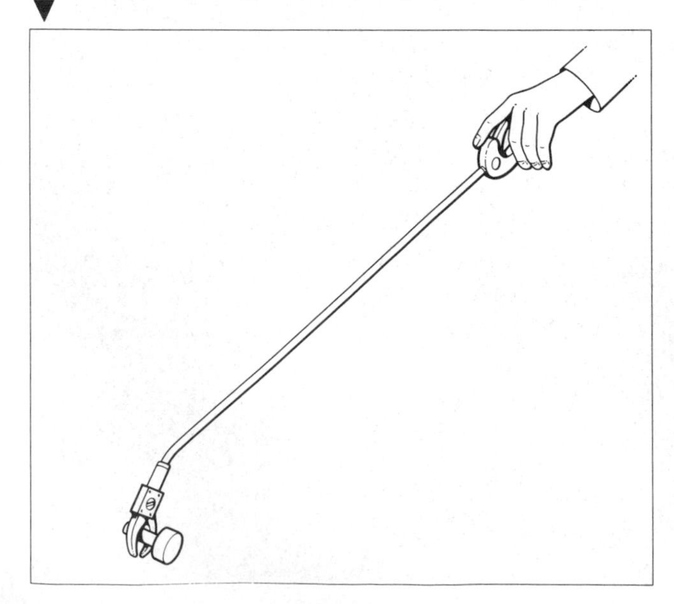

Experiments with scaler and ratemeter

Radioactive radiation in large doses is dangerous to health and the use of radioactive materials in schools and colleges is governed by safety regulations issued by the Department of Education and Science. It is assumed that the following experiments will be carried out by competent persons using the appropriate techniques. Sources of radioactive material for schools are 'sealed sources' so that the actual material cannot be handled or be lost. They are stored in lead containers. The sources are handled with long tongs to keep them away from the body, figure 39.25.

If a scaler is connected to a G–M tube and power supply, and the voltage across the tube adjusted to the correct value, when the scaler is switched on it will be observed that counts are recorded, although the tube is not near a known source of radioactivity. This is due to **background radiation** such as cosmic rays, traces of radioactive material in the atmosphere and even radioactivity in our own bodies. This background radiation must be measured and taken account of when making observations. The background count may well vary in different parts of the laboratory.

Another thing that will be noticed is that the count rate is not constant. It varies from time to time, showing the random nature of the radioactive changes taking place. The same thing can be observed if the G–M tube is connected to an amplifier and loudspeaker. The clicks, indicating the passage of radiation through the tube, occur irregularly.

The way in which the different radiations are absorbed by different materials can easily be demonstrated. It is convenient to use sources which are now known to emit one type of radiation only, although this was not possible for the early experimenters who often had to deal with all three radiations at the same time.

The tube is placed a fixed distance from the source, and sheets of the absorbing material are placed between the two, figure 39.26. The count rate (i.e. the

fig. 39.26 Measuring the differences in the radiation absorption of various materials
▼

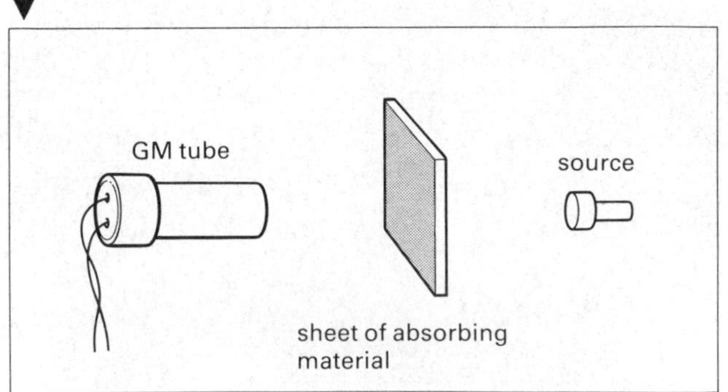

number of counts per second, found by taking the total count over, say, two minutes) is observed with and without the material, and for different thicknesses of the material. The results will confirm the absorptivities of the α-, β- and γ-radiation in table 39.1.

- α-particles are stopped by a sheet of paper.
- β-particles are stopped by a few millimetres of aluminium or plastic.
- γ-radiation is very penetrating and is only reduced in intensity by several centimetres of lead.

A fundamental property of radioactive decay can be studied using an ionisation chamber, amplifier and milliammeter which together act in the same way as a ratemeter, giving a meter reading which measures the activity of the source, figure 39.27. A convenient source is a compound of thorium which can be supplied in a sealed plastic bottle. The thorium decays to a radioactive gas, thoron, which in turn decays, emitting α-particles. The thoron is pumped into the ionisation chamber through a system of valves by squeezing the plastic bottle.

The ionisation current is a measure of the number of α-particles present in the chamber and this in turn depends on the quantity of thoron present. The reading of the milliammeter is noted every five seconds and the reading plotted against the time. The graph, figure 39.28, shows the activity decreases with time.

The **activity** of the source is measured by the number of atoms disintegrating per second. The activity of the source decreases because, as the atoms of the original sample disintegrate, there will gradually be fewer and fewer atoms left to disintegrate and produce α-particles. The time for half the total number of atoms originally present to disintegrate is called the **half-life** of the material. During the half-life period the activity of the material is reduced to half its value because there are only half the number of atoms present.

From the graph, the half-life of thoron is 55 seconds. After another half-life period there will only be a quarter of the original atoms left, after a third half-life period there will be an eighth of the number remaining, and so on. This type of decrease in numbers is an **exponential decay**.

The half-life periods of the radioactive elements vary enormously, from thousands of millions of years

table 39.2 Widely different half-life periods

Element	Half-life periods	
uranium-238	4.5×10^{9}	years
radium-226	1620	years
thorium-234	24	days
radium-214	20	minutes
polonium-212	3×10^{-7}	seconds

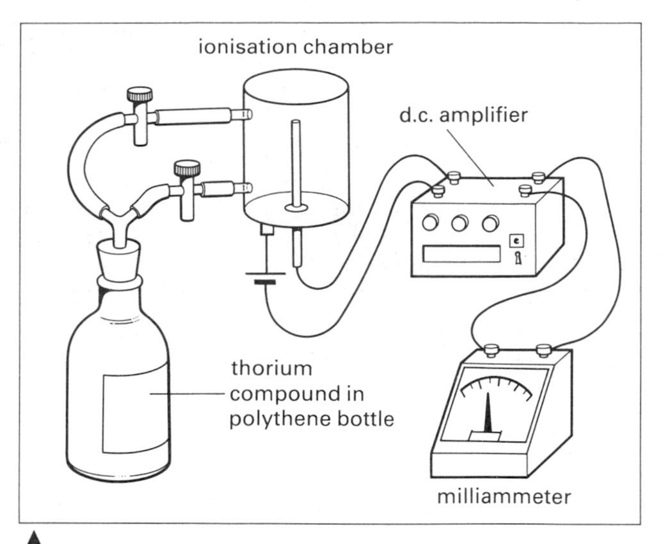

fig. 39.27 Measuring the decay of thoron

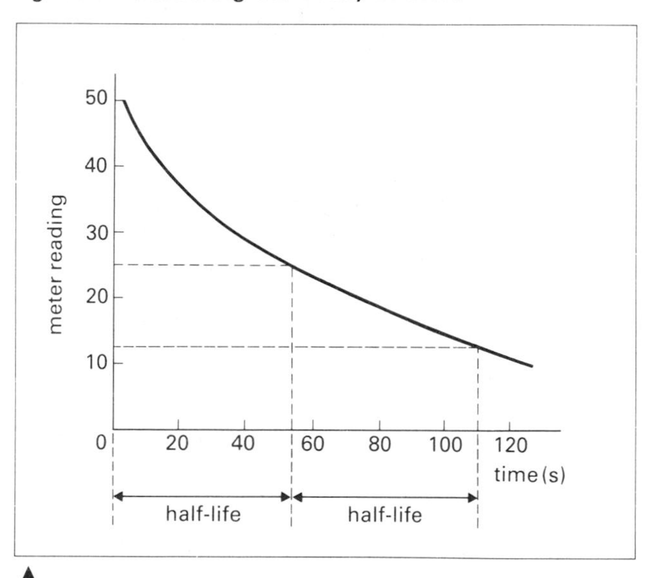

fig. 39.28 Radioactive decay curve for thoron

fig. 39.29 Exponential decay

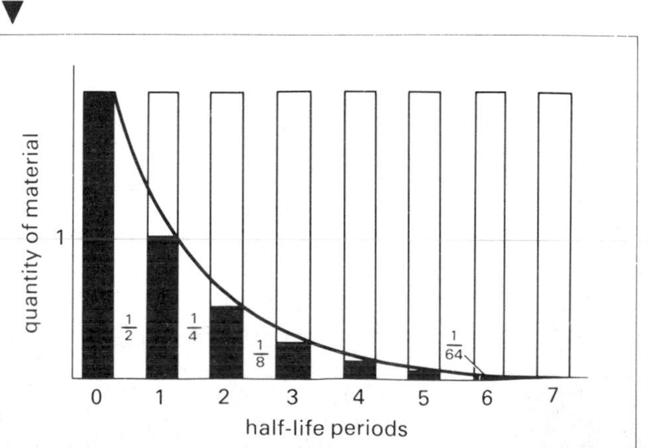

to a few millionths of a second. Table 39.2 gives some examples of these variations. Some artificially-produced radioactive elements may once have existed in the earth, but because of their comparatively short half-lives have completely disappeared.

The idea behind the half-life period is a statistical one. What we are saying is that, in one half-life period, half the atoms present will disintegrate. We cannot say anything about the behaviour of any particular atom. Indeed, the idea of a half-life period, like all statistical laws, only holds good if we are dealing with a very large number of atoms.

EXAMPLE

A Geiger counter is used to measure the decay of a radioactive substance at a place where the background count is 20 counts/minute. The readings are given in the table below. Estimate the half-life.

Time (hours)	0	6	8	10.5	20
Count-rate (counts/minute)	120	70	60	50	30

The corrected count-rate, making allowance for the background radiation, is found by subtracting twenty from each of the readings. The corrected count-rate is then plotted against the time. Figure 39.30 shows the result.

Lines drawn across the graph at count-rates of 50, 25 and 12.5 will give one, two and three half-life periods respectively. If there is any difference between the values they should be averaged.

The half-life of this particular radioactive substance is 6 hours.

fig. 39.30

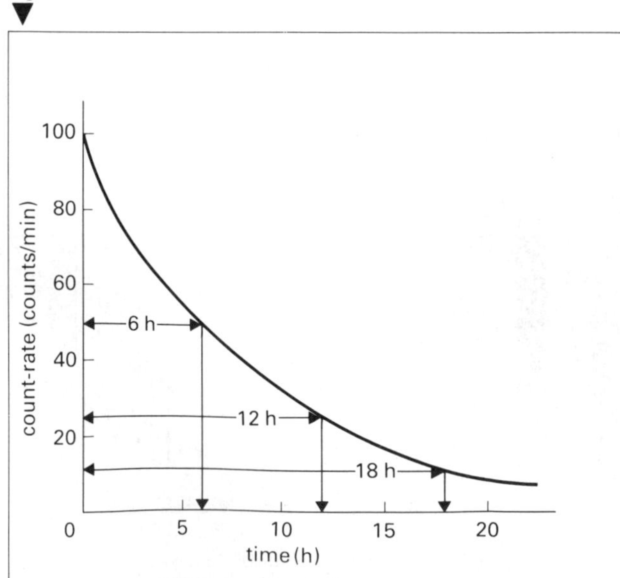

Changes in the nucleus when α- and β-particles are emitted

Radioactive changes take place in the nucleus. The nuclei of the atoms of the heavy elements are unstable. The emission of particles and energy means that the new nucleus so formed is more stable.

When a radioactive nucleus emits an α-particle the new **daughter** nucleus will have two protons and two neutrons less than the **parent** nucleus, figure 39.31.

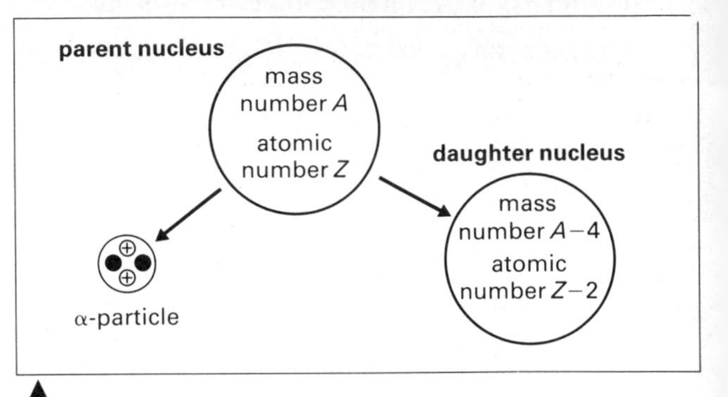

fig. 39.31 Emission of an α-particle

The emission of an α-particle means a reduction of 2 in the atomic number and a reduction of 4 in the mass number. In general terms:

$$^{A}_{Z}\mathbf{X} \rightarrow {}^{4}_{2}\mathbf{He} + {}^{A-4}_{Z-2}\mathbf{Y}$$

parent nucleus — alpha particle — daughter nucleus

To take a specific case, uranium-235 decays by emission of an α-particle into thorium-231:

$$^{235}_{92}\mathrm{U} \rightarrow {}^{4}_{2}\mathrm{He} + {}^{231}_{90}\mathrm{Th}$$

Notice that the numbers at the tops of the symbols balance on both sides of the equation (conservation of mass) and the numbers at the bases of the symbols balance on both sides of the equation (conservation of charge).

The emission of a β-particle (electron) is explained

fig. 39.32 Emission of a β-particle

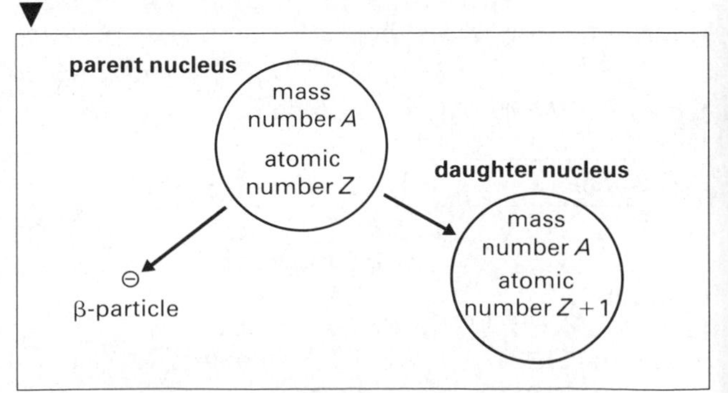

as follows. There are no electrons in the nucleus, but we can regard a neutron as being made up of a proton and an electron. The positive and negative charges would cancel out and the mass of the neutron is slightly greater than that of a proton, so the idea makes sense. During β-emission a neutron changes to a proton and an electron:

$$ {}_0^1\mathrm{n} \rightarrow {}_1^1\mathrm{p} + {}_{-1}^{0}\mathrm{e} $$

The proton stays in the nucleus, and the electron is emitted. The effect of this is to give the daughter nucleus an extra proton, but the mass number remains the same because there is still the same number of nucleons in the nucleus, figure 39.32.

The general equation for the emission of a β-particle is

$$ {}_Z^A\mathrm{X} \rightarrow {}_{-1}^{0}\mathrm{e} + {}_{Z+1}^{A}\mathrm{Y} $$

To take a specific case, radium-228 decays by emitting a β-particle to actinium-228:

$$ {}_{88}^{228}\mathrm{Ra} \rightarrow {}_{-1}^{0}\mathrm{e} + {}_{89}^{228}\mathrm{Ac} $$

table 39.3 Effect of emission on Z and A

Radiation	Atomic number (Z)	Mass number (A)
α-emission	decreases by 2	decreases by 4
β-emission	increases by 1	unchanged
γ-emission	unchanged	unchanged

Gamma radiation is high-frequency electromagnetic radiation and when it is emitted there is no change in the number or nature of the nucleons, so the mass number and the atomic number remain the same. The γ-radiation is usually emitted at the same time as the α- and β-particle emissions and represents the excess energy of the daughter nucleus as it settles down into a more stable condition.

Radioactive decay series

When a parent nucleus decays to form a new daughter nucleus it often happens that the daughter nucleus itself is radioactive and decays. Thus a whole series

fig. 39.33 The radioactive thorium series

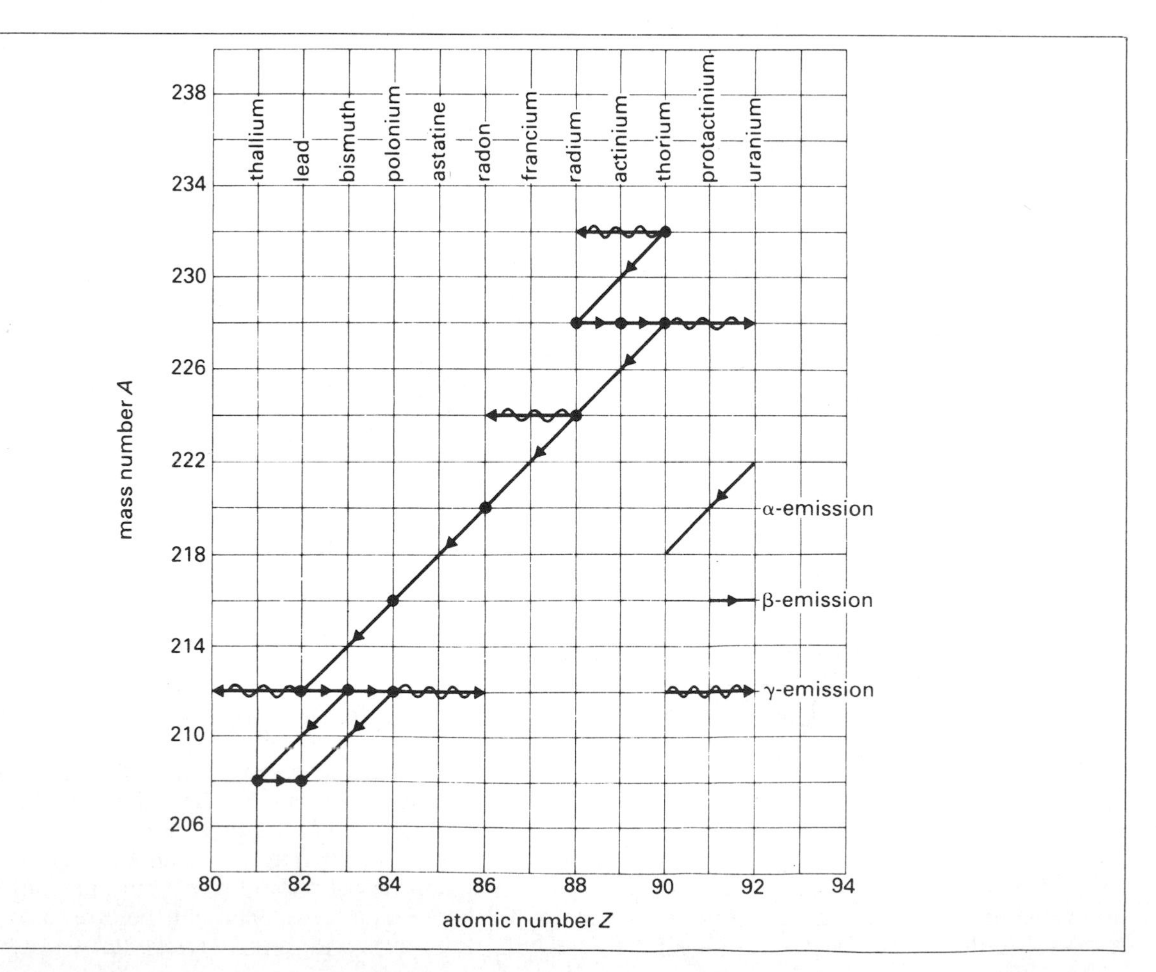

of new elements can be produced, each one decaying until finally a stable nuclear arrangement is arrived at.

Three naturally occurring series are known:

Uranium series starting with $^{238}_{92}U$ and ending with the lead isotope $^{206}_{82}Pb$.
Actinium series starting with $^{235}_{92}U$ and ending with the lead isotope $^{207}_{82}Pb$.
Thorium series starting with $^{232}_{90}Th$ and ending with the lead isotope $^{208}_{82}Pb$.

The thorium decay series is given in detail in figure 39.33. The emission of an α-particle moves the element two places to the left and four places down the table, while the emission of a β-particle moves the element one place to the right. The γ-radiation is shown by a wavy line with an arrowhead.

Artificial radioactivity

The naturally-occurring radioactive elements have large atomic numbers and large mass numbers. Normally stable elements with low atomic numbers can be made radioactive by bombardment with nuclear particles. Two examples of nuclear transformations were discussed in Chapter 27 where the bombarding particles were alpha-particles and protons. The simplest way to produce artificially-radioactive atoms is to allow them to absorb neutrons produced by the chain reaction in a nuclear reactor. For example, a cobalt atom absorbing a neutron would become an isotope of cobalt (same atomic number, different mass number) and, having an extra neutron, would be unstable and radioactive. It is called a **radioisotope**.

$$\underset{\text{normal cobalt atom}}{^{59}_{27}Co} + \underset{\text{neutron}}{^{1}_{0}n} \rightarrow \underset{\text{radioactive cobalt atom}}{^{60}_{27}Co}$$

Cobalt-60 is radioactive with a half-life of 5.3 years. It decays by β-emission to nickel:

$$^{60}_{27}Co \rightarrow {^{60}_{28}Ni} + {^{0}_{-1}e}$$

Radioactive isotopes of many elements are produced commercially, among them sodium-24, phosphorus-32, sulphur-35 and iodine-53.

Uses of radioisotopes

Radioisotopes have a great number of uses in industry, agriculture, medicine and research. The γ-radiation which many of them emit can be used in place of X-rays. Welding faults in steel pipes can be detected by photographing the weld, using cobalt-60 as a source of radiation. Gamma radiation can also be used for the treatment of cancer cells in the body. Thickness and density measurements can be made based on the quantity of β-radiation absorbed as it passes through the material. If there is any difference between the readings of the two detectors this shows on the gauge, which also indicates whether the production material is thicker or thinner than the standard sample. The reading can be fed back to the machinery producing the material so that correcting adjustment is automatic, figure 39.34.

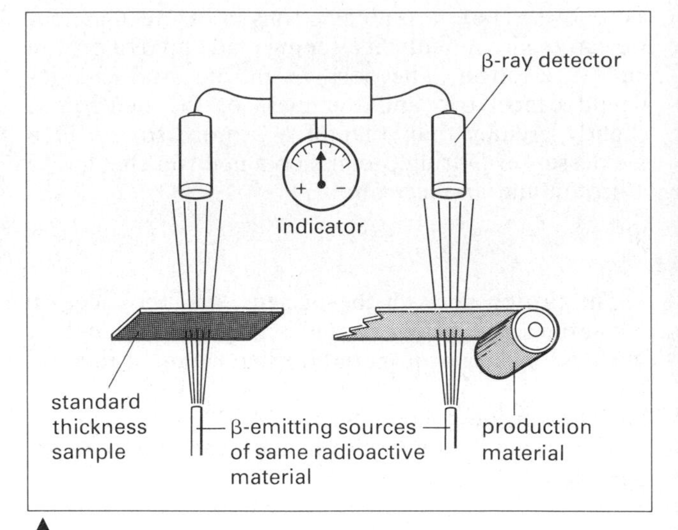

▲ fig. 39.34 Using radioisotopes to make thickness measurements

The radioisotopes can be used as **tracers** to follow the path taken by a material passing through a series of operations. The radioactive atoms are 'labelled' atoms; we can find out where they are by using a Geiger-counter. Very small amounts are required because the detectors are so sensitive, and if the half-life is reasonably short the tracer element will disappear by natural decay. In ten half-lives the activity will fall to 2^{-10}, i.e. $\frac{1}{1024}$ of its original value. The half-life of sodium-24 is 15 hours. At the end of a week its activity will have fallen to $\frac{1}{2500}$ of its original value.

Examples of the uses of radioisotopes are shown on the opposite page.

Biological effects of radiation

All forms of radioactive radiation and X-rays can damage body tissues. One of the problems is that we have no sense organs in our bodies which detect radiation, so that we could be absorbing dangerous quantities of radiation without our knowing what was happening to us. Many of the effects of radiation are long-term and so the damage done would not become

Some uses of radioactive isotopes

Inspecting an aircraft engine with the help of gamma radiography. The source is iridium 192. Minute flaws can be located quickly and accurately, and their seriousness assessed

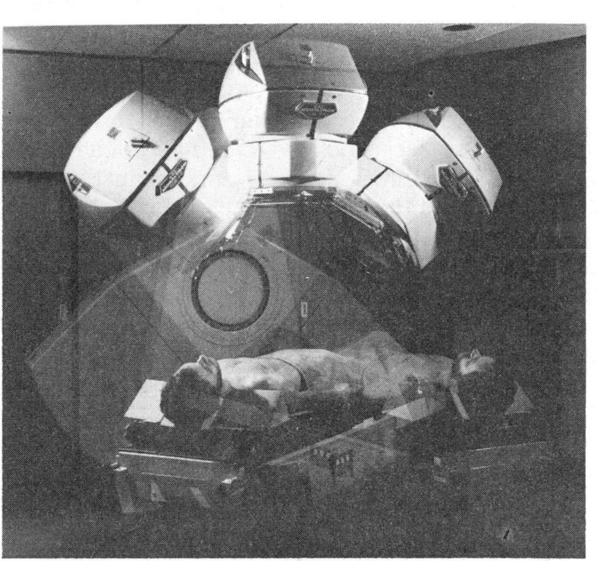

The Mobaltron MS 90 uses a minicomputer to control the shape of the irradiated tissue volume in three dimensions, as the patient undergoes treatment. The source is cobalt-60

This portable gamma-ray spectrometer will simultaneously measure uranium, thorium and potassium concentrations

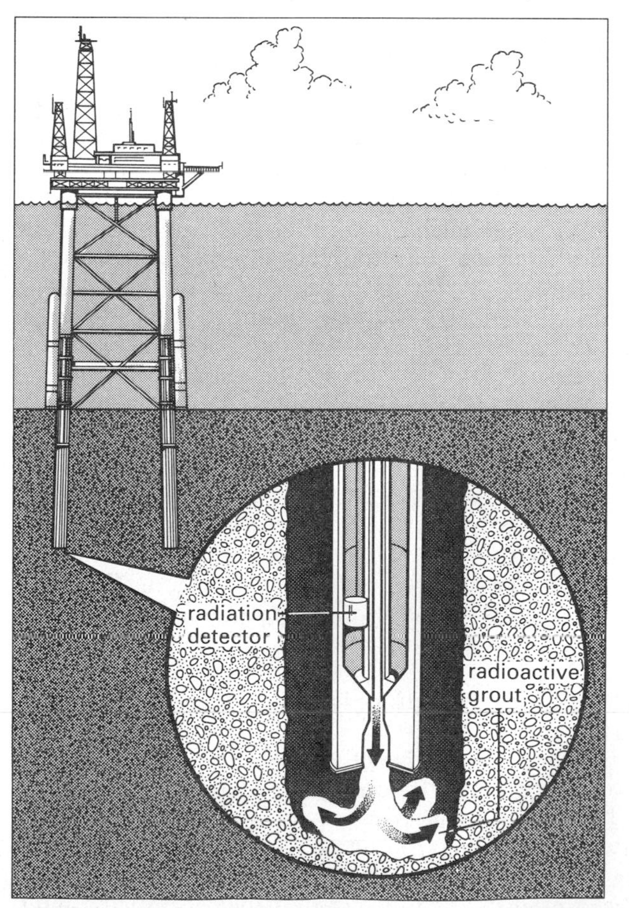

When building oil rig foundations, a radioactive tracer can be added to the concrete grout so that its position and density can be accurately monitored

evident for a number of years. Radioactive substances are dangerous and their handling is strictly controlled.

The most serious effects of radiation are on cells which are dividing to form new tissues. The cells in the bone marrow are producing new blood cells. The ovaries and testes are producing new eggs and sperm. The lining of the alimentary canal is constantly being renewed. Excessive exposure to radiation can cause anaemia and leukemia. Even if the reproductive cells are not destroyed the chromosomes they contain may be damaged. The chromosomes are made up of genes which carry the hereditary information that determines the characteristics of the offspring. A change in a gene is called a **mutation** and this will affect the particular characteristic associated with the gene for all future generations.

Strict safety precautions have to be taken by all who work in nuclear power stations and in the handling and transport of radioactive isotopes.

Radioactive dating

The decay of radioactive nuclei acts like a clock ticking off time in the rocks of the earth's crust, because the nuclei disintegrate at a constant rate. We know that during a half-life period half the atoms originally present have disintegrated. Let us assume that a uranium ore contained pure uranium-238 when the earth was 'created'. Uranium-238 decays through a series of radioactive products to finish as lead-206. So in any uranium ore the proportion of lead-206 to uranium-238 increases with time. The half-life of uranium-238 is 4500×10^6 years, so if we found a uranium ore which contained equal amounts of U-238 and Pb-206 we could say that the decay process had been going on for 4500 million years. The present ratio U-238/Pb-206 in rocks indicates a decay process which has been going on for about 5000 million years and this is taken as a reasonable estimate for the age of the earth. The decay of other radioactive materials gives a similar result.

Green plants absorb carbon dioxide from the atmosphere and convert it to sugar, starch and then to cellulose during the process of **photosynthesis**. Following figure 39.35 atmospheric carbon dioxide contains atoms of two isotopes of carbon, C-12 and C-14. C-14 is formed by the bombardment of nitrogen atoms by cosmic-ray neutrons:

$$^{14}_{7}N + ^{1}_{0}n \rightarrow ^{1}_{1}H + ^{14}_{6}C$$

Carbon-14 is radioactive and decays by β-emission to nitrogen. The ratio C-14/C-12 atoms in the atmosphere is constant because an equilibrium has been reached in which the rate of decay of the C-14 atoms is balanced by their rate of formation. In living plants the C-14/C-12 ratio is constant because of the interchange of carbon dioxide with the atmosphere. But when the plant dies no new carbon is taken in, the C-14 goes on decaying and so the ratio C-14/C-12 gets less. The half-life of C-14 is 5590 years, so we can calculate how long the decay process has been going on and determine the age of the material. If the ratio C-14/C-12 has fallen to a quarter of its original atmospheric value, the decay process must have been going on for 2×5590 years.

All forms of radioactive dating have limitations. For example, C-14 dating can only be carried out on materials containing carbon which has been absorbed when they were alive, e.g. wood, cotton, paper, papyrus, etc. The quantity of C-14 is very small and the activity difficult to determine. It is less than the normal background count, so that causes problems. Material can be dated up to about 40 000 years. After 70 000 years virtually no carbon-14 exists.

fig. 39.35 The process by which we can date material using C-14 dating

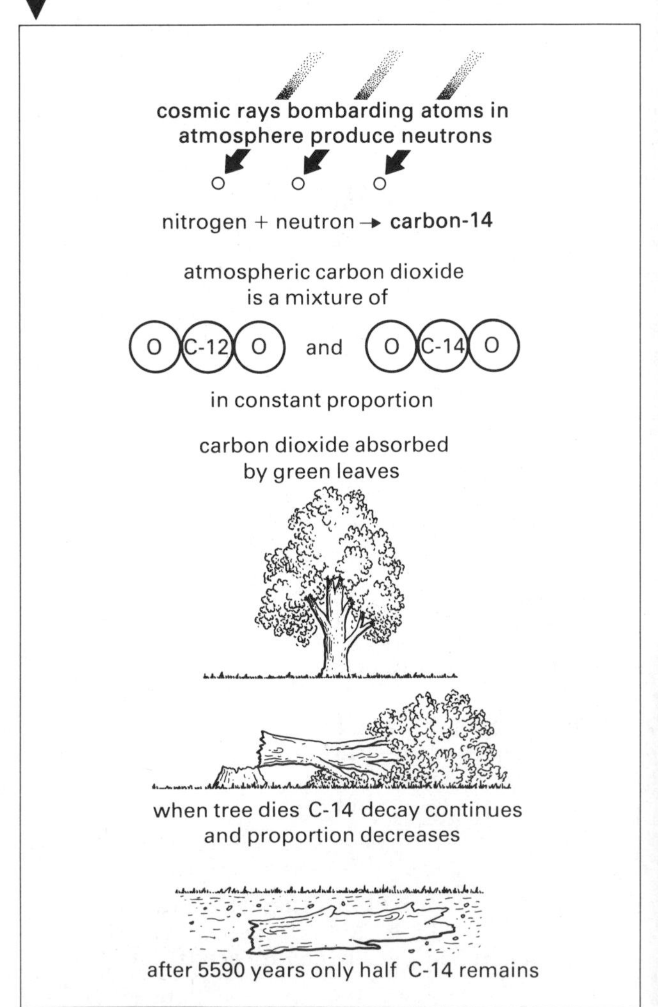

EXERCISE 39

In questions 1–5 select the most suitable answer.

1 Which of the following is *not* a correct statement about X-rays?
A They are a stream of particles of small mass.
B They can cause damage to body tissue.
C They pass easily through sheets of cardboard.
D They are formed when fast moving electrons are stopped.
E They are a form of electromagnetic radiation.

2 An alpha particle consists of
A 2 protons and 2 electrons
B 4 protons
C 2 protons and 2 neutrons
D 4 neutrons
E 2 neutrons and 2 electrons

3 A radioactive source emits one type of radiation only. On investigation it is found that the rays will penetrate quite well through 1 mm of aluminium, but will not penetrate 1 cm of lead. The radiation emitted will most probably be
A beta rays
B alpha particles
C cathode rays
D infrared rays
E gamma rays

4 A radioactive substance has a half-life of 1 minute. How long does it take for the activity of a sample of this substance to fall to one-sixteenth of its original value?
A 2 minutes
B 3 minutes
C 4 minutes
D 8 minutes
E 16 minutes

5 When a $^{228}_{90}Th$ nucleus undergoes radioactive decay by the emission of an alpha particle the new nucleus formed is
A $^{228}_{89}Ac$
B $^{228}_{90}Ac$
C $^{224}_{88}Ra$
D $^{224}_{87}Fr$
E $^{228}_{88}Ra$

6 Figure 39.36 shows an X-ray tube.
a (i) What are the parts labelled E, F and G? (ii) What material is used for the target? (iii) Why is this material suitable? (iv) Out of which window, H or I, are the rays emitted?
b What is the reason for the fins shown on figure 39.36?
c X-rays are described as being of long or short wavelength. Which type is the more penetrating?
d Name *two* uses of X-rays.

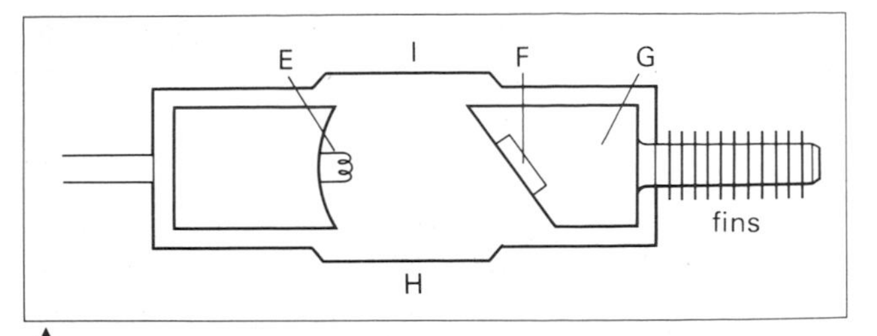

fig. 39.36

7 Figure 39.37 shows three types of radiation A, B and C being emitted by a radioactive source.
a Name each of the radiations A, B and C.
b What would happen to each radiation if the magnetic field was removed?
c If the following materials were placed directly over the hole at X, in the path of the rays, state which of the above radiations, if any, would be absorbed by (i) one sheet of tissue paper, (ii) one sheet of this book paper, (iii) an aluminium sheet 1 mm thick, (iv) a lead sheet 1 mm thick.

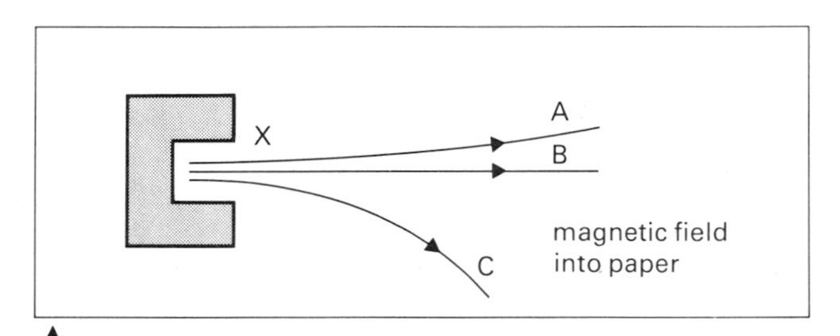

fig. 39.37

8 a Why are nuclei represented by the symbols $^{214}_{82}X$ and $^{210}_{82}Y$ isotopes?
b In what way do the nuclei differ?
c How many orbiting electrons does each nucleus have?
d If $^{214}_{82}X$ is radioactive and decays with the emission of a β-particle, write down the symbol for nucleus Z which is formed.

9 A radioactive substance emits radiation which is counted. As time goes by the radiation decreases. Some typical results are given in the table.

Time (s)	20	40	60	80	100	120	140	160
Activity (per s)	150	124	104	85	70	58	48	40

a Draw a graph of the results (plot time on the *x*-axis).
b What is the time for the activity to fall from: (i) 140 to 70? (ii) 70 to 35?
c What do you notice about these times?
d What name is given to this time?

10 Figure 39.38 shows a section through a Geiger–Müller tube (used for detecting alpha and beta radiation), and its connected electrical circuit.

fig. 39.38

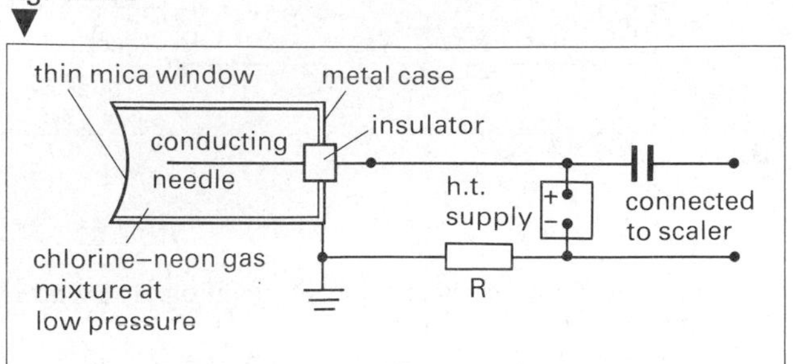

a Why is it important that the tube has a thin mica end?
b Explain how the tube detects a particle such as a beta particle.
c Alpha particles may be written $^{4}_{2}He$. What product is formed by the alpha particle decay of $^{213}_{84}Po$? Choose from: $^{212}_{85}At$; $^{209}_{83}Bi$; $^{209}_{82}Pb$; $^{217}_{86}Rn$.

Time (seconds)	Counts per second
10	50
20	38
30	34
40	32
50	28
60	25
70	22
80	19
90	16
100	14
110	13
120	11
130	10
140	8
150	7

11 a The table above shows the results of an experiment to measure the radiations from a particular radioactive source over a period of time. Allowance has been made for the background count. From the table, show how the half-life of this radioactive substance may be calculated, starting at
(i) 50 counts per second,
(ii) 28 counts per second.
b Sodium-24 has a half-life of 14 hours. What fraction of a sample of sodium-24 will have decayed after 42 hours?

12 The piston rings of a car are made radioactive before fitting. The engine is assembled and run; wear of the rings takes place and tiny quantities of the radioactive isotope which the piston rings contain find their way into the engine oil. Here they can be detected from the outside so that the rate of wear of the piston rings can be measured.

Answer the following questions about the experiment described above.
a What is meant by the word *radioactive*?
b What is meant by the word *isotope*?
c State *three* safety precautions necessary when radioactive materials are being used.
d Which type of radiation, alpha, beta or gamma, do you consider to be most suitable for use in the experiment? Give a reason for your choice.
e Draw a labelled diagram of a piece of apparatus which you could use to detect this radiation.
f Would you use an isotope with a half-life of 50 seconds, 50 minutes, 50 days or 50 years? Give a reason for your answer.
g Outline how you think the wear of a piston ring would have been detected without the use of radioactive materials.

13 a A radium nucleus may be represented by the symbol $^{226}_{88}Ra$. What is indicated by (i) the number 226? (ii) the number 88?
b Radium can emit an α-particle producing a nucleus X, represented by the following equation:

$$^{226}_{88}Ra \rightarrow X + ^{4}_{2}He$$

(i) What is the charge on $^{4}_{2}He$? (ii) Complete the equation by putting in the correct numbers for X. (iii) What does $^{4}_{2}He$ indicate?
c Instead of losing an α-particle, a radium nucleus can lose a β-particle, as shown by the following equation:

$$^{228}_{88}Ra \rightarrow Y + \beta\text{-particle}$$

(i) What is the charge on the β-particle? (ii) Complete the equation by putting in the correct numbers for Y (the new nucleus produced).
d What is the difference between $^{226}_{88}Ra$ and $^{228}_{88}Ra$?

14 A nucleus which is radioactive has a half-life of 1000 years. Explain what is meant by 'a half-life of 1000 years'.
b Gamma radiation may be emitted from a radioactive nucleus. Give *two* ways in which this type of radiation differs from alpha and beta emission.
c Describe *three* precautions which should be taken when using or transporting radioactive materials.
d What are 'isotopes'?
e Give examples of *two* uses to which radioactive isotopes may be put.

ELECTRICAL AND ELECTRONIC GRAPHICAL SYMBOLS

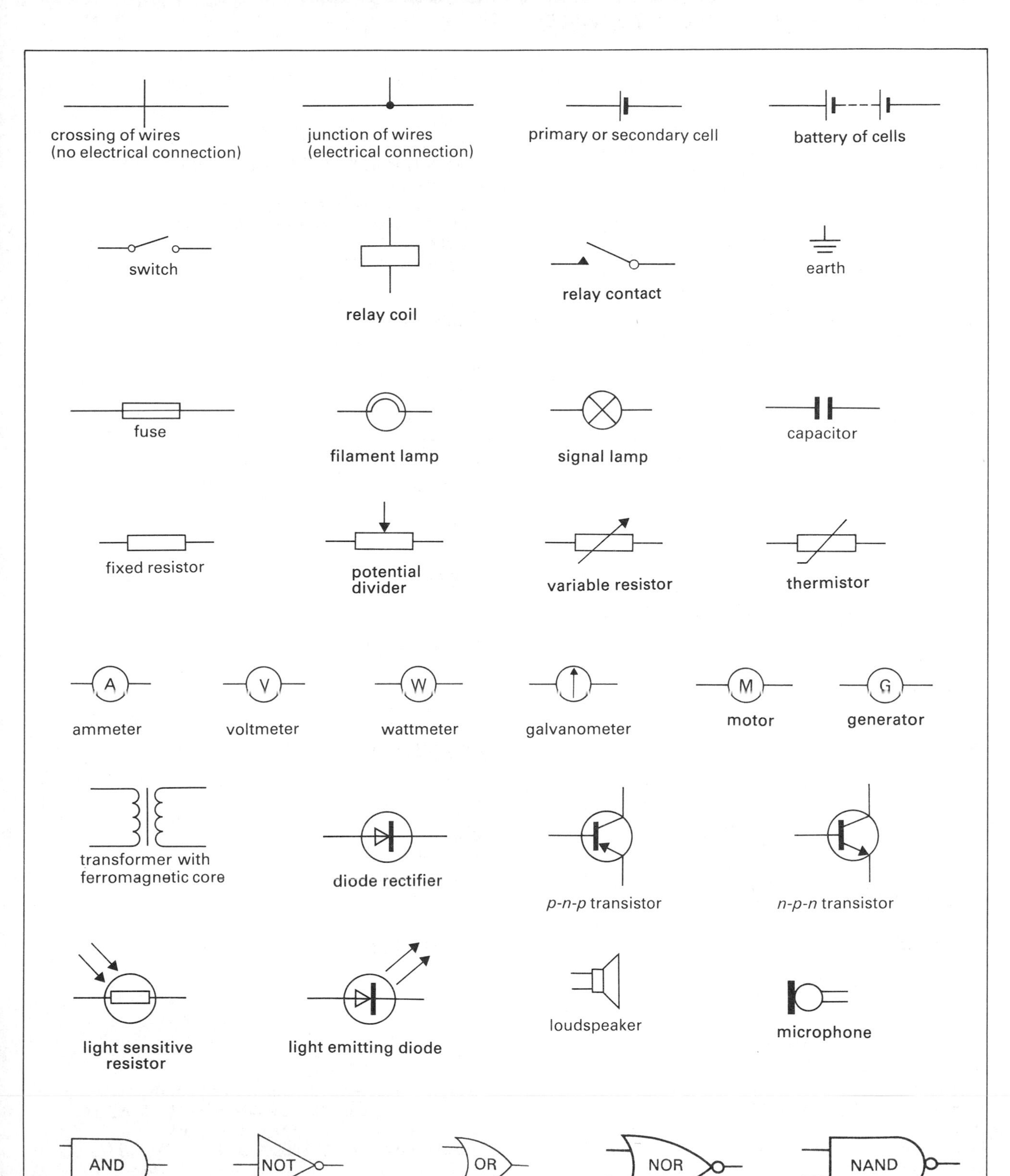

ANSWERS TO MULTIPLE CHOICE AND NUMERICAL QUESTIONS

EXERCISE 1

1 A
2 C
3 C
4 D
5 **a** 17.5 m **b** 65 **c** 2.5 m/s
6 **a** AB 2 m/s **b** CD 1 m/s **c** BC and DE, 250 s **d** EF **e** 100 m **f** 100 m **g** 500 s
7 **c** 16 cm/s **d** 19 cm/s
8 **b**(i) 5 s (ii) 1.325 m/s **d** 0.28 m/s^2
9 **b**(ii) 7 cm per 10 ticks (iii) 35 cm/s **d** 40 cm, 37 cm, 46 cm **e**(i) 2 s (ii) 23 cm/s
10 **a** 10 m/s **b** 20 m/s **c** 20 m

EXERCISE 2

1 D
2 A
3 E
4 **a** 2 *ma* **b** 3 *ma* **c** 6 *ma*
5 4 N
6 **a**(ii) 24 mm/N **b**(i) 4.5 N (ii) 45 m/s^2
7 **a**(iii) 10 cm (iv) 5 N **b** B 0.5 N, C 2 N, D 1 N
9 **a**(i) 15 m/s^2 (ii) 37.5 m/s^2

EXERCISE 3

1 E
2 A
3 B
4 C
7 **a** 1.33×10^6 kg m/s **b** 1.11×10^4 N
8 **f** 0.08 N
9 **a** 2.5 m/s^2 **b** 1.25 N **c** 5 kg m/s
11 **b** 0.5 m/s **c** 132 kg m/s **d** 1.1 g
12 1.5 m/s
13 **b**(i) 45 cm/s (ii) 0.16 s (iii) 15 cm/s

EXERCISE 4

1 B
2 C
3 A
4 C
5 D
6 **a** 60 000 J **b** 500 W
7 **a** 15 000 J **b** 50 W
8 **a** 1000 J **b** 100 W **c** 1000 J
10 **b** 10 J
11 **a** 15 550 N **b**(i) 1000 N (ii) 16 550 N (iii) 250 W

EXERCISE 5

1 A
2 C
3 **a** 1 **b**(i) 1 (ii) ½ **d** 20 N **e** 20 J **f** 40 J
4 **b** 120 N, 10 m
5 **d**(i) 1500 J (ii) 1800 J (iii) 83.3%
6 **b**(i) 10 cm (ii) 10 **c**(i) 8 (ii) 80% **e** 8 N
7 410 N **c** 800 J
8 **a** 5 **b** 6 **c** 83.3%

EXERCISE 6

1 C
2 A
3 D
4 **c** 500 N **d** 50 N
5 100 N
6 **b** 50 N
8 400 N
9 **b** 450 N
10 150 N
11 **b** 816.7 N, 283.3 N

EXERCISE 7

1 B
2 C
5 **a** 45 N **b** 3 m **c** 15 N
7 **b** 55 N **c** 35 N
8 **c**(i) 1:4 (iii) 500 J (iv) 50 J
9 **a** 71.5° **b** 63.4°

EXERCISE 8

1 B
2 E
3 D
4 C
5 **a** 26 ml **b** 42 ml **c** 16 ml **d** 2.5 g/cm^3
6 **a** 20 cm^3 **b** 0.2 cm^3 **c** 9 g/cm^3
7 2493 kg/m^3

EXERCISE 9

1 B
2 B
3 A
4 **a** 32 500 N/m^2 **b** 300 000 N/m^2
5 **b** 250 500 N/m^2
6 **a** 6×10^6 J **b** 3×10^4 N/m^2
7 **c**(i) atmospheric (ii) greater than atmospheric (iii) atmospheric **f**(i) 2500 N/m^2 (ii) 0.25 m
8 **a** 5×10^5 N/m^2 **b** 500 N **c** 0.5 cm **d** 5
10 **a**(ii) 500 N **b**(i) 250 000 N/m^2 (ii) 50 N

EXERCISE 10

1 B
2 D

3 A
4 D
7 **a** 64 000 N
8 **a** a **b** 6 m^2 **c** 6000 N/m^2 **g** 4×10^5 N/m^2

EXERCISE 11

1 E
2 C
3 A
5 0.0024 m
9 **c** 5.008 m

EXERCISE 12

1 D
2 D
3 E
4 D
10 **b** 900 °C

EXERCISE 13

1 C
2 C
3 E
4 D
5 **a** 4500 J **b** 4200 J
6 40 000 J/g
7 653.3 J/s
8 63 000 J, 350 W
9 **c**(i) 120 000 J (ii) 1000 J/s (iii) 2000 J/s
10 **a** 50 min **b**(i) 5 h 20 min, 2 h (ii) 4.2×10^7 J

EXERCISE 14

1 D
2 D
3 E
4 C
5 E
6 C
8 30 J
12 **a** 200 000 J/kg
13 **a** 100 °C **c** 8000 J/min **e** 216 000 J

EXERCISE 15

1 A
2 E
3 E
4 C
6 **d** 15 °C
8 **a**(ii) 48 °C **c**(i) 4500 J (ii) 5 °C (iii) 900 J/kg °C

EXERCISE 16

1 B
2 C
3 B
4 D

EXERCISE 17

1 A
2 A
3 B
4 D
5 E
7 **c** 6 cm^3
9 **c** 2.14×10^5 Pa
10 **c** 1.14×10^5 N/m^2
13 **b**(i) 115 cm of mercury (ii) 115 cm of mercury, 11.7 cm

EXERCISE 18

1 E
2 B
3 C
4 A

EXERCISE 19

1 C
2 D
3 C
4 A
5 C
11 **a**(i) 10 cm (ii) 40 cm **c**(ii) 20 cm
15 **b** 32 cm, 32 cm, 16 cm

EXERCISE 20

1 E
2 C
3 D
4 A
5 D
6 D
7 **c** 0.067 m
8 **d**(iii) 3/2
11 **b** 54.74° **c** 120°

EXERCISE 21

1 C
2 C
3 E
4 E
5 D
9 **a** 1 cm **b** 5 cm **c** 7 cm **d** 35 cm **e** 7
10 **b** 30 cm, 15 cm

EXERCISE 22

1 E
2 A
3 D
4 E
5 A
6 **a** 20 dioptres **c** 2 dioptres

EXERCISE 23

1 C
2 D
3 E
4 C
8 **a** lens B **c** 40 cm from A
10 **b**(i) 60 cm from the lens
11 **i**(i) 3 cm (iv) 6 cm
13 **d** 5×

EXERCISE 24

1 B
2 D
3 D
4 A
5 E
6 A
7 **a** 50 cm **b** 2 Hz **c** 100 cm/s
8 **e**(i) 4.5 Hz (ii) 3.33 cm (iii) 15 cm/s
9 **a** 1.5 **b** 4×10^{-7} m **c** 5×10^{14} Hz
11 **c**(i) 0.5 m (ii) 330 m/s

EXERCISE 25

1 E
2 C
3 B
4 A
5 C
7 **b** 0.7 m

EXERCISE 26

1 D
2 B
3 C
4 C
5 **a** 85 m
7 **a**(i) 825 m **b**(i) 3000 m/s

EXERCISE 27

1 D
2 A
3 B
4 C
5 B
6 D
8 **b** 1 **c** 12 **d** ${}^{4}_{2}He$
10 (iii) P and S

EXERCISE 28

1 D
2 E
3 A
4 A

EXERCISE 29

1 B
2 D
3 B
7 55° E of N

EXERCISE 30

1 E
2 E
3 A
4 B

EXERCISE 31

1 C
2 D
3 B
4 B
5 B
6 **a** 2 V **c** 3 V
8 **c**(i) 12 V (ii) 24 V
11 **a** 4.5 V **b** 1.5 V **c** 6 V

EXERCISE 32

1 B
2 E

EXERCISE 33

1 B
2 A
3 E
4 A

EXERCISE 34

1 E
2 C
3 E
4 B
5 C
8 **a**(i) 11 Ω (ii) 1 Ω **b** 1 Ω
9 **a** 6 **b** 2 V, 6 **e** 8 V **f** 4 V **g** 0.67 A **h** 1.35 A **i** 3 Ω
10 **a** 6 Ω **b** 2 Ω **c** 2 A **d** 0.67 A **e** 0.67 A
11 **a**(i) 0.2 A (ii) 0.4 A **b**(i) 2.4 A (ii) 1.33 A
13 **c** 7 Ω

EXERCISE 35

1 D
2 D
3 D
4 D
5 **b** 3 A
6 50 kWh
7 **a** 4800 J **b** 80 A
8 **a**(i) 6.25 A **a**(ii) 8 mA **b** 12 **c** 1.8 MJ
9 **b** 72p
10 **a**(i) 20 V (ii) 0.15 A (iii) 0.15 A **b** 129.6p
12 **a** 0.05 A **b** 3 C **c** 150 J **d** 2.5 W
14 **a** 0.25 A **b** 960 Ω

EXERCISE 36

1 A
2 D
3 B
4 A
9 **b** 2750 rev/min **c** 2900 rev/min **d** 10 A **e** 0 rev/min **f** 13 A **i** 480 W

EXERCISE 37

1 B
2 A
3 C
4 A
5 D
6 D
8 **c** 75 turns
9 **a**(v) 3 A, 0.25 A
10 **a**(i) 4 A (ii) 40:1 (iii) 0.1 A **b**(i) 1.5 Ω **b**(ii) 234 V **b**(iii) 58.5 Ω (iv) 936 J

EXERCISE 38

1 E
2 D
3 D
4 C
8 **b**(iii) 55

EXERCISE 39

1 A
2 C
3 A
4 C
5 C
8 **c** 82 **d** $^{214}_{83}Z$
9 **b** (i) 75 s (ii) 75 s
11 **a** 50 s **b** 7/8
13 **b**(i) $+2e$ (ii) $^{222}_{86}X$ **c**(i) $-e$ (ii) $^{228}_{89}Y$

Acknowledgements

The author and publisher thank the following for supplying prints and for permission to reproduce copyright material.

Babcock Power Ltd 39.4
Barnaby's Picture Library 16.1, 24.7
Lady Blackett and Royal Society 39.19, 39.20
Bourns Electronics Ltd 34.23
British Petroleum Co. PLC 9.7
Central Electricity Generating Board 4.16
Electricity Council 35.2
Ferranti PLC 37.23
Fibreglass Ltd 15.5, 15.8
GEC Witton Kramer Ltd 33.9
Griffin and George Ltd 1.12, 1.13, 1.16, 2.24, 3.6, 24.19, 24.20, 24.21, 24.22, 30.14, 31.2, 34.19, 34.23
Philip Harris Ltd 30.17
Kodansha Ltd (UK agent: Philip Harris Biological Ltd) 1.1, 1.17, 2.27
Lansing Ltd 4.21
London Transport Executive 7.12
Mullard Ltd 38.13, 38.19, 38.20
Negretti Automation 10.8, 12.6, 12.8
North of Scotland Hydro Electric Board 4.17
Novosti/Camera Press Ltd 23.11
Pergamon Press Ltd 39.21
PKG Electronics 38.36, 38.37, 38.38
Rex Features Ltd 2.15, 3.1, 4.8, 4.9, 4.10, 7.11, 14.20
St Bartholomew's Hospital 39.5
George H. Scholes PLC 35.15
Science and Engineering Research Council, Rutherford Appleton Laboratory 39.18, 39.22
Science Museum, London 38.17
Sharp Electronics (UK) Ltd page 198
Space Frontiers Ltd 2.21, 32.10
Stothert and Pitt PLC 5.7
The Telegraph Colour Library front cover
University of Dundee 15.22
United Kingdom Atomic Energy Authority 27.21, 30.16, 38.63, 38.64, 39.15, page 347
Yerkes Observatory 23.6

INDEX